Toxicological Assessment of Combined Chemicals in the Environment

Toxicological Assessment of Combined Chemicals in the Environment

Edited by

De-Sheng Pei and Yiyun Liu
Chongqing Medical University
Chongqing, China

WILEY

Published by John Wiley & Sons, Inc., Hoboken, New Jersey.

Published simultaneously in Canada.

Library of Congress Cataloging-in-Publication Data has been applied for.

Hardback: 9781394158324
ePDF: 9781394158331
epub: 9781394158348
oBook: 9781394158355

Cover Design: Wiley
Cover Image: © Hybrid_Graphics/Shutterstock

Dedication

This book is in the memory of my parents, Mr. Chao Pei (1942–2020) and Ms. Jinhua Wang (1943–2014). This book is dedicated to the spirit of collaboration, recognizing that the protection of our environment is a shared responsibility that transcends borders and cultures.

De-Sheng Pei PhD
Professor of Chongqing Medical University

This book is dedicated to the invaluable mentors and steadfast supporters who have illuminated my journey in the realm of scientific research.

Yiyun Liu PhD
Assistant Researcher of Chongqing Medical University

Contents

Foreword

During the early stages of my research, I struggled to fully explain the toxicity of fluoride in drinking water on human health. It was only later that I realized the necessity of incorporating combined toxicology concepts to properly understand the problem. I would like to congratulate Prof. De-Sheng Pei, Dr. Yiyun Liu, and the contributing researchers for their arduous task of compiling this excellent monograph on combined toxicology.

Unlike toxicological research on single compounds, combined toxicology focuses on adverse reactions that reflect actual environmental exposure conditions, gaining increasing attention from researchers globally. The challenges of combined toxicity experiments, analysis, and predictive models indicate that this field is in its infancy and requires national and international cooperation to address these issues effectively.

This book begins with a comprehensive assessment of combined toxicological philosophy, elucidating toxicokinetic and toxicodynamic processes. The authors pay particular attention to developing essential routes for understanding the fate and mechanisms of toxicants and the detection techniques in combined chemical matrices. They propose carefully chosen experimental models and highlight the importance of advanced computational and "omics" technologies in researching chosen toxicant groups. Future research directions and knowledge transfer avenues are succinctly presented.

The concepts of combined toxicology also provide crucial information and expertise for regulatory authorities, decision-makers, and others to develop programs and policies that limit human exposure to toxic substances, thereby preventing or reducing the risk of disease and other adverse health outcomes.

Therefore, I believe this book will be an invaluable resource for students, researchers, scientists, and policymakers dedicated to managing and protecting the environment. It will also serve as a valuable reference for emerging scientists in the field of combined toxicology.

Rohan Weerasooriya, PhD
Fellow of the National Academy of Sciences, Sri Lanka
Professor, National Institute of Fundamental Studies, Kandy, Sri Lanka
Professor, China-Sri Lanka Joint Research and Demonstration Center for Water Technology,
University of Peradeniya, Ministry of Water Supply
Distinguished Professor, Hefei University of Technology, China

Preface

Toxicology is known as the study of poisons and intoxication, which allows for the evaluation of dangers posed by industrial chemicals, plant protection products, biocides, medicinal medications and equipment, and consumer goods to humans, animals, and the environment. With the development of science and technology, the types and quantities of pollutants in water, atmosphere, and soil are increased. Many of the same or different kinds of pollutants exist together bringing harm to the environment. Pollutants in the environment can interact physically or chemically to generate new toxic substances. These substances, which can have stronger or weaker toxicity, can affect the balance of the ecosystem. The toxicity study of a single substance is not suitable for analyzing a variety of environmental pollution effects, but the combined toxicity effect can more accurately and truly reflect the biological toxicity of pollutants. Therefore, this book aims to focus on the research methods of combined toxicity, the progress of research on the combined toxicity of environmental pollutants, and propose the challenges of establishing relevant technical means and models in future research strategies

Introductory Chapter 1 proposes the concept of combined toxicology, including its background, key terminologies, and significance for assessing risks from environmental contaminants. Although it can provide a more realistic understanding of the potential environmental risks and health effects of exposure to multiple chemicals, it also poses some challenges for research due to the complexity of chemical mixtures, lack of standardized methods, and limited availability of data. Chapter 2 provides information on the toxicokinetics of chemical mixtures, such as organic halides, insecticides, and plasticizers, and understands the degradation, oxidation, and distribution processes of the mixture in the environment. The application of toxicokinetic evaluation models and methods can help predict the harmful effects of chemical mixtures on organisms and provide guidance for health risks. Chapter 3 provides an overview of the toxicological effects of several major chemical contaminants present in the environment, using chemical mixtures as an entry point and the common model animals as the main research objects. The toxicological effects of the major environmental chemical pollutants on the target organs of the heart, nerves, blood vessels, liver, and kidneys are analyzed. We hope to let readers understand the importance and practical significance of combined toxicity research by clarifying the causes and hazards of compound mixed exposure in Section 1.

It is an important research subject and premise of toxicology to understand the mixing mode and the fate of the environmental behavior of compounds before the biological toxicity test. Therefore, in Section 2, we will introduce the main sources and enrichment mechanisms of different chemical mixtures, and clarify the biological exposure pathways of different compounds. Chapter 4 comprehensively investigates the sources, pathways, and mechanisms involved in the formation of chemical mixtures containing heavy metals and pesticides in soil environments, as well as those comprising heavy metals, microplastics, and nanoparticles in various environmental media. Chapter 5 interpretes how the interaction between the multi-chemicals in the mixture form changed their toxicity to aquatic vertebrates and invertebrates and provided valuable insights into critical impacts on the ecological hazard of their combinations. Chapter 6 highlights the overall geochemical behavior of volatile organic pollutants (VOCs) mixtures. The concentration, dispersion, and disappearance of pollutants caused by spatial motion are defined as migration. At the same time, pollutants will be converted into new compounds, which may cause greater harm to the environment. Their migration and transformation are not only affected by their solubility, adsorption constant, volatility, and other properties but also limited by environmental conditions such as photolysis and microbial hydrolysis. Chapter 7 emphasizes analytical methods for identifying, characterizing, and monitoring classes of hazardous compounds in intricate matrices and techniques that integrate chemical analyses, which may make it easier to identify chemical mixes that cause a combined risk.

The use of an experimental model is a common method of toxicological research. Chapter 8 introduces the toxicological model of zebrafish, analyzes the research cases, and proposes the parameters that can be used to characterize the reproductive function of zebrafish and the corresponding detection methods. This chapter provides a reference for model organism selection and model establishment. At the same time, based on the different toxicological effects of different compounds, Chapter 9 describes the combined molecular toxicity of heavy metals in cardiovascular, nervous, reproductive, liver, immune, and olfactory systems from the perspective of population investigation and experimental testing. Chapter 10 highlights pesticide pollution has been linked to a wide range of human diseases and disorders including neurodevelopmental disorders, neurodegenerative diseases, cancers, neurotoxicity, and congenital defects. Pesticides alter the body's antioxidant defensive mechanisms and increase the accumulation of reactive oxygen species leading to oxidative stress. Pesticides also inhibit cytochrome P450 enzymes competitively binding at the enzyme's active site and interfering with the electron transfer chain. Chapter 11 reviews the developmental toxicity and endocrine effects caused by mixtures of persistent organic pollutants (POPs) in epidemiological and experimental studies. Meanwhile, it discusses the molecular toxicity mechanisms of POPs, including epigenetic changes, hormone receptors, oxidative stress, cell death-related signaling pathways, and mitochondrial dysfunction. Chapters 12 and 13 provide an overview of the "state of the art" literature data on the effects and molecular toxicity mechanisms of pharmaceuticals and personal care products (PPCPs) and phthalates (PAEs), with special emphasis on their mixture scenarios. Chapter 14 highlights the impact of microplastic mixtures on molecular toxicity mechanisms. By considering the varying toxicological effects of different pollutants, it explores the combined toxicity mechanisms of heavy metals, persistent organic pollutants, pathogens, engineered nanoparticles, and other pollutants with microplastics. Chapter 15 provides a useful summary of the combined toxicity and molecular mechanisms of exposure to flame retardants (FRs) mixtures, presenting the scientific challenges and potential research directions. Chapter 16 discusses the challenges of adverse outcome pathway (AOP) as a theoretical framework and method for obtaining and analyzing toxicological data and suggests that omics could be employed as a more specific tool to differentiate between mode and mechanism of action.

Section 4 introduces the computer mathematical modeling method for hazard assessment of chemical mixtures. This is a fast and promising risk assessment method, which can predict the physical and chemical properties related to the fate of compounds and predict the toxicological endpoint. Chapter 17 delves into the vital realm of combined toxicity prediction and offers a comprehensive insight into mathematical models for combined toxicity prediction. Chapter 18 reviews the quantitative structure-activity relationships (QSARs) models for predicting endocrine disrupting activities and acute toxicities, as well as the QSAR models based on machine learning method, biomolecular interaction networks, toxicokinetic–toxicodynamic studies, high-throughput transcriptomics approach, and geospatial modeling approach.

Although the combined toxicity of chemical mixtures has attracted more and more attention, there are still many challenges to overcome. Therefore, Chapter 19 emphasizes the mechanistic combined toxicity of the legacy and emerging environmental pollutants, both in vivo and in vitro models, and discusses the challenges of applying experimental, analytical, and predictive models in combined toxicity assessment. Chapter 20 highlights that these mixtures can exhibit diverse interactions, including additive, synergistic, or antagonistic effects, that cannot be precisely anticipated by isolating the constituent chemicals. Employing the procedure of toxicological assessment is a thorough approach to scrutinizing the possible dangers and perils posed by chemicals. This new technology is expected to effectively assess the hazards of simultaneous exposure to chemical substances and propose mitigation strategies.

Chapter 21 provides final thoughts and concluding remarks. The increasing pollution rate has resulted in numerous chemical pollutants in complex mixtures, severely threatening human and environmental health. This book systematically examines combined toxicity, environmental behavior, sampling and detection techniques, experimental programs, toxicity mechanisms, and model analysis strategies. However, challenges persist due to vast chemical combinations, exposure scenarios, and mixture interaction complexities. This book is a comprehensive resource for research on combined toxicology. It highlights current progress in the field and underscores the need for continued research, collaboration, and the application of scientific findings to practical solutions.

Prof. De-Sheng Pei and Dr. Yiyun Liu
School of Public Health
Chongqing Medical University
Chongqing
China

Acknowledgments

The editors are sincerely thankful to all contributing authors, who worked hard to improve the quality of this book.

This work was supported by the Chongqing Technology Innovation and Application Development Sichuan-Chongqing Special Key Project (CSTB2024TIAD-CYKJCXX0017), Scientific Research Foundation of Chongqing Medical University (R4014 and R4021), Chongqing Postdoctoral Innovation Mentor Studio (X7928 D.S.P.), China and Chongqing Postdoctoral Science Foundation (2022TQ0394, CSTB2022NSCQ-BHX0641), and Chongqing Special Funding for Postdoctoral Research Projects (2022CQBSHTB1006 and X7947).

List of Contributors

Abdul Wahab Hussain
College of Earth and Environmental Sciences
University of the Punjab
Lahore
Pakistan

Ali Omar Jimale
School of Public Health
Chongqing Medical University
Chongqing
China

Anam Ashraf
Landscape Lab Xishuangbanna Tropical Botanical Garden
Chinese Academy of Sciences
Menglun
Mengla
Yunnan
China

Asad Ullah Saeed
Pakistan Council for Scientific and Industrial Research
Lahore
Pakistan

Bao-Fu Zhang
School of Public Health
Chongqing Medical University
Chongqing
China

Chunjiao Lu
Engineering Research Center of Key Technique for
Biotherapy of Guangdong Province
Shantou University Medical College
Shantou
China
Department of Public Health
School of Medicine, Guangxi University of Science
and Technology
Liuzhou
Guangxi
China

Dejun Huang
Gansu Key Laboratory of Biomonitoring and
Bioremediation for Environmental Pollution
School of Life Sciences
Lanzhou University
Lanzhou
China

De-Sheng Pei
School of Public Health
Chongqing Medical University
Chongqing
China

Fasiha Javaid
School of Public Health
Chongqing Medical University
Chongqing
China
Quaid-i-Azam University
Islamabad
Pakistan

Hossam El Din H. Abdelhafez
Mammalian and Aquatic Toxicology Department
Central Agricultural Pesticides Laboratory
Agricultural Research Center
Dokki
Giza
Egypt

Jun Wang
Joint Laboratory of Guangdong Province and Hong Kong Region on Marine Bioresource Conservation and Exploitation
College of Marine Sciences
South China Agricultural University
Guangzhou
China

Karunanithi Vidhya
PG & Research Department of Physics
Sri Sarada College for Women (Autonomous)
Salem
Tamil Nadu
India

Kousar Parveen
The Department of Environmental Sciences
The Women University Multan
Pakistan

Kusheng Wu
Shantou University Medical College
Shantou
China

Licheng Peng
Key Laboratory of Ecological Prewarning
Protection and Restoration of Bohai Sea
Ministry of Natural Resources
School of Life Sciences
Shandong University
Qingdao
China

Magdy Moheb El-Dein Saad
Department of Food Toxicology and Contaminants
Institute of Food Industries & Nutrition
National Research Centre
Dokki
Cairo
Egypt

Manoharan Saravanan
Department of Environmental Science
Periyar University
Department of Zoology
Padmavani Arts & Science College for Women (Autonomous)
Salem
Tamil Nadu
India

Marriya Sultan
Chongqing Institute of Green and Intelligent Technology
Chinese Academy of Sciences
Chongqing School
University of Chinese Academy of Sciences
Chongqing
China

Maryam Mumtaz
Department of Pathology
University of Veterinary and Animal Sciences
Lahore
Pakistan

Mehvish Mumtaz
College of Earth and Environmental Sciences
University of the Punjab
Lahore
Pakistan

Mohamed Bedair M. Ahmed
Department of Food Toxicology and Contaminants
Institute of Food Industries & Nutrition
National Research Centre
Dokki
Cairo
Egypt

Muhammad Azher Hassan
Tianjin Key Lab of Indoor Air Environmental Quality Control
School of Environmental Science and Engineering
Tianjin University
Tianjin
China

Muhammad Fahad Sardar
Key Laboratory of Ecological Prewarning
Protection and Restoration of Bohai Sea
Ministry of Natural Resources
School of Life Sciences
Shandong University
Qingdao
China

Muhammad Faheem
Department of Civil Infrastructure and Environmental Engineering
Khalifa University of Science and Technology
Abu Dhabi
United Arab Emirates

Muhammad Junaid
Joint Laboratory of Guangdong Province and Hong Kong Region on Marine Bioresource Conservation and Exploitation
College of Marine Sciences
South China Agricultural University
Guangzhou
China
Guangdong Provincial Key Laboratory of Utilization and Conservation of Food and Medicinal Resources in Northern Region
Shaoguan University
Shaoguan
China

Na Li
Key Laboratory of Drinking Water Science and Technology
Research Center for Eco-Environmental Sciences
Chinese Academy of Sciences
Beijing
China

Naima Hamid
Faculty of Science and Marine Environment
University Malaysia Terengganu
Kuala Nerus
Terengganu
Malaysia

Predrag Ilić
PSRI Institute for Protection and Ecology of the Republic of Srpska
Vidovdanska
Banja Luka
Republic of Srpska
Bosnia and Herzegovina

Qiang Yue
Guangdong Provincial Key Laboratory of Utilization and Conservation of Food and Medicinal Resources in Northern Region
Shaoguan University
Shaoguan
China

Ruijia Zhang
School of Environmental Science and Engineering
Hainan University
Haikou
China

Shakeel Ahmad
School of Environment
Tsinghua University
Beijing
China

Tariq Mehmood
Helmholtz Centre for Environmental Research - UFZ
Department of Environmental Engineering
Permoserstr
Leipzig
Germany

Tauseef Ahmad
Key Laboratory of Ecological Prewarning
Protection and Restoration of Bohai Sea
Ministry of Natural Resources
School of Life Sciences
Shandong University
Qingdao
China

Thodhal Yoganandham Suman
Department of Environmental Engineering
School of Smart and Green Engineering
Changwon National University
Changwon
Gyeongsangnamdo
Republic of Korea

Tiangang Luan
Guangdong Provincial Laboratory of Chemistry and Fine Chemical Engineering Jieyang Center
Jieyang
China
Smart Medical Innovation Technology Center
Guangdong University of Technology
Guangzhou
China
School of Environmental and Chemical Engineering
Wuyi University
Jiangmen
China
State key Laboratory of Biocontrol
School of Life Sciences, Sun Yat-sen University
Guangzhou
China

Ting Yang
Engineering Research Center of Key Technique for Biotherapy of Guangdong Province
Shantou University Medical College
Shantou
China

Wenlong Huang
Shantou University Medical College
Shantou
China

Xiang Ge
School of Environmental Science and Engineering
Institute of Environmental Health and Pollution Control
Guangdong University of Technology
Guangdong
China

Xiangsheng Hong
National Engineering Research Center of Industrial Wastewater Detoxication and Resource Recovery
Research Center for Eco-Environmental Sciences
Chinese Academy of Sciences
Beijing
China

Xiaojun Yang
Engineering Research Center of Key Technique for Biotherapy of Guangdong Province
Shantou University Medical College
Shantou
China

Xin Meng
Engineering Research Center of Key Technique for Biotherapy of Guangdong Province
Shantou University Medical College
Shantou
China

Xinyan Li
School of Biomedical and Pharmaceutical Sciences
Guangdong University of Technology
Guangzhou
China
Guangdong Provincial Laboratory of Chemistry and Fine Chemical Engineering Jieyang Center
Jieyang
China
Smart Medical Innovation Technology Center
Guangdong University of Technology
Guangzhou
China

Xuan Ma
Gansu Key Laboratory of Biomonitoring and Bioremediation for Environmental Pollution
School of Life Sciences
Lanzhou University
Lanzhou
China

Ya Liu
School of Environment
Hangzhou Institute for Advanced Study
University of Chinese Academy of Sciences
Hangzhou
China

Yin Liu
School of Environment
Hangzhou Institute for Advanced Study
University of Chinese Academy of Sciences
Hangzhou
China

Yingjie Xia
Division of Life Science
Hong Kong University of Science and Technology
Clear Water Bay
NT
Hong Kong
China

Yingxin Yu
School of Environmental Science and Engineering
Institute of Environmental Health and Pollution Control
Guangdong University of Technology
Guangdong
China

Yiyun Liu
School of Public Health
Chongqing Medical University
Chongqing
China

Yuequn Chen
Shantou University Medical College
Shantou
China

1

Introduction to Combined Toxicology, Background, Key Terminologies, and Significance

Yingjie Xia[a,*]

[a] *Division of Life Science, Hong Kong University of Science and Technology, Clear Water Bay, NT, Hong Kong, China*

*Corresponding author.
E-mail: yxiaas@connect.ust.hk

1.1 Introduction

Toxicology is the study of harmful effects of chemicals on living organisms. Traditionally, toxicology has focused on studying the effects of individual chemicals in isolation. However, in reality, human beings are exposed to multiple chemicals simultaneously, and effects or the toxicity of these chemical mixtures can differ from those of individual chemicals. This has led to the emergence of a new subfield of toxicology known as combination toxicology. Investigations into the toxicological characteristics of compound mixtures have been pursued over several decades. Initial research predominantly focused on simple combinations, such as binary combinations. Next, investigations expanded to include both explicitly defined mixtures consisting of more than two substances, and complex, undefined mixtures. Recently, researchers realized that the interpretation of toxicological effects is complicated due to the indeterminate composition of the mixture and the potential for significant variation in that composition. Nonetheless, this complexity is frequently overlooked in favor of a more simplistic understanding. In addition to the composition of the mixture, there are many factors that significantly affect its toxicity: the dosage level, dosage rate, the order of exposure, the method of administration, and the potential interaction among the components, all of which contribute significantly to the final toxicological response. Therefore, the assessment of combined toxicology would provide a realistic approach to evaluate the health risks of chemical cocktails that coexisted in the environment. In this chapter, we would like to provide the background of the development of combination toxicology, key terminologies, and its significance for assessing risks from environmental contaminants.

1.1.1 History and Drivers of Combined Toxicology

The advent of the chemical and industrial revolutions during the mid-nineteenth centuries led to extensive emissions of naturally present chemicals into the environment, alongside the manufacturing and discharge of novel substances that were previously nonexistent. This led directly to the emergence of several branches of toxicology, such as pharmacology, occupational toxicology, pesticide toxicology, and environmental toxicology. The formal concept of toxicology as an academic discipline was recognized around 1959, although the toxicological studies had been conducted throughout the early twentieth century [1, 2].

However, early toxicologic research were devoted to exploring the toxicity of individual chemicals, which makes it impossible to explain the exposure toxicity of complex mixtures of chemicals or pollutants at a time in the real world. In the environment, organisms are exposed to complex mixtures of chemicals from diverse sources such as industrial emissions, agricultural runoff, and consumer products. This has led to the growing recognition of the importance of studying chemical interactions and combined effects of multiple toxicants, known as combined toxicology. The early discussion of combined

Toxicological Assessment of Combined Chemicals in the Environment, First Edition. Edited by De-Sheng Pei and Yiyun Liu.

toxicity was defined as synergistic or antagonistic action, when the effect of the combination is weaker or greater than that suggested by the component toxic effects [3]. The study of combined toxicology can be traced back to the early twentieth century from studies in pharmacology on the combined toxicity of two drugs [4, 5]. However, it was not until the late twentieth century that scientists started systematically investigating mixture effects beyond the simple addition of individual component toxicities. Hence, research about combined toxicity has intensified over the past decades [6–8].

The intensive study of combined toxicology was driven primarily by two factors: first, rise of environmentalism and enactment of regulations. For instance, the Clean Air Act (CAA) revealed widespread pollution from complex chemical mixtures discharged into the air and water. Under the CAA, US Environmental Protection Agency (EPA) initiated to consider the combined effects of chemicals from various sources and has been mandated to incorporate information on air pollutants [9]. Under the Clean Water Act, the US EPA suggested that the cumulative effects of individual chemicals and chemical combinations should be addressed using approaches such as single-chemical testing, whole-effluent toxicity (WET) testing, and bioassays [10]. Measuring and interpreting the combined impacts of these contaminant cocktails posed new challenges for toxicologists. Second, advances in analytical chemistry allowed more accurate measurement of very-low-concentration mixtures in the parts-per-billion or parts-per-trillion range that organisms commonly encounter. These detection capabilities highlighted the need to study interactive toxic effects below normal individual "no observed adverse effect levels."

Nowadays, combination toxicology considers how chemical mixtures interact in biological systems and impact health. Examining chemical interactions, especially at concentrations that are relevant to the environment, represents a crucial stride in enhancing our comprehension of the effects of mixtures on human well-being and the environment. A major goal is to understand and predict the toxicity of environmental contaminant mixtures based on the individual constituent chemicals. This field has gained more and more attention due to uncertainties surrounding the additive, synergistic, or antagonistic effects that can occur when organisms are exposed to chemical mixtures.

1.1.2 Combined Toxicology on Environmental Pollutants

Environmental pollution is a major public health concern as various pollutants from industrial activities, vehicles, waste disposal, and agricultural sources are released into the environment, as shown in Figure 1.1. Exposure to these pollutants, which include heavy metals, pesticides, particulate matter (PM), and volatile organic compounds (VOCs), has been associated with increased risks of cardiovascular disease [11], neurological disorders [12], reproductive problems [13], respiratory disorders [14],

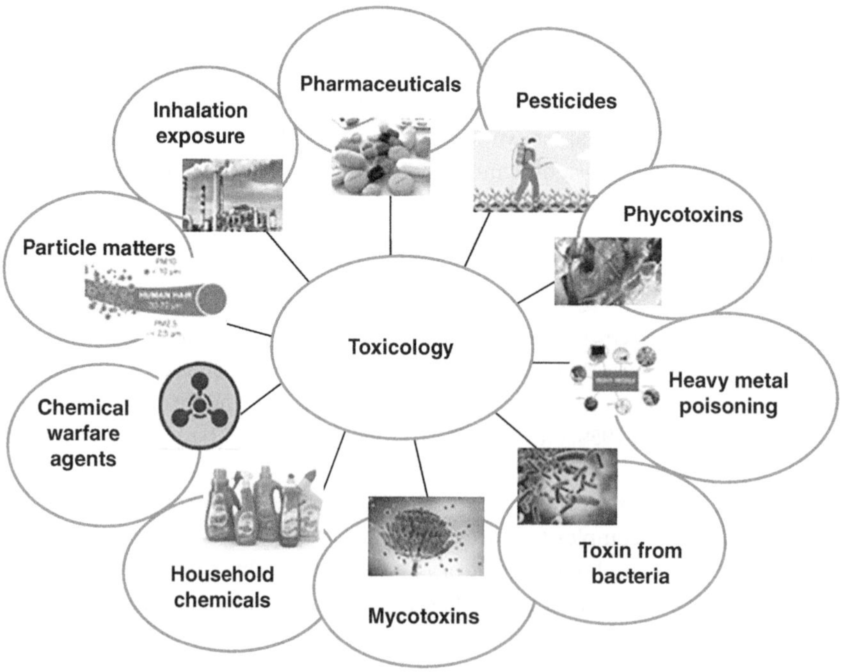

Figure 1.1 The scope of environmental toxicology [22, 23].

immunological diseases [15], and developmental disorders [16]. The most concerning pollutants include ozone (O_3), VOCs, nitrogen oxides (NO_x), PM, heavy metals, and pesticides such as organophosphates and pyrethroids [17–21]. Unfortunately, most toxicological research has focused on studying the effects of individual pollutants in isolation, which does not reflect real-world exposures. In reality, populations are exposed to mixtures of pollutants that may interact and produce synergistic toxic effects that are greater than the sum of individual components. Therefore, understanding the combined toxicology of real-world environmental pollutant mixtures is crucial for fully assessing health risks and guiding regulatory policies.

Here, the combined toxicology of major environmental pollutant mixtures has been classified according to the resources, including air pollution, pesticide mixtures, heavy metal–pesticide mixtures, and electronic waste mixtures. In addition, the advances and challenges in evaluating the combined toxicology of major environmental pollutant mixtures are discussed.

1.1.2.1 Air Pollution Mixtures

Air pollution is a complex mixture resulting largely from motor vehicle emissions, power generation, and industrial activities. Major hazardous components include particulate matter (PM2.5 and PM10), ground-level O_3, NOx, CO, and VOCs such as benzene and formaldehyde [24]. Epidemiological studies clearly link air pollution exposure to increased cardiovascular and respiratory mortality [25–27]. Controlled exposure studies in animals and humans demonstrate inflammation [28], oxidative stress [29], and pulmonary dysfunction [30] as underlying mechanisms. Interactions between air pollutants may enhance toxicity. For example, co-exposure to O_3 and PM2.5 in mice shows synergistic effects on inflammation and tissue damage compared to individual exposure [31]. This synergy likely results from oxidative stress caused by O_3 rendering lung tissues more susceptible to physical injury from PM2.5. Combined exposure to O_3 and PM2.5 also produces greater inflammatory effects and oxidative stress in rats than individual chemicals [32]. However, air pollution mixtures consist of numerous chemical compounds and vary with time. These compounds can interact with one another in the atmosphere, forming new compounds or changing their properties. The large number of chemicals and their interactions make it difficult to identify and quantify the toxic effects.

1.1.2.2 Pesticide Mixtures

Pesticides, such as organophosphates, carbamates, pyrethroids, and neonicotinoids, are used ubiquitously in agriculture. Given their acute neurotoxicity, most studies have focused on the neurological effects of individual pesticides [33]. However, several pesticides are applied together and interact to increase neurotoxicity in animal studies. In salmon, mixtures of organophosphates and carbamates inhibit cholinesterase enzymes additively compared with individual pesticides [34]. It was reported that formamidine pesticides, such as chlordimeform and amitraz, effectively synergize the toxic actions of certain pyrethroid and neonicotinoid insecticides in some insect species, which is based on the ability to activate the octopamine receptor [35]. Beyond acute toxicity, combined lifelong exposure to multiple pesticides may contribute to chronic neurodegenerative disease [36, 37]. The ability of pesticide mixtures to induce oxidative stress, inflammation, and mitochondrial dysfunction is likely an underlying mechanism for neurological damage [38–40]. Analyzing the role of pesticide mixtures is critical for understanding chronic neurodegenerative risks among agricultural populations.

1.1.2.3 Heavy Metal–Pesticide Mixtures

Heavy metals and pesticides are major food and water contaminants that can interact and increase toxicity. The combined toxicity of heavy metal–pesticide mixtures can have detrimental effects on ecosystems. The presence of heavy metals can affect the degradation and bioavailability of pesticides, altering their environmental fate and persistence. Furthermore, heavy metals can interact with pesticides in soil, leading to increased mobility and potential groundwater contamination. These mixtures can have adverse effects on various organ systems, including the nervous, respiratory, and immune systems [41, 42]. For instance, the combination of Cr and fenitrothion elevated testicular toxicity compared with that of single exposure in rats, which was mediated through the increased oxidative stress, androgenic hormone changes, and reduction of free radical scavenging capacity [43]. The combined Cd and chlorpyrifos could synergistically induce hepatic toxicity through the increased levels of cholesterol and triglyceride in the cells [44, 45]. Overall, these studies demonstrate the importance of considering metal–pesticide mixtures for accurately assessing health risks. More research is still needed on the toxicological interactions between heavy metals and the wide variety of pesticides and herbicides used in agriculture.

1.1.2.4 Electronic Waste Mixtures

Rapid generation of electronic waste (e-waste) containing complex chemical mixtures is an emerging environmental threat [46, 47]. E-waste is defined as those electronic devices which are broken, unwanted, or discarded by their users. Recycling e-waste to recover valuable metals using crude techniques in developing countries exposes workers and communities to mixtures of metals, flame retardants, and plastic additives. Developmental neurotoxicity has been recognized as a serious concern, but studies on the toxic effects and underlying mechanisms in humans are scarce [48]. Toxicological studies show that e-waste mixtures generate systemic DNA damage, oxidative stress, upregulation of inflammatory cytokines, and endocrine disruption at lower doses [49, 50]. More research is urgently needed on e-waste recycling mixtures to develop proper remediation strategies.

1.1.3 The Challenges in Analyzing the Combined Toxicity of Environmental Pollutant Mixtures

Tons of pollutants can be released into the environment, and the mixture could be much more complicated. Besides, the components of a mixture may be altered depending on previous use, the mixture, route of release into the environment, and time in the environment. Hence, it is not easy to assess the toxicity of environmental mixtures. Here are some key challenges in analyzing the combined toxicity of complex environmental mixtures: (1) Very large number of chemicals – environmental mixtures may contain hundreds or thousands of compounds, making it unfeasible to test every combination. This creates uncertainties in predicting interactions. (2) Complex interactions – chemicals can interact through multiple mechanisms like addition, synergism, or antagonism. These interactions are difficult to model without extensive data [51]. (3) Unknown mixture components – many chemicals in mixtures have not been identified, limiting the ability to predict toxicity. Analytical chemistry techniques are still developing to comprehensively profile mixtures. (4) Variability in mixtures – the composition of mixtures in different locations and over time is highly variable [52]. This makes it difficult to generalize results across studies. (5) Route-specific effects – toxicity can depend on the route of exposure (inhalation, ingestion, dermal) [53]. Most studies only assess one route, creating uncertainties in extrapolating real-world exposures. (6) Low-dose effects – mixtures may exert effects at very low doses of individual components that are missed by traditional testing [54, 55] and require more sensitive models like in vitro assays. (7) Inter-individual variability – factors like genetics, age, and health status affect susceptibility to mixtures, complicating determinations of "safe" levels [56]. (8) Limited experimental models – Current in vitro and animal models have limited ability to capture complex processes and low-dose effects relevant to mixtures. (9) Regulatory challenges – Policies focused on individual chemicals need to evolve to account for cumulative risks from mixtures. Lack of data is a barrier. Overall, addressing these challenges requires a combination of computational modeling, high-throughput screening methods, analytical chemistry advancements, and evolving regulatory guidance. Significant interdisciplinary research is still needed in this emerging area.

1.1.4 The Methodology to Analyze the Combined Toxicity of Environmental Pollutant Mixtures

How to assess the combined toxicity of environmental mixtures is an intractable and challenging question. At present, some high-throughput screening methods have been applied to analyze the toxicity of complex environmental mixtures.

1. Cell-based reporter gene assays: Cells are engineered to express a reporter gene like luciferase when specific pathways are activated, allowing rapid screening of mixtures for effects like endocrine disruption or oxidative stress. For example, the aryl hydrocarbon receptor (AhR)-mediated activity has been identified as one of the key targets in the toxicity of polycyclic aromatic hydrocarbons (PAHs). Hence, a human gene reporter AZ-AhR cell line was developed to evaluate the AhR-mediated activities of a lot of environmental PAHs and further perform the risk assessment of PAHs [57]. Similarly, the *CALUX* reporter gene was introduced to the human osteosarcoma cell line to investigate and characterize the endocrine-disrupting potencies of the chemical mixtures, such as the crude extracts of indoor dust [58]. However, these assays have certain limitations, e.g. single-target assessment, that reporter gene assays typically focus on a specific molecular target or pathway; lack of metabolic considerations; and absence of tissue-specific responses.
2. Microfluidic organ-on-a-chip models: Miniaturized systems recreate organ structures and functions to model the toxicity of mixtures at the tissue level. This model can combine tissue engineering and micro-manufacturing to simulate the critical physiological functions and environment of the human organs, which can be applied to assess the combined toxicity of chemical mixtures. For example, the gut-on-a-chip system has been established for drug screening because of the critical role of the gut in drug administration and absorption [59]. This application provides new directions to evaluate the toxicity of environmental pollutants [60]. The drawback of this methodology lies in the fact that the simulation of an individual organ fails to account for crucial biological phenomena that transpire within the intricate network of physiological processes in the human body. Consequently, this oversimplification significantly constrains the scope of inferences that can be derived.
3. Zebrafish embryos: Transparent zebrafish embryos can be used for rapid developmental toxicity screening of mixtures to assess their effects. The zebrafish embryo model is sensitive to quantifying the combined effects of chemical mixtures of

water contaminants [61]. For example, a transgenic zebrafish strain tg (cyp19a1b-GFP) has been developed to assess the endocrine disruption of five different oestrogenic chemicals [62], which was also predicted to serve as an available model to determine the combined toxicity of other pollutant mixtures.

4. Quantitative high-throughput screening (qHTS): Robotic systems conduct dose–response testing for mixtures across a wide concentration range, allowing the generation of large datasets for computational modeling [63].
5. High-content imaging: Automated microscopy combined with fluorescent dyes/tags allows multiple cellular endpoints such as oxidative stress, apoptosis, or mitochondrial damage to be rapidly assayed.
6. OMICS approaches: OMICS approaches techniques like transcriptomics, proteomics, and metabolomics profile global changes across thousands of biomolecules to identify mechanisms and biomarkers.
7. Computational modeling: Algorithms predict interactions between untested chemicals in mixtures using quantitative structure–activity relationship (QSAR) models.

Most studies have only examined interactions between a handful of chemicals, while real-world exposures involve hundreds of pollutants. High-throughput screening approaches combining computational toxicology with in vitro bioassays show promise for comprehensively profiling environmental mixtures. Advanced analytical chemistry techniques like nontargeted gas chromatography–mass spectrometry (GC/MS) and chromatography tandem mass spectrometry (LC/MS) can help exhaustively identify mixture components as a starting point. Overall, substantial research is still needed on developing biomarkers, nonanimal testing models, and computational approaches to predict synergies between untested chemicals. Moving forward, understanding the health risks of complex environmental mixtures requires an interdisciplinary approach combining environmental chemistry, toxicology, and public health expertise. Regulatory policies also need to evolve to account for cumulative risks from pollutant mixtures rather than individual chemicals. Implementing these science-based reforms is critical for protecting public health from complex environmental exposures.

1.2 Key Terminologies

Combined toxicology, also known as mixture toxicology or mixture risk assessment, is a multidisciplinary field that investigates the combined effects of multiple chemical substances on living organisms. Here, we aim to provide a comprehensive overview of key terminologies commonly used in combined toxicology research. It discusses essential concepts and definitions related to this field, such as mixture toxicity, interaction types, does–responses, and exposure assessment.

1.2.1 Mixture Toxicity

The first term is "mixture toxicity," which refers to the combined toxic effects of multiple chemicals. Combined toxicology focuses on understanding the potential interactions and cumulative effects that arise from exposure to mixtures of chemical substances. Mixture toxicity refers to the combined effects of two or more chemical substances in a mixture that differ from the effects of individual chemicals when tested alone. This term encompasses both synergistic (enhanced effects) and antagonistic (reduced effects) interactions between the components of the mixture.

1.2.2 Types of Combined Action

Studies on mixture toxicity or combined toxicity are often accompanied by statements about the type of combined action, which requires an understanding of the basic concepts of the combined toxicology, i.e. dose-addition, effect, or response-addition, and interaction such as synergism, potentiation, and antagonism. The main interaction types of the chemical mixtures include additivity, synergism, antagonism, and potentiation, which result in distinct effects of the mixtures, as shown in Table 1.1.

1.2.2.1 Additivity

Additivity refers to an event where the combined effect of two or more chemicals is equal to the sum of their individual effects. The chemicals act independently of each other to produce a combined effect that is predictable based on their individual effects. According to the way of action among the mixtures, additivity can be further divided into two types, including dose-addition and response-addition. Dose-addition refers to that mixture components contribute to toxicity as if they acted through a common mechanism, and toxic units are summed. The concept of dose-addition originally

Table 1.1 The interaction types of chemical mixtures.

Type of interaction	Toxic effects of chemical A (%)	Toxic effects of chemical B (%)	Combined effect of chemical A + B (%)	Notes
Additivity	20	30	50	A combination of two or more chemicals is the sum of the expected individual responses
Antagonism	20	30	<50	Two or more chemicals in combination have an overall effect that is less than the sum of their individual effects
Potentiation	0	30	50	A chemical that does not have a specific toxic effect makes another chemical more toxic
Synergism	20	30	>50	Combined two or more chemicals have an overall effect that is more than the sum of their individual effects

stated that when noninteracting chemicals are combined in a mixture, their behavior resembles that of diluted substances, and as a result, their effects can be quantitatively related using potency factors [64]. From a risk assessment perspective, the practical implication of dose-addition is that a mixture comprising two or more chemicals, each present at a concentration equating to half of its toxic threshold, is expected to induce a quantifiable toxic effect. The notable case to this concept is the approach to dealing with the carcinogenicity of dioxin [2,3,7,8-tetrachlorodibenzo-*p*-dioxin (TCDD)]-like chemicals since they share the same target organ via the same mechanism of action with 2,3,7,8-TCDD, the designated reference compound [65]. In contrast, response-addition assumes that the mixture components act through different pathways to produce an overall effect. Each contributes independently to the total response [66].

1.2.2.2 Nonadditivity

However, nonadditivity refers to a situation where the combined effect of two or more chemicals is different from the sum of their individual effects. Nonadditivity can be further divided into three categories: synergism, antagonism, and potentiation.

1.2.2.3 Synergism

Synergism refers to a situation where the combined effect of two or more chemicals is greater than the sum of their individual effects. In other words, the chemicals act together to produce a more potent effect than they would individually. Synergism is a fundamental principle in various fields, including biology, chemistry, drug development, and toxicology. Synergism can occur when two or more chemicals act on the same biochemical pathway or target organ. For example, if chemical A and chemical B both inhibit the same enzyme, their combined effect may be greater than the sum of their individual effects, e.g. the synergistic inhibition of berberine and palmatine on acetylcholinesterase activity [67]. Research on synergism has made significant advances in recent years, driven by advancements in technology, computational modeling, and interdisciplinary collaboration. One area where synergism has been extensively studied is in the field of pharmacology. Synergistic drug combinations have the potential to enhance therapeutic efficacy while minimizing side effects. Researchers have employed various experimental and computational approaches to identify and optimize synergistic drug combinations, leading to improved treatment strategies for various diseases, including cancer, infectious diseases, and neurological disorders. In parallel, the combined toxicity of chemical mixtures has been identified with a synergistic effect. Quantifying synergism can be challenging and depends on the specific context. In pharmacology, for instance, researchers often use the combination index (CI) or combination synergy index (CSI) to measure the degree of synergism between drugs. These indices are calculated based on the observed effects of individual drugs and their combination, taking into account factors such as dose–response curves and the nature of the interaction (e.g. additive, synergistic, or antagonistic). Various mathematical models, such as the Chou–Talalay method, have been developed to facilitate the calculation and interpretation of synergistic effects in drug combinations [68, 69]. According to the Chou–Talalay theory, the calculated values of CI are as follows: CI close to 1 = additive effect; $CI > 1$ = antagonistic effect; and $CI < 1$ = synergistic effect [70].

1.2.2.4 Antagonism

Antagonism refers to a situation where the combined effect of two or more chemicals is less than the sum of their individual effects. In other words, the chemicals cancel out each other's effects, resulting in a weaker effect than would be expected based on their individual effects. Antagonism can occur when two or more chemicals act on different biochemical pathways or target organs. For example, if chemical A and chemical B both cause liver damage but by different mechanisms, their combined effect may be less than the sum of their individual effects. Antagonism can also occur when one chemical counteracts the effects of another chemical, leading to reduced toxicity. Antagonism plays a significant role in combined toxicology, where the effects of exposure to multiple toxic substances are evaluated. The interaction between selenium and mercury is a typical environmental example of antagonism. Studies have shown that selenium decreases the assimilation of methylmercury in the body and reduces its toxicity. Selenium is thought to bind to or "tie up" the mercury so it cannot be taken up as easily. This binding prevents the mercury from interacting with important selenium-containing enzymes in the body that contain sulfur. These sulfur enzymes are involved in many of mercury's harmful effects. By holding onto the mercury, selenium protects these enzymes and lowers mercury's ability to harm the body. Hence, selenium is identified as an "antagonist" to mercury toxicity [71, 72]. This antagonism between mercury and selenium helps limit the damage from methylmercury exposure.

1.2.2.5 Potentiation

The definition of potentiation is similar to synergism, which refers to the fact that the combined effect of two or more chemicals is greater than the sum of their individual effects. However, potentiation is mostly applied to describe that a chemical that does not have a specific toxic effect makes another chemical more toxic. While potentiation and synergism are sometimes used interchangeably, they refer to distinct phenomena when describing chemical interactions. Some researchers define mixture toxicity as synergistic specifically when one constituent is present at a toxic concentration alone, but other components at subtoxic levels contribute to overall harmfulness [73]. In this case, the substances in subtoxic amounts would be inactive independently, yet boost the effects of the primary toxicant as part of the combination. Such compounds enhancing toxicity despite individual inefficacy would be labeled synergists. The key distinction is that potentiation involves one chemical altering the metabolic processing of another in a mixture, changing its inherent activity. Specifically, potentiation occurs when one partner impacts how the other is broken down or transformed inside an organism. This interaction at the biochemical level leads to magnified adverse outcomes greater than merely additive toxicities from each ingredient separately. Studies have reported that heavy metals, in particular, often have potentiated toxicity in mixtures [74]. On the other statement, this term is often used in the enhancement of allosteric effect. That is the positive allosteric modulators (PAMs) bind to certain agonists of receptors to change that receptor's response to stimuli. For example, galantamine, which acts as a PAM of nicotinic acetylcholine receptors, can potentiate the acetylcholine-induced microglial phagocytotic ability. However, galantamine alone does not influence this ability [75].

1.3 Significance

The study of combined toxicology is significant for several reasons. First, it helps us understand the potential health effects of exposure to multiple chemicals, which is more representative of real-world scenarios than exposure to individual chemicals in isolation. Second, it can help us identify situations where the combined effect of chemicals is greater than expected, which can have important implications for risk assessment and regulation. Third, it can help us identify situations where the combined effect of chemicals is less than expected, which can be useful for developing strategies to mitigate the harmful effects of chemical mixtures.

One example of the significance of combined toxicology is the study of endocrine-disrupting chemicals (EDCs). EDCs are chemicals that interfere with the normal functioning of the endocrine system, which is responsible for regulating many physiological processes in the body, including growth, development, reproduction, and metabolism. EDCs can mimic or block the effects of hormones, leading to a wide range of health effects, including developmental abnormalities, reproductive disorders, and cancer. EDCs are often found in mixtures, and the effects of these mixtures on human health are not well understood. However, studies have shown that the combined effects of EDCs can be different from the effects of individual EDCs. Another example of the significance of combined toxicology is the study of mixtures of pesticides. Pesticides are used to control pests in agriculture, but they can also have harmful effects on human health. The effects of pesticide mixtures can be difficult to predict because different pesticides can have different modes of action and target different organs or biochemical pathways in the body. However, studies have shown that the combined effects of pesticide mixtures can be greater than the effects of individual pesticides.

The significance of combined toxicology also extends to the regulation of chemicals. The traditional approach to chemical regulation has focused on individual chemicals, with little consideration given to the effects of chemical mixtures. However, the emerging field of combined toxicology suggests that this approach may not be sufficient to protect public health. Therefore, the study of combined toxicology can help inform the development of more effective regulatory frameworks that take into account the potential health effects of chemical mixtures.

Despite its importance, studying combined toxicology poses several challenges. One of the main challenges is the complexity of chemical mixtures. Living organisms are exposed to a wide range of chemicals, and the number of possible chemical combinations is vast. Additionally, the effects of chemical mixtures can be influenced by a wide range of factors, such as the dose, the timing of exposure, and the route of exposure. Therefore, studying the effects of chemical mixtures requires a multidisciplinary approach that combines chemistry, biology, and toxicology. Another challenge in studying combined toxicology is the lack of standardized methods for assessing the toxicity of chemical mixtures. Traditional toxicology studies have focused on assessing the toxicity of individual chemicals, and methods for assessing the toxicity of chemical mixtures are still in development. Additionally, the effects of chemical mixtures can be influenced by factors that are not well understood, such as the interactions between chemicals in the body, the effects of metabolites, and the effects of genetic variability. Finally, the study of combined toxicology is limited by the availability of data on chemical mixtures. Many chemicals have not been tested for their potential to interact with other chemicals, and the effects of many chemical mixtures on human health are not well understood. Additionally, the effects of chemical mixtures can be influenced by factors that are difficult to measure, such as the effects of chronic exposure or the effects of exposure to low doses over long periods of time.

In conclusion, the study of combined toxicology is an important subfield of toxicology that helps us understand the effects of chemical mixtures on living organisms. Mixture toxicity can be described in terms of additivity, synergism, and antagonism, and the significance of the study of combined toxicology lies in its ability to provide a more realistic understanding of the potential health effects of exposure to multiple chemicals and its implications for risk assessment and regulation. Despite its importance, studying combined toxicology poses several challenges, including the complexity of chemical mixtures, the lack of standardized methods for assessing the toxicity of chemical mixtures, and the limited availability of data on chemical mixtures. Therefore, future research in combined toxicology should focus on developing new methods for assessing the toxicity of chemical mixtures, improving our understanding of the factors that influence the effects of chemical mixtures, and generating more data on the effects of chemical mixtures on human health.

References

1 National Research Council (US) Committee on toxicology. The National Research Council's Committee on Toxicology: The First 50 Years 1947–1997. National Academies Press (US), Washington (DC), 1997.

2 Doull J. Toxicology comes of age. Annu Rev Pharmacol Toxicol. 2001, 41: 1–21.

3 Yang RS. Introduction to the toxicology of chemical mixtures. In: Yang RS (Ed.), Toxicology of Chemical Mixtures. Academic Press, 1994, pp. 1–10.

4 de Zwart D, Slooff W. Toxicity of mixtures of heavy metals and petrochemicals to Xenopus laevis. Bull Environ Contam Toxicol. 1987, 38(2): 345–351.

5 Diehl H, Gäting J, Habermann D. An experimental design for the simulation of combined-acting pollutants. Ecotoxicol Environ Saf. 1985, 9(3): 369–372.

6 Carpenter DO, Arcaro K, Spink DC. Understanding the human health effects of chemical mixtures. Environ Health Perspect. 2002, 110 (Suppl 1): 25–42.

7 Feron VJ, Cassee FR, Groten JP, van Vliet PW, van Zorge JA. International issues on human health effects of exposure to chemical mixtures. Environ Health Perspect. 2002, 110 (Suppl 6): 893–839.

8 Seed J, Brown RP, Olin SS, Foran JA. Chemical mixtures: current risk assessment methodologies and future directions. Regul Toxicol Pharmacol. 1995, 22(1): 76–94.

9 United States Environmental Protection Agency. The Plain English Guide to the Clean Air Act. EPA 400-K-93-001. U.S. Environmental Protection Agency, Washington, DC, 1993.

10 United States Environmental Protection Agency. Guidance for Water Quality Decisions: The TMDL Process. EPA 440/4-91-001. U.S. Environmental Protection Agency, Washington, DC, 1991.

11 Cosselman KE, Navas-acien A, Kaufman JD. Environmental factors in cardiovascular disease. Nat Rev Cardiol. 2015, 12: 627–642.

12 Aleya L, Uddin MS. Environmental pollutants and the risk of neurological disorders. Environ Sci Pollut Res Int. 2020, 27(36): 44657–44658.

13 Amir S, Shah STA, Mamoulakis C, Docea AO, Kalantzi OI, Zachariou A, Calina D, Carvalho F, Sofikitis N, Makrigiannakis A, Tsatsakis A. Endocrine disruptors acting on estrogen and androgen pathways cause reproductive disorders through multiple mechanisms: a review. Int J Environ Res Public Health. 2021, 18(4): 1464.

14 Vooicu G, Bechir A, Arghir OC. Chemical air pollution and the prevalence of respiratory disorders and diseases. J Environ Protect Ecol. 2012, 13: 1357.

15 Suzuki T, Hidaka T, Kumagai Y, Yamamoto M. Environmental pollutants and the immune response. Nat Immunol. 2020, 21(12): 1486–1495.

16 Ha S. Air pollution and neurological development in children. Dev Med Child Neurol. 2021, 63(4): 374–381.

17 Manisalidis I, Stavropoulou E, Stavropoulos A, Bezirtzoglou E. Environmental and health impacts of air pollution: a review. Front Public Health. 2020, 8: 14.

18 Mostafalou S, Abdollahi M. Pesticides: an update of human exposure and toxicity. Arch Toxicol. 2017, 91(2): 549–599.

19 Jaishankar M, Tseten T, Anbalagan N, Mathew BB, Beeregowda KN. Toxicity, mechanism and health effects of some heavy metals. Interdiscip Toxicol. 2014, 7(2): 60–72.

20 Hegazy A, Zaher M, Abd El-hafez M, Morsy R, Saleh R. Relation between exposure to pesticides and occurrence of lymphoid neoplasm. BMC Public Health. 2012, 12: 1–3.

21 Kim KH, Kabir E, Kabir S. A review on the human health impact of airborne particulate matter. Environ Int. 2015, 74: 136–143.

22 Gallo MA, Doull J. History and scope of toxicology. In: Klassen, CD (Ed.), Cassarett and Doull's Toxicology. The Basic Science of Poisons (Fifth Edition). McGraw-Hill, New York, 1996, pp. 3–11.

23 Sinha M, Sachan DK, Parthasarathi R. Artificial intelligence in clinical toxicology. In: Artificial Intelligence in Medicine. Springer International Publishing, Cham, 2021, pp. 1–15.

24 Kampa M, Castanas E. Human health effects of air pollution. Environ Pollut. 2008, 151(2): 362–367.

25 Brook RD. Cardiovascular effects of air pollution. Clinical Sci. 2008, 115: 175–187.

26 Al-Kindi SG, Brook RD, Biswal S, Rajagopalan S. Environmental determinants of cardiovascular disease: lessons learned from air pollution. Nat Rev Cardiol. 2020, 17(10): 656–672.

27 Hoek G, Krishnan RM, Beelen R, Peters A, Ostro B, Brunekreef B, Kaufman JD. Long-term air pollution exposure and cardio-respiratory mortality: a review. Environ Health. 2013, 12(1): 43.

28 Arias-Pérez RD, Taborda NA, Gómez DM, Narvaez JF, Porras J, Hernandez JC. Inflammatory effects of particulate matter air pollution. Environ Sci Pollut Res Int. 2020, 27(34): 42390–42404.

29 Lodovici M, Bigagli E. Oxidative stress and air pollution exposure. J Toxicol. 2011, 2011: 487074.

30 Sunyer J. Urban air pollution and chronic obstructive pulmonary disease: a review. Eur Respir J. 2001, 17(5): 1024–1033.

31 Lian Z, Qi H, Liu X, Zhang Y, Xu R, Yang X, Zeng Y, Li J. Ambient ozone, and urban PM2.5 co-exposure, aggravate allergic asthma via transient receptor potential vanilloid 1-mediated neurogenic inflammation. Ecotoxicol Environ Saf. 2022, 243: 114000.

32 Wang G, Zhao J, Jiang R, Song W. Rat lung response to ozone and fine particulate matter (PM2.5) exposures. Environ Toxicol. 2015, 30(3): 343–356.

33 Richardson JR, Fitsanakis V, Westerink RHS, Kanthasamy AG. Neurotoxicity of pesticides. Acta Neuropathol. 2019, 138(3): 343–362.

34 Scholz NL, Truelove NK, Labenia JS, Baldwin DH, Collier TK. Dose-additive inhibition of chinook salmon acetylcholinesterase activity by mixtures of organophosphate and carbamate insecticides. Environ Toxicol Chem. 2006, 25(5): 1200–1207.

35 Ahmed MA, Matsumura F. Synergistic actions of formamidine insecticides on the activity of pyrethroids and neonicotinoids against Aedes aegypti (Diptera: Culicidae). J Med Entomol. 2012, 49(6): 1405–1410.

36 Baltazar MT, Dinis-Oliveira RJ, de Lourdes Bastos M, Tsatsakis AM, Duarte JA, Carvalho F. Pesticides exposure as etiological factors of Parkinson's disease and other neurodegenerative diseases – a mechanistic approach. Toxicol Lett. 2014, 230(2): 85–103.

37 Kamel F, Hoppin JA. Association of pesticide exposure with neurologic dysfunction and disease. Environ Health Perspect. 2004, 112(9): 950–958.

38 Sharma H, Zhang P, Barber DS, Liu B. Organochlorine pesticides dieldrin and lindane induce cooperative toxicity in dopaminergic neurons: role of oxidative stress. Neurotoxicology. 2010, 31(2): 215–222.

39 Min N, Park H, Hong T, An G, Song G, Lim W. Developmental toxicity of prometryn induces mitochondrial dysfunction, oxidative stress, and failure of organogenesis in zebrafish (*Danio rerio*). J Hazard Mater. 2023, 443(Pt A): 130202.

40 Mokarizadeh A, Faryabi MR, Rezvanfar MA, Abdollahi M. A comprehensive review of pesticides and the immune dysregulation: mechanisms, evidence and consequences. Toxicol Mech Methods. 2015, 25(4): 258–278.

41 Wallace DR, Djordjevic AB. Heavy metal and pesticide exposure: a mixture of potential toxicity and carcinogenicity. Curr Opin Toxicol. 2020, 19: 72–79.

42 Alengebawy A, Abdelkhalek ST, Qureshi SR, Wang MQ. Heavy metals and pesticides toxicity in agricultural soil and plants: ecological risks and human health implications. Toxics. 2021, 9(3): 42.

43 El-Demerdash FM, Jebur AB, Nasr HM, Hamid HM. Modulatory effect of *Turnera diffusa* against testicular toxicity induced by fenitrothion and/or hexavalent chromium in rats. Environ Toxicol. 2019, 34(3): 330–339.

44 He W, Guo W, Qian Y, Zhang S, Ren D, Liu S. Synergistic hepatotoxicity by cadmium and chlorpyrifos: disordered hepatic lipid homeostasis. Mol Med Rep. 2015, 12(1): 303–308.

45 Chen L, Qu G, Sun X, Zhang S, Wang L, Sang N, Du Y, Liu J, Liu S. Characterization of the interaction between cadmium and chlorpyrifos with integrative techniques in incurring synergistic hepatoxicity. PLoS One. 2013, 8(3): e59553.

46 Saha L, Kumar V, Tiwari J, Rawat S, Singh J, Bauddh K. Electronic waste and their leachates impact on human health and environment: global ecological threat and management. Environ Technol Innov. 2021, 24: 102049.

47 Perkins DN, Brune Drisse MN, Nxele T, Sly PD. E-waste: a global hazard. Ann Glob Health. 2014, 80(4): 286–295.

48 Chen A, Dietrich KN, Huo X, Ho SM. Developmental neurotoxicants in e-waste: an emerging health concern. Environ Health Perspect. 2011, 119(4): 431–438.

49 Tahir M, Rehman MYA, Malik RN. Heavy metal-associated oxidative stress and glutathione S-transferase polymorphisms among E-waste workers in Pakistan. Environ Geochem Health. 2021, 43(11): 4441–4458.

50 Issah I, Arko-Mensah J, Agyekum TP, Dwomoh D, Fobil JN. Electronic waste exposure and DNA damage: a systematic review and meta-analysis. Rev Environ Health. 2021, 38(1): 15–31.

51 Groten JP, Feron VJ, Sühnel J. Toxicology of simple and complex mixtures. Trends Pharmacol Sci. 2001, 22(6): 316–322.

52 Castro EN, Lomonte B, del Carmen Gutiérrez M, Alagón A, Gutiérrez JM. Intraspecies variation in the venom of the rattlesnake *Crotalus simus* from Mexico: different expression of crotoxin results in highly variable toxicity in the venoms of three subspecies. J Proteomics. 2013, 87: 103–121.

53 Shakeel M, Jabeen F, Shabbir S, Asghar MS, Khan MS, Chaudhry AS. Toxicity of nano-titanium dioxide (TiO2-NP) through various routes of exposure: a review. Biol Trace Elem Res. 2016, 172(1): 1–36.

54 Cavieres MF, Jaeger J, Porter W. Developmental toxicity of a commercial herbicide mixture in mice: I. Effects on embryo implantation and litter size. Environ Health Perspect. 2002, 110(11): 1081–1085.

55 Welshons WV, Thayer KA, Judy BM, Taylor JA, Curran EM, vom Saal FS. Large effects from small exposures. I. Mechanisms for endocrine-disrupting chemicals with estrogenic activity. Environ Health Perspect. 2003, 111(8): 994–1006.

56 Miura N. Individual susceptibility to cadmium toxicity and metallothionein gene polymorphisms: with references to current status of occupational cadmium exposure. Ind Health. 2009, 47(5): 487–494.

57 Vondráček J, Pěnčíková K, Neča J, Ciganek M, Grycová A, Dvořák Z, Machala M. Assessment of the aryl hydrocarbon receptor-mediated activities of polycyclic aromatic hydrocarbons in a human cell-based reporter gene assay. Environ Pollut. 2017, 220(Pt A): 307–316.

58 Suzuki G, Tue NM, Malarvannan G, Sudaryanto A, Takahashi S, Tanabe S, Sakai S, Brouwer A, Uramaru N, Kitamura S, Takigami H. Similarities in the endocrine-disrupting potencies of indoor dust and flame retardants by using human osteosarcoma (U2OS) cell-based reporter gene assays. Environ Sci Technol. 2013, 47(6): 2898–2908.

59 Beaurivage C, Naumovska E, Chang YX, Elstak ED, Nicolas A, Wouters H, van Moolenbroek G, Lanz HL, Trietsch SJ, Joore J, Vulto P, Janssen RAJ, Erdmann KS, Stallen J, Kurek D. Development of a gut-on-a-chip model for high throughput disease modeling and drug discovery. Int J Mol Sci. 2019, 20(22): 5661.

60 Yang S, Chen Z, Cheng Y, Liu T, Lihong Yin, Pu Y, Liang G. Environmental toxicology wars: organ-on-a-chip for assessing the toxicity of environmental pollutants. Environ Pollut. 2021, 268(Pt B): 115861.

61 Schmidt S, Busch W, Altenburger R, Küster E. Mixture toxicity of water contaminants-effect analysis using the zebrafish embryo assay (*Danio rerio*). Chemosphere. 2016, 152: 503–512.

62 Petersen K, Fetter E, Kah O, Brion F, Scholz S, Tollefsen KE. Transgenic (cyp19a1b-GFP) zebrafish embryos as a tool for assessing combined effects of oestrogenic chemicals. Aquat Toxicol. 2013, 138–139: 88–97.

63 Shukla SJ, Huang R, Austin CP, Xia M. The future of toxicity testing: a focus on in vitro methods using a quantitative high-throughput screening platform. Drug Discov Today. 2010, 15(23-24): 997–1007.

64 Greco WR, Bravo G, Parsons JC. The search for synergy: a critical review from a response surface perspective. Pharmacol Rev. 1995, 47(2): 331–385.

65 Borgert CJ, Quill TF, McCarty LS, Mason AM. Can mode of action predict mixture toxicity for risk assessment? Toxicol Appl Pharmacol. 2004, 201(2): 85–96.

66 Bliss CI. The toxicity of poisons applied jointly. Ann Appl Biol. 1939, 26: 585–615.

67 Mak S, Luk WW, Cui W, Hu S, Tsim KW, Han Y. Synergistic inhibition on acetylcholinesterase by the combination of berberine and palmatine originally isolated from Chinese medicinal herbs. J Mol Neurosci. 2014, 53(3): 511–516.

68 Chou TC, Motzer RJ, Tong Y, Bosl GJ. Computerized quantitation of synergism and antagonism of taxol, topotecan, and cisplatin against human teratocarcinoma cell growth: a rational approach to clinical protocol design. J Natl Cancer Inst. 1994, 86(20): 1517–1524.

69 Ashton JC. Drug combination studies and their synergy quantification using the Chou-Talalay method – letter. Cancer Res. 2015, 75(11): 2400.

70 Chou TC. Drug combination studies and their synergy quantification using the Chou–Talalay method. Cancer Res. 2010, 70(2): 440–446.

71 Luque-Garcia JL, Cabezas-Sanchez P, Anunciação DS, Camara C. Analytical and bioanalytical approaches to unravel the selenium-mercury antagonism: a review. Anal Chim Acta. 2013, 801: 1–13.

72 Sørmo EG, Ciesielski TM, Øverjordet IB, Lierhagen S, Eggen GS, Berg T, Jenssen BM. Selenium moderates mercury toxicity in free-ranging freshwater fish. Environ Sci Technol. 2011, 45(15): 6561–6566.

73 Heys KA, Shore RF, Pereira MG, Jones KC, Martin FL. Risk assessment of environmental mixture effects. RSC Adv. 2016, 6: 47844–47857.

74 Utgikar VP, Chaudhary N, Koeniger A, Tabak HH, Haines JR, Govind R. Toxicity of metals and metal mixtures: analysis of concentration and time dependence for zinc and copper. Water Res. 2004, 38(17): 3651–3658.

75 Takata K, Kitamura Y, Saeki M, Terada M, Kagitani S, Kitamura R, Fujikawa Y, Maelicke A, Tomimoto H, Taniguchi T, Shimohama S. Galantamine-induced amyloid-{beta} clearance mediated via stimulation of microglial nicotinic acetylcholine receptors. J Biol Chem. 2010, 285(51): 40180–40191.

2

Toxicokinetics of Chemical Mixture Exposure in the Environment

*Abdul Wahab Hussain[a,**], Mehvish Mumtaz[a*,**], Asad Ullah Saeed[b], Maryam Mumtaz[c]*

[a] *College of Earth and Environmental Sciences, University of the Punjab, Lahore, Pakistan*
[b] *Pakistan Council for Scientific and Industrial Research, Lahore, Pakistan*
[c] *Department of Pathology, University of Veterinary and Animal Sciences, Lahore, Pakistan*

*Corresponding author.
**First two authors shared equal contributions
E-mail: mehvish.cees@pu.edu.pk; mehvish.env@gmail.com (M.M.)

2.1 Unveiling Toxicokinetics

Toxicokinetics is a specialized branch of toxicology that deals with absorption, distribution, and elimination processes [1]. Considering the toxicological principles, nil exposure leads to no danger. Toxicokinetics is information about the kinetics of absorption, distribution, metabolism, and excretion of a xenobiotic in circumstances of toxicity evaluation. Toxicokinetics is becoming increasingly important in assessing toxicity and interpreting toxicological data. It is a valuable tool in discovering and developing new products [1]. Toxicokinetics is the terminal point that informs about the penetration into and fate within the body of a toxic substance, including the possible appearance of metabolites [2].

2.2 Toxicokinetics and Chemical Mixture Assessment

When assessing chemical mixtures, a comprehensive analysis of exposure helps to accurately determine the cumulative impact of compounds. To initiate chemical risk management, it is crucial to understand the extent and duration of exposure [3]. Several factors greatly influence the course of action of drugs. The drug's extent of absorption or volume of distribution determines whether the chemical effects are perilous, harmless, or advantageous. Additional factors that can influence the toxicity of the chemical, apart from the amount, include the method of ingestion, the duration and frequency of exposure, variations among distinct species, and differences between various species (Figure 2.1) [4].

2.2.1 Approaches Adopted for Chemical Mixture Assessment

2.2.1.1 A Toxicity-driven Approach

The concepts are derived from the mechanism of action, wherein a combination of mixtures is identified to have significantly increased toxicity due to the additive or synergistic method of mixture toxicity.

2.2.1.2 An Exposure-driven Approach

This approach kindles for parallel exposure in which the components that may result in similar directions are examined for amplified toxicity. The reason is that they are mutually supportive. It is impossible to signal out one of these methods as optimal, and a combo is preferable. If possible, the method starts with the concentration addition for mixture toxicity. Consequently, risk description may signal toxic effects that amplify and add to any component's toxicity [5].

Toxicological Assessment of Combined Chemicals in the Environment, First Edition. Edited by De-Sheng Pei and Yiyun Liu.

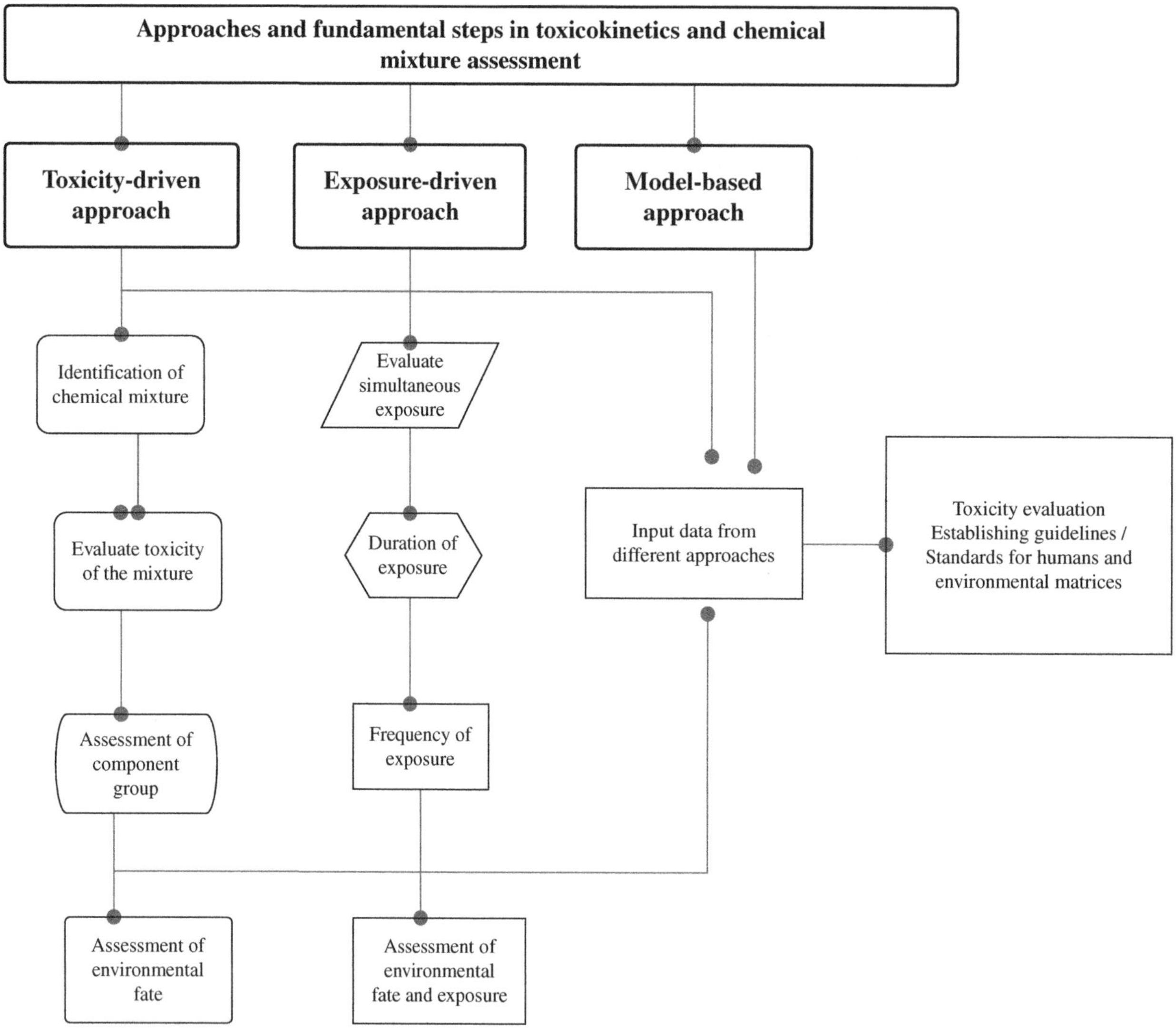

Figure 2.1 Steps/approaches adopted in chemical mixture assessment.

Two exposure assessment methods are used to evaluate the effects of chemical mixtures and to assess environmental risk exposure to both humans and surroundings.

- The concepts are derived from the mechanism of action, wherein a combination of mixtures is identified to have significantly increased toxicity due to the additive or synergistic method of mixture toxicity. The direct approach includes evaluating exposure at the point of interaction or updating by assessing chemical density in people or environment (i.e. food, air, and water).
- An indirect approach uses illustration and predicting techniques to estimate exposure levels. Indirect contact evaluation starts with the emission data, foreseeing the environmental chemicals' fate and concluding concentrations in the different ecological ecosystems [3].

2.2.1.3 Modeling-based Approach

Physiologically based toxicokinetic (PBTK) modeling is particularly well-suited to calculate tissue doses of chemicals and their metabolites in diverse species across various exposure scenarios. The compartments in PBTK models correspond to discrete tissues or sets of tissues with good volumes and blood flow rates and test chemical metabolism pathways. Relevant biochemical and physicochemical parameters for metabolism and tissue solubility are explicitly incorporated into the descriptions of each tissue compartment in PBTK models. Routes of administration are considered in their relationship to the overall physiological structure. Variances in exposure scenarios are accounted for in the temporal sequence of the dose input terms [6].

2.2.2 The Fundamental Steps Involved in Chemical Mixture Assessment

Exposure is determined by factors impacting the chemical's bioavailability, such as binding strength to soil or sediment particles. Chemical-to-chemical interactions can have an impact on availability. The individual behavior of creatures, including humans, may also influence exposure. Techniques for determining orientation primarily comprise residue analysis in exposure media such as air, water, soil, food, or tissues of exposed organisms. Furthermore, these monitoring systems typically focus on single compounds rather than mixtures. These long-term and large-scale data sets can give critical information for validating emission scenario results and environmental fate models [7].

2.2.2.1 Determination of the Composition of the Mixture

The composition of the mixture might change both in time and space. Measurements taken at the source (point of emission) may aid in determining (variations in) the mixture composition [8].

2.2.2.2 Component Group Selection (Optional)

Different (groups of) components within a mixture might be identified. These components can be classified based on the features that influence their environmental fate. The material's volatility, solubility, density qualities, and the nature of the hazard provided by the substance define its conduct and final 'fate' (toxicity, flammability, reactivity, explosive, corrosive, etc.). Aggregation of property data is relevant to each category. This step entails gathering properties beneficial to predict the destiny of various (groups of) components in the environment [9].

2.2.2.3 Models of Environmental Fate and Exposure

Each group's environmental fate is assessed. Fate models can forecast the ecological destiny of mixture components at various spatial scales. Such models can anticipate air, water, soil, and sediment distribution. Environmental fate and exposure models effectively integrate information on chemicals, their partitioning and degrading behavior, ecological scenarios, and emissions to create a picture of chemical distribution and fluxes in the multimedia environment [10].

2.3.2.4 Evaluation of Environmental and Human Vulnerability

Finally, concentrations for each of the (groups of) mixture components can be determined in various exposure media (inhaled air, ingested water, food items) or environmental compartments (soil, sediment, air, and surface or groundwater). This information, however, may need to be completed: only a portion of the total concentration in an ambient compartment is often biologically available for uptake by organisms. Furthermore, species behaviors and individual behavior may influence the nature of exposure. Third, life-stage-specific factors may be crucial in determining vulnerability to mixtures; this element has been researched extensively for human exposure, but it is also relevant to ecological evaluation for some taxa [11].

2.3.2.5 Constructing Standards

Toxicology data from animal experiments and epidemiological research are utilized to develop exposure guidelines. The approach used to create a policy is determined by the type of chemical, the duration, and the frequency of exposure. Understanding the difference between an experimental dose (mg/kg) and an environmental concentration (mg/m^3 or ppm) is also critical. To help make safe conclusions, exposure guidelines are supplied as concentrations that can be compared to concentrations detected by air monitoring equipment. The data collected from different approaches in the past may help to construct the guidelines and standards for different chemicals that are present in the environment and may harm humans and the environment directly and indirectly (Figure 2.1).

2.3 List of Toxic Chemicals/Pollutants Present in the Environment

Any substance, regardless of its physical state, has the potential to become hazardous when it is outside its natural environment or when it is present in higher-than-normal quantities. However, chemical pollution is considered a pollutant only when it accumulates to a level that causes harm to the ecosystem. Currently, there is an extensive array of industrial chemicals available in the market that pose a significant threat to the environment and its inhabitants [12]. A noticeable list of

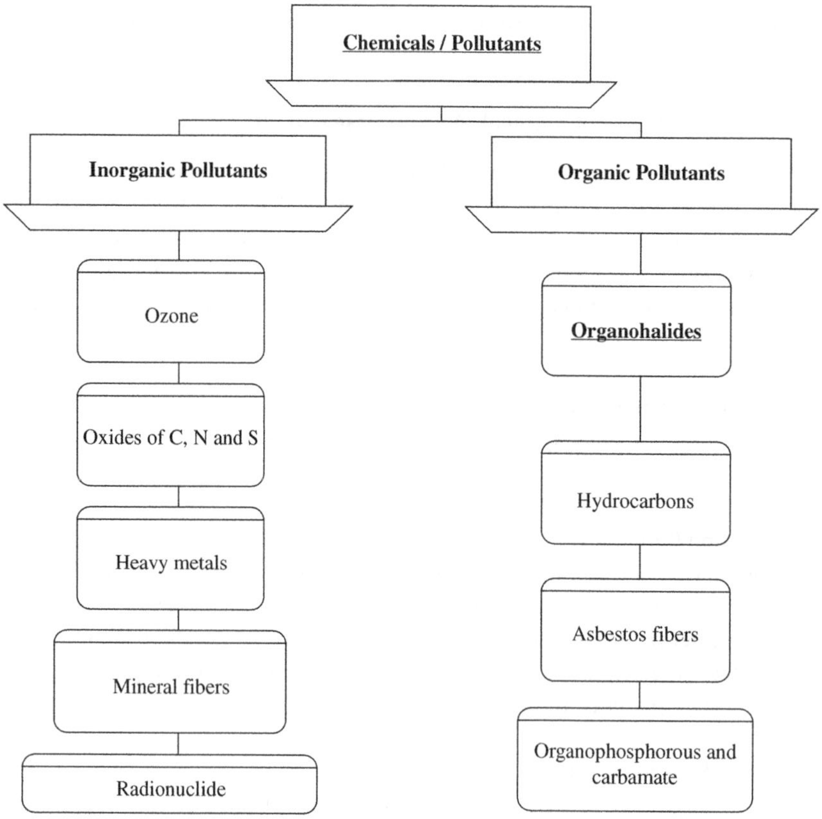

Figure 2.2 The list of compounds present in the environment.

pollutants is present in any environment as a mixture (Figure 2.2). Here, we will interpret the course of action, i.e. the toxic behavior of organic halogen compounds and mixtures when living organisms are exposed [13].

2.4 Toxicokinetics of Selected Organohalide Compounds/Mixtures

Organohalides are classified as halogenated organic compounds or chemicals widely manufactured and used for industrial and agricultural purposes. At least 5000 organohalides originating from synthetic or natural resources have been discovered. Organohalides are a constant risk to the health of individuals and the environment because of their inherent stability, bioaccumulation, long-distance migration, toxicity, and even carcinogenicity. The United States Environmental Protection Agency (USEPA) enlisted 126 important pollutants that contain 69 organic halides. The Stockholm convention list of persistent organic pollutants (POPs) had 34 halogenated organic compounds – the most persistent environmental toxicants because of their distinct physical and chemical properties. From the long list of organohalides so far, certain compounds were selected and toxic risks were discussed. The selected list is given in Figure 2.3 [14].

2.4.1 Chlorinated Organic Compounds

Several of the highest priority pollutants, such as dichlorodiphenyltrichloroethane (DDT), polychlorinated biphenyls (PCBs), polychlorinated dibenzo-*p*-dioxins (PCDDs), and polychlorinated dibenzofurans, are among the chlorinated organic compounds (COCs) that have been widely employed in numerous industrial processes involving high temperatures and chlorine, such as combustion in municipal solid waste incinerators, metallurgical operations, chlorine gas bleaching of pulp, and industrial chemical systems [15].

DDT is one of the most well-known and researched chlorinated hydrocarbon pollutants. DDT is still present in the environment and life-forms, apart from a ban on its manufacture in the United States established in the early 1970s. DDTs

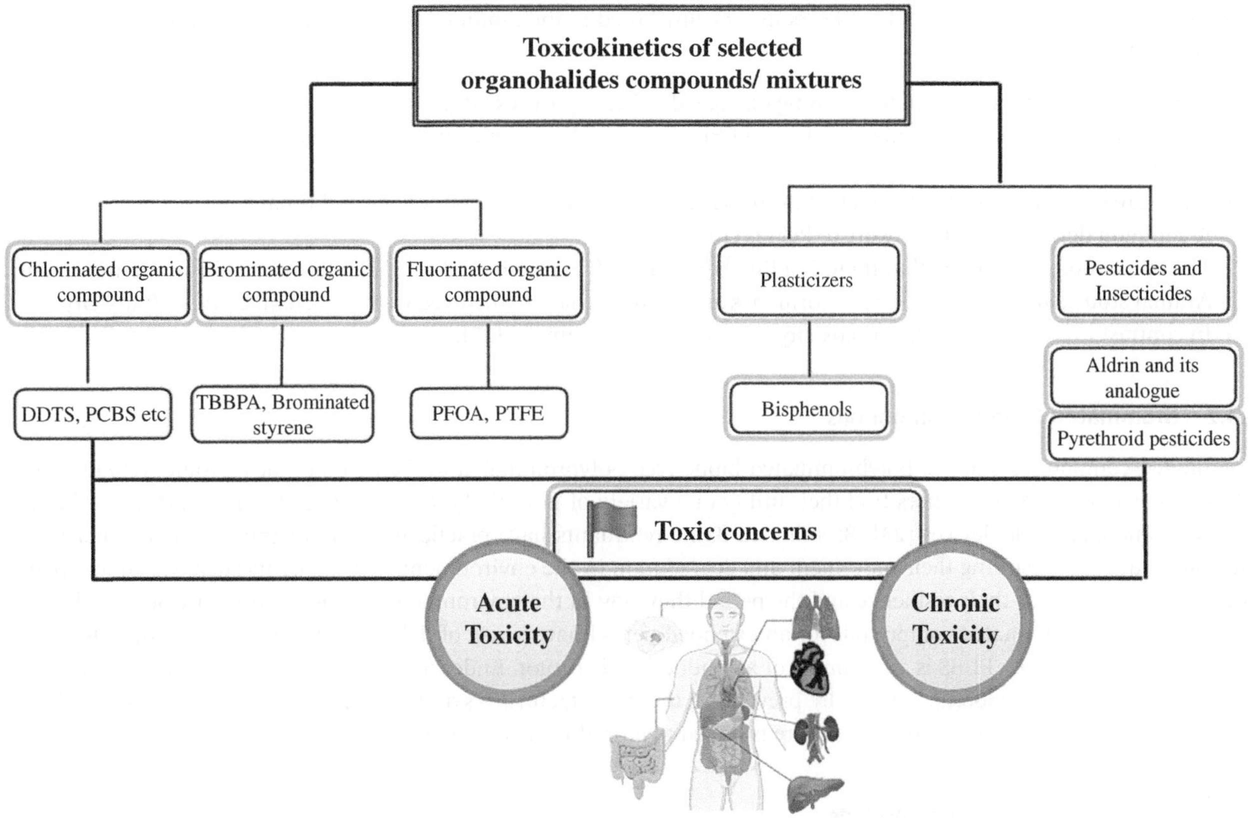

Figure 2.3 List of selected toxic compounds.

lipophilic nature allows it to accumulate and remain in lipid-rich biota tissues, and its biomagnification in the food chain is a significant concern. DDT, a highly persistent and bioaccumulative insecticide, is transformed in organisms into the even more stable metabolites dichlorodiphenyldichloroethylene (DDE) and dichlorodiphenyl dichloroethane (DDD). Depending on the dominant mechanisms, such impacts can be channel-specific. DDT is classified as a B2 carcinogen by the EPA. This implies that DDT has been shown in laboratory animals to cause cancer, but there is insufficient evidence that it may cause cancer in humans [16]. DDT is metabolized in the body into various breakdown products. When no more DDT exposures exist, the breakdown products are progressively eliminated from the body at about 1% of the accumulated DDT daily. Furthermore, when fat reserves are utilized during periods of starvation, DDT breakdown products are released into the bloodstream, where they are harmful to the liver and brain system. Within a population, its quantity in human breast milk is analyzed to assess DDT's influencing exposure. As DDT stalks in fat, it is likely in the population's blood and breast milk samples globally. However, there are no grounds that tell us that DDTs stalked in the body would influence human infant birth weight. DDT can be eliminated in the urine, stool, or breast milk once accumulated. DDT exposure can be detected in the blood and fat, where its presence is anticipated [16].

Polychlorinated biphenyls (PCBs) are a class of legacy chemicals that are still dangerous to human and environmental health [17]. They were created through bulk chlorination of biphenyl, resulting in complicated chemical combinations containing 140–150 of the 209 potential PCB congeners. PCBs are chemically and thermally stable and hence stay in the environment. Like POPs, PCBs bioconcentrate in living creatures and biomagnify in terrestrial and marine food chains due to their lipid solubility. Due to growing environmental and human health concerns, their industrial manufacturing was outlawed in the late 1970s. Recent studies, however, show that PCBs are erroneously produced as part of some industrial processes, such as the creation of paint pigments and adhesives [18].

PCBs are converted to phenols (through arene oxide intermediates) by the microsomal monooxygenase system catalyzed by cytochrome P-450, which can then be conjugated or further hydroxylated to generate a catechol. They can covalently bind to nucleophilic biological macromolecules (e.g. protein, DNA, RNA) and trigger DNA strand breaks and repair, contributing to PCB toxicity. Furthermore, arene oxide intermediates can be conjugated with glutathione and converted to generate methyl sulfonyl metabolites, which have been found in human serum and tissue samples and laboratory animals. Some of the harmful effects of PCBs may be attributed to the binding of methyl sulfonyl metabolites to specific proteins [19].

- PCBs disturb the reproductive functioning on considerable exposures [20].
- PCBs exposure has been recognized as an agent causing neurobehavioral and developmental impairments in neonates in the uterus [21].
- There are solid grounds that tell us that exposure to PCBs causes cutaneous effects in humans.
- A common dermal sign of exposure to PCB is chloracne.
- The epidemiology suggests that there is a link between PCB exposure and thyroidal hormone poisoning in humans. Animal studies indicate that thyroid hormones are the ones that are involved in the mechanism of PCB toxicity.
- In contrast to animals, over-hepatotoxicity is uncommon in humans [22].

2.4.2 Brominated Organic Compounds

Brominated compounds such as polybrominated biphenyls, polybrominated diphenyl ether, hexabromocyclododecane, tetrabromobisphenol A, and others find their utility in a variety of household items such as clothing, textiles for homes, furniture, and electronic devices [23]. Brominated flame retardants have practicality in industries and consumer items; they raised queries regarding their movement and degradation in the environmental components such as soil and plants and animals because of their resilience and the period they stay in the environment. Pregnant women who have been in contact with polybrominated compounds for an extended period have a risk of delivering babies with abnormalities and developmental disorders. PBBS is an example of an endocrine disruptor. Endocrine disruptors are a group of compounds that interact with the endocrine system by preventing the manufacturing synthesis, binding, or transport of vital hormones. Increased endocrine disruptor exposure is linked to reproductive disorders [23].

2.4.3 Fluorinated Organic Compounds

Per and polyfluoroalkyl substances (PFASs) have been manufactured since the 1940s. They have distinctive physicochemical properties; thus, they are commercialized and used in industrial and commercial applications such as foam surfactants [24–26]. These partially fluorinated compounds are widespread in the environment [27]. Indeed, once it enters the environment, transformation products are generated and a few of them transform into persistent long-chain carboxylates and sulfonates [25, 28–30]. Besides these transformation products, the list of synthetic PFASs is also getting extensive. To date, more than 4000 PFASs have been reported [31]. Long-chain PFASs (by definition, perfluoroalkyl carboxylic acids [PFCAs] with eight or more carbon (C) atoms or perfluoroalkane sulfonates [PFSAs] with six or more carbon atoms) [25], like perfluorooctane sulfonate (PFOS) and perfluorooctanoic acid (PFOA), are now declared as persistent, bioaccumulative, and toxic compounds; they rapidly spread into aquatic environment[32–34], causing adverse effects on humans [32, 35, 36]. Over time, questions were raised (carcinogenic effects, persistence), and most of the long-chain compounds were banned. The PFOS and PFOA were also listed in the Stockholm Convention [37]. Besides the phase-out initiative around the world (Figure 2.4), the substitution of long chains with short chains was also ongoing. The leading companies around the globe (Chemours, Dynax, Solvay, Daikin, Arkema, etc.) focus on telomerization technology. The predominant C8 chemistry was replaced, therefore manufacturing the C6 telomere-based products. Owing to the extensive use of telomerization technology, precursors are also widespread in the environment [38, 39]. There are some known precursors reported in the literature that are transformed into PFOS. Few telomere-based precursors exist that are transformed into PFCAs (particularly PFOA) through biodegradation [30, 40], and ample evidence was found regarding health effects [32, 35, 41]. Nevertheless, these phase-out initiatives and substitutions are still under debate (Madrid, Helsingør, and Zürich statements) [42–45] and questionable [46]. In this chapter, toxicity of two compounds (i.e. PFOS and PFOA) is considered.

2.4.3.1 Perfluorooctanoic Acid

PFOA is an octanoic acid whose derivatives are produced by fluorine substitution in aliphatic hydrocarbons [47]. Its salts are used for industrial products or commercial usage. It is also resistant to biodegradation and biotransformation. It is found in the general population and occupationally exposed individuals [48]. According to a study on occupational workers exposed to the compounds, PFOA's half-life in humans is 4.37 years. It has a detrimental effect on human health upon accumulation [49]. The toxicokinetics of PFOA is unknown. In experimental animals, it has low toxicity. It is well absorbed

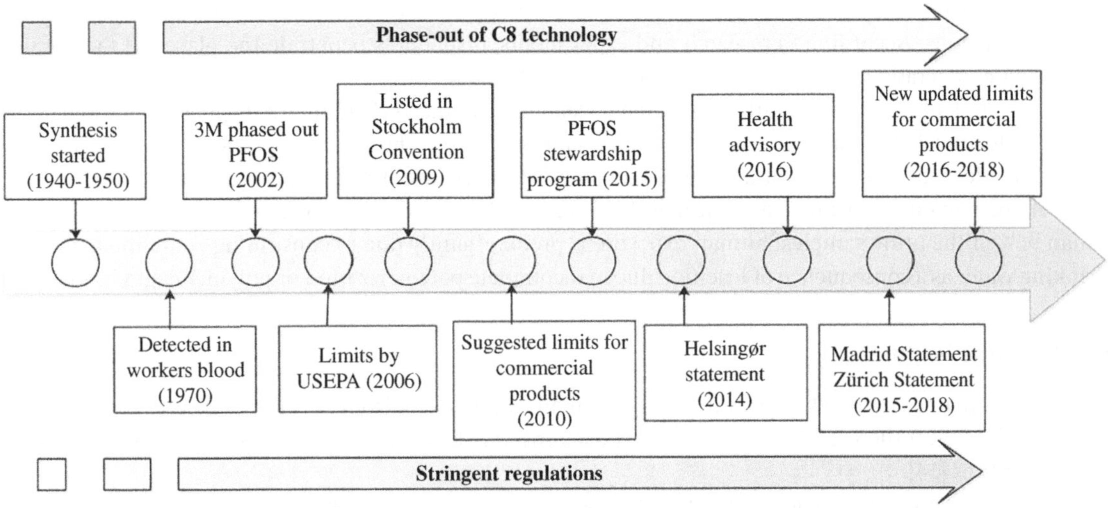

Figure 2.4 Global concerns and initiatives regarding PFASs.

through oral and respiratory routes and, to a lesser extent, by the dermal way. Once absorbed, it is distributed to the liver and plasma and, to a lesser extent, to the kidneys and lungs. It is excreted in urine and feces. The half-life of PFOA is different in different sexes and different species due to additional renal clearance. In rats, renal clearance is regulated by the sex hormone, i.e. testosterone. It is excreted by active urine secretion, and anion transporters are involved in this function. PFOA is also passed through fecal excretion. Inhibition of the transport mechanism leads to its elimination and detoxification in the human body [50].

2.4.3.2 Perfluorooctane Sulfonic Acid

PFOS is a fluorinated negatively charged ion with molecular formula C8F17SO2Y where Y=OH negative ion or other anions. PFOS is a persistent and bioaccumulated chemical having a uniquely specific course of action. These are surprisingly stable compounds and cannot be easily degraded under extreme circumstances like high temperatures and pressure. PFOS chemicals are likely to bind with proteins when they enter the living body, thereby their accumulative potential is related to their ability to bind with proteins. PFOS are hydrophobic and lipophobic surfactants that find a handsome utility in the industry. According to the UN Stockholm Convention, a wide variety of utilities have been reported in textiles, leather products, metallic plating, food packaging, firefighting foams, floor polishes, denture cleansers, coatings, etc. To extrapolate its toxicokinetics in humans and top mammals, we have to hire biofouling and modeling approaches in different species because bioaccumulation is associated with environmental pollutants. Now, we will discuss the toxicokinetics of PFOS on different organs of mammals. The liver is the most affected organ by biofouling of PFOS as the liver is majorly the site of detoxification and metabolism. They cause various dysfunctionalities in the liver like hepatic steatosis, hepatomegaly, hepatocellular hyperplasia, and oxidative damage to hepatocytes. Other experimentations have shown that lying vulnerable to PFOS results in motor behavior, learning memory, and cognition dysfunction. PFOS can cause aerobic damage in nerve cells. The in vivo and in vitro studies have pointed out that PFOS can disrupt immune cells' expansion, polarity, and routine functionality. Too much vulnerability to PFOS can disrupt the anatomy and function of thyroid glands and thyroid hormones. Laying open to PFOS can lead to myocardial infarction, disturbed heart rate, induction of cellular mortality, and death of cardiomyocytes.

2.4.4 Plasticizers (Bisphenols)

Plasticizers are minimal molecular weight polymers that enhance the space between crystalline polymer chains, making them more malleable; hence, harder plasticizers are used in materials to soften and make them more pliable. Some plasticizers are more important or influential than others at this level [51].

Bisphenols (BP) are one of the most massively produced anthropogenic substances in the world. They find their utility in manufacturing polycarbonate plastic and epoxy resin varnishes for food and beverage containers. Phenols are also present in medical and dental instruments, building materials, thermal receipt paper, and children's toys, with a half-life of 5 hours. Consumed BPA is promptly converted into the liver and excreted in bile or urine. Despite its fast metabolism and excretion, BPA is pernicious. According to research and experiments, bisphenols penetrate the placental membrane and accumulate in neonates at a ratio higher than in the maternal bloodstream [52].

This could be due to the underdeveloped detoxification defenses leading to reduced clearance of bisphenols from the fetal compartments, as grounds provided by the investigations in pregnant mammals. The fetus is significantly jeopardized by the disruptive endocrine effects of bisphenols and other alien compounds because it is going through vital development processes and stages of organ maturation and endocrine axis development. Resilient and persistent in our environment and present in more than 92% of the urine samples, human exposure is predominantly due to consuming contaminated food, beverages, and drinking water as a consequence of leaching due to incomplete polymerization or polymer degradation [53].

2.4.5 Pyrethroid Pesticide Mixtures

Pesticides are chemical compounds used to kill pests that affect humans. Pesticide poisoning is a significant cause of death in developing countries. Every year there are 3 million cases of poisoning and 20000 deaths; most (about 99%) of the deaths occur in third-world countries [54].

Pyrethroids are a group of synthetic pesticides. Pyrethroid pesticides are commonly applied insecticides for residential purposes in the United States. Survey data have shown recurrent pyrethroids in childcare facilities and residential areas. Their presence is a cause of concern as most children spend their time indoors. Parathyroid contains 1–3 chiral carbons and is divided into two groups depending on the presence or absence of cyano group at alpha carbon of alcohol. Both groups are neurotoxic to mammalian species. Pyrethroids cause excessive saliva secretion/choreoathetosis syndrome [55]. Pyrethroids are detrimental, as chronic exposure can cause endocrine dysfunction, hepatic function loss, and respiratory damage. Oxidative stress, lipid peroxidation, and allergy may also cause toxicity [56]. Long-term exposure to permethrin at different concentrations and alternative reproductive hormones leads to significant elevation of inflammatory indicators, as well as inhibition of testicular tissue functions [57]. The toxicokinetic behavior of organic pollutants is shown in Table 2.1.

Table 2.1 The toxicokinetic behavior of selected organohalides.

Pollutants	Salient characteristics	General utility	Toxicokinetic behavior	Ref.
DDT	Environmental hydrocarbon pollutants	A component widely used in insecticides	Increases the risk of cancer	[58, 59]
	Bioaccumulative due to lipophilic nature	Used to combat mosquitos' proliferation	Disrupts the liver and brain anatomy and functionality	
	Highly persistent chemical		Causes skin diseases and eye irritation	
PCBs	Chemically and thermally stable compounds	Component in flame retardants	Disturbs reproductive functioning	[60, 61]
	Bioaccumulative chemicals	Also used widely in electrical appliances	Causes neurobehavioral and developmental impairments	
		Also used in ink and coating fluids	Causes thyroidal gland dysfunctioning	
PBBs	Resilient organic compound	Used in the textile industry, electronic devices, and flame retardants	Endocrine disruptor	[62, 63]
	Halogenated		Cause reproductive disorders	
	environmental pollutant		Disrupt normal functionality of gland-secreting hormones and increase the risk of cancerous division	
PFOA	Stable organic pollutants resistant to biodegradation and biotransformation	Widely used in industrial products	Causes detrimental effects on the human body's reproductive system and developmental effects on the organ development in the fetus	[49]
PFOS	Persistent and bioaccumulative chemicals	Used in floor polishing	Causes dysfunction in motor behavior	[64, 65]
	Hydrophobic and lipophilic in nature	Used in denture cleansers add electroplating	Damages the liver anatomy and functionality	
			Causes myocardial diseases	

Pollutants	Salient characteristics	General utility	Toxicokinetic behavior	Ref.
Bisphenol A	Pernicious compound Resilient and persistent	Used in the manufacturing of epoxy resins and food packaging	Disrupts the endocrine functionality Causes abnormality in the fetal developmental stages	[66, 67]
Pyrethroid	Synthetic pesticides and poisonous compounds	Commonly used in residential insecticides	Neurotoxic to mammalian species and leads to choreoathetosis syndrome Causes liver impairment Results in defective detoxification of the xenobiotics and disrupts glycogenesis	[68, 69]
Aldrin and dieldrin	Organochlorine pesticide and carcinogenic	Used vastly in the field of agriculture	Increase the risk of cancer Cause poisoning in humans Exposure leads to gastrointestinal damage	[70, 71]
Atrazine and its analogue	Herbicide and respiration inhibitor	Widely used as pesticides	Disrupt the endocrine functionality Lead to impairment in the liver and kidney	[4, 72]
Neonicotinoid insecticides and their analogue	Persistent in nature	Mostly used in agricultural fields	Involved in cytotoxicity, genotoxicity, and impaired immune functioning; researches indicate its involvement in increasing the risk of hepatocarcinogenicity Directly influence the glandular secretions and lead to syndromes	[73, 74]

2.4.6 Organochlorine Pesticides and Their Analogues

Compounds with one or more covalent bonds and chlorine atoms are known as organochlorines. These exist in various compositions and chemical properties. Organochlorines have more density than water as they have more atomic mass [75]. These can be formed by using hydrogen, chlorine, and chlorinating agents. Living beings can get it through their skin, lungs, and gut wall. Skin easily allows cycloidei, hexachlorocyclohexane, endosomes, and lindane to pass through. Their absorption gets more significant with fat and fat solvents. They are evaporative, so most of the part is found in fat tissue and discharged from urinary tracts and lipophilic compounds from maternal milk (Figure 2.5). These can cause hyper-excitable states in the brain, causing tremors, convulsions, etc. More significant usage of organochlorine pesticides pollutes the environment because of their nonbiodegradable nature. They have highly venomous effects on all phyla, including human beings. Finally, using organochlorine pesticides should be reduced to enhance the health conditions of humans and the environment [76].

2.4.7 Aldrin and Its Analogues

The far-ranging insecticides, primarily contact, stomach, and inhalation poisons, are aldrin and dieldrin, which are chlorinated cycloidei [77]. Aldrin is being transmuted to dieldrin, which is principally observed as one of the most tenacious among all pesticides. For about 20 years, these insecticides were used significantly in the field of agriculture until their utility was suspended by EPA (1974). Still, the use of them in controlling termites was retained. The applications of aldrin and dieldrin are being outlawed in many countries. Still, despite the fact these insecticides were persistently being mass-produced in several European countries until 1978, they are still used throughout the world [71].

Major motor convulsions from aldrin and dieldrin cause human poisoning. They were followed by other principal effects comprising malaise, incoordination, headache, dizziness, and gastrointestinal disturbances. The episode of seizures may or may not accompany other poisoning side effects. Acute intoxication is nearly reported, involving fatalities, from ingestion of aldrin and dieldrin.

Aldrin and dieldrin are carcinogenic in mice, leading to increased liver tumors. By epoxidation of the double bond, aldrin is readily converted to 6,7-epoxide dieldrin. This type of reaction is a microsomal oxidation reaction in the soil,

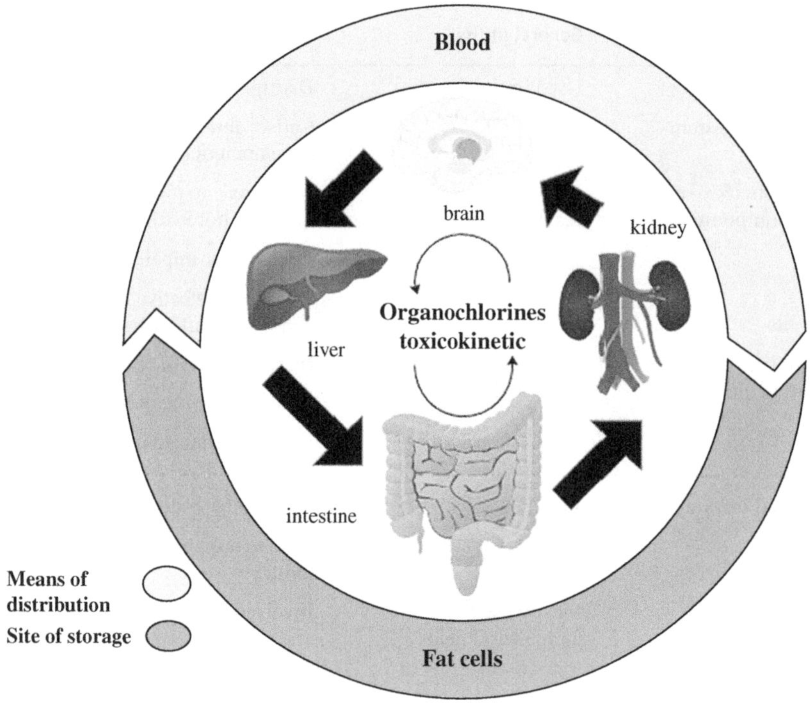

Figure 2.5 The toxicokinetic behavior of organochlorines.

plants, and animals. This reaction is more toxic than its parent compound form as it also yields photo dieldrin. Hydroxy degradation products of aldrin are usually affixed and conjugated before being excreted [78]. When dieldrin was administered as a single dose to rats and mice by gavage, 50–70% was eliminated in 1 week. Most radioactivity was detected in the feces – 10 times more than in the urine. Aldrin and dieldrin are piled up in adipose tissue. For workers exposed to aldrin and dieldrin, concentrations in adipose tissue and plasma may vary slightly from 5.67 ppm and 0.0185 ppm, respectively. Dieldrin has been found in the fatty tissue of the general mass in various countries, where it is usually deposited (Figure 2.5) [79].

2.4.8 Atrazine and Its Analogues

Chloro-*s*-triazines are a class of herbicides structurally similar but differ only in specific *s*-triazine 4- and 6-*N*-alkyl substituents. They all work by blocking mitochondrial respiration. Currently used in agriculture are atrazine, cyanazine, propazine, simazine, and terbuthylazine. Terbuthylazine is the leading triazine herbicide in the EU, although atrazine was the most widely used pesticide in the United States in 1997 and is currently the second most commonly used pesticide [80]. Atrazine has been the subject of extensive investigation and concern in the United States. It has been used for an extended period in the United States and evaluated by the USEPA, with the most recent assessment being completed in 2019. Current metabolites or degradation products in drinking water have put terbuthylazine on the agenda of the European Food Safety Authority (EFSA), leading to the creation of an EFSA Panel on Plant Protection Products and their Residues (PPR) Scientific Opinion [81].

Atrazine herbicide toxicokinetic research on humans has yet to be conducted for apparent reasons, and regulatory animal (rat) studies are the primary source of toxicokinetic knowledge. Due to study designs centered on clarifying mass balance or an elementary parent and metabolite extraction as a function of time, further in-depth toxicokinetic data are only sometimes available. As a result, it is rarely possible to interpret more accurate toxicokinetic data, such as the parent or primary metabolite's clearance or half-life. This might help integrate and interpret species differences or extrapolate to the human situation in regulatory risk assessment [72].

The World Health Organization (WHO) and central regulatory agencies' assessment and regulatory materials were used to compile the data [82]. Atrazine absorbs rapidly in the human body, ultimately disrupting the endocrine system by distributing erythrocytes, liver, and kidney. Atrazine is usually disposed of through a series of metabolic pathways; that is why atrazine has a long path. The excretion rate of atrazine is less than half (50%) within a day in urine, which gradually decreases in the case of excretion through feces [83].

The rate of absorption of cyanazine is high, and it does not target any of the specific organs or tissues. The part of the body being affected depends on the pathway through which this chemical is exposed to the body. The processes involved in the metabolism of cyanazine or detoxifying the harmful impacts of this chemical affect a chain of reactions indicating the extensive metabolism of cyanine [84]. Propazine also has a rapid absorption rate and no specific target and can affect any part of the body exposed to it. It also involves a series of reactions during its metabolism showing extensive metabolism; 82–95% of propazine can be excreted within 48 hours through urine, and this rate gradually decreases when there is excretion through feces [84].

Simazine, a herbicide, also absorbs rapidly in animals, especially humans. It can cause many health effects but is mainly distributed in the spleen, liver, and kidney. It has an extensive metabolism, and the excretion rate is usually higher in urine than in the excretion through the fecal pathway [85]. Terbuthylazine has rapid absorption, but the extent rate is usually less than 60%. This herbicide is mainly distributed in the kidneys and liver and can be transported through the blood, causing many health impacts in animals like humans. It also has a series of metabolic pathways, and the excretion rate is 75% within 24 hours through bile, and this rate gradually decreases when there is excretion through urine and feces.

The toxicokinetics of chloro-*s*-triazines are comparable, with absorption, distribution, biotransformation, and excretion via urine and feces, with bile being a primary excretory route. In humans, biomarkers in bodily fluids can be used to estimate exposure, with desethylatrazine and diamino chlorotriazine being the significant metabolites of occupational exposure to atrazine. Additionally, dealkylated metabolites of triazine herbicides can be formed and excreted in the urine, providing class exposure information. However, measuring the intact pesticide in blood products is rare [86].

2.4.9 Neonicotinoid Insecticides and Their Analogues

Neonicotinoid insecticides are extensively used in urban and agricultural fields around the globe. They have been considered noble replacements for more toxic chemicals like organophosphate due to their perceived finite probability of affecting the environment and human health. This review investigates neonicotinoids' environmental influence and toxicity, their metabolites, and their potential exposure harms. Neonicotinoids are prevalent in the environment, drinking water, and food, with lesser exposure usually documented below tolerable daily intake standards. Toxicological data from animal studies indicate that cytotoxicity, genotoxicity, impaired immune response, and reduced growth and reproduction are possibly at low concentrations. Also, limited data from ecological or cross-sectional epidemiological studies suggest that the associated health hazards, both acute and chronic diseases, which may range from acute respiratory, cardiovascular, and neurological symptoms to more severe oxidative genetic damage and congenital disabilities are eminent [87]. Because neonicotinoids are vastly used and have the potential for cumulative chronic exposure, these insecticides represent novel risks. There is a dire need for further studies to explore the hazards humans can face. The presence of actinoids in the food has led to an increased focus on combined toxicity with other insecticides. It has also been reported that they disrupt the central nervous system. At lower concentrations, they can stimulate the central nervous system, but at higher concentrations, receptor blockade, paralysis, and death may occur. Neonicotinoids are more toxic to insects than vertebrates. Nevertheless, harmful impacts have also been seen in the nontarget insects residing on land and water. Moreover, honeybees have exhibited neurodevelopmental toxicity and immunotoxicity. According to USEPA, neonicotinoid insecticides have toxic effects on mammals' liver, kidneys, thyroid, and immune systems [88].

The toxicity assessment of neonicotinoid insecticides was done based on cell viability assays and combined threshold analysis. The effect was found more on Sf-9 cells (culture cells) than the impact on SK-N-SH cells. A combination of pesticides is more effective than using a single pesticide. Based on the evidence, there is a limited adverse effect if neonicotinoids are combined. The negative effects include reproductive, sperm, and thyroid abnormalities. The liver causes the detoxification of neonicotinoids. The single-neonicotinoid treatment can increase the liver's weight, edema, and hepatocarcinogenicity. The mixing of acetamiprid causes a change in the morphology of the liver resulting in vacuolar degeneration and dilation of vessels and sinusoids in mice. The combined effect was more than using a single pesticide of the same concentration. There is a wide use of neonicotinoid combinations, and these mixtures are being detected in human bodies using them. They can target various organs of mammals. The mechanism of their toxicity warrants investigation of a metabolic pathway, biological response, and impact on CYP450s (cytochrome cells). It has become a premarket requirement for pesticide manufacturers due to relevant data. The hazards were systematically assessed at a realistic environmental concentration of exposure, providing practical guidelines to prevent adverse effects of combining neonicotinoids with other pesticides in humans.

2.5 Conclusions

Toxicokinetics is a crucial field that significantly contributes to the evaluation of risks, assessment of dose–response, and analysis of exposure frequency for chemical substances and mixtures. The utilization of toxicokinetic evaluation models and methodologies has facilitated the identification of the harmful impacts of chemical mixtures such as DDTs, PCBs, PBBs, PFOS, PFAS, and others on living beings. Conducting toxicokinetic studies on these aforementioned compounds enables us to establish guidelines and gain a deeper understanding of the direction in which these chemicals are taking us. To reduce the magnitude of the risk, it is possible to monitor the frequency and dosage of the chemicals. Harmful chemical mixtures should be avoided or replaced with nontoxic or less hazardous alternatives whenever possible. The presence of a combination of chemical components, notably persistent organic compounds, significantly raises the likelihood of chronic illnesses, neurodevelopmental abnormalities, and cancer in humans, specifically. Moreover, the interplay among the harmful chemical combinations can lead to diverse reactions, which vary based on disparities in chemical characteristics and the body's response mechanisms.

References

1 Dixit R, Riviere J, Krishnan K, Andersen ME. Toxicokinetics and physiologically based toxicokinetics in toxicology and risk assessment. J Toxicol Environ Health B Crit Rev. 2003, 6(1): 1–40.
2 Humpage A. Toxin types, toxicokinetics and toxicodynamics. Adv Exp Med Biol. 2008, 619: 383–415.
3 Spurgeon DJ, Jones OA, Dorne JL, Svendsen C, Swain S, Stürzenbaum SR. Systems toxicology approaches for understanding the joint effects of environmental chemical mixtures. Sci Total Environ. 2010, 408(18): 3725–3734.
4 Abass K, Pelkonen O, Rautio A. Chloro-s-triazines-toxicokinetic, toxicodynamic, human exposure, and regulatory considerations. Curr Drug Metab. 2021, 22(8): 645–656.
5 McCarty LS, Borgert CJ. Review of the toxicity of chemical mixtures: theory, policy, and regulatory practice. Regul Toxicol Pharmacol. 2006, 45(2): 119–143.
6 Haddad S, Charest-Tardif G, Tardif R, Krishnan K. Validation of a physiological modeling framework for simulating the toxicokinetics of chemicals in mixtures. Toxicol Appl Pharmacol. 2000, 167(3): 199–209.
7 Krishnan K. Chapter 47 - Physiologically based pharmacokinetic models in the risk assessment of developmental neurotoxicants. In: Slikker W, Paule MG, C Wang (Eds.), Handbook of Developmental Neurotoxicology (Second Edition). Academic Press, 2018, pp. 539–557.
8 Punt A. Toxicokinetics in risk evaluations. Chem Res Toxicol. 2018, 31(5): 285–286.
9 Fardel O, Kolasa E, Le Vee M. Environmental chemicals as substrates, inhibitors or inducers of drug transporters: implication for toxicokinetics, toxicity and pharmacokinetics. Expert Opin Drug Metab Toxicol. 2012, 8(1): 29–46.
10 Becka M, Urfer W. Statistical aspects of inhalation toxicokinetics. Environ Ecol Stat. 1996, 3: 51–64.
11 van Birgelen AP, van den Berg M. Toxicokinetics. Food Addit Contam. 2000, 17(4): 267–273.
12 He H, Li Y, Shen R, Shim H, Zeng Y, Zhao S, Lu Q, Mai B, Wang S. Environmental occurrence and remediation of emerging organohalides: a review. Environ Pollut. 2021, 290: 118060.
13 Viau C. Biological monitoring of exposure to mixtures. Toxicol Lett. 2002, 134 (1–3): 9–16.
14 Sarigiannis DA, Karakitsios S, Dominguez-Romero E, Papadaki K, Brochot C, Kumar V, Schuhmacher M, Sy M, Mielke H, Greiner M, Mengelers M, Scheringer M. Physiology-based toxicokinetic modelling in the frame of the European Human Biomonitoring Initiative. Environ Res. 2019, 172: 216–230.
15 Persson, Y. Chlorinated organic pollutants in soil and groundwater at chlorophenol-contaminated sawmill sites. 2007, Kemi.
16 Lotufo GR, Landrum PF, Gedeon ML, Tigue EA, Herche LR. Comparative toxicity and toxicokinetics of DDT and its major metabolites in freshwater amphipods. Environ Toxicol Chem. 2000, 19(2): 368–379.
17 Mackay DPD, Shiu WY, Ma KC. Monoaromatic Hydrocarbons, Chlorobenzenes, and PCBs. Lewis Publishers, 1992.
18 Drouillard KG, Norstrom RJ. Dietary absorption efficiencies and toxicokinetics of polychlorinated biphenyls in ring doves following exposure to Aroclor® mixtures. Environ Toxicol Chem. 2000. 19(11): 2707–2714.
19 Kania-Korwel I, Lehmler HJ. Toxicokinetics of chiral polychlorinated biphenyls across different species--a review. Environ Sci Pollut Res Int. 2016, 23(3): 2058–2080.
20 Cogliano VJ. Assessing the cancer risk from environmental PCBs. Environ Health Perspect. 1998, 106(6): 317–323.

21 Golden R, Doull J, Waddell W, Mandel J. Potential human cancer risks from exposure to PCBs: a tale of two evaluations. Crit Rev Toxicol. 2003, 33(5): 543–580.

22 McFarland VA, Clarke JU. Environmental occurrence, abundance, and potential toxicity of polychlorinated biphenyl congeners: considerations for a congener-specific analysis. Environ Health Perspect. 1989, 81: 225–239.

23 Hakk H, Letcher RJ. Metabolism in the toxicokinetics and fate of brominated flame retardants--a review. Environ Int. 2003, 29(6): 801–828.

24 Qin Z, McNee D, Gleisner H, Raab A, Kyeremeh K, Jaspars M, Krupp E, Deng H, Feldmann J. Fluorine speciation analysis using reverse phase liquid chromatography coupled off-line to continuum source molecular absorption spectrometry (CS-MAS): identification and quantification of novel fluorinated organic compounds in environmental and biological samples. Anal Chem. 2012, 84(14): 6213–6219.

25 Buck RC, Franklin J, Berger U, Conder JM, Cousins IT, de Voogt P, Jensen AA, Kannan K, Mabury SA, van Leeuwen SP. Perfluoroalkyl and polyfluoroalkyl substances in the environment: terminology, classification, and origins. Integr Environ Assess Manag. 2011, 7(4): 513–541.

26 Wang Z, DeWitt JC, Higgins CP, Cousins IT. A never-ending story of per- and polyfluoroalkyl substances (PFASs)? Environ Sci Technol. 2017, 51(5): 2508–2518.

27 Ye F, Tokumura M, Islam MS, Zushi Y, Oh J, Masunaga S. Spatial distribution and importance of potential perfluoroalkyl acid precursors in urban rivers and sewage treatment plant effluent--case study of Tama River, Japan. Water Res. 2014, 67: 77–85.

28 Butt CM, Muir DC, Mabury SA. Biotransformation pathways of fluorotelomer-based polyfluoroalkyl substances: a review. Environ Toxicol Chem. 2014, 33(2): 243–267.

29 Prevedouros K, Cousins IT, Buck RC, Korzeniowski SH. Sources, fate and transport of perfluorocarboxylates. Environ Sci Technol. 2006, 40(1): 32–44.

30 Liu J, Mejia Avendaño S. Microbial degradation of polyfluoroalkyl chemicals in the environment: a review. Environ Int. 2013, 61: 98–114.

31 Barzen-Hanson KA, Roberts SC, Choyke S, Oetjen K, McAlees A, Riddell N, McCrindle R, Ferguson PL, Higgins CP, Field JA. Discovery of 40 classes of per- and polyfluoroalkyl substances in historical Aqueous Film-Forming Foams (AFFFs) and AFFF-impacted groundwater. Environ Sci Technol. 2017, 51(4): 2047–2057.

32 Sunderland EM, Hu XC, Dassuncao C, Tokranov AK, Wagner CC, Allen JG. A review of the pathways of human exposure to poly- and perfluoroalkyl substances (PFASs) and present understanding of health effects. J Expo Sci Environ Epidemiol. 2019, 29(2): 131–147.

33 Dauchy X. Per- and polyfluoroalkyl substances (PFASs) in drinking water: current state of the science. Curr Opin Environ Sci Health. 2019, 7: 8–12.

34 Hu XC, Andrews DQ, Lindstrom AB, Bruton TA, Schaider LA, Grandjean P, Lohmann R, Carignan CC, Blum A, Balan SA, Higgins CP, Sunderland EM. Detection of Poly- and Perfluoroalkyl Substances (PFASs) in U.S. drinking water linked to industrial sites, military fire training areas, and wastewater treatment plants. Environ Sci Technol Lett. 2016, 3(10): 344–350.

35 Zhang S, Tan R, Pan R, Xiong J, Tian Y, Wu J, Chen L. Association of perfluoroalkyl and polyfluoroalkyl substances with premature ovarian insufficiency in Chinese women. J Clin Endocrinol Metab. 2018, 103(7): 2543–2551.

36 Di Nisio A, Sabovic I, Valente U, Tescari S, Rocca MS, Guidolin D, Dall'Acqua S, Acquasaliente L, Pozzi N, Plebani M, Garolla A, Foresta C. Endocrine disruption of androgenic activity by perfluoroalkyl substances: clinical and experimental evidence. J Clin Endocrinol Metab. 2019, 104(4): 1259–1271.

37 Stockholm Conference of the Parties decided to list perfluorooctane sulfonic acid (PFOS), its salts and perfluorooctane sulfonyl fluoride (PFOSF) in Annex B to the Stockholm Convention (decision SC-4/17). 2009. Available from: http://chm.pops.int/Implementation/PFOS/Overview/tabid/5221/Default.aspx.

38 Houtz EF, Higgins CP, Field JA, Sedlak DL. Persistence of perfluoroalkyl acid precursors in AFFF-impacted groundwater and soil. Environ Sci Technol. 2013, 47(15): 8187–8195.

39 Kuroda K, Murakami M, Oguma K, Takada H, Takizawa S. Investigating sources and pathways of perfluoroalkyl acids (PFAAs) in aquifers in Tokyo using multiple tracers. Sci Total Environ. 2014, 488-489: 51–60.

40 Paul AG, Jones KC, Sweetman AJ. A first global production, emission, and environmental inventory for perfluorooctane sulfonate. Environ Sci Technol. 2009, 43(2): 386–392.

41 Xiao F, Simcik MF, Halbach TR, Gulliver JS. Perfluorooctane sulfonate (PFOS) and perfluorooctanoate (PFOA) in soils and groundwater of a U.S. metropolitan area: migration and implications for human exposure. Water Res. 2015, 72: 64–74.

42 Scheringer M, Trier X, Cousins IT, de Voogt P, Fletcher T, Wang Z, Webster TF. Helsingør statement on poly- and perfluorinated alkyl substances (PFASs). Chemosphere. 2014, 114: 337–339.

43 Blum A, Balan SA, Scheringer M, Trier X, Goldenman G, Cousins IT, Diamond M, Fletcher T, Higgins C, Lindeman AE, Peaslee G, de Voogt P, Wang Z, Weber R. The Madrid statement on Poly- and Perfluoroalkyl Substances (PFASs). Environ Health Perspect. 2015, 123(5): A107–A111.

44 Ritscher A, Wang Z, Scheringer M, Boucher JM, Ahrens L, Berger U, Bintein S, Bopp SK, Borg D, Buser AM, Cousins I, DeWitt J, Fletcher T, Green C, Herzke D, Higgins C, Huang J, Hung H, Knepper T, Lau CS, Leinala E, Lindstrom AB, Liu J, Miller M, Ohno K, Perkola N, Shi Y, Småstuen Haug L, Trier X, Valsecchi S, van der Jagt K, Vierke L. Zürich statement on future actions on Per- and Polyfluoroalkyl Substances (PFASs). Environ Health Perspect. 2018, 126(8): 84502.

45 Nakayama SF, Yoshikane M, Onoda, Y, Nishihama Y, Iwai-Shimada M, Takagi M, Kobayashi Y, Isobe, T. Worldwide trends in tracing poly- and perfluoroalkyl substances (PFAS) in the environment. TrAC: Trends Analyt Chem. 2019, 121: 115410.

46 Land M, De WCA, Bignert A, Cousins LT, Herzke D, Johansson JH, Martin JW. What is the effect of phasing out long-chain per- and polyfluoroalkyl substances on the concentrations of perfluoroalkyl acids and their precursors in the environment? A systematic review. Environ Evid. 2018, 7(1): 4.

47 Satbhai K, Vogs C, Crago J. Comparative toxicokinetics and toxicity of PFOA and its replacement GenX in the early stages of zebrafish. Chemosphere. 2022, 308(Pt 1): 136131.

48 Goeden HM, Greene CW, Jacobus JA. A transgenerational toxicokinetic model and its use in derivation of Minnesota PFOA water guidance. J Expo Sci Environ Epidemiol. 2019, 29(2): 183–195.

49 Kudo N, Kawashima Y. Toxicity and toxicokinetics of perfluorooctanoic acid in humans and animals. J Toxicol Sci. 2003, 28(2): 49–57.

50 Conley JM, Lambright CS, Evans N, Medlock-Kakaley E, Dixon A, Hill D, McCord J, Strynar MJ, Ford J, Gray LE Jr. Cumulative maternal and neonatal effects of combined exposure to a mixture of perfluorooctanoic acid (PFOA) and perfluorooctane sulfonic acid (PFOS) during pregnancy in the Sprague-Dawley rat. Environ Int. 2022, 170: 107631.

51 Pérez-Albaladejo E, Fernandes D, Lacorte S, Porte C. Comparative toxicity, oxidative stress and endocrine disruption potential of plasticizers in JEG-3 human placental cells. Toxicol In Vitro. 2017, 38: 41–48.

52 Chen MY, Ike M, Fujita M. Acute toxicity, mutagenicity, and estrogenicity of bisphenol-A and other bisphenols. Environ Toxicol. 2002, 17(1): 80–86.

53 Waidyanatha S, Gibbs S, South N, Smith JP, Mutlu E, Burback B, Cao Y, Rider CV. Toxicokinetics of the plasticizer, N-butylbenzenesulfonamide, in plasma and brain following oral exposure in rodents: route, species, and sex comparison. Toxicol Rep. 2020, 7: 711–722.

54 Michael L, Jason B, Craig W, Bruce H, Debra D. Challenges in regulating pesticide mixtures. Ecol Soc. 2004. 9(6): 1.

55 Abou El-Magd SA, Sabik LM, Shoukry AJLSJ. Pyrethroid toxic effects on some hormonal profile and biochemical markers among workers in pyrethroid insecticides company. Life Sci J. 2011, 8(1): 311–22.

56 Deneer JW. Toxicity of mixtures of pesticides in aquatic systems. Pest Manag Sci. 2000, 56(6): 516–520.

57 Starr JM, Scollon EJ, Hughes MF, Ross DG, Graham SE, Crofton KM, Wolansky MJ, Devito MJ, Tornero-Velez R. Environmentally relevant mixtures in cumulative assessments: an acute study of toxicokinetics and effects on motor activity in rats exposed to a mixture of pyrethroids. Toxicol Sci. 2012, 130(2): 309–318.

58 Filipak Neto F, Zanata SM, Silva de Assis HC, Nakao LS, Randi MA, Oliveira Ribeiro CA. Toxic effects of DDT and methyl mercury on the hepatocytes from Hoplias malabaricus. Toxicol In Vitro. 2008, 22(7): 1705–1713.

59 Hellou J, Lebeuf M, Rudi M. Review on DDT and metabolites in birds and mammals of aquatic ecosystems. Environ Rev. 2013. 21(1): 53–69.

60 Ross G. The public health implications of polychlorinated biphenyls (PCBs) in the environment. Ecotoxicol Environ Saf. 2004, 59(3): 275–291.

61 Kimbrough RD. Polychlorinated biphenyls (PCBs) and human health: an update. Crit Rev Toxicol. 1995, 25(2): 133–163.

62 Darnerud PO. Toxic effects of brominated flame retardants in man and in wildlife. Environ Int. 2003, 29(6): 841–853.

63 Robertson LW, Parkinson A, Bandiera S, Lambert I, Merrill J, Safe SH. PCBs and PBBs: biologic and toxic effects on C57BL/6J and DBA/2J inbred mice. Toxicology. 1984, 31 (3–4): 191–206.

64 Tarazona JV, Rodríguez C, Alonso E, Sáez M, González F, San Andrés MD, Jiménez B, San Andrés MI. Toxicokinetics of perfluorooctane sulfonate in rabbits under environmentally realistic exposure conditions and comparative assessment between mammals and birds. Toxicol Lett. 2016, 241: 200–206.

65 Tarazona JV, Rodríguez C, Alonso E, Sáez M, González F, San Andrés MD, Jiménez B, San Andrés MI. Toxicokinetics of perfluorooctane sulfonate in birds under environmentally realistic exposure conditions and development of a kinetic predictive model. Toxicol Lett. 2015, 232(2): 363–368.

66 Gingrich J, Pu Y, Ehrhardt R, Karthikraj R, Kannan K, Veiga-Lopez A. Toxicokinetics of bisphenol A, bisphenol S, and bisphenol F in a pregnancy sheep model. Chemosphere. 2019, 220: 185–194.

67 Tominaga T, Negishi T, Hirooka H, Miyachi A, Inoue A, Hayasaka I, Yoshikawa Y. Toxicokinetics of bisphenol A in rats, monkeys and chimpanzees by the LC-MS/MS method. Toxicology. 2006, 226 (2–3): 208–217.

68 Bradbury SP, Coats J. Toxicokinetics and toxicodynamics of pyrethroid insecticides in fish. Environ Toxicol Chem. 1989, 8(5): 373–380.

69 Harwood AD, You J, Lydy MJ. Temperature as a toxicity identification evaluation tool for pyrethroid insecticides: toxicokinetic confirmation. Environ Toxicol Chem. 2009, 28(5): 1051–1058.

70 Deck AT, Reinke EN, McCain WC. Wildlife toxicity assessment for aldrin and dieldrin. Wildlife Toxicity Assessments for Chemicals of Military Concern. Elsevier, 2015, pp. 367–384.

71 National Research Council (US) Committee on Toxicology. An Assessment of the Health Risks of Seven Pesticides Used for Termite Control. National Academies Press (US), Washington (DC), 1982.

72 Belden JB, Lydy MJ. Impact of atrazine on organophosphate insecticide toxicity. Environ Toxicol Chem. 2000, 19(9): 2266–2274.

73 Li H, Zhang Q, Su H, You J, Wang WX. High tolerance and delayed responses of daphnia magna to neonicotinoid insecticide imidacloprid: toxicokinetic and toxicodynamic modeling. Environ Sci Technol. 2021, 55(1): 458–467.

74 Frew JA, Brown JT, Fitzsimmons PN, Hoffman AD, Sadilek M, Grue CE, Nichols JW. Toxicokinetics of the neonicotinoid insecticide imidacloprid in rainbow trout (Oncorhynchus mykiss). Comp Biochem Physiol C Toxicol Pharmacol. 2018, 205: 34–42.

75 Mrema EJ, Rubino FM, Brambilla G, Moretto A, Tsatsakis AM, Colosio C. Persistent organochlorinated pesticides and mechanisms of their toxicity. Toxicology. 2013, 307: 74–88.

76 Jayaraj R, Megha P, Sreedev P. Organochlorine pesticides, their toxic effects on living organisms and their fate in the environment. Interdiscip Toxicol. 2016, 9 (3–4): 90–100.

77 Mumtaz MM, Tully DB, El-Masri HA, De Rosa CT. Gene induction studies and toxicity of chemical mixtures. Environ Health Perspect. 2002, 110 (Suppl 6): 947–956.

78 McLachlan JA. Functional toxicology: a new approach to detect biologically active xenobiotics. Environ Health Perspect. 1993, 101(5): 386–387.

79 Cleveland FP. A summary of work on aldrin and dieldrin toxicity at the Kettering Laboratory. Arch Environ Health. 1966, 13(2): 195–198.

80 Pape-Lindstrom PA, Lydy MJ. Synergistic toxicity of atrazine and organophosphate insecticides contravenes the response addition mixture model. Environ Toxicol Chem. 1997, 16(11): 2415–2420.

81 Anderson TD, Lydy MJ. Increased toxicity to invertebrates associated with a mixture of atrazine and organophosphate insecticides. Environ Toxicol Chem. 2002, 21(7): 1507–1514.

82 Howe GE, Gillis R, Mowbray RC. Effect of chemical synergy and larval stage on the toxicity of atrazine and alachlor to amphibian larvae. Environ Toxicol Chem. 1998, 17(3): 519–525.

83 Wacksman MN, Maul JD, Lydy MJ. Impact of atrazine on chlorpyrifos toxicity in four aquatic vertebrates. Arch Environ Contam Toxicol. 2006, 51(4): 681–689.

84 Cedergreen N. Quantifying synergy: a systematic review of mixture toxicity studies within environmental toxicology. PLoS One. 2014, 9(5): e96580.

85 Shefali KR, Sankhla MS, Kumar R, Sonone SS. Impact of pesticide toxicity in aquatic environment. Biointerface Res Appl Chem. 2021, 11(3): 10131–10140.

86 Solomon KR, Williams WM, Mackay D, Purdy J, Giddings JM, Giesy JP. Properties and uses of chlorpyrifos in the United States. Rev Environ Contam Toxicol. 2014, 231: 13–34.

87 Sánchez-Bayo F, Tennekes HA. Time-cumulative toxicity of neonicotinoids: experimental evidence and implications for environmental risk assessments. Int J Environ Res Public Health. 2020, 17(5): 1629.

88 Zhao GP, Yang FW, Li JW, Xing HZ, Ren FZ, Pang GF, Li YX. Toxicities of neonicotinoid-containing pesticide mixtures on nontarget organisms. Environ Toxicol Chem. 2020, 39(10): 1884–1893.

3

Toxicodynamics of Chemical Mixtures Exposure in the Organism Body

Chunjiao Lu[a,b], *Ting Yang*[a], *Xin Meng*[a], *Xiaojun Yang*[a,*]

[a] *Engineering Research Center of Key Technique for Biotherapy of Guangdong Province, Shantou University Medical College, Shantou, China*
[b] *Department of Public Health, School of Medicine, Guangxi University of Science and Technology, Liuzhou, Guangxi, China*

*Corresponding author.
E-mail: yangx@stu.edu.cn (X.Y.)

3.1 Chemical Mixtures in Environment

Humans and the environment are exposed to a huge number of chemicals and compounds. Humans are exposed to a wide variety of chemicals through respiration, skin, and mouth. Mixtures of substances are the primary types of combined exposures and are divided into two categories: intentional and unintentional [1]. Intentional mixtures refer to chemical components in most known mixtures (e.g. food additives, pesticide formulations, and paint), while unintentional mixtures include coincidences, emissions, and environmental mixtures [2]. Coincidental mixtures refer to multi-source mixtures, which result in increased component complexity. The emission mixtures are defined as substances that are released in a single transpose during production, transportation, etc. Environmental mixtures are defined as the combination of chemical components from different origins that occur in their respective exposed environments [3].

Chemicals usually exist in the environment in the form of mixtures. Organic chemicals and heavy metals are two kinds of common pollutants in the environment. Organic pollutants refer to pollutants composed of natural organic substances in the form of biodegradable synthetic organic substances such as carbohydrates, proteins, amino acids, and fats. Organic pollutants can be divided into two main categories: natural and synthetic. Organic chemical pollutants are highly toxic, persistent, and bioaccumulative, and mainly include volatile halogenated hydrocarbons (VHCs), volatile organic compounds (VOCs), and polycyclic aromatic hydrocarbons (PAHs).

3.1.1 Organic Chemicals

3.1.1.1 Volatile Halogenated Hydrocarbons

VHCs consist of a diverse range of organic compounds, most of which are characterized by environmental persistence, bioaccumulation, and biotoxicity [4]. VHCs are the main components of VOCs in the atmosphere [5, 6]. As an important greenhouse gas, they have the dual role of influencing tropospheric climate and destroying the stratospheric ozone layer [5, 6]. Moreover, VHC is a compound in which the hydrogen atom(s) of a hydrocarbon molecule is replaced by a halogen atom(s) [7]. Toxic and hazardous VHCs including carbon tetrachloride; 1,1,1-trichloroethane; o-dichlorobenzene; chlorobenzene; tetrachloroethylene; trichloroethylene; dichloromethane; chloroform; dichloromethane, etc., which are important chemical solvents, organic synthesis raw materials, or intermediates [8]. Various VHCs produced by plastic petroleum and landfill leachate flow into urban rivers, altering the character of the water environment and causing potential adverse effects on humans [9].

Toxicological Assessment of Combined Chemicals in the Environment, First Edition. Edited by De-Sheng Pei and Yiyun Liu.

3.1.1.2 Volatile Organic Compounds

VOCs are organic compounds found in a variety of products that can easily evaporate and enter the environment under normal conditions [10]. VOCs have higher volatility and fluidity, and they are not easily degraded and can be transported over long distances in the environment [10]. The most common VOCs are aromatic hydrocarbons and halogenated hydrocarbons, such as toluene, xylene, benzene, ethylbenzene, trichloroethylene, vinyl chloride, and so on. VOCs are strongly associated with the development of cancer [10]. They are classified into anthropogenic sources and natural sources (Figure 3.1). Anthropogenic sources refer to VOCs generated during human processes, such as food extraction, the use of pesticides and fertilizers, chlorination, transportation, hydrocarbon fuel combustion, septic tanks, textile cleaning, the pharmaceutical industry, the printing industry, etc. [11, 12]. Natural sources include plant emissions, anaerobic bog processes, and forest fires caused by natural causes [11, 12].

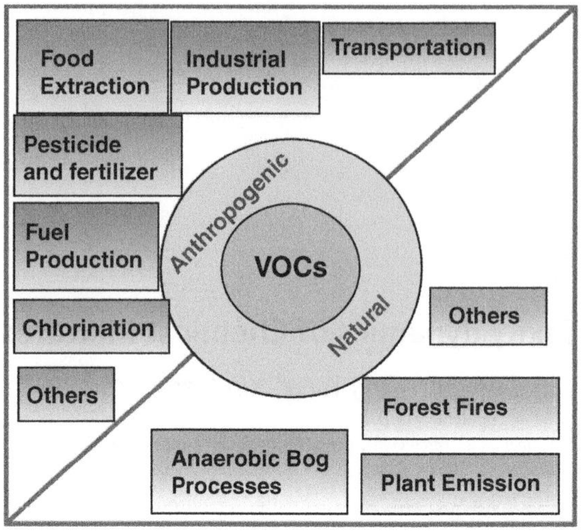

Figure 3.1 The sources of VOCs.

3.1.1.3 Polycyclic Aromatic Hydrocarbons

PAHs are organic compounds with primarily colorless, white, or pale-yellow solids. PAHs are generally classified into low molecular PAHs (LMW) and high molecular PAHs (HMW) [13]. Moreover, the toxicity of 2–3 rings LMW, such as fluorene, naphthalene, and phenanthrene, is much more toxic than that of 4–7 rings HMW [13]. PAHs usually enter the environment in different ways (Figure 3.2); some are produced during biological processes, and others are products of incomplete combustion from artificial combustion sources (vehicle emissions and cigarette smoke) or natural combustion sources (forest and bush fires) [14, 15]. Thus, PAHs are widespread in the environment, including air, sediment, water, and soil [15].

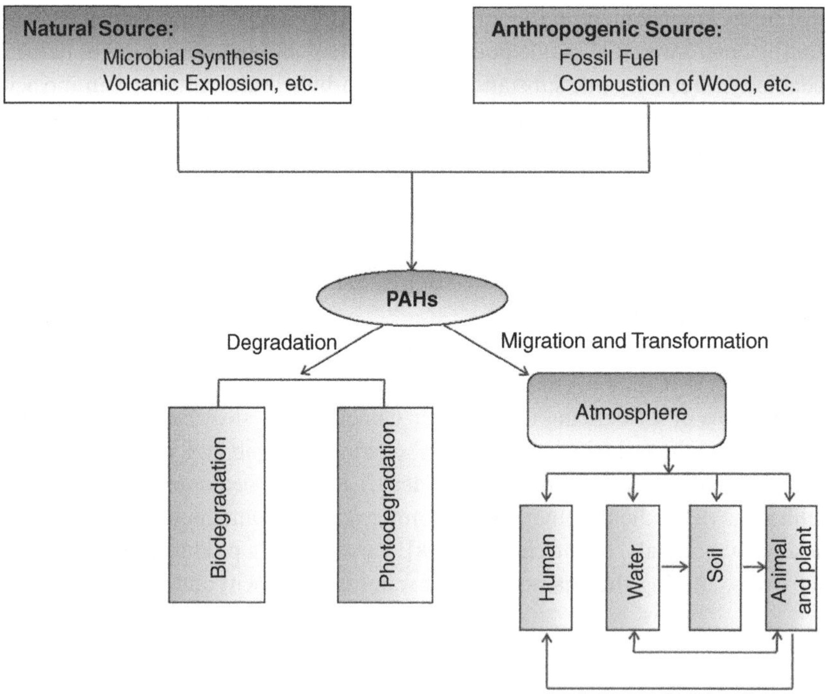

Figure 3.2 PAHs generation, migration, and transformation in the environment.

3.1.2 Heavy Metals

Heavy metals are naturally occurring elements with relatively high atomic weights and a minimum density of five times that of water [16]. Heavy metals are widely used in households, agriculture, medicine, and industry, resulting in their wide distribution in the environment [16]. Human exposure to heavy metals can cause severe health problems that worsen over time. It can affect a variety of major organs such as the brain, liver, kidneys, and heart [17]. Most heavy metals are highly toxic and they have many exposure routes, including inhalation, ingestion, and dermal absorption, and exposure to heavy metal contaminants can cause adverse health effects. Various toxic heavy metals, such as lead (Pb), chromium (Cr), mercury (Hg), nickel (Ni), cadmium (Cd), etc., are commonly used in various industrial processes and end up in natural resources such as soil, oceans, rivers, and land [18]. The environmental pollution caused by heavy metals has a severe impact on animals, plants, and humans [19, 20].

3.1.2.1 Lead (Pb)

Pb (Figure 3.3) is one of the oldest dynamic metals on earth and also is a powerful non-biodegradable public health toxin [21]. It has been roughly estimated that 90% of the mined Pb is consumed in various applications in the industrial sector, such as the production of various metal alloys; the preparation of paints, inks, and petroleum products; etc. [21, 22]. Human exposure to Pb can occur in a variety of ways, such as kitchen utensils containing Pb used for cooking, ingestion of contaminated edible items, drinking contaminated water, inhaling contaminated gases, or through skin contact, etc. [23, 24]. Research has suggested that Pb, as a potential toxin, can disrupt the normal functioning of human biochemical and physiological processes [24].

3.1.2.2 Chromium (Cr)

Cr (Figure 3.4) is very frequently used in various industries, such as chemical industry, mining, chrome-plating and tanneries, etc. [25, 26]. Cr is widely used in metallurgical industry, chemical industry, refractory industry, and foundries [27]. In the metallurgical industry, Cr is used in the production of metals such as chrome alloys and Cr [28]. In the chemical industry, Cr is used for leather tanning, metal corrosion inhibition, wood preservation, and textile dyes [28]. Cr exists in a variety of oxidation states, among which trivalent Cr (Cr(III)) and hexavalent Cr (Cr(VI)) are most commonly noticed. Cr has a wide range of functions, such as lipid, protein, and carbohydrate metabolism. Cr is one of the major micronutrients. Excessive exposure to the heavy metal Cr can also cause serious environmental problems, affecting human, animal, and plant health.

3.1.2.3 Mercury (Hg)

Hg is a volatile element with an atomic number of 80 (Figure 3.5). It has the characteristics of a metal and a liquid at room temperature, with silvery white, shimmering, and faint blue tint [29]. Hg is widely distributed in the atmosphere in gaseous form [30]. Hg can cause adverse effects on the environment and human health and is considered a global pollutant. It exists in the environment in inorganic, organic, and elemental forms [31]. The main sources of Hg pollution include mineral extraction, application of Hg containing pesticides and fertilizers, mining activities, fossil fuels burning, industrial activities, and

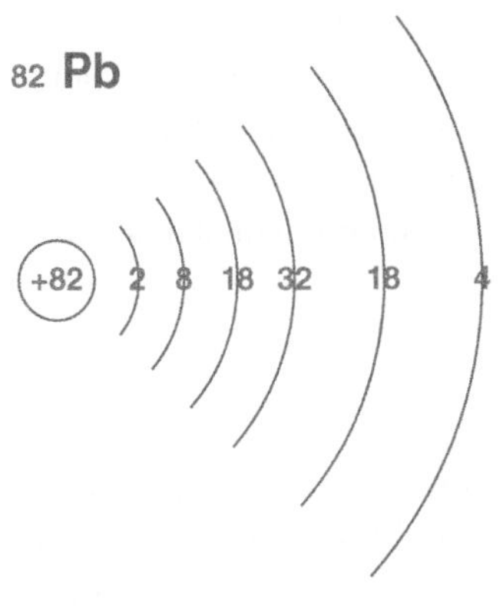

Figure 3.3 Schematic diagram of the atomic structure of Pb.

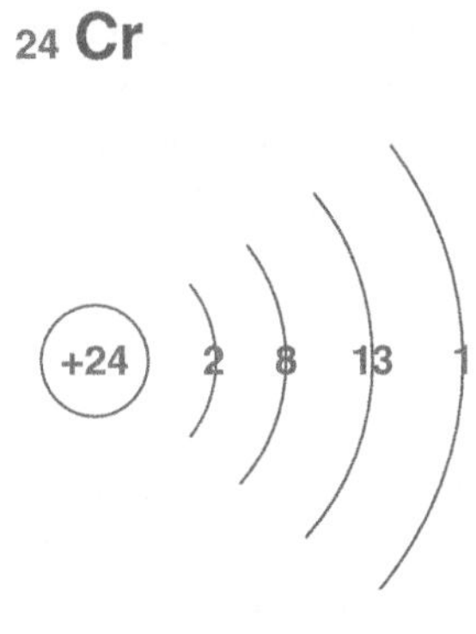

Figure 3.4 Schematic diagram of the atomic structure of Cr.

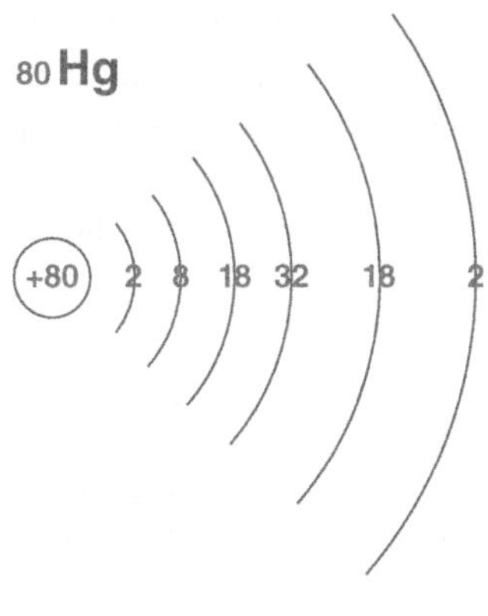

Figure 3.5 Schematic diagram of the atomic structure of Hg.

waste discharge [32]. Some studies reported that coal-fired power plants alone account for 65% of environmental Hg deposition emissions [33]. Due to its unique properties (highly volatile, persistent), Hg is easily absorbed by plant systems and accumulated in the food chain, eventually entering the human body and affecting human health [30].

3.1.2.4 Nickel (Ni)

Ni (Figure 3.6) is the 24th most abundant element in the earth's crust [34]. At low concentrations, Ni is an essential nutrient trace element for animals [35, 36]. However, at high concentrations, Ni is a toxic element [36, 37]. Ni is a natural component of soil and water [38], and concentrations of Ni in soil and surface water are below 100 and 0.005 mg/kg, respectively [39, 40]. Ni is released into the environment through various anthropogenic activities, such as fossil fuel burning, smelting, industry, phosphate fertilizer, electroplating industry, vehicle emissions, and sludge water [41]. The widespread use of Ni has increased its concentration in the environment and increased its exposure to humans as well as animals and plants through environmental pollution and occupational exposure [42]. The body is exposed to Ni mainly through inhalation, ingestion, and skin touch [43]. The extensive use of Ni can have adverse effects on the environment and flora and fauna.

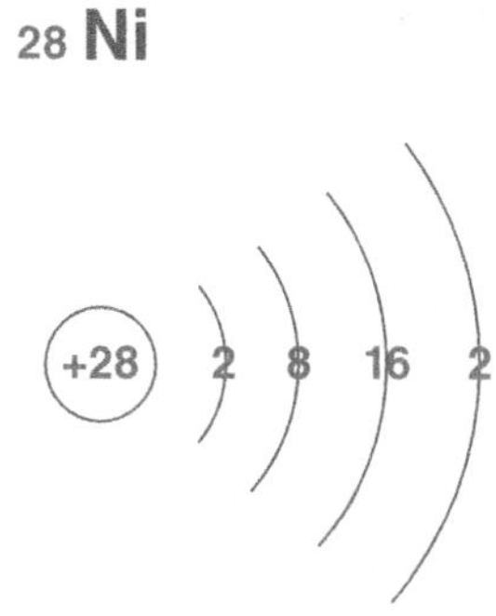

Figure 3.6 Schematic diagram of the atomic structure of Ni.

3.1.2.5 Cadmium (Cd)

Cd (Figure 3.7) is a naturally occurring metal that sits between zinc (Zn) and Hg in the periodic table. The chemical behavior of Cd is similar to Zn [44]. It usually exists as a divalent cation. Cd occurs in the earth's crust at about 0.1 parts per million, which is usually present as an impurity in Zn or Pb deposits, and therefore being produced primarily as a by-product of Zn or Pb smelting [44]. Cd is one of the most toxic trace elements in the environment; it can harm the health of animals and humans [45]. Cd is widely used in TV screens, batteries, paints, and cosmetics [44, 46]. Humans are mainly exposed to Cd through inhalation or ingestion. Exposure to Cd usually occurs through the ingestion of contaminated food or water and can cause adverse health effects [44, 46].

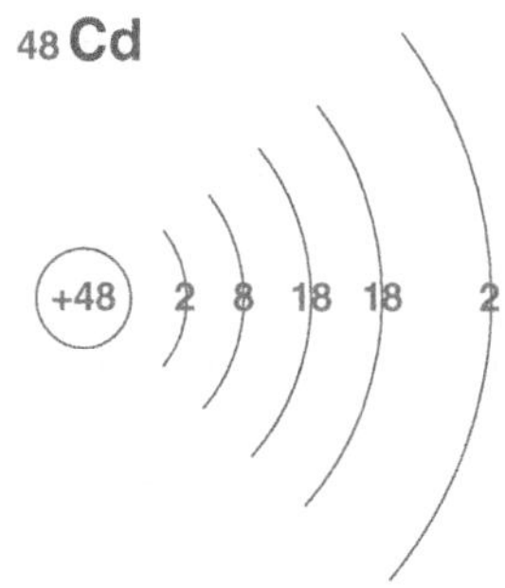

Figure 3.7. Schematic diagram of the atomic structure of Cd.

3.2 Model Organism in Toxicology

Model organisms play an important role in life science research. In addition, they have implications for the treatment of certain human diseases. To date, most of the major discoveries in life science published in important journals have been made using model organisms. Common model organisms include *Escherichia coli* and *Haemophilus influenzae* in prokaryotes; *Saccharomyces cerevisiae* in eukaryotes (unicellular); *Caenorhabditis elegans, Drosophila melanogaster*, and *Arabidopsis thaliana* in eukaryotes (multicellular); *Danio rerio* in ichthyoids; and *Mus musculus* in mammalian classes. A model organism has the following advantages in the study of life science: (1) it can represent a certain kind of large group in the biological world; (2) harmless to human body and the environment; (3) easy to obtain, breed, and raise in the laboratory; (4) clear genetic background; and (5) easy to carry out experimental operations. Modern life science research mainly focuses on nematode worms, fruit flies, zebrafish, mice, and *A. thaliana*. Next, we will focus our discussion on the model organisms – zebrafish and mice.

3.2.1 Danio Rerio

D. rerio (zebrafish, Figure 3.8), also known as blue-striped fish and striped fish, is a common tropical freshwater fish originating from India, Bangladesh, and other places. The main advantages of zebrafish are as follows: (1) small size (the body length of zebrafish is about 3–4 cm), (2) strong reproductive ability, (3) easy breeding, (4) transparent embryos, easy to visual observation, (5) embryo development rapidly, (6) easy to inject the foreign genes into the embryo, and (7) homology with human genes up to 87% [47–49]. In addition, as an overall experimental animal model, zebrafish can evaluate the therapeutic effect and toxic side effects of drugs. As a model animal, zebrafish not only has a short screening

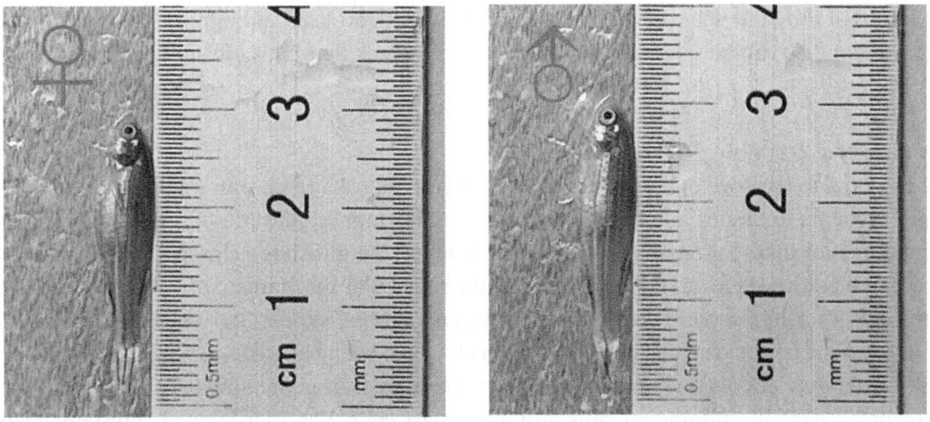

Figure 3.8 Zebrafish diagram.

cycle, low experimental cost, easy operation, and observation, but also has the characteristics of high efficiency, rapid speed, and large scale [49]. It is one of the important model vertebrates in current life science research [47].

Research on zebrafish toxicology basically follows the following steps: (1) to detect the type and concentration of toxicants, (2) to assess the biological phenotype of zebrafish and obtain relevant data, (3) to interpret the obtained data and observed phenotypes, and (4) to confirm the data and phenotypes in step 3 [50] This toxicological research process enables the interpretation and confirmation of biological phenotypes and molecular mechanisms for subsequent studies. At present, zebrafish, as a model animal, has been widely used in studies of cardiodevelopmental toxicity, neurodevelopmental toxicity, hepatotoxicity, and reproductive toxicity.

3.2.1.1 Neurotoxicity in Zebrafish

The nervous system of zebrafish embryo begins to develop six hours after fertilization. Like other vertebrates, the central nervous system of zebrafish is derived from the ectodermal epithelium on the dorsal side of the embryo, also known as the neural plate. The gastrula terminal nerve plates of the zebrafish converge to form neural tubes 9–10 hours after fertilization; 16 hours after fertilization, the neural tube is subdivided into different regions, some of which develop into ventricles; by 24 hours, the brain has almost completed morphogenesis, dividing the brain into forebrain, diencephalon, telencephalon, midbrain, posterior brain, and forming the spinal cord [51]; and 48 hours after fertilization, the brain ventricle forms, and the zebrafish embryos begin to respond to tactile stimuli. Conventional assessment of neurodevelopmental toxicity in zebrafish includes the following components: (1) cholinesterase activity, (2) caudal axonal morphology, (3) brain cells, (4) forebrain morphology, (5) motor neuron pathway, (6) neurogenesis-related genes, (7) motor axons, etc. Glutamic, glycine, and γ-aminobutyric acids are three major neurotransmitters in the central nervous system, and the *VGLUT, GlyT2*, and *GAD* genes are markers of glutaminergic neurons, glutamatergic neurons, and aminobutyric neurons, respectively [52]. Transgenic zebrafish strains – *Tg (vglut2a:GFP)*, *Tg(glyt2:Gal4)*, and *Tg(gad1b:GFP)* – play important role in the toxicological study of nervous system development [52].

3.2.1.2 Cardiotoxicity in Zebrafish

The heart of a zebrafish is a fist-like asymmetrically developed organ. At 16 hours after fertilization, the precursor cells of the heart begin to differentiate and transfer to the midline of the embryo [53]. At 18 hours after fertilization, myocardial myosin light chain 2 promotes the fusion of atrial and ventricular precursors into simple linear cardiac tubes [53]. At 24 hours after fertilization, ventricular and atrial structures appear [53]. At 24–48 hours after fertilization, the cardiac loops form, and the linear cardiac ducts gradually curve into S-shaped rings along the atrioventricular boundary [54]. At 36 hours after fertilization, each ventricle of the heart contract orderly and markedly [54]; and 72 hours after fertilization, ion channels are formed on the cell membrane of zebrafish myocardium, and the heart has the ability to complete a series of metabolic processes [54].

Phenotypes of cardiodevelopmental toxicity include tachycardia, bradycardia, arrhythmia, cardiac injury, pericardial effusion, apoptosis of cardiac cells, epicardial epithelial cells, and so on. These toxic phenotypes were visualized in zebrafish embryos 48 hours after fertilization. At present, transgenic zebrafish has been widely used in toxicology research. The green fluorescence signal of transgenic zebrafish *Tg (cmlc2:GFP)* specifically marks the heart tissue, and the green fluorescence signal can be expressed at the early stage of embryonic development and lasts throughout the heart development process [55]. This transgenic zebrafish can be used for the study of compounds that have effects on cardiac circulation and heart valve formation [55].

3.2.1.3 Vascular Toxicity in Zebrafish

Blood vessels are essential to the development and survival of vertebrates and play an important role throughout life, providing oxygen and nutrients to tissues and organs, transporting hormones, and providing a staging ground for immune responses and signaling molecules for organ and tissue development. Zebrafish is a closed blood circulation system, and its blood vessel structure, blood vessel formation process, and the molecular mechanism of blood vessel formation are highly similar to human and other higher vertebrates [56]. Transgenic strains expressing green fluorescent protein (GFP) or enhanced green fluorescent protein (eGFP) in endothelial cells have been constructed under the control of specific gene promoters [57, 58]. Zebrafish endogenous promoter-driven vascular transgenic strain *Tg (fli1a:GFP)* or *Tg (kdrl:eGFP)* expressed fluorescent proteins in endothelial cells and angioblast cells. Vascular-labeled transgenic zebrafish has been widely used in various fields field of vascular research [59, 60]. The evaluation of vasodevelopmental toxicity usually focuses on bleeding spots and abnormal angiogenesis in different parts of zebrafish body. The construction of transgenic zebrafish strains is helpful to observe vascular defects and circulation abnormalities.

3.2.1.4 Hepatotoxicity and Nephrotoxicity in Zebrafish

Zebrafish liver begins to develop at about 28 hours after fertilization, and its growth factors and gene expressions are similar to human beings [56, 61]. At 78 hours, liver development is complete and acquire the basic functions [56, 61]. The liver originates from the endodermal progenitor cells at the edge of the blastoderm and goes through germination and growth stages. The germination stage of liver development occurs at 24–50 hours after fertilization. During the subsequent growth period, the liver of zebrafish becomes larger, its shape changes, and hepatocyte differentiation occurs [61]. At present, some studies have found that many compounds can cause hepatodevelopmental toxicity, such as APAP, TA, and PQ can cause swelling, degeneration, necrosis, and intracellular cytoplasmic vacuole of zebrafish hepatocytes [62]. When the testing of toxicity for chemical compounds and drugs, some compounds/drugs are not toxic, but substance produced after liver metabolism will be toxic to the organism [63].

The embryonic kidney of zebrafish has high homology with that of human. The kidney development of zebrafish begins 14 hours after fertilization and is completed within 72 hours [64]. The green fluorescence of transgenic zebrafish strain *Tg (wt1b:GFP)* marked the glomeruli, anterior renal tubules, and exocrine glands of zebrafish embryos. The transgenic strain could be used to screen compounds for renal developmental toxicity. Transgenic zebrafish strains *Tg (wt1b:GFP)*, *Tg (cdh17:GFP)*, *Tg (pod:GFP)*, etc., can meet the toxicity screening of different renal development sites.

3.2.2 Mus Musculus

Mice belongs to vertebrate phylum, mammalia (Class), rodentia (Order), and murinae (subfamily). At present, *Mus musculus* is a widely used animal model in the field of life science research. Compared with zebrafish, the life cycle of mice is relatively slow and difficult to operate. However, as mammals, mice are extremely closely related to humans because of their exalted position on the evolutionary tree. In mammalian experimental animal models, mice have been recognized as the best model mammals due to their small size, convenient feeding and management, easy control, rapid reproduction, and clear quality control standards, and they have a large number of inbred lines, mutant lines, and closed groups. Mice and humans share many basic biological and behavioral processes. Thus, using mice as model organisms helps to study human susceptibility to heart disease, neurological disease, cancer, liver and kidney disease, and reproductive system disease. In addition, transgenic mice also play an important role in toxicological research. At present, the toxicology studies of transgenic mice mainly included: carcinogenesis, mutagenesis, reproduction, poison metabolism, etc.

3.2.2.1 Neurotoxicity in Mice

Mammals have highly developed nervous systems that rapidly coordinate their internal environment and respond to external pressures. The nervous system consists of the brain, the spinal cord, and the peripheral nervous system, which are connected to them and spread throughout the body. It is divided into two main parts: the central nervous system and the peripheral nervous system. The toxicity of central nervous system and peripheral nervous system may cause reversible or irreversible damage to the nervous system and eventually lead to various neurological diseases. Neurotoxicity is usually identified by several indicators of observational model animals, such as neurobehavioral, biochemical, clinical, and histopathological analyses [65]. The developmental process and nervous system structure and function of mice are similar to that of human beings, and the research of neurodevelopmental toxicity is more comparable to that of human beings.

3.2.2.2 Cardiotoxicity in Mice

Mice are very useful model animals for studying the development of cardiovascular systems. The heart, as the first organ of mouse embryo formation, is part of a synergistic process of multicellular lineage. The development of the mouse heart and blood vessels starts from the progenitor cells in the mesoderm (lateral plate mesoderm)—cardiac progenitor cells and vasogenic cells, which are activated by different induction signals and begin to differentiate [66]. The development of heart of mice is an extremely complex process; it goes through the formation of the original cardiac tube and complex morphogenesis, and finally forms the heart with a four-chambered structure. Cardiac development in mice is mainly divided into the following processes: (1) induction and specialization of cardiogenic cells, (2) cardiac tube formation, (3) cardiac cyclization and asymmetric development, and (4) ventricular specialization and growth. In mice, the expression of genes related to cardiac development, the activities of key enzymes, markers, and cardiac phenotypes were detected to further explore the cardiac developmental toxicity under the action of compounds. In toxicology, genetically modified animal models are mainly used in the study of chemical drugs, compounds, and so on. Transgenic mice have been constructed and widely used in toxicity studies. Transgenic mice may be more sensitive than wild-type mice in cardiotoxicity tests. For example, mice modified with heart-specific *CYP2E1* gene can be used to research potential cardiotoxic drugs and environmental compounds [67].

3.2.2.3 Hepatotoxicity and Nephrotoxicity in Mice

The liver is the main organ involved in digestion in animals. Its main functions are metabolism, storage, and detoxification. Various toxic and harmful factors and substances, such as drugs, compounds, viruses, hypoxia, endotoxins, and so on, can target liver cells and cause liver cell damage. Liver injury can induce liver cell apoptosis, necrosis, and other characteristics of the disease [68]. Some poisons undergo a series of biochemical transformations in organs such as the liver, which can be transformed into non-toxic or less toxic substances that are easy to excrete. However, some substances will become more toxic, difficult to excretion, and then residual in the body, causing more serious adverse effects. In addition, the kidney, as the main organ for the excretion of metabolites in animals, should be analyzed for toxicity monitoring of drugs and compounds. Through the pathological analysis of the kidney and the detection of blood creatinine, urea nitrogen, and other changes, the effects of compounds on kidney functioning can be studied [69, 70].

3.3 Toxicodynamics of Chemical Mixtures

Environmental pollution has become one of the main concerns of human beings. Organic chemicals and heavy metals are common environmental pollutants. In the real world, organic chemicals and heavy metals often exist in mixtures. Interactions between chemical mixtures (organic chemicals and heavy metals) affect their biological uptake, accumulation, and toxicity. Studies have suggested that exposure to chemical mixtures can have harmful effects on organisms [71]. Even at low doses, the heart, kidneys, and brain were also affected [71]. Toxicodynamics is widely used in toxicological research. It refers to the process of toxicant-target interaction and its subsequent effects at the suborganism or organism level, including physiological processes [72]. Toxicodynamics is helpful to better understand the toxic effects of pollutants. Therefore, we will focus on the toxicity kinetics of chemical mixtures' exposure to an organism's body (zebrafish and mice) in the following sections.

3.3.1 Toxicodynamics of Neurotoxicity by Chemical Mixtures

Exposure to VHCs (carbon tetrachloride) significantly increased the activity of lipid peroxides and thione oxidase, which induced neuronal cell damage through the production of oxygen-free radicals [73]. Wang et al. exposed mice to VOC mixtures (benzene, toluene, xylene, and formaldehyde) and found that VOCs exposure caused behavioral disorders in the Morris water maze and the Grip strength test [74]. In addition, the levels of oxidative stress markers in the brain, reactive oxygen species (ROS), MDA, and glutamate (Glu), were significantly elevated [74]. The levels of acetylcholinesterase (AChE), choline acetyltransferase (ChAT), and acetylcholine (ACh), and the expression of NMDA receptor were significantly decreased [74]. After zebrafish were exposed to VOCs, the expression levels of stress response-related genes, DMA methylation pattern, and behavior were affected, among which the most obvious was the change in zebrafish behavior [75].

The neurotoxicity induced by PAHs has been extensively studied in rodents, and its mechanism is as follows: (1) the binding of PAHs to the aryl hydrocarbon receptor (AHR), (2) AHR-dependent modulation of the transcription of N-methyl-D-aspartate glutamate receptor (NMDAR) subunits, (3) NMDAR-mediated loss of neuronal activity and decreased long-term enhancement (LTP), and (4) impaired learning and memory [76–79]. A zebrafish model was used to evaluate the developmental toxicity of PAHs (BaP). It was found that exposure to BaP would lead to overactivity of larval photomotor response (LPR) phenotype, affect zebrafish behavior, and cause neurotoxicity [80, 81]. DNA methylation has been found to play a role in BaP-induced neurotoxicity. BaP treatment, DNA methyltransferase (DNMTs), and brain-derived neurotrophic factor (BDNF) IV hypermethylation were increased in the cerebral cortex of mice, while aspirin-combined therapy restored the activation of DNMT and hypermethylation of BDNF IV [82].

Pb was found to cause memory loss and neuronal damage in adult mice [83]. The treatment of Pb significantly increased malondialdehyde levels and the activities of acetylcholinesterase (AchE) and monoamine oxidase-A and -B, and inhibited the levels of norepinephrine, serotonin, and 5-hydroxyindoleacetic acid [83]. Pb exposure can affect the neural behavior of zebrafish and disrupt the mRNA levels of neural development-related genes (*gap43, dat, drd1b, syn2a, th*), neurotoxic effects-related genes (*gfap, c-fos*), and stress response–related genes (*mt1, hsp70, tap, hsp90*) [84]. By interfering with one carbon metabolism, Cr(III) and Cr(VI) can affect histone methylation and DNA methylation, regulate the expression of neuron-related genes, and thus lead to neurotoxicity [85]. Methylmercury is highly neurotoxic and can lead to a variety of neurodegenerative diseases [86]. Methylmercury exposure induces excessive ROS accumulation in neuro-associated cells and inhibits cell activity [86]. Methylmercuric chloride was found to accumulate in the cerebral cortex, subcutaneous layer, brainstem, and cerebellum of mice after 1 – 9 days of oral administration [87]. Zebrafish exposure to methylmercury induces the accumulation of Hg in brain tissue and abnormal expression levels of proteins associated with oxidative phosphorylation and mitochondrial dysfunction [88]. In addition, the mTOR signaling pathway in zebrafish was activated by methylmercury exposure [88].

Zebrafish larvae exposed to $NiCl_2$ were observed to have delayed hatching, decreased heart rate, and behavioral abnormalities [89]. Adult zebrafish exposed to $NiCl_2$ were observed to have reduced anxious behavior and aggression [89]. In zebrafish, early exposure to Ni can lead to morphological changes, resulting in avoidance response disorders and movement disorders [89]. In adult zebrafish, exposure to Ni can lead to anxiety response, impaired memory, and reduced aggressive behavior [89]. Ni treatment may induce neurotoxicity by inducing oxidative stress in the brain. Reduced glutathione and increased malondialdehyde were observed in the brains of mice treated with Ni [90]. Ni exposure inhibited δ-aminolevulinate dehydratase and upregulated brain metallothionein [90]. Cd is a potent neurotoxic heavy metal, which can cause oxidative stress and membrane disturbance in the brain. After Cd treatment, the AChE level of mice was inhibited, the level of oxidative stress markers was significantly increased, and the enzymatic antioxidants and non-enzymatic antioxidants in the brain were inhibited [91]. Cd exposure can induce early developmental impairments in zebrafish, and the neurobehavioral patterns of zebrafish, such as movement and response to ambient signals, are found to be disturbed [92]. In addition, Cd exposure to zebrafish can cause significant neurodevelopmental disorders and neuroimmune responses [92]. In zebrafish larvae exposed to Cd, cell cycle arrest and cell apoptosis together lead to downregulation of neuron density, thus leading to neurodevelopmental toxicity [92].

Heavy metals share several neurotoxic pathways, such as acting as a competing ion for Ca^{2+} (MeHg, Pb, Cd), interfering with the Na^+/K^+-ATPase pump (MeHg, Cd), affecting glutamate uptake and clearance (MeHg, Pb), disrupting NMDA receptor expression (MeHg, Pb), and causing an imbalance between antioxidants and ROS [93]. Metal mixtures may have a greater cumulative toxicity inducing neurotoxicity than single metal elements. It has been found that mixtures of Cd, Pb, and arsenic upregulate oxidative stress levels in neurons by upregulating Ca^{2+} and stimulating the c-Jun terminal kinase (JNK), extracellular signal-regulated kinase (ERK) and mitogen-activated protein kinase 3 (MAPK3) signaling pathways [94]. After heavy metal mixtures (Pb, Cd, Hg) were treated in rats, oxidative stress, intracellular free calcium, and apoptosis were significantly increased in the brain, and learning memory and sensory perception were impaired [95].

3.3.2 Toxicodynamics of Cardiotoxicity by Chemical Mixtures

Cardiotoxicity is characterized by damage to the heart muscle and electrophysiological dysfunction resulting in inefficient pumping activity of the heart and reduced circulation of blood [96]. Environmental pollutants, heavy metals, oxidants, etc., may cause cardiac toxicity [97, 98]. Exposure to VHCs (carbon tetrachloride) could induce cardiotoxicity in mice with increased levels of lipid peroxidation (thiobarbituric acid reactive substances), protein carbonyl level, and antioxidant markers (superoxide dismutase [SOD], catalase, glutathione peroxidase, glutathione, and vitamin C) in the heart tissue [99]. VOCs are the main indoor air pollutants, among which benzene, toluene, xylene, and formaldehyde are more common VOCs. It has been suggested that benzene exacerbates heart failure by promoting endothelial activation and neutrophil recruitment [100]. PAHs are harmful factors in inducing heart disease. After oral administration of PAHs in mice, the protein expressions of transforming growth factor β1 and its downstream transcription factors Smad3 and Smad4 were significantly downregulated, which caused the loss of cardiac fiber [101]. In addition, the downregulation of phosphatidylinositol (PI) 3-kinase and AKT induced the increased expression of caspase3, caspase9, and BAX, and inhibited the expression of bcl2, leading to cell apoptosis [101]. However, loss of myocardial fibers and apoptosis can lead to decreased heart/weight ratio [101]. BaP (benzo-a-pyrene), BeP, and BkF are relatively high molecular weight PAHs, all of which target the cardiovascular system [102]. Their potential role is ligand-activated aryl hydrocarbon receptor (AHR2) and the tissue-specific induction of the cytochrome p450 metabolic pathway (CYP1A) [102]. Zebrafish exposed to BaP caused AHR2-dependent bradycardia, pericardial edema, and myocardial CYP1A immunofluorescence [102]. BkF exposure induces pericardial edema, ring defects, and atrioventricular valvular erythrocyte reflux [102]. When exposed to BaP, the level of CYP1A+ signal in the vascular endothelium of the head and trunk of zebrafish was lower [102].

Pb exposure causes cardiac mitochondrial dysfunction, increases malondialdehyde (MDA) levels, and decreases mitochondrial SOD activity [96, 103]. The pathological changes in mitochondrial energy metabolism are closely related to myocardial cell apoptosis dysfunction and heart failure [96]. Cr treatment of mice induces increased levels of lipid peroxidation in the heart, lungs, testes, and spleen, while inhibiting levels of nonprotein thiols (NPSH) [104]. Exposure to the heavy metal Hg is associated with an increased risk of cardiovascular disease and cardiotoxicity. Hg treatment increased the expression and activity levels of soluble epoxide hydrolase (sEH) and decreased the levels of 14,15-, and 11,12-epoxyeicosatrienoic acids (EETs) in the heart of mice [105]. Acute Hg toxicity regulates the expression of cytochrome P450s (CYP) and sEH, leading to the decrease of cardioprotective EETs [105].

As one of heavy metal pollutants, Ni exposure can induce mitochondrial apoptosis and upregulate the mRNA expression levels of caspase-3, caspase-9, p53, bax, cytochrome c, LC3, atg1, TNF-α, COX 2, IL-1β, and IL-2 [106]. The mRNA expression of bcl2, p62, and mTOR was inhibited [106]. At the same time, Ni exposure could induce oxidative stress, increase the contents of glutathione disulfide (GSSG) and MDA, and decrease the activity of catalase (CAT) and total superoxide dismutase (T-SOD) [106]. Cd exposure resulted in focal necrosis, myofibril disarray, perivascular and interstitial fibrosis, and disorganized sarcomere structures in cardiac tissues in mice [107]. In addition, Cd can induce apoptosis of cardiomyocytes and increase the expression of matrix metalloproteinase (MMP-2, MMP-14) in cardiac tissue. In arterial tissue, Cd exposure can damage the intimal and medial layers of the aorta and reduce the activity of cultured aortic smooth muscle cells in vitro [107]. Cd-exposed mice inhibited the activities of antioxidant enzymes, that is, the activities of SOD, CAT, glutathione-S-transferase (GST), glutathione reductase (GR), glutathione peroxidase (GPx), and glucose-6-phosphate dehydrogenase (G6PD) were decreased [108]. Glutathione (GSH) and total mercaptan contents were decreased, while glutathione oxide (GSSG) levels, lipid peroxidation end products, protein carbonyl content, and DNA fragmentation were increased [108]. The action of two or more heavy metals on the same organism is much more complex than that of a single heavy metal. It was found that the pathological phenotype of the heart in rats exposed to a mixture of heavy metals (equal concentrations of Ni, Cd, Co and Hg chloride, and Pb acetate) was myocardial inflammation and a trend toward increased cardiovascular biomarkers (troponin T, C-reactive protein (CRP), and BNP) [109].

3.3.3 Toxicodynamics of Vascular Toxicity by Chemical Mixtures

Previous studies have shown that prepregnancy exposure to PAHs resulted in a significant decrease in the number of small-diameter intraplacental vessels in mice and affected fetal growth [110]. Both 2,3,7,8-tetrachlorodibenzo-p-dioxin (TCDD) and BaP exposures caused dose-dependent AhR-related malformations in zebrafish [111]. Coexposure of Bap and alpha-naphthoflavone (ANF) can cause significant changes in vascular morphology and increase the blood vessel luminal diameter of zebrafish [111]. Individual exposure to TCDD, BaP, and ANF can induce the mRNA expression level of CYP1A [111]. Principal component analysis showed that cardiac malformations and vascular dilatation were positively correlated with CYP1A [111].

Pb exposure has been found to cause damage to nervous system microvasculature in mice [112]. Pb exposure induced Pb accumulation in brain capillaries and inhibited synaptosomal constitutive nitric oxide synthase (NOS) [112]. Short-term exposure to Ni hydroxide-nanoparticles (NH-NPs) in mice caused acute endothelial destruction and altered vasoconstriction and diastole function [113]. Ni exposure has been shown to cause endothelial dysfunction in mice through oxidative stress–dependent inhibition of endothelial nitric oxide synthase (eNOS) dimerization [114]. Hg, Cd, and other heavy metals have a high affinity for sulfhydryl groups (-SH), which inactivate many enzymatic reactions, amino acids, and sulfur-containing antioxidants (NAC, GSH, ALA), thereby interfering with oxidative defenses and inducing oxidative stress [115]. The toxic effects of Hg on the blood vessels include the following: (1) inflammatory responses, (2) thrombosis, (3) endothelial dysfunction, (4) dyslipidemia, (5) vascular smooth muscle dysfunction, (6) oxidative stress, (7) mitochondrial dysfunction, etc. [115]. Cd exposure is known to adversely affect all vital organ systems in the body, especially the vascular system [116]. It can induce cardiovascular diseases, such as hypertension, increased oxidative stress, endothelial dysfunction, and atherosclerosis [116]. Cd exposure increased arterial blood pressure in mice and reduced the response of blood vessels to vasoactive agents [117]. In addition, Cd-treated mice were found to have increased superoxide production in the thoracic aorta, increased urinary nitrate/nitrite, increased plasma protein carbonyl, increased plasma and tissue MDA contents, and decreased blood glutathione (GSH) [117].

3.3.4 Toxicodynamics of Hepatotoxicity and Nephrotoxicity by Chemical Mixtures

Exposure to VHCs (carbon tetrachloride) induced liver injury in mice, and the levels of alanine aminotransferase (ALT), aspartate aminotransferase (AST), alkaline phosphatase (ALP), MDA, and serum nitric oxide were significantly increased compared with the control. Total protein, serum catalase, tissue SOD, and GSH levels were inhibited [118]. For mice exposed to VHCs (carbon tetrachloride), serum levels of creatinine and blood urea were significantly increased [119]. The levels of MDA, glutathione, catalase, and SOD were significantly elevated in renal tissue [119]. VHCs (carbon tetrachloride) treatment induced liver and kidney toxicity in mice. Exposure to a VOCs' mixture can cause immune system disruption and liver and kidney damage. The AST, ALT, ALP, and creatinine (CREA) were significantly upregulated in serum by the VOCs' mixture exposure in mice [120].

PAHs (BaP) exposure was found to alter the metabolism of glycerophospholipids, glycerolipids, and fatty acid metabolism in the mouse liver, with increasing triacylglycerol (TG), PI, and phosphatidylcholine (PC), and decreases in lysophosphatidylcholine (LysoPC)phosphatidyl ethanolamine, lysophosphatidylethanolamine, free fatty acids, and eosin compounds [121]. Exposure to 3-methylcholanthracene (3MC) significantly increased lipid droplets and triglyceride (TG) levels in the liver of mice, and inhibited the expression levels of acyl-coA carboxylase (Acc), fatty acid synthetase (Fas), and stearoyl-CoA desaturase 1 (Scd1)-related genes [122]. In addition, low doses of 3MC activate aromatics receptors that negatively regulate lipid synthesis in the liver [122]. Zebrafish exposed to phenylene (PHE), a common PAHs in the aquatic environment, have been found to have deteriorated liver morphology and increased liver cell damage [123]. The activities of GST and SOD were significantly different when zebrafish were exposed to PHE, and the expression of related genes in the liver was also significantly different [123].

Oral administration of Pb nitrate in male mice induced significant increases in the levels of hepatic AST, ALT, ALP, acid phosphatase, cholesterol, and lipid peroxidation [124]. In addition, the liver protein levels of Pb-exposed mice were inhibited, and the REDOX status of the liver was affected, such as the levels of antioxidants (SOD, catalase, and glutathione) were significantly reduced [124]. Pb acetate treatment increased the contents of serum glutamic oxaloacetic transaminase (SGOT), serum glutamic pyruvate transaminase (SGPT), alkaline phosphatase (ALP), and MDA, and decreased the contents of antioxidant enzyme (SOD, GPx) in the liver of mice [125]. The study suggests that inhibition of antioxidant enzymes may increase the amount of free radicals in liver tissue, leading to liver damage in mice [125]. Pb exposure can also cause severe nephrotoxicity. The exposure of mice to Pb caused a significant accumulation of the toxic metal in the kidneys and elicited inflammation with leukocyte infiltrations near the glomerulus [126]. Serum biochemical analysis showed that Pb exposure altered renal function markers in mice [126]. Pb-induced nephrotoxicity is associated with significant downregulation of renal concentrations of phosphorus, calcium, selenium, and copper [126].

Some studies have suggested that Cr exposure can induce liver toxicity through oxidative stress [127]. After Cr exposure, the liver tissue structure is abnormal in mice, liver cells are damaged, and liver inflammation is induced [127]. In addition, RNA-seq transcriptome showed that oxidative stress, apoptosis, and inflammation-related signaling pathways were upregulated in mice exposed to Cr, and KEGG pathway analysis found that the activation of NF-kB signaling pathways were significantly upregulated [127]. Cr may induce liver injury in mice by mediating inflammatory response through the NF-kB signaling pathway [127]. Cr is also toxic to the zebrafish. Exposure to Cr can cause hepatocyte apoptosis and change the distribution of liver elements [128]. In Cr-treated zebrafish, production of ROS, increase in reduced glutathione content, and increase in MDA content along with enhanced catalase activity were evident [129]. Cr exposure can induce not only liver toxicity but also kidney toxicity. Intraperitoneal injection of potassium dichromate in mice resulted in an abnormal increase of lipid peroxidation levels, inhibition of non-protein sulfhydryl levels, catalase activity, and SOD activity in the liver and kidney [130]. After Cr treatment, liver and kidney indexes were increased, such as the serum levels of ALT, AST, ALP, MDA, and creatinine were increased [131]. Histopathology of the liver revealed the distortion of hepatic cords, necrosis, damaged glomerulus, and Bowman's capsule [131].

Hg-induced hepatotoxicity was triggered by intrinsic apoptosis pathway, oxidative stress, dysactivity of nuclear receptors, and nuclear kinases (including Gsk3, which regulates the Wnt signaling pathway), gluconeogenesis, and lipogenesis, resulting in mitochondrial dysfunction, endocrine disruption, and metabolic disruption [132]. HgCl2 exposure can cause pathological injury, oxidative stress, and immunotoxicity of liver tissue in zebrafish [133] The toxicity of Hg to animals and humans has been confirmed. As the main site of Hg deposition, the kidney is also one of the main target organs of Hg toxicity. It was found that mice treated with HgCl(2) induced the increase of MDA content and ACP activity, and inhibited the activities of LDH and ALP [134]. Hg exposure can also cause pathological changes in the kidney, such as glomerular, proximal, and distal tubular degeneration [134].

Ni is a kind of harmful pollutant, excessive Ni exposure can cause toxicity to human and animal organs. NiCl2 was found to cause mitochondrial damage in mouse liver. NiCl2 inhibits mitochondrial biogenesis by reducing protein and mRNA expression levels of PGC-1α, TFAM, and NRF1 [135]. In addition, mitochondrial fusion proteins (Mfn1, Mfn2) were significantly reduced, and mitochondrial fission proteins (Drip1, Fis1) were upregulated [135]. The detection of mitochondrial autophagy showed that $NiCl_2$ could promote the aggregation of mitochondrial PINK1 and the recruitment of Parkin, and the expressions of mitochondrial autophagy receptor proteins Bnip3 and FUNDC1 were increased in the liver of NiCl2-treated mice [135]. Ni treatment induces ROS production and oxidative stress, upregulates lipid peroxidation levels in the liver, and depletes intracellular glutathione levels [136]. While Ni can induce liver toxicity, its harm to kidney toxicity cannot be ignored. It was found that after Ni treatment, serum creatinine, and the urea and uric acid levels of mice were significantly upregulated, and epithelial cell necrosis and inflammatory cell infiltration were observed in renal tissue [137].

Cd as a heavy metal contaminant exists widely in the environment, it has high toxicity to human health and organs. Studies have suggested that Cd treatment can accumulate in the liver and cause oxidative stress and inflammatory responses [138]. The levels of AST, ALT, LDH, and ALP in the serum of mice treated with Cd were significantly upregulated [138]. In addition, SOD, GSH, GST, and CAT levels in mouse liver were inhibited, while MDA content was significantly upregulated [138]. In renal tissue, Cd treatment causes its accumulation in the kidneys and leads to nephrotoxicity by mediating oxidative stress, inflammatory responses, and apoptosis [139]. After Cd treatment, serum urea and creatinine levels of mice were significantly increased, and renal tissue lesions were significant [139]. In addition, MDA, nitric oxide, p65, TNF-α, and IL-1β were increased; the activities of glutathione, catalase, and SOD were inhibited by Cd [139]. It was found that exposure to low doses of a mixture of heavy metals Pb, Cd, and Hg affected the liver and kidneys of rats [95]. CYP450 and PON1 were elevated in the liver and KIM1 in the kidney after heavy metal mixtures [95]. Exposure to heavy metal mixtures caused histopathological lesions in the liver and kidneys [95].

3.4 Conclusions

This chapter discusses the kinetics of the toxic effects of chemical mixtures exposed in organisms. It begins with an introduction to the main contaminants (organic compounds and heavy metals) in the environment, in terms of the morphology of the compounds/heavy metals, their production, and the main sources of their corresponding contamination. It contributes to a comprehensive understanding of the properties of chemical mixtures and their possible toxicity mechanisms. This is followed by an introduction to the occurrence of toxicity in the target organs (heart, nerves, liver, kidneys, etc.) of zebrafish and mice. It is helpful to understand the main toxicity mode of model biological target organs and to study the toxic effects of compounds on the body. Finally, focusing on the target organs (heart, nerve/brain, blood vessels, liver, kidney, etc.), we analyzed the toxicodynamics of environmental pollutants on the target organs of zebrafish/mice. Zebrafish and mice as the main model animals, with the heart, nerves, blood vessels, and liver/kidney as the main target organs, to analyze the toxicodynamics of different chemical contaminants, which will help subsequent in-depth studies on the toxic mechanisms of chemical mixtures.

References

1 Sexton K, Hattis D. Assessing cumulative health risks from exposure to environmental mixtures – three fundamental questions. Environ Health Perspect. 2007, 115(5): 825–832.

2 Kienzler A, Bopp SK, van der Linden S, Berggren E, Worth A. Regulatory assessment of chemical mixtures: requirements, current approaches and future perspectives. Regul Toxicol Pharmacol. 2016, 80: 321–334.

3 Ford LC, Jang S, Chen Z, Zhou YH, Gallins PJ, Wright FA, Chiu WA, Rusyn I. A population-based human in vitro approach to quantify inter-individual variability in responses to chemical mixtures. Toxics. 2022, 10(8): 441.

4 Song C, Liu X, Song Y, Liu R, Gao H, Han L, Peng J. Key blackening and stinking pollutants in Dongsha River of Beijing: spatial distribution and source identification. J Environ Manage. 2017, 200: 335–346.

5 Montzka SA, Dlugokencky EJ, Butler JH. Non-CO2 greenhouse gases and climate change. Nature. 2011, 476(7358): 43–50.

6 Liu Y, Yvon-Lewis SA, Thornton DCO, Campbell L, Bianchi T. Spatial distribution of brominated very short-lived substances in the eastern Pacific. J Geophys Res Oceans. 2013, 118(5): 2318–2328.

7 Brown SG, Eberly S, Paatero P, Norris GA. Methods for estimating uncertainty in PMF solutions: examples with ambient air and water quality data and guidance on reporting PMF results. Sci Total Environ. 2015, 518–519: 626–635.

8 Wang J, Wu S, Yang Q, Gu Y, Wang P, Li Z, Li L. Performance and mechanism of the in situ restoration effect on VHCs in the polluted river water based on the orthogonal experiment: photosynthetic fluorescence characteristics and microbial community analysis. Environ Sci Pollut Res Int. 2022, 29(28): 43004–43018.

9 Wang H, Wang J, Bo G, Wu S, Luo L. Degradation of pollutants in polluted river water using Ti/IrO2-Ta2O5 coating electrode and evaluation of electrode characteristics. J Clean Prod. 2020, 273: 123019.

10 David E, Niculescu VC. Volatile Organic Compounds (VOCs) as environmental pollutants: occurrence and mitigation using nanomaterials. Int J Environ Res Public Health. 2021, 18(24): 13147.

11 Pandey P, Yadav R. A review on volatile organic compounds (VOCs) as environmental pollutants: fate and distribution. Int J Plant Environ. 2018, 4(2): 14–26.

12 Liang HM, Liao CM. Modeling VOC-odor exposure risk in livestock buildings. Chemosphere. 2007, 68(4): 781–789.

13 Dai C, Han Y, Duan Y, Lai X, Fu R, Liu S, Leong KH, Tu Y, Zhou L. Review on the contamination and remediation of polycyclic aromatic hydrocarbons (PAHs) in coastal soil and sediments. Environ Res. 2022, 205: 112423.

14 Abdel-Shafy HI, Mansour MSM. A review on polycyclic aromatic hydrocarbons: source, environmental impact, effect on human health and remediation. Egypt J Pet. 2016, 25(1): 107–123.

15 Armstrong B, Hutchinson E, Unwin J, Fletcher T. Lung cancer risk after exposure to polycyclic aromatic hydrocarbons: a review and meta-analysis. Environ Health Perspect. 2004, 112(9): 970–978.

16 Al Osman M, Yang F, Massey IY. Exposure routes and health effects of heavy metals on children. Biometals. 2019, 32(4): 563–573.

17 Hamdy S, Shaban AM, Aziz Y, Mahmoud A, Moeme LA, Ibrahim W, Gad NS. Ameliorative role of *Jania rubens* alga against toxicity of heavy metal polluted water in male rats. 2019, 2(2): 38–46.

18 Jaishankar M, Tseten T, Anbalagan N, Mathew BB, Beeregowda KN. Toxicity, mechanism and health effects of some heavy metals. Interdiscip Toxicol. 2014, 7(2): 60–72.

19 Yin J, Wang L, Chen Y, Zhang D, Hegazy AM, Zhang X. A comparison of accumulation and depuration effect of dissolved hexavalent chromium (Cr6+) in head and muscle of bighead carp (Aristichthys nobilis) and assessment of the potential health risk for consumers. Food Chem. 2019, 286: 388–394.

20 Mohinuzzaman M, Saadat AHM, Islam SMN, Tareq SM. Health risk assessment of chromium-accumulated fish and vegetables at Gulshan Lake of Bangladesh: a case study. Pollution. 2018, 4(3): 459–469.

21 Mani MS, Kabekkodu SP, Joshi MB, Dsouza HS. Ecogenetics of lead toxicity and its influence on risk assessment. Hum Exp Toxicol. 2019, 38(9): 1031–1059.

22 Tong S, von Schirnding YE, Prapamontol T. Environmental lead exposure: a public health problem of global dimensions. Bull World Health Organ. 2000, 78(9): 1068–1077.

23 Singh N, Kumar A, Gupta VK, Sharma B. Biochemical and molecular bases of lead-induced toxicity in mammalian systems and possible mitigations. Chem Res Toxicol. 2018, 31(10): 1009–1021.

24 Hegazy AMS, Fouad UA. Evaluation of lead hepatotoxicity; histological, histochemical and ultrastructural study. Forensic Med Anat Res. 2014, 02(03): 70–79.

25 Liu J, Xue J, Wei X, Su H, Xu R. Optimization of Cr^{6+} removal by bacillus subtilis strain SZMC 6179 J from chromium-containing soil. Indian J Microbiol. 2020, 60(4): 430–435.

26 Elahi A, Ajaz M, Rehman A, Vuilleumier S, Hussain SZ. Isolation, characterization, and multiple heavy metal-resistant and hexavalent chromium-reducing Microbacterium testaceum B-HS2 from tannery effluent. J King Saud Univ Sci. 2019, 31(4): 1437–1444.

27 Rahman Z, Singh VP. The relative impact of toxic heavy metals (THMs) (arsenic (As), cadmium (Cd), chromium (Cr)(VI), mercury (Hg), and lead (Pb)) on the total environment: an overview. Environ Monit Assess. 2019, 191(7): 419.

28 Dhal B, Thatoi HN, Das NN, Pandey BD. Chemical and microbial remediation of hexavalent chromium from contaminated soil and mining/metallurgical solid waste: a review. J Hazard Mater. 2013, 250-251: 272–291.

29 Beckers F, Rinklebe J. Cycling of mercury in the environment: sources, fate, and human health implications: a review. Crit Rev Environ Sci Technol. 2017, 47(9): 693–794.

30 Singh AD, Khanna K, Kour J, Dhiman S, Bhardwaj T, Devi K, Sharma N, Kumar P, Kapoor N, Sharma P, Arora P, Sharma A, Bhardwaj R. Critical review on biogeochemical dynamics of mercury (Hg) and its abatement strategies. Chemosphere. 2023, 319: 137917.

31 Ganguly J, Kulshreshtha D, Jog M. Mercury and movement disorders: the toxic legacy continues. Can J Neurol Sci. 2022, 49(4): 493–501.

32 Teng D, Mao K, Ali W, Xu G, Huang G, Niazi NK, Feng X, Zhang H. Describing the toxicity and sources and the remediation technologies for mercury-contaminated soil. RSC Adv. 2020, 10(39): 23221–23232.

33 Yang L, Zhang Y, Wang F, Luo Z, Guo S, Strähle U. Toxicity of mercury: molecular evidence. Chemosphere. 2020, 245: 125586.

34 Zambelli B, Uversky VN, Ciurli S. Nickel impact on human health: an intrinsic disorder perspective. Biochim Biophys Acta. 2016, 1864(12): 1714–1731.

35 Zdrojewicz Z, Popowicz E, Winiarski J. Nikiel – rola w organizmie człowieka i działanie toksyczne [Nickel – role in human organism and toxic effects]. Pol Merkur Lekarski. 2016, 41(242): 115–118.

36 Guo H, Liu H, Wu H, Cui H, Fang J, Zuo Z, Deng J, Li Y, Wang X, Zhao L. Nickel carcinogenesis mechanism: DNA damage. Int J Mol Sci. 2019, 20(19): 4690.

37 Shahzad B, Tanveer M, Rehman A, Cheema SA, Fahad S, Rehman S, Sharma A. Nickel; whether toxic or essential for plants and environment – a review. Plant Physiol Biochem. 2018, 132: 641–651.

38 Hedfi A, Mahmoudi E, Boufahja F, Beyrem H, Aïssa P. Effects of increasing levels of nickel contamination on structure of offshore nematode communities in experimental microcosms. Bull Environ Contam Toxicol. 2007, 79(3): 345–349.

39 McIlveen WD, Negusanti JJ. Nickel in the terrestrial environment. Sci Total Environ. 1994, 148(2-3): 109–138.

40 Mcgrath SP, Smith S. Chromium and nickel in heavy metals in soils. Blackie & Sons, 1990.

41 Hassan MU, Chattha MU, Khan I, Chattha MB, Aamer M, Nawaz M, Ali A, Khan MAU, Khan TA. Nickel toxicity in plants: reasons, toxic effects, tolerance mechanisms, and remediation possibilities-a review. Environ Sci Pollut Res Int. 2019, 26(13): 12673–12688.

42 Doll R. Nickel exposure: a human health hazard. IARC Sci Publ. 1984, (53): 3–21.

43 Zeinali T, Salmani F, Naseri K. Dietary Intake of cadmium, chromium, copper, nickel, and lead through the consumption of meat, liver, and kidney and assessment of human health risk in Birjand, Southeast of Iran. Biol Trace Elem Res. 2019, 191(2): 338–347.

44 Bernhoft RA. Cadmium toxicity and treatment. Sci World J. 2013, 2013: 394652.

45 Recatalá L, Sánchez J, Arbelo C, Sacristán D. Testing the validity of a Cd soil quality standard in representative Mediterranean agricultural soils under an accumulator crop. Sci Total Environ. 2010, 409(1): 9–18.

46 Bhattacharya S. The role of probiotics in the amelioration of cadmium toxicity. Biol Trace Elem Res. 2020, 197(2): 440–444.

47 Howe K, Clark MD, Torroja CF, Torrance J, Berthelot C, Muffato M, Collins JE, Humphray S, McLaren K, Matthews L, McLaren S, Sealy I, Caccamo M, Churcher C, Scott C, Barrett JC, Koch R, Rauch GJ, White S, Chow W, Kilian B, Quintais

LT, Guerra-Assunção JA, Zhou Y, Gu Y, Yen J, Vogel JH, Eyre T, Redmond S, Banerjee R, Chi J, Fu B, Langley E, Maguire SF, Laird GK, Lloyd D, Kenyon E, Donaldson S, Sehra H, Almeida-King J, Loveland J, Trevanion S, Jones M, Quail M, Willey D, Hunt A, Burton J, Sims S, McLay K, Plumb B, Davis J, Clee C, Oliver K, Clark R, Riddle C, Elliot D, Threadgold G, Harden G, Ware D, Begum S, Mortimore B, Kerry G, Heath P, Phillimore B, Tracey A, Corby N, Dunn M, Johnson C, Wood J, Clark S, Pelan S, Griffiths G, Smith M, Glithero R, Howden P, Barker N, Lloyd C, Stevens C, Harley J, Holt K, Panagiotidis G, Lovell J, Beasley H, Henderson C, Gordon D, Auger K, Wright D, Collins J, Raisen C, Dyer L, Leung K, Robertson L, Ambridge K, Leongamornlert D, McGuire S, Gilderthorp R, Griffiths C, Manthravadi D, Nichol S, Barker G, Whitehead S, Kay M, Brown J, Murnane C, Gray E, Humphries M, Sycamore N, Barker D, Saunders D, Wallis J, Babbage A, Hammond S, Mashreghi-Mohammadi M, Barr L, Martin S, Wray P, Ellington A, Matthews N, Ellwood M, Woodmansey R, Clark G, Cooper J, Tromans A, Grafham D, Skuce C, Pandian R, Andrews R, Harrison E, Kimberley A, Garnett J, Fosker N, Hall R, Garner P, Kelly D, Bird C, Palmer S, Gehring I, Berger A, Dooley CM, Ersan-Ürün Z, Eser C, Geiger H, Geisler M, Karotki L, Kirn A, Konantz J, Konantz M, Oberländer M, Rudolph-Geiger S, Teucke M, Lanz C, Raddatz G, Osoegawa K, Zhu B, Rapp A, Widaa S, Langford C, Yang F, Schuster SC, Carter NP, Harrow J, Ning Z, Herrero J, Searle SM, Enright A, Geisler R, Plasterk RH, Lee C, Westerfield M, de Jong PJ, Zon LI, Postlethwait JH, Nüsslein-Volhard C, Hubbard TJ, Roest Crollius H, Rogers J, Stemple DL. The zebrafish reference genome sequence and its relationship to the human genome. Nature. 2013, 496(7446): 498–503.

48 Lu C, Luo J, Liu Y, Yang X. The oxidative stress responses caused by phthalate acid esters increases mRNA abundance of base excision repair (BER) genes in vivo and in vitro. Ecotoxicol Environ Saf. 2021, 208: 111525.

49 Ma YB, Lu CJ, Junaid M, Jia PP, Yang L, Zhang JH, Pei DS. Potential adverse outcome pathway (AOP) of silver nanoparticles mediated reproductive toxicity in zebrafish. Chemosphere. 2018, 207: 320–328.

50 Fitzgerald JA, Könemann S, Krümpelmann L, Županič A, Vom Berg C. Approaches to test the neurotoxicity of environmental contaminants in the zebrafish model: from behavior to molecular mechanisms. Environ Toxicol Chem. 2021, 40(4): 989–1006.

51 Wilson SW, Ross LS, Parrett T, Easter SS Jr. The development of a simple scaffold of axon tracts in the brain of the embryonic zebrafish, Brachydanio rerio. Development. 1990, 108(1): 121–145.

52 Satou C, Kimura Y, Hirata H, Suster ML, Kawakami K, Higashijima S. Transgenic tools to characterize neuronal properties of discrete populations of zebrafish neurons. Development. 2013, 140(18): 3927–3931.

53 Yelon D, Horne SA, Stainier DY. Restricted expression of cardiac myosin genes reveals regulated aspects of heart tube assembly in zebrafish. Dev Biol. 1999, 214(1): 23–37.

54 Rottbauer W, Baker K, Wo ZG, Mohideen MA, Cantiello HF, Fishman MC. Growth and function of the embryonic heart depend upon the cardiac-specific L-type calcium channel alpha1 subunit. Dev Cell. 2001, 1(2): 265–275.

55 Kim YS, Kim MJ, Koo TH, Kim JD, Koun S, Ham HJ, Lee YM, Rhee M, Yeo SY, Huh TL. Histone deacetylase is required for the activation of Wnt/β-catenin signaling crucial for heart valve formation in zebrafish embryos. Biochem Biophys Res Commun. 2012, 423(1): 140–146.

56 Isogai S, Horiguchi M, Weinstein BM. The vascular anatomy of the developing zebrafish: an atlas of embryonic and early larval development. Dev Biol. 2001, 230(2): 278–301.

57 Yamaguchi TP, Dumont DJ, Conlon RA, Breitman ML, Rossant J. flk-1, an flt-related receptor tyrosine kinase is an early marker for endothelial cell precursors. Development. 1993, 118(2): 489–498.

58 Shalaby F, Rossant J, Yamaguchi TP, Gertsenstein M, Wu XF, Breitman ML, Schuh AC. Failure of blood-island formation and vasculogenesis in Flk-1-deficient mice. Nature. 1995, 376(6535): 62–66.

59 Eichmann A, Corbel C, Nataf V, Vaigot P, Bréant C, Le Douarin NM. Ligand-dependent development of the endothelial and hemopoietic lineages from embryonic mesodermal cells expressing vascular endothelial growth factor receptor 2. Proc Natl Acad Sci U S A. 1997, 94(10): 5141–5146.

60 Liao W, Bisgrove BW, Sawyer H, Hug B, Bell B, Peters K, Grunwald DJ, Stainier DY. The zebrafish gene cloche acts upstream of a flk-1 homologue to regulate endothelial cell differentiation. Development. 1997, 124(2): 381–389.

61 Tao T, Peng J. Liver development in zebrafish (Danio rerio). J Genet Genomics. 2009, 36(6): 325–334.

62 Driessen M, Kienhuis AS, Pennings JL, Pronk TE, van de Brandhof EJ, Roodbergen M, Spaink HP, van de Water B, van der Ven LT. Exploring the zebrafish embryo as an alternative model for the evaluation of liver toxicity by histopathology and expression profiling. Arch Toxicol. 2013, 87(5): 807–823.

63 Vliegenthart AD, Tucker CS, Del Pozo J, Dear JW. Zebrafish as model organisms for studying drug-induced liver injury. Br J Clin Pharmacol. 2014, 78(6): 1217–1227.

64 Ding YJ, Chen YH. Developmental nephrotoxicity of aristolochic acid in a zebrafish model. Toxicol Appl Pharmacol. 2012, 261(1): 59–65.

65 Usanmaz SE, Akarsu ES, Vural N. Neurotoxic effects of acute and subacute formaldehyde exposures in mice. Environ Toxicol Pharmacol. 2002, 11(2): 93–100.

66 Bushdid PB, Osinska H, Waclaw RR, Molkentin JD, Yutzey KE. NFATc3 and NFATc4 are required for cardiac development and mitochondrial function. Circ Res. 2003, 92(12): 1305–1313.

67 Ghanayem BI, Hoffler U. Investigation of xenobiotics metabolism, genotoxicity, and carcinogenicity using Cyp2e1(-/-) mice. Curr Drug Metab. 2007, 8(7): 728–749.

68 Cai W, Wu J, Hong L, Xu Y, Tang Q, Shi C. Oxidative injury and hepatocyte apoptosis in total parenteral nutrition-associated liver dysfunction. J Pediatr Surg. 2006, 41(10): 1663–1668.

69 Ni W, Zhang Y, Yin Z. The protective mechanism of Klotho gene-modified bone marrow mesenchymal stem cells on acute kidney injury induced by rhabdomyolysis. Regen Ther. 2021, 18: 255–267.

70 Xu XM, Cai GY, Bu R, Wang WJ, Bai XY, Sun XF, Chen XM. Beneficial effects of caloric restriction on chronic kidney disease in rodent models: a meta-analysis and systematic review. PLoS One. 2015, 10(12): e0144442.

71 Rana MN, Tangpong J, Rahman MM. Toxicodynamics of lead, cadmium, mercury and arsenic- induced kidney toxicity and treatment strategy: a mini review. Toxicol Rep. 2018, 5: 704–713.

72 Escher BI, Ashauer R, Dyer S, Hermens JL, Lee JH, Leslie HA, Mayer P, Meador JP, Warne MS. Crucial role of mechanisms and modes of toxic action for understanding tissue residue toxicity and internal effect concentrations of organic chemicals. Integr Environ Assess Manag. 2011, 7(1): 28–49.

73 Vohra BP, Hui X. Taurine protects against carbon tetrachloride toxicity in the cultured neurons and in vivo. Arch Physiol Biochem. 2001, 109(1): 90–94.

74 Wang F, Li C, Liu W, Jin Y. Potential mechanisms of neurobehavioral disturbances in mice caused by sub-chronic exposure to low-dose VOCs. Inhal Toxicol. 2014, 26(4): 250–258.

75 Chatterjee N, Kim C, Im J, Kim S, Choi J. Mixture and individual effects of benzene, toluene, and formaldehyde in zebrafish (Danio rerio) development: metabolomics, epigenetics, and behavioral approaches. Environ Toxicol Pharmacol. 2023, 97: 104031.

76 Chepelev NL, Moffat ID, Bowers WJ, Yauk CL. Neurotoxicity may be an overlooked consequence of benzo[a]pyrene exposure that is relevant to human health risk assessment. Mutat Res Rev Mutat Res. 2015, 764: 64–89.

77 Lyu Y, Ren XK, Zhang HF, Tian FJ, Mu JB, Zheng JP. Sub-chronic administration of benzo[a]pyrene disrupts hippocampal long-term potentiation via inhibiting CaMK II/PKC/PKA-ERK-CREB signaling in rats. Environ Toxicol. 2020, 35(9): 961–970.

78 Yang K, Jiang X, Su Q, Wang J, Li C, Xia Y, Cheng S, Qin Q, Cao X, Chen C, Tu B. Disruption of glutamate neurotransmitter transmission is modulated by SNAP-25 in benzo[a]pyrene-induced neurotoxic effects. Toxicology. 2017, 384: 11–22.

79 Wang J, Li CL, Tu BJ, Yang K, Mo TT, Zhang RY, Cheng SQ, Chen CZ, Jiang XJ, Han TL, Peng B, Baker PN, Xia YY. Integrated epigenetics, transcriptomics, and metabolomics to analyze the mechanisms of benzo[a]pyrene neurotoxicity in the hippocampus. Toxicol Sci. 2018, 166(1): 65–81.

80 Knecht AL, Truong L, Simonich MT, Tanguay RL. Developmental benzo[a]pyrene (B[a]P) exposure impacts larval behavior and impairs adult learning in zebrafish. Neurotoxicol Teratol. 2017, 59: 27–34.

81 Knecht AL, Truong L, Marvel SW, Reif DM, Garcia A, Lu C, Simonich MT, Teeguarden JG, Tanguay RL. Transgenerational inheritance of neurobehavioral and physiological deficits from developmental exposure to benzo[a]pyrene in zebrafish. Toxicol Appl Pharmacol. 2017, 329: 148–157.

82 Li Y, Cao J, Hao Z, Liu A, Li X, Li H, Xia N, Wang Z, Zhang Z, Bai J, Zhang H. Aspirin ameliorates the cognition impairment in mice following benzo[a]pyrene treatment via down-regulating BDNF IV methylation. Neurotoxicology. 2022, 89: 20–30.

83 Xu Y, Li G, Han C, Sun L, Zhao R, Cui S. Protective effects of Hippophae rhamnoides L. juice on lead-induced neurotoxicity in mice. Biol Pharm Bull. 2005, 28(3): 490–494.

84 Li X, Kong H, Ji X, Gao Y, Jin M. Zebrafish behavioral phenomics applied for phenotyping aquatic neurotoxicity induced by lead contaminants of environmentally relevant level. Chemosphere. 2019, 224: 445–454.

85 Xu Y, Wang L, Zhu J, Jiang P, Zhang Z, Li L, Wu Q. Chromium induced neurotoxicity by altering metabolism in zebrafish larvae. Ecotoxicol Environ Saf. 2021, 228: 112983.

86 Kaur P, Aschner M, Syversen T. Glutathione modulation influences methyl mercury induced neurotoxicity in primary cell cultures of neurons and astrocytes. Neurotoxicology. 2006, 27(4): 492–500.

87 Vandewater LJ, Racz WJ, Norris AR, Buncel E. Methylmercury distribution, metabolism, and neurotoxicity in the mouse brain. Can J Physiol Pharmacol. 1983, 61(12): 1487–1493.

88 Rasinger JD, Lundebye AK, Penglase SJ, Ellingsen S, Amlund H. Methylmercury induced neurotoxicity and the influence of selenium in the brains of adult zebrafish (Danio rerio). Int J Mol Sci. 2017, 18(4): 725.

89 Nabinger DD, Altenhofen S, Bitencourt PER, Nery LR, Leite CE, Vianna MRMR, Bonan CD. Nickel exposure alters behavioral parameters in larval and adult zebrafish. Sci Total Environ. 2018, 624: 1623–1633.

90 Šulinskienė J, Bernotienė R, Baranauskienė D, Naginienė R, Stanevičienė I, Kašauskas A, Ivanov L. Effect of zinc on the oxidative stress biomarkers in the brain of nickel-treated mice. Oxid Med Cell Longev. 2019, 2019: 8549727.

91 Dong C. Protective effect of proanthocyanidins in cadmium-induced neurotoxicity in mice. Drug Res (Stuttg). 2015, 65(10): 555–560.

92 Xu Y, Zhao H, Wang Z, Gao H, Liu J, Li K, Song Z, Yuan C, Lan X, Pan C, Zhang S. Developmental exposure to environmental levels of cadmium induces neurotoxicity and activates microglia in zebrafish larvae: from the perspectives of neurobehavior and neuroimaging. Chemosphere. 2022, 291(Pt 1): 132802.

93 Karri V, Schuhmacher M, Kumar V. Heavy metals (Pb, Cd, As and MeHg) as risk factors for cognitive dysfunction: a general review of metal mixture mechanism in brain. Environ Toxicol Pharmacol. 2016, 48: 203–213.

94 Rai A, Maurya SK, Khare P, Srivastava A, Bandyopadhyay S. Characterization of developmental neurotoxicity of As, Cd, and Pb mixture: synergistic action of metal mixture in glial and neuronal functions. Toxicol Sci. 2010, 118(2): 586–601.

95 Zhou F, Yin G, Gao Y, Liu D, Xie J, Ouyang L, Fan Y, Yu H, Zha Z, Wang K, Shao L, Feng C, Fan G. Toxicity assessment due to prenatal and lactational exposure to lead, cadmium and mercury mixtures. Environ Int. 2019, 133(Pt B): 105192.

96 Chen Z, Huo X, Chen G, Luo X, Xu X. Lead (Pb) exposure and heart failure risk. Environ Sci Pollut Res Int. 2021, 28(23): 28833–28847.

97 Pai VB, Nahata MC. Cardiotoxicity of chemotherapeutic agents: incidence, treatment and prevention. Drug Saf. 2000, 22(4): 263–302.

98 Taleb A, Ahmad KA, Ihsan AU, Qu J, Lin N, Hezam K, Koju N, Hui L, Qilong D. Antioxidant effects and mechanism of silymarin in oxidative stress-induced cardiovascular diseases. Biomed Pharmacother. 2018, 102: 689–698.

99 Hamed H, Gargouri M, Bellassoued K, Ghannoudi Z, Elfeki A, Gargouri A. Cardiopreventive effects of camel milk against carbon tetrachloride-induced oxidative stress, biochemical and histological alterations in mice. Arch Physiol Biochem. 2018, 124(3):253–260.

100 Zelko IN, Dassanayaka S, Malovichko MV, Howard CM, Garrett LF, Uchida S, Brittian KR, Conklin DJ, Jones SP, Srivastava S. Chronic benzene exposure aggravates pressure overload-induced cardiac dysfunction. Toxicol Sci. 2021, 185(1): 64–76.

101 Zhang S, Ou K, Huang J, Fang L, Wang C. In utero exposure to mixed PAHs causes heart mass reduction in adult male mice. Ecotoxicol Environ Saf. 2021, 225: 112804.

102 Incardona JP, Linbo TL, Scholz NL. Cardiac toxicity of 5-ring polycyclic aromatic hydrocarbons is differentially dependent on the aryl hydrocarbon receptor 2 isoform during zebrafish development. Toxicol Appl Pharmacol. 2011, 257(2): 242–249.

103 Davuljigari CB, Gottipolu RR. Late-life cardiac injury in rats following early life exposure to lead: reversal effect of nutrient metal mixture. Cardiovasc Toxicol. 2020, 20(3): 249–260.

104 Boşgelmez Iİ, Güvendik G. Beneficial effects of N-acetyl-L-cysteine or taurine pre- or post-treatments in the heart, spleen, lung, and testis of hexavalent chromium-exposed mice. Biol Trace Elem Res. 2019, 190(2): 437–445.

105 Amara IE, Elshenawy OH, Abdelrady M, El-Kadi AO. Acute mercury toxicity modulates cytochrome P450, soluble epoxide hydrolase and their associated arachidonic acid metabolites in C57Bl/6 mouse heart. Toxicol Lett. 2014, 226(1): 53–62.

106 Liu Y, Ma W, Liu Q, Liu P, Qiao S, Xu L, Sun Y, Gai X, Zhang Z. Decreased thioredoxin reductase 3 expression promotes nickel-induced damage to cardiac tissue via activating oxidative stress-induced apoptosis and inflammation. Environ Toxicol. 2023, 38(2): 436–450.

107 Chou SH, Lin HC, Chen SW, Tai YT, Jung SM, Ko FH, Pang JS, Chu PH. Cadmium exposure induces histological damage and cytotoxicity in the cardiovascular system of mice. Food Chem Toxicol. 2023, 175: 113740.

108 Manna P, Sinha M, Sil PC. Amelioration of cadmium-induced cardiac impairment by taurine. Chem Biol Interact. 2008, 174(2): 88–97.

109 Nabil-Adam A, Shreadah MA. Ameliorative role of Ulva extract against heavy metal mixture-induced cardiovascular through oxidative/antioxidant pathways and inflammatory biomarkers. Environ Sci Pollut Res Int. 2021, 28(21): 27006–27024.

110 Rennie MY, Detmar J, Whiteley KJ, Yang J, Jurisicova A, Adamson SL, Sled JG. Vessel tortuousity and reduced vascularization in the fetoplacental arterial tree after maternal exposure to polycyclic aromatic hydrocarbons. Am J Physiol Heart Circ Physiol. 2011, 300(2): H675–684.

111 Bugiak BJ, Weber LP. Phenotypic anchoring of gene expression after developmental exposure to aryl hydrocarbon receptor ligands in zebrafish. Aquat Toxicol. 2010, 99(3): 423–437.

112 Garcia-Arenas G, Claudio L, Perez-Severiano F, Rios C. Lead acetate exposure inhibits nitric oxide synthase activity in capillary and synaptosomal fractions of mouse brain. Toxicol Sci. 1999, 50(2): 244–248.

113 Cuevas AK, Liberda EN, Gillespie PA, Allina J, Chen LC. Inhaled nickel nanoparticles alter vascular reactivity in C57BL/6 mice. Inhal Toxicol. 2010, 22 Suppl 2(0 2): 100–106.

114 Ying Z, Xu X, Chen M, Liu D, Zhong M, Chen LC, Sun Q, Rajagopalan S. A synergistic vascular effect of airborne particulate matter and nickel in a mouse model. Toxicol Sci. 2013, 135(1): 72–80.

115 Houston MC. The role of mercury and cadmium heavy metals in vascular disease, hypertension, coronary heart disease, and myocardial infarction. Altern Ther Health Med. 2007, 13(2): S128–133.

116 Oliveira TF, Batista PR, Leal MA, Campagnaro BP, Nogueira BV, Vassallo DV, Meyrelles SS, Padilha AS. Chronic cadmium exposure accelerates the development of atherosclerosis and induces vascular dysfunction in the aorta of ApoE-/- Mice. Biol Trace Elem Res. 2019, 187(1): 163–171.

117 Donpunha W, Kukongviriyapan U, Sompamit K, Pakdeechote P, Kukongviriyapan V, Pannangpetch P. Protective effect of ascorbic acid on cadmium-induced hypertension and vascular dysfunction in mice. Biometals. 2011, 24(1): 105–115.

118 Said ES, Mohammed AH, Ali HM, Babiker AY, Alnughaymishi R, Althaqeel NZ, Ahmed AS. Evaluation of hepatoprotective effect of Nebivolol and sodium copper Chlorophyllin on CCL4-induced hepatotoxicity in mice. Eur Rev Med Pharmacol Sci. 2022, 26(5): 1717–1728.

119 Ebaid H, Al-Tamimi J, Hassan I, Alhazza I, Al-Khalifa M. Antioxidant bioactivity of Samsum ant (Pachycondyla sennaarensis) venom protects against CCL_4-induced nephrotoxicity in mice. Oxid Med Cell Longev. 2014, 2014: 763061.

120 Wang F, Liu F, Liu H, Chen W, Si X, Ma X. Effects of immunological and hematological parameter in mice exposed to mixture of volatile organic compounds. Inhal Toxicol. 2016, 28(4): 164–169.

121 Li F, Xiang B, Jin Y, Li C, Ren S, Wu Y, Li J, Luo Q. Hepatotoxic effects of inhalation exposure to polycyclic aromatic hydrocarbons on lipid metabolism of C57BL/6 mice. Environ Int. 2020, 134: 105000.

122 Jin Y, Miao W, Lin X, Wu T, Shen H, Chen S, Li Y, Pan Q, Fu Z. Sub-chronically exposing mice to a polycyclic aromatic hydrocarbon increases lipid accumulation in their livers. Environ Toxicol Pharmacol. 2014, 38(2): 353–363.

123 Mai Y, Peng S, Li H, Lai Z. Histological, biochemical and transcriptomic analyses reveal liver damage in zebrafish (Danio rerio) exposed to phenanthrene. Comp Biochem Physiol C Toxicol Pharmacol. 2019, 225: 108582.

124 Sharma A, Sharma V, Kansal L. Amelioration of lead-induced hepatotoxicity by Allium sativum extracts in Swiss albino mice. Libyan J Med. 2010, 5(1): 4621.

125 Wardani G, Farida N, Andayani R, Kuntoro M, Sudjarwo SA. The potency of red seaweed (eucheuma cottonii) extracts as hepatoprotector on lead acetate-induced hepatotoxicity in mice. Pharmacognosy Res. 2017, 9(3): 282–286.

126 Gluhcheva Y, Pashkunova-Martic I, Schaier M, Vladov I, Stoykova S, Petrova E, Pavlova E, Dorkov P, Helbich TH, Keppler B, Ivanova J. Comparative effects of deferiprone and salinomycin on lead-induced disturbance in the homeostasis of intrarenal essential elements in mice. Int J Mol Sci. 2022, 23(8): 4368.

127 Shen J, Kom MC, Huang H, Fu G, Xie Y, Gao Y, Tang Y, Yan J, Jin L. Role of NF-κB signaling pathway in hexavalent chromium-induced hepatotoxicity. Environ Toxicol. 2023, 38(6): 1361–1371.

128 Shaw P, Mondal P, Dey Bhowmik A, Bandyopadhyay A, Sudarshan M, Chakraborty A, Chattopadhyay A. Environmentally relevant hexavalent chromium disrupts elemental homeostasis and induces apoptosis in zebrafish liver. Bull Environ Contam Toxicol. 2022, 108(4): 716–724.

129 Shaw P, Mondal P, Bandyopadhyay A, Chattopadhyay A. Environmentally relevant concentration of chromium activates Nrf2 and alters transcription of related XME genes in liver of zebrafish. Chemosphere. 2019, 214: 35–46.

130 Boşgelmez İİ, Güvendik G. N-acetyl-L-cysteine protects liver and kidney against chromium(VI)-induced oxidative stress in mice. Biol Trace Elem Res. 2017, 178(1): 44–53.

131 Nauroze T, Ali S, Kanwal L, Ara C, Akbar Mughal T, Andleeb S. Ameliorative effect of Nigella sativa conjugated silver nanoparticles against chromium-induced hepatotoxicity and renal toxicity in mice. Saudi J Biol Sci. 2023, 30(3): 103571.

132 Ung CY, Lam SH, Hlaing MM, Winata CL, Korzh S, Mathavan S, Gong Z. Mercury-induced hepatotoxicity in zebrafish: in vivo mechanistic insights from transcriptome analysis, phenotype anchoring and targeted gene expression validation. BMC Genomics. 2010, 11: 212.

133 Chen QL, Sun YL, Liu ZH, Li YW. Sex-dependent effects of subacute mercuric chloride exposure on histology, antioxidant status and immune-related gene expression in the liver of adult zebrafish (Danio rerio). Chemosphere. 2017, 188: 1–9.

134 Sharma MK, Sharma A, Kumar A, Kumar M. Evaluation of protective efficacy of Spirulina fusiformis against mercury-induced nephrotoxicity in Swiss albino mice. Food Chem Toxicol. 2007, 45(6): 879–887.

135 Guo H, Wei L, Wang Y, Cui H, Deng H, Zhu Y, Deng J, Geng Y, Ouyang P, Lai W, Du Z, Ni X, Yin H, Fang J, Zuo Z. Nickel induces hepatotoxicity by mitochondrial biogenesis, mitochondrial dynamics, and mitophagy dysfunction. Environ Toxicol. 2023, 38(5): 1185–1195.

136 Liu CM, Zheng GH, Ming QL, Chao C, Sun JM. Sesamin protects mouse liver against nickel-induced oxidative DNA damage and apoptosis by the PI3K-Akt pathway. J Agric Food Chem. 2013, 61(5): 1146–1154.

137 Kadi IE, Dahdouh F. Vitamin C pretreatment protects from nickel-induced acute nephrotoxicity in mice. Arh Hig Rada Toksikol. 2016, 67(3): 210–215.

138 Yang Z, He Y, Wang H, Zhang Q. Protective effect of melatonin against chronic cadmium-induced hepatotoxicity by suppressing oxidative stress, inflammation, and apoptosis in mice. Ecotoxicol Environ Saf. 2021, 228: 112947.

139 Algefare AI. Renoprotective and oxidative stress-modulating effects of taxifolin against cadmium-induced nephrotoxicity in mice. Life (Basel). 2022, 12(8): 1150.

4

Principal Sources, Fate, and Mechanism of Chemical Mixtures in the Environment

Bao-Fu Zhang[a], *Yiyun Liu*[a], *De-Sheng Pei*[a,*]

[a] *School of Public Health, Chongqing Medical University, Chongqing, China*

*Corresponding author.
E-mail: peids@cqmu.edu.cn (D.S.P.)

4.1 Introduction

Medical studies have demonstrated that genetics alone cannot fully explain up to two-thirds of the risk of chronic diseases in humans. This suggests that environmental factors and the interaction between genes and the environment play a significant role [1]. Furthermore, according to the Lancet Committee on Pollution and Health, approximately 16% of premature deaths worldwide are attributed to environmental pollutants [2]. While studies on the toxicity of individual pollutants have been extensively conducted, the reality is that environmental conditions involve a complex interplay of multiple pollutants with structural interactions. This interaction often leads to the emergence of a new class of pollutants, which pose significant risks to both human health and the environment.

Most of the chemical mixtures come from human daily activities and can enter the environment through various mediations. When the residue is released into the environment, all possible pollutants will go through physical, chemical, and biological actions to change the composition of the original pollutants, degrade the pollutants, and reduce the degree of pollution [3]. However, some chemicals can be completely degraded and reduce the number of heterophytes and the formation of new pollutants, and other chemicals can combine with other pollutants to form new pollutants in the process of degradation [4] (Figure 4.1). These compounds acquire new properties in the process of transformation. Compared with the parent compounds, these transformed products show more obvious toxicity and stronger persistence, and this chemical transformation follows the biotic or abiotic transformation process [5].

4.2 MP Mixtures

Numerous studies have confirmed that microplastics (MPs) are widely present in various water bodies such as oceans, rivers, lakes, beaches, and agricultural soils and have become an environmental issue of global concern [6–8]. MPs are plastic particles ≤5 mm, which are composed of primary and secondary MPs. [9, 10]. The primary MPs refer to the plastic particles produced directly, while the secondary MPs refer to the plastic fragments, which are decomposed under ultraviolet light, weathering, high temperature, oxidation, freeze–thaw cycle, and biological action [11, 12]. MPs have the characteristics of species, size, strong adsorption capacity, easy migration, hydrophobicity, and persistence and can be widely distributed in various environmental media. The soil environment connects the atmosphere, water, rock, and other environmental systems, so the soil environment is considered to be an important medium of MPs pollution [13].

MPs in soil mainly come from two sources: human activities, such as fertilization, plastic film mulching, and household waste, and natural factors such as deposition and plastic degradation. The abundance, type, size, and shape of MPs are

Toxicological Assessment of Combined Chemicals in the Environment, First Edition. Edited by De-Sheng Pei and Yiyun Liu.

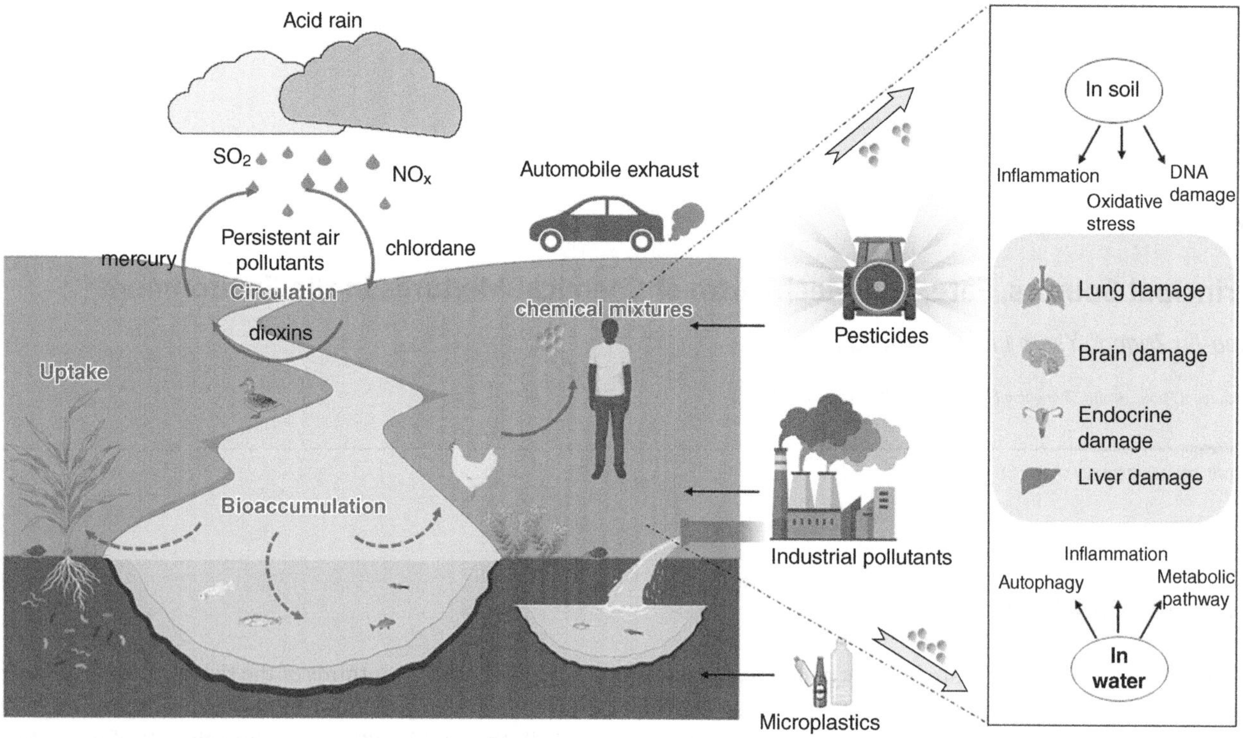

Figure 4.1 Principal sources, fate, and mechanism of chemical mixtures in the environment.

closely related to their source. Smaller particles of MPs can directly enter the human circulatory system through the food chain, and there is a possibility of cancer, a serious threat to human health, which even affects the entire ecosystem [14, 15]. In the long-term process of natural evolution and biodegradation, MPs can not only leach and release harmful substances such as organic or inorganic plasticizers and polymers but also adsorb heavy metals (HMs), organic pollutants, and antibiotics on their surface and then further migrate and accumulate into the surrounding environmental media and organisms through energy flow, material flow, and biological accumulation [16, 17]. The study further revealed the adsorption mechanism of MPs and other pollutants. The adsorption mechanism mainly includes hydrophobic interaction [18, 19], electrostatic interaction [20], van der Waals interaction [21], surface complexation cooperation [22], π–π bond, hydrogen bond [23], monolayer cover [24], hole filling mechanism [25], and intraparticle diffusion [26]. The results show that the adsorption capacity of MPs is directly proportional to hydrophobicity, and the more hydrophobic the MPs are, the stronger the adsorption capacity for most pollutants is. The sources of adsorbed organic pollutants in soil may be pesticides, wastewater, and herbicides [27]. Other studies have shown that polycyclic aromatic hydrocarbons (PAHs), polychlorinated biphenyls (PCBs), glyphosate, and amoxicillin (AMX) can be adsorbed on soil MPs [28, 29]. MPs affect the migration, toxicity, bioaccumulation, and degradation of organic pollutants through adsorption and desorption [30]. Some studies have pointed out that MPs have synergistic toxicity to the body after absorbing organic pollutants such as bisphenol A (BPA). This study compared the toxicity of polystyrene MPs with different particle sizes to colon cancer Caco-2 cells before and after the adsorption of BPA [31]. It was found that nanosized MPs were easier to adsorb BPA because of their large specific surface area, resulting in increased mitochondrial depolarization and oxidative stress, resulting in enhanced cytotoxicity [31]. As pigments and stabilizers in plastic products, HMs can be absorbed by MPs and accumulated in plants and animals in soil. Previous studies have shown that simultaneous exposure to MPs and cadmium reduces the growth rate and mortality of earthworms and aggravates oxidative damage, and MPs may increase the bioaccessibility of HM ions in the soil environment and increase the retention of MPs particles in earthworms [32]. Zhang et al. [33] showed that individual or combined exposure to PS-MPs (10 mg/L) and 3,6-dibromocarbazole (3,6-DBCZ; 0.5 mg/L) could significantly increase the rate of zebrafish embryo deformity. Furthermore, exposure to 3,6-DBCZ or PS-MPs increased reactive oxygen species (ROS) levels in zebrafish embryos. Comparably, the levels of oxidative stress in zebrafish embryos were significantly reduced with the combined exposure of 3,6-DBCZ and PS-MPs. These observations suggest that the combined exposure of 3,6-DBCZ and

PS-MPs has an antagonistic effect on oxidative stress and apoptosis. Animal husbandry uses antibiotics to fight disease, and animal feces containing antibiotics are excreted into the soil, which are selectively enriched in MPs, which act as carriers to promote their migration [34]. Antibiotics and HMs are more easily absorbed by MPs, which leads to the larger the particle size of MPs, the stronger the weathering effect, or the higher the content of antibiotics and HMs in MPs when MPs come from the soil with longer cultivation years [35].

In addition, the surface of MPs that have existed in the environment for a long time will undergo varying degrees of weathering and aging. The aging process can change the surface properties of MPs (size, surface area, porosity, surface polarity), thereby changing the adsorption capacity of MPs. The research of Wu et al. [36] shows that aged PP particles have a looser material structure and larger amorphous domains to form small-sized plastic particles, and the aging process will help to produce plastic particles on the surface of the particles. More available adsorption sites can improve the adsorption capacity of PP for triclosan (TCS). Zhang et al. [37] used an artificial aging mode to study the impact of the aging process on the adsorption of 9-NAnt on MPs. The results showed that the adsorption capacity of PS decreased at most. This is because the aging process will introduce oxygen-containing functional groups and enhance the surface polarity of MPs, thereby hindering the adsorption process of hydrophobic organic pollutants.

MPs/NPs in the environment are formed by the fragmentation of plastic macromolecular polymers. Due to its high persistence, environmental MPs/NPs are difficult to completely eliminate by degradation, leading to increased environmental accumulation of MPs/NPs [38]. External pressure (such as freezing and thawing of plastics in an aqueous environment) may also lead to the mechanical degradation of plastics [39, 40]. But to break down most plastics, more complex reactions mediated by free radicals triggered by solar radiation are needed [41]. In recent years, MPs/NPs pollution has been reported in both freshwater and ocean environments. The sources of MPs in freshwater primarily include the discharge of wastewater from sewage treatment plants, industrial and domestic sources, and runoff from soil and road surfaces. In contrast, land-based plastic waste is considered the primary source of marine MPs. Recently, it has been reported that these plastic fragments can be used as carriers to absorb harmful chemicals, through their interaction with polymers, and then transferred to different nutritional levels in the food chain. Studies have shown that MPs can be used as carriers of persistent organic pollutants (POPs) in organisms. In the presence of MPs, the accumulation of organic pollutants in organisms is also increased [42, 43]. Combined exposure to MPs will affect the survival rate, predation ability, biochemical indexes, histopathological damage, and gene expression. Avio et al. [44] found that combined exposure to MPs and pyrene significantly decreased the activity of acetylcholinesterase (AChE) in the gills of mussels, and the same phenomenon was also found in goby. Further studies showed that phenanthrene or fluoranthene combined with MPs had a high level of histopathological damage to catfish or mussels [45, 46]. When the adsorbed chemicals and plastics reach the intestines of aquatic organisms, endocrine-disrupting chemicals (EDCs) are used to disrupt the endocrine system, resulting in an imbalance of hormone levels in the organism [47, 48]. It also affects organisms, such as crustaceans and other aquatic fish, disrupting their hormonal regulation and interfering with biochemical pathways, leading to metabolic changes [49]. Rainieri et al. [50] found that PE MPs (125–25 μm) adsorbed organic pollutants (PCBs, perfluorinated compounds, and methylmercury) had stronger compound toxicological effects on zebrafish than single MP exposure or organic exposure, resulting in higher vacuolization in zebrafish liver. In addition, the compound pollutants formed by the combination of MPs and HMs can affect the oxidative stress of aquatic organisms or produce synergistic toxicity with superoxide dismutase, AChE, and catalase [51].

4.3 Pesticide Mixtures

Pesticides are important agricultural chemicals that are widely used around the world and are of great significance in promoting agricultural development. While pesticides in the soil are adsorbed by soil solid matter, water-soluble pesticides can penetrate into groundwater with precipitation, or migrate laterally and diffuse to surrounding water bodies through surface runoff. Fat-soluble pesticides are easily adsorbed by soil and are absorbed by crop roots due to poor mobility, causing chronic harm to organisms high in the food chain. Many pesticide combinations can cause changes in metabolic pathways, alter lipid metabolism, cause oxidative stress, cause homeostasis disorders, disrupt endocrine glands, cause malformations, affect fetal growth, and may even sometimes cause carcinogenic mutagenesis when altering DNA [52]. In some related reports, organophosphorus and carbamate pesticides inhibit the activity of AChE through an additive concentration mode to aggravate neurotoxicity [53]. In in vivo experimental studies, these two types of pesticides can interact with other biochemical targets in the organism. For example, oxidative desulfurization and carboxylesterase can combine with these two types of pesticides and covalently combine to phosphorylate them, which can effectively prevent AChE from being inhibited, thereby degrading pesticides and playing a detoxification role [54–56]. If carboxylesterase is inhibited, the toxic effects of pesticides are enhanced [57]. In addition, reports have found that 90% of pesticides used on crops such as certain cereals and corn can alter the metabolism of glucose, lipids, amino acids, and AChE. The combination of phenylacetylglycine and hippuric acid has been observed to disrupt intestinal microbiota, which can also be excreted through the urine of affected individuals. Metabolic changes in pathways such as amino acid metabolism in urea cycle disorders may occur when patients are exposed to pesticide mixtures containing atrazine, ticks, and endosulfan [58]. Previous studies have shown that the use of single and combined herbicides and insecticides poses serious threats to some taxa, but no changes in metamorphosis were observed. Pesticides kill 99% of amphibian larvae, causing free radicals to produce ROS, causing inflammation, and disrupting metabolic pathways. Some data on exposure to combinatorial pesticides, either alone or in combination, include organophosphates [59, 60], organochlorines, pyrethroids, and carbamates, showing how these pesticides induce liver and brain abnormalities in different model organisms, resulting in metabolic and neurological systemic disease. Used alone or in combination, these pesticides can cause mitochondrial damage due to depletion of antioxidants, production of ROS, and upregulation of exogenous metabolic enzymes, thereby triggering lipid peroxidation in the cells of exposed individuals [61–63]. This releases inflammatory cytokines that affect insulin receptor phosphorylation, leading to impaired insulin signaling [64, 65].

The combined toxicity of organophosphorus insecticides (OIs) and copper was studied in the Pacific oyster *Crassostrea gigas* [66]. The results showed that compared with the use of pesticides or copper alone, developmental abnormalities occurred during combined exposure. The number of larvae and the expression levels of genes related to xenobiotic transformation (e.g. glutathione S-transferase [GST]) increased. Furthermore, simultaneous exposure to cadmium and dimethoate (a pesticide) alters many enzyme activities in the freshwater snail *Galba truncatula* [67], which results in increased activities of aspartate aminotransferase, alanine aminotransferase, alkaline phosphatase, lactate dehydrogenase, glutathione peroxidase and catalase, AChE, and glucose-6-phosphate dehydrogenase and reduced glycogen activity.

4.4 HM and Metalloid Metal Mixtures

HMs and metalloid metals are naturally produced and found in the whole crust, but with the acceleration of industrialization and urbanization, high concentrations of sewage containing HMs are discharged into the aquatic environment, resulting in metal pollution. Microorganisms can accumulate in higher organisms through food webs, thus affecting human health [68]. HMs naturally exist in aquatic environments in low concentrations. HMs can absorb other environmental pollutants, such as MPs, creating various mixtures that are then absorbed and accumulated by aquatic organisms [69, 70]. With the development of characterization methods and the deepening of toxicological research, the toxicological effects, and ecological health risks under the coexistence of many kinds of HMs, the effects of exposure to mixtures of HMs in aquatic organisms may vary depending on the type and quantity of metals, species, exposure time, and surrounding environment [71–73]. Several studies have reported that simultaneous exposure to multiple metals can lead to competition for cell binding sites, changes in absorption pathways, and changes in detoxification mechanisms, resulting in synergistic toxicity [72].

According to various studies, the combined impact of POPs and HMs results in more toxicity than that of POPs or HMs alone. In the amphipod *Hyalella azteca*, combined exposure to two PAHs (phenanthrene or phenanthrenequinone) and copper led to increased mortality and ROS levels compared to copper exposure alone. The accumulation of copper was increased in the presence of phenanthrenequinone, but was not changed in the presence of phenanthrene [74]. The combined toxicity of a pesticide mixture and copper was studied in the Pacific oyster *C. gigas* [66].

Nanoparticles could induce synergistic or addictive toxic effects by acting as carriers for HMs in aquatic invertebrates. For example, the interactional effects of titanium dioxide nanoparticles ($nTiO_2$) and inorganic mercury (IHg) or methyl mercury (MeHg) were assessed in water fleas in freshwater [75]. The results showed higher adsorption of IHg on $nTiO_2$ than MeHg and an increase in the total mercury burden for IHg and MeHg depending on the concentration of $nTiO_2$ during direct waterborne and $nTiO_2$ pre-treated two hours exposure, indicating that $nTiO_2$ might combine with mercury and thus increase uptake of mercury by water flea. Also, combined exposure to carbon nanoparticles and chromium showed synergistic activation of electron transport system activity in the clam *Ruditapes decussatus*, but had no effects on neurotoxicity or antioxidant enzyme activities. In contrast, combination exposure changed the activities of antioxidant enzymes (SOD, GPx) with no effect on other parameters in the clam *Ruditapes philippinarum* [76]. These varied effects demonstrate that species-specific responses can be induced under even the same exposure conditions.

4.5 Nanoparticle Mixtures

At all stages of production, transportation, consumption, and understanding, nanoparticles may enter the natural environment, especially the water environment, through spillage, leakage, waste, and discharge [77]. The toxicity and reactivity of nanomaterials with different core chemical compositions are different; for example, the toxicity of carbon nanomaterials in the water environment is much lower than that of metal or metal oxide nanoparticles [78]. The ability of nanoparticles to absorb pollutants plays a decisive role in the toxicity of nanoparticles or pollutants. The high specific surface area of nanomaterials makes it easy to adsorb other pollutants in the water environment, such as pesticides, HMS, POPs, and natural organic poisons, as carriers for the transport of these pollutants, causing pollutants to enter organisms in the form of free molecules (or ions) and/or particulate pollutant complexes [79]. Furthermore, it affects its distribution in the environment and its bioavailability and produces a variety of compound toxic effects on organisms.

Carbon nanotubes (CNTs) have a strong affinity for pollutants [80, 81]. The outer surface, interstitial pore space, inner cavity, and peripheral grooves are four possible adsorption sites, most of which are accessible and allow the adsorption of contaminants [82], thereby promoting the bioaccumulation of pollutants in the target organism. For example, the combined effect of single-walled carbon nanotubes (SWCNTs) and metal Cd co-exposure was discussed, and it was found that SWCNTs were fed after adsorbing HMs, which increased the mortality of *Daphnia magna* [83]. Xu et al. [84] applied graphitic carbon nitride nanosheets (g-C3N4 NSs) for the alleviation of Cd toxicity in soybeans. The g-C3N4 NSs supplementation significantly improved plant growth and reduced oxidative damage in the Cd-intoxicated soybean seedlings through hydroponic culture.

POPs are ubiquitous in the water environment and can be absorbed by fish [85]. Phenanthrene is a kind of PAHs. It has been found that SWCNTs promotes the bioaccumulation of phenanthrene in *Oryzias latipes*, and the bioaccumulation and release of phenanthrene depend on the uptake and digestion of SWCNT [83]. Multiwalled carbon nanotubes (MWCNTs) increased the inhibition of photosynthesis of *Chlorella vulgaris* by five times [79]. MWCNTs can also aggravate the immobilization of *Spirulina platensis* induced by phenanthrene [86]. Similarly, a mixture of MWCNTs and 2 BDE-47-tetrabromodiphenyl ether (BDE-47) attenuated the developmental inhibition, oxidative stress, apoptosis, and DNA damage induced by BDE-47 in the early developmental stage of zebrafish [87]. Previous studies have also shown that oxidized MWCNTs enhance lead-induced acute lead toxicity of *Tilapia niloticus*, leading to the proliferation and displacement of epithelial structures and epithelial cells, leading to severe histopathological changes [88].

Metal oxide nanoparticles are another common class of nanoparticles, which can also increase the toxicity of pollutants through the "Trojan Horse effect." It is reported that there is an additive effect between $nTiO_2$ and dioxins (TCDD). In the experiment on marine shells, Canesi et al. found that $nTiO_2$ and NPs can promote the accumulation of TCDD in Mediterranean mussels, which may be due to the adsorption effect of $nTiO_2$ and NPs on TCDD, which can be used as a carrier of TCDD to promote the transport of TCDD, showing the "Troy effect" [89]. $NTiO_2$ increases the bioavailability and toxicity of BPA, leading to endocrine disorders and reproductive disorders [90]. Nano-Al_2O_3 and As(V) synergistically increased the mortality of *Ceriodaphnia dubia* as reported by Fang et al. [91]. Mice co-exposed to zinc oxide nanoparticles and $Pb(Ac)_2$ showed more lead deposition in organs and higher levels of liver ROS, proinflammatory cytokines, and liver injury [92].

4.6 Persistent Organic Pollutants

POPs are ubiquitous environmental pollutants that can adversely affect living organisms and are highly resistant to environmental degradation [93, 94]. POPs leach into water during use, coexisting with other chemical mixtures like HMs and MPs, leading to toxic changes.

When HMs and POPs coexist in the same environmental medium, POPs can pass through the cell membrane due to their lipid affinity, causing structural changes in the membrane and membrane potential, thereby increasing the membrane fluidity [95]. Increased membrane fluidity improves the bioavailability of HMs with low lipid affinity. The electron-rich functional groups of POPs act as electron donors and interact with metal ions. Phenolic hydroxyl groups, amine groups, and carboxyl groups are common functional groups in organic matter. Different electron-donating abilities lead to differences in the complexing abilities of the organic pollutants they constitute. At the same time, the greater the number of electron-rich functional groups, the stronger the complexing ability of organic matter [96]. Cation–π interactions between PAHs and HMS form PAH-HMS complexes. Lipophilic organometallic complexes are more likely to be membrane permeable to HMs, thus promoting the bioaccumulation and toxicity of HMs [97]. According to various studies, the combined effects of POPs and HMs are more toxic than POPs or HMs alone. Exposure of the amphipod *Hyalella azteca* to both phenanthrene or phenanthrenequinone and copper resulted in greater mortality and ROS levels than exposure to copper alone. The accumulation of copper increased in the presence of phenanthrenequinone, but there was no change in the accumulation of copper in the presence of phenanthrenequinone [98]. However, not all POP and HM interactions result in increased toxicity; in some cases, the result may be reduced or no change in the toxic effects of HMs on aquatic invertebrates. In chironomids, benthic invertebrates exposed to both cypermethrin and lead; the LC_{50} increased compared with each pollutant alone, showing an antagonistic effect [74].

When MPs and POPs coexist in the same environmental medium, their interaction mainly relies on the adsorption capacity of MPs. Research by Wang et al. [18] pointed out that due to the different functional groups of POPs, the adsorption level of perfluorooctanesulfonamide (FOSA) on PE is higher than that of perfluorooctanesulfonate (PFOS). The molecular composition and structure of MPs, specifically the surface oxygen–containing functional groups (carbonyl oxygen) and the spatial arrangement of the rubber domain, also vary. As a result, the adsorption of PFOS and FOSA on PE and polyvinyl chloride (PVC) is stronger than that of PS. Chen et al. [25] also pointed out that due to the different physical and chemical properties and molecular structures of PE and PVC, the adsorption of PVC on tris(2-chloroethyl phosphate) ester (TCEP) has a higher affinity than PE, but PVC has a lower affinity for tri-*n*-butyl phosphate (TnBP) than PE. Since PVC has a higher adsorption capacity for TCS, in the presence of TCS, in this case, the toxic effect of PVC on *Skeletonema costatum* is less than that of PE and PS [99]. Interacting with PE MPs, glyphosate acid inhibited large leeches by 53.3%, while glyphosate exposure alone inhibited large leeches by 18.3% [100]; 13 μm PA alleviated the toxicity of nonylphenol (NP) to *Chlorella pyrenoidosa* at 48 hours, while 150 μm PA reduced the toxicity of NP at 96 hours [101]. PS with a particle size of 5 μm will not significantly affect the toxic effect of triphenyltin chloride (TPTCl) on *Chlorella pyrenoidosa*, while PS with a nanoparticle size (550 nm) will increase the toxic effect of TPTCl [102]. Furthermore, nanoscale PS (50 and 500 nm) caused stronger accumulation in marine rotifers compared with micron-sized PS (6 μm) [103].

4.7 Antibiotic Mixtures

In recent years, environmental pollution and ecological risks caused by antibiotics as emerging contaminants have gained widespread attention. Antibiotics have the property of inhibiting the growth and inactivating specific pathogenic microorganisms, leading to their frequent use in treating diseases in humans and animals and in promoting animal growth. Antibiotics are widely used in humans and animals. About 30–90% of antibiotics persist in the environment as prototypes or metabolites, contributing to a vicious cycle of transformation and bioaccumulation that threatens ecosystems [104]. Antibiotics have been detected in various environmental media, including surface water, sediment, soil, groundwater, and drinking water [105]. Antibiotic residues in the environment, due to their exogenous nature and high biological activity, can affect the population size and structure of various microorganisms and other organisms at different trophic levels. They can even kill or inhibit organisms that maintain ecosystem stability, seriously affecting the ecosystem's balance [106]. Additionally, the selective pressure of antibiotic residues can lead to the transfer of drug-resistant genes to human pathogens, causing cell death, metabolic dysfunction, and other problems [107].

When antibiotics coexist with HMs, many antibiotics, such as the widely used quinolones and tetracycline (TC) antibiotics, contain carboxyl, carbonyl, or piperazine groups, which can act as potential electron donors to coordinate HMs [108]; in the case of sulfonamide antibiotics, the type and quantity of functional groups are small, and nitrogen-containing functional groups are mainly involved in coordination [109]. Although TC antibiotics have many functional groups and are theoretically prone to complex reactions with HM ions, an excessive number of functional groups. It will bring steric hindrance and reduce the complexing ability [110]. Some studies have shown that the reasons why TC antibiotics easily form complexes with HMs include good electronic transitions, the ionic radii of HM ions, and the stability of chelate rings [111].

The type of HM ions will also affect the complexation of antibiotics and HMs. Studies [112] have shown that the higher the charge number of HM ions, the greater the possibility of complexing with antibiotics. K^+ Monovalent HM ions such as monovalent HM ions are unable to complex with antibiotics. Cu^{2+} and Fe^{3+} have strong complexing abilities, but their complexing effects with antibiotics are equivalent. The essence of complexing between HM ions and antibiotics is that HM ions share functional group atoms with antibiotics. Thus, increasing the electronegativity (electron-withdrawing ability) of HM ions will amplify the complexing effect.

Currently, few studies have investigated the joint toxicity of antibiotics and HMs. Existing studies have shown that compound pollution of antibiotics and HMs has a joint toxic effect on organisms. Combined exposure to fluoroquinolones

(FQ), ciprofloxacin (CIP) and enrofloxacin (ENR), and metals (Cd and Cu) at environmentally relevant concentrations (0.2 μM) has been reported to be genotoxic to zebrafish embryos Including DNA damage and apoptosis. Compared with single exposure to either FQ or metals, combined exposure caused less ROS overproduction but higher genotoxicity [113].

Antibiotics and MNPs have similar sources and migration pathways, and they inevitably coexist in the aquatic environment, forming composite pollution. The hydrophobicity of antibiotics and the properties related to their hydrophobicity (log K_{ow} and ionization constant pKa) play an important role in the adsorption process [114]. Due to their large specific surface area, strong hydrophobicity, and high mobility, MNPs can accumulate various toxins and chemical pollutants in the environment and serve as carriers for long-distance transport of pollutants. Due to the specific structures and properties of antibiotics and MNPs, there are large differences in the adsorption capacity and the ability to form complexes. The adsorption process of antibiotics and MNPs is affected by hydrogen bonds, hydrophobic interactions, van der Waals forces, and ion exchange. The combination of the two causes harm to organisms through bioaccumulation. Some specific functional groups in antibiotics contribute to the formation of hydrogen bonds. For example, the carbonyl group (a proton acceptor group) in AMX, TC, and CIP and the amide groups in polyamide (PA) (proton donor groups) can form hydrogen bonds [115, 116]. Yang et al. [117] found that as the pH value increased, the hydrogen ions on CIP decreased, and the adsorption amount on PS decreased. Therefore, it was speculated that CIP and PS were adsorbed through hydrogen bonds. At the same time, hydrogen bonding has also been proven to be the main reason for the adsorption of sulfamethoxazole (SMX) on PA, PS, PVC, polyethylene (PE), polypropylene (PP), and polyethylene terephthalate (PET) mechanism [20]. Most MNPs have abundant alkyl groups and strong hydrophobicity, and the adsorption process is dominated by hydrophobic interactions [30]. Studies have shown that hydrophobic interaction dominates the adsorption of AMX, TC, CIP, trimethoprim (TMP), tylosin (TYL), SMX, sulfamethazine, and sulfadiazine (SDZ) on MNPs to a certain extent, and antibiotics with higher *n*-octanol/water partition coefficients (log K_{ow}) have a stronger affinity for MNPs [22, 115, 118–121]. Both PS and TC have benzene rings in their structures, so they are mainly adsorbed together through hydrophobic interactions [122]. Fu et al. [123] also found that aged PA, PVC, and PET can adsorb sulfonamide antibiotics through hydrophobic interactions. Van der Waals forces are also the most common way for MNPs to adsorb antibiotics, mainly consisting of π–π and electrostatic interactions. For example, PE only adsorbs CIP, TMP, and SDZ through van der Waals forces, and CIP in freshwater systems increases its adsorption capacity on the surface of MNPs through electrostatic attraction [116]. Oxytetracycline (OTC) adsorption to PS is predominantly driven by cation exchange [24]. PE adsorbs CIP through hydrophobic and electrostatic interactions [124]. Some MNPs can simultaneously adsorb antibiotics through hydrogen bonding and π–π stacking, allowing the two to stably combine to form a complex [125].

Numerous antibiotics and MNPs in aquatic environments can exert toxic effects on aquatic organisms and are transferred and enriched through the food chain, posing further threats to human health [126–128]. A large amount of literature shows that the compound contamination of antibiotics-MNPs may cause changes in the molecular tissue, cells, and behavior of organisms, leading to damage to organisms. The research results of Zhang et al. [129] showed that compared with the intestinal villi of crucian carp (*Carassius auratus*) in the control group, exposure to roxithromycin (ROX) resulted in vacuolization of intestinal epithelial cells and damage to intestinal cilia, while in aged MNPs (5 and 50 μm) and ROX combined exposure group, the phenomena of intestinal cell vacuolation, ciliary damage, and lymphocyte infiltration were more serious. The carrier effect of aged MNPs is enhanced by changes in their functional groups and specific surface area, which can increase the accumulation of ROX in crucian carp. Therefore, it can be inferred that the combined effect of PS/aging PS and ROX may show a synergistic relationship. Huang et al. [130] also obtained similar results using combined exposure of aged PS and SMX. Compared with SMX alone, the AChE activity in the combined exposure group of aged PS and SMX was significantly increased. Therefore, aged PS alleviated SMX-induced neurotoxicity. However, some studies have shown that sulfamethazine (SMZ) will have a significant impact on the stability and antioxidant properties of *Oryzias melastigma*'s intestinal flora, while PS (100 nm) only slightly changes the composition of the intestinal flora. However, combining PS with SMZ can reduce the overall intestinal toxicity of SMZ. It is speculated that the combined effect between PS and SMZ may be an antagonistic relationship, thereby reducing the intestinal toxicity of SMZ to medaka fish [131].

4.8 Conclusions

Chemical mixtures are pervasive in our environment. Their complex interactions hinder remediation and ecological risk assessment, which complicates conventional treatment. Research on compound pollution is limited and superficial. To progress, we must clarify the formation mechanisms of chemical mixtures. This includes bioavailability, toxicokinetics, biotransformation, and toxicodynamics of single and mixed chemicals. This knowledge will improve compound pollutant remediation. Furthermore, toxicity experiments are limited to specific organisms, lacking systematic approaches. There is potential to expand research. Future studies could assess the overall toxicity of organisms in soil, water, and air and investigate the combined toxicity of mixtures. Current research focuses on binary mixtures, but real-world pollution involves multiple substances. Multiple compound pollution is a significant concern for future research. Environmental complexity necessitates considering compound pollution to align research with real-world conditions. This comprehensive approach will advance our understanding of compound pollution challenges.

Acknowledgments

The authors are grateful for the support from the Chongqing Technology Innovation and Application Development Sichuan-Chongqing Special Key Project (CSTB2024TIAD-CYKJCXX0017), High-level Talents Project of Chongqing Medical University (No. R4014), and Chongqing Postdoctoral Innovation Mentor Studio (X7928 D.S.P.).

Competing Interests

The authors declare no competing interests.

References

1 Rappaport SM, Smith MT. Epidemiology. Environment and disease risks. Science. 2010, 330: 460–461.
2 Landrigan PJ, Fuller R, Acosta NJR, Adeyi O, Arnold R, Basu NN, Baldé AB, Bertollini R, Bose-O'Reilly S, Boufford JI, Breysse PN, Chiles T, Mahidol C, Coll-Seck AM, Cropper ML, Fobil J, Fuster V, Greenstone M, Haines A, Hanrahan D, Hunter D, Khare M, Krupnick A, Lanphear B, Lohani B, Martin K, Mathiasen KV, McTeer MA, Murray CJL, Ndahimananjara JD, Perera F, Potočnik J, Preker AS, Ramesh J, Rockström J, Salinas C, Samson LD, Sandilya K, Sly PD, Smith KR, Steiner A, Stewart RB, Suk WA, van Schayck OCP, Yadama GN, Yumkella K, Zhong M. The Lancet Commission on pollution and health. Lancet Lond Engl. 2018, 391: 462–512.
3 Vischetti C, Casucci C, De Bernardi A, Monaci E, Tiano L, Marcheggiani F, Ciani M, Comitini F, Marini E, Taskin E, Puglisi E. Sub-lethal effects of pesticides on the DNA of soil organisms as early ecotoxicological biomarkers. Front Microbiol. 2020, 11: 1892.
4 K'oreje K, Okoth M, Van Langenhove H, Demeestere K. Occurrence and point-of-use treatment of contaminants of emerging concern in groundwater of the Nzoia River basin, Kenya. Environ Pollut. 2022, 297: 118725.
5 Reymann T, Kerner M, Kümmerer K. Assessment of the biotic and abiotic elimination processes of five micropollutants during cultivation of the green microalgae *Acutodesmus obliquus*. Bioresour Technol Rep. 2020, 11: 100512.
6 Guo X, Wang J. The chemical behaviors of microplastics in marine environment: A review. Mar Pollut Bull. 2019, 142: 1–14.
7 Liu M, Lu S, Song Y, Lei L, Hu J, Lv W, Zhou W, Cao C, Shi H, Yang X, He D. Microplastic and mesoplastic pollution in farmland soils in suburbs of Shanghai, China. Environ Pollut. 2018, 242: 855–862.
8 Bai Y, Zang C, Gu M, Gu C, Shao H, Guan Y, Wang X, Zhou X, Shan Y, Feng K. Sewage sludge as an initial fertility driver for rapid improvement of mudflat salt-soils. Sci Total Environ. 2017, 578: 47–55.
9 Thompson RC, Olsen Y, Mitchell RP, Davis A, Rowland SJ, John AW, McGonigle D, Russell AE. Lost at sea: where is all the plastic? Science. 2004, 304(5672): 838.
10 Blettler MCM, Ulla MA, Rabuffetti AP, Garello N. Plastic pollution in freshwater ecosystems: macro-, meso-, and microplastic debris in a floodplain lake. Environ Monit Assess. 2017, 189(11): 581.
11 Yang X, An C, Feng Q, Boufadel M, Ji W. Aggregation of microplastics and clay particles in the nearshore environment: characteristics, influencing factors, and implications. Water Res. 2022, 224: 119077.
12 Qi R, Jones D L, Li Z, Liu Q, Yan C. Behavior of microplastics and plastic film residues in the soil environment: a critical review. Sci Total Environ. 2020, 703: 134722.
13 Delangiz N, Aliyar S, Pashapoor N, Nobaharan K, Asgari Lajayer B, Rodríguez-Couto S. Can polymer-degrading microorganisms solve the bottleneck of plastics' environmental challenges? Chemosphere. 2022, 294: 133709.

14 Cole M, Lindeque P, Fileman E, Halsband C, Goodhead R, Moger J, Galloway TS. Microplastic ingestion by zooplankton. Environ Sci Technol. 2013, 47: 6646–6655.

15 Fendall LS, Sewell MA. Contributing to marine pollution by washing your face: microplastics in facial cleansers. Mar Pollut Bull. 2009, 58: 1225–1228.

16 Yang H, Yumeng Y, Yu Y, Yinglin H, Fu B, Wang J. Distribution, sources, migration, influence and analytical methods of microplastics in soil ecosystems. Ecotoxicol Environ Saf. 2022, 243: 114009.

17 Riaz S, Nasreen S, Burhan Z, Shafique S, Alvi SA, Khan MA. Microplastics assessment in Arabian Sea fishes: accumulation, characterization, and method development. Braz J Biol. 2023;84: e270694. Published 2023 Feb 13. doi:10.1590/1519-6984.270694.

18 Wang F, Shih KM, Li XY. The partition behavior of perfluorooctanesulfonate (PFOS) and perfluorooctanesulfonamide (FOSA) on microplastics. Chemosphere. 2015, 119: 841–847.

19 Wang J, Liu X, Liu G, Zhang Z, Wu H, Cui B, Bai J, Zhang W. Size effect of polystyrene microplastics on sorption of phenanthrene and nitrobenzene. Ecotoxicol Environ Saf. 2019, 173: 331–338.

20 Guo X, Chen C, Wang J. Sorption of sulfamethoxazole onto six types of microplastics. Chemosphere. 2019, 228: 300–308.

21 Vaid M, Sarma K, Gupta A. Microplastic pollution in aquatic environments with special emphasis on riverine systems: current understanding and way forward. J Environ Manage. 2021, 293: 112860.

22 Guo X, Pang J, Chen S, Jia H. Sorption properties of tylosin on four different microplastics. Chemosphere. 2018, 209: 240–245.

23 Elizalde-Velázquez A, Subbiah S, Anderson TA, Green MJ, Zhao X, Cañas-Carrell JE. Sorption of three common nonsteroidal anti-inflammatory drugs (NSAIDs) to microplastics. Sci Total Environ. 2020, 715: 136974.

24 Zhang H, Wang J, Zhou B, Zhou Y, Dai Z, Zhou Q, Chriestie P, Luo Y. Enhanced adsorption of oxytetracycline to weathered microplastic polystyrene: kinetics, isotherms and influencing factors. Environ Pollut. 2018, 243: 1550–1557.

25 Wang W, Wang J. Comparative evaluation of sorption kinetics and isotherms of pyrene onto microplastics. Chemosphere. 2018, 193: 567–573.

26 Xu P-C, Guo J, Ma D, Ge W, Zhou Z-F, Chai C. [Sorption of polybrominated diphenyl ethers by virgin and aged microplastics]. Huan Jing Ke Xue Huanjing Kexue. 2020, 41: 1329–1337.

27 Wang J, Liu X, Li Y, Powell T, Wang X, Wang G, Zhang P. Microplastics as contaminants in the soil environment: a mini-review. Sci Total Environ. 2019, 691: 848–857.

28 Fan P, Yu H, Xi B, Tan W. A review on the occurrence and influence of biodegradable microplastics in soil ecosystems: are biodegradable plastics substitute or threat? Environ Int. 2022, 163: 107244.

29 Wang J, Coffin S, Sun C, Schlenk D, Gan J. Negligible effects of microplastics on animal fitness and HOC bioaccumulation in earthworm *Eisenia fetida* in soil Environ. Pollut. 2019, 249: 776–784.

30 Koelmans AA, Bakir A, Burton GA, Janssen CR. Microplastic as a vector for chemicals in the aquatic environment: critical review and model-supported reinterpretation of empirical studies. Environ Sci Technol. 2016, 50: 3315–3326.

31 Wang Q, Bai J, Ning B, Fan L, Sun T, Fang Y, Wu J, Li S, Duan C, Zhang Y, Liang J, Gao Z. Effects of bisphenol A and nanoscale and microscale polystyrene plastic exposure on particle uptake and toxicity in human Caco-2 cells. Chemosphere. 2020, 254: 126788.

32 Zhou Y, Liu X, Wang J. Ecotoxicological effects of microplastics and cadmium on the earthworm *Eisenia fetida*. J Hazard Mater. 2020, 392: 122273.

33 Zhang J, Bai Y, Meng H, Zhu Y, Yue H, Li B, Wang J, Wang J, Zhu L, Du Z. Combined toxic effects of polystyrene microplastics and 3,6-dibromocarbazole on zebrafish (*Danio rerio*) embryos. Sci Total Environ. 2024, 913: 169787.

34 Khalid N, Aqeel M, Noman A, Khan SM, Akhter N. Interactions and effects of microplastics with heavy metals in aquatic and terrestrial environments. Environ Pollut. 2021, 290(12): 118104.

35 Lu X-M, Lu P-Z, Liu X-P. Fate and abundance of antibiotic resistance genes on microplastics in facility vegetable soil. Sci Total Environ. 2020, 709: 136276.

36 Wu X, Liu P, Huang H, Gao S. Adsorption of triclosan onto different aged polypropylene microplastics: critical effect of cations. Sci Total Environ. 2020, 717: 137033.

37 Wu P, Cai Z, Jin H, Tang Y. Adsorption mechanisms of five bisphenol analogues on PVC microplastics. Sci Total Environ. 2019, 650: 671–678.

38 Horton AA, Walton A, Spurgeon DJ, Lahive E, Svendsen C. Microplastics in freshwater and terrestrial environments: evaluating the current understanding to identify the knowledge gaps and future research priorities. Sci Total Environ. 2017, 586: 127–141.

39 Kumar M, Chen H, Sarsaiya S, Qin S, Liu H, Awasthi MK, Kumar S, Singh L, Zhang Z, Bolan NS, Pandey A, Varjani S and Taherzadeh MJ. Current research trends on micro- and nano-plastics as an emerging threat to global environment: a review. J Hazard Mater. 2021, 409: 124967.

40 Liu Q, Chen C, Li M, Ke J, Huang Y, Bian Y, Guo S, Wu Y, Han Y, Liu M. Neurodevelopmental toxicity of polystyrene nanoplastics in *Caenorhabditis elegans* and the regulating effect of presenilin. ACS Omega 2020, 5: 33170–33177.

41 Liu K, Wang X, Wei N, Song Z, Li D. Accurate quantification and transport estimation of suspended atmospheric microplastics in megacities: implications for human health. Environ Int. 2019, 132: 105127.

42 Chua E M, Shimeta J, Nugegoda D, Morrison PD, Clarke BO. Assimilation of polybrominated diphenyl ethers from microplastics by the marine amphipod, *Allorchestes compressa*. Environ Sci Technol. 2014, 48: 8127–8134.

43 Wardrop P, Shimeta J, Nugegoda D, Morrison PD, Miranda A, Tang M, Clarke BO. Chemical pollutants sorbed to ingested microbeads from personal care products accumulate in fish. Environ Sci Technol. 2016, 50: 4037–4044.

44 Oliveira M, Ribeiro A, Hylland K, Guilhermino L. Single and combined effects of microplastics and pyrene on juveniles (0+ group) of the common goby *Pomatoschistus microps* (Teleostei, Gobiidae). Ecol Indic. 2013, 34: 641–647.

45 Karami A, Goh Y-M, Jahromi MF, Lazorchak JM, Abdullah M, Courtenay SC. Diploid and triploid African catfish (*Clarias gariepinus*) differ in biomarker responses to the pesticide chlorpyrifos. Sci Total Environ. 2016, 557–558: 204–211.

46 Paul-Pont I, Lacroix C, González Fernández C, Hégaret H, Lambert C, Le Goïc N, Frère L, Cassone A-L, Sussarellu R, Fabioux C, Guyomarch J, Albentosa M, Huvet A and Soudant P. Exposure of marine mussels Mytilus spp. to polystyrene microplastics: toxicity and influence on fluoranthene bioaccumulation. Environ Pollut. 2016, 216: 724–737.

47 Andreas, Hadibarata T, Sathishkumar P, Prasetia H, Hikmat, Pusfitasari ED, Tasfiyati AN, Muzdalifah D, Waluyo J, Randy A, Ramadhaningtyas DP, Zuas O, Sari AA. Microplastic contamination in the Skipjack Tuna (*Euthynnus affinis*) collected from Southern Coast of Java, Indonesia. Chemosphere. 2021, 276:130185.

48 Amir S, Shah STA, Mamoulakis C, Docea AO, Kalantzi O-I, Zachariou A, Calina D, Carvalho F, Sofikitis N, Makrigiannakis A, Tsatsakis A. Endocrine disruptors acting on estrogen and androgen pathways cause reproductive disorders through multiple mechanisms: a review. Int J Environ Res Public Health. 2021, 18: 1464.

49 Johnson P, Loganathan C, Krishnan V, Sakayanathan P, Raji V, Vijayan S, Sathishkumar P, Murugesan K, Palvannan T. Plant extract as environmental-friendly green catalyst for the reduction of hexavalent chromium in tannery effluent. Environ Technol. 2018, 39: 1376–1383.

50 Rainieri S, Conlledo N, Larsen BK, Granby K, Barranco A. Combined effects of microplastics and chemical contaminants on the organ toxicity of zebrafish (*Danio rerio*). Environ Res. 2018, 162: 135–143.

51 Santos D, Félix L, Luzio A, Parra S, Cabecinha E, Bellas J, Monteiro S M. Toxicological effects induced on early life stages of zebrafish (*Danio rerio*) after an acute exposure to microplastics alone or co-exposed with copper. Chemosphere 2020, 261: 127748.

52 Bellinger DC. Children's cognitive health: the influence of environmental chemical exposures. Altern Ther Health Med. 2007, 13: S140–144.

53 Scholz NL, Truelove NK, Labenia JS, Baldwin DH, Collier TK. Dose-additive inhibition of chinook salmon acetylcholinesterase activity by mixtures of organophosphate and carbamate insecticides. Environ Toxicol Chem. 2006, 25: 1200–1207.

54 Jokanović M. Biotransformation of organophosphorus compounds. Toxicology. 2001, 166: 139–160.

55 Pohanka M. Acetylcholinesterase inhibitors: a patent review (2008–present). Expert Opin Ther Pat. 2012, 22: 871–886.

56 Laetz CA, Baldwin DH, Collier TK, Hebert V, Stark JD, Scholz NL. The synergistic toxicity of pesticide mixtures: implications for risk assessment and the conservation of endangered Pacific salmon Environ. Health Perspect. 2009, 117: 348–353.

57 Barata C, Solayan A, Porte C. Role of B-esterases in assessing toxicity of organophosphorus (chlorpyrifos, malathion) and carbamate (carbofuran) pesticides to Daphnia magna. Aquat Toxicol Amst Neth. 2004, 66: 125–139.

58 Sathishkumar P, Palvannnan T, Rajesh RV, Boopathy R. Effect of pesticide exposure in erythrocyte membrane bound acetylcholinesterase. New Biotechnol. 2009, 25: S370–371.

59 Holmes E, Li JV, Athanasiou T, Ashrafian H, Nicholson JK. Understanding the role of gut microbiome-host metabolic signal disruption in health and disease. Trends Microbiol. 2011, 19: 349–359.

60 Tsuchiyama T, Ito Y, Oya N, Nomasa K, Sato H, Minato K, Kitamori K, Oshima S, Minematsu A, Niwa K, Katsuhara M, Fukatsu K, Miyazaki H, Ebara T, Kamijima M. Quantitative analysis of organophosphate pesticides and dialkylphosphates in duplicate diet samples to identify potential sources of measured urinary dialkylphosphates in Japanese women. Environ Pollut. 2022, 298: 118799.

61 Zhang Q, Yao Y, Wang Y, Zhang Q, Cheng Z, Li Y, Yang X, Wang L, Sun H. Plant accumulation and transformation of brominated and organophosphate flame retardants: a review. Environ Pollut. 2021, 288: 117742.

62 Gardes T, Portet-Koltalo F, Debret M, Copard Y. Historical and post-ban releases of organochlorine pesticides recorded in sediment deposits in an agricultural watershed, France. Environ Pollut. 2021, 288: 117769.

63 He H, Li Y, Shen R, Shim H, Zeng Y, Zhao S, Lu Q, Mai B, Wang S. Environmental occurrence and remediation of emerging organohalides: a review. Environ Pollut. 2021, 290: 118060.

64 Pereira PCG, Parente CET, Carvalho GO, Torres JPM, Meire RO, Dorneles PR, Malm O. A review on pesticides in flower production: a push to reduce human exposure and environmental contamination. Environ Pollut. 2021, 289: 117817.

65 Sabarwal A, Kumar K, Singh RP. Hazardous effects of chemical pesticides on human health-Cancer and other associated disorders. Environ Toxicol Pharmacol. 2018, 63: 103–114.

66 Mai H, Cachot J, Clérandeau C, Martin C, Mazzela N, Gonzalez P, Morin B. An environmentally realistic pesticide and copper mixture impacts embryonic development and DNA integrity of the Pacific oyster, Crassostrea gigas. Environ Sci Pollut Res Int. 2020, 27(4): 3600–3611.

67 Dai Y, Sun S, Cao R, Zhang H, Chen J, Geng N. Residual levels and health risk assessment of trace metals in Chinese resident diet. J Environ Sci (China). 2024, 136: 451–459.

68 Jeong H, Byeon E, Kim DH, Maszczyk P, Lee JS. Heavy metals and metalloid in aquatic invertebrates: a review of single/mixed forms, combination with other pollutants, and environmental factors. Mar Pollut Bull. 2023, 191: 114959.

69 Gu Y-G, Lin Q, Huang H-H, Wang L-G, Ning J-J, Du F-Y. Heavy metals in fish tissues/stomach contents in four marine wild commercially valuable fish species from the western continental shelf of South China Sea. Mar Pollut Bull. 2017, 114: 1125–1129.

70 Hao Z, Chen L, Wang C, Zou X, Zheng F, Feng W, Zhang D, Peng L. Heavy metal distribution and bioaccumulation ability in marine organisms from coastal regions of Hainan and Zhoushan, China. Chemosphere. 2019, 226: 340–350.

71 Nys C, Versieren L, Cordery KI, Blust R, Smolders E, De Schamphelaere KAC. Systematic evaluation of chronic metal-mixture toxicity to three species and implications for risk assessment. Environ Sci Technol. 2017, 51: 4615–4623.

72 Rebolledo UA, Páez-Osuna F, Fernández R. Single and mixture toxicity of As, Cd, Cr, Cu, Fe, Hg, Ni, Pb, and Zn to the rotifer *Proales similis* under different salinities. Environ Pollut. 2021, 271: 116357.

73 Koppel DJ, Adams MS, King CK, Jolley DF. Chronic toxicity of an environmentally relevant and equitoxic ratio of five metals to two Antarctic marine microalgae shows complex mixture interactivity. Environ Pollut. 2018, 242: 1319–1330.

74 Gauthier PT, Norwood WP, Prepas EE, Pyle GG. Metal-polycyclic aromatic hydrocarbon mixture toxicity in Hyalella azteca. 2. Metal accumulation and oxidative stress as interactive co-toxic mechanisms. Environ Sci Technol. 2015, 49(19): 11780–11788.

75 Li M, Slaveykova VI. Dual role of titanium dioxide nanoparticles in the accumulation of inorganic and methyl mercury by crustacean Daphnia magna through waterborne and dietary exposure. Environ Pollut. 2022, 295: 118619.

76 Sturla Lompré J, De Marchi L, Pinto J, Soares AMVM, Pretti C, Chielini F, Pereira E, Freitas R. Effects of carbon nanoparticles and chromium combined exposure in native (*Ruditapes decussatus*) and invasive (*Ruditapes philippinarum*) clams. Nanomater Basel Switz. 2023, 13: 690.

77 Mackevica A, Olsson ME, Hansen SF. The release of silver nanoparticles from commercial toothbrushes. J Hazard Mater. 2017, 322: 270–275.

78 Klaper RD. The known and unknown about the environmental safety of nanomaterials in commerce. Small. 2020, 16: e2000690.

79 Shen M, Xia X, Wang F, Zhang P, Zhao X. Influences of multiwalled carbon nanotubes and plant residue chars on bioaccumulation of polycyclic aromatic hydrocarbons by *Chironomus plumosus* larvae in sediment. Environ Toxicol Chem. 2012, 31: 202–209.

80 Schwab F, Bucheli TD, Camenzuli L, Magrez A, Knauer K, Sigg L, Nowack B. Diuron sorbed to carbon nanotubes exhibits enhanced toxicity to Chlorella vulgaris. Environ Sci Technol. 2013, 47: 7012–7019.

81 Yang XY, Edelmann RE, Oris JT. Suspended C60 nanoparticles protect against short-term UV and fluoranthene photo-induced toxicity, but cause long-term cellular damage in Daphnia magna. Aquat Toxicol Amst Neth. 2010, 100: 202–210.

82 Zhang S, Shao T, Kose HS, Karanfil T. Adsorption of aromatic compounds by carbonaceous adsorbents: a comparative study on granular activated carbon, activated carbon fiber, and carbon nanotubes. Environ Sci Technol. 2010, 44: 6377–6383.

83 Wang X, Qu R, Liu J, Wei Z, Wang L, Yang S, Huang Q, Wang Z. Effect of different carbon nanotubes on cadmium toxicity to Daphnia magna: the role of catalyst impurities and adsorption capacity. Environ Pollut. 2016, 208(Pt B): 732–738.

84 Xu K, Guo Y, Xing C, Fu R, Zou B, Liu R, Cai L, Yan J, Wu XL, Cai M. Graphitic carbon nitride nanosheets mitigate cadmium toxicity in Glycine max L. by promoting cadmium retention in root and improving photosynthetic performance. J Environ Sci (China). 2024, 139: 543–555.
85 Sun R, Sun Y, Li Q X, Zheng X, Luo X, Mai B. Polycyclic aromatic hydrocarbons in sediments and marine organisms: implications of anthropogenic effects on the coastal environment. Sci Total Environ. 2018, 640–641: 264–272.
86 Su Y, Yan X, Pu Y, Xiao F, Wang D, Yang M. Risks of single-walled carbon nanotubes acting as contaminants-carriers: potential release of phenanthrene in Japanese medaka (Oryzias latipes). Environ Sci Technol. 2013, 47: 4704–4710.
87 Zindler F, Glomstad B, Altin D, Liu J, Jenssen BM, Booth AM. Phenanthrene bioavailability and toxicity to daphnia magna in the presence of carbon nanotubes with different physicochemical properties. Environ Sci Technol. 2016, 50: 12446–12454.
88 Wang W, Zhao X, Ren X, Duan X. Antagonistic effects of multi-walled carbon nanotubes and BDE-47 in zebrafish (*Danio rerio*): Oxidative stress, apoptosis and DNA damage. Aquat Toxicol Amst Neth. 2020, 225: 105546.
89 Barbieri E, Campos-Garcia J, Martinez DST, da Silva JRMC, Alves OL, Rezende KFO. Histopathological effects on gills of Nile tilapia (Oreochromis niloticus, Linnaeus, 1758) exposed to Pb and carbon nanotubes. Microsc Microanal. 2016, 22: 1162–1169.
90 Canesi L, Frenzilli G, Balbi T, Bernardeschi M, Ciacci C, Corsolini S, Della Torre C, Fabbri R, Faleri C, Focardi S, Guidi P, Kočan A, Marcomini A, Mariottini M, Nigro M, Pozo-Gallardo K, Rocco L, Scarcelli V, Smerilli A, Corsi I. Interactive effects of n-TiO_2 and 2,3,7,8-TCDD on the marine bivalve Mytilus galloprovincialis. Aquat Toxicol Amst Neth. 2014, 153: 53–65.
91 Fang Q, Shi Q, Guo Y, Hua J, Wang X, Zhou B. Enhanced bioconcentration of bisphenol a in the presence of nano-TiO2 can lead to adverse reproductive outcomes in zebrafish. Environ Sci Technol. 2016, 50: 1005–1013.
92 Wang D, Hu J, Forthaus BE, Wang J. Synergistic toxic effect of nano-Al2O3 and As(V) on Ceriodaphnia dubia. Environ Pollut. 2011, 159: 3003–3008.
93 Jia J, Li F, Zhai S, Zhou H, Liu S, Jiang G, Yan B. Susceptibility of overweight mice to liver injury as a result of the ZnO nanoparticle-enhanced liver deposition of Pb2. Environ Sci Technol. 2017, 51: 1775–84.
94 Tang Y, Rong J, Guan X, Zha S, Shi W, Han Y, Du X, Wu F, Huang W, Liu G. Immunotoxicity of microplastics and two persistent organic pollutants alone or in combination to a bivalve species. Environ Pollut. 2020, 258: 113845.
95 Duttagupta S, Mukherjee A, Bhattacharya A, Bhattacharya J. Wide exposure of persistent organic pollutants (PoPs) in natural waters and sediments of the densely populated Western Bengal basin, India. Sci Total Environ. 2020, 717: 137187.
96 Gautam RK, Sharma SK, Mahiya S, Chattopadhyaya MC. CHAPTER 1: Contamination of Heavy Metals in Aquatic Media: Transport, Toxicity and Technologies for Remediation. RSC Publisher, 2014.
97 Xiangfeng H, Yongjiao X, Kaiming P, Lijun LU, Jia L. The progress of antibiotics removal performance under the complexion effect of metal ions. Environ Chem. 2016, 35: 133–140.
98 Liang X, Zhu L, Zhuang S. Sorption of polycyclic aromatic hydrocarbons to soils enhanced by heavy metals: perspective of molecular interactions. J Soil Sediments. 2016, 16: 1509–1518.
99 Mehler WT, Du J, Lydy MJ, You J. Joint toxicity of a pyrethroid insecticide, cypermethrin, and a heavy metal, lead, to the benthic invertebrate Chironomus dilutus. Environ Toxicol Chem. 2011, 30(12): 2838–2845.
100 Zhu Z-L, Wang S-C, Zhao F-F, Wang S-G, Liu F-F, Liu G-Z. Joint toxicity of microplastics with triclosan to marine microalgae Skeletonema costatum. Environ Pollut. 2019, 246: 509–517.
101 Zocchi M, Sommaruga R. Microplastics modify the toxicity of glyphosate on Daphnia magna. Sci Total Environ. 2019, 697: 134194.
102 Yang W, Gao X, Wu Y, Wan L, Tan L, Yuan S, Ding H, Zhang W. The combined toxicity influence of microplastics and nonylphenol on microalgae Chlorella pyrenoidosa. Ecotoxicol Environ Saf. 2020, 195: 110484.
103 Yi X, Chi T, Li Z, Wang J, Yu M, Wu M, Zhou H. Combined effect of polystyrene plastics and triphenyltin chloride on the green algae Chlorella pyrenoidosa. Environ Sci Pollut Res Int. 2019, 26: 15011–15018.
104 Miao S, Zhang Y, Men C, Mao Y, Zuo J. A combined evaluation of the characteristics and antibiotic resistance induction potential of antibiotic wastewater during the treatment process. J Environ Sci (China). 2024, 138: 626–636.
105 Bhatt P, Bhandari G, Bilal M. Occurrence, toxicity impacts and mitigation of emerging micropollutants in the aquatic environments: recent tendencies and perspectives. J Environ Chem Eng. 2022, 10: 107598.
106 Yi X, Lin C, Ong EJL, Wang M, Zhou Z. Occurrence and distribution of trace levels of antibiotics in surface waters and soils driven by non-point source pollution and anthropogenic pressure. Chemosphere. 2019, 216: 213–223.

107 Grenni P, Ancona V, Caracciolo AB. Ecological effects of antibiotics on natural ecosystems: a review. Microchem J. 2018, 136: 25–39.

108 Wang J, Xu S, Zhao K, Song G, Zhao S, Liu R. Risk control of antibiotics, antibiotic resistance genes (ARGs) and antibiotic resistant bacteria (ARB) during sewage sludge treatment and disposal: a review. Sci Total Environ. 2023, 877: 162772.

109 Zhang Y, Cai X, Lang X, Qiao X, Li X, Chen J. Insights into aquatic toxicities of the antibiotics oxytetracycline and ciprofloxacin in the presence of metal: complexation versus mixture. Environ Pollut. 2012, 166: 48–56.

110 Tommasino JB, Renaud FNR, Luneau D, Pilet G. Multi-biofunctional complexes combining antiseptic copper(II) with antibiotic sulfonamide ligands: structural, redox and antibacterial study. Polyhedron. 2011, 30: 1663–1670.

111 Lambs L, Reverend D L, Kozlowski H, Berthon G. Metal ion-tetracycline interactions in biological fluids. 9. Circular dichroism spectra of calcium and magnesium complexes with tetracycline, oxytetracycline, doxycycline, and chlortetracycline and discussion of their binding modes. Inorg Chem. 1988, 27: 3001–3012.

112 Pulicharla R, Hegde K, Brar SK, Surampalli RY. Tetracyclines metal complexation: significance and fate of mutual existence in the environment. Environ Pollut. 2017, 221: 1–14.

113 Chang PH, Li Z, Jean JS, Jiang W-T. Desorption of tetracycline from montmorillonite by aluminum, calcium, and sodium: an indication of intercalation stability. Int J Environ Sci Technol. 2014, 11: 633–644.

114 Jia D, You X, Tang M, Lyu Y, Hu J, Sun W. Single and combined genotoxicity of metals and fluoroquinolones to zebrafish embryos at environmentally relevant concentrations. Aquat Toxicol Amst Neth. 2023, 258: 106495.

115 Hüffer T, Hofmann T. Sorption of non-polar organic compounds by micro-sized plastic particles in aqueous solution. Environ Pollut. 2016, 214: 194–201.

116 Li J, Zhang K, Zhang H. Adsorption of antibiotics on microplastics. Environ Pollut. 2018, 237: 460–467.

117 Antony A, Fudianto R, Cox S, Leslie G. Assessing the oxidative degradation of polyamide reverse osmosis membrane – accelerated ageing with hypochlorite exposure. J Membr Sci. 2010, 347: 159–164.

118 Changfu Y, Jiani G, Yidi Y, Yijin L, Yiyao L, Yu F. Interface behavior changes of weathered polystyrene with ciprofloxacin in seawater environment. Environ Res. 2022, 212: 113132.

119 Xu B, Liu F, Brookes PC, Xu J. The sorption kinetics and isotherms of sulfamethoxazole with polyethylene microplastics. Mar Pollut Bull. 2018, 131: 191–196.

120 Ding L, Mao R, Ma S, Guo X, Zhu L. High temperature depended on the ageing mechanism of microplastics under different environmental conditions and its effect on the distribution of organic pollutants. Water Res. 2020, 174: 115634.

121 Liu P, Qian L, Wang H, Zhan X, Lu K, Gu C, Gao S. New insights into aging behavior of microplastics accelerated by advanced oxidation processes. Environ Sci Technol. 2019, 53: 3579–3588.

122 Xu B, Liu F, Brookes PC, Xu J. Microplastics play a minor role in tetracycline sorption in the presence of dissolved organic matter. Environ Pollut. 2018, 240: 87–94.

123 Fu J, Li Y, Peng L, Gao W, Wang G. Distinct chemical adsorption behaviors of sulfanilamide as a model antibiotic onto weathered microplastics in complex systems. Colloids Surf Physicochem Eng Asp. 2022, 648: 129337.

124 Atugoda T, Wijesekara H, Werellagama DRIB, Jinadasa KBSN, Vithanage M. Adsorptive interaction of antibiotic ciprofloxacin on polyethylene microplastics: implications for vector transport in water. Environ Technol Innov. 2020, 19: 100971.

125 Zhou Z, Sun Y, Wang Y, Yu F, Ma J. Adsorption behavior of Cu(II) and Cr(VI) on aged microplastics in antibiotics-heavy metals coexisting system. Chemosphere. 2022, 291(Pt 1): 132794.

126 Barboza LGA, Dick Vethaak A, Lavorante BRBO, Lundebye A-K, Guilhermino L. Marine microplastic debris: an emerging issue for food security, food safety and human health. Mar Pollut Bull. 2018, 133: 336–348.

127 Barboza LGA, Vieira LR, Guilhermino L. Single and combined effects of microplastics and mercury on juveniles of the European seabass (*Dicentrarchus labrax*): changes in behavioural responses and reduction of swimming velocity and resistance time. Environ Pollut. 2018, 236: 1014–1019.

128 Wright SL, Kelly FJ. Plastic and human health: a micro issue? Environ Sci Technol. 2017, 51: 6634–6647.

129 Zhang P, Lu G, Sun Y, Zhang J, Liu J, Yan Z. Aged microplastics change the toxicological mechanism of roxithromycin on Carassius auratus: size-dependent interaction and potential long-term effects. Environ Int. 2022, 169: 107540.

130 Huang Y, Ding J, Zhang G, Liu S, Zou H, Wang Z, Zhu W, Geng J. Interactive effects of microplastics and selected pharmaceuticals on red tilapia: role of microplastic aging. Sci Total Environ. 2021, 752: 142256.

131 Zhang YT, Chen H, He S, Wang F, Liu Y, Chen M, Yao G, Huang Y, Chen R, Xie L, Mu J. Subchronic toxicity of dietary sulfamethazine and nanoplastics in marine medaka (Oryzias melastigma): Insights from the gut microbiota and intestinal oxidative status. Ecotoxicol Environ Saf. 2021, 226: 112820.

5

Experimental Designs and Sampling Strategies for Combined Toxicity Studies Based on Concentrations

Manoharan Saravanan[a,b,], Karunanithi Vidhya[c]*

[a] *Department of Environmental Science, Periyar University, Salem 636 011, Tamil Nadu, India*
[b] *Department of Zoology, Padmavani Arts & Science College for Women (Autonomous), Salem – 636 011, Tamil Nadu, India*
[c] *PG & Research Department of Physics, Sri Sarada College for Women (Autonomous), Salem – 636 016, Tamil Nadu, India*

*Corresponding author.
E-mail: msktox@gmail.com (M.S.)

5.1 Introduction

Every day, many chemical compounds are simply discarded into various segments of the environment (aquatic, terrestrial, and atmospheric ecosystem), which have been prepared and manufactured in large quantities for the day-to-day life of human and animal welfare in terms of medical, agricultural, industrial, commercial, and aquacultural purposes at the global level. Once these anthropogenic chemical compounds reached the environment in single and mixture form, the chemicals mixed themselves and combined with other xenobiotic compounds (organic and inorganic) due to their interaction and reactions with environmental factors. In addition to this, different types of environmental pollutants usually coexist in the natural environment [1]. However, due to consumer exposure (water, air, and food), occupational exposure, and environmental exposure, animals, humans, and ecosystems are continually exposed to very complex mixtures of chemicals (unintentional exposure to unknown chemical mixtures). The chemical nature such as bioaccumulative, nonbiodegradable, and toxic to aquatic biota and modes of actions of chemical substances may cause ecotoxicological impacts solely and/or combinedly on nontarget living organisms. To date, there are numerous reports on the toxicity of individual chemicals in living organisms compared to the combined toxicity of mixtures of chemicals. Consequently, in recent years, there has been a growing awareness and concern about the combined toxicity of mixtures of chemicals [2].

The combined toxicity of chemical mixtures is difficult to predict due to interactions between chemicals. Toxicologists are developing innovative designs, methodologies, and instruments to study the combined effects of toxicants on ecosystems, animals, and humans. In addition, more research has been conducted to identify the fate and occurrence (μg/L to ng/L) ratio, degree of toxicity, and mechanism of toxicity using different methodologies and new experimental designs. This research has increased in the last couple of decades due to the need to assess risks from chemical mixtures (ranging from simple to complex). To address the challenges posed by the combined toxicity of chemicals, various regulatory frameworks and methodologies have been developed, guided by organizations, such as the Organisation for Economic Co-operation and Development (OECD), European Food Safety Authority (EFSA), and US Environmental Protection Agency (USEPA) [3]. However, there is an urgent need to generate new knowledge and scientific advancements, driving the development and harmonization of methodologies to assess exposure and hazards of chemical mixtures in different forms. In this sense, this chapter deals with available technologies, methodologies, and approaches used to predict the level of occurrence of chemical compounds in mixtures and aims to highlight the mixture effects of multiple chemical stressors on biological systems, with particular focus on challenges associated with future research work on combined toxicity.

Toxicological Assessment of Combined Chemicals in the Environment, First Edition. Edited by De-Sheng Pei and Yiyun Liu.

5.2 Need for Risk Assessment Studies

Combined toxicological assessment presents a necessary and feasible approach to evaluating the cumulative interactions of chemicals and their corresponding health hazards to organisms and ecosystems [4, 5]. Ecological and health assessments of xenobiotic chemicals should account for the combined toxicological effects of pollutants [1, 6]. The hazard of chemical mixtures can be assessed either by testing the whole mixture or based on the concentration and effect information of the individual components of the mixture. However, it is not feasible to test all possible combinations of chemicals in mixtures [7]. In toxicology studies, the toxicity of chemicals to ecosystems, animals, and humans is determined by dose–response, concentration–response, and time–response relationships. Even though some of the chemical compounds exist in small amounts, such as pesticides, pharmaceuticals, polycyclic aromatic hydrocarbons (PAHs), and nanomaterials, in the range of microgram to nanogram levels [8], they produce more impact on the environment and living organisms. Importantly, studies on mixture exposure of pesticides and pharmaceutical drugs, heavy metals and agrochemicals, nanomaterials and pesticides, pharmaceutical drugs, and microplastics (MPs) are still scarce. However, in most assessments, only single chemicals are considered, with no general guidelines as to when combinations of chemicals should be carried out. Recent ecotoxicological studies report that multicomponent chemical mixtures are ubiquitous in the environment, existing as parent compounds, metabolites, and transformation products. These different forms can induce varying effects on biomolecules, tissues, organs, and biological systems [5, 9, 10]. Therefore, identifying agents causing toxicity in complex mixtures is of great importance in diagnostic assessment and is recognized as a major concern in the environmental risk assessment of chemicals. Over the past years, understanding the combined toxicity of chemical mixtures has been a challenging task for the scientific community [5, 11]. The data based on individual pollutants do not reflect real field situations, which may lead to the underestimation of chemical mixtures in aquatic vertebrate populations [1]. Hence, it is necessary to evaluate different concentrations of emerging pollutants in mixture form in the environment to environmental organisms under acute and chronic toxicity studies.

5.3 Risk Assessment of Mixture Toxicity

The best way to address the mixture toxicity problem is to conduct toxicological studies using various biological models. The integration of dose- or concentration-based effect screening through in vivo/in vitro bioassays and in silico mathematical modeling provides a holistic approach for integrated toxicological assessment, which is of critical importance to prioritize toxic chemical mixtures for safer chemical research [5]. Chemicals with common modes of action will act jointly to produce combination effects that are larger than the effects of each mixture component applied singly. These effects can be described by dose/concentration addition (CA). In toxicological studies, the sample size is of primary importance. Experiments combining toxicological studies usually describe the combined effect of two or more chemicals as either stronger (synergistic, potentiating, supra-additive) or weaker (antagonistic, inhibitive, subadditive, infra-additive) than would be expected based on dose/concentration or response addition alone. Interactions may therefore vary with dose level, route, timing, duration of exposure (including component persistence), and biological target. Human health and animal health risk assessment of combined exposure to multiple chemicals uses the same steps as single-substance risk assessment, namely problem formulation, exposure assessment, hazard assessment, and risk characterization. The main unique feature of combined risk assessment is the assessment of combined exposure, toxicity, and risk [12]. In combined toxicological studies, the hazard of chemical mixtures can be assessed as a whole, after fractionation, or based on the individual components of the mixture. In this case, there are some principles used to understand the induction of toxicity classified as (1) whole-mixture approaches, (2) component-based approaches, (3) grouping of mixture components based on structural similarities, and (4) grouping by toxicological or biological responses/effects [12].

5.4 Experimental Designs and Methodologies for Mixture Toxicity

Generally, several anthropogenic chemical classes (pesticides, heavy and trace metals, pharmaceutical drugs, personal care products, phthalates [PAEs], MPs, nanomaterials, flame retardants, diesel exhaust fumes, and tobacco smoke) occur in the environment. Multicomponent mixtures ubiquitously exist in the environment as parent compounds, metabolites, and transformation products [10]. When organisms are subjected to short- and long-term exposure to these chemical mixtures, substantial amounts can readily permeate the body through various pathways, including inhalation, ingestion, dermal contact, and dietary intake. Once inside, these chemical substances disperse and bioaccumulate in different body parts, potentially leading to a range of adverse health effects. These chemical mixtures engage in biochemical interactions with various molecules, including enzymes and protein receptors. They can modulate gene expression levels, either amplifying or suppressing them, and alter the concentration of intracellular ions. Such changes can ultimately disrupt the entire metabolic mechanism, underscoring the profound impact these mixtures can have on biological systems. This highlights the critical need for comprehensive studies and effective strategies to mitigate the potential risks associated with exposure to chemical mixtures [13] (Figure 5.1).

Experimental designs and methodologies offer a comprehensive understanding of studies investigating the combined toxicity of diverse toxicants. These studies emphasize screening various biological responses in target and nontarget organisms [5]. Consequently, it has been reported that the molecular interaction at the target sites and the adverse combined effects of multichemical exposures are quantitatively predicted with the classical mathematical models such as CA and independent addition (IA), and the combination index (CI) is also used to evaluate the nature of the possible toxicological interactions among mixture chemicals [5]. Moreover, these CA, IA, and CI models are commonly used as important tools for mixture toxicity in predictive accuracy and to define toxicological interactions in mixtures of environmental contaminants.

Generally, the basic objective of the experimental design is to understand the effect of the factors used in experiments on different chemical mixtures and the interaction mechanisms between concentrations/levels and combinations/factors [14]. However, in previous studies, different experimental designs have been used particularly depending upon the research objectives, the chemical composition of the mixture, and their exposure concentrations. The most common experimental designs included (1) Taguchi orthogonal design, (2) factorial design, (3) combination design, and (4) uniform design [13, 15] (Table 5.1). Among these, factorial design is one of the popular experimental designs that can be used to screen and identify responses with individual and multiple combinations [16]. However, this design has a few limitations in that it can

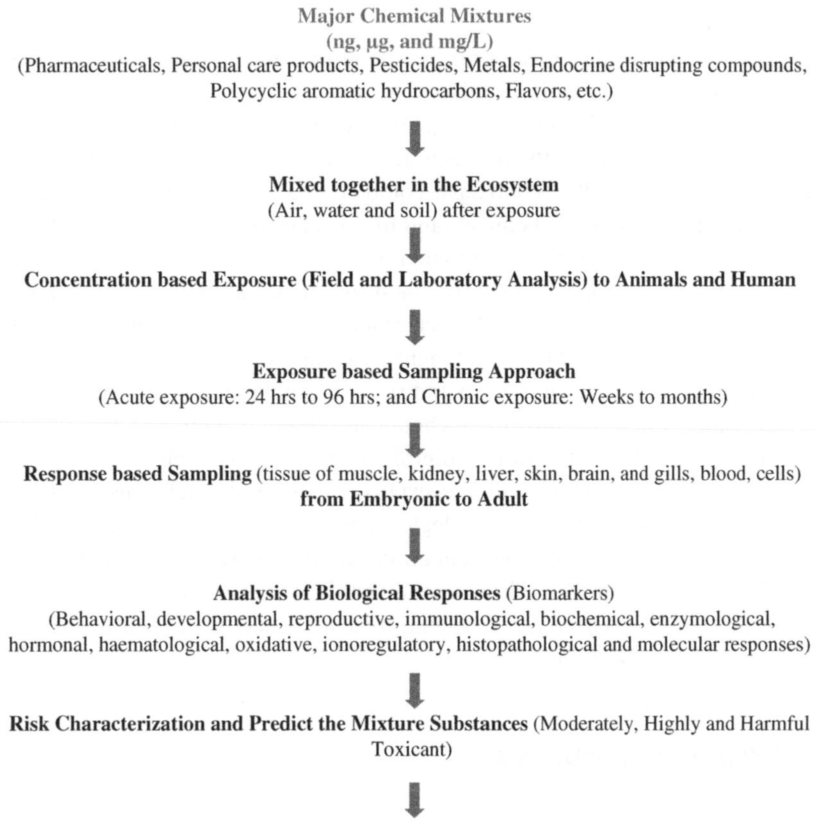

Figure 5.1 Common experimental design on combined toxicity study.

Table 5.1 Commonly used experimental designs for the combined toxicity studies of mixture chemicals.

No.	Experimental design	Formula	Characteristics
1.	Factorial design (FD)	FD_{125} (5^3); where levels = 5, combinations = 125, mixture = 3.	It can be applied only to binary and ternary mixtures with maximum exposure levels of 2 or 3.
2.	Combinational/Taguchi orthogonal (OD)	OD_{25} (5^3); where levels = 5, combinations = 25, mixture = 3	It can increase the number of experiments (n) exponentially with the levels/concentrations (y), robust, time-consuming, and not cost-effective.
3.	Uniform design ray/diagonal ray design (UD/UD-ray)	U_5 (5^3); where levels = 5, combinations = 5, mixture = 3	It alternates the complete combination of all possible experimental parameters with minimum trials, based on equitoxic ratio, cost-effectiveness, and the most used experimental design in toxicological studies.

Source: Adapted from Ref. [5]/with permission of Elsevier.

only be amenable to smaller sample sizes, such as binary and ternary mixtures with concentration levels of two or three [15, 17]. Conversely, combinational and Taguchi orthogonal experimental designs can be used with large datasets, but the number of experiments increases exponentially with the levels/concentrations [18]. In contrast, the uniform design is an experimental design that works on large sample data and predicts the combined toxicity with minimal experiments [19]. This experimental design is the most cost-effective and widely used in toxicological studies [20]. However, a rational experimental design is very important for predicting and evaluating the mixture toxicity of chemicals. Conversely, from the published studies, uniform design and combination design are the most commonly used designs for determining the toxic effects of emerging pollutant mixtures [10, 21].

In addition to chemical factors, nonchemical factors such as temperature, salinity and pH of the water environment, and exposure time also play an important role in determining mixture toxicities [11] due to the transformation and degradation of these compounds and also the formation of byproducts [22]. For instance, the temperature increased the lethal toxicity of pesticide mixtures to *Culex pipiens* by changing the interaction from additive action to synergism [23], and changes in the acid–base condition of water quinolone antibiotics to algae could be slowed down or reduced the growth rate on combined toxicity [24]. Therefore, the concentration of chemicals and timing of exposure play a vital role in toxicological studies because the responses exerted by organisms vary owing to their size and age.

Assessing combined exposure requires the application of the classical steps of risk assessment, namely problem formulation, exposure assessment, hazard assessment, and risk characterization [25]. Two different approaches are used for risk characterization: the whole mixture approach and the component-based approach; the selection of the approach depends on the level of characterization of the mixture composition and the exposure and toxicity data available [12]. The whole mixture approach essentially considers the whole mixture of chemicals present in a product or environmental media sample as a single chemical substance. Recently, the Scientific Committee of EFSA has published two guidance documents to harmonize methodologies for the risk assessment of combined exposure to multiple chemicals mentioned as "Harmonised methodologies for the human health, animal health, and ecological RA of combined exposure to multiple chemicals" [12] and "Scientific criteria for grouping chemicals into assessment groups for human risk assessment of combined exposure to multiple chemicals" [26]. In recent years, human welfare, animal health, and ecological risk assessment of the effects of combined exposure to chemical mixtures has been a challenging topic for academia, regulatory agencies, risk managers, industry, and nongovernmental organizations [27].

5.5 Concentration of Mixture Substances

Indeed, it is important to discuss available methods for assessing the combined risk from exposure to multiple chemicals (with or without a common mode of action [MOA]) via all relevant routes and pathways. In toxicology studies, once external chemical compounds enter the body through various routes such as respiration, food, or water, the primary processes of absorption, distribution, metabolism, and excretion (ADME) occur. It induces toxic impacts on the physiological responses of the organisms when they exceed the acceptable daily intake (ADI). Based on the MOA of the substance on organs, tissues, and cell organelles, threshold limit value (TLV), effective concentration (EC), no-observed-adverse-effect level (NOAEL), lowest-observed-adverse-effect level (LOAEL), and lowest-observed-effect levels (LOEL) can be determined. For instance, once xenobiotic chemical compounds enter organisms internally, they tend to bioaccumulate in vital organs (toxicokinetic process) and induce adverse effects at the molecular and cellular levels. Furthermore, they may lead to abnormal physiological responses (toxicodynamic processes) in terms of adverse outcome pathways in the organisms (Figure 5.2). The concentrations of various components in the mixtures designed are usually the ECs of individual components [28]. By specifying the toxicological endpoint and exposure time, the toxicity of a mixture is closely correlated with the concentration ratio of various components in the mixture as well as its total concentration [11]. However, most experimental designs used in mixture toxicity studies are based on a fixed concentration ratio, such as 50% or 10% EC ratio [29, 30] and no/lowest observed effect concentration ratio [31, 32].

5.6 Concentration-based Strategy

When designing mixture toxicity studies, understanding the selection of exposure concentration is paramount to achieving the expected combined effects [33, 34]. Over the last decade, various strategies have been devised and applied using diverse concentration gradients for different chemical exposure groups [14, 15]. The concentration effect is a principal factor for determining the risk assessment in chemical mixtures [31]. After determining the concentration levels, the combined outcome may also depend on the concentration ratio of the mixture components. Moreover, the exposure assessment should be performed based on the environment-relevant concentrations (ERCs) of the chemicals rather than using high doses/concentrations that do not exist in the natural environment [35]. Two key concepts exist about exposure concentration for chemical mixtures, i.e. nominal concentration and EC [36]. Nominal concentration is defined as "the amount of a compound that is expected to be present in a sample," whereas the EC refers to "the concentration at which the maximum effects can be observed is called effective concentration" [35]. Exposure effects largely depend upon the chemicals' fate, sorption effects, degradation, and transformation (metabolites) processes [14]. Among all the ECs, EC_{50} is considered the

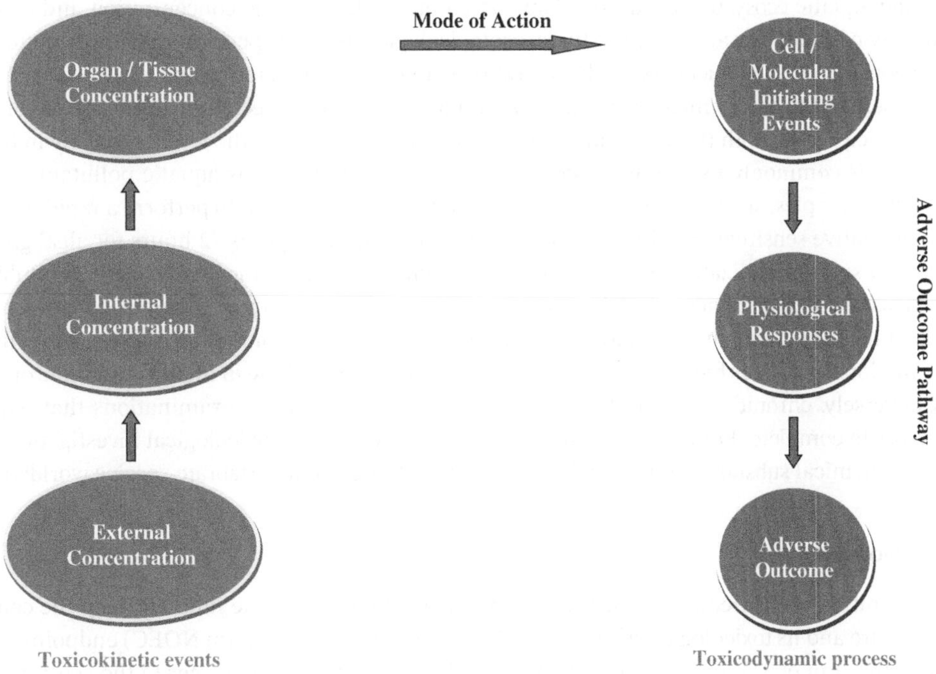

Figure 5.2 Schematic illustration of each single chemical and/or combined chemical compound induces adverse effects on target/nontarget organisms via toxicokinetic and toxicodynamic processes during short- and long-term exposures. *Source:* Adapted from Ref. [26].

graded or the optimum dose–response curve parameter that represents the concentration of a chemical required to obtain a 50% maximum effect. It is calculated using several statistical analyses, such as hillslope analysis, logit and probit analyses, and Weibull analysis [21, 31]. All mathematical models, such as IA, CA, and CI, are also dependent on the dose-receptor-based (EC_{50}) relationship, which is further used as a baseline for predicting mixture toxicities [31, 34]. In conclusion, the concentration should be the amount that ultimately reaches the site of action and executes the chemical kinetics and toxicodynamics after exposure [37, 38] (Figure 5.2). Therefore, concentration selection requires a high degree of awareness that ultimately assists in overcoming problems during the interpretation of results.

5.6.1 Lethal Concentration 50 (LC_{50}) and Effective Concentration 50 (EC_{50})

The combined effects of acute and chronic toxicity depended on the concentration of the mixture. Most of the combined toxicity studies of mixture substances have focused on the LC_{50} values of the chemical substances. For instance, 10%, 20%, and/or 50% of LC_{50} values were used to investigate certain types of mixture toxicity studies in aquatic organisms. For instance, Jeon et al. [8] studied the combined toxicity of 3,5,6-trichloro-2-pyridinol (TCP) and 2-(bromomethyl)naphthalene (2-BMN) using the LC_{10} (0.51) and LC_{50} (0.64 mg/L) values on 2-BMN LC_{10} (1.5 mg/L) and LC_{50} (2.3 mg/L) values. Similar to LC_{50}, many combined toxicity of mixture substances have been done using EC_{50} values. The concentration effect is a principal factor for determining the risk assessment in chemical mixtures [31]. All mathematical models, such as IA, CA, and CI, are also dependent on the dose-receptor-based (EC_{50}) relationship, which is further used as a baseline for predicting mixture toxicities [34]. Quaternary antibiotic mixtures were exposed based on the EC_{50} value and investigated the steroid synthesis in H295R cells [36]. Similarly, PAEs and metal co-exposure (DBP, DEHP, Cu) in the mixture concentration gradient were also dependent on the EC_{50} value [39].

5.7 Ecotoxicity Tests: Acute (Short-term Exposure) and Chronic Test (Long-term Exposure)

In general, the major chemical products like household products, pesticides, metals, EDCs, pharmaceutical drugs and personal care products, nanomaterials, and micro- and nano-sized plastics (parent compounds and their metabolites) adversely impact the aquatic ecosystems and organisms based on the level of the concentration and existing period. To determine the toxicity of chemicals and their mixtures, ecotoxicological tests are performed within environmental hazard assessment frameworks. In toxicological assays, bioassays play a vital role in determining the toxicity of a specific compound, which is present in the environment at a concentration high enough to affect flora and fauna. It can be divided into acute and chronic toxicity tests. Both these tests are commonly executed to predict the level of toxicity in living organisms. An acute toxicity test is commonly used to evaluate the potential threat of various aquatic pollutants. The acute test is characterized by fewer samples, short duration, and low cost and is frequently used to perform a rapid toxicity assessment or to determine the relative sensitivity of different species [40]. Typically, it requires 72 hours for algal growth inhibition, 48 hours for *Daphnia* sp. immobilization, 96 hours for fish toxicity, and 7 days for Lemna growth inhibition. The most important advantages of conducting acute toxicity tests on organisms (bacteria, algae, daphnia, earthworms, and fish) are that they have short life cycles and show speedy responses to any kind of environmental change. Nevertheless, they still perform similar functions in various biochemical mechanisms, comparable to those in higher vertebrate organisms, including mammals. Conversely, chronic toxicity testing is more complex, with lengthy examinations that require more time from months to years to complete. Both tests have been extensively used in the toxicological investigations of single, mixture, and coexisting chemical substances on animal models of vertebrate and invertebrate species worldwide (Table 5.2).

5.7.1 Toxic Unit Models

The model of toxic units (TUs) is frequently used in ecotoxicology. It represents the ratio between the concentration of a component in a mixture and its toxicological acute (e.g. LC_{50}) or chronic (e.g. long-term NOEC) endpoint. The toxic unit of a mixture (TUm) is the sum of TUs of individual chemicals. The TU model does not refer to the ecosystem, but only to a specific organism representative of a group of organisms ecologically or taxonomically relevant to the aquatic ecosystem (such as algae, daphnia, and fish). The TU concept can be used to quantify the toxicity of a mixture (assuming the dose/concentration addition principle) based on its composition [12].

Table 5.2 Combined toxicity studies on the most commonly used animal models during 2014–2024.

S. No.	Compound name	Mixture concentration	Species	Exposure period	Endpoints studied	Ref.
1.	Ultraviolet-aged polystyrene (UV-PS) and tetrabromobisphenol A (TBBPA)	UV-PS and TBBPA: 1 μg/L	*Caenorhabditis elegans*	24 h	Altered morphology, intestinal damage, oxidative stress, and expression of *gst-4::GFP*	[97]
2.	Cadmium (Cd) and diclofenac (DCF)	Cd and DCF: 50 μg/L Cd + 20 μg/L DCF 500 μg/L Cd + 20 μg/L DCF	Crayfish (*Procambarus clarkii*)	7 and 14 days	Antioxidant enzyme activity, tissue accumulation (hepatopancreas > gill > intestine > muscle), histopathology, and gut microbiome	[83]
3.	MPs, Perfluorooctane sulfonate (PFOS), and 6:2 chlorinated polyfluorinated ether sulfonate (F-53B)	10 μg/L PFOS/F-53B + 50 μg/L MPs	Zebrafish	14 days	Liver oxidative stress, immunoinflammatory, and energy metabolism disorders and also impact on 16S rRNA gene in gut microbiota	[101]
4.	Polyethylene microplastics (PEMPs) and nickel oxide nanoparticles (nano-NiO)	Nano-NiO (200mg/kg) + PEMPs: 0.1%, 1%, and 10% sizes (13, 50, and 500 μm)	Earthworm (*Eisenia andrei*)	28 days	Induction of oxidative stress and enzyme responses such as SOD, POD, CAT, MDA, AChE, Na^+/K^+-ATPase, and cellulase	[70]
5.	*Fungicides:*trifloxystrobin (TRI) and fluopyram (FLU)	TRI and FLU: 0.2, 0.3, 0.45, 0.675, and 1.0125 mg/L	Zebrafish embryos (*Danio rerio*)	7 days	Developmental toxicity, inhibited skeletal development, decreased heart rate, reduced melanin, pericardial cysts, and yolk cysts, increased frequency of voluntary movements, alterations on *Bmpr-1a* and *Bmpr-1ba* gene receptors	[102]
6.	Polystyrene microplastics (PSMPs) and 3,6-dibromocarbazole (3,6-DBCZ)	MPs and 3,6-DBCZ: 10 and 0.5 mg/L	Zebrafish embryos	48, 72, and 96 h post-fertilization (hpf)	Malformation, bioaccumulation, induced antagonistic effect on oxidative stress, DNA damage, and apoptosis	[6]
7.	Methyl tert-butyl ether (MTBE) and Rubber dust (RP)	MTBE + RP: 2.5 μL/L + 5 mg/L 2.5 μL/L + 10 mg/L 5 μL/L + 5 mg/L 5 μL/L + 10 mg/L	Nile tilapia (*Oreochromis niloticus*)	28 days	Increased response of enzymes (AST, ALT, LDH, CPK, ALP, and GGT) in plasma and antioxidants (SOD, CAT, GPx, GR, and MDA) in hepatocytes. Alterations in plasma glucose, triglycerides, cholesterol, creatinine, and impaired growth performance	[88]
8.	Microplastics (MPs) and polycyclic aromatic hydrocarbons (PAHs)	MP: 10 mg/L PAH: 2.28 μg/L	Neotropical fish (*Astyanax lacustris*)	96 h	Muscle glycogen	[85]
9.	3,5,6-trichloro-2-pyridinol (TCP) and 2-(bromomethyl) naphthalene (2-BMN)	TCP LC_{10} and LC_{50} values: 0.51 and 0.64 mg/L, 2-BMN LC_{10} and LC_{50} values: 1.5 and 2.3 mg/L	Zebrafish embryos	96 hpf	Developmental toxicity: high mortality, strong inhibition of hatching, severe morphological changes, downregulation of *cyp1a, myl7, cacna1c, edn1*, and *vmhc* expression, and upregulation of *nppa* genes	[8]

(Continued)

Table 5.2 (Continued)

S. No.	Compound name	Mixture concentration	Species	Exposure period	Endpoints studied	Ref.
10.	Microsized polystyrene (mPS, 5 μm) and Chlorpyrifos (CPF)	500 μg/L of mPS + 20 μg/L and 500 μg/L of mPS + 200 μg/L of CPF	Zebrafish	14 days	Accumulation and histopathological alterations in the gut and liver, hyperactive swimming behavior, decreased AChE activity, and severe oxidative stress	[78]
11.	Microfiber (MF) and Bead microplastics (MBs)	MF: 10 and 100 fibers/mL MB:10 and 100 beads/mL	Goldfish (*Carassius auratus*)	120 h	Higher levels of bioaccumulation, concentration of H_2O_2, antioxidant responses related to mRNAs, cell (DNA) damage, aminotransferase (ALT), and aspartate aminotransferase (AST) in blood plasma and tail DNA	[87]
12.	Cadmium (Cd) and Nitrate (NO_3)	Cd^{2+}: 0, 0.2, and 2 mg/L NO_3: 0 and 80 mg/L	Juvenile Japanese flounder (*Paralichthys olivaceus*)	30 days	Alterations on intestinal morphology and immune response, and microbiota	[84]
13.	Triclosan (TCS), Caffeine (CAF), Nanoplastics (NPs), and Microplastics (MPs)	TCS, CAF, NPs and MPs: 10, 50, 100, 200, and 400 mg/L	*Daphnia magna*	24 and 48 h	Increased mortality rate	[22]
14.	Polystyrene microplastics (PSMPs) and arsenic (AsIII)	PSMP + As(III): 10 mg/L and 5 mg/L	Zebrafish embryo	4, 72, and 120 hpf	Developmental toxicity, Accumulation, antioxidant enzyme activity, and changes in metabolic pathway	[71]
15.	Benz[a]anthracene (BaA) and polystyrene MPs (PSMPs)	1 ppm (BaA) + 0.2 μm PSPMs), 5 ppm + 1 μm, and 10 ppm + 10 μm	Zebrafish embryo	96 hpf	Induced cardiotoxicity (cardiac defect, ROS levels, and cell death)	[99]
16.	*Pesticides:* Imidacloprid (IMI), Cycloxaprid (CYC), and Tebuconazole (TBZ)	IMI + TBZ: 1.54 + 0.29 mg/L CYC + TBZ: 0.96 + 0.29 mg/L	*Daphnia magna*	48 h and 21 days	Acute and chronic toxicity impaired developmental and reproductive and induced antioxidant responses	[91]
17.	*Nanoparticles:* nMOx, aluminum oxide (nAl_2O_3), and cerium oxide ($nCeO_2$)	nMOx NPs: 1 mg/L nAl_2O_3: 1 mg/L and $nCeO_2$: 1 mg/L	Zebrafish embryos	96 h	Bioaccumulation and antioxidative stress response	[100]
18.	*Metal and pesticides:* Cadmium (Cd), Acetamiprid (ACT), Carbendazim (CAR), Azoxystrobin (AZO), Chlorpyrifos (CHL), and Bifenthrin (BIF)	Cd + ACT + CAR + AZO + CHL + BIF: 0.021 mg/L	Zebrafish larvae	96 h	Induced synergistic effects	[1]

19.	*Antibiotics:* Sulfamethoxazole (SMZ) and Polystyrene microplastics (PSMPs)	SMZ + PSMPs: 1 + 10 μg/L 10 + 10 μg/L 100 + 10 μg/L	Zebrafish embryos and Juvenile	96 hpf	Decrease in fetal movement and heartbeat rate, and enhanced the mortality rate and malformation rate in embryos and alteration in VTG, E2, testosterone (T), and T_3 in juveniles	[74]
20.	*Pesticides:* Fenvalerate and Tebuconazole	Fenvalerate + Tebuconazole: 3.00×10^{-5} mg/L 3.00×10^{-4} mg/L 3.00×10^{-3} mg/L	Zebrafish larvae and adult	2 days	Upregulation of behavioral responses	[75]
21.	*Flavors:* Benzyl alcohol (BEA), phenethyl alcohol (PHA) and cinnamaldehyde (CID)	100, 200, and 400 μL	*Caenorhabditis elegans*		High toxicity sensitivities and mortality	[96]
22.	*Pesticides:* Chlorpyrifos (CPF), profenofos (PRF) and triazophos (TAZ)	CPF + PRF: 2.1 + 4.4 μg/L CPF + TAZ: 2.9 + 1.07 μg/L PRF + TAZ: 9.8 + 1.64 μg/L	Silver perch (*Bidyanus bidyanus)*	48 h	Significant mortality rate and brain acetylcholinesterase (AChE) inhibition	[90]
23.	*Pharmaceuticals:* Metformin and Ranitidine	Metformin: 14.3 mg/L + Ranitidine: 247.3 mg/L	*Daphnia similis*		Synergistic effects on developmental and behavioral responses	[92]
24.	*Plasticizers:* (Bisphenol A and DBP)	BPA + DBP: 10 + 50 mg/L	Zebrafish embryos	12 h	Molecular action at the proteome level, Integrated toxicity analysis predicted that proteins function in common regulatory networks that are associated with developmental and metabolic disorders	[76]
25.	*Phthalates:* DEHP and DEP	DEHP: 1, 10 μM + DEP: 1, 10 μM	*Caenorhabditis elegans*	48 h	Gene expression: upregulated *fasn-1, pod-2, fat-5, acs-6* and *sbp-1, vtg*, and stress response genes, *ced-1 wah-1, daf-21, gst-4* upregulated; downregulated *ctl-1, cdf-2, hsp-16.1, hsp-16.48*, and *sip-1* genes	[95]
26.	*Plasticizers:* acetyl tributyl citrate (ATBC), dibutyl phthalate (DBP) and di-(2-ethylhexyl) phthalate (DEHP)	50 μg/L	Zebrafish	3 months	Reproductive toxicity: Decrease in the GSI, impairment of fecundity, disruption of oogenesis, spermatogenesis. Male zebrafish were more sensitive to combined exposure	[81]

(Continued)

Table 5.2 (Continued)

S. No.	Compound name	Mixture concentration	Species	Exposure period	Endpoints studied	Ref.
27.	*Antibiotics:* Caffeine, Ibuprofen, Carbamazepine and Tamoxifen	Caf, Ibu, Cbz, Tmx: low mix: 0.05 μM of Caf, Ibu, Cbz and 0.003 μM of Tmx, and high mix: 5 μM of Caf, Ibu, Cbz and 0.3 μM of Tmx	Zebrafish	96 h	Alterations in gene expression and enzymatic activity. Significant upregulation of *cyp1a and vtg*	[82]
28.	Carbofuran and Nitric acid-treated multiwalled carbon nanotubes (HNO_3-MWCNTs)	Carbofuran: 1, 0.5, 2.0, 4.0, and 8.0 mg/L + HNO_3-MWCNTs: 1.0 mg/L	Nile tilapia (*Oreochromis niloticus*)	24 h	Histopathological alterations in the gills	[98]
29.	Phthalates (DBP) and Metal (Pb)	Pb: 0, 0.01 and 1 mg/L DBP: 0, 0.005 and 0.5 mg/L	Zebrafish embryos	120 h	Neurological and apoptosis-related gene expressions: Co-exposure to Pb + DBP upregulated *NR1A, NR2A, NR2 D* (neurological gene) *c-fos* apoptosis-related gene and downregulated *bcl-2* gene	[77]
30.	Phthalates (DBP, DEHP) and Metal (Cu)	DBP and Cu were 0.25, 0.5, 0.625, 0.75, 0.875 and a (a was 390 mg/L for DBP) and 7 mg/L for Cu(II)	*Daphnia magna*	14 h	Antagonism effects were found in the joint toxicity of Cu(II) combined with DBP or DEHP	[39]
31.	Phthalates (DBP, DEHP) and Metal (Cu)	Cu 50, 100 μg/L + DBP 50, 500 μg/L; Cu 50, 100 μg/L + DEHP 50, 500 μg/L	Goldfish (*Carassius auratus*)	7 and 21 days	Antioxidant response (SOD, CAT, MDA, and GSH) and antagonistic effects using Integrated biomarker response index	[86]
32.	*Pesticides:* Endosulfan (ED) and Lambda-cyhalothrin (LC)	ED: 1.1 μg/L + LC: 0.7 μg/L	South American omnivorous fish (*Piaractus mesopotamicus*)	96 h	Increase in the liver-somatic index, LPO in liver, kidney, and brain, hematological changes, and decrease in liver AST, ALT, and ALP activities, antioxidant enzyme induction in gills, liver, kidney, brain, and muscle	[89]
33.	*Phthalates:* BBP, DBP, DEHP, DIDP, DINP, and DNOP	BBP, DBP, DEHP, DIDP, DINP, DNOP: 1.5, 0.06, and 0.01 ppm for a 1:1 mixture	Marine medaka fish (*Oryzias melastigma*) and Zebrafish	72 h	Developmental toxicity and enhanced estrogenic activity	[80]
34.	*Phthalates:* DBP and 17a-ethynylestradiol (EE2)	EE2: 0.005 or 0.020 mg/L + DBP: 100 or 500 mg/L	Zebrafish	21 days	Decreased VTG and AOX protein levels.	[79]

5.8 Sampling of Biomarkers

In the assessment of polluted water bodies and aquatic animal health, biomarkers are widely used as early diagnostic tools [41]. Typically, biomarkers are extensively analyzed for diagnosing the structural and functional status of the animal exposed to environmental toxicants and for monitoring the aquatic ecosystem [42, 43]. They are often sensitive to environmental or physiological changes, are easily measurable, and provide an integrated measure of the physiological changes in organisms [44, 45].

5.8.1 Antioxidant Enzyme Activity

In living organisms, environmental toxicants can lead to the production of reactive oxygen species (ROS), such as hydrogen peroxide, superoxide, and/or hydroxyl radicals through direct and/or indirect mechanisms [46]. Oxidative stress results from increased ROS production and reduced antioxidant defense, or a combination of both mechanisms [47]. Furthermore, oxidative stress induces severe molecular damage and leads to the dysfunction of multiple organs and systems. Consequently, analysis of antioxidant enzymes could serve as biomarkers of oxidative stress in many organisms. Malondialdehyde (MDA) is a main lipid oxidation product formed by the decomposition of unsaturated fatty acid peroxides, which are normally degenerated by ROS [48]. The MDA content can directly reflect the degree of oxidative damage induced by toxicants. Glutathione (GSH) is an important low-molecular-weight nonenzymatic antioxidant involved in cellular metabolism, ROS detoxification, and xenobiotic excretion [47]. GSH acts directly as a reductant or as a substrate for antioxidant enzymes and ensures the reduction of oxidants and the neutralization of peroxides [49]. The antioxidant defense system performs an important function against the generated ROS. Superoxide dismutase (SOD) primarily catalyzes the transformation of superoxide anion to hydrogen peroxide (H_2O_2), whereas catalase (CAT) converts superoxide anion to H_2O and O_2 [47]. Therefore, SOD and CAT activities are the first lines of defense against oxidative stress, and these enzymes protect the cells against lipid peroxidation (LPO), protein oxidation, and DNA damage caused by ROS [48]. These antioxidant parameters are routinely measured and used as valuable biomarkers to assess the impacts of environmental toxicants [50, 51].

5.8.2 Hormonal Responses

In toxicological studies, hormonal responses are often used to determine the effects of endocrine-disrupting chemicals (EDCs) on the endocrine systems of fish [52]. Cortisol is a major osmoregulatory corticosteroid hormone that plays a vital role in mediating stress-related responses by mobilizing energy substrates in vertebrates exposed to xenobiotic compounds [53]. Therefore, cortisol is commonly examined as an excellent indicator of primary stress response in fish [54]. The gonadosomatic index (GSI) and hepatosomatic index (HSI) are often evaluated to determine environmental risks in fish health [55, 56]. The natural hormone 17β-estradiol (E2) is one of the essential sex steroids, involved in many physiological functions in the hypothalamus–pituitary–gonadal (HPG) axis of fish [57]. Furthermore, the androgenic steroid hormone 11-ketotestosterone (11-KT) in male fish plays a pivotal role in the development of secondary sexual characteristics and the regulation of reproductive behaviors [58]. Vitellogenin is a yolk precursor protein, which is mainly produced in the liver by E2 secreted from the ovary of female fish. Even though most juvenile and/or male fish are not producing VTG naturally, many studies have reported that immature or male fish are capable of synthesizing VTG when exposed to exogenous estrogenic sources [59, 60]. Therefore, the measurement of VTG synthesis can be used as a biomarker even in immature fish [61]. Moreover, measurements of the thyroid hormones (THs), namely thyroid stimulating hormone (TSH), tri-iodothyronine (T_3), and thyroxine (T_4) endpoints, were routinely used for evaluating TH homeostasis in fish and also consistently used as potential biomarkers in fish [62, 63].

5.8.3 Hematological, Biochemical, and Immunological Parameters

Blood is commonly used in medical and toxicological research and also in the environmental biomonitoring field as a promising indicator of physiological and pathological changes. In this, the hematological parameters, such as hemoglobin (Hb), hematocrit (Hct), red blood cell (RBC), white blood cell (WBC), and other hematological indices, namely mean cellular volume (MCV), mean cellular hemoglobin (MCH), and mean cellular hemoglobin concentration (MCHC), are frequently examined to assess the toxic stress induced by aquatic pollutants. Alterations of these parameters are considered an anemic condition in different animals [42]. Similarly, environmental pollutants can exert toxic effects at cellular or molecular levels that induce significant changes in the biochemical parameters of aquatic vertebrate organisms. In this, glucose, protein, cholesterol, albumin, globulin, and triglycerides in plasma/serum; glycogen content in gill and muscle; and enzymes such as alanine aminotransferase (ALT), aspartate aminotransferase (AST), acid phosphatase (ACP), alkaline phosphatase (ALP), lactate dehydrogenase (LDH), and AChE activities are widely used as biomarkers to assess the toxic stress, integrity of the immune system, and tissue damage [43, 64]. Moreover, gill Na^+/K^+-ATPase plays an important role in the maintenance of ions (electrolytes) between extracellular and intracellular milieus [65]. In immunological studies, the levels of alternative complement activity (ACH_{50}), lysozymes, and total immunoglobulin (Ig) have been suggested to predict the health status of aquatic organisms, since changes in these parameters can reflect disease/stressful conditions [66].

5.8.4 Bioaccumulation and Histopathological Responses

It is known that the aquatic environment can act as a final sink for many environmental pollutants that are likely to be quickly absorbed by aquatic animals and accumulate in their vital organs (gill, liver, kidney, and muscle) at a considerable level through the aquatic food chain. In fish, gills are the first organ in the absorption of xenobiotic toxicants when they contact them. The liver and kidneys are the main sites of accumulation and storage of aquatic pollutants and play a vital role in the detoxification and excretion of toxicants [67]. Therefore, tissue-specific accumulation of chemical compounds has been proposed as a key indicator of chronic exposure in many toxicological studies [68]. Hence, accumulation and histological studies allow identification of cellular changes that occur in the vital organs such as the gills, liver, and kidney.

5.9 Animal Models for Combined Toxicity

There are many organisms used as an animal model for various toxicological investigations to determine the ecotoxicological impacts of single, mixed, and coexisting chemical substances. Risk assessment of combined exposure to multiple chemicals should be monitored and investigated in every segment of the environment (aquatic, terrestrial, and atmospheric environment). Among these, the aquatic ecosystem is more susceptible to anthropogenic chemical substances, and it is a place of final sink, where the chemicals pose a potential threat to aquatic biota. Furthermore, aquatic organisms are usually challenged by many pollutants, contaminants, and toxicants, and exposure to multicomponent mixtures may induce substantial toxic impacts. Typically, fish, mussels, frogs, shrimps, prawns, crabs, diatoms, daphnia, bacteria, etc. are extensively used as aquatic animal models for toxicological investigations. Among the aquatic organisms, fish species have been most commonly used for assessing the effects of environmental pollution on aquatic ecosystems. In this regard, different experimental animal models, designs, and concentration-based samplings for combined toxicity studies are given in Table 5.2.

5.9.1 Aquatic Vertebrates: Fish Species

It is known that the fish models are increasingly used in many toxicological early phases of many chemical substances, including pesticide and pharmaceutical development and toxicity evaluation [69]. Fish play an important role as a bioindicator for water pollution by responding with great sensitivity to changes in their normal physiological functions such as behavioral, molecular, cellular, biochemical, hormonal, and pathophysiological responses [43]. Therefore, fish species have been extensively used as animal models for many toxicological investigations since 80% of physiological functions are very close to human beings. Among animal models used for mixture effect evaluation, studies on fish are essentially needed to screen the combined toxicity of chemicals and monitor the aquatic ecosystem.

5.9.1.1 Zebrafish

It is well depicted that the zebrafish (*Danio rerio*) have several advantages compared to other vertebrate animal models; therefore, it has become largely known as an excellent model organism for studies of vertebrate biology, behavior, embryonal development, genetics, physiology, immunity, human diseases and drug screening, and nutrition. Therefore, zebrafish is increasingly used as an experimental model in toxicological examinations to predict the effects on human and also animal responses to aquatic pollutants. Globally, MPs and nanoplastics (NPs) are emerging as hotspot pollutants in both freshwater and marine water ecosystems. As different environmental factors, such as temperature, light, salinity, pH, and force of waves, induce fractions on plastic materials and also changes in physiochemical properties of plastics, where they are inevitably undergoing aging, there is a decrease in their particle size into macroplastics, MPs, and NPs. In recent times, there have been lots of reports established on the occurrence and ecotoxicity of MPs and NPs in many countries. However, their fractions and combinations with other major and minor pollutants (pesticides, metals, pharmaceuticals, EDCs, etc.) cause major issues in the normal physiological functions of living organisms. In recent times, the co-occurrence of heavy metals and MPs and NPs is an emerging issue that has attracted considerable attention, worldwide [70].

To investigate the combined toxicity of binary exposure of polystyrene MPs (PSMPs) and arsenic (As(III)), PSMPs mixed with As(III) in the range of 10 and 5 mg/L, respectively, are exposed to zebrafish embryos. In this study, the mixture of PSMPs and As(III) induced an impaired development of embryos due to heavy accumulation, altered antioxidant enzyme activities, and metabolic pathway by the production of oxidative stress in embryos at 4, 72, and 120 hpf [71]. In other mixture toxicity studies, the most commonly used fungicides, namely trifloxystrobin and fluopyram, were exposed to zebrafish embryos at different concentrations levels (0.2, 0.3, 0.45, 0.675, and 1.0125 mg/L) for 7 days binary exposure. In this study, significant levels of developmental toxicity (inhibited skeletal development, decreased heart rate, reduced melanin, pericardial cysts, yolk cysts, and increased frequency of voluntary movements) and alterations in *Bmpr-1a* and *Bmpr-1ba* gene receptors on *D. rerio* were observed [72]. Similarly, 10 mg/L of PSMPs and 0.5 mg/L of 3,6-dibromocarbazole (3,6-DBCZ) were combined and exposed to zebrafish embryos. At the end of 48, 72, and 96 hpf, developmental defects and oxidative stress (DNA damage and apoptosis) were observed due to the bioaccumulation of both chemicals [72]. Furthermore, Jian et al. [73] found lower activity in SOD and nitric oxide synthase (TNOS) and also higher levels of transcription abundances in *ndufb8, cat1* and *gstr1, il-8, cc-chem*, and *nrf2 mRNA* levels when exposed to PFOS + MPs group and F-53B + MPs in zebrafish embryos under 14-day treatment.

Jeon et al. [8] studied the combined toxicity of TCP and 2-BMN on zebrafish embryos under 96 hpf. In this study, the LC_{10} (0.51 mg/L) and LC_{50} (0.64 mg/L) values of TCP, respectively, mixed with the LC_{10} (1.5 mg/L) and LC_{50} (2.3 mg/L) values of 2-BMN caused high mortality, strong inhibition of hatching, severe morphological changes, and also downregulation of *cyp1a, myl7, cacna1c, edn1*, and *vmhc* expression and upregulation of *nppa* genes. Furthermore, when exposed to 1 and 10 μg/L, 10 and 10 μg/L, and 100 and 10 μg/L of PSMPs and antibiotic drug sulfamethoxazole (SMZ), respectively, they jointly induced significant decrease of fetal movement and higher heartbeat rate and enhanced mortality rate and malformation rate in embryos and also significant alterations on vitellogenin (VTG), 17β-Estradiol (E2), testosterone (T), and triiodothyronine (T_3) activities in juvenile zebrafish [74]. In aquatic environments, pesticide residues are mainly present as mixtures of varying complexity, which may lead to compound pollution of organisms. Therefore, it is of great practical importance to evaluate the combined toxic effects of pesticides [75]. According to this, the senary mixture of cadmium and five pesticides (acetamiprid, carbendazim, azoxystrobin, chlorpyrifos, and bifenthrin) had significantly induced synergistic effects on zebrafish larvae when exposed to 0.021 mg/L at 96 hours [1], and severe behavioral abnormalities were also noticed on the zebrafish larvae and adult species when treated with the binary mixture exposure of fenvalerate and tebuconazole at the end of 2 days [75].

Another study on zebrafish embryos exposed to bisphenol A and DBP with the combined mixture ratio of 10 and 50 mg/L, respectively, induced molecular action at the proteome level. This integrated toxicity analysis predicted that proteins function in common regulatory networks that are associated with developmental and metabolic disorders [76]. Furthermore, the zebrafish embryos treated with the binary concentrations of 0.005 and 0.5 mg/L of PAEs (DBP) and 0.01 and 1 mg/L of the metal lead (Pb) intoxicated on neurological and apoptosis-related gene expressions like the co-exposure to Pb + DBP upregulated *NR1A, NR2A*, and *NR2D* (neurological gene) *c-fos* apoptosis-related genes and downregulated *bcl-2* gene [77].

Moreover, the 14-day binary exposure of microsized polystyrene (mPS, 5 μm) and chlorpyrifos (CPF) in the range of 500 μg/L of mPS and 20 μg/L of CPF and 500 μg/L of mPS and 200 μg/L of CPF induced significant histopathological alterations in gut and liver due to higher level of accumulation, hyperactive swimming behaviors, decrease in acetylcholinesterase (AChE) activity, and severe oxidative stress in adult zebrafish [78]. Similarly, during 21 days, binary mixture of PAEs (DBP) at 100 or 500 mg/L and 17α-ethynylestradiol (EE2) at the concentration of 0.005 or 0.020 mg/L induced antagonistic effects on VTG and AOX protein levels in zebrafish [79]. However, different types of PAEs, namely BBP, DBP, DEHP, DIDP, DINP, and DNOP, used 1.5, 0.06, and 0.01 ppm for 1:1 mixture (quinary exposure for 72 hours) and exhibited enhanced estrogenic activity and developmental toxicity in zebrafish and the marine medaka fish *Oryzias melastigma* [80]. The long-term ternary exposure of 50 μg/L concentration of DEHP, DBP, and ATBC on zebrafish has exhibited significant reproductive toxicity, decreased GSI activity, impaired fecundity, and disruption of oogenesis and spermatogenesis [81]. Furthermore, Aguirre-Martínez et al. [82] observed that the zebrafish treated with the quaternary mixture of pharmaceutical drugs, namely caffeine, ibuprofen, carbamazepine, and tamoxifen (low mix: 0.05 μM of Caf, Ibu, and Cbz and 0.003 μM of Tmx and high mix: 5 μM of Caf, Ibu, and Cbz and 0.3 μM of Tmx), significantly altered gene expression (upregulation of *cyp1a* and *vtg*) and enzymatic activity at the end of 96 hours.

5.9.1.2 *Procambarus clarkii*

The Crayfish *Procambarus clarkii* is widely used in ecotoxicological studies as a potential bioindicator. Very recently, the Crayfish treated the binary mixture of Cd and DCF mixed with 50 and 20 μg/L and 500 with 20 μg/L, respectively, has induced significant changes in the antioxidant enzyme activities and higher amount of tissue accumulation in hepatopancreas > gill > intestine > muscle and also caused histopathological changes at the end of 7- and 14-day exposures [83]. The authors suggested that the mixture of both Cd and DCF exposure tends to accumulate in the vital organs and cause oxidative stress due to the histopathological conditions in the organs as combined toxic effects.

5.9.1.3 *Paralichthys olivaceus*

Paralichthys olivaceus is also commonly used to investigate the toxic effects of xenobiotic compounds. Recently, Liu et al. [84] reported that the binary mixture of cadmium (0, 0.2, and 2 mg/L) and nitrate (0 and 80 mg/L) treated with the Juvenile Japanese flounder fish (*Paralichthys olivaceus*) has shown significant alterations in intestinal morphology, immune response, and microbiota under 30 days of exposure. The results of this study suggested that the significant changes might be due to the combined toxicity of Cd and NO_3.

5.9.1.4 *Astyanax lacustris*

Astyanax lacustris is Neotropical fish, considered one of the ideal animals for toxicological studies. A combined toxicity study was conducted in the fish *A. lacustris* exposed to the binary mixture of MPs and PAHs at the concentrations of 10 mg/L and 2.28 μg/L, respectively. In this work, the muscle glycogen was significantly depleted under the short-term exposure of 96 hours for the toxic virulence of MPs and PAHs [85].

5.9.1.5 *Carassius auratus*

The goldfish *Carassius auratus* is an unconventional model organism, which is largely used for medical and toxicological research work. A binary mixture toxicity study on phthalate (DBP at 50 and 500 μg/L) and the metal copper (Cu at 50 and 100 μg/L) and 50 and 100 μg/L of Cu with 50 and 500 μg/L of DEHP has shown a significant alteration in the antioxidant responses (SOD, CAT, MDA, and GSH) via antagonistic effects on *C. auratus* when exposed to 7 and 21 days due to their combined toxic effects [86]. Similarly, a higher level of bioaccumulation pattern in intestine, gill, and liver; significant changes in antioxidant (SOD and CAT)-related gene expressions; cell (DNA) damage and mRNAs; Caspase-3 (*casp-3*) mRNA expression; and also the enzymes such as ALT and AST were found in the goldfish *C. auratus* exposed to MPs like 500 μm fiber-type microplastic (MF; 10 and 100 fibers/L) and two sizes (0.2 and 1.0 μm) of microbeads (MB; 10 and 100 beads) under in vivo and in vitro treatments for 120 hours [87]. The authors confirmed that binary exposure to both fiber- and bead-type MPs has induced higher levels of oxidative stress and bioaccumulation in the vital organs of *C. auratus.*

5.9.1.6 *Oreochromis niloticus*

The fish species Nile tilapia (*Oreochromis niloticus*) is a commonly used freshwater model for many scientific studies. The methyl tert-butyl ether (MTBE) from gas stations, gasoline spills, and its residues from vehicles, roads, parking lots, and the release of unburned petrol from ships can contribute to water pollution [88]. In the case of rubber powder (RP), it is more extensively derived from the vehicular traffic on roads and tires and tend to erode over time discharging particulate matter into the air, which can be easily transported to aquatic ecosystems. To determine the toxic potential of both MTBE and RP, Banaee et al. [88] recently investigated combined toxic effects on *O. niloticus* exposed to mixture concentrations of MTBE and RP like 2.5 μL/L of MTBE mixed with 5 mg/L of RP, 2.5 μL/L with 10 mg/L, 5 μL/L with 5 mg/L, and 5 μL/L with 10 mg/L, respectively. In this study, the authors have reported a significant increase in the enzymes such as AST, ALT, LDH, CPK, ALP, and GGT activity in blood plasma and higher levels of antioxidants (SOD, CAT, GPx, GR, and MDA) in hepatocytes. Furthermore, they found a significant alteration in the biochemical responses such as plasma glucose, triglycerides, cholesterol, and creatinine and impaired growth performance. The results of this study envisage that the mixture of MTBE and RP can produce toxicological effects on fish species even when exposed to lower concentrations under long-term exposures.

5.9.1.7 *Piaractus mesopotamicus*

The *P. mesopotamicus* is a widely farmed edible South American omnivorous fish, which is commonly used for toxicological investigations. The binary exposure of endosulfan (ED) and lambda-cyhalothrin (LC) has significantly altered many biological responses in *P. mesopotamicus* exposed to the mixture concentrations of 1.1 μg/L of ED and 0.7 μg/L of LC at the end of 96 hours. In this study, the liver-somatic index, hematological changes, and LPO levels were significantly increased in the liver, kidney, and brain of exposed fish, whereas a significant decrease in liver transaminases' (AST, ALT, and ALP) activities and antioxidant enzymes (GST, GR, GPx, and CAT) were recorded in gills, liver, kidney, brain, and muscle [89].

5.9.1.8 *bidyanus*

A severe mortality rate and significantly lowered AChE activity in the brain were observed in the silver perch fish (*Bidyanus bidyanus*) when exposed to different combinedly mixed pesticides such as chlorpyrifos, profenofos, and triazophos under 24 hours binary exposures [90].

5.9.2 Aquatic Invertebrates

5.9.2.1 *Daphnia magna*

Daphnia magna is a crucial model organism, which is widely used next to fish species in toxicological investigations. Recently, Pashaei et al. [22] found that the acute toxicity of triclosan (TCS), caffeine (CAF), NPs, MPs, and their mixtures on *D. magna* caused higher mortality rates with increasing concentration and duration of exposure to substances. In this study, a TCS concentration of 200 μg/L with 400 mg/L NPs resulted in 50% mortality after 24 hours and 90% mortality after 48 hours, and the highest TCS concentration of 400 μg/L resulted in 100% mortality at both time intervals with the concentration of NPs 400 mg/L. Like this, in the combinations of TCS and MPs, at TCS concentrations of 50 μg/L with 200 mg/L of MPs, a 50% mortality rate was observed after 24 hours, and this increased to 60% after 48 hours. A TCS concentration of 100 μg/L resulted in 100% mortality at both 24 and 48 hours when mixed with 200 mg/L of MPs. CAF concentration of 200 μg/L with a concentration of MPs 400 mg/L resulted in 60% mortality at 24 hours and 70% at 48 hours. The highest CAF concentration of 400 μg/L resulted in 50% mortality after 24 hours and 90% mortality after 48 hours when mixed with the concentration of MPs 400 mg/L. In another study, the mixtures of pesticides namely imidacloprid (IMI) and tebuconazole (TBZ) at the concentrations of 1.54 mg/L mixed with 0.29 mg/L of TBZ and 0.96 mg/L of cycloxaprid (CYC) with 0.29 mg/L of TBZ showed significant developmental and reproductive impairments, and elevated antioxidant responses were noticed in the *D. magna* after 48 hours and 21 days of combined exposures [91]. Furthermore, Huang et al. [39] reported that the mixture concentrations of PAEs (DBP, DEHP) and Cu (0.25, 0.5, 0.625, 0.75, and 0.875 mg/L for DBP; 390 mg/L for DBP; and 7mg/L for Cu(II)) induced significant antagonism effects on the *D. magna* under 14 hours of Cu(II) combined with DBP or DEHP.

5.9.2.2 *Daphnia similis*

Daphnia similis is a planktonic crustacean like *D. magna* and is also used as an animal model. Synergistic effects on developmental and behavioral responses were found in *D. similis* exposed to the binary mixture of pharmaceutical drugs, namely metformin and ranitidine, at the combined concentrations of 14.3 and 247.3 mg/L, respectively, under 48 hours acute exposure[92].

5.9.3 Terrestrial Species

Terrestrial animals are seldom used in toxicological studies due to ethical problems. Very limited species, such as rabbits, rats, mice, and monkeys, are predominantly used in medical fields to discover new medicines for human and animal welfare. Among them, laboratory rats have served as the traditional animal model of choice for research and regulatory developmental and reproductive toxicity testing conducted to support human health hazard identification and risk assessment [93]. Moreover, the animals living in the soil, namely *Caenorhabditis elegans, Eisenia andrei,* and some other earthworm species, have been more generally studied in ecotoxicological research works due to similar responses, low cost, and easy handling. However, aquatic animals have been more largely used as animal models for toxicological studies when compared to other organisms. In this case, the common terrestrial animal models, such as *C. elegans* and *E. Andrei*, used for analyzing the combined toxic effects of mixed chemical substances are discussed in the following.

5.9.3.1 *Caenorhabditis elegans*

Caenorhabditis elegans is one of the common animal models for toxicological investigations – a free-living transparent nematode about 1 mm in length that lives in temperate soil environments [94]. It has been extensively used as an animal model for the ecotoxicological effects of various types of anthropogenic chemical substances including the combined toxicity studies. In recent years, Pradhan et al. [95] have found an upregulated gene expression of *fasn-1, pod-2, fat-5, acs-6, sbp-1*, and *vtg* and the stress response genes, *ced-1, wah-1, daf-21*, and *gst-4*. In contrast to this, the genes such as *ctl-1, cdf-2, hsp-16.1, hsp-16.48*, and *sip-1* were significantly downregulated due to the binary toxicity of 1 and 10 μM of DEHP and DEP exposed to *C. elegans* for 48 hours binary exposure. Like this, a binary mixture exposure of three flavor chemicals such as benzyl alcohol (BEA), phenethyl alcohol (PHA), and cinnamaldehyde (CID) at 100, 200, and 400 μL concentrations, respectively, exhibited higher toxicity and mortality rates in *C. elegans* due to combined toxic effects of BEA, PHA, and CID [96]. Very recently, the binary mixture of ultraviolet-aged polystyrene (UV-PS) and tetrabromobisphenol A (TBBPA) at the concentration of 1 μg/L each significantly altered the morphology of *C. elegans* and induced expression of *gst-4:GFP* gene, intestinal damage, and oxidative stress under the short-term 24 hours exposure. The results obtained from this study report that the combined exposure of 1 μg/L of UV-PS and TBBPA produced a higher level of mixture toxicity in *C. elegans* [97].

5.9.3.2 *Eisenia andrei*

Even though there are many reports available on the toxicological effects of xenobiotic chemical compounds using *E. andrei* as an animal model, there is limited literature on the combined toxic effects of a mixture of chemical substances. Recently, Yang et al. [70] found that the ecotoxicological impacts of the interaction of nickel oxide nanoparticle (nano-NiO) (200 mg/kg) and polyethylene (PE) MPs with different concentrations (0.1%, 1%, and 10%) and sizes (13, 50, and 500 μm) were exposed to the earthworms (*E. andrei*) for 28 days. In this study, the combined exposure of PEMPs and nano-NiO showed significant increment in the antioxidants such as SOD, CAT, and MDA and also the enzyme Na^+/K^+-ATPase and cellulase with increasing concentrations after 28 days of exposure.

5.10 Conclusions

To date, most toxicological studies have focused on individual chemical compounds rather than complex mixtures found in the environment. However, in recent years, scientific communities have established new guidelines and revised existing methodologies to predict the ecological impacts of chemical mixtures and co-occurring substances on various environmental compartments and nontarget organisms. This chapter reviews recent studies on the combined toxicity of chemical mixtures in aquatic and terrestrial environments based on environmentally relevant and nominal concentrations. Most studies have investigated the toxic effects of combined exposures to chemical compounds using different fish species, *C. elegans*, and earthworms under short-term and long-term treatments. Results show that low levels of chemical mixtures can affect organisms from the embryonic stage, as they are particularly sensitive to xenobiotic chemicals. Embryonic exposure to toxicants can lead to population declines in subsequent generations due to endocrine disruption (e.g. the decline of vulture populations due to diclofenac toxicity). Juvenile and adult organisms are also affected by the combined toxicity of multiple chemicals. Since actual concentrations of chemical mixtures in the environment are difficult to predict, they may impact entire ecosystems, including animals and human health. Therefore, more attention should be paid to assessing the combined toxicity of chemical compounds by developing new technologies and experimental designs. Further studies are needed to elucidate the synergistic/antagonistic effects of multiple chemicals and the combined effects of environmental contaminants using environmentally relevant concentrations. To understand the ecotoxicological impacts of chemical mixtures and exposures, appropriate risk management measures are essential to mitigate adverse effects and protect ecosystems and organisms. More research is needed to better understand their potential to produce additive, synergistic, or antagonistic toxicities in the environment.

References

1 Hu G, Wang H, Wan Y, Zhou L, Wang Q, Wang M. Combined toxicities of cadmium and five agrochemicals to the larval zebrafish (*Danio rerio*). Sci Rep. 2022, 12: 16045.

2 Białk-Bielinska A, Mulkiewicz E, Stokowski M, Stolte S, Stepnowski P. Acute aquatic toxicity assessment of six anti-cancer drugs and one metabolite using biotest battery - biological effects and stability under test conditions. Chemosphere. 2017, 189: 689–698.

3 Rotter S, Beronius A, Boobis AR, Hanberg A, van Klaveren J, Luijten M, Machera K, Nikolopoulou D, van der Voet H, Zilliacus J, Solecki R. Overview on legislation and scientific approaches for risk assessment of combined exposure to multiple chemicals: the potential EuroMix contribution. Crit Rev Toxicol. 2018, 48: 796–781.

4 Yang RSH. 1.24-toxicology and risk assessment of chemical mixtures and multiple stressors. In: McQueen CA (Ed.), Comprehensive Toxicology (Third edition). Elsevier, Oxford, 2018, pp. 489–518.

5 Hamid N, Junaid M, Pei DS. Combined toxicity of endocrine-disrupting chemicals: a review. Ecotoxicol Environ Saf. 2021, 215: 112136.

6 Zhang J, Bai Y, Meng H, Zhu Y, Yue Z, Li B, Wang J, Wang J, Zhu L, Du Z. Combined toxic effects of polystyrene microplastics and 3,6-dibromocarbazole on zebrafish (*Danio rerio*) embryos. Sci Total Environ. 2024, 913: 169787.

7 Bopp SK, Kienzler A, Richarz A-N, van der Linden, SC, Paini A, Parissis N, Worth AP. Regulatory assessment and risk management of chemical mixtures: challenges and ways forward. Crit Rev Toxicol. 2019, 49(2): 174–189.

8 Jeon HJ, Cho Y, Kim K, Kim C, Lee SE. Combined toxicity of 3,5,6-trichloro-2-pyridinol and 2-(bromomethyl) naphthalene in the early stages of zebrafish (*Danio rerio*) embryos: abnormal heart development at lower concentrations via differential expression of heart forming-related genes. Environ Pollut. 2023, 325: 121450.

9 Vasquez MI, Lambrianides A, Schneider M, Kummerer K, Fatta-Kassinos D. Environmental side effects of pharmaceutical cocktails: what we know and what we should know. J Hazard Mater. 2014, 279: 169–189.

10 Liu L, Wu W, Zhang J, Lv P, Xu L, Yan Y. Progress of research on the toxicology of antibiotic pollution in aquatic organisms. Acta Ecol Sin. 2018, 38: 36–41.

11 Ding TT, Liu SS, Wang ZJ, Huang P, Gu, ZW, Tao MT. A novel equal frequency sampling of factor levels (EFSFL) method is applied to identify the dominant factor inducing the combined toxicities of 13 factors. Environ Int. 2023, 175: 107940.

12 EFSA Scientific Committee, More SJ, Bampidis V, Benford D, Bennekou SH, Bragard C, Halldorsson TI, Hernandez-Jerez AF, Koutsoumanis K, Naegeli H, Schlatter JR. Guidance on harmonised methodologies for human health, animal health and ecological risk assessment of combined exposure to multiple chemicals. EFSA J. 2019, 17: e05634.

13 Escher BI, Fenner K. Recent advances in environmental risk assessment of transformation products. Environ Sci Technol. 2011, 45: 3835–3847.

14 Ginebreda A, Kuzmanovic M, Guasch H, de Alda ML, Lopez-Doval JC, Munoz I, Ricart M, Romani AM, Sabater S, Barcelo D. Assessment of multi-chemical pollution in aquatic ecosystems using toxic units: compound prioritization, mixture characterization and relationships with biological descriptors. Sci Total Environ. 2014, 468-469: 715–723.

15 Dou RN, Liu SS, Mo LY, Liu HL, Deng FC. A novel direct equipartition ray design (EquRay) procedure for toxicity interaction between ionic liquid and dichlorvos. Environ Sci Pollut Res. 2011, 18: 734–742.

16 Softeland L, Eide I, Olsvik PA. Factorial design applied for multiple endpoint toxicity evaluation in Atlantic salmon (*Salmo salar* L.) hepatocytes. Toxicol Vitr. 2009, 23: 1455–1464.

17 Lin Z, Ping Z, Kong D, Yin K, Cai Z. The ratios of individual chemicals in a mixture determine the degree of joint effect: the climax hypothesis. Arch Environ Contam Toxicol. 2005, 49: 1–8.

18 Chen J, Sheui RG. Using Taguchi's method and orthogonal function approximation to design optimal manipulated trajectory in batch processes. Ind Eng Chem Res. 2002, 41: 2226–2237.

19 Li T, Liu SS, Qu R, Liu HL. Global concentration additivity and prediction of mixture toxicities, taking nitrobenzene derivatives as an example. Ecotoxicol Environ Saf. 2017, 144: 475–481.

20 Liu C, Li G, Mo L, Hou M, Zhang J. Alteration in concentration-response curves of four N-alkylpyridinium chloride by exposure concentration, time and in their mixtures by uniform design. Sci Total Environ. 2020, 712: 136493.

21 Miller MF, Goodson WH, Manjili MH, Kleinstreuer N, Bisson WH, Lowe L. Low-dose mixture hypothesis of carcinogenesis workshop: scientific underpinnings and research recommendations. Environ Health Perspect. 2017, 125: 163–169.

22 Pashaei R, Dzingeleviciene R, Putna-Nimane I, Overlinge D, Błaszczyk A, Walker TR. Acute toxicity of triclosan, caffeine, nanoplastics, microplastics, and their mixtures on *Daphnia magna*. Mar Poll Bull. 2023, 192: 115113.

23 Delnat V, Tran TT, Janssens L, Stoks R. Daily temperature variation magnifies the toxicity of a mixture consisting of a chemical pesticide and a biopesticide in a vector mosquito. Sci Total Environ. 2019, 659: 33–40.

24 Ren ZX, Xu HH, Wang YW, Li YF, Han S, Ren JB. Combined toxicity characteristics and regulation of residual quinolone antibiotics in water environment. Chemosphere. 2021, 263: 128301.

25 EFSA Scientific Committee, Dorne JLC, Hogstrand C, Dujardin B, Kass GE, Liem AD, Tarazona J, Machera K, Robinson T, Manini P, Benfenati E. EFSA International Workshop on RA of Combined Exposure to Multiple Chemicals. EFSA Support Publ. 2022, 19: 7422E.

26 EFSA Scientific Committee, More SJ, Bampidis V, Benford D, Bragard C, Hernandez-Jerez A, Bennekou SH, Halldorsson TI, Koutsoumanis KP, Lambre C, Machera K, Naegeli H, Nielsen SS, Schlatter JR, Schrenk D, Silano V, Turck D, Younes M, Benfenati E, Crepet A, Te Biesebeek JD, Testai E, Dujardin B, Dorne JLCM, Hogstr and C. Guidance Document on Scientific criteria for grouping chemicals into assessment groups for human risk assessment of combined exposure to multiple chemicals. EFSA J. 2021, 19(12): 7033.

27 Cattaneo I, Kalian AD, Di Nicola MR, Dujardin B, Levorato S, Mohimont L, Nathanail AV, Carnessechi E, Astuto MC, Tarazona JV, Kass GEN, Djien Liem AK, Robinson T, Manini P, Hogstrand C, Price PS, Dorne JLCM. Risk assessment of combined exposure to multiple chemicals at the European food safety authority: principles, guidance documents, applications and future challenges. Toxins. 2023, 15: 40.

28 Liu SS, Xiao QF, Zhang J, Yu M. Uniform design ray in the assessment of combined toxicities of multi-component mixtures. Sci Bull. 2016, 61: 52-58.

29 Sanchez-Aceves LM, Perez-Alvarez I, Gomez-Olivan LM, Islas-Flores H, Barcelo D. Developmental alterations, teratogenic effects, and oxidative disruption induced by ibuprofen, aluminum, and their binary mixture on *Danio rerio*. Environ Pollut.2021, 291: 118078.

30 Pikula K, Johari SA, Santos-Oliveira R, Golokhvast K. Individual and binary mixture toxicity of five nanoparticles in marine microalga *Heterosigma akashiwo*. Int J Mol Sci. 2022, 23: 990.

31 Ge HL, Liu SS, Su BX, Qin LT. Predicting synergistic toxicity of heavy metals and ionic liquids on photobacterium Q67. J Hazard Mater. 2014, 268: 77–83.

32 Bisesi JH, Sweet LE, Peter V, Klaine SJ. Effects of an antidepressant mixture on the brain serotonin and predation behavior of hybrid striped bass. Environ Toxicol Chem. 2016, 35: 938–945.

33 Cleuvers M. Mixture toxicity of the anti-inflammatory drugs diclofenac, ibuprofen, naproxen, and acetylsalicylic acid. Ecotoxicol Environ Saf. 2004, 59: 309–315.

34 Dalla Bona M, Di Leva V, De Liguoro M. The sensitivity of *Daphnia magna* and *Daphnia curvirostris* to 10 veterinary antibacterials and to some of their binary mixtures. Chemosphere. 2024, 115: 67–74.

35 Kabir ER, Rahman MS, Rahman I. A review on endocrine disruptors and their possible impacts on human health. Environ Toxicol Pharmacol. 2015, 40: 241–258.

36 Hadrup N, Taxvig C, Pedersen M, Nellemann C, Hass U, Vinggaard AM. Concentration addition, independent action and generalized concentration addition models for mixture effect prediction of sex hormone synthesis in vitro. PLoS One. 2013,8: e70490.

37 Magos L. In: Yang RSH (Eds.). Toxicology of Chemical Mixtures Case Studies, Mechanisms, and Novel Approaches. Academic Press, San Diego, 1994, pp. 720.

38 Pal A, Gin KYH, Lin AYC, Reinhard M. Impacts of emerging organic contaminants on freshwater resources: review of recent occurrences, sources, fate and effects. Sci Total Environ. 2010, 408: 6062–6069.

39 Huang B, Li D, Yang Y. Joint toxicity of two phthalates with waterborne copper to *Daphnia magna* and *Photobacterium phosphoreum*. Bull Environ Contam Toxicol. 2016, 97: 380–386.

40 Lomba L, Lapena D, Ros N, Aso E, Cannavo M, Errazquin D, Giner B. Ecotoxicological study of six drugs in *Aliivibrio fischeri, Daphnia magna* and *Raphidocelis subcapitata*. Environ Sci Pollut Res. 2020, 27: 9891–9900.

41 Cajaraville MP, Bebianno MJ, Blasco J, Porte C, Sarasquete C, Viarengo A. The use of biomarkers to assess the impact of pollution in coastal environments of the Iberian Peninsula: a practical approach. Sci Total Environ. 2000, 247 (2-3): 295–311.

42 Saravanan M, Karthika S, Malarvizhi A, Ramesh M. Ecotoxicological impacts of clofibric acid and diclofenac in common carp (*Cyprinus carpio*) fingerlings: hematological, biochemical, ionoregulatory and enzymological responses. J Hazard Mater. 2011, 195: 188–194.

43 van der Oost R, Beyer J, Vermeulen N. Fish bioaccumulation and biomarkers in environmental risk assessment: a review. Environ Toxicol Pharmacol. 2023, 13: 57–149.

44 Saravanan M, Ramesh M. Short and long-term effects of clofibric acid and diclofenac on certain biochemical and ionoregulatory responses in an Indian major carp. *Cirrhinus mrigala*. Chemosphere. 2013a, 93(2): 388–396.

45 Saravanan M, Ramesh M, Petkam R. Alterations in certain enzymological parameters of an Indian major carp, *Cirrhinus mrigala* exposed to short and long-term exposure of clofibric acid and diclofenac. Fish Physiol Biochem. 2013b, 39(6): 1431–1440.

46 Livingstone DR. Contaminant-stimulated reactive oxygen species production and oxidative damage in aquatic organisms. Mar Pollut Bull. 2001, 42: 656–666.

47 Lesser MP. Oxidative stress in marine environments: biochemistry and physiological ecology. Annu Rev Physiol. 2006, 68: 253–278.

48 Lushchak VI. Environmentally induced oxidative stress in aquatic animals. Aquat Toxicol. 2011, 101: 13–30.

49 Di Giulio RT, Washburn PC, Wenning RJ, Winston GW, Jewell CS. Biochemical responses in aquatic animal: a review of determinants of oxidative stress. Environ Toxicol Chem. 1989, 8: 1103–1123.

50 Haque MN, Eom HJ, Rhee JS. Waterborne phenanthrene modulates immune, biochemical, and antioxidant parameters in the bloods of juvenile olive flounder. Toxicol Environ Health Sci. 2018, 10: 194–202.

51 Jung JH, Moon YS, Kim BM, Lee YM, Kim M, Rhee JS, Comparative analysis of distinctive transcriptome profiles with biochemical evidence in bisphenol S- and benzo[a]pyrene-exposed liver tissues of the olive flounder *Paralichthys olivaceus*. PLoS One. 2018, 13: e0196425.

52 Kime DE. Classical and nonclassical reproductive steroids in fish. Rev Fish Biol Fish. 1993, 3: 160–180.

53 Mommsen TP, Vijayan MM, Moon T. Cortisol in teleosts: dynamics, mechanisms of action, and metabolic regulation. Rev Fish Biol Fish. 1999, 9: 211–268.

54 Wendelaar Bonga SE. The stress response in fish. Physiol Rev. 1997, 77: 591–625.

55 Pait AS, Nelson JO. Vitellogenesis in male *Fundulus heteroclitus* (Killifish) induced by selected estrogenic compounds. Aquat Toxicol. 2003, 64: 331–342.

56 Louiz I, Ben-Attia M, Ben-Hassinea OK. Gonadosomatic index and gonad histopathology of *Gobius niger* (Gobiidea, Teleost) from Bizerta lagoon (Tunisia): evidence of reproduction disturbance. Fish Res. 2009, 100: 266–273.

57 Yadetie F, Arukwe A, Goksoyr A, Male R. Induction of hepatic estrogen receptor in juvenile Atlantic salmon in vivo by the environmental estrogen, 4-nonylphenol. Sci Total Environ. 1999, 233: 201–210.

58 Trudeau VL, Peter RE, Soley BD. Testosterone and estradiol potentiate the serum gonadatropin response to gonadotropin releasing hormone in goldfish. Biol Reprod. 1991, 48: 951–960.

59 Mommsen TP, Walsh PJ. Vitellogenesis and oocyte assembly. In: Hoar WS, Randall DJ (Eds.), The Physiology of Developing Fish. Vol. 2. Academic, San Diego, CA, USA, 1988, pp. 1–59.

60 Sumpter JP, Jobling S. Vitellogenesis as a biomarker for estrogenic contamination of the aquatic environment. Environ Health Perspect. 1995, 103: 173–178.

61 Christiansen LB, Pedersen KL, Korsgaard B, Bjerregaard P. Estrogenicity of xenobiotics in rainbow trout (*Oncorhynchus mykiss*) using in vivo synthesis of vitellogenin as a biomarker. Mar Environ Res. 1998, 46(1-5): 137–140.

62 Zaccaroni A, Gamberoni M, Mandrioli L, Sirri R, Mordenti O, Scaravelli D, Sarli G, Parmeggiani A. Thyroid hormones as a potential early biomarker of exposure to 4-nonylphenol in adult male shubunkins (*Carassius auratus*). Sci Total Environ.2009, 407: 3301–3306.

63 Saravanan M, Hur JH, Arul N, Ramesh M. Toxicological effects of clofibric acid and diclofenac on plasma thyroid hormones of an Indian major carp, *Cirrhinus mrigala* during short and long-term exposures. Environ Toxicol Pharmacol. 2014, 38(3): 948–958.

64 Nemcsok J, Benedeczky I. Effect of sublethal concentrations of phenol on some enzyme activities and blood sugar level of carp *Cyprinus carpio* (L.). Environ Monit Assess. 1990, 14: 377–383.

65 McCormick SD. Effects of growth hormone and insulin-like growth factor I on salinity tolerance and gill Na^+/K^+-ATPase in Atlantic salmon (*Salmo salar*): interaction with cortisol. Gen Comp Endocrinol. 1996, 101: 3–11.

66 Saravanan M, Lee DH, Eom HJ, Nam SE, Rhee JS. Long-term exposure to waterborne nonylphenol alters reproductive physiological parameters in economically important marine fish. Comp Biochem Physiol C: Toxicol Pharmacol. 2019, 216: 10–18.

67 Malarvizhi A, Saravanan M, Hur JH, Poopal RK, Ramesh M. Accumulation of cadmium and antioxidant and hormonal responses in the Indian major carp *Cirrhinus mrigala* during acute and sublethal exposure. Water Air Soil Pollut. 2017, 228: 310.

68 Wang WX, Rainbow PS. Comparative approaches to understand metal bioaccumulation in aquatic animals. Comp Biochem Physiol C Toxicol Pharmacol. 2008, 148: 315–323.

69 Powers DA. Fish as model system. Science. 1989, 246: 352–358.

70 Yang Z, Tian X, Shu W, Yang Y, Xu J, Kan S. Combined toxicity of polyethylene microplastics and nickel oxide nanoparticle on earthworm (*Eisenia andrei*): oxidative stress responses, bioavailability and joint effect. Environ Sci Pollut Res. 2024, 31: 34910–34921.

71 Zhang Z, Li X, Li J, Pan Y, Zhuang Z, Zhang X, Chen C, Liu Y, Zhang L, Luo Y, Liu F, Fan H, Luo Q, Chen X. The combined toxic effects of polystyrene microplastics and different forms of arsenic on the zebrafish embryos (*Danio rerio*). Sci Total Environ. 2023, 887: 164017.

72 Zhang J, Bai Y, Meng H, Zhu Y, Yue Z, Li B, Wang J, Wang J, Zhu L, Du Z. Combined toxic effects of polystyrene microplastics and 3,6-dibromocarbazole on zebrafish (*Danio rerio*) embryos. Sci Total Environ. 2024, 913: 169787.

73 Jian M, Chen X, Liu S, Liu, Y, Liu Y, Wang Q, Tu W. Combined exposure with microplastics increases the toxic effects of PFOS and its alternative F-53B in adult zebrafish. Sci Total Environ. 2024, 920: 170948.

74 Lu J, Wu J, Gong L, Cheng Y, Yuan Q, He Y. Combined toxicity of polystyrene microplastics and sulfamethoxazole on zebrafish embryos. Environ Sci Pollut Res Int. 2022, 29(13): 19273–19282.

75 Yao C, Huang L, Li C, Nie D, Chen Y, Guo X, Cao N, Li X, Pang S. Exposure to fenvalerate and tebuconazole exhibits combined acute toxicity in zebrafish and behavioral abnormalities in larvae. Front Environ Sci. 2022, 10: 975634.

76 Dong X, Qiu X, Meng S, Xu H, Wu X, Yang M. Proteomic profile and toxicity pathway analysis in zebrafish embryos exposed to bisphenol A and di-n-butyl phthalate at environmentally relevant levels. Chemosphere. 2018, 193: 313–320.

77 Mao G, Zhou Z, Chen Y, Wang W, Wu X, Feng W, Cobbina SJ, Huang J, Zhang Z, Xu H, Yang L, Wu X. Neurological toxicity of individual and mixtures of low dose arsenic, mono and di (n-butyl) phthalates on sub-chronic exposure to mice. Biol Trace Elem Res. 2016, 170: 183–193.

78 Huang Z, Xiao X, Wang D, Zhong Y, Ding Q, You J. Joint effects of micro-sized polystyrene and chlorpyrifos on zebrafish based on multiple endpoints and gut microbial effects. J Environ Sci. 2023, 126: 184–197.

79 Xu N, Chen P, Liu L, Zeng Y, Zhou H, Li S. Effects of combined exposure to 17α-ethynylestradiol and dibutyl phthalate on the growth and reproduction of adult male zebrafish (*Danio rerio*). Ecotoxicol Environ Saf. 2014, 107: 61–70.

80 Chen X, Xu S, Tan T, Lee S, Cheng S, Lee F, Xu S, Ho K. Toxicity and estrogenic endocrine disrupting activity of phthalates and their mixtures. Int J Environ Res Public Health. 2014, 11: 3156–3168.

81 Gao HT, Xu R, Cao WX, Qian LL, Wang M, Lu L, Xu Q, Yu SQ. Effects of six priority controlled phthalate esters with long-term low-dose integrated exposure on male reproductive toxicity in rats. Food Chem Toxicol. 2017, 101: 94–104.

82 Aguirre-Martinez GV, Reinardy HC, Martin-Diaz ML, Henry TB. Response of gene expression in zebrafish exposed to pharmaceutical mixtures: implications for environmental risk. Ecotoxicol Environ Saf. 2017, 142: 471–479.

83 Jiang H, Li R, Zhao M, Peng X, Sun M, Liu C, Liu G, Xue H. Toxic effects of combined exposure to cadmium and diclofenac on freshwater crayfish (*Procambarus clarkii*): Insights from antioxidant enzyme activity, histopathology, and gut microbiome. Aquat Toxicol. 2024, 268: 106844.

84 Liu P, Wan Y, Zhang Z, Ji Q, Lian J, Yang C, Wang X, Qin B, Zhu L, Yu J. Toxic effects of combined exposure to cadmium and nitrate on intestinal morphology, immune response, and microbiota in juvenile Japanese flounder (*Paralichthys olivaceus*). Aquat Toxicol. 2023, 264: 106704.

85 Campos JM, Wintruff LTT, Souza-Bastos LR, Pont GD, Dolatto RG, Westphal GCC, Grassi MT, Ostrensky A, Sadauskas-Henrique H. Osmoregulatory responses in the neotropical fish species *Astyanax lacustris*, exposed to single and combined microplastics, polycyclic aromatic hydrocarbons, and their mixture. Aquat Toxicol. 2023, 263: 106693.

86 Huang B, Feng M, Li D, Yang Y. Antagonistic joint toxicity assessment of two current-use phthalates with waterborne copper in liver of *Carassius auratus* using biochemical biomarkers. Ecotoxicol Environ Saf. 2015, 116: 107–112.

87 Kim JA, Kim MJ, Park YS, Kang CK, Kim JM. Effects of microfiber and bead microplastic exposure in the goldfish *Carassius auratus*: bioaccumulation, antioxidant responses, and cell damage. Aquat Toxicol. 2023, 263: 106684.

88 Banaee M, Badr AA, Multisanti CR, Haghi BN, Faggio C. The toxicity effects of the individual and combined exposure of methyl tert-butyl ether (MTBE) and tire rubber powder (RP) on Nile tilapia fish (*Oreochromis niloticus*). Comp Biochem Physiol C. 2023, 274: 109759.

89 Bacchetta C, Rossi A, Ale A, Campana M, Parma MJ, Cazenave J. Combined toxicological effects of pesticides: a fish multi-biomarker approach. Ecol Indic. 2014, 36: 532–538.

90 Arora S, Kumar A. Mixture toxicity assessment of selected insecticides to silver perch fingerling, *Bidyanus*. Ecotoxicol Environ Saf. 2021, 226: 112790.

91 Man Y, Sun T, Wu C, Liu X, He M. Evaluating the impact of individual and combined toxicity of imidacloprid, cycloxaprid, and tebuconazole on *Daphnia magna*. Toxics. 2023, 11: 428.

92 Godoy AA, de Oliveira AC, Silva JGM, de Jesus Azevedo CC, Domingues I, Nogueira AJA, Kummrow F. Single and mixture toxicity of four pharmaceuticals of environmental concern to aquatic organisms, including a behavioral assessment. Chemosphere. 2019, 235: 373–382.

93 Manservisi F, Marquillas CB, Buscaroli A, Huff J, Lauriola M, Mandrioli D, Manservigi M, Panzacchi S, Silbergeld EK, Belpoggi F. An integrated experimental design for the assessment of multiple toxicological endpoints in rat bioassays. Environ Health Perspect. 2017, 125(3): 289–295.

94 Wood WB. The Nematode Caenorhabditis elegans. Cold Spring Harbor Laboratory Press, Cold Spring Harbor, 1988, p. 1.

95 Pradhan A, Olsson PE, Jass J. Di(2-ethylhexyl) phthalate and diethyl phthalate disrupt lipid metabolism, reduce fecundity and shortens lifespan of *Caenorhabditis elegans*. Chemosphere. 2018, 190: 375–382.

96 Lu S, Liu SS, Huang P, Wang ZJ, Wang Y. Study on the combined toxicities and quantitative characterization of toxicity sensitivities of three flavor chemicals and their mixtures to *Caenorhabditis elegans*. ACS Omega. 2021, 6: 35745–35756.

97 Chen H, Chen X, Ding P, Gu Y, Jiang Y, Li X, Hu G, Li L, Wang C, Yu J, Li H. Photoaging enhances combined toxicity of microplastics and tetrabromobisphenol A by inducing intestinal damage and oxidative stress in *Caenorhabditis elegans*. Sci Total Environ. 2024, 912: 169259.

98 Campos-Garcia J, Martinez DST, Rezende KFO, da Silva JRMC, Alves OL, Barbieri E. Histopathological alterations in the gills of Nile tilapia exposed to carbofuran and multiwalled carbon nanotubes. Ecotoxicol Environ Saf. 2016, 133: 481–488.

99 Sim Y, Cho HJ, Lee JS, Lee WS, Kim H, Jeong J. Combined effects of microplastics and benz[a]anthracene on cardiotoxicity in zebrafish (*Danio rerio*) larvae: size matters. Chemosphere. 2023, 330: 138723.

100 Bhagat J, Zang L, Kaneco S, Nishimura N, Shimada Y. Combined exposure to nanoplastics and metal oxide nanoparticles inhibits efflux pumps and causes oxidative stress in zebrafish embryos. Sci Total Environ. 2022, 835: 155436.

101 Jian M, Chen X, Liu S, Liu Y, Liu Y, Wang Q, Tu W. Combined exposure with microplastics increases the toxic effects of PFOS and its alternative F-53B in adult zebrafish. Sci Total Environ. 2024, 920: 170948.

102 Zhang T, Yuan J, Guo Y, Wang, Li QX, Zhang J, Xie J, Miao W, Fan Y. Combined toxicity of trifloxystrobin and fluopyram to zebrafish embryos and the effect on bone development. Aquat Toxicol. 2024, 268: 106834.

6

Migration and Transformation of Chemical Pollutants as Mixtures

*Xiang Ge, Yingxin Yu**

School of Environmental Science and Engineering, Institute of Environmental Health and Pollution Control, Guangdong University of Technology, Guangdong, China

*Corresponding author.
E-mail: yuyingxin@gdut.edu.cn (Y.Y.)

6.1 Introduction

When pollutants enter the environment, their migration is defined as concentration, dispersion, and disappearance induced by spatial movement (Figure 6.1). Pollutants will be transformed in the meantime to produce new compounds, which may pose more hazards to the environment. Many organic pollutants are highly toxic and challenging to break down in natural environments. Their migration and transformation are influenced not only by the properties of themselves (solubility, adsorption constant, volatility, etc.) but also by environmental conditions (photolysis, microbial hydrolysis, etc.).

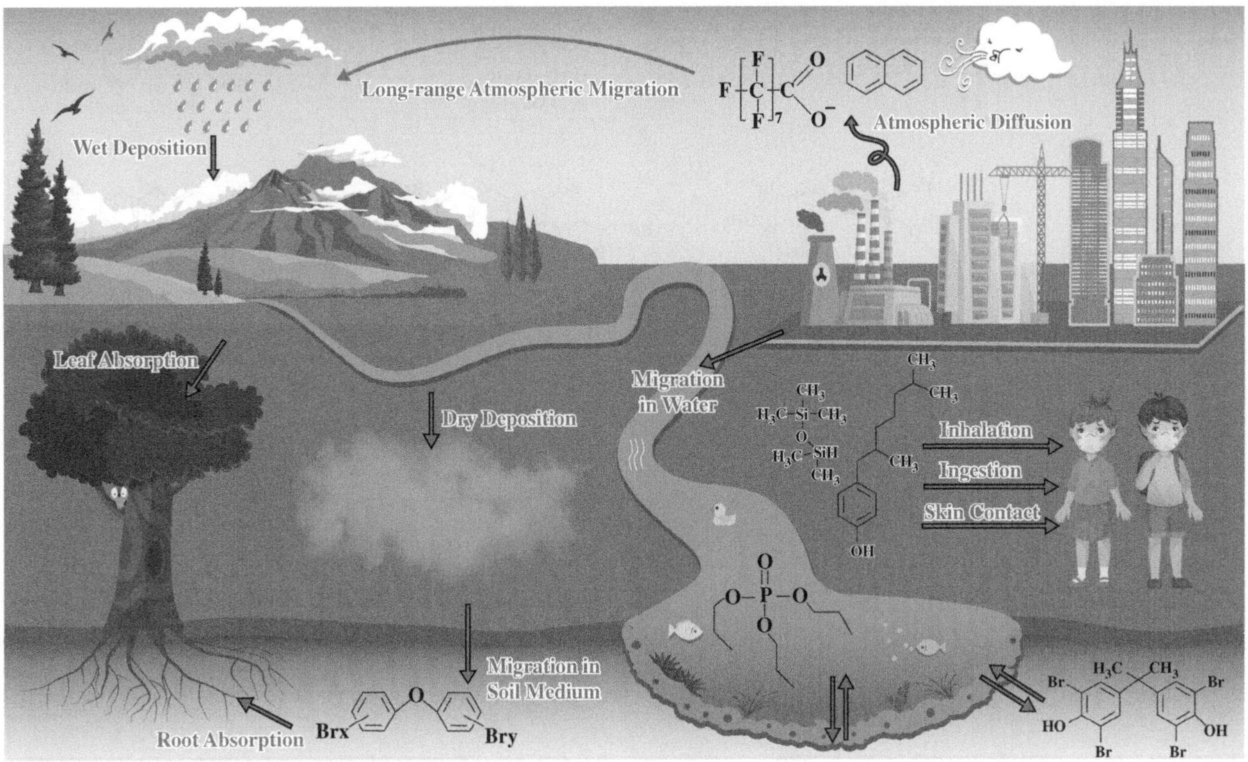

Figure 6.1 The migration of POPs into the environment.

Toxicological Assessment of Combined Chemicals in the Environment, First Edition. Edited by De-Sheng Pei and Yiyun Liu.

6.2 Volatile Organic Compounds

Volatile organic compounds (VOCs) are defined by the World Health Organization as organic compounds with boiling points between 50°C and 260°C, which include alkanes, alkenes, aromatic hydrocarbons, esters, ethers, ketones, and aldehydes (Figure 6.2) [1]. VOCs mainly exist in the atmosphere and the concentration, migration, and transformation of VOCs in the atmosphere are closely related to the distribution of emission sources, emission fluxes, topography, and meteorologic conditions [2, 3]. VOCs are emitted by a variety of sources into the environment, including petrochemical industry, coking, chemical and pharmaceutical manufacture, food, dry cleaning, tire manufacturing, tobacco, textiles, electronics, and electrical industries [4, 5]. It is worth emphasizing that the rapid expansion of the petrochemical industry in China has made it not only a pillar industry of the national economy but also one of the important industries that require joint pollution prevention and management [6].

Specifically, the VOCs emitted by petrochemical parks are mainly alkanes and olefins, with average hourly concentrations of 45.03 and 13.08 ppb, accounting for 44.0% and 29.1% of total volatile organic compounds (TVOCs), respectively. Compared with nonpetrochemical parks, there is a significant difference in the composition of VOCs [7]. It is interesting that the VOCs emitted from petrochemical parks also exhibit different characteristic pollutants due to different processes. For example, in the refining area, alkanes account for about 70%, while in the aromatic hydrocarbon production area, 80% of the VOCs are aromatic hydrocarbons [8]. The VOCs in the olefin production area are mainly composed of olefins and aromatic hydrocarbons [8]. In the sewage treatment area, 28% are alkanes, 43% are aromatic hydrocarbons, and 21% are halogenated alkanes [8]. In addition, in the monitoring of petrochemical parks, cities, and background points, the average emission concentrations were 45.2 ± 28.4, 17.6 ± 8.1, and 7.7 ± 3.7 ppbv, respectively. Among them, VOC emissions in petrochemical parks were intensive, with significant emissions of alkanes and olefins [9].

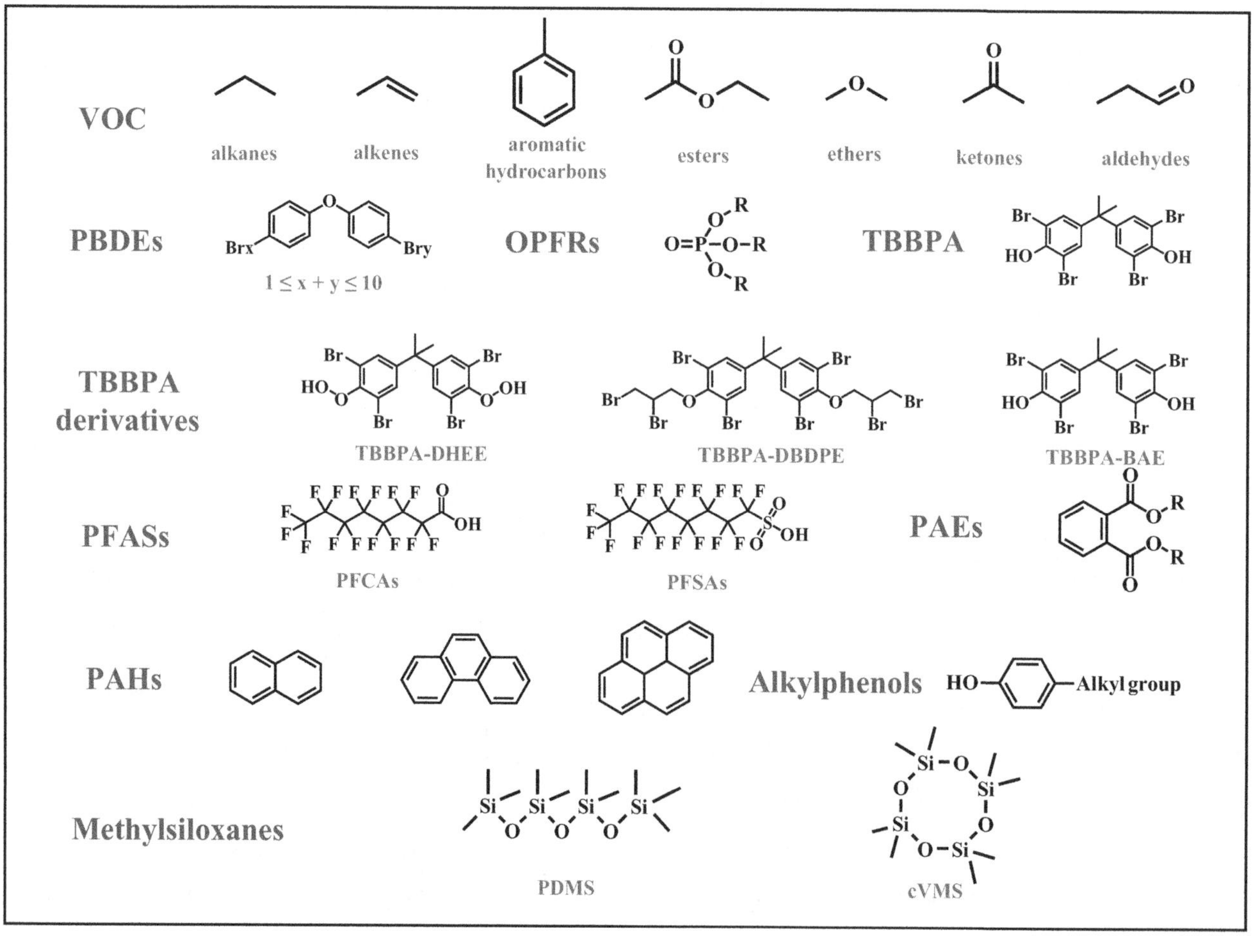

Figure 6.2 The molecular structure of the chemicals mentioned in this chapter.

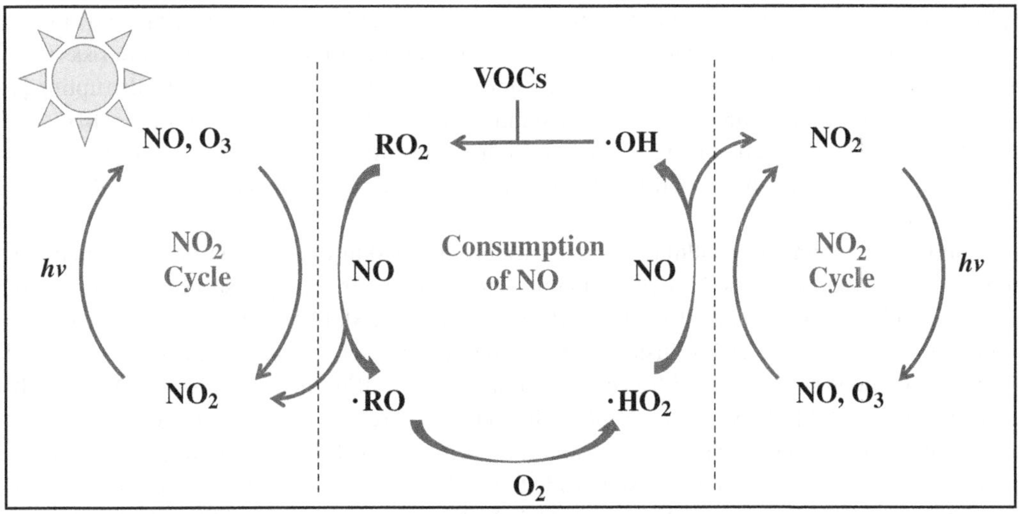

Figure 6.3 The mechanism of VOCs leading to ozone accumulation in the atmosphere.

Upon emission to the atmosphere, VOCs are not only susceptible to transport via atmospheric flows but also capable of migrating among different environmental compartments. A study calculated concentration weighted trajectory (CWT) to determine the VOCs across the provincial atmosphere and found that the VOCs emitted by petrochemical parks during VOCs pollution events exacerbate the pollution in the atmosphere [2]. In addition, research on VOCs monitoring and species migration models in petrochemical parks has shown that long-distance VOCs exhibit mass transfer and accumulation in atmospheric diffusion under the influence of terrain [3]. Backward trajectory clustering and potential source contribution function (PSCF) models have also been used to study the transport of VOCs in the atmosphere. The results indicate that the regional transport of VOCs' air masses is an important component of regional atmospheric VOCs that cannot be ignored [10]. Besides, VOCs in the atmosphere also migrate into the soil. Research has conducted sedimentation calculations using particle gas distribution and sedimentation resistance models and found that the deposition rates of VOCs' emissions from the petrochemical industry in the Bohai Economic Rim, Yangtze River Economic Belt, and Pearl River Delta are 0.24–1.53%, 0.24–1.57%, and 0.28–1.75%, respectively [11].

During migration, VOCs undergo a series of chemical reactions to generate pollutants such as secondary organic aerosol (SOA) and O_3 in the atmosphere (Figure 6.3). For the form of SOA, when the VOC concentration in the atmosphere is supersaturated, it will condense on the surface of particulate matter to form SOA. When in a subsaturated state, VOC can form SOA on the surface of particles through chemical absorption or adsorption. In addition, VOC undergoes an oxidation reaction to generate low VOCs, which then undergo gas/particle distribution to form SOA [12, 13].

At the beginning of O_3 generation process, NO_2 can produce an excited triple-state oxygen atom ($O(^3P)$) and NO. Then, $O(^3P)$ would combine with O_2 to form O_3, although O_3 could react with NO to generate NO_2 and O_2, to form a closed loop throughout the entire reaction process. However, whether O_3 is in a photostable state is determined by the NO_2 photolysis rate and the $[NO_2]/[NO]$ ratio in the atmosphere. After numerous VOCs are discharged, the peroxy radicals (RO_2) generated by the reaction between OH radicals and VOCs have stronger oxidizing properties than those of O_3. The RO_2 would immediately react with NO to generate NO_2 and •RO, and then, •RO would react with O_2 to generate $•HO_2$, which could also quickly convert NO to NO_2 and generate OH radicals. OH radicals continue to react with VOCs, causing O_3 in the atmosphere to accumulate continuously [14].

6.3 Polybrominated Diphenyl Ethers

Polybrominated diphenyl ethers (PBDEs) are a very popular type of traditional organic brominated flame retardants that are widely used in commercial products such as electronics, furniture and textiles, plastic products, and building materials (Figure 6.2) [15]. There are 209 isomers, differing in the number and position of bromine atoms on two benzene rings. Due to their lack of chemical bonding, they can easily migrate from the materials into the surrounding environment [16]. As typical POPs, PBDEs have characteristics including long-range transportability, bioaccumulation, and environmental persistence [17]. PBDEs can bioaccumulate and biomagnify in the food web, eventually bringing health risks to human beings [18]. In terms of toxicity, PBDEs are reprotoxic, growth and developmental toxic, and endocrine disrupting [19]. For commercial use, PBDEs are divided into three categories, including Penta-BDEs, Octa-BDEs, and Deca-BDEs [20]. Although Penta-BDEs and Octa-BDEs were listed as POPs by the Stockholm Convention in 2009 and the same for Deca-BDE in 2017, Deca-BDEs have not been completely banned in China and the amount of products containing Deca-BDEs produced worldwide is enormous [21].

PBDEs are ubiquitous environmental pollutants and are widely detected in various environmental matrices such as air, water, soil, and sediment [22]. Because soil is rich in organic matter, PBDEs are easily adsorbed in particles and aged, making it suitable for the soil to become a sink of PBDEs in the environment [23]. PBDEs were often detected in high concentrations in industrial areas (Figure 6.4). For instance, in the high-concentration manufacturing base of Weifang, China, PBDE was commonly found in surface soil with a concentration of 1.22×10^3ng/g, especially for BDE 209, which accounts for 95.9% and 44.4% of the total PBDEs in LG and BH industrial parks, respectively [24]. In addition, large quantities of PBDEs could be released into the soil via e-waste dismantling activities [17]. In an e-waste dismantling park in Guiyu, China, considerable PBDEs were detected in the soil inside the park (1.54×10^3–3.10×10^5 ng/g), whereas the contamination level was relatively lower in the surrounding areas (11.6–3.60×10^4 ng/g) [25]. In addition to direct source emissions, there is also a distribution of PBDEs in petrochemical areas, which may come from the leakage of PBDEs from products and the migration of PBDEs [24].

In the environment, PBDEs could be transported from polluted regions to remote areas through long-range atmospheric transport. For example, low concentrations of PBDEs (0.02–0.30 ng/g) have been detected in the Tibetan Plateau with PBDEs emitted nowhere [26]. The main reason is dry/wet deposition, where PBDEs could adsorb on atmospheric particles and then deposit in the soil far from pollution sources [27]. The spatial distribution of PBDEs in the soil might also be influenced by the direction of the prevailing wind [25, 28], and the wind blows in different directions in the winter and summer, which may affect how chemicals are distributed (Figure 6.4). In addition, the frequency and amount of rainfall could affect the concentration of pollutants [28]. PBDE concentrations declined with soil depth [28]. This may be due to the degradation of PBDEs in deeper soil, or the low migration ability of PBDEs in surface soil towards deeper soil depths. Furthermore, PBDEs could be partitioned among various environmental matrices. When PBDEs enter the water, some of them could evaporate into the air again due to the temperature rise, while others migrate to distant places with the flow of water, ultimately accumulating in sediments [29]. The adsorption/desorption of PBDEs in sediment is influenced by the structure of sediment particles and organic matter content. PBDEs adsorbed on sediment may be desorbed and released back into water once the conditions of the sediment/water interface system change [30].

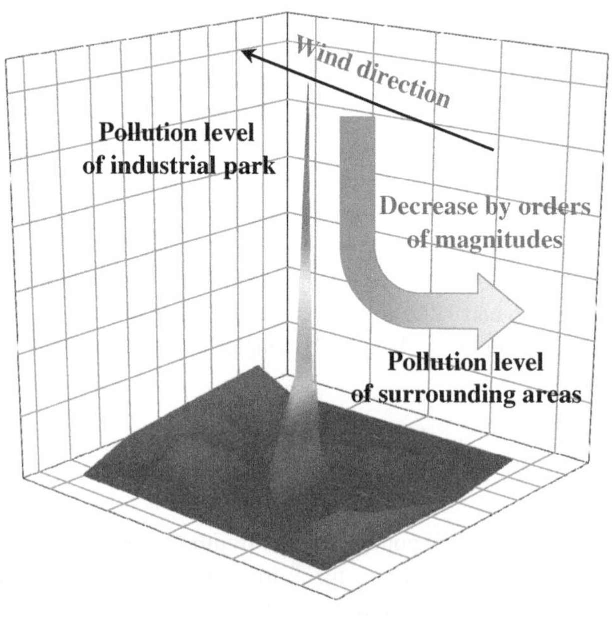

Figure 6.4 Comparison of pollution situation in industrial parks and their surrounding areas.

Many studies showed that photolysis and biodegradation are the two pathways of PBDEs for transformation in the environment [31]. Photolysis mainly occurs in the atmosphere and the surface of soil. The first and most important step is the successive removal of bromine atoms to produce low-brominated diphenyl ethers. The second step is intramolecular cyclization followed by dehydrobromination to produce highly toxic polybrominated dibenzo-*p*-dioxins/furans (PBDD/Fs) [32]. In addition, the rate of photolysis could be affected by different matrices to which PBDEs adhere. A study investigated the degradation of BDE 209 in different matrices (toluene, sand, silica gel, soil, and sediment) under simulated solar irradiation and found that degradation was fastest on sand and silica gel, while it was relatively slow in soil and sediment [33]. However, there was no difference in the debromination sequence of PBDE congeners under different substrates or different light conditions [33].

In sediment, anaerobic microbiota play a significant role in the transformation [34]. The main mechanism is catalytic reduction and debromination, but the difficulty depends on the substitution position of bromine on the benzene ring. Zeng et al. [35] used three different microbial strains, i.e. *Desulfitobacterium hafniense, Dehalobacter restrictus*, and ANAS195. It is found that the anaerobic degradation products vary depending on different bacterial cultures. For example, in the ANAS195 culture, BDE-28 was produced in minor concentrations, whereas in the *D. restrictus* and *D. hafniense* cultures, BDE-17 immediately dibrominated further to BDE-4. In general, the anaerobic degradation products of Deca-BDEs could be congeners ranging from Hexa-BDEs to Nona-BDEs [36], whereas the anaerobic degradation of an Octa-BDEs and a Penta-BDEs mixture usually produces two hepta-BDEs congeners and two tetra-BDEs congeners, respectively [35].

6.4 Tetrabromobisphenol A and Its Derivatives

Tetrabromobisphenol A (TBBPA) is not only used as an additive, but commonly as a reactive flame retardant. The "—OH" on the benzene ring provides convenience for TBBPA to bond with other monomers to produce various electronic and electrical materials, including epoxy resin, phenolic resin, and polycarbonate (Figure 6.2) [37]. In addition, different substituents replace "H" atom of "—OH" to generate a series of TBBPA derivatives, such as tetrabromobisphenol A dihydroxyethyl ether (TBBPA-DHEE), tetrabromobisphenol A bis(allyl ether) (TBBPA-BAE), and tetrabromobisphenol A bis(2,3-dibromopropyl) ether (TBBPA-BDBPE), which can be added to the material to increase fire resistance. With the gradual withdrawal of traditional flame retardants from the market, TBBPA and its derivatives are currently widely used. However, they can leak into the environment during the use and recycling of materials and accumulate continuously. Thus, TBBPA and its derivatives have attracted widespread attention due to their difficult degradation and potential toxicity, especially flame-retardant properties.

Overall, printed circuit board manufacturing and electronic waste (e-waste) dismantling activities emit a large amount of TBBPA and its derivatives. TBBPA was detected in production wastes (fine solid waste, rising water, effluent, and sludge) and environmental matrices (dust and PM_{10}). It was found that fine solid waste has the highest contribution to the environmental emission of TBBPA, suggesting the need for strict control of its production and disposal [38]. Inside an e-waste dismantling park, the TBBPA content in surface soil exhibited a high level, far exceeding that around the park [39]. Although there are few reports of TBBPA derivatives in the environment, TBBPA-DBDPE has been detected in air, sediment, soil, and sewage samples from the Pearl River Delta and an adjacent e-waste area [40]. It is worth noting that with the replacement of flame retardants, TBBPA derivatives would accumulate more in the environment, and their half-lives in water and soil are higher than those of TBBPA [41].

TBBPA is detected in multiple environmental media and even in places without emission sources, indicating that it has undergone migration. At an e-waste dismantling park, Ge et al. found that the proportion of TBBPA in the mixed pollutants in the soil inside and outside the park was basically the same. At the same time, using principal component analysis further revealed that TBBPA had undergone long-distance migration [39]. In addition, Li et al. found that the concentration of TBBPA in the rivers near the e-waste dismantling park was lower in the upstream than that in the downstream, following the trend of upstream<downstream<close to the park [42]. TBBPA has low vapor pressure (Vp, 1.41×10^{-7} Torr) and high octanol–water partition coefficient (log K_{ow}, 4.50–6.53) [41], which makes it easier for TBBPA to adsorb on the particulate matter then settle into dust and soil in the atmosphere, as well as to bind with organic matter in the sediment of the river. As for TBBPA derivatives, there is no relevant research. However, the lower Vp and higher log K_{ow} than those of TBBPA would reduce the possibility of long-distance migration of TBBPA derivatives, leading to a higher accumulation of pollution at the emission source.

The most common transformation pathway of TBBPA is debromination. In various environmental matrices around a chemical industrial park, which mainly produces brominated flame retardants, tribromobisphenol A (TriBBPA), dibromobisphenol A (DBBPA), and monobromobisphenol A (MBBPA) were widely detected [43]. A cluster of mixed bromine/chloride transformation products of TBBPA was found in the soil around the electronic waste dismantling park. Although Ge et al. believed that such transformation products were generated during the thermal treatment of printed circuit boards, and the occurrence in surroundings with no emission sources was attributed to the migration inside the park, there was no direct evidence to deny that the bromine/chloride radical substitution reaction could take place under natural conditions [39, 44]. Besides debromination, β-scission is also one of the ways of TBBPA transformation, which produces various phenolic substances, including 2,6-dibromo-4-isopropylphenol, 2,6-dibromo-4-hydroxyl-phenol, 2,6-dibromophenol, 4-isopropylene-2,6-dibromophenol, and 4-(2-hydroxyisopropyl)-2,6-dibromophenol [45]. As for TBBPA derivatives, TBBPA-BAE, TBBPA-BDBPE, and TBBPA-BHEE can break single-sided ether bonds for preliminary degradation [46]. Especially TBBPA-DBDPE can be converted into TBBPA-BAE through debromination, and both can completely break two ether bonds to generate TBBPA [45].

6.5 Per- and Polyfluoroalkyl Substances

Per- and polyfluoroalkyl substances (PFASs) are a class of organically synthesized compounds, in which the hydrogen atom attached to the carbon atom is completely or partially substituted with a fluorine atom (Figure 6.2) [47]. There are two major subcategories of PFASs, i.e. ionic PFASs (i-PFASs) and neutral PFASs (n-PFASs). The former mainly include perfluoroalkyl carboxylic acids (PFCAs) and perfluoroalkyl sulfonates (PFSAs); the latter mainly include fluorotelomer alcohols (FTOHs) and perfluoroalkyl sulfonamides (FOSAs) [48]. Due to high chemical bonding energy and large electronegativity of C—F bonds, PFASs have high chemical stability, thermal stability, high surface activity, and hydrophobic and oleophobic properties [49]. Thus, they are widely used in textile treatment, firefighting foam extinguishing agents, metal electroplating, surfactants, cleaning agents, cosmetic products, and other industrial and consumer products [50]. Currently, more than 4700 precursors of PFASs and related substances have been recorded [51]. Ecotoxicological studies found that PFASs may have immunotoxicity, neurotoxicity, developmental toxicity, reproductive toxicity, endocrine disruption, and other toxic effects, as well as potential carcinogenic effects [27]. Although some traditional PFASs have been listed as POPs, some new substitutes are gradually expanding the market [52].

PFASs are not readily degradable in the natural environment and can persist in the environment with bioconcentration ability. Until now, PFASs have been detected globally in surface water, groundwater, air, soil, sediment, wildlife, and even human urine [53, 54]. i-PFASs are mainly found in the aqueous environment and sediment, whereas most n-PFASs are volatile and are mainly found in the atmospheric environment [55]. The concentration of n-PFASs in the atmosphere of Switzerland and Arctic was 980 and 3.8pg/m^3, respectively [56, 57]. The total concentration of PFCAs and PFSA in the Yellow Sea seawater was 1.1 $\pm$ 6.75ng/L, which was significantly higher than that of PFCAs and PFSAs in the East China Sea (1.93 $\pm$ 2.90ng/L) and the South China Sea (2.84 $\pm$ 2.06 ng/L) [58]. It has been shown that the i-PFAS concentrations in the surface water and sediments of the Bohai Sea ranged from 19.3 to 967 ng/L and 0.70 to 4.13 ng/g, respectively [59]. The PFAS concentrations in Chinese residential soils ranged from 244 to 13 564 pg/g, which indicated the wide variation of PFAS contamination in China [52].

There are two sources of PFASs in the environment: one is the direct release of PFASs into the environment through the production, use, transportation, and disposal of perfluorinated products; the other is environmental behaviors such as long-range atmospheric migration and deposition [60, 61]. For example, PFASs in water bodies and the atmosphere can enter soil and water through rain and snow deposition, while PFASs in soil can migrate to surface water, groundwater, and the atmosphere through volatilization, diffusion, leaching, and stormwater runoff [62]. Therefore, soil is a potential long-term environmental pollution source of PFASs. In the atmosphere, PFASs are transported over long distances by direct binding with aerosols and particulate matter [63]. Compared with long-chain PFASs, short-chain PFASs have higher vapor pressure and are more likely to enter the atmosphere from surface water and other matrices than long-chain ones. And short-chain PFASs could quickly be transported over long distances via oceanic microsurface atmospheric aerosols in oceanic waters [63]. In water and soil environments, short-chain PFASs are more water-soluble and have a smaller mud-to-water partition ratio than long-chain ones, so they are weaker for soil adsorption and thus easily migrate with water flow [64]. In the soil–plant system, PFASs from the soil enter the plant in two processes. The first is the diffusion process of

PFASs from the soil particles to the solution, and then, the enrichment of PFASs from the solution by plant roots takes place [65].

In the environment, FTOHs are considered a precursor compound for the transformation reaction of perfluorinated compounds [66]. FTOHs are accompanied by a certain degree of abiotic degradation (photolysis, thermal degradation) and biotic degradation during the migration transformation process, eventually generating perfluorooctanoic acid (PFOA) and other PFCAs with long half-lives [67]. Wallington et al. [68] found that FTOHs can be oxidized in the atmospheric troposphere by OH radicals via a three-dimensional global atmospheric chemistry model. Plumee et al. [69] found that *N*-ethyl perfluorooctane sulfonamidoethanol (N-EtFOSE) could eventually produce FOSAs and PFOA through radical oxidation and dealkylation by sunlight irradiation of N-EtFOSE dissolved in aqueous hydrogen peroxide solution. In addition, it was found that FTOH could be degraded by aerobic microorganisms, and PFCAs such as PFOA were detected in the degradation products [70]. However, there are still many new PFASs whose conversion mechanisms and products have not been studied yet, and their environmental and human risks are also unknown. They should continue to be monitored by relevant institutions of POPs.

6.6 Organophosphorus Flame Retardants

Organophosphorus flame retardants (OPFRs) are a class of phosphoric acid esters containing alkyl, chlorinated alkyl, and aryl groups (Figure 6.2) [71, 72]. OPFRs are used as additives without forming chemical bonds with the materials, making OPFRs easily leak into the environment via volatilization, wear, leaching, deposition, and infiltration dissolution [73, 74]. The high vapor pressures may lead to potential air pollution. For example, tris(chloropropyl)phosphate (TCPP) and triethylphosphate (TEP), with vapor pressures of 0.75 mmHg (20°C) and 0.29mmHg (25°C), respectively, are likely to escape into the air [75]. Yue et al. found that the concentrations of TCPP and TEP in an e-waste dismantling park were 1.01×10^7 pg/m^3 and 7.11×10^4 pg/m^3, respectively [76]. In the atmosphere, reaction with the OH radicals is expected to be the dominant atmospheric transformation process for OPFRs [77]. The atmospheric lifetimes for OPFRs are estimated to be less than 1.3 days [78]. However, it was found that particle-bound OPFRs are highly persistent in the atmosphere about OH radical oxidation, which demonstrated that OPFRs could undergo medium- or long-range transport (LRT) [78]. Aryl and alkyl OPFRs with high molecular weights are highly lipophilic and have a strong affinity to sediments and soil, whereas chlorinated alkyl OPFRs have been shown to have better water solubility and pose a continuing threat to aquatic animals [75, 79].

Compared with other flame retardant, such as PBDEs and TBBPA, OPFRs have less persistence and biological enrichment, but as their substitutes, huge amount of use inevitably leads to OPFRs occurrence in various environmental matrices, including the atmosphere, dust, sediment, soil, and water [80, 81]. A study measured the OPFRs in the atmosphere from East Asia to the high Arctic as 230–2900 pg/m^3 and the concentration from East Asia toward the Indian Ocean to the Antarctic as 120–1700 pg/m^3 [82]. Concentrations of the OPFRs in the Bohai and Yellow Seas ranged from 8.12 to 98.04 ng/L and tended to decrease from nearshore to offshore [83]. OPFRs were also detected in the sludge of six municipal wastewater treatment plants in the Pearl River Delta (96.7–1312.9 ng/g) and in the sediments of Taihu Lake (3.38–14.3 ng/g) [84, 85]. Especially, in the soil from an e-waste dismantling park in Guiyu, China, a high level of OPFRs was detected at 1.20×10^4 ng/g, and the surrounding areas were also contaminated with 256 ng/g [25]. All studies showed that OPFRs have prevalent sources and have been distributed around the world. In particular, pollution levels are much higher in areas engaged in flame-retardant manufacturing or e-waste dismantling than in other areas [25].

Atmospheric OPFRs could enter soil through dry/wet deposition, and soil–air partitioning is a complicated process influenced by many factors, among which local source emission intensity is the most critical [71]. In heavily polluted areas, soil is a secondary pollution source. OPFRs may volatilize from soil to air, although the direction of soil–air exchange is also influenced by the physicochemical properties of OPFRs, temperature, and relative humidity (RH) [71, 81, 86]. OPFRs with log K_{oa} >9 tend to deposit from the air into the soil [87]. Conversely, TMP (log K_{oa}: 5.88), TEP (log K_{oa}: 6.63), and TCIPP (log K_{oa}: 8.20) tend to volatilize from soil to air [87]. High temperatures and RH could decrease the soil–air partition coefficient values, thus leading to the volatilization of OPFRs [86]. OPFRs in the soil could transfer into groundwater and surface water via partitioning, leaching, and runoff [88]. The transferability of OPFRs from soil to groundwater depends on their solubility and soil properties such as dissolved organic matter (DOM) [89]. High DOM contents could fix OPFRs in soils, and high temperatures may promote the leach of OPFRs [71, 89]. Also, OPFRs could diffuse to deeper soil through leaching and gravity until reaching the underground layer, probably posing harm to human drinking water [83].

The transformation of OPFRs in the environment includes photolysis and hydrolysis. According to whether the active substances are produced with the photosensitizer, photolysis could be divided into direct and indirect ways [90–92]. Because of the low light penetration into the soil, direct photolysis is typically limited to the top 0.4mm of soil [93]. Cristale et al. found that only the aryl OPFRs and TBOEP had slight spectral absorption at 220 and 400 nm, whereas the chlorinated and alkyl OPFRs did not absorb the whole spectrum. This indicated that the direct photolysis of OPFRs was generally hard to occur due to the lack of chromophore in the structure [94]. There are three possible indirect photolysis pathways of OPFRs in the soil surface: (1) •OH addition to the phenyl ring; (2) •OH addition to the central phosphorus atom, followed by the cleavage of an alkoxy or a phenoxy group from the phosphoric center; and (3) the reaction equations are described as follows:

$$RH + \bullet OH \rightarrow R\bullet + H_2O$$

$$R\bullet + O_2 \rightarrow RO_2\bullet$$

$$RO_2\bullet + RO_2\bullet \rightarrow R - H{=}O + ROH$$

$$RO_2\bullet + RO_2\bullet \rightarrow 2RO\bullet + O_2$$

$$RO\bullet + R'H \rightarrow R'\bullet + ROH$$

$$RO\bullet \rightarrow \text{decomposition products}$$

There are two hydrolysis pathways of OPFRs: (1) the nucleophile attacks the central phosphorus atom, which is positively charged and (2) the nucleophile can also attack the carbon atom with a positive charge. Generally, the hydrolysis rates of OPFRs are in the order of alkyl OPFRs>chlorinated alkyl OPFRs>aryl OPFRs [95]. Rapid degradation of OPFRs is observed in the presence of Fe(III) (hydr)oxide minerals, which the half-lives would reduce to <10 days. When no minerals are present, OPFRs tend to hydrolyze under alkaline conditions [95, 96]. In fact, the toxicities of the degradation products of tris(2-chloroethyl)phosphate (TCEP) and triphenyl phosphate (TPhP) are even greater than those of their parent compounds [97, 98]. Therefore, the study on the degradation of OPFRs is conducive to a better understanding of the transformation behavior and ecological toxicity of OPFRs in environmental matrices [83].

6.7 Phthalic Acid Esters

Phthalic acid esters (PAEs) are compounds synthesized by double esterification from phthalic anhydride and specific alcohols (Figure 6.2) [99]. The excellent plasticity makes PAEs used in various materials for product manufacturing, such as medical devices, food packaging, children's toys, and personal care products [100]. Due to noncovalently binding to materials, PAEs are prone to migration into the environment via volatilization and exudation [101]. The environmental protection agencies of countries including the United States, Canada, China, and the European Union have classified some PAE compounds as priority pollutants for control [102]. Many studies investigate that PAEs as endocrine-disrupting compounds could induce reproductive, genetical, and developmental abnormalities [103]. PAEs with high boiling points and high lipophilicity are difficult to migrate in the environment but are likely to accumulate at the emission source [99, 100]. Therefore, PAEs with the characteristics of POPs should be given more attention.

At present, PAEs are detected widely in different environmental matrices. In China, almost all surface water contains PAEs [101]. In Taihu Lake, the total concentrations of PAEs (ΣPAEs) spanned in a range of 740–13 000 ng/L [104], and in Chaohu Lake, 467–17 953 ng/L of ΣPAEs was detected [105]. Several other studies pointed out the existence of PAEs in the surface water of Russia (57.4 ng/L), Spain (254–1759 ng/L), Germany (120–8800 ng/L), and the United States (1562–3698 ng/L) [101]. In addition, it indicates that PAEs were detected in agricultural soil around the world, such as butyl benzyl phthalate (BBP) in China (0.001−1232 μg/g); dibutyl phthalate (DBP) in Paris, France (0.15–1.09 μg/g); DBP in the United Kingdom (0.042–0.099 μg/g), dimethyl phthalate (DMP) in Novi Sad, Serbia Anzali wetland, and Iran (4.18–20.71 μg/g) [101]. Specifically, di(2-ethylhexyl)phthalate (DEHP) is the most abundant compound in the environment [100]. In China, the occurrence of PAEs in surface water has been widely in southern cities (0.087–0.63 μg/L) than in northern cities (3.54–101.1 μg/L), with the highest concentration of 101 μg/L DEHP found in Haihe river [106, 107]. The concentration of PAEs in soil was found in cotton fields of South Xinjiang, with DEHP ranging from 103.5 to149 μg/g, which was due to the extensive use of plastic films [100]. Besides, a study in Tianjin showed that DEHP was dominant in both PM2.5 and PM10 in the air [105].

The TaPL3 model, which was developed specifically for estimating environmental persistence and LRT potential is set to mimic a closed system, from which no advective losses occur (via air or water flows) [108]. Evaluative models indicate that phthalates have moderate long-range atmospheric transport potentials [109]. Also, a study tested seawater and found that PAEs gradually decreased from south to north along the coast of Norway [109]. This proves that PAEs can migrate not only in the atmosphere but also in water. In addition, PAEs could migrate among different environmental matrices. The study regarding uptake and metabolism by plants showed that PAEs migrate through soil and reach groundwater [110]. The fate and migration of PAEs in the environment mainly depend on their physicochemical properties, especially log K_{ow}, log K_{oa}, log K_{aw} (air–water partition coefficient), log K_{oc} (organic carbon partition coefficient), and vapor pressure [111]. Due to the low volatility of PAEs, their participation in the atmospheric environment would be limited [100]. PAEs mainly exist in the form of aerosols in the air and can be transferred to the aquatic environment and soil through dry/wet deposition [112].

In the environment, the transformation of PAEs is mainly carried out in two ways, i.e. nonbiodegradation and biodegradation. Biodegradation is the main way to decomposition, whereas the efficiency of nonbiodegradation including hydrolysis and photolysis is relatively lower [99]. In specific, hydrolysis is a slow process and the products include acid and alcohol [99]. Photodegradation occurs through the absorption of UV light from sunlight, which is mediated by either direct or indirect mechanisms [99]. The direct photolysis mechanism refers to the energy transition of PAEs absorbing photons to the excited singlet state, which undergoes a reaction and is directly converted into products, while the indirect photolysis mechanism refers to the reaction in which certain substances present in the environment absorb light energy in an excited state and then induce a series of PAEs to participate [113]. Complete biodegradation of PAEs requires primary and secondary steps. The primary step refers to beta-oxidation, desertification or dealkylation, and trans-esterification or demethylation [112]. PAEs with a long alkyl chain are converted to a short alkyl chain via beta-oxidation [112]. Then, the shorter chain is oxidized to phthalic acid (PA) through either de-esterification of diethyl phthalate (DEP) or trans-esterification of DMP [112]. For DEP, the hydrolysis of each ethyl group occurs in the de-esterification, producing mono-ethyl phthalate (MEP) and finally PA [112]. For DMP, trans-esterification produces mono-methyl phthalate (MMP) and finally PA [112]. In the secondary degradation, PA is metabolized and mineralized under aerobic and anaerobic conditions [112]. Under aerobic conditions, PA finally oxidized to pyruvate and oxaloacetate [112]. Under anaerobic conditions, PA is converted to benzoate via decarboxylation [112].

6.8 Polycyclic Aromatic Hydrocarbons

Polycyclic aromatic hydrocarbons (PAHs) are composed of two or more fused benzene rings (Figure 6.2) [114, 115], which are lipophilic, semi-volatile, hydrophobic, and highly toxic [116–118]. According to the number of rings, they can be divided into low-molecular-weight PAHs and high-molecular-weight PAHs [114, 119]. PAHs could be transported over long distances through aerosol migration, and research has detected their presence in remote areas such as the Arctic [120]. The hydrophobicity, electrochemical stability, durability, anti-biodegradability, and carcinogenicity of PAHs would increase with the ring number and chemical structure change, whereas volatility would also decrease with the increase in its molecular weight [117, 118, 121]. Natural sources like forest fires and volcanic eruptions, as well as anthropogenic sources like coal burning, vehicle exhaust emissions, and industrial incineration, are the main sources of PAHs [122, 123].

PAHs in the environment can be divided into natural sources and anthropogenic sources, and multiple types of PAHs often co-exist in environmental matrices [119, 124]. Volcanic eruptions and forest fires were considered the main natural sources of atmospheric PAHs [114, 125]. The anthropogenic sources of PAHs include incomplete combustion of fossil fuels, other biomass, and various organic compounds [125, 126]. Therefore, metal smelting, petrochemical industry, waste incineration, and electronic waste treatment were all considered the main industries for PAH emissions [124]. PAHs were mainly formed from unsaturated hydrocarbons under anoxic conditions. Anoxic combustion and pyrolysis were the two main mechanisms leading to the formation of PAHs [117, 127].

Nowadays, numerous researches have been conducted on PAHs. Oh et al. [128] detected 16 types of PAHs in industrial wastewater, sludge, and sediment, with a total concentration of 2.10 μg/g. Chen et al. [122] collected total suspended particulate matter, PM_{10}, and $PM_{2.5}$ samples in the electronic waste dismantling park, and the PAHs content measured in these three samples was 2.60–16.1 $\mu g/m^3$, 2.20–15.1 $\mu g/m^3$, and 1.90–14.6 $\mu g/m^3$, respectively. Zhi et al. [129] detected the presence of low-molecular-weight PAHs in the water of Panyang Lake, with concentrations ranging from 5.56 to 266.1 ng/L. Qi et al. conducted a concentration assessment and source analysis on 16 PAHs in indoor dust samples from 45 private residences and 36 public buildings in China [125], and the results showed that the total concentration range of PAHs was 1.00–470 μg/g. The analysis of this research indicates that PAHs in indoor dust samples from urban and rural areas mainly come from vehicle traffic, coal combustion, and indoor cooking or heating. In addition, the research has found a significant correlation between the concentration of PAHs in indoor dust and geographical location [125]. The contents of PAHs in indoor dust in the northern region were higher than those in the southern region, which was related to the coal-fired heating in the northern region. From the West to the East, the use of oil and fossil fuels, economic development, and increased population density have led to an increase in anthropogenic PAH emissions.

PAHs mainly adsorbed onto particulate matter in the atmosphere and then migrated to soil or water via dry/wet deposition [130]. Owing to their high lipophilicity and semi-volatility, PAHs could bind tightly to soil organic matter and minerals, but the close binding with soil organic matter makes the biodegradation of PAHs more challenging [118, 130]. PAHs could be removed from soil through a variety of processes, including volatilization, photo-oxidation, abiotic losses, and plant uptake, leaching to groundwater, irreversible sorption to soil organic matter, and microbial decomposition [118]. Due to the hydrophobicity, it is simple for PAHs to combine with suspended particles in water and they can also deposit together with suspended particles in sediment [117, 131]. When the quantities of PAHs in water decline, the sediment could turn into a secondary source and release pollutants back into the water to achieve a steady state between water and sediment [131].

Moreover, PAHs can also form oxygenated PAHs, nitrated PAHs, hydroxylated PAHs, halogenated PAHs, and other derivatives [126, 130, 132]. The oxygenated PAHs were formed by the replacement of one or more hydrogen atoms with oxygen-containing functional groups in the molecular structure of PAHs [130]. It can originate from incomplete combustion or other pathways such as photochemical reactions, photo-oxidation, or in vivo transformation [130]. The formation pathways of other derivatives were mostly like the oxygenated PAHs [126, 133]. Compared with the PAHs, the typical characteristics of PAHs derivatives are greater toxicity, higher molecular weight, and lower vapor pressure, which make them easier to adsorb in particulate matter than PAHs [126].

6.9 Alkylphenols

Alkylphenols are a class of compounds synthesized by the alkylation of phenols, and the main commercially available ones include propylphenol, butylphenol, pentylphenol, heptylphenol, octylphenol, nonylphenol, decylphenol, and related long-chain alkylphenols (Figure 6.2) [134]. Alkylphenols are endocrine disruptors, which are of strong estrogen activity, and continuously accumulate after exposure to human beings [135]. Alkylphenols are used in many fields, such as plastics production, industrial detergents, packaging materials, and medical devices [136]. This inevitably contaminant the air, water, soil, and organisms through leakage and migration; nevertheless, studies regarding the occurrence of alkylphenols in the environment have rarely been reported [137].

Alkylphenols exist in water in two ways, i.e. being dissolved in water and being adsorbed on suspended particles (or sediments). Water emitted from wastewater treatment plants and fishery and agricultural activities are considered the main sources of rivers [138]. The sludge content released from 11 sewage treatment plants in the United States is 5.4–887 mg/kg, which indicates that Alkylphenol in sewage is a major source of soil alkylphenol pollution [139]. Soil is an important storage sink and re-release source for alkylphenols, and large amounts of alkylphenols could aggregate when they enter the soil with activated sludge and water for agricultural purposes. In some countries with long sludge agricultural history, the nonylphenol concentration in soils is relatively high, such as Sweden and Canada where sludge was applied on land with approximately 2 mg/kg [140]. The concentration of alkylphenols in the air was easily affected by temperature; 4-tert-butylphenol (4-t-BP), 4-tert-octylphenol (4-long-distance), and 4-nonylphenol (4-NP) were detected in homes, offices, and outdoor locations in Tokyo, Japan. Their concentrations and detection frequencies were higher in indoor air with higher temperatures than in outdoor air with lower temperatures [141]. It is clearly shown that the water is more prone to be contaminated by alkylphenols than the air and soil due to their high solubility.

In water, although alkylphenols are easily accumulated in sediments, the structure of sediment and external factors such as temperature, hydraulic disturbance, and pH all could affect the re-release of alkylphenols from sediment [142]. Adsorption is an exothermic process, and according to the principle of equilibrium movement, an increase in temperature is not conducive to adsorption. The release effect can be seen as the reverse process of adsorption, and an increase in temperature is beneficial for desorption. In addition, alkylphenols are semi-volatile and capable of evaporating from water or soil into the atmospheric environment. Then, they could either be removed by dry and wet deposition [143] or photolyzed through the reaction with OH radicals [144]. The research shows that the removal amount of alkylphenol in the Santa Ana River (SAR) is about 50%. The alkylphenols attenuation might be due to sorption and biotransformation, and phototransformation may also play a minor role [145]. The research on the photodegradation of alkylphenol needs further research in the future. As for soil, the adsorption–desorption process of soil particles for alkylphenol is very important [146], which is significantly affected by the organic matter content, although other properties such as clay content and pH should not be ignored [147].

The primary transformation pathway of alkylphenols in the environment is degradation [148], and the degradation rate of alkylphenols in water is lower than that in soil [149]. In water, naturally, dissolved oxygen, organic matter, bicarbonate, nitrate, and ferric iron could promote photolysis of alkylphenol in aqueous solution. Patrick Mazellier found that the degradation rate of 4-t-OP and 4-OP in pure aqueous solution depends on the oxygen concentration, and the degradation rate increases with the increase in oxygen concentration, and 4-tert-octylcatechol has been identified as one of the degradation products. The later product may arise from the oxidation of 4-tert-octylcatechol by hydrogen peroxide or a subsequent reaction of hydroxyl radicals with 4-tert-octylcatechol [144]. Moreover, alkylphenols in soil could be degraded within 7 days under aerobic conditions [150]. The degree of degradation is related to the oxygen content and microbial activity, whereas they hardly degrade under anaerobic conditions [150]. Meanwhile, another study suggested a significant negative correlation between the aerobic degradation rate and the size of alkyl groups [147]. In addition, it was shown that Fungi have a higher degradation ability of 4-t-BP than that of bacteria. Specifically, strain SB-4 (*Penicillium pinophilum*) showed 76.8% and strain SB-7 (*Trichoderma asperellum*) showed 49.2% degradation of 4-t-BP. Apart from 4-t-BP, strain SB-7 could degrade other alkylphenols such as 4-sec-butylphenol, 4-t-OP, and nonylphenol [151]. At present, there is little research on the degradation of alkylphenol in the environment, so alkylphenol should be paid more attention.

6.10 Methylsiloxanes

Methylsiloxanes are a class of organosilicon compounds composed of alternating silicon and oxygen atoms, with the methyl group on the silicon side [152]. For structural form, polydimethylsiloxanes (PDMS) are linear, whereas cyclic volatile methylsiloxanes (cVMS) are cyclic (Figure 6.2) [153]. Methylsiloxanes possess exceptional physicochemical properties such as low surface tension, high thermal stability, good lubricity, and hydrophobicity [154]. Thus, these chemicals have been used for decades in a wide range of industrial and consumer products, including cosmetics, personal care products, construction and decorative materials, detergents, plastics, and paper [155]. Of these, octamethylcyclotetrasiloxane (D4), decamethylcyclopentasiloxane (D5), and dodecamethylcyclohexasiloxane (D6) have been classified as high-yield chemicals by the United States Environmental Protection Agency because of their widespread use [156]. Methylsiloxanes are toxic and may be potentially harmful to human health. Toxicity tests have shown that high concentrations of D4 and D5 can cause a number of effects in mice, such as enlarged livers, smaller ovaries, and enlarged uterus [157].

The issues of environmental contamination by methylsiloxanes have received increasing attention in recent decades. It appears in various matrices such as atmosphere, soil, water, sediments, and organisms. For example, Kierkegaard et al. [158] reported the presence of methylsiloxanes in Swedish air samples with concentrations ranging from 1.8 to 8.0 ng/m^3 for D4 and 2.3 to 3.7 ng/m^3 for D5. Sanchez-Brunete et al. [159] detected D5 and D6 in the range of 9–57 ng/g and 6–27 ng/g in agricultural soils from Spain. Moreover, D4, D5, and D6 in sediments collected from Dongting Lake, China, were detected by Zhang et al. [160] with concentrations ranging from 3.98 to 360.29 ng/g, 13.57 to 304.52 ng/g, and 4.63 to 332.14 ng/g, respectively. All data show that due to mass production and use, methylsiloxanes have become widespread in various environmental matrices, which may bring potential health risks to human beings and cannot be ignored.

In the environment, methylsiloxanes could enter into the atmosphere via volatilization and ultimately be transformed. In the atmosphere, methylsiloxanes react chemically with OH radicals to form silanols. Silanols could be easily removed from the atmosphere by wet deposition due to their high polarity [161]. In addition, studies have shown that in the atmospheric troposphere, D4 and D5 could involve heterogeneous reactions on the surface of aerosols, which may be the main pathway for the removal of cVMS from the atmosphere [162].

In addition, the use of daily chemical products and personal care products would contribute to the contamination of methylsiloxanes in water through the discharge of municipal wastewater. However, it was revealed that there is still a certain amount of methylsiloxanes that could enter the natural water environment, although the removal efficiency of methylsiloxanes by wastewater treatment plants is high [163]. Meanwhile, some of the methylsiloxanes would undergo transformation via hydrolysis. For example, the cVMS could open into straight chains and then continue to be converted to monomeric units in the aqueous environment. Kozerski's work shows that D4 and D5 are readily hydrolyzed in the pH ranging from 4 to 9. Both acids and bases have a catalytic effect and the rate of hydrolysis is influenced by pH and temperature [164].

The main sources of methylsiloxanes in the soil are the leakage of sewage, the land use of sludge, landfills, and so on [162]. In soil, many factors influence the degradation and volatilization of methylsiloxanes, such as humidity, temperature, and clay content [165]. The study by Xu et al. found that hydrolysis was faster than volatilization in dry soil, while volatilization was faster in moist soil [165]. Furthermore, by tracing the degradation of methylsiloxanes, Buch et al. [166] found that linear compound L5 produced a mixture of linear oligomers L2–L8 and cyclic oligomers D4–D7; cyclic model compounds produced a mixture of D4–D8 and oligomeric linear siloxane diols.

Especially, halogenated methylsiloxanes have begun to receive attention from researchers. At some fields such as e-waste dismantling parks and oil and water treatment stations in oil extraction plants, halogenated methylsiloxanes were widely detected in water and soil around the vicinity of the plants with low concentrations of bromo-methylsiloxanes and chloro-cVMS, respectively [167, 168]. These halogenated transformation products are derived from bromination or chlorination reactions, in which the presence of a halogen source is very important. The high levels of chlorine and bromine in the waste from the e-waste dismantling park, whereas the presence of electro-oxidation units in the Victory Oilfield, China, may also release free halogen compounds [167, 168]. Furthermore, there may be more ways for these substances to be transformed in the environment, and more research should be conducted to gain a deeper understanding.

References

1 Wang H, Sun S, Nie L, Zhang Z, Li W, Hao Z. A review of whole-process control of industrial volatile organic compounds in China. J Environ Sci. 2023;123: 127–139.

2 Liu Y, Wang H, Jing S, Peng Y, Gao Y, Yan R, Wang Q, Lou S, Cheng T, Huang C. Strong regional transport of volatile organic compounds (VOCs) during wintertime in Shanghai megacity of China. Atmos Environ. 2021, 244: 117940.

3 Zhang T, Kang W, Ge X, Lin Q, Chen Q, Yu Y, An T. Explication on distribution patterns of volatile organic compounds in petro-chemistry and oil refineries of China using a species-transport model and health risk assessment. Sci Total Environ. 2023, 863: 160707.

4 Li B, Ho SSH, Li X, Guo L, Chen A, Hu L, Yang Y, Chen D, Lin A, Fang X. A comprehensive review on anthropogenic volatile organic compounds (VOCs) emission estimates in China: comparison and outlook. Environ Int. 2021, 156: 106710.

5 Yang Y, Liu B, Hua J, Yang T, Dai Q, Wu J, Feng Y, Hopke PK. Global review of source apportionment of volatile organic compounds based on highly time-resolved data from 2015 to 2021. Environ Int. 2022, 165: 107330.

6 Fang L, Hao R, Xie X, Li G, Wang H. Characteristics of VOCs emissions from circulating water of typical petrochemical enterprises and their impact on surroundings. Atmosphere. 2022, 13(12): 1985.

7 Yang Y, Guo W, Sun J, Chen Q, Meng X, Wang L, Tao H, Yang L. Characteristics of volatile organic compounds and secondary organic aerosol pollution in different functional areas of petrochemical industrial cities in Northwest China. Sci Total Environ. 2023, 858: 159903.

8 Mo Z, Shao M, Lu S, Qu H, Zhou M, Sun J, Gou B. Process-specific emission characteristics of volatile organic compounds (VOCs) from petrochemical facilities in the Yangtze River Delta, China. Sci Total Environ. 2015, 533: 422–431.

9 Mu J, Zhang Y, Xia Z, Fan G, Zhao M, Sun X, Liu Y, Chen T, Shen H, Zhang Z, Zhang H, Pan G, Wang W, Xue L. Two-year online measurements of volatile organic compounds (VOCs) at four sites in a Chinese city: significant impact of petrochemical industry. Sci Total Environ. 2023, 858: 159951.

10 Yao D, Tang G, Wang Y, Yang Y, Wang L, Chen T, He H, Wang Y. Significant contribution of spring northwest transport to volatile organic compounds in Beijing. J Environ Sci (China) 2021, 104: 169–181.

11 Dai L, Meng J, Zhao X, Li Q, Shi B, Wu M, Zhang Q, Su G, Hu J, Shu X. High-spatial-resolution VOCs emission from the petrochemical industries and its differential regional effect on soil in typical economic zones of China. Sci Total Environ. 2022, 827: 154318.

12 Pankow JF. An absorption model of gas/particle partitioning of organic compounds in the atmosphere. Atmos Environ. 1994, 28: 182–188.

13 Pandis SN, Harley RA, Cass GR, Seinfeld JH. Secondary organic aerosol formation and transport. Atmos Environ. 1992, 26A: 2269–2285.

14 Carter W P L. Development of ozone reactivity scales for volatile organic compounds. J Air Waste Manag Assoc. 1994, 44: 881–899.

15 Li W-L, Liu L-Y, Zhang Z-F, Song W-W, Huo C-Y, Qiao L-N, Ma W-L, Li Y-F. Brominated flame retardants in the surrounding soil of two manufacturing plants in China: Occurrence, composition profiles and spatial distribution. Environ Pollut. 2016, 213: 1–7.

16 Chen Y, Zhang A, Li H, Peng Y, Lou X, Liu M, Hu J, Liu C, Wei B, Jin J. Concentrations and distributions of polybrominated diphenyl ethers (PBDEs) in surface soils and tree bark in Inner Mongolia, northern China, and the risks posed to humans. Chemosphere. 2020, 247: 125950.

17 Wu Z, Han W, Yang X, Li Y, Wang Y. The occurrence of polybrominated diphenyl ether (PBDE) contamination in soil, water/sediment, and air. Environ Sci Pollut Res. 2019, 26(23): 23219–23241.

18 Ma X, Zhang H, Yao Z, Zhao X, Wang L, Wang Z, Chen J, Chen J. Bioaccumulation and trophic transfer of polybrominated diphenyl ethers (PBDEs) in a marine food web from Liaodong Bay, North China. Mar Pollut Bull. 2013, 74(1): 110–115.

19 Zhang S, Xu X, Wu Y, Ge J, Li W, Huo X. Polybrominated diphenyl ethers in residential and agricultural soils from an electronic waste polluted region in South China: Distribution, compositional profile, and sources. Chemosphere. 2014, 102: 55–60.

20 de Wit CA. An overview of brominated flame retardants in the environment. Chemosphere. 2002, 46(5): 583–624.

21 Chokwe TB, Magubane MN, Abafe OA, Okonkwo JO, Sibiya IV. Levels distributions, and ecological risk assessments of polybrominated diphenyl ethers and alternative flame retardants in river sediments from Vaal River, South Africa. Environ Sci Pollut Res. 2019, 26(7): 7156–7163.

22 Klinčić D, Dvoršćak M, Jagić K, Mendaš G, Herceg Romanić S., Levels and distribution of polybrominated diphenyl ethers in humans and environmental compartments: a comprehensive review of the last five years of research. Environ Sci Pollut Res. 2020, 27(6): 5744–5758.

23 Cheng Z, Wang Y, Wang S, Luo C, Li J, Chaemfa C, Jiang H, Zhang G. The influence of land use on the concentration and vertical distribution of PBDEs in soils of an e-waste recycling region of South China. Environ Pollut. 2014, 191: 126–131.

24 Ren H, Su P, Kang W, Ge X, Ma S, Shen G, Chen Q, Yu Y, An T. Heterologous spatial distribution of soil polycyclic aromatic hydrocarbons and the primary influencing factors in three industrial parks. Environ Pollut. 2022, 310: 119912.

25 Ge X, Ma S, Zhang X, Yang Y, Li G, Yu Y. Halogenated and organophosphorous flame retardants in surface soils from an e-waste dismantling park and its surrounding area: Distributions, sources, and human health risks. Environ Int. 2020, 139: 105741.

26 Sun Y, Yuan G-L, Li J, Li J-C, Wang G-H. Polybrominated diphenyl ethers in surface soils near the Changwengluozha Glacier of Central Tibetan Plateau, China. Sci Total Environ. 2015, 511: 399–406.

27 Jiang Y, Yuan L, Lin Q, Ma S, Yu Y. Polybrominated diphenyl ethers in the environment and human external and internal exposure in China: a review. Sci Total Environ. 2019, 696: 133902.

28 Wu Z, Han W, Xie M, Han M, Li Y, Wang Y. Occurrence and distribution of polybrominated diphenyl ethers in soils from an e-waste recycling area in northern China. Ecotoxicol Environ Saf. 2019, 167: 467–475.

29 Lee HJ, Jeong HJ, Jang YL, Kim GB. Distribution, accumulation, and potential risk of polybrominated diphenyl ethers in the marine environment receiving effluents from a sewage treatment plant. Marine Pollut Bulln. 2018, 129(1): 364–369.

30 Huang Y, Zhang D, Yang Y, Zeng X, Ran Y. Distribution and partitioning of polybrominated diphenyl ethers in sediments from the Pearl River Delta and Guiyu, South China. Environ Pollut. 2018, 235: 104–112.

31 Chen D, Hale RC, Letcher RJ. Photochemical and microbial transformation of emerging flame retardants: cause for concern? Environ Toxicol Chem. 2015, 34(4): 687–699.

32 Wang R, Tang T, Xie J, Tao X, Huang K, Zou M, Yin H, Dang Z, Lu G. Debromination of polybrominated diphenyl ethers (PBDEs) and their conversion to polybrominated dibenzofurans (PBDFs) by UV light: mechanisms and pathways. J Hazard Mater. 2018, 354: 1–7.

33 Söderström G, Sellström U, de Wit CA, Tysklind M., Photolytic debromination of decabromodiphenyl ether (BDE 209). Environ Sci Technol. 2004, 38(1): 127–132.

34 Qiu M, Chen X, Deng D, Guo J, Sun G, Mai B, Xu M. Effects of electron donors on anaerobic microbial debromination of polybrominated diphenyl ethers (PBDEs). Biodegradation. 2012, 23(3): 351–361.

35 Zeng X, Massey Simonich SL, Robrock KR, Korytár P, Alvarez-Cohen L, Barofsky DF. Application of a congener-specific debromination model to study photodebromination, anaerobic microbial debromination, and FE0 reduction of polybrominated diphenyl ethers. Environ Toxicol Chem. 2010, 29(4): 770–778.

36 Tokarz JA III, Ahn M-Y, Leng J, Filley TR, Nies L. Reductive debromination of polybrominated diphenyl ethers in anaerobic sediment and a biomimetic system. Environ Sci Technol. 2008, 42(4): 1157–1164.

37 Covaci A, Harrad S, Abdallah MA, Ali N, Law RJ, Herzke D, de Wit CA. Novel brominated flame retardants: a review of their analysis, environmental fate and behaviour. Environ Int. 2011, 37(2): 532–556.

38 Zhou X, Guo J, Zhang W, Zhou P, Deng J, Lin K. Tetrabromobisphenol A contamination and emission in printed circuit board production and implications for human exposure. J Hazard Mater. 2014, 273: 27–35.

39 Ge X, Ma S, Huo Y, Yang Y, Luo X, Yu Y, An T. Mixed bromine/chlorine transformation products of tetrabromobisphenol A: Potential specific molecular markers in e-waste dismantling areas. J Hazard Mater. 2022, 423(Pt A): 127126.

40 Shi T, Chen SJ, Luo XJ, Zhang XL, Tang CM, Luo Y, Ma YJ, Wu JP, Peng XZ, Mai BX. Occurrence of brominated flame retardants other than polybrominated diphenyl ethers in environmental and biota samples from southern China. Chemosphere. 2009, 74(7): 910–916.

41 Sunday OE, Bin H, Guanghua M, Yao C, Zhengjia Z, Xian Q, Xiangyang W, Weiwei F. Review of the environmental occurrence, analytical techniques, degradation and toxicity of TBBPA and its derivatives. Environ Res. 2022, 206: 112594.

42 Li H, La Guardia MJ, Liu H, Hale RC, Mainor TM, Harvey E, Sheng G, Fu J, Peng P. Brominated and organophosphate flame retardants along a sediment transect encompassing the Guiyu, China e-waste recycling zone. Sci Total Environ. 2019, 646: 58–67.

43 Lan J, Shen Z, Gao W, Liu A. Occurrence of bisphenol-A and its brominated derivatives in tributary and estuary of Xiaoqing River adjacent to Bohai Sea, China. Mar Pollut Bull. 2019, 149: 110551.

44 Ge X, Gao Y, Yang Y, Chen G, Ma S, Hu B, Yu Y, An T. Mixed bromine/chlorine transformation products of tetrabromobisphenol A formed in the combustion of printed circuit boards: Emission characteristics and transformation pathways. Sci Total Environ. 2023, 859: 160104.

45 Liu A, Zhao Z, Qu G, Shen Z, Shi J, Jiang G. Transformation/degradation of tetrabromobisphenol A and its derivatives: a review of the metabolism and metabolites. Environ Pollut. 2018, 243(Pt B): 1141–1153.

46 Liu A, Shi J, Qu G, Hu L, Ma Q, Song M, Jing C, Jiang G. Identification of emerging brominated chemicals as the transformation products of tetrabromobisphenol A (TBBPA) derivatives in soil. Environ Sci Technol. 2017, 51(10): 5434–5444.

47 Ding G, Xue H, Zhang J, Cui F, He X. Occurrence and distribution of perfluoroalkyl substances (PFASs) in sediments of the Dalian Bay, China. Marine Pollut Bull. 2018, 127: 285–288.

48 Wang Q, Zhao Z, Ruan Y, Hua X, Chen H, Wang Y, Jin L, Tsui MMP, Yao Y, Lam PKS, Sun H. Occurrence and seasonal distribution of legacy and emerging per- and polyfluoroalkyl substances (PFASs) in different environmental compartments from areas around ski resorts in northern China. J Hazard Mater. 2021, 407: 124400.

49 Sznajder-Katarzynska K, Surma M, Cieslik I. A review of Perfluoroalkyl Acids (PFAAs) in terms of sources, applications, human exposure, dietary intake, toxicity, legal regulation, and methods of determination. J Chem. 2019, 2019: 1–20.

50 Lindstrom AB, Strynar MJ, Libelo EL. Polyfluorinated compounds: past, present, and future. Environ Sci Technol. 2011, 45(19): 7954–7961.

51 Feng H, Ruan Y, Zhang K, Lam PKS. Current analytical methodologies and gaps for per- and polyfluoroalkyl substances determination in the marine environment. Trac-Trends Anal Chem. 2019, 121: 115372.

52 Li J, He J, Niu Z, Zhang Y. Legacy per- and polyfluoroalkyl substances (PFASs) and alternatives (short-chain analogues, F-53B, GenX and FC-98) in residential soils of China: present implications of replacing legacy PFASs. Environ Int. 2020, 135: 105419.

53 Zhang C, Lei M, Liu X, Zhou Z, Liu M, Chen H, Yang W, Wang X. Presence and inputs of legacy and novel per- and polyfluoroalkyl substances from rivers and drainage outlets to Liaodong Bay, China. Reg Stud Marine Sci. 2022, 56: 102684.

54 Cao X, Wang C, Lu Y, Zhang M, Khan K, Song S, Wang P, Wang C. Occurrence, sources and health risk of polyfluoroalkyl substances (PFASs) in soil, water and sediment from a drinking water source area. Ecotoxicol Environ Saf. 2019, 174: 208–217.

55 Yao Y, Chang S, Zhao Y, Tang J, Sun H, Xie Z. Per- and poly-fluoroalkyl substances (PFASs) in the urban, industrial, and background atmosphere of Northeastern China coast around the Bohai Sea: Occurrence, partitioning, and seasonal variation. Atmos Environ. 2017, 167: 150–158.

56 Mueller C. E, Gerecke A. C, Bogdal C, Wang Z, Scheringer M, Hungerbuehler K. Atmospheric fate of poly- and perfluorinated alkyl substances (PFASs): I. Day-night patterns of air concentrations in summer in Zurich, Switzerland. Environ Pollut. 2012, 169: 196–203.

57 Wong F, Shoeib M, Katsoyiannis A, Eckhardt S, Stohl A, Bohlin-Nizzetto P, Li H, Fellin P, Su Y, Hung H. Assessing temporal trends and source regions of per- and polyfluoroalkyl substances (PFASs) in air under the Arctic Monitoring and Assessment Programme (AMAP). Atmos Environ. 2018, 172: 65–73.

58 Lee J-W, Lee H-K, Lim J-E, Moon H-B. Legacy and emerging per- and polyfluoroalkyl substances (PFASs) in the coastal environment of Korea: occurrence, spatial distribution, and bioaccumulation potential. Chemosphere 2020, 251: 126633.

59 Zhao Z, Cheng X, Hua X, Jiang B, Tian C, Tang J, Li Q, Sun H, Lin T, Liao Y, Zhang G. Emerging and legacy per- and polyfluoroalkyl substances in water, sediment, and air of the Bohai Sea and its surrounding rivers. Environ Pollut. 2020, 263: 114391.

60 Paul AG, Jones KC, Sweetman AJ. A first global production, emission, and environmental inventory for perfluorooctane sulfonate. Environ Sci Technol. 2009, 43(2): 386–392.

61 Ellis DA, Martin JW, De Silva AO, Mabury SA, Hurley MD, Andersen MPS, Wallington, TJ. Degradation of fluorotelomer alcohols: a likely atmospheric source of perfluorinated carboxylic acids. Environ Sci Technol. 2004, 38(12): 3316–3321.

62 Sorengard M, Bergstrom S, McCleaf P, Wiberg K, Ahrens L. Long-distance transport of per- and polyfluoroalkyl substances (PFAS) in a Swedish drinking water aquifer. Environ Pollut. 2022, 311: 119981.

63 Giesy JP, Kannan K. Perfluorochemical surfactants in the environment. Environ Sci Technol. 2002, 36(7): 146A–152A.

64 Neuwald IJ, Huebner D, Wiegand HL, Valkov V, Borchers U, Noedler K, Scheurer M, Hale SE, Arp HPH, Zahn D. Ultra-short-chain PFASs in the sources of German drinking water: prevalent, overlooked, difficult to remove, and unregulated. Environ Sci Technol. 2022, 56(10): 6380–6390.

65 Zhao S, Fang S, Zhu L, Liu L, Liu Z, Zhang Y. Mutual impacts of wheat (Triticum aestivum L.) and earthworms (Eisenia fetida) on the bioavailability of perfluoroalkyl substances (PFASs) in soil. Environ Pollut. 2014, 184: 495–501.

66 Pabon M, Corpart J. M. Fluorinated surfactants: synthesis, properties, effluent treatment. J Fluorine Chem. 2002, 114(2): 149–156.

67 Weber EJ, Tebes-Stevens C, Washington JW, Gladstone R. Development of a PFAS reaction library: identifying plausible transformation pathways in environmental and biological systems. Environ Sci Process Impact. 2022, 24(5): 689–653.

68 Wallington TJ, Hurley MD, Xia J, Wuebbles DJ, Sillman S, Ito A, Penner JE, Ellis DA, Martin J, Mabury SA, Nielsen OJ, Andersen MPS. Formation of C7F15COOH (PFOA) and other perfluorocarboxylic acids during the atmospheric oxidation of 8: 2 fluorotelomer alcohol. Environ Sci Technol. 2006, 40(3): 924–930.

69 Plumlee MH, McNeill K, Reinhard M. Indirect photolysis of perfluorochemicals: hydroxyl radical-initiated oxidation of n-ethyl perfluorooctane sulfonamido acetate (N-EtFOSAA) and other perfluoroalkanesulfonamides. Environ Sci Technol. 2009, 43(10): 3662–3668.

70 Li F, Su Q, Zhou Z, Liao X, Zou J, Yuan B, Sun W. Anaerobic biodegradation of 8: 2 fluorotelomer alcohol in anaerobic activated sludge: Metabolic products and pathways. Chemosphere. 2018, 200: 124–132.

71 Tian YX, Chen HY, Ma J, Liu QY, Qu YJ, Zhao WH. A critical review on sources and environmental behavior of organophosphorus flame retardants in the soil: Current knowledge and future perspectives. J Hazard Mater. 2023, 452: 131161.

72 Bekele TG, Zhao H, Yang J, Chegen RG, Chen J, Mekonen S, Qadeer A. A review of environmental occurrence, analysis, bioaccumulation, and toxicity of organophosphate esters. Environ Sci Pollut Res. 2021, 28(36): 49507–49528.

73 Wei G-L, Li D-Q, Zhuo M-N, Liao Y-S, Xie Z-Y, Guo T-L, Li J-J, Zhang S-Y, Liang Z-Q. Organophosphorus flame retardants and plasticizers: Sources, occurrence, toxicity and human exposure. Environ Pollut. 2015, 196: 29–46.

74 Sundkvist A. M, Olofsson U, Haglund P. Organophosphorus flame retardants and plasticizers in marine and fresh water biota and in human milk. J Environ Monit. 2010, 12(4): 943–951.

75 van der Veen I, de Boer J. Phosphorus flame retardants: properties, production, environmental occurrence, toxicity and analysis. Chemosphere. 2012, 88(10): 1119–1153.

76 Yue C, Ma S, Liu R, Yang Y, Li G, Yu Y, An T. Pollution profiles and human health risk assessment of atmospheric organophosphorus esters in an e-waste dismantling park and its surrounding area. Sci Total Environ. 2022, 806(Pt 3): 151206.

77 Aschmann SM, Long WD, Atkinson R. Temperature-dependent rate constants for the gas-phase reactions of OH radicals with 1,3,5-trimethylbenzene, triethyl phosphate, and a series of alkylphosphonates. J Phys Chem A. 2006, 110(23): 7393–7400.

78 Liu Y, Liggio J, Harner T, Jantunen L, Shoeib M, Li S-M. Heterogeneous OH initiated oxidation: a possible explanation for the persistence of organophosphate flame retardants in air. Environ Sci Technol. 2014, 48(2): 1041–1048.

79 Reemtsma T, Quintana JB, Rodil R, Garcı´a-López M, Rodrı´guez I. Organophosphorus flame retardants and plasticizers in water and air I. Occurrence and fate. TrAC Trends Anal Chem. 2008, 27(9): 727–737.

80 Yang J, Zhao Y, Li M, Du M, Li X, Li Y., A review of a class of emerging contaminants: the classification, distribution, intensity of consumption, synthesis routes, environmental effects and expectation of pollution abatement to Organophosphate Flame Retardants (OPFRs). Int J Mol Sci. 2019, 20(12): 2874.

81 Zhang QY, Wang Y, Zhang C, Yao YM, Wang L, Sun HW. A review of organophosphate esters in soil: Implications for the potential source, transfer, and transformation mechanism. Environ Res. 2022, 204: 112122.

82 Möller A, Sturm R, Xie Z, Cai M, He J, Ebinghaus R. Organophosphorus flame retardants and plasticizers in airborne particles over the northern pacific and Indian Ocean toward the polar regions: evidence for global occurrence. Environ Sci Technol. 2012, 46(6): 3127–3134.

83 Zhong M, Tang J, Mi L, Li F, Wang R, Huang G, Wu H. Occurrence and spatial distribution of organophosphorus flame retardants and plasticizers in the Bohai and Yellow Seas, China. Marine Pollut Bull. 2017, 121(1): 331–338.

84 Zeng X, He L, Cao S, Ma S, Yu Z, Gui H, Sheng G, Fu J. Occurrence and distribution of organophosphate flame retardants/plasticizers in wastewater treatment plant sludges from the Pearl River Delta, China. Environ Toxicol Chem. 2014, 33(8): 1720–1725.

85 Cao S, Zeng X, Song H, Li H, Yu Z, Sheng G, Fu J. Levels and distributions of organophosphate flame retardants and plasticizers in sediment from Taihu Lake, China. Environ Toxicol Chem. 2012, 31(7): 1478–1484.

86 Deng Y, Xu W, Zeng Q, Sun F, Wang F, Li Y. Effects of temperature and relative humidity on soil–air partition coefficients of organophosphate flame retardants and polybrominated diphenyl ethers. Chemosphere. 2022, 291: 132716.

87 Li W, Wang Y, Kannan K. Occurrence, distribution and human exposure to 20 organophosphate esters in air, soil, pine needles, river water, and dust samples collected around an airport in New York state, United States. Environ Int. 2019, 131: 105054.

88 Mihajlović I, Fries E., Atmospheric deposition of chlorinated organophosphate flame retardants (OFR) onto soils. Atmos Environ. 2012, 56: 177–183.

89 Liao R, Jiang J, Li Y, Gan Z, Su S, Ding S, Li Z, Hou L. Distribution and leaching behavior of organophosphorus and brominated flame retardants in soil in Chengdu. Environ Sci: Proces Impact. 2020, 22(5): 1295–1305.

90 Andreozzi R, Raffaele M, Nicklas P., Pharmaceuticals in STP effluents and their solar photodegradation in aquatic environment. Chemosphere. 2003, 50(10): 1319–1330.

91 Wang Y, Roddick FA, Fan L. Direct and indirect photolysis of seven micropollutants in secondary effluent from a wastewater lagoon. Chemosphere. 2017, 185: 297–308.

92 Zepp RG, Cline DM. Rates of direct photolysis in aquatic environment. Environ Sci Technol. 1977, 11(4): 359–366.

93 Su L, Caywood LM, Sivey JD, Dai N. Sunlight photolysis of safener benoxacor and herbicide metolachlor as mixtures on simulated soil surfaces. Environ Sci Technol. 2019, 53(12): 6784–6793.

94 Cristale J, Dantas RF, De Luca A, Sans C, Esplugas S, Lacorte S. Role of oxygen and DOM in sunlight induced photodegradation of organophosphorous flame retardants in river water. J hazard Mater. 2017, 323: 242–249.

95 Su G, Letcher RJ, Yu H. Organophosphate flame retardants and plasticizers in aqueous solution: pH-dependent hydrolysis, kinetics, and pathways. Environ Sci Technol. 2016, 50(15): 8103–8111.

96 Fang Y, Kim E, Strathmann TJ. Mineral- and base-catalyzed hydrolysis of organophosphate flame retardants: potential major fate-controlling sink in soil and aquatic environments. Environ Sci Technol. 2018, 52(4): 1997–2006.

97 Su G, Crump D, Letcher R, Kennedy SW. Rapid in vitro metabolism of the flame retardant triphenyl phosphate and effects on cytotoxicity and mRNA expression in chicken embryonic hepatocytes. Environ Sci Technol. 2014, 48(22): 13511–13519.

98 Takahashi S, Miura K, Abe K, Kera Y. Complete detoxification of tris(2-chloroethyl) phosphate by two bacterial strains: Sphingobium sp. strain TCM1 and Xanthobacter autotrophicus strain GJ10. J Biosci Bioeng. 2012, 114(3): 306–311.

99 Amritha PS, Vinod V, Harathi PB. A critical review on extraction and analytical methods of phthalates in water and beverages. J Chromatogr A. 2022, 1675: 463175.

100 Gao D, Li Z, Wang H, Liang H. An overview of phthalate acid ester pollution in China over the last decade: environmental occurrence and human exposure. Sci Total Environ. 2018, 645: 1400–1409.

101 Wang Z, Ma J, Wang T, Qin C, Hu X, Mosa A, Ling W. Environmental health risks induced by interaction between phthalic acid esters (PAEs) and biological macromolecules: a review. Chemosphere. 2023, 328: 138578.

102 Lü H, Mo C-H, Zhao H-M, Xiang L, Katsoyiannis A, Li Y-W, Cai Q-Y, Wong M-H. Soil contamination and sources of phthalates and its health risk in China: a review. Environ Res. 2018, 164: 417–429.

103 Ghosh, S, Sahu, M. Phthalate pollution and remediation strategies: a review. J Hazard Mater Adv. 2022, 6: 100065.

104 Gao X, Li J, Wang X, Zhou J, Fan B, Li W, Liu Z. Exposure and ecological risk of phthalate esters in the Taihu Lake basin, China. Ecotoxicol Environ Saf. 2019, 171: 564–570.

105 He W, Qin N, Kong X, Liu W, He Q, Ouyang H, Yang C, Jiang Y, Wang Q, Yang B, Xu F. Spatio-temporal distributions and the ecological and health risks of phthalate esters (PAEs) in the surface water of a large, shallow Chinese lake. Sci Total Environ. 2013, 461-462: 672–680.

106 Chi J. Phthalate acid esters in Potamogeton crispus L. from Haihe River, China. Chemosphere. 2009, 77(1): 48–52.

107 Zeng F, Cui K, Xie Z, Liu M, Li Y, Lin Y, Zeng Z, Li F. Occurrence of phthalate esters in water and sediment of urban lakes in a subtropical city, Guangzhou, South China. Environ Int. 2008, 34(3): 372–380.

108 Cousins IT, Mackay D, Parkerton TF. In Physical-Chemical Properties and Evaluative Fate Modelling of Phthalate Esters, Springer, Berlin, 2003, pp. 57–84.

109 Xie Z, Ebinghaus R, Temme C, Lohmann R, Caba A, Ruck W. Occurrence and air—sea exchange of phthalates in the arctic. Environ Sci Technol. 2007, 41(13): 4555–4560.

110 Sun J, Wu X, Gan J. Uptake and metabolism of phthalate esters by edible plants. Environ Sci Technol. 2015, 49(14): 8471–8478.

111 Kashyap D, Agarwal T. Concentration and factors affecting the distribution of phthalates in the air and dust: a global scenario. Sci Total Environ. 2018, 635: 817–827.

112 Tran H-T, Nguyen M-K, Hoang H-G, Hutchison JM, Vu CT. Composting and green technologies for remediation of phthalate (PAE)-contaminated soil: current status and future perspectives. Chemosphere. 2022, 307: 135989.

113 Feng J-R, Deng Q-X, Ni H-G. Photodegradation of phthalic acid esters under simulated sunlight: mechanism, kinetics, and toxicity change. Chemosphere. 2022, 299: 134475.

114 Anam M, Mariaas C. A review on polycyclic aromatic hydrocarbon (PAH) emissions from energy generation. Environ Sci Technol. 2000, 34: 15.

115 Cao S, Hu J, Wu Q, Wei X, Ma G, Yu H. Prediction study on the distribution of polycyclic aromatic hydrocarbons and their halogenated derivatives in the atmospheric particulate phase. Ecotoxicol Environ Saf. 2022, 245: 114111.

116 Tsibart AS, Gennadiev AN. Polycyclic aromatic hydrocarbons in soils: sources, behavior, and indication significance (a review). Eurasian Soil Sci. 2013, 46(7): 728–741.

117 Li Y, Duan X, Polycyclic aromatic hydrocarbons in sediments of China Sea. Environ Sci Pollut Res. 2015, 22(20): 15432–15442.

118 Mazarji M, Minkina T, Sushkova S, Mandzhieva S, Bidhendi GN, Barakhov A, Bhatnagar A. Effect of nanomaterials on remediation of polycyclic aromatic hydrocarbons-contaminated soils: a review. J Environ Manag. 2021, 284: 112023.

119 Abdel-Shafy HI, Mansour MSM. A review on polycyclic aromatic hydrocarbons: source, environmental impact, effect on human health and remediation. Egyptian J Petrol. 2016, 25(1): 107–123.

120 Wu X, Sun W, Huai B, Wang L, Han C, Wang Y, Mi W. Seasonal variation and sources of atmospheric polycyclic aromatic hydrocarbons in a background site on the Tibetan Plateau. J Environ Sci. 2023, 125: 524–532.

121 Chen HJ, Ma ST, Yu YX, Liu RR, Li GY, Huang HB, An TC. Seasonal profiles of atmospheric PAHs in an e-waste dismantling area and their associated health risk considering bioaccessible PAHs in the human lung. Sci Total Environ. 2019, 683: 371–379.

122 Kariyawasam T, Doran GS, Howitt JA, Prenzler PD. Polycyclic aromatic hydrocarbon contamination in soils and sediments: sustainable approaches for extraction and remediation. Chemosphere. 2022, 291(Pt 3): 132981.

123 Guo J, Chen J, Wang J. Sedimentary records of polycyclic aromatic hydrocarbons in China: a comparison to the worldwide. Crit Rev Environ Sci Technol. 2017, 47(17): 1612–1667.

124 Yang L, Zhang H, Zhang X, Xing W, Wang Y, Bai P, Zhang L, Hayakawa, K, Toriba A, Tang N. Exposure to Atmospheric Particulate Matter-Bound Polycyclic Aromatic Hydrocarbons and Their Health Effects: A Review. International Journal of Environmental Research and Public Health 2021, 18(4): 2177.

125 Qi H, Li W. L, Zhu NZ, Ma WL, Liu LY, Zhang F, Li YF. Concentrations and sources of polycyclic aromatic hydrocarbons in indoor dust in China. Sci Total Environ. 2014, 491-492: 100–107.

126 Yadav IC, Devi NL, Singh VK, Li J, Zhang G. Concentrations, sources and health risk of nitrated- and oxygenated-polycyclic aromatic hydrocarbon in urban indoor air and dust from four cities of Nepal. Sci Total Environ. 2018, 643: 1013–1023.

127 Imam A, Kumar Suman S, Kanaujia PK, Ray A. Biological machinery for polycyclic aromatic hydrocarbons degradation: a review. Bioresour Technol. 2022, 343: 126121.

128 Oh JY, Choi SD, Kwon HO, Lee SE. Leaching of polycyclic aromatic hydrocarbons (PAHs) from industrial wastewater sludge by ultrasonic treatment. Ultrason Sonochem. 2016, 33: 61–66.

129 Zhi H, Zhao Z, Zhang L. The fate of polycyclic aromatic hydrocarbons (PAHs) and organochlorine pesticides (OCPs) in water from Poyang Lake, the largest freshwater lake in China. Chemosphere. 2015, 119: 1134–1140.

130 Almouallem W, Michel J, Dorge S, Joyeux C, Trouve G, Le Nouen D. A comparative study of the sorption of O-PAHs and PAHs onto soils to understand their transport in soils and groundwater. J Environ Sci. 2023, 124: 61–75.

131 Kong J, Ma T, Cao X, Li W, Zhu F, He H, Sun C, Yang S, Li S, Xian Q. Occurrence, partition behavior, source and ecological risk assessment of nitro-PAHs in the sediment and water of Taige Canal, China. J Environ Sci. 2023, 124: 782–793.

132 Walgraeve C, Demeestere K, Dewulf J, Zimmermann R, Van Langenhove H. Oxygenated polycyclic aromatic hydrocarbons in atmospheric particulate matter: molecular characterization and occurrence. Atmos Environ. 2010, 44(15): 1831–1846.

133 Sun C, Qu L, Wu L, Wu X, Sun R, Li Y. Advances in analysis of nitrated polycyclic aromatic hydrocarbons in various matrices. Trends Anal Chem. 2020, 127: 115878.

134 Priac A, Morin-Crini N, Druart C, Gavoille S, Bradu C, Lagarrigue C, Torri G, Winterton P, Crini G. Alkylphenol and alkylphenol polyethoxylates in water and wastewater: a review of options for their elimination. Arab J Chem. 2017, 10: S3749–S3773.

135 Omar TFT, Ahmad A, Aris AZ, Yusoff FM. Endocrine disrupting compounds (EDCs) in environmental matrices: review of analytical strategies for pharmaceuticals, estrogenic hormones, and alkylphenol compounds. TrAC Trends Anal Chem. 2016, 85: 241–259.

136 Crini G, Cosentino C, Bradu C, Fourmentin M, Torri G, Ruzimuradov O, Alaton IA, Tomei MC, Derco J, Barhoumi M, Prosen H, Malinović BN, Vrabel' M, Huq MM, Soltan J, Lichtfouse E, Morin-Crini N. Innovative technologies to remove alkylphenols from wastewater: a review. Environ Chem Lett. 2022, 20(4): 2597–2628.

137 Metcalfe CD, Bayen S, Desrosiers M, Munoz G, Sauve S, Yargeau V. An introduction to the sources, fate, occurrence and effects of endocrine disrupting chemicals released into the environment. Environ Res. 2022, 207: 112658.

138 Boitsov S, Mjos SA, Meier S. Identification of estrogen-like alkylphenols in produced water from offshore oil installations. Mar Environ Res. 2007, 64(5): 651–665.

139 La Guardia MJ, Hale RC, Harvey E, Mainor TM. Alkylphenol ethoxylate degradation products in Land-applied sewage sludge (biosolids). Environ Sci Technol. 2001, 35(24): 4798–4804.

140 Liber K, Knuth ML, Stay FS. An integrated evaluation of the persistence and effects of 4-nonylphenol in an experimental littoral ecosystem. Environ Toxicol Chem. 1999, 18(3): 357–362.

141 Saito I, Onuki A, Seto H. Indoor air pollution by alkylphenols in Tokyo. Indoor Air. 2004, 14(5): 325–332.

142 Means JC, Wood SG, Hassett JJ, Banwart WL. Sorption of polynuclear aromatic hydrocarbons by sediments and soils. Environ Sci Technol. 1980, 14(12): 1524–1528.

143 Fries E. Occurrence of 4-Nonylphenol in rain and snow. Atmos Environ. 2004, 38(13): 2013–2016.

144 Mazellier P, Leverd J. Transformation of 4-tert-octylphenol by UV irradiation and by an H2O2/UV process in aqueous solution. Photochem Photobiol Sci. 2003, 2(9): 946–953.

145 Lin AY, Plumlee MH, Reinhard M. Natural attenuation of pharmaceuticals and alkylphenol polyethoxylate metabolites during river transport: photochemical and biological transformation. Environ Toxicol Chem. 2006, 25(6): 1458–1464.

146 Stumpe B, Marschner B. Dissolved organic carbon from sewage sludge and manure can affect estrogen sorption and mineralization in soils. Environ Pollut. 2010, 158(1): 148–154.

147 Shibata A, Inoue Y, Katayama A. Aerobic and anaerobic biodegradation of phenol derivatives in various paddy soils. Sci Total Environ. 2006, 367(2-3): 979–987.

148 Loffredo E, Senesi N. Fate of anthropogenic organic pollutants in soils with emphasis on adsorption/desorption processes of endocrine disruptor compounds. Pure Appl Chem. 2006, 78(5): 947-961.

149 Salgueiro-Gonzalez N, Turnes-Carou I, Besada V, Muniategui-Lorenzo S, Lopez-Mahia P, Prada-Rodriguez D. Occurrence, distribution and bioaccumulation of endocrine disrupting compounds in water, sediment and biota samples from a European river basin. Sci Total Environ. 2015, 529: 121–30.

150 Toyama T, Momotani N, Ogata Y, Miyamori Y, Inoue D, Sei K, Mori K, Kikuchi S, Ike M. Isolation and characterization of 4-tert-butylphenol-utilizing Sphingobium fuliginis strains from Phragmites australis rhizosphere sediment. Appl Environ Microbiol. 2010, 76(20): 6733–6740.

151 Chang Y-C, Reddy MV, Umemoto H, Kondo S, Choi D. Biodegradation of alkylphenols by rhizosphere microorganisms isolated from the roots of Hosta undulata. J Environ Chem Eng. 2020, 8(3): 103771.

152 Allen RBKP, Chandra G. Organosilicon Materials: The Handbook of Environmental Chemistry/Anthropogenic Compounds (part H). Springer-Verlag, Berlin, Germany, 1997;Vol. 3.

153 Tran TM, Kannan K. Occurrence of cyclic and linear siloxanes in indoor air from Albany, New York, USA, and its implications for inhalation exposure. Sci Total Environ. 2015, 511: 138–144.
154 Horii Y, Kannan K. Survey of organosilicone compounds, including cyclic and linear siloxanes, in personal-care and household products. Arch Environ Contam Toxicol. 2008, 55(4): 701–710.
155 Lu Y, Yuan T, Wang W, Kannan K. Concentrations and assessment of exposure to siloxanes and synthetic musks in personal care products from China. Environ Pollut. 2011, 159(12): 3522–3528.
156 Non-confidential Inventory Update Reporting Production Volume Information. Toxic Substances Control Act (TSCA). United States Environmental Protection Agency, 2002.
157 Mckim JM, Wilga PC, Kolesar GB, Supratim C, Ajay M, Dochterman LW, Breen JG, Andrew P, Mast RW, Meeks RGJTS. Evaluation of octamethylcyclotetrasiloxane (D4) as an inducer of rat hepatic microsomal cytochrome P450, UDP-glucuronosyltransferase, and epoxide hydrolase: a 28-day inhalation study. Toxicol Sci. 1998, 41(1): 29–41.
158 Kierkegaard A, McLachlan MS. Determination of linear and cyclic volatile methylsiloxanes in air at a regional background site in Sweden. Atmos Environ. 2013, 80: 322–329.
159 Sanchez-Brunete C, Miguel E, Albero B, Tadeo JL. Determination of cyclic and linear siloxanes in soil samples by ultrasonic-assisted extraction and gas chromatography-mass spectrometry. J Chromatogr A. 2010, 1217(45): 7024–7030.
160 Zhang Y, Shen M, Tian Y, Zeng G. Cyclic volatile methylsiloxanes in sediment, soil, and surface water from Dongting Lake, China. J Soil Sediment. 2018, 18(5): 2063–2071.
161 Janechek NJ, Hansen KM, Stanier CO. Comprehensive atmospheric modeling of reactive cyclic siloxanes and their oxidation products. Atmos Chem Phys. 2017, 17(13): 8357–8370.
162 Rucker C, Kummerer K. Environmental chemistry of organosiloxanes. Chem Rev. 2015, 115(1): 466–524.
163 Van Egmond R, Sparham C, Hastie C, Gore D, Chowdhury N. Monitoring and modelling of siloxanes in a sewage treatment plant in the UK. Chemosphere. 2013, 93(5): 757–765.
164 Kozerski GE, Durham JA. Abstracts, 38th Central Regional Meeting of the American Chemical Society. Frankenmuth, MI, May 16-20, 2006.
165 Xu S. Hydrolysis of poly(dimethylsiloxanes) on clay minerals as influenced by exchangeable cations and moisture. Environ Sci Technol. 1998, 32(20): 3162–3168.
166 Buch RR. Rearrangement of poly (dimethylsiloxane) fluids on soil. Environ Sci Technol. 1979, 13(6): 676–679.
167 Liu N, Zhang J, He X, Xu L, Cai Y. Occurrence and fate of chlorinated methylsiloxanes in surrounding aqueous systems of Shengli oilfield, China. J Environ Sci (China). 2023, 125: 332–339.
168 Xu L, Huang Z, Zhang Q, Xiang X, Zhang S, Cai Y. Methylsiloxanes and their brominated products in one e-waste recycling area in China: emission, environmental distribution, and elimination. Environ Sci Technol. 2020, 54(7): 4267–4274.

7

Analytical Techniques Used to Detect Chemical Mixtures in the Environment

Thodhal Yoganandham Suman[a,b,*]

[a] *Department of Environmental Engineering, Changwon National University, Changwon, Gyeongsangnamdo, Republic of Korea*
[b] *School of Smart and Green Engineering, Changwon National University, Changwon, Gyeongsangnamdo, Republic of Korea*

*Corresponding author.
E-mail: sumancas2010@gmail.com

7.1 Introduction

In recent decades, a growing number of organic compounds, collectively termed emerging organic pollutants (EOPs), have garnered significant attention due to their ubiquitous presence in the environment and potential for adverse ecological and human health effects. Released in vast quantities from various anthropogenic activities, these toxic compounds permeate our modern world, profoundly impacting diverse environmental matrices. Waterways, unfortunately, serve as major sinks for EOPs, accumulating these contaminants from a multitude of sources and posing risks to aquatic organisms and human populations reliant on these water resources.

The fate and transport of EOPs in the environment are complex and are influenced by a delicate interplay of factors. These include inherent chemical and physical properties of each compound, their susceptibility to transformation via chemical reactions or microbial degradation, and prevailing environmental conditions such as water availability, temperature, oxygen, and light [1]. These transport mechanisms, coupled with their persistence, contribute to the global distribution of synthetic organic compounds, encompassing a wide range of chemicals including pesticides, industrial byproducts, pharmaceuticals, personal care products, and their degradation products. This widespread contamination raises concerns about potential bioaccumulation in food webs, ultimately leading to human exposure through the consumption of contaminated food and water, with potential consequences for human health.

The significant contribution of environmental factors to human health is increasingly recognized, with studies attributing up to two-thirds of chronic disease risk to gene–environment interactions [2]. This is further emphasized by the Lancet Commission on Pollution and Health, which estimates that pollution contributes to a staggering 16% of all premature deaths globally [3]. These alarming figures underscore the urgent need for comprehensive research to unravel the complex relationship between environmental quality, chemical exposure, and human health outcomes. While concerted national and international regulatory efforts have led to a decline in the concentrations of certain legacy pollutants [4], a persistent challenge arises from the continuous introduction of novel chemicals into the environment.

This challenge is exemplified by the exponential growth in registered compounds within the Chemical Abstract Service Registry, surging from 20 million in 2002 to 156 million by 2019. The sheer scale of this chemical expansion often outpaces our capacity to conduct thorough toxicological assessments, potentially leaving knowledge gaps concerning their long-term environmental and health impacts. Furthermore, the replacement of phased-out chemicals frequently results in the introduction of substitute substances with comparable properties, including toxicity profiles. This regrettable trend is exemplified by flame retardants, plasticizers, chlorinated paraffins, and polyfluoroalkyl substance replacements that often mirror the hazardous characteristics of their predecessors. While many contemporary agricultural pesticides, industrial chemicals, pharmaceuticals, and personal care products are designed for shorter environmental persistence, their widespread use and continuous release result in persistent environmental burdens. Moreover, the degradation products of these chemicals can exhibit unexpected behavior, with some being more stable, persistent, and even more toxic than their parent

Toxicological Assessment of Combined Chemicals in the Environment, First Edition. Edited by De-Sheng Pei and Yiyun Liu.

compounds [5, 6]. While degradation often leads to increased water solubility and potentially reduced hazard [6], there are documented cases where transformation products exhibit greater toxicity than the original compound [7]. These complexities highlight the need for comprehensive risk assessments that consider not only the parent compound but also its potential transformation products throughout its environmental lifecycle.

7.2 Navigating the Complexities of Mixture Toxicity: A Persistent Challenge in Environmental Health

Historically, environmental contamination has often been attributed to specific classes of industrial chemicals, leading to a somewhat siloed approach to risk assessment. However, a growing body of evidence highlights the critical need to shift toward a more holistic understanding of chemical exposures, recognizing that we are continuously exposed to complex mixtures of chemicals, many of which remain unidentified and uncharacterized. This paradigm shift is particularly relevant in the context of globalized manufacturing, with large-scale production increasingly shifting toward developing nations that may face challenges in implementing robust chemical regulations and infrastructure for wastewater treatment [3]. These regions may consequently experience disproportionately higher exposures to complex chemical mixtures, underscoring the need for equitable environmental health protections on a global scale.

The concept of the exposome has emerged as a valuable framework for addressing the complexities of chemical mixtures and their implications for human and ecological health [8]. Critically, chemicals can exert combined effects even at concentrations below their individual effect thresholds or analytical detection limits [9], highlighting the limitations of traditional risk assessment approaches that typically focus on single chemicals in isolation.

While the principles of concentration addition (for similarly acting chemicals) and independent action (for dissimilarly acting chemicals) provide a useful starting point for understanding mixture toxicity [10], extrapolating these concepts to complex environmental mixtures presents significant challenges. At environmentally relevant concentrations, both models tend to converge, suggesting a generally additive relationship between concentration and effect [11, 12]. While synergistic interactions (where the combined effect is greater than the sum of individual effects) can occur, they are typically observed at higher concentrations involving a limited number of components [11] and are less likely to be the dominant driver of toxicity in complex environmental mixtures.

A major challenge in assessing the risks posed by environmental mixtures lies in the significant proportion of unidentified chemicals present. While assuming an additive model for low-concentration mixtures with diverse modes of action may be reasonable, the contribution of these unknown entities to overall toxicity remains largely unknown. This knowledge gap underscores the need for advanced analytical techniques capable of characterizing complex mixtures and for the development of more sophisticated toxicological approaches that can better predict the combined effects of multiple stressors. Bioanalytical tools show promise in addressing these challenges by providing sensitive and biologically relevant endpoints for assessing mixture effects in complex environmental samples.

7.3 Advancing Environmental Monitoring: Toward a Holistic Understanding of Chemical Mixtures

The evolving landscape of chemical pollution necessitates a paradigm shift in environmental monitoring strategies. Traditional approaches relying on discrete, active sampling of individual environmental compartments are increasingly being replaced by multimedia sampling and passive sampling techniques, enabling a more comprehensive assessment of complex environmental mixtures (Figure 7.1) [13].

Passive samplers, in particular, offer several advantages for characterizing chemical mixtures. These devices allow for the integrated collection of multiple chemicals over time, simplifying sampling logistics and providing a more representative picture of environmental exposure. Importantly, extracts from passive samplers can be directly coupled with bioassays to assess the overall toxicity of the bioavailable fraction of the environmental mixture [16–18]. This effect-directed analysis (EDA) approach, integrating passive sampling with bioassays, has gained traction as a powerful tool for identifying bioactive chemicals within complex environmental matrices [19–21].

Among the various passive sampling devices, silicone wristbands (SWBs) have emerged as a promising tool for personal exposure monitoring [22, 23]. The use of SWBs for assessing human exposure to a wide range of chemicals has grown

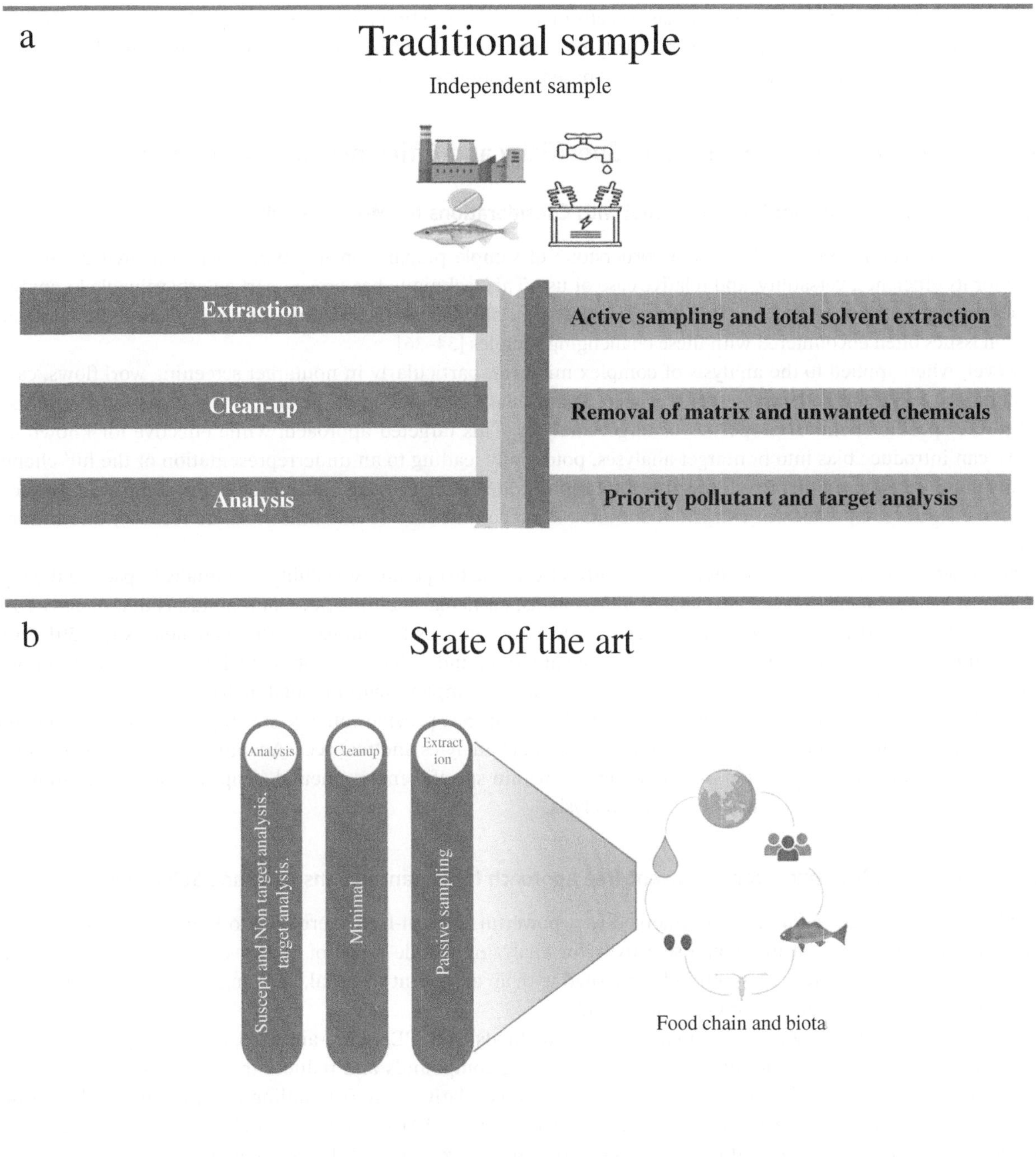

Figure 7.1 Sampling and (bio)analytical methods for complex environmental mixtures [14, 15].

rapidly, encompassing both targeted analyses of specific compounds [24–28] and nontargeted approaches seeking to characterize the entirety of chemicals present [25, 29–32]. High-resolution mass spectrometry (HRMS) analysis of SWB extracts has revealed a remarkable diversity of chemicals absorbed during wear, providing valuable insights into personal exposomes. Furthermore, SWBs have demonstrated utility in ecological and animal studies, enabling comparative exposure assessments across species. Notably, a study comparing human and dog exposomes using SWBs revealed remarkably similar chemical profiles [33].

Passive samplers, including SWBs, offer a cost-effective, user-friendly, and versatile approach to reconstructing the chemical profile of environmental mixtures. However, challenges remain in terms of sample preparation and data processing, particularly for suspect screening and nontarget analysis workflows, which can be time-consuming and complex. The following section will delve into advanced mass spectrometry and extraction techniques that hold potential for enhancing analytical throughput and improving the specificity of chemical characterization in environmental samples, ultimately advancing our ability to assess and manage the risks posed by complex chemical mixtures.

7.4 Sample Preparation: Enhancing Specificity and Efficiency in Mixture Analysis

7.4.1 Solid-phase Extraction: A Versatile Tool with Considerations for Mixture Analysis

Solid-phase extraction (SPE) has become a cornerstone of sample preparation in environmental analytical chemistry, lauded for its efficiency, versatility, and relative ease of use. This technique has proven particularly valuable in analyzing complex matrices such as oil and gas (O&G)-produced waters, effectively removing interfering compounds and mitigating emulsion issues often encountered with these challenging samples [34–36].

However, when applied to the analysis of complex mixtures, particularly in nontarget screening workflows, careful consideration of SPE method parameters is crucial. Traditional SPE relies on the selection of sorbents and solvents tailored to the specific chemical properties of target analytes. This targeted approach, while effective for known compounds, can introduce bias into nontarget analyses, potentially leading to an underrepresentation of the full chemical diversity present in the sample. To address this limitation, strategies employing multiple sorbents with varying selectivities, coupled with comprehensive extraction solvents, are recommended to maximize analyte recovery and minimize bias [37].

Furthermore, manual SPE procedures are inherently susceptible to operator variability, potentially impacting the reproducibility of extraction efficiencies. Automated SPE systems, particularly online SPE configurations that directly couple extraction with analytical instrumentation, offer significant advantages for mitigating these challenges [37–39]. Online SPE minimizes sample handling, reduces solvent consumption, and enhances reproducibility, making it an attractive approach for high-throughput analysis of emerging pollutants in complex environmental matrices [37–39].

While standardized SPE methods, such as US EPA Method 3535A, provide valuable starting points for environmental analysis [40], they often require optimization for specific target analytes and matrices. Adapting existing EPA methods to incorporate relevant emerging contaminants of concern within specific environmental samples holds great promise for improving the accuracy and efficiency of mixture analysis.

7.4.2 Solid-phase Microextraction: A Solvent-free Approach for Enhanced Sensitivity and Selectivity

Solid-phase microextraction (SPME) has emerged as a powerful, solvent-free alternative to traditional extraction techniques, offering enhanced sensitivity and selectivity for analyzing a wide range of analytes in complex environmental matrices [36]. This technique utilizes a thin fiber coated with an adsorbent material, allowing for the direct extraction of target analytes from liquid samples or headspace gas [36].

SPME presents several advantages over conventional methods like SPE, purge-and-trap, and headspace analysis. Its solvent-free nature minimizes the introduction of interfering compounds and reduces the risk of analyte loss during solvent evaporation steps [36]. This attribute, coupled with its relatively short handling time, minimizes background noise and enhances method sensitivity. Moreover, the selectivity of SPME can be fine-tuned by selecting fibers coated with specific adsorbent materials tailored to the physicochemical properties of the target analytes.

SPME has been successfully employed in a wide array of environmental applications, including the analysis of volatile and semivolatile organic compounds in complex samples such as fire debris, where it has proven effective for detecting gasoline components [41]. Notably, recent advancements have led to the development of online SPME systems, directly coupled with analytical instrumentation such as liquid chromatography (LC) [42]. This hyphenated approach eliminates the need for offline sample preparation, further enhancing analytical efficiency and minimizing the potential for analyte loss. The efficacy of SPME/LC has been demonstrated in studies investigating the presence of surfactants like Triton X-100 and alkylphenol ethoxylates in environmental samples [42].

Furthermore, the coupling of SPME with HRMS, particularly gas chromatography-time-of-flight MS (GC-ToF MS), has enabled the comprehensive characterization of complex environmental mixtures [43]. Researchers utilizing this powerful combination have successfully identified and quantified over 1 400 chemicals in environmental samples, encompassing diverse chemical classes including cycloalkanes, alkanes, aromatic hydrocarbons, and their alkylated derivatives [43].

7.4.3 Liquid–Liquid Extraction: Balancing Simplicity and Selectivity Challenges in Complex Matrices

Liquid–liquid extraction (LLE) remains a widely employed technique for sample preparation due to its relative simplicity, ease of implementation, and adaptability to various analytes [44]. It has been particularly favored for extracting organic compounds from O&G-produced waters prior to gas chromatography analysis [45–48]. This preference stems partly from the LLE's ability to reduce the concentration of hydrophobic components susceptible to biotransformation, thus aiding in sample preservation. Additionally, modified LLE methods targeting hydrophilic analytes have been developed, enabling the selective extraction of water-soluble compounds like polyethylene glycol (PEG), which are prevalent in flowback and produced waters [49, 50].

However, LLE's inherent limitations, particularly its lower selectivity compared to techniques like SPE, pose challenges when analyzing complex environmental mixtures [36]. This lack of selectivity often leads to the co-extraction of numerous interfering compounds, resulting in increased background noise and potentially compromising sensitivity and accurate quantification of target analytes. While LLE can effectively remove nonpolar compounds, its application is limited when analyzing water-soluble components. Furthermore, the technique is often labor-intensive and challenging to automate, limiting its throughput for large sample sets.

A significant drawback of LLE, especially pronounced in matrices like O&G-produced waters, is the frequent formation of emulsions during extraction [36]. These emulsions arise from the presence of high concentrations of surfactant-like substances, which are estimated to constitute approximately 20% of the organic carbon in these waters [51, 52]. Surfactants disrupt the clear separation of the aqueous and organic phases, leading to the formation of a troublesome third, intermediate phase that hinders accurate analyte quantification.

Despite these challenges, several strategies can be employed to mitigate emulsion formation and improve LLE performance. Adding salt to the aqueous phase can alter its polarity, forcing analytes with low solubility into the organic phase and facilitating phase separation [36]. Additionally, passing the extract through silica columns can help to break emulsions and remove interfering compounds [53]. While these modifications can improve LLE's efficacy, alternative techniques like SPE or SPME might be more suitable for analyzing complex environmental mixtures where high selectivity and sensitivity are paramount.

7.5 Expanding the Analytical Toolbox: Emerging Extraction Techniques for Complex Environmental Matrices

While traditional extraction methods like LLE and SPE remain valuable, emerging techniques offer distinct advantages for analyzing complex environmental mixtures, particularly for trace-level contaminants. These novel approaches aim to overcome the limitations of conventional methods, such as high solvent consumption, limited selectivity, and challenges in extracting specific analyte classes.

7.5.1 Liquid–Liquid–Liquid Microextraction: Enhancing Sensitivity and Minimizing Contamination

Liquid–liquid–liquid microextraction (LLLME) represents a miniaturized and highly sensitive alternative to conventional LLE, particularly well-suited for analyzing trace-level polar contaminants in aqueous matrices. In contrast to SPME, where analytes partition directly into a coated fiber, LLLME utilizes a three-phase system [54]. Target analytes are first extracted from the aqueous sample into a small volume of organic solvent, followed by back-extraction into a final aqueous phase, typically contained within a disposable hollow fiber. This approach minimizes solvent consumption, enhances analyte enrichment, and reduces the risk of carryover and contamination due to the disposable nature of the extraction device. For instance, LLLME has proven highly effective for extracting sulfamides from water samples at nanomolar concentrations, showcasing its potential for analyzing pharmaceuticals and personal care products in environmental samples [54].

7.5.2 Molecularly Imprinted Polymers: Tailoring Selectivity for Challenging Analytes

Molecularly imprinted polymers (MIPs) offer a highly selective approach for extracting specific analytes or classes of compounds from complex environmental matrices. These tailor-made polymers are synthesized using a template molecule, resulting in a material with specific binding cavities designed to selectively capture the target analyte. While promising,

MIPs face challenges related to template bleeding, which can limit their reusability and complicate accurate quantification, particularly at environmentally relevant concentrations [55]. To address this, researchers are exploring strategies like using dummy templates structurally similar to the target analyte but with distinct chromatographic properties, minimizing interference during analysis. For example, using 6-ketoestradiol as a dummy template for 17β-estradiol extraction from river water demonstrated comparable recoveries to SPE while achieving lower detection limits, highlighting the potential of MIPs for analyzing trace-level hormones in complex environmental samples [56].

7.6 Analytical Techniques for Characterizing Organic Pollutants: Advancing Beyond Traditional Approaches

Characterizing the diverse array of semivolatile organic compounds (SVOCs) present in environmental matrices requires sophisticated analytical techniques capable of resolving complex mixtures and providing detailed chemical information. While techniques like Fourier-transform infrared (FTIR) spectroscopy and nuclear magnetic resonance (NMR) spectroscopy offer valuable insights into functional group composition and structural properties [57–59], gas chromatography-mass spectrometry (GC-MS) and liquid chromatography-mass spectrometry (LC-MS) have emerged as the gold standards for comprehensive chemical characterization.

Figure 7.2 illustrates the strengths and limitations of various analytical approaches for analyzing environmental organic pollutants across three key dimensions: time/size resolution, chemical resolution, and completeness. While nonchromatographic methods like FTIR and NMR provide valuable information on compound classes and functional groups, they often lack the resolution and sensitivity needed for a detailed characterization of complex mixtures [57]. For instance, FTIR analysis successfully identified various functional groups (CH, CO) corresponding to different chemical classes (aliphatic aldehydes, ketones, carboxylic acids, hydrocarbons) in sediments and rural samples [58]. Similarly, NMR spectroscopy

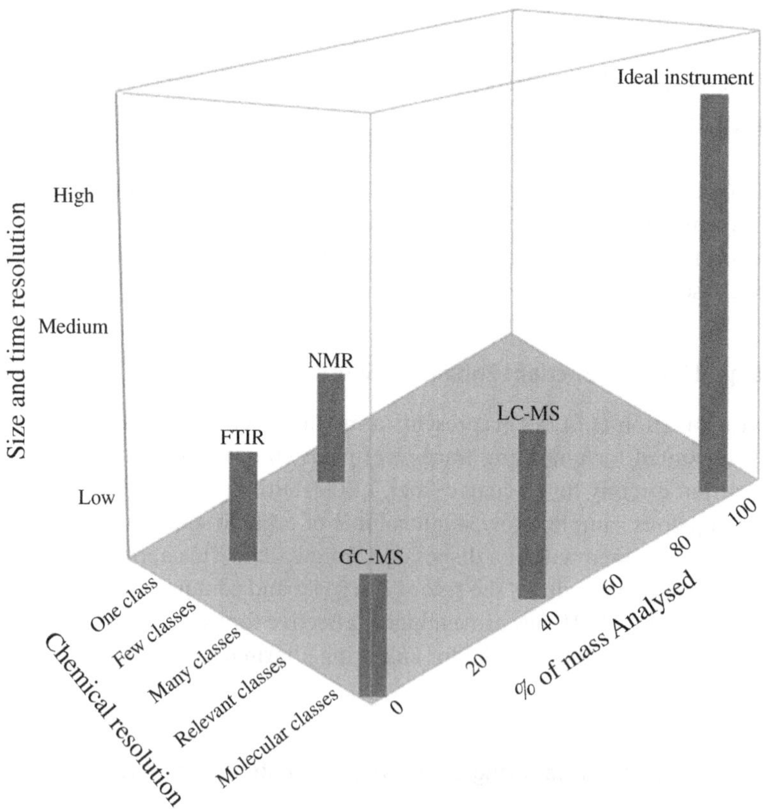

Figure 7.2 Three-dimensional representation of some determination techniques currently used for the analysis of the organic content.

elucidated structural characteristics of organic compounds, including cyclopropane, halogenated alkanes, and methacrylonitrile, in wastewater samples [59]. However, these techniques often struggle to resolve individual compounds within complex mixtures, limiting their ability to provide comprehensive compositional information.

In contrast, chromatographic techniques coupled with MS (GC-MS and LC-MS) excel in separating and identifying individual components within complex mixtures, offering superior chemical resolution and sensitivity. GC-MS is particularly well-suited for analyzing volatile and semivolatile compounds, while LC-MS extends the analytical capabilities to encompass a wider range of polarities and molecular weights, encompassing less volatile and thermally labile compounds. This enhanced separation power, combined with the high sensitivity and selectivity of MS for detection and identification, makes GC-MS and LC-MS indispensable tools for characterizing the vast diversity of organic pollutants present in environmental samples.

7.6.1 Advancing Chemical Characterization: A Comparative Analysis of MS Detectors for Environmental Analysis

Mass spectrometry (MS) detectors play a crucial role in identifying and quantifying organic pollutants in environmental samples. This section compares the strengths and limitations of common MS detectors, highlighting their suitability for analyzing complex mixtures like those found in O&G-produced waters.

7.6.1.1 Single Quadrupole MS: Simplicity and Limitations

Single quadrupole MS detectors, often coupled with GC or LC, are widely employed due to their simplicity, reliability, and cost-effectiveness [60–66]. However, their limited mass resolution hinders the accurate identification of compounds within complex matrices [66]. This limitation often leads to ambiguous compound assignments and potential false positives, especially when analyzing samples without extensive pretreatment to remove interferences. While full-scan mode is commonly employed in standard EPA methods like 8260B (volatiles) and 8270C (semivolatiles) [66], this approach may prove insufficient for characterizing the complexity of O&G-produced waters. Selective ion monitoring (SIM) enhances sensitivity and reduces background noise by targeting specific ions; however, it requires prior knowledge of the sample composition and may miss unexpected compounds.

7.6.1.2 Time-of-flight MS: Enhanced Resolution for Complex Matrices

Time-of-flight (ToF) MS analyzers offer significantly higher mass resolution (1–10 ppm) and accuracy compared to single quadrupole systems, enabling more confident compound identification within complex mixtures [67]. This capability has proven invaluable in identifying surfactants like polyethylene glycols (PEGs) and polypropylene glycols (PPGs) in O&G-produced waters [35, 51, 52, 68, 69].

7.6.1.3 Ion Trap MS: Versatility and Sensitivity with Resolution Trade-offs

Ion trap MS analyzers, including the Orbitrap, offer high sensitivity and versatility. However, their resolving power declines with increasing ion populations within the trap [70]. The Orbitrap, when coupled with LC, provides high resolution and mass accuracy comparable to ToF systems, making it suitable for analyzing polyvalent ions often encountered in O&G-produced waters [71, 72]. While the Fourier-transform (FT) mode offers superior resolution, tandem MS (MS/MS) analysis necessitates the use of the lower resolution MSI mode (Table 7.1).

Alternatively, FT ion cyclotron resonance mass spectrometry (FT-ICR-MS) provides exceptional resolution and has shown promise in characterizing complex mixtures like shale oil and identifying halogenated byproducts in O&G wastewaters [73]

7.6.2 HRMS Screening Methods – Target and Suspect

HRMS has emerged as a transformative technology in environmental analytical chemistry, offering unparalleled capabilities for identifying and quantifying organic contaminants in complex matrices. HRMS achieves this by accurately determining both the mass and isotopic fingerprint of individual molecules, allowing for confident identification even in the presence of interfering compounds. This section will explore the diverse applications of HRMS in environmental analysis, focusing on targeted and suspect, screening approaches. Recent advances in HRMS, particularly HRMS/MS instrumentation, have significantly enhanced the capabilities of environmental analytical chemistry. While LC coupled with tandem

Table 7.1 Analytical methods for the determination of emerging organic contaminants in environmental matrices.

Sample	Compounds	Extraction technique	Suggested analytical instrument	Approach	Ref.
Water samples	Pharmaceuticals and personal care products	Solid-phase extraction	LCMS	Suspect and nontarget screening	Park et al. (2018)
Wetland (water samples)	Synthetic organic Contaminants	Solid-phase extraction	LCMS	Suspect and nontarget screening	Kang et al. (2020)
Domestic and industrial effluents	Pharmaceuticals, pesticides, poly-/perfluorinated alkyl substances, 2 organophosphate flame retardants	Solid-phase extraction	LCMS	Suspect and nontarget screening	Choi et al. (2021)
Seawater in the Arctic region	Pharmaceutical and personal care products	Solid-phase extraction	LCMS	Suspect and nontarget screening	Choi et al. (2020)
Streams in agricultural watersheds	Herbicides, insecticides, fungicides	Solid-phase extraction	Ultrahigh performance liquid chromatograph	Suspect and nontarget screening	Tian et al. (2021)
Drinking water and source water	Haloacetaldehydes, haloketones, haloacetamides, haloacetonitriles, halonitromethanes, and iodinated-trihalomethanes	Liquid–liquid extraction	GC-(TOF)MS 200-Rtx	Target and nontarget	Kimura et al. (2019)

mass spectrometry (LC-MS/MS) has become the gold standard for targeted screening of organic contaminants in water, especially at low concentrations [74], HRMS/MS offers superior selectivity and accuracy for both targeted and suspect screening workflows.

7.6.2.1 Targeted Screening: Precise Quantification of Known Contaminants

Targeted HRMS screening focuses on detecting and quantifying a predefined list of known compounds. This approach is essential for monitoring regulated contaminants, assessing compliance with environmental guidelines, and tracking the fate of specific compounds in the environment. To achieve reliable and accurate results, targeted HRMS screening requires the use of authentic standards. Standards allow analysts to optimize instrument parameters, such as chromatographic separation (e.g. retention time) and mass spectrometer settings (e.g. MS/MS transitions), to ensure optimal sensitivity and selectivity for the target compounds. By comparing the retention time, accurate mass, isotopic pattern, and MS/MS fragmentation patterns of the sample signals to those of the authentic standard, analysts can confidently confirm the presence of the target compounds. Calibration curves generated using a series of known concentrations of the authentic standards allow for the accurate quantification of the target compounds in the samples. While triple quadrupole MS systems can be used for targeted analysis, HRMS offers superior selectivity and confidence in compound identification, especially in complex environmental samples.

7.6.2.2 Suspect Screening: Expanding the Scope of Analysis

Suspect screening utilizes HRMS data to identify compounds in a sample based on a predefined list of potential candidates. This approach is particularly valuable for characterizing emerging contaminants, transformation products, or compounds for which authentic standards are not readily available [74, 75]. Databases like PubChem, ChemSpider, and STOFF-IDENT contain vast libraries of chemical structures and properties, including accurate mass, molecular formula, and predicted fragmentation patterns. Published studies on similar environmental samples or industrial processes can provide valuable insights into potential contaminants of concern. Software programs can predict compound properties, such as retention time and MS/MS fragments, based on chemical structure, aiding in the creation of comprehensive suspect lists. R Statistical Software offers powerful tools for data management and statistical analysis in this context [76, 77].

7.7 Outlook: Toward Holistic Environmental Monitoring with Advanced Analytical Tools

The rapid evolution of analytical techniques, particularly HRMS, is leading to an unprecedented increase in the detection of chemicals in environmental samples. This growing body of data necessitates a paradigm shift in environmental monitoring, moving beyond traditional single-compound analyses toward a more holistic understanding of complex chemical mixtures and their combined effects on ecosystems and human health.

While standardized monitoring methods remain crucial for establishing robust environmental baselines and assessing regulatory compliance, the sheer number of chemicals present in the environment, often at trace levels, poses a significant analytical challenge. Robust and sensitive methods for sampling, extraction, detection, and data processing are essential to ensure reliable identification and quantification of both parent compounds and their bioactive transformation products.

The integration of HRMS data with bioassay results and comprehensive metadata holds immense promise for revolutionizing environmental monitoring. This convergence of "big data" approaches will enable the application of powerful statistical and machine learning techniques to unravel complex relationships between chemical exposures and biological responses. By elucidating these connections, we can move toward a more comprehensive understanding of mixture effects, a critical aspect often overlooked in current risk assessment frameworks.

Given the ubiquitous presence of chemical mixtures in the environment, it is imperative to transition from a single-compound regulatory paradigm to one that explicitly considers the combined effects of multiple stressors. Advanced analytical techniques, coupled with sophisticated data analysis tools, provide a powerful toolkit for unraveling the complexity of environmental exposures and informing more effective risk management strategies for safeguarding ecosystem and human health.

Acknowledgments

The authors are thankful for the postdoctoral scholarship program of Changwon National University (to T.Y. Suman), and they are grateful for providing necessary facilities from the Changwon National University, Changwon, Gyeongsangnam-do, South Korea.

Conflicts of Interest

The authors declare no conflict of interest.

References

1 Walker CH, Sibly RM, Peakall DB. Principles of Ecotoxicology. CRC press, Boca Raton, FL, 2005.

2 Rappaport SM, Smith MT. Environment and disease risks. Science. 2010, 330(6003): 460–461.

3 Landrigan PJ, Fuller R, Acosta NJ, Adeyi O, Arnold R, Baldé AB, Bertollini R, Bose-O'Reilly S, Boufford JI, Breysse PN, Chiles T. The Lancet commission on pollution and health. Lancet. 2018, 391(10119): 462–512.

4 Hites RA. Statistical approach for assessing the Stockholm convention's effectiveness: great lakes atmospheric data. Environ Sci Technol. 2019, 53(15): 8585–8590.

5 Fenner K, Canonica S, Wackett LP, Elsner M. Evaluating pesticide degradation in the environment: blind spots and emerging opportunities. Science. 2013, 341(6147): 752–758.

6 Suman TY, Kim SY, Yeom DH, Jeon J. Transformation products of emerging pollutants explored using non-target screening: perspective in the transformation pathway and toxicity mechanism – a review. Toxics. 2022, 10(2): 54.

7 Escher BI, Fenner K. Recent advances in environmental risk assessment of transformation products. Environ Sci Technol. 2011, 45(9): 3835–3847.

8 Wild CP. Complementing the genome with an "exposome": the outstanding challenge of environmental exposure measurement in molecular epidemiology. Cancer Epidemiol Biomarkers Prev. 2005, 14(8): 1847–1850.

9 Kortenkamp A, Faust M. Regulate to reduce chemical mixture risk. Science. 2018, 361(6399): 224–226.

10 Kortenkamp A. Low dose mixture effects of endocrine disrupters and their implications for regulatory thresholds in chemical risk assessment. Curr Opin Pharmacol. 2014, 19: 105–111.

11 Cedergreen N. Quantifying synergy: a systematic review of mixture toxicity studies within environmental toxicology. PloS One. 2014, 9(5): e96580.

12 Boobis A, Budinsky R, Collie S, Crofton K, Embry M, Felter S, Hertzberg R, Kopp D, Mihlan G, Mumtaz M, Price P. Critical analysis of literature on low-dose synergy for use in screening chemical mixtures for risk assessment. Crit Rev Toxicol. 2011, 41(5): 369–383.

13 Escher BI, Henneberger L, König M, Schlichting R, Fischer FC. Cytotoxicity burst? Differentiating specific from nonspecific effects in Tox21 in vitro reporter gene assays. Environ Health Perspect. 2020, 128(7): 077007.

14 Kim K, Choi Y, Mok S, Moon HB, Jeon J. Optimization of the QuEChERS method for multi-residue analysis of pharmaceuticals and pesticides in aquaculture products. Food Chem. 2023, 399: 133958.

15 Jeon H, Yeo S, Park EA, Kang D, Shen K, Kim M, Lee I, Jeon J, Moon B, Ji K, Kim S, Kho Y. Identification and quantification of Acetyl Tributyl Citrate (ATBC) metabolites using human liver microsomes and human urine. Chemosphere. 2024, 363: 142840.

16 Jahnke A, Witt G, Schäfer S, Haase N, Escher BI. Combining passive sampling with toxicological characterization of complex mixtures of pollutants from the aquatic environment. Adv Biochem Eng Biotechnol. 2016, 157: 225–261.

17 Claessens M, Monteyne E, Wille K, Vanhaecke L, Roose P, Janssen CR. Passive sampling reversed: coupling passive field sampling with passive lab dosing to assess the ecotoxicity of mixtures present in the marine environment. Mar Pollut Bull. 2015, 93(1-2): 9–19.

18 Smith KE, Jeong Y. Passive sampling and dosing of aquatic organic contaminant mixtures for ecotoxicological analyses. Environ Sci Technol. 2021, 55(14): 9538–9547.

19 Sonavane M, Schollee JE, Hidasi AO, Creusot N, Brion F, Suter MJ, Hollender J, Aït-Aïssa S. An integrative approach combining passive sampling, bioassays, and effect-directed analysis to assess the impact of wastewater effluent. Environ Toxicol Chem. 2018, 37(8): 2079–2088.

20 Bergmann AJ, North PE, Vasquez L, Bello H, del Carmen Gastañaga Ruiz M, Anderson KA. Multi-class chemical exposure in rural Peru using silicone wristbands. J Expo Sci Environ Epidemiol. 2017, 27(6): 560–568.

21 Zwart N, Jonker W, Ten Broek R, de Boer J, Somsen G, Kool J, Hamers T, Houtman CJ, Lamoree MH. Identification of mutagenic and endocrine disrupting compounds in surface water and wastewater treatment plant effluents using high-resolution effect-directed analysis. Water Res. 2020, 168: 115204.

22 O'Connell SG, Kincl LD, Anderson KA. Silicone wristbands as personal passive samplers. Environmental Sci Technol. 2014, 48(6): 3327–3335.

23 Anderson KA, Points GL, Donald CE, Dixon HM, Scott RP, Wilson G, Tidwell LG, Hoffman PD, Herbstman JB, O'Connell SG. Preparation and performance features of wristband samplers and considerations for chemical exposure assessment. J Expo Sci Environ Epidemiol. 2017, 27(6): 551–559.

24 Hammel SC, Hoffman K, Webster TF, Anderson KA, Stapleton HM. Measuring personal exposure to organophosphate flame retardants using silicone wristbands and hand wipes. Environ Sci Technol. 2016, 50(8): 4483–4491.

25 Aerts R, Joly L, Szternfeld P, Tsilikas K, De Cremer K, Castelain P, Aerts JM, Van Orshoven J, Somers B, Hendrickx M, Andjelkovic M. Silicone wristband passive samplers yield highly individualized pesticide residue exposure profiles. Environ Sci Technol. 2018, 52(1): 298–307.

26 Romanak KA, Wang S, Stubbings WA, Hendryx M, Venier M, Salamova A. Analysis of brominated and chlorinated flame retardants, organophosphate esters, and polycyclic aromatic hydrocarbons in silicone wristbands used as personal passive samplers. J Chromatogr. 2019, 1588: 41–47.

27 Kile ML, Scott RP, O'Connell SG, Lipscomb S, MacDonald M, McClelland M, Anderson KA. Using silicone wristbands to evaluate preschool children's exposure to flame retardants. Environ Res. 2016, 147: 365–372.

28 Travis SC, Aga DS, Queirolo EI, Olson JR, Daleiro M, Kordas K. Catching flame retardants and pesticides in silicone wristbands: evidence of exposure to current and legacy pollutants in Uruguayan children. Sci Total Environ. 2020, 740: 140136.

29 Doherty BT, Pearce JL, Anderson KA, Karagas MR, Romano ME. Assessment of multipollutant exposures during pregnancy using silicone wristbands. Front Public Health. 2020, 8: 570.

30 Bergmann AJ, Tanguay RL, Anderson KA. Using passive sampling and zebrafish to identify developmental toxicants in complex mixtures. Environ Toxicol Chem. 2017, 36(9): 2290–2298.

31 Manzano CA, Dodder NG, Hoh E, Morales R. Patterns of personal exposure to urban pollutants using personal passive samplers and GC × GC/ToF–MS. Environ Sci Technol. 2018, 53(2): 614–624.

32 Koelmel JP, Lin EZ, Nichols A, Guo P, Zhou Y, Godri Pollitt KJ. Head, shoulders, knees, and toes: placement of wearable passive samplers alters exposure profiles observed. Environ Sci Technol. 2021, 55(6): 3796–3806.

33 Wise CF, Hammel SC, Herkert N, Ma J, Motsinger-Reif A, Stapleton HM, Breen M. Comparative exposure assessment using silicone passive samplers indicates that domestic dogs are sentinels to support human health research. Environ Sci Technol. 2020, 54(12): 7409–7419.

34 Regnery J, Coday BD, Riley SM, Cath TY. Solid-phase extraction followed by gas chromatography-mass spectrometry for the quantitative analysis of semivolatile hydrocarbons in hydraulic fracturing wastewaters. Analytic Met. 2016, 8(9): 2058–2068.

35 Cluff MA, Hartsock A, MacRae JD, Carter K, Mouser, PJ. Temporal changes in microbial ecology and geochemistry in produced water from hydraulically fractured Marcellus shale gas wells. Environ Sci Technol. 2014, 48(11): 6508–6517.

36 Majors ER. Sample Preparation Fundamentals for Chromatography. Agilent Technologies. Inc., 13 November 2013.

37 Haun J, Leonhardt J, Portner C, Hetzel T, Tuerk J, Teutenberg T, Schmidt TC. Online and splitless NanoLC × CapillaryLC with quadrupole/time-of-flight mass spectrometric detection for comprehensive screening analysis of complex samples. Analytic Chem. 2013, 85(21): 10083–10090.

38 Axel M, Ewelina K, Jenny-Maria B, Leif K. An online SPE LC-MS/MS method for the analysis of antibiotics in environmental water. Environ Sci Pollut Res. 2017, 24(9): 8692–8699.

39 Heffernan AL, Thompson K, Eaglesham G, Vijayasarathy S, Mueller JF, Sly PD, Gomez MJ. Rapid, automated online SPE-LC-QTRAP-MS/MS method for the simultaneous analysis of 14 phthalate metabolites and 5 bisphenol analogues in human urine. Talanta. 2016, 151: 224–233.

40 Oppenheimer J, Eaton A, Badruzzaman M, Haghani AW, Jacangelo, JG. Occurrence and suitability of sucralose as an indicator compound of wastewater loading to surface waters in urbanized regions. Water Res. 2011, 45(13): 4019–4027.

41 Snyder LR. Classification off the solvent properties of common liquids. J Chromatogr Sci. 1978, 16(6): 223–234.

42 Aulakh JS, Malik AK, Kaur V, Schmitt-Kopplin, P. A review on solid phase microextraction – high performance liquid chromatography (SPME-HPLC) analysis of pesticides. Crit Rev Anal Chem. 2005, 35(1): 71–85.

43 Khan NA, Engle M, Dungan B, Holguin FO, Xu P, Carroll KC. Volatile-organic molecular characterization of shale-oil produced water from the Permian Basin. Chemosphere. 2016, 148: 126–136.

44 De Nicola F, Graña EC, Aboal JR, Carballeira A, Fernández JÁ, Mahía PL, Rodríguez, DP, Lorenzo SM. PAH detection in Quercus robur leaves and Pinus pinaster needles: a fast method for biomonitoring purpose. Talanta. 2016, 153: 130–137.

45 Jin R, Zheng M, Yang H, Yang L, Wu X, Xu Y, Liu G. Gas–particle phase partitioning and particle size distribution of chlorinated and brominated polycyclic aromatic hydrocarbons in haze. Environ Pollut. 2017, 231(Pt 2): 1601–1608.

46 Hladik ML, Focazio MJ, Engle M. Discharges of produced waters from oil and gas extraction via wastewater treatment plants are sources of disinfection by-products to receiving streams. Scihe Total Environ. 2014, 466: 1085–1093.

47 Llewellyn GT, Dorman F, Westland JL, Yoxtheimer D, Grieve P, Sowers T, Humston-Fulmer E, Brantley, S. Evaluating a groundwater supply contamination incident attributed to Marcellus shale gas development. Proc Natl Acad Sci USA. 2015, 112(20): 6325–6330.

48 Maguire-Boyle, SJ, Barron AR. Organic compounds in produced waters from shale gas wells. Environ Sci Process Impacts. 2014, 16(10): 2237–2248.

49 Ho TS, Pedersen-Bjergaard S, Rasmussen, KE. Recovery, enrichment and selectivity in liquid-phase microextraction: comparison with conventional liquid–liquid extraction. J Chromatogr A. 2002, 963(1-2): 3–17.

50 Crawford Scientific, CHROMacademy, Crawford Sci(2016). Available from:http://www.chromacademy.com/ (accessed 09 January 2016).

51 Thurman EM, Ferrer I, Rosenblum J, Linden K, Ryan JN. Identification of polypropylene glycols and polyethylene glycol carboxylates in flowback and produced water from hydraulic fracturing. J Hazard Mater. 2017, 323: 11–17.

52 Thurman EM, Ferrer I, Blotevogel J, Borch T. Analysis of hydraulic fracturing flowback and produced waters using accurate mass: identification of ethoxylated surfactants. Anal Chem. 2014, 86(19): 9653–9661.

53 de Villiers A, Venter P, Pasch H. Recent advances and trends in the liquid-chromatography–mass spectrometry analysis of flavonoids. J Chromatogr A. 2016, 1430: 16–78.

54 McClure EL, Wong CS. Solid phase microextraction of macrolide, trimethoprim, and sulfonamide antibiotics in wastewaters. J Chromatogr A. 2007, 1169(1-2): 53–62.

55 Meng Z, Chen W, Mulchandani A. Removal of estrogenic pollutants from contaminated water using molecularly imprinted polymers. Environ Sci Technol. 2005, 39(22): 8958–8962.

56 Watabe Y, Kubo T, Nishikawa T, Fujita T, Kaya K, Hosoya K. Fully automated liquid chromatography–mass spectrometry determination of 17β-estradiol in river water. J Chromatogr A. 2006, 1120(1-2): 252–259.

57 Hallquist M, Wenger JC, Baltensperger U, Rudich Y, Simpson D, Claeys M, Dommen J, Donahue NM, George C, Goldstein AH, Hamilton JF. The formation, properties and impact of secondary organic aerosol: current and emerging issues. Atmos Chem Phys. 2009, 9(14): 5155–5236.

58 Song Z, Jones KW, Marinkovic N, Xiao XM, Feng H, Tchouparova E. Characterization of organic contaminants in New York/New Jersey harbor sediments using FTIR-ATR and synchrotron FTIR. CLEAN–Soil Air Water. 2011, 39(12): 1041–1049.

59 Farounbi AI, Mensah PK, Olawode EO, Ngqwala NP. 1H-NMR determination of organic compounds in municipal wastewaters and the receiving surface waters in Eastern Cape Province of South Africa. Molecules. 2020, 25(3): 713.

60 Park N, Choi Y, Kim D, Kim K, Jeon J. Prioritization of highly exposable pharmaceuticals via a suspect/non-target screening approach: a case study for Yeongsan River, Korea. Sci Total Environ. 2018, 639: 570–579.

61 Kang D, Doudrick K, Park N, Choi Y, Kim K, Jeon J. Identification of transformation products to characterize the ability of a natural wetland to degrade synthetic organic pollutants. Water Res. 2020, 187: 116425.

62 Choi Y, Lee JH, Kim K, Mun H, Park N, Jeon J. Identification, quantification, and prioritization of new emerging pollutants in domestic and industrial effluents, Korea: Application of LC-HRMS based suspect and non-target screening. J Hazard Mater. 2021, 402: 123706.

63 Choi Y, Kim K, Kim D, Moon HB, Jeon J. Ny-Ålesund-oriented organic pollutants in sewage effluent and receiving seawater in the Arctic region of Kongsfjorden. Environ Pollut. 2020, 258: 113792.

64 Tian Z, Wark DA, Bogue K, James CA. Suspect and non-target screening of contaminants of emerging concern in streams in agricultural watersheds. Sci Total Environ. 2021, 795: 148826.

65 Kimura SY, Cuthbertson AA, Byer JD, Richardson SD. The DBP exposome: development of a new method to simultaneously quantify priority disinfection by-products and comprehensively identify unknowns. Water Res. 2019, 148: 324–333.

66 Dawson PH. Quadrupole Mass Spectrometry and its Applications. Elsevier, 2013.

67 Chernushevich IV, Loboda AV, Thomson BA. An introduction to quadrupole–time-of-flight mass spectrometry. J Mass Spectrom. 2001, 36(8): 849–865.

68 Lester Y, Ferrer I, Thurman EM, Sitterley KA, Korak JA, Aiken G, Linden, KG. Characterization of hydraulic fracturing flowback water in Colorado: implications for water treatment. Sci Total Environ. 2015, 512: 637–644.

69 Coday BD, Miller-Robbie L, Beaudry EG, Munakata-Marr J, Cath TY. Life cycle and economic assessments of engineered osmosis and osmotic dilution for desalination of Haynesville shale pit water. Desalination. 2015, 369: 188–200.

70 Adams RP. Identification of Essential oils by Ion Trap Mass Spectroscopy. Academic Press, 2012.

71 Zubarev RA, Makarov A. Orbitrap mass spectrometry. Anal Chem. 2013, 85(11): 5288–5296.

72 Getzinger GJ, O'Connor MP, Hoelzer K, Drollette BD, Karatum O, Deshusses MA, Ferguson PL, Elsner M, Plata DL. Natural gas residual fluids: sources, endpoints, and organic chemical composition after centralized waste treatment in Pennsylvania. Environ Sci Technol. 2015, 49(14): 8347–8355.

73 Luek JL, Schmitt-Kopplin P, Mouser PJ, Petty WT, Richardson SD, Gonsior M. Halogenated organic compounds identified in hydraulic fracturing wastewaters using ultrahigh resolution mass spectrometry. Environ Sci Technol. 2017, 51(10): 5377–5385.

74 Rutter AP, Snyder DC, Stone EA, Shelton B, DeMinter J, Schauer JJ. Preliminary assessment of the anthropogenic and biogenic contributions to secondary organic aerosols at two industrial cities in the upper Midwest. Atmos Environ. 2014, 84: 307–313.

75 Odabasi M, Dumanoglu Y, Kara M, Altiok H, Elbir T, Bayram, A. Polychlorinated naphthalene (PCN) emissions from scrap processing steel plants with electric-arc furnaces. Sci Total Environ. 2017, 574: 1305–1312.

76 Zhang Y, Li R, Fang J, Wang C, Cai Z. Simultaneous determination of eighteen nitro-polyaromatic hydrocarbons in PM2. 5 by atmospheric pressure gas chromatography-tandem mass spectrometry. Chemosphere. 2018, 198: 303–310.

77 Cazier F, Genevray P, Dewaele D, Nouali H, Verdin A, Ledoux F, Hachimi A, Courcot L, Billet S, Bouhsina S, Shirali P. Characterisation and seasonal variations of particles in the atmosphere of rural, urban and industrial areas: organic compounds. J Environ Sci. 2016, 44: 45–56.

8

Common Toxicological Experimental Models

Xuan Ma[a], *Dejun Huang*[a,*]

[a] *Gansu Key Laboratory of Biomonitoring and Bioremediation for Environmental Pollution, School of Life Sciences, Lanzhou University, Lanzhou, China*

*Corresponding author.
E-mail: huangdj@lzu.edu.cn

Zebrafish (*Danio rerio*) are a tropical freshwater species that has been widely used as a model animal in laboratory conditions [1]. Zebrafish have a short reproductive cycle, are transparent during early developmental stages, and function similarly to the human reproductive system. Zebrafish can be assessed for potentially toxic compounds to the reproductive system since it is highly sensitive to environmental factors [2]. These compounds commonly have a significant negative impact on the structure and physiological functioning of the reproductive organ, germ cell development, and fertility. Because zebrafish and humans have highly similar genetic functions, the toxicity of combined contaminants can be initially assessed via zebrafish. Furthermore, zebrafish can perform genetic manipulation, which provides an important means to study the underlying mechanisms of reproductive toxicity [3].

Zebrafish show discrepant anatomical structures of reproductive organs between males and females (Figure 8.1). The reproductive system is composed of the gonads that produce gametes, the gametes that perform reproductive functions, and a strictly regulated neuroendocrine system [4]. In general, reproductive functional abnormalities in males are mainly caused by factors such as abnormal development of sexual organs and poor sperm quality and quantity [5]. While in females, these factors include abnormal ovarian development, abnormal ovulation, and poor egg quality. The intersection of male and female factors includes hormonal imbalance and abnormal reproductive behavior [6, 7].

8.1 Male Reproductive Function

Male zebrafish possesses a pair of testes located on the dorsal sides of the ventral cavity. The testes are divided into several tubular tissues using septa. Within each seminiferous tubule, supporting cells (Sertoli cells) line the wall of the tubule and function primarily to support testicular morphogenesis and spermatogenesis, while testicular interstitial cells (Leydig cells) are the main location of androgen production [8]. The morphological structure and histological

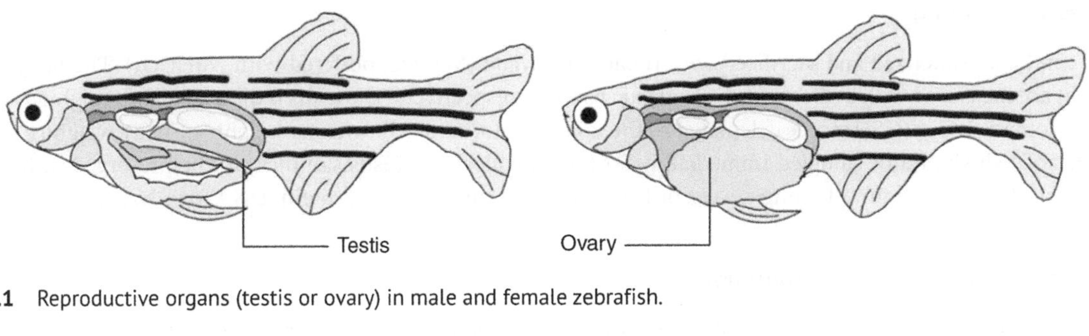

Figure 8.1 Reproductive organs (testis or ovary) in male and female zebrafish.

Toxicological Assessment of Combined Chemicals in the Environment, First Edition. Edited by De-Sheng Pei and Yiyun Liu.

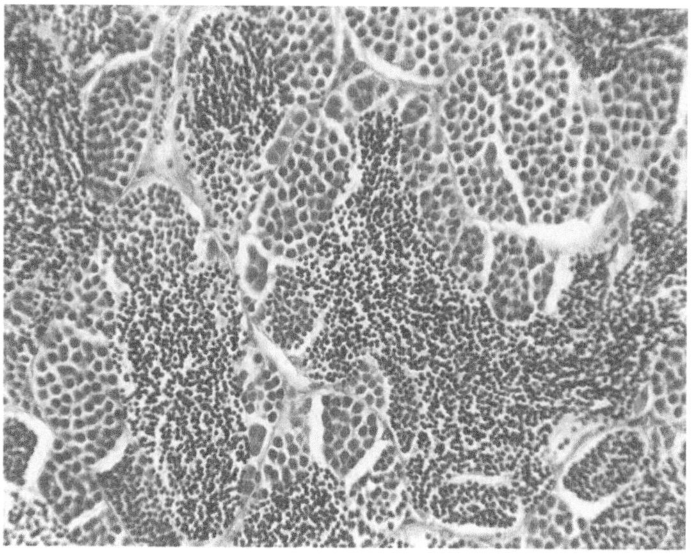

Figure 8.2 A representative image of HE staining of zebrafish testis.

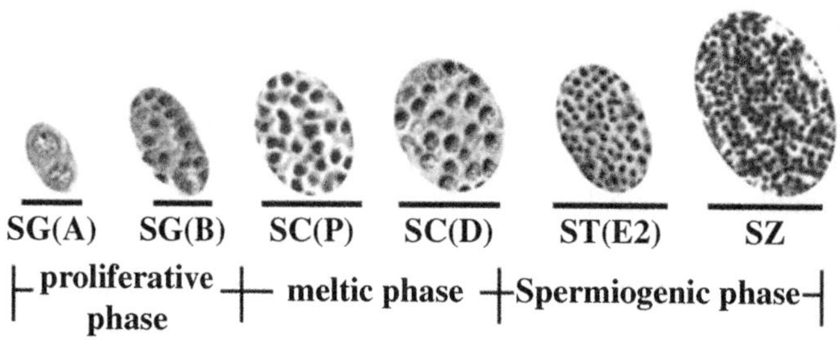

Figure 8.3 Different germ cells during the spermatogenic process in zebrafish. SG(A), A-type spermatogonia; SG(B), B-type spermatogonia; SC(P), spermatocytes at pachytene; SC(D), spermatocytes at diplotene; ST(E2), intermediate spermatids; SZ, spermatozoa.

analysis of fish testis could be performed by hematoxylin–eosin (H&E) staining (Figure 8.2) showing the different stages of germ cells [9, 10] (Figure 8.3). The potential reproductive damage caused by combined contaminants includes abnormal gonad development, reduced number and quality of germ cells, fertilization and hatching success, and the viability of larval and juvenile [11, 12].

The frequently used toxicological experimental assays for the male reproductive function of zebrafish can include the following contents:

8.1.1 Testis Sampling

Zebrafish can be anesthetized and sacrificed with tricaine mesylate (MS-222, buffered with $NaHCO_3$). The body size and body weight are measured with a ruler and an electronic balance, respectively. The testis is dissected and weighted, and then the gonado-somatic index (GSI) can be calculated as the following formula: GSI = testis weight (g)/body weight(g) [13].

The testes of fish should be sampled immediately for sperm collection. Testis samples can be preserved in fixative for histological studies and immunostaining assays, or frozen in liquid nitrogen for protein, DNA, or RNA analyses.

8.1.2 Histological Study (H&E Staining)

Testicular tissue is fixed via paraformaldehyde fix solution (PFA) and treated with a gradient of 50 to 100% ethanol for 30 minutes, respectively. The tissue is embedded in paraffin and sectioned at 4–7 μm. Finally, the slides are stained with H&E to observe the structural changes in the testicular tissue.

8.1.3 Sperm Collection

Male zebrafish are anesthetized in a solution of tricaine mesylate (MS-222, buffered with $NaHCO_3$) and placed under a stereo microscope. The male genital area needs to be dried to avoid water activation of sperm [5]. Sperm are collected by gently pressing the abdomen with a 10 μL glass microcapillary. Sperm volume can be measured through the microcapillary before transferring the sperm into Hank's solution kept on ice.

Alternatively, the abdominal cavity can be opened by dissection, both testes are removed and crushed with forceps. The sperm are released into 300 mOsm/kg (0.137 M NaCl, 5.55 mM glucose, 5.4 mM KCl, 4.2 mM $NaHCO_3$, 1.3 mM $CaCl_2$, 1.0 mM $MgSO_4$, and 0.25 mM Na_2HPO_4, pH 7.2) frozen Hanks' balanced salt solution (HBSS) to maintain sperm viability and avoid sperm activation. The sperm solutions need to be kept on ice until all sperm are collected. All sperm analyses should be performed within 15 minutes of the collection of the sperm samples [14].

8.1.4 Sperm Motility Measurement

Activate 1 μL of diluted sperm with 10–20 μL of ultrapure water on a special slide coated with 1% polyvinyl alcohol to avoid sperm adhering to the glass slide. The coverslips are placed in the testing chamber and a computer-assisted sperm analyzer (CASA) is used to measure sperm motility parameters, including sperm viability (%), the velocity of the average path (VAP), straight-line velocity (VSL), and curve line velocity (VCL) [15]. At least 100 sperm cells should be assessed for each parameter.

The parameters of sperm motility should be assessed when the sperm are just activated (t0) and at 30 s after sperm activation (t30). For each ejaculation, sperm motility is measured at least in triplicate, and the reproducibility of the measured sperm motility parameters is calculated [5]. Sperm longevity is measured by recording the time (seconds) of activation to resting for ≥ 80% of the sperm in the viewfinder.

8.1.5 Sperm Viability and Quantitative Analysis

Sperm viability and quantity are estimated by the Dual Fluorescence Sperm Cell Counter. The sperm is diluted in Hank's solution in order to have the optimum sperm concentration for counting. Sperm viability is measured by acridine orange (AO) staining, which is membrane permeable and marks live sperm in green, and propidium iodide (PI) staining, which marks apoptotic or dead sperm in red. In each sample, at least 1000 spermatozoa should be analyzed [5].

8.1.6 Determination of Cell Viability

Plasma membrane integrity is a marker of sperm fertilization capacity and reflects the functional state of the sperm membrane. A structurally intact and functionally activated plasma membrane is required for physiological activities such as sperm cell metabolism, attachment, and crossing of the egg cell. Dual fluorescent staining using PI and SYBR-14 is the most commonly used method for sperm plasma membrane integrity analysis.

Sperm viability can be determined by the integrity of the sperm plasma membrane. The sperm is diluted to the appropriate concentration in 300 mOsm HBSS and incubated with SYBR-14 (final concentration 0.1 μM) for 5 minutes at room temperature, followed by PI (final concentration 12 μM) for 5 minutes. SYBR-14 staining is used first to show plasma membrane integrity based on its ability to freely cross living sperm cells with intact plasma membranes and bind to intranuclear nucleic acids. In contrast, PI is rejected by living sperm cells with intact plasma membranes (both motile and nonmotile sperm) and readily enters broken or dead sperm cells, indicating plasma membrane damage. Double staining is used to distinguish between the structural integrity of the plasma membrane and its absence. The status of the plasma membrane can then be detected by fluorescence microscopy/flow cytometry.

Samples are assayed by flow cytometry immediately after incubation. A total of 10 000 cells should be evaluated per sample. SYBR-14 and PI filters are excited at 488 nm and green and red fluorescence is detected at 535 nm and 595 nm emission wavelengths, respectively.

Flow results are plotted in frequency to determine the percentage of positive populations for PI and SYBR-14 staining. Cell populations in the lower right quadrant have intact membrane structures, whereas cell populations in the upper right and upper left quadrants represent damaged cells [16].

Fluorescence microscopic observations are listed in Table 8.1.

Table 8.1 Fluorescence microscopic observation of sperm plasma membrane integrity analysis results.

Classification	Staining results	Significance
I	Bright green fluorescence of the entire sperm head	Plasma membrane intact
II	Red fluorescence of the whole or part of the sperm head	Plasma membrane completely damaged
III	Orange fluorescence throughout the sperm head	Plasma membrane damage

Figure 8.4 Zebrafish sperm treated with the Halomax-SCD kit. The sperm with intact DNA maintain compact circular light spots, while sperm with fragmented DNA have a distinctly diffuse halo [5].

8.1.7 Sperm DNA Fragmentation Assay

The fragmentation of sperm DNA is measured by using the sperm chromatin dispersion (SCD) Kit. Sperm are diluted in Hank's solution to the appropriate concentration and the manufacturer's instructions are followed. To test the DNA integrity of sperm, slides are made and stained with Midori Green Advance dye and viewed using a fluorescent microscope. The sperm with intact DNA show a compact halo of small fluorescent spots, whereas sperm with fragmented DNA show a fuzzy halo of large fluorescent spots (Figure 8.4). A minimum of 200 sperm cells in each sample are counted [5].

8.1.8 Measurement of Reproductive Behavior

Prior to the measurement of reproductive behavioral, females and males are kept separately for 5 days prior to the experiment to ensure successful reproduction. One male fish from the treated group and one untreated female are placed in a breeding tank separated by a plastic partition at the night before mating. After 10 hours of acclimatization in the dark, the partition is removed. After 5 minutes of light stimulation, 20 minutes of breeding behavior is recorded with a digital camera set 50 cm above the breeding tank.

Four behavioral parameters are generally observed and assessed: (1) frequency of meeting (male meets the female at a distance of less than 0.5 cm), (2) visits of both sexes in the spawning area (male successfully lures the female to the spawning area), (3) duration of both sexes in the spawning area, and (4) visits of each sex in the spawning area [13].

8.1.9 Sperm *in Vitro* Fertilization Analysis

The two chambers of the 12-well plate are connected by a linking tube and filled with fresh culture water until the level of the liquid is the same on both sides. The eggs are then placed in one of the chambers and the sperm are added to the other well, along with an equal volume of fresh culture water to the egg. The sperm and eggs are allowed to contact and are incubated together for 30 minutes at room temperature [17, 18].

Afterward, the eggs are transferred to clean Petri dishes containing fresh culture water and incubated in an incubator at 28 °C for 3 hours. Viable embryos continue to develop; however, unfertilized eggs undergo apoptosis or atrophy. Dead eggs are distinguished from viable eggs under the stereo microscope. Fertilized eggs are counted and the fertilization rates are calculated using the formula: $N_v/N_t \times 100\%$ (Nv: number of viable eggs; Nt: number of total eggs) to assess the fertilization capacity of the sperm [19].

8.1.10 Sperm *in Vivo* Fertilization Analysis

At the end of the treatment, an experimental individual is mated with a control individual of the opposite sex. Females and males are separated by a plastic divider which is removed when the lights are switched on to promote mating. Then, all eggs in the bottom layer of the reproductive tank are collected and randomly grouped. The number of fertilized and unfertilized eggs is calculated to quantify the fertilization rate (the number of fertilized eggs/total eggs oviposited) at 2 hpf, the mortality rate (the number of dead eggs/ total eggs oviposited) at 24 hpf, and the hatching rate (the number of hatched eggs/total fertilized eggs) at 72 hpf. The experiment is repeated three times [20].

8.1.11 Enzyme-linked Immunosorbent Assay

Enzyme-linked immunosorbent assay (ELISA) is a specifically labeled immunoassay and is regarded as the gold standard of immunoassays. This immunological test is very sensitive and can be used to detect and quantify substances, including antibodies, antigens, proteins, glycoproteins, and hormones.

For the general protocol, diluted standards and samples are added to the reaction wells and incubated for 1 hour, followed by the addition of the detection antibody, which is incubated for 1 hour, the addition of the secondary antibody, which is incubated for 1 hour, and the addition of the substrate solution, which is incubated for 30 minutes. The reaction is then stopped by adding the termination solution. The absorbance is read using a spectrophotometer or microplate reader to calculate the concentration of the substance in the sample [21].

8.1.12 Spermatogonial Cell Culture

Testes of 2–3-month-old zebrafish are dissected and separated in collagenase solution (0.2% collagenase and 0.002% DNase in PBS) at 28.5 °C for 1 hour [22]. The dissociated tissue is passed through a discontinuous Percoll gradient (20–40%) method to purify zebrafish spermatogonia. Zebrafish spermatogonia can be found in fractions 25%, 30%, 33%, and 36%, most of which are observed in fractions 25% and 30%. The suspension of cells obtained is then transferred to individual wells of a 12-well plate to be cultured (cells from 1 to 1.5 fish/well).

Modified StemPro®-34 SFM culture medium supplemented with minimal essential medium (MEM) vitamin solution, MEM nonessential amino acid solution, 5% KnockOut™ serum replacement, 0.4% BSA, 10 μg/mL insulin, 2 mM l-glutamine, 100 μg/mL transferrin, 30 nM sodium selenite, 3 mg/mL D-(+)-glucose, 10 μg/mL d-biotin, 1 μL/mL DL-lactic acid, 60 μM putrescine, 50 μM 2-mercaptoethanol, 30 ng/mL β-estradiol, 100 μM ascorbic acid, 60 ng/mL progesterone, 2 μM retinol, 40 ng/mL human epidermal growth factor (EGF), 1% FBS, 1% fish serum, and 30% conditioned medium were used [23].

The cells are cultured at 28.5 °C in an atmosphere of 3% CO_2 and the medium is refreshed every 3–4 days.

8.1.13 Testis Tissue Culture *in Vitro*

For short-term cultures (18 hours for steroid release levels and gene expression analysis), the testes can be immersed directly in the medium. Whereas for long-term exposure (7 days for morphology and BrdU/EdU assay analysis), testes should be placed on nitrocellulose membranes on agarose cylinders and cultured in 24-well flat-bottomed plates with 1 mL of medium [24].

The pair of testes are removed from the zebrafish by dissection and immersed in 0.5% v/v commercial bleach in PBS containing 1 mM $MgCl_2$ and $CaCl_2$ for 2 minutes. Wash with buffer for 2 minutes, and the two testes are placed in control and treatment cultures, respectively. During the culture, testis tissue explants are placed on a 0.25 cm^2 nitrocellulose membrane (25 μm thick, 0.22 μm pore size), which themselves are placed on 750 μL agarose cylinder (1.5% w/v in pH 7.4 Ringer's solution) in 1 mL medium in 24-well flat-bottom plates [25] (Figure 8.5). The agarose cylinders are prepared using 48-well flat-bottom plates as models. The membranes and cylinders need to be soaked overnight at 4 °C in basal medium before culture.

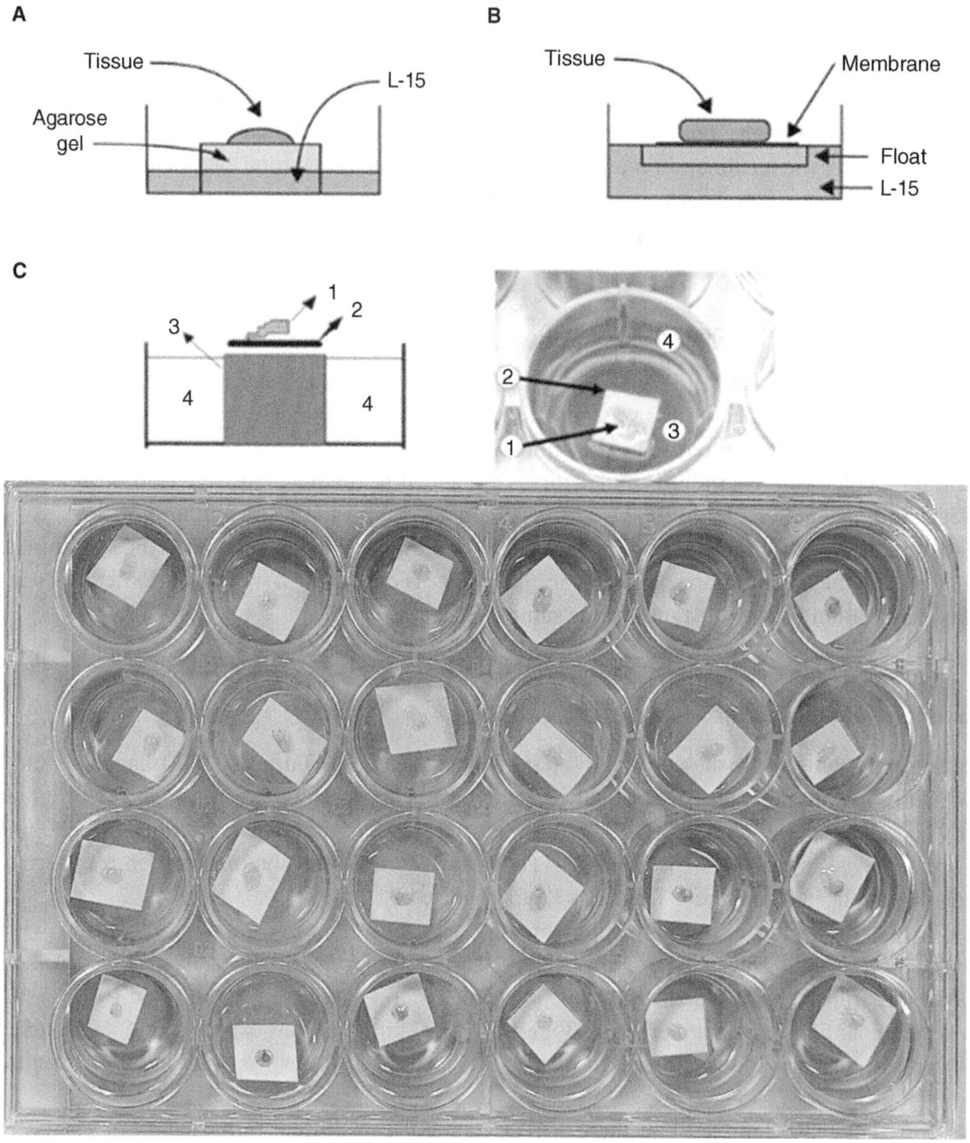

Figure 8.5 The fish testicular organ culture system. (A) The classical agarose gel model. (B) The Japanese eel testis culture model. (C) The zebrafish testis primary culture model (1, testis tissue; 2, nitrocellulose membrane; 3, agarose cylinder; 4, 1 mL medium) [25].

The basal culture medium is Leibovitz-15 supplemented with 0.5% w/v bovine serum albumin fraction V, 10 mM HEPES, 0.4 mg/L amphotericin B, 10 nM retinoic acid, 200 μg/mL streptomycin, 200 U/mL penicillin, and pH 7.4. Cultures are incubated in a moist air atmosphere at 25 °C for 7 days, the medium is changed on day 3. To measure the amount of androgens released from testicular tissue, the medium can be renewed by directly transferring the nitrocellulose membrane supporting the tissue explants into fresh culture wells [25].

8.2 Female Reproductive Function

Female zebrafish have been shown to be able to spawn frequently after sexual maturity, with a spawn ranging from single to several hundred eggs. The eggs laid by zebrafish are translucent and usually larger compared to other fish, being approximately 0.7 mm in diameter. Healthy reproductive organs, uninterrupted steroid production, and normal courtship behavior are the key criteria for successful reproduction [26].

In vertebrates, the ovary is the place where eggs are developed and produced. The reproductive organs of the female zebrafish have a pair of ovaries located on either side of the abdominal wall. The ovarian walls are lined with thin epithelium.

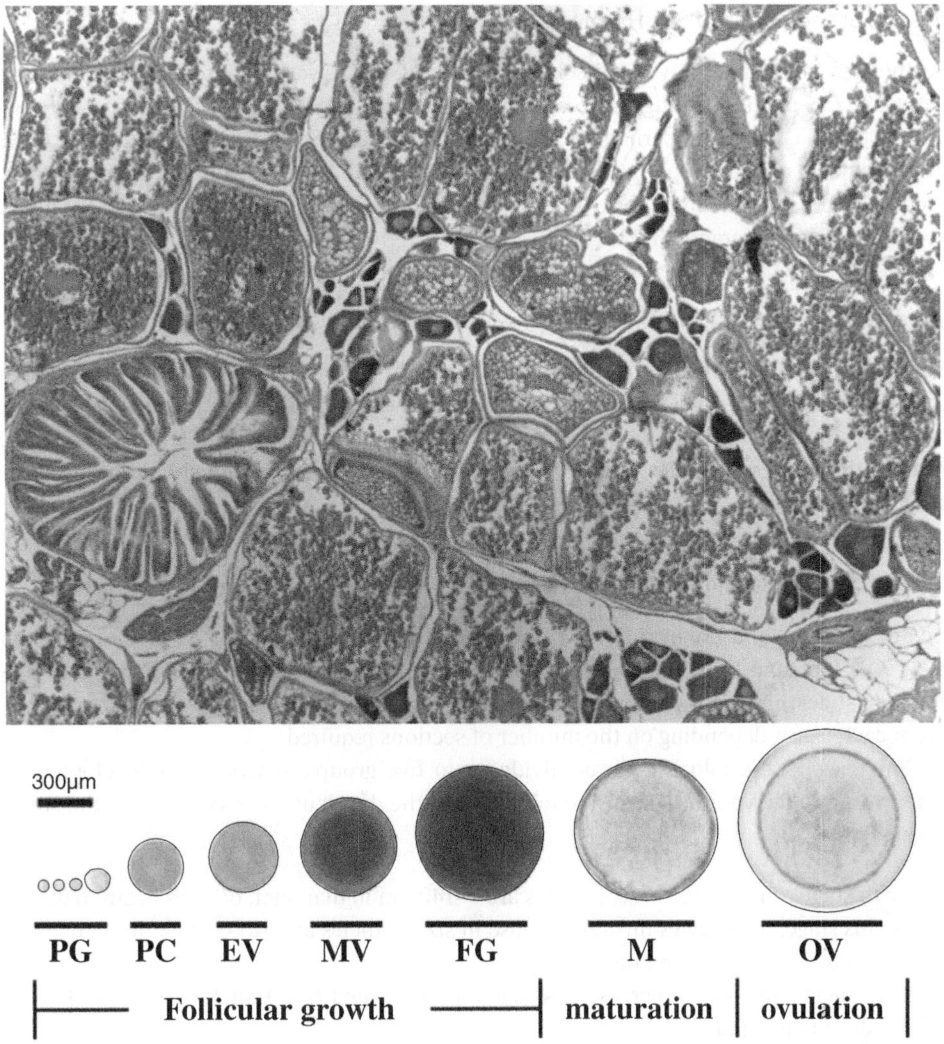

Figure 8.6 A representative HE staining image of zebrafish ovary (upper, provided by the authors), and follicle development (lower [10]), which includes follicular growth, maturation, and ovulation. PV, previtellogenic stage; EV, early vitellogenic stage; M, maturation; OV, ovulation.

The oogenic cells, oocytes, and follicles in mature ovaries are surrounded by mesenchymal tissue and somatic cells to form a sac-like structure.

Ovarian development in zebrafish consists of four stages (Figure 8.6): the primary oocyte stage, where smaller spherical cells with less cytoplasm are observed; the cortico-alveolar stage, in which the oocyte becomes progressively larger in size and the interior is filled with cortical alveoli; the vitellogenic stage, characterized by the beginning of yolk granules and vesicles in the oocyte; and finally, the maturation stage, in which the cytoplasm can be observed to be filled with yolk granules.

In zebrafish, the oocyte is surrounded by an inner layer of granulosa cells and an outer follicular layer of membrane cells. Ovulation occurs after the follicular layer is ruptured; a process induced mainly by sex hormones. The GSI, sex hormone level, follicle development, egg quality, and the expression of gonad-related genes are complex biological traits for determinants of reproductive function in zebrafish [27].

8.2.1 Measurement of the GSI

The GSI is widely used as a criterion for assessing ovarian development and function in many animal models. Firstly, the fish are anesthetized, dried with paper towels, and weighed by an electronic balance; then, the abdomen is dissected, and the ovaries are removed and weighed; Finally, the GSI is calculated as follows: GSI = ovary weight (g)/body weight (g) [28].

8.2.2 Measurement of the Levels of Sex Hormones

Zebrafish are anesthetized in tricaine mesylate dissolved in water at 4 °C. Ovaries are removed by dissection and weighed and homogenized using 75% methanol (1 μg tissue in 1 μl solution) or 1 × PBS (1 μg tissue in 9 μl solution). The samples are centrifuged at 2000g at 4 °C and the supernatants are collected. The above process is repeated three times and the supernatants are then combined, vacuum lyophilized, and stored at −80 °C. The levels of sex hormones, including follicle-stimulating hormone (FSH), progesterone (P), luteinizing hormone (LH), and estradiol, (E2) can be measured by high-performance liquid chromatography (HPLC) or ELISA.

8.2.3 Collection of Eggs

The eggs are collected by gently squeezing the abdomen of sexually mature females. Briefly, adult female zebrafish are anesthetized with tricaine mesylate, dried with paper towels, and placed in a plastic petri dish. The eggs are pushed out by gently pressing a finger along the ventral side of the fish from front to back. A metal spatula is used to gently scrape the eggs off the fish. Eggs from several females may be combined in the experiment and should be used immediately after extrusion (in vitro fertilization) [19].

8.2.4 Histology and Classification of Follicles

The zebrafish are anesthetized, and the ovaries are removed and fixed overnight at 4 °C in 4% paraformaldehyde. After dehydration in 50%, 70%, 90%, and 100% ethanol, and 100% xylene, the samples are embedded in paraffin. Sections 4–7 μm thick are collected onto slides, air dried, stained with H&E, and masked with coverslips. Multiple sections are collected at 5–10 μm intervals in each ovary, depending on the number of sections required.

Under the light microscope, the follicles can be divided into five groups, representing developing, transitional, and mature follicles, according to the volume size of the follicles and the distribution of contents in the surrounding components [10, 29] (Figure 8.6).

Stage I: primary growth stage, small oocyte size, follicles are 7–140 μm in diameter, oocytes begin to develop;
Stage II: cortical alveolus stage, the oocytes further increase in size, follicles are 140–340 μm in diameter, and many small vesicles appear in the cortical area around the oocyte;
Stage III: vitellogenesis stage, the oocytes increase in size at an accelerated rate, follicles are 340–690 μm in diameter and the yolk granules accumulate rapidly;
Stage IV: oocyte maturation stage, the oocytes are increasing in size, follicles are 690–730 μm in diameter, and the yolk granules fill the cytoplasm of the oocyte;
Stage V: mature egg stage, the follicles increase in diameter to 730–750 nm and the oocyte matures with a cell filled with coarse yolk granules;

Generally, 30–80 follicles should be counted from each section, depending on the size of the section [28].

8.2.5 Mating

The treated females are placed in the mating tank in a 1:1 ratio with the normally reared males. Early the next morning, the divider in the middle of the reproductive tank is removed and the male and female zebrafish are allowed to mate for 1 hour. It is possible to calculate indicators including the number of eggs laid by the female, fertilization rate, mortality, and hatching rate [30].

8.2.6 Egg Quality

The most straightforward standard for determining egg quality is embryo survival. Eggs in good quality are defined as a batch of embryos with a very high survival rate (>90%) at 48 hpf, whereas poor quality eggs are those with a very low survival rate (<40%) at 48 hpf.

Female zebrafish occasionally produce nonviable eggs, which are the result of the absorption of mature eggs. Nonviable eggs can be distinguished by their opaque and asymmetrical appearance. The production of eggs (viable or nonviable) indicates that spawning has occurred (i.e. reproductive conditions have stimulated the female to generate eggs); however, only viable eggs can be used as an assessment of fecundity.

In addition, to assess egg quality, fertilized eggs can be placed in a Petri dish containing E3 embryo medium, photographed, and observed under a stereo microscope, and the diameter of the egg and yolk measured randomly [7].

8.2.7 The Expression of Gonad-related Genes (Applicable to Testicular Tissue)

Total RNA is isolated from ovarian/testis tissue by using the Trizol method. The integrity of RNA samples (intact rRNA 28S/18S) is checked by 1% agarose gel electrophoresis. The quantity and quality of the RNA samples were determined using micro spectrophotometry. cDNAs are synthesized from 3 μg total RNA using a commercial reverse transcription kit. Quantitative real-time PCR (qRT-PCR) is performed to evaluate the gene expression profiles by a real-time PCR instrument. The relative transcript levels of target genes are determined using the $2^{-\Delta\Delta Ct}$ method, using a particular reference gene as a normalization [31].

References

1 Hoo JY, Kumari Y, Shaikh MF, Hue SM, Goh BH. Zebrafish: a versatile animal model for fertility research. Biomed Res Int. 2016, 2016(1): 20169732780.
2 Li HQ, Liu YA, Chen QQ, Jin LB, Peng RY. Research progress of zebrafish model in aquatic ecotoxicology. Water-Sui. 2023, 15(9): 1735.
3 Shao XT, Xiao DD, Yang ZY, Jiang LL, Li Y, Wang Y, Ding YL. Frontier of toxicology studies in zebrafish model. J Appl Toxicol. 2024, 44(4): 488–500.
4 Zohar Y. Fish reproductive biology – reflecting on five decades of fundamental and translational research. Gen Comp Endocr. 2021, 300(5): 113544.
5 Cattelan S, Gasparini C. Male sperm storage impairs sperm quality in the zebrafish. Sci Rep. 2021, 11(1): 16689.
6 Mitchell K, Zhang WS, Lu C, Tao B, Chen L, Hu W, Trudeau VL. Targeted mutation of secretogranin-2 disrupts sexual behavior and reproduction in zebrafish. Proc Natl Acad Sci U S A. 2020, 117(23): 12772–12783.
7 Massey MD, Fredericks MK, Malloy D, Arif S, Hutchings JA. Differential reproductive plasticity under thermal variability in a freshwater fish (*Danio rerio*). Proc Biol Sci. 2022, 289(1982): 20220751.
8 Qian PP, Kang JH, Liu D, Xie GC. Single cell transcriptome sequencing of zebrafish testis revealed novel spermatogenesis marker genes and stronger Leydig-germ cell paracrine interactions. Front Genet. 2022, 13: 851719.
9 Leal MC, Cardoso ER, Nobrega RH, Batlouni SR, Bogerd J, Franca LR, Schulz RW. Histological and stereological evaluation of zebrafish (Danio rerio) spermatogenesis with an emphasis on spermatogonial generations. Biol Reprod. 2009, 81(1): 177–187.
10 Chu L, Li J, Liu Y, Cheng CH. Gonadotropin signaling in zebrafish ovary and testis development: insights from gene knockout study. Mol Endocrinol. 2015, 29(12): 1743–1758.
11 Shaliutina O, Materiienko A, Shaliutina-Kolesova A, Gazo I. Using fish spermatozoa in toxicity tests: a potential toxicology tool. Aquaculture. 2021, 539(5): 736647.
12 Riesco MF, Valcarce DG, Martinez-Vazquez JM, Robles V. Effect of low sperm quality on progeny: a study on zebrafish as model species. Sci Rep. 2019, 9(1): 11192.
13 Lan XR, Li YW, Chen QL, Shen YJ, Liu ZH. Tributyltin impaired spermatogenesis and reproductive behavior in male zebrafish. Aquat Toxicol. 2020, 224: 105503.
14 Cheng Y, Franek R, Rodina M, Xin M, Cosson J, Zhang S, Linhart O. Optimization of sperm management and fertilization in zebrafish (Danio rerio (Hamilton)). Animals (Basel). 2021, 11(6): 1558.
15 Kvarnemo C, Green L, Svensson O, Lindstrom K, Schold S, Griful-Dones M, Havenhand JN, Leder EH. Molecular, behavioural and morphological comparisons of sperm adaptations in a fish with alternative reproductive tactics. Evol Appl. 2023, 16(2): 338–353.
16 Yang H, Daly J, Tiersch TR. Determination of sperm concentration using flow cytometry with simultaneous analysis of sperm plasma membrane integrity in zebrafish Danio rerio. Cytometry A. 2016, 89(4): 350–356.
17 Takemoto K, Nishimura T, Kawasaki T, Imai Y, Levy K, Hart N, Olaya I, Burgess SM, Elkouby YM, Tanaka M, Sakai N. In vitro storage of functional sperm at room temperature in zebrafish and medaka. Zebrafish. 2023, 20(6): 229–235.

18 Nayak R, Franek R, Sindelka R, Psenicka M. Enhancement of zebrafish sperm production via a large body-sized surrogate with germ cell transplantation. Commun Biol. 2023, 6(1): 412.

19 Chen Y, Wang H, Wang F, Chen C, Zhang P, Song D, Luo T, Xu H, Zeng X. Sperm motility modulated by Trpv1 regulates zebrafish fertilization. Theriogenology. 2020, 151: 41–51.

20 Tanaka S, Zmora N, Levavi-Sivan B, Zohar Y. Chemogenetic depletion of hypophysiotropic GnRH neurons does not affect fertility in mature female zebrafish. Int J Mol Sci. 2022, 23(10): 5596.

21 Zhang W, Sheng N, Wang M, Zhang H, Dai J. Zebrafish reproductive toxicity induced by chronic perfluorononanoate exposure. Aquat Toxicol. 2016, 175: 269–276.

22 Wong TT, Collodi P. Dorsomorphin promotes survival and germline competence of zebrafish spermatogonial stem cells in culture. PLoS One. 2013, 8(8): e71332.

23 Xie X, Nóbrega R, Psenicka M. Spermatogonial stem cells in fish: characterization, isolation, enrichment, and recent advances of in vitro culture systems. Biomolecules. 2020, 10(4): 644.

24 Tovo-Neto A, Martinez ERM, Melo AG, Doretto LB, Butzge AJ, Rodrigues MS, Nakajima RT, Habibi HR, Nobrega RH. Cortisol directly stimulates spermatogonial differentiation, meiosis, and spermiogenesis in zebrafish (Danio rerio) testicular explants. Biomolecules. 2020, 10(3): 429.

25 Leal MC, De Waal PP, Garcia-Lopez A, Chen SX, Bogerd J, Schulz RW. Zebrafish primary testis tissue culture: an approach to study testis function ex vivo. Gen Comp Endocrinol. 2009, 162(2): 134–138.

26 Delbes G, Blázquez M, Fernandino JI, Grigorova P, Hales BF, Metcalfe C, Navarro-Martín L, Parent L, Robaire B, Rwigemera A, Van Der Kraak G, Wade M, Marlatt V. Effects of endocrine disrupting chemicals on gonad development: mechanistic insights from fish and mammals. Environ Res. 2022, 204: 112040.

27 Li JZ, Ge W. Zebrafish as a model for studying ovarian development: recent advances from targeted gene knockout studies. Mol Cell Endocrinol. 2020, 507: 110778.

28 Liu C, Yue S, Solarz J, Lee J, Li L. Improving the sexual activity and reproduction of female zebrafish with high testosterone levels. Sci Rep. 2021, 11(1): 3822.

29 Wu K, Zhai Y, Qin M, Zhao C, Ai N, He J, Ge W. Genetic evidence for differential functions of figla and nobox in zebrafish ovarian differentiation and folliculogenesis. Commun Biol. 2023, 6(1): 1185.

30 Anjusha M, Prakash S, Kathirvelpandian A, Sakthivel M, Kalaiselvi N, Suresh E. Estimation of repeatability and phenotypic correlation of reproductive traits in zebrafish. Indian J Fish. 2023, 70(1): 31–35.

31 Hu Y, Ma X, Liu R, Mushtaq I, Qi Y, Yuan C, Huang D. 2,4-Dichlorophenol increases primordial germ cell numbers via ESR2a-dependent pathway in zebrafish larvae. Environ Sci Technol. 2022, 56(19): 13878–13887.

9

Combined Molecular Toxicity Mechanism of Heavy Metals Mixtures

Kusheng Wu[a,], Yuequn Chen[a], Wenlong Huang[a]*

[a] *Shantou University Medical College, Shantou, China*

*Corresponding author.
E-mail: kswu@stu.edu.cn

9.1 Introduction: Heavy Metal Exposures Are Everywhere

With the acceleration of industrialization, the metal pollutants discharged into the environment are increasing, and the impact of environmental pollution on human health has gradually attracted people's attention. Heavy metals are widely distributed in the environment and have persistent bioaccumulation. As reported, the main sources of heavy metals in soil were natural sources and human activities, such as mineral resource development, metal processing and smelting, chemical production, factory discharge, and sewage irrigation [1]. The anthropogenic sources including industrialization, urbanization, and intensive agriculture are the primary reasons for local anomaly patterns of heavy metals pollution [2]. The anthropogenic activities associated with mining and smelting emissions are likely to be major sources of Cu, Zn, and Pb from the sediments [3]. The source of heavy metals in Jing River Basin indicated that human activities have significant impacts on the distribution and accumulation of Pb and As, while Hg was attributed to farmland irrigation, and Cr, Ni, Cu, Zn, and Cd contents in soil were affected by natural source [4]. A study also showed that the main sources of pollution in the Fenghe River Basin are urban construction and transportation, the electronics industry, machinery manufacturing, and tourism [5]. One report focusing on the source of heavy metals in PM2.5 showed that Ni, Cu, Zn, Cd, and Pb mainly came from vehicle emissions and metallurgical activities [6]. Both Cr and Mn were mainly from vehicle emissions and road dust, while As was mainly from coal combustion in the air [6]. The bioaccumulation routes of heavy metals are ingestion and inhalation [7], and they could accumulate in some vital body organs, such as the liver, heart, kidney, and brain, and disturb normal biological functioning [8].

9.2 Combined Molecular Toxicity of Heavy Metal Mixtures in the Cardiovascular System

Cardiovascular disease (CVD) is a group of diseases, including atherosclerosis and hypertension, which affect the heart and blood vessels, eventually leading to cardiac failure [9]. Oxidative stress and vascular inflammation have been implicated in the development of CVD [10, 11]. Arterial stiffness, associated with CVD, results from decreased elastin and increased collagen, with irreversible, non-enzymatic cross-linking of collagen contributing to early vascular aging (EVA) [12]. Mitochondrial dysfunction is an identified source of oxidative damage in vascular smooth muscle cells (VSMC) [12] and cardiomyocytes [13]. Increased fibroblastic activity is linked to heart disease, where extracellular matrix (ECM) remodels, and increases collagen deposits in the myocardium, leading to fibrosis [12–14].

Toxicological Assessment of Combined Chemicals in the Environment, First Edition. Edited by De-Sheng Pei and Yiyun Liu.

Excessive toxic heavy metals in the environment can enter the human body through the respiratory tract, digestive tract, skin contact, and other ways, affecting the structure and function of blood vessels, including atherosclerosis, vascular remodeling, endothelial dysfunction, platelet adhesion and aggregation, thrombosis, and ultimately lead to vascular disease.

These metals usually induce free radical production and mainly interact with protein sulfhydryl groups, altering their activities. The biochemical basis for these effects is often attributed to the inhibition of several ATPases, including Na-K-ATPase (NKA) and Ca-ATPase [15, 16], angiotensin-converting enzyme (ACE), and to the production of oxidative damage (free radicals by activation of NADPH oxidase). Metals with vascular functions promote endothelial dysfunction, hypertension, and vascular diseases through these effects. The fact that the endothelium is affected by low doses or concentrations of heavy metals necessitates the need to improve the recognition of the mechanisms by which these metals promote the development of CVD.

Previous epidemiological studies (Table 9.1) from Bangladesh ($N = 959$ adults, median urine arsenic levels 357.9 μg/g) [17] and Mexico ($N = 199$ children, mean urine arsenic levels 74.3 μg/L) [18] mostly focused on carotid intima-media thickness (IMT) as a subclinical atherosclerosis endpoint and generally reported positive cross-sectional associations. In 2402 American Indian men and women participating in the *Strong Heart Study* (baseline urine arsenic median levels 9.20 μg/g), the prospective association of arsenic and carotid IMT was consistently positive [19]. Urine arsenic, however, was not prospectively associated with plaque presence in the *Strong Heart Study*. In a Multi-Ethnic Study of Atherosclerosis (MESA) pilot study using preliminary data from 310 participants, urine arsenic levels (median 3.08 μg/g) were not associated with carotid IMT or coronary artery calcification [20].

Table 9.1 Population studies about the effects of metals on the cardiovascular system.

Region	Heavy metals	Levels and source of sample	Study design	Population	Main findings	Ref.
Bangladesh	Arsenic	Median urine arsenic level 357.9 μg/g	Cross-sectional study	$N = 959$ Adults	Arsenic exposure was positively correlated with cIMT.	[17]
Mexico	Arsenic	Mean urine arsenic level 74.3 μg/L	Cross-sectional study	$N = 199$ Children (3–14 years old)	Arsenic exposure and plasma ADMA levels were positively associated with cIMT.	[18]
American	Arsenic	Baseline urine arsenic median level 9.20 μg/g	Prospective cohort study	$N = 2\,402$ American Indian men and women	Consistently positive prospective association of arsenic and carotid IMT.	[19]
American	Arsenic	Median urine arsenic level 3.08 μg/g	Multi-ethnic study	$N = 6\,814$ American multiethnic adults	No identified association with carotid IMT nor with coronary artery calcification.	[20]
Austria	Cadmium	Serum	Cross-sectional study	$N = 195$ Young women	Positive association with carotid IMT.	[21]
Sweden	Cadmium	Blood	Cohort study	$N = 4\,639$ Middle-aged	Plasma cadmium levels were associated with carotid plaque development.	[22]
Sweden	Cadmium	Blood	Cohort study (mean follow-up 5.4 years)	$N = 599$ Women (64 years old)	Cadmium in blood and urine was associated with the formation and progress of carotid plaque.	[23]
American	Cadmium	Urine cadmium (geometric mean 0.27 μg/g creatinine)	cross-sectional survey	$N = 1\,857$ Adults	Cadmium in urine was positively associated with self-reported heart attack/myocardial infarction in models, adjusted for urine barium cobalt levels.	[24]

Region	Heavy metals	Levels and source of sample	Study design	Population	Main findings	Ref.
American	Cadmium	Blood and urine	Cohort study (mean follow-up 7.4 years)	$N = 26\ 056$ Adults (20–85 years old)	Cadmium and antimony in urine were associated with an increased risk of CVD mortality in models after adjustment for the other metals as covariates.	[25]
Spanish	Cadmium and antimony	Urine	Hortega Follow-Up Study	$N = 1\ 171$ Adults (15–85 years old)	Cadmium and antimony were associated with CVD incidence endpoints, in models adjusted for other metals.	[26]
American	Tungsten	Urine tungsten (mean 0.14 μg/L creatinine)	Cohort study	$N = 8\ 614$ Adults (18 –74 years old)	Tungsten in urine was positively associated with stroke in models adjusted for urine molybdenum and cobalt	[27]
Aragon	Arsenic, cadmium, and titanium	Median arsenic, cadmium, and titanium levels 1.83, 0.27, and 9.8 μg/g creatinine, respectively	Cohort study	$N = 1\ 783$ Workers	Increased arsenic, cadmium, and titanium exposures were positively associated with the presence of subclinical atherosclerosis in mid-life working adults	[28]

cIMT: carotid intima-media thickness; CVD: cardiovascular disease.

Cross-sectional studies showed that increased serum cadmium levels have been positively associated with carotid IMT in 195 young healthy women from Austria [21]. In Sweden, blood cadmium levels were associated with carotid plaque development in 4639 middle-aged men and women [22]. Consistently, blood and urine cadmium were associated with carotid plaque formation and increased after a mean of 5.4 years of follow-up in 599 64-year-old women from the same cohort in Sweden [23].

In the National Health and Nutrition Examination Survey 2007 to 2008 ($N = 1857$), urine cadmium (geometric mean: 0.27 μg/g creatinine) was positively associated with self-reported heart attack/myocardial infarction in models adjusted for urine barium cobalt, molybdenum, tungsten, and other heavy metals [24]. In a recently published National Health and Nutrition Examination Survey from 1999 to 2014 ($N = 26\ 056$), urine cadmium and antimony were associated with increased risk of CVD mortality in models after adjustment for the other metals as covariates [25]. Consistent with the Hortega Study, a representative sample of the Spanish population, reported associations of cadmium and antimony with CVD incidence endpoints in models adjusted for other metals [26]. In the National Health and Nutrition Examination Survey 1999 to 2010 (N = 8614), urine tungsten (mean: 0.14 μg/L creatinine) was positively associated with stroke in models adjusted for urine molybdenum and cobalt [27].

Maria Grau-Perez et al. evaluated the individual and joint association of a panel of nonessential metals measured in urine with subclinical atherosclerosis measured in multiple vascular territories, including the carotid, the femoral, and coronary. Increased arsenic, cadmium, and titanium exposures were positively associated with the presence of subclinical atherosclerosis in mid-life working adults [28]. These associations were driven by arsenic and cadmium in carotid, cadmium, and titanium in femoral, and titanium in coronary territories and remained mostly unchanged after additional adjustment for the other relevant metals. In Bayesian Kernel Machine Regression (BKMR) analysis, they also found suggestive evidence of a synergistic interaction between arsenic and the cadmium and titanium mixture component. In secondary analyses, titanium, cadmium, and antimony were positively associated with alternative definitions of elevated coronary calcium. Other metals, including chromium, uranium, vanadium, and tungsten, were not associated with subclinical atherosclerosis measures in the study population (Table 9.1).

Vascular endothelium is an important target of cadmium toxicity. Cadmium exposure mainly damages vascular endothelial cells through oxidative stress, proinflammatory effects, endocrine interference, and other mechanisms [29–32]. As an important vascular protective factor, NO can inhibit platelet adhesion and aggregation, enhance vasodilation function by stimulating VSMCs to produce cyclic guanosine monophosphate (cGMP), reduce the interaction between leukocytes and endothelial cells, reduce the proliferation of smooth muscle cells, and treat atherosclerosis. It is produced by uncoupled endothelial nitric oxide synthase (eNOS), which catalyzes L-arginine under the action of the cofactor tetrahydrobiopterin (BH4) [33]. Some studies have shown that acute cadmium exposure increases the local release of angiotensin II, thereby increasing the activities of cyclooxygenase-2 (COX-2) and nicotinamide adenine dinucleotide phosphate (NADPH) oxidase, and the production of free radicals. Superoxide anions interact with NO to form peroxynitrite, thus reducing the bioavailability of NO [31, 34]. In addition, cadmium exposure also inhibits antioxidant enzymes including glutathione, thioredoxin, superoxide dismutase, and catalase to reduce vascular antioxidant capacity [35, 36], thereby promoting vascular oxidative stress. Studies have shown that cadmium can also directly lead to the release of many inflammatory factors in blood vessels, such as tumor necrosis factor-α (TNF-α) and interleukin-6 (IL-6), IL-8, IL-1β. It can trigger vascular inflammatory reactions and vascular permeability, destroy the integrity of the endothelial barrier, and promote vascular atherosclerosis [37, 38]. After cadmium exposure, human umbilical vein endothelial cells will cause the release of mitochondrial reactive oxygen species (ROS). Mitochondrial ROS mediates the activation of Nod-like receptor protein 3 (NLRP3) inflammatory bodies. NLRP3 inflammatory bodies also cause the activation of caspase-1. Caspase-1 induces DNA damage, leads to membrane pore formation, and cell swelling, and ultimately leads to endothelial cell apoptosis [29]. Cadmium has been considered as a metal hormone [39], which can induce changes in the structure and function of the vascular system through endocrine interference mechanisms [30]. As we all know, steroid hormones have a great impact on vascular function at all stages of body growth [40], low dose $CdCl_2$ exposure (0–10 μmol/L) can downregulate the expression of androgen receptor (AR) in human umbilical vein endothelial cells and upregulate estrogen receptor β and *Cyp19a1* gene expression level, to interfere with AR signal transduction, affect the proliferation of vascular endothelial cells, interfere with angiogenesis, repair, and other functions, and promote atherosclerosis [30].

Lead can promote the production of ROS in blood vessels through the Fenton type Haber Weiss reaction [41], and at the same time, lead can cause a relative lack of antioxidant enzymes such as catalase and glutathione peroxidase [42], thus inducing oxidative stress in blood vessels. Vascular oxidative stress induced by lead-containing compounds can reduce the bioavailability of NO and cause vascular endothelial dysfunction [43, 44]. Research shows that the increase of lead ions in blood will lead to the increase of homocysteine levels in blood [45, 46]. Hyperhomocysteinemia can cause vascular injury by promoting the proliferation of VSMCs, changing the elasticity of the vascular wall, inhibiting NO synthesis, and promoting oxidative stress. Studies have shown that the vascular inflammation caused by lead-containing compounds exposure is related to the activation of nuclear factor κB (NF-κB) [47]. NF-κB is a universal transcription factor of many proinflammatory cytokines, chemokines, and adhesion molecules. Oxidative stress can promote the activation of NF-κB, causing vascular inflammation, vascular fibrosis, and vascular cell apoptosis. After exposure to lead-containing compounds, vasoconstriction, and diastolic function decrease, which increases the risk of hypertension. In addition to the above oxidative stress, related mechanisms also involve a variety of possible mechanisms [48]. Studies have shown that the increase in vascular contraction caused by lead is to some extent caused by the increase of protein kinase C (PKC) activity. PKC, a member of the serine-threonine kinase family, regulates cell growth, vascular contraction, blood flow, and cell permeability [49]. In addition, lead can also promote vasoconstriction by affecting hormone production. Studies have shown that lead exposure increases the production of adrenaline, angiotensin II, and endothelin, thereby increasing vasoconstriction [41, 47]. Fibroblast growth factor (bFGF) is synthesized and stored by endothelial cells, which can promote the migration and proliferation of adjacent endothelial cells to start the process of vascular repair. Previous studies have shown that lead can inhibit the binding of bFGF and heparan sulfate proteoglycan (HSPG) on the surface of endothelial cells. However, because this inhibition is not obvious, it is unlikely that lead's direct inhibition of the binding of bFGF and HSPG is one of the main mechanisms of its inhibition of endothelial cell proliferation [50]. In addition, heparan sulfate can promote the binding of bFGF and its receptor. This study shows that lead can inhibit the proliferation of vascular endothelial cells by inhibiting the synthesis of heparan sulfate to reduce the response of endogenous bFGF.

Mercury and its compounds damage the vascular system mainly through oxidative stress. Studies have shown that even very low levels of chronic mercury exposure ($HgCl_2$, 6 nmol/L) can promote endothelial cell dysfunction through oxidative stress, inflammatory response, and other mechanisms. The increase of the local release of angiotensin II-induced by $HgCl_2$ may lead to the increase of COX-2 and NADPH oxidase activities, thus increasing the release of ROS and inducing oxidative stress [51]. In addition, mercury compounds can also inhibit the expression of nitric oxide synthase gene and reduce the production of NO [52].

Research shows that Cr(VI) remarkably inhibited the viability of HUVECs with a dose-dependent manner, Cr(VI) increased the expression of inflammatory factors (IL-1β, VCAM-1, ICAM-1, and TNF-α) and NLRP3 inflammasome on mRNA level [53]. ROS generation was increased in the Cr (VI) treated HUVECs. Furthermore, Cr(VI) induced the apoptosis of HUVECs by activating the intrinsic apoptosis pathway. Meanwhile, Cr(VI) induced oxidative stress response in HUVECs by JNK and p38 MAPK signaling pathways. These results are useful for Cr (VI) toxicity in HUVECs.

As exists ubiquitously in four possible oxidation states (3−, 0, 3+, and 5+) in the environment, trivalent arsenite [As (III)] and pentavalent arsenate [As (V)] are the predominant inorganic arsenic (iAs) species in natural water, whereas arsenite is more prevalent in groundwater. Arsenic trioxide (ATO), a trivalent As is released into the air and water through natural and industrial processes. It is the most produced form of iAs that may generate arsenite in alkaline solutions. The data from mouse models and an established endothelial monolayer model demonstrate that ATO-induced vascular endothelial dysfunction is mediated by increased CAPN-1 activity [54]. This increased activity is associated with a rapid rise in intracellular Ca^{2+}, increased CAPN-1 levels, and membrane localization of CAPN-1, which leads to increased proteolytic disorganization of VE-cadherin. The present data support the notion that therapeutics targeting CAPN-1 may be useful in attenuating atherogenesis induced by As exposure.

The functioning of the vascular system involves complex interactions among multiple cell types, with each one utilizing a myriad of cellular signaling pathways that allow the vascular system to respond and adapt to ever-changing environmental conditions. Recent studies have provided new insights into the mechanisms by which metals can influence vascular function. Expression of the ZIP8 metal ion transporter (*Slc39a8* gene) appears to be a key factor contributing to the selective toxicity of Cd in the endothelial cells of organs such as the testes and kidneys [32]. At the cellular level, metals such as Cd and As have profound effects on the process of angiogenesis. These effects involve alterations in the function and expression of cell adhesion molecules such as VE-cadherin and PECAM-1, although the specific signaling pathways that mediate these actions have yet to be elucidated [32]. Further research is needed, especially in the area of metal-induced hypertension, to determine the significance and the mechanism of the adverse effects.

Heavy metal contamination poses a significant threat to public health, with numerous studies linking it to detrimental health outcomes. To more accurately quantify the health implications of heavy metal exposure, there is an urgent need to enhance nationwide surveillance of heavy metal concentrations within the general population. While the focus has traditionally been on the association between heavy metal pollution and neurodevelopmental disorders, cancers, and hepatic pathologies, emerging evidence suggests a correlation between such exposure and the development of various vascular diseases. Investigating the mechanisms underlying vascular damage induced by heavy metals is crucial. The research can facilitate the advancement of clinical interventions for heavy metal-induced vascular diseases and inform the development of preventive strategies aimed at mitigating the risk of vascular diseases within the broader population.

9.3 Combined Molecular Toxicity of Heavy Metal Mixtures in the Nervous System

The effects of lead, mercury, cadmium, arsenic, manganese, antimony, and other metallic elements on the nervous system of organisms have been confirmed by numerous studies [55]. Some epidemiological studies have also reported the relationship between metal exposure and neurodevelopmental disorders (Table 9.2).

9.3.1 Relationship Between Lead Levels and Neurotoxicity

As a common heavy metal element, many scholars have conducted extensive research on the neurotoxic effects of lead (Pb). Eubig et al. reported that elevated lead levels may stimulate the inappropriate release of neurotransmitters that interfere with brain function [56]. Parajuli et al. reported a significant negative association between increased lead concentration in cord blood and the neurodevelopmental status of newborns, suggesting that prenatal exposure to high levels of lead may delay utero neurodevelopment [57]. Two epidemiological studies have shown a positive association between elevated lead levels and the risk of several neurodevelopmental disorders, such as ASD and ADHD [55, 58]. A follow-up study of more than 30 years has also revealed the long-term neurotoxicity of heavy metal lead, namely that childhood lead exposure may impair the development of normal brain structures and reduce intellectual performance [59].

A growing number of clinical studies have also shown that early exposure to Pb may lead to neurobehavioral disorders such as memory and learning disabilities [60]. In addition, Wang et al. explored the immediate and long-term effects of early-life exposure to environmental levels of lead on the central nervous system, and the cellular and molecular mechanisms underlying that lead to abnormal neurobehavior by using the zebrafish model. This study found that exposure to Pb in zebrafish larvae (2–120 hpf) may cause disruption of early neurodevelopment and the prolonged modulation of neuromorphic and neurotransmitter systems, leading to neurobehavioral disorders in juveniles and adults [61]. Several potential mechanisms for Pb neurotoxicity have been proposed, including disruption of normal neurotransmitter release, cytotoxicity, energy metabolism disorders, and epigenetic changes [62, 63]. In addition, Ca^{2+} dysregulation and apoptosis have been extensively documented as major mediators of Pb-induced neurotoxicity [64, 64]. Using SD rats as experimental models, Li et al. further demonstrated that the IP3R-Ca2+-ASK1-p38 signaling pathway mediated PB-induced hippocampal neuronal apoptosis. PAS-Na ameliorated these changes within a specific dose range [64, 64]. As a key factor in the regulation of apoptosis, ASK1 is activated through a common Ca^{2+}-dependent mechanism [65]. Studies have shown that Ca^{2+} plays an important role in the development of many neurodegenerative diseases [66]. Calcium release is primarily mediated by

Table 9.2 A summary of neurotoxic effects caused by heavy metal exposure.

Heavy metals	Species	Symptoms of neurotoxicity	Possible mechanisms	Ref.
Lead (Pb)	Human	a. ASD and ADHD b. Reduced intelligence c. Neurodevelopmental disorders d. Memory and learning disabilities	a. Inappropriate release of neurotransmitters b. Early neurodevelopment and the prolonged modulation of neuromorphic and neurotransmitter systems c. Disruption of normal neurotransmitter release, cytotoxicity, energy metabolism disorder and epigenetic changes, etc. d. Ca^{2+} dysregulation (IP3R-Ca^{2+}-ASK1-p38) and apoptosis in the hippocampal neurons	[55, 58–64]
	Zebrafish	a. Neurobehavioral disorders		
	SD rats	a. The damage to hippocampus neuron		
Manganese (Mn)	Humans	a. Clinical symptoms of psychiatry and fine motor, attention, cognitive and intellectual deficits in children	a. Neurotransmitter systems including dopaminergic, cholinergic, and GABAergic b. Interfering with disease-associated proteins, such as synaptic proteins and amyloid proteins c. Oxidative stress, generation of free radicals, and apoptosis d. Ca^{2+} homeostasis e. MAPK signaling pathway, cell apoptosis, and DNA damage	[67, 69–71]
	Zebrafish	a. Neurotoxic effect		
Cadmium (Cd)	Humans	a. Cognitive behavior, dyslexia, behavioral problems, decreased attention, decreased learning and memory, olfactory impairment, etc.	a. Dysregulation of neural stem cells (NSCs) in proliferation and differentiation b. Cell cycle arrest and apoptosis together led to a significant decrease c. Interferes with the release of neurotransmitters and the expression and function of neurotransmitter receptors d. Interferes with the normal function of calcium ions and affects the metabolism of copper and zinc e. Affects neuron number and neurogenesis f. Activates mitochondrial apoptosis and autophagy pathways g. Activates astrocytes and neuro-microglia, causes neuroinflammatory reactions	[72, 74–84]
	Animals	a. Impaired hippocampal-dependent learning and memory abilities, as well as olfactory memory impairments b. Neurodevelopmental disorders and neuroimmune responses		
Antimony (Sb)	Zebrafish	a. Alters the developmental and neurobehavioral disorder	Affects the neurotransmitter system and oxidative stress	[87]

Heavy metals	Species	Symptoms of neurotoxicity	Possible mechanisms	Ref.
Combined exposure to heavy metals	Humans	a. Co, Zn and Pb in urine were significantly correlated with dyslexia in children b. Mn, Pb, and Cr in adolescents were associated with lower IQ scores c. Levels of metals and elements during pregnancy (such as arsenic, cadmium, copper, mercury, manganese, magnesium, and lead) are associated with ASD and ADHD in children	a. Density of dendritic spines in the CA1 and DG regions of the hippocampus was decreased b. HDAC2 plays an important regulatory role in the neurotoxicity caused by co-exposure of Pb and Cd c. Co-exposure to Pb+Cd+As+Hg induced dose-dependent reductions in dendrite length and branching, spine density, and mature phenotype in primary hippocampal neurons, as well as stimulated neurite growth in NGF-differentiated PC12 cells d. Combined exposure to heavy metals arsenic (As), lead (Pb), and cadmium (Cd) changes dopaminergic neurotransmission e. Neuronal apoptosis and changes in active oxide and [Ca^{2+}](i) levels	[76, 92–97, 99]
	Sprague-Dawley rats	a. Developmental manganese, lead, and barren cage exposure have adverse long-term neurocognitive, behavioral, and monoamine effects b. Single or combined exposure to Pb/Cd could impair the learning and memory ability of rats c. Low levels of combined exposure to Hg, Pb, and Cd can cause synergistic neurotoxic damage		
	Others	a. Zn, Cu, and Cd, alone and combined, interfered in the antioxidant defenses and induced neurotoxicity in tadpoles		

ryanodine receptors (RyRs), which play a dominant role in mediating learning, memory, and cognitive molecular mechanisms. So, in addition to increased Ca^{2+} concentration and apoptosis in the hippocampal neurons, it may also be involved in increased reactive oxygen species and upregulation of RyR3 expression [64, 64].

9.3.2 Relationship Between Manganese Level and Neurotoxicity

Manganese (Mn) is one of the most abundant elements in the earth's crust and mainly exists in the form of organic and inorganic compounds in rocks, soil, and water. Mn is also an essential metal ion for the human body and is essential for normal growth, development, and cellular homeostasis. Studies have shown that high levels of Mn in the central nervous system can cause clinical symptoms of psychiatry [67] and fine motor, attention, cognitive, and intellectual deficits in children. Dietary intake of high Mn content can also produce certain toxicity to the human body. For example, studies have shown that children exposed to a certain level of Mn for 3 years have significantly lower IQ levels and do less well in school [68].

Although the toxicity of manganese has been reported in many studies in recent years, the underlying mechanism remains unclear. Mn neurotoxicity also affects several neurotransmitter systems including dopaminergic, cholinergic, and GABAergic [69]. Recent studies have also suggested that exposure to Mn at the environmental level may cause neurodegeneration by interfering with disease-associated proteins, such as synaptic proteins and amyloid proteins [70]. Excess manganese is known to cause oxidative stress, generation of free radicals, and apoptosis, which were influenced by mitochondrial integrity. So, these changes can be dependent on intracellular Ca^{2+} homeostasis. Moreover, exposure to Mn can promote the expression of related genes such as cell apoptosis and DNA damage by inducing the activation of the MAPK signaling pathway in zebrafish, thus showing the neurotoxic effect of Mn [71].

9.3.3 The Relationship Between Cadmium Levels and Neurotoxicity

Chronic exposure to cadmium at a certain level will cause certain toxic effects on the body, such as kidney, liver, and bone damage and neurotoxicity [72]. In addition, Cd can increase the permeability of the blood-brain barrier and permeate and accumulate in the brain [73]. There is growing epidemiological evidence that Cd exposure is associated with neurotoxicity, such as cognitive behavior, dyslexia, behavioral problems, decreased attention, decreased learning and memory, olfactory impairment, etc. [74–77].

Exposure to Cd in adult animals has been shown to lead to sustained hippocampal-dependent learning and memory impairments, as well as olfactory memory deficits, which suggests dysregulation of neural stem cell (NSC) proliferation and differentiation [78–80]. These studies suggest that Cd may affect neurogenesis and lead to neurological dysfunction such as learning and memory. In addition, Song et al. showed that Cd has a toxic effect on NSCs differentiation at the single-cell level through the data of RNA-SEQ, which provides a reference for an in-depth understanding of the potential molecular mechanism of Cd in neurogenesis [81]. Xu et al. also found that Cd^{2+} induced significant neurodevelopmental disorders and neuroimmune responses, while cell cycle arrest and apoptosis together led to a significant decrease in the density of neurons [82].

To sum up, the neurotoxicity caused by chronic cadmium exposure has gradually attracted people's attention, and the mechanisms involved in neurotoxicity mainly include the following: cadmium interferes with the release of neurotransmitters and the expression and function of neurotransmitter receptors [81]; Cadmium interferes with the normal function of calcium ions and affects the metabolism of copper and zinc [83]; Cadmium affects neuron number and neurogenesis; Cadmium induces ROS and activates mitochondrial apoptosis and autophagy pathways [72, 84]; Cadmium can also activate astrocytes and microglia and cause neuroinflammatory reactions [82].

9.3.4 The Relationship Between Antimony Levels and Neurotoxicity

With the rapid development of mining and manufacturing, antimony (Sb) is gradually recognized. Sb mainly exists in the form of a trivalent (III) or pentavalent (V) oxidation state in nature, and almost all its compounds are toxic to the human body [85]. The main ways Sb enters the human body are through the food chain, skin contact, and respiration [86]. Previous studies have shown that Sb and all its compounds have toxic effects on the immune system, digestive system cardiovascular system, etc. However, few studies have mentioned the adverse effects of antimony on the nervous system. In 2021, Xia et al. reported on the neurotoxic effects of Sb, showing that antimony can alter the developmental and neurotoxicity of zebrafish embryos by affecting the neurotransmitter system and oxidative stress, thereby altering behavior [87].

9.3.5 The Relationship Between Combined Exposure to Heavy Metals and Neurotoxicity

Several previous studies have explored the association between co-exposure to multiple metals and neurodevelopmental disorders [88–90]. Huang et al. used the polymetallic exposure model and found that Co, Zn, and Pb in urine were significantly correlated with dyslexia in children [76]. Stein et al. demonstrated that middle childhood is a developmental window for vulnerability to adverse neurodevelopmental consequences of exposure to environmental toxicants [91]. Bauer et al. also pointed out that higher levels of Mn, Pb, and Cr in adolescents were associated with lower IQ scores, especially lower levels of Cu [92]. Levels of metals and elements during pregnancy (such as arsenic, cadmium, copper, mercury, manganese, magnesium, and lead) are associated with ASD and ADHD in children, and even population levels of these compounds may have negative effects on neurodevelopment [93]. These results suggest that polymetallic co-exposure may lead to neurological dysfunction in the body to some extent, and this neurodevelopmental disorder is particularly common in children.

Although a growing body of epidemiological evidence explores the relationship between co-exposure to multiple metals and neurotoxicity, the mechanism of neurotoxicity associated with co-exposure to heavy metals remains unclear. Study shows that developmental manganese, lead, and barren cage exposure causes adverse long-term neurocognitive, behavioral, and monoamine effects in Sprague–Dawley rats [94], suggesting an interaction between Mn and Pb. Zhou et al. [95] verified the neurotoxicity of single and joint exposure to Pb and Cd through in vivo and in vitro experiments and found that single or combined exposure to Pb/Cd could inhibit the neurite growth of PC12 cells and impair the learning and memory ability of rats. Also, the density of dendritic spines in the CA1 and DG regions of the hippocampus was decreased. Interestingly, the co-exposure of Pb/Cd aggravated this nerve damage. This phenomenon has also been observed in other studies. In the study by Zhou et al, HDAC2 plays an important regulatory role in the neurotoxicity caused by co-exposure of Pb and Cd. Previously, Zhou et al. also reported for the first time that co-exposure to Pb + Cd + As + Hg induced dose-dependent reductions in dendrite length and branching, spine density, and mature phenotype in primary hippocampal neurons, as well as stimulated neurite growth in NGF-differentiated PC12 cells [96]. In recent years, other studies have shown that combined exposure to heavy metals As, Pb, and Cd changes dopaminergic neurotransmission, and this may help explain the neurobehavioral disorders induced by metal mixtures [97]. Similar to the mechanism of neurotoxicity caused by single metal exposure, low levels of combined exposure to Hg, Pb, and Cd can cause synergistic neurotoxic damage, which may be related to neuronal apoptosis and changes in active oxide and $[Ca^{2+}](i)$ levels [98]. The above studies suggest that joint exposure to heavy metals can produce neurotoxic effects on the body and may have a synergistic effect to aggravate such nerve damage. This phenomenon is also found in other studies. Zn, Cu, and Cd, alone and combined, interfered with the antioxidant defenses and induced neurotoxicity in *Lithobates catesbeianus* tadpoles [99]. However, compared to individual metals, combined metals are more toxic, particularly at low doses [100, 101].

The mechanism involved in combined exposure may be like the neurotoxic mechanism caused by single metals, but there are still few studies on the neurotoxic mechanism of combined exposure. Due to the important role of metal exposure in neurodevelopment, more studies are needed to explore the association between mono-metal and polymetallic exposure and neurotoxicity, especially their complex pathogenic mechanisms.

9.4 Combined Molecular Toxicity of Heavy Metal Mixtures in the Male Reproductive System

Reproduction is a necessary process for biological reproduction, a process that requires the transmission of genetic material to the next generation through germ cells and has a crucial role in the life process. Approximately 15% of modern couples suffer from infertility [102], more than 20 to 30% of which are caused by male reproductive disorders, the detailed mechanisms of which are largely unknown [103].

The reproductive performance of most species is adversely affected by harmful heavy metals such as lead, cadmium, mercury, arsenic, zinc, and copper [104]. Environmental and occupational exposure to heavy metals has resulted in various deleterious damages to biological systems, with infertility being one of them [104].

9.4.1 Male Reproductive System

Starting in the late twentieth century, scientists have expressed concerns about the declining semen quality in men. In the 2010s, it was clear that there had been a significant, steady decline in sperm count and semen volume. In a 2017 Hebrew University meta-analysis, decreases in sperm concentration of up to 52.4% and 59.3% in sperm count were noted from 1973 to 2011 [105]. Two other studies presented at the 2018 American Society for Reproductive Medicine (ASRM) scientific congress had similar findings: reduced sperm counts and motility during the 2000s [106]. A 2012 paper published by French researchers and Yeshiva University's Institute for Public Health Sciences in the Journal of Human Reproduction studied French males from 1989 to 2005 and concluded that sperm counts and the proportion of normal, motile sperm fell by 32.2% and 8.1%, respectively [107]. Various studies in Finland, Southern Tunisia, and Argentina also showed a decline in sperm count, motility, morphology, and seminal volume. Males from India had a 30.3% decline in sperm count, a 22.9% decline in sperm motility, and a 51% decrease in morphology over a decade. According to a research report, the semen of 30 male workers who worked in mining enterprises in several regions of Russia who worked for more than 5 years and were married for more than 1 year without giving birth (infertile marriage) was investigated [108]. The results showed that only 33.5% of the men had normal sperm pattern parameters.

Male fertility is determined by the health of the reproductive organs and the quality of reproductive cells. The process of spermatogenesis requires the processes of spermatogonia accretion, spermatocyte meiosis, and spermatocyte degeneration. The effects of healthy testicular aging and testicular premature aging may lead to spermatogenesis disorders, low sperm motility, sperm malformations, and abnormal sperm acquisition, and finally, result in poor reproductive capacity.

Heavy metals alter several reproductive functions in both male and female animals, such as sperm count, motility, spermatogenesis, hormone imbalance, follicular atresia, and delayed oocyte maturation [109]. As endocrine-disrupting compounds (EDCs), heavy metals are not only found at higher concentrations in blood or urine, but also in semen [110, 111]. Continual exposure to them has been identified as an influential factor in male fertility and spermatogenesis [112]. Heavy metals affect male reproductive function by damaging reproductive organs, interfering with reproductive hormone levels, or directly affecting sperm quality [113]. It has been found that the toxicity of inorganic arsenic (iAs^{3+}/iAs^{5+}), cadmium

(Cd^{2+}), lead (Pb^{2+}), and mercury (Hg^{2+}/CH_3Hg^{2+}) could damage the epididymis, which plays a key role in sperm maturation [114]. Researchers in the field of heavy metals even suggest that semen could be considered an early biomarker of environmental exposure to Cu, Zn, and Cr [111].

9.4.2 Arsenic Exposure and Male Reproductive System

Arsenic is ubiquitous in the environment in organic or inorganic states. Water with arsenic above the World Health Organization guideline of 10 ppb has been reported in many countries, especially in Asia [115]. Meanwhile, arsenic is one of the important poisons of PM2.5 air pollution [116]. Current epidemiological and animal toxicology studies suggest that arsenic exposure can directly disrupt androgenic steroidogenesis. Arsenic affects mitochondrial enzymes and interrupts energy production. High concentrations of arsenic may inhibit the sensitivity of gonadotroph cells to GnRH by increasing the level of plasma glucocorticoids, thereby reducing the secretion of gonadotropins, and ultimately affecting testicular development and interfering with spermatogenesis [117]. At the same time, arsenic exposure could suppress levels of sex hormones in the blood by altering the expression of steroids [118]. An epidemiological survey of 1040 volunteers showed that Arsenic exposure is inversely associated with sex hormones, which are thought to be anti-androgenic. However, Leydig cell exposure experiments in vitro confirmed that low doses of arsenic in the environment could stimulate the synthesis of steroid hormones [119]. In another epidemiological survey involving 1020 volunteers, higher urinary arsenic was associated with increased levels of oxidative stress markers, which were also associated with impaired semen quality [120]. The mediation analysis indicated that 8-isoPGF 2a partially mediated the negative association between urinary cadmium and total forward movements.

9.4.3 Mercury Exposure and Male Reproductive System

Mercury (Hg) pollution to the environment and its impact on human health are currently global issues [121]. Some anthropogenic factors, such as mining, coal combustion, chemical, and industry cement production, will contaminate the biosphere. When Hg is released into the groundwater or the other natural environment, microorganisms are responsible for the biotransformation of the metal leading to the formation of methylmercury (MeHg) which will accumulate in food chains, freshwaters, and ecosystems, leading to Hg exposure in human [122, 123].

Mercury can cross the blood–testis barrier, and mercury exposure could directly damage sperm [124]. Inorganic mercury exposure also affects androgen secretion, damages testicular cells, and consequently impairs spermatogenesis, and testosterone synthesis [124]. Mercury chloride exposure disrupts the reproductive system and induces testicular immunosuppression and fibrosis through the inhibition of the CD74 signaling pathway [124]. Mercury can change the shape and movement of sperm, reducing their quantity and quality. Among men exposed to mercury, decreased erection, sexual performance, and ejaculation quality were found [125].

Exposure of mice to Hg confirmed adverse results on male fertility, spermatogenesis, and testicular morphology, whereas MeHg and inorganic Hg, to an extent, decreased the synthesis of DNA in spermatogonia [126]. Researchers also found that Hg will decrease the mobility and number of spermatozoids. Impairment of the tissue structure of the testes and a decline in the level of testosterone in the serum of male mice were noticed [127]. Hg exposure could also cause dysfunction in Sertoli cells [128]. It could nurture a pro-oxidative microenvironment and foster the formation of ROS leading to abnormal sperm functionality and viability, testicular apoptosis, erroneous protamination, oxidative DNA damage, and thus male infertility [129, 130].

9.4.4 Chromium, Cadmium Exposure and Male Reproductive System

Chromium(VI) is a strong oxidizing agent due to its unstable electronic conformation and is commonly found in the metal and coating industries. In the testis, chromium(VI) induces complete disruption of spermatogenesis by degrading seminiferous tubules, chromatin fragmentation, mitochondrial dysregulation, and disruption of the blood–testis barrier [131]. Furthermore, it reduces the activity of antioxidant enzymes and the concentration of non-enzymatic antioxidants in the testes and semen. The severe oxidative state promoted by chromium(VI) ultimately leads to decreased sperm quality and male fertility [132, 133]. However, the toxicity of chromium(VI) could be reversed by supplementing with antioxidant compounds such as vitamin C (0.5 g/L; 1.0 g/L; 2.0 g/L) [134].

Cadmium (Cd) has an exceedingly long biological half-life of 20–40 years and it accumulates in different tissues over time, particularly in male reproductive organs [135]. Indeed, testes are reported to be more sensitive to Cd than other important organs. The toxic effects on male reproductive function can be directly produced by the action on the reproductive organs and sperm quality, or, indirectly, by disturbing reproductive hormone levels [136]. *In vivo* exposure to Cd, will decrease sperm count and impair sperm quality, increase the number of dead and abnormal sperm, as well as reduce testis and epididymis weight [137, 138]. It primarily targets Sertoli cells and impairs spermatogenesis by compromising the adhesion between Sertoli cells or Sertoli–germ cells. Cd will lead to adhesion loss between Sertoli cells and germ cells, and trigger Sertoli cell–germ cell co-culture disruption and alterations in Sertoli cell morphology [139]. However, other cell types in the testes, such as peritubular cells, remain unaffected, which highlights the specific targets of Cd.

9.5 Combined Molecular Toxicity of Heavy Metal Mixtures in the Female Reproductive System

Exposure to heavy metals is also reported to be associated with multiple dysfunctions such as endocrine toxicity, infertility, and spontaneous abortions [140]. An estimated 15% of couples worldwide experience infertility [141]. The effects of heavy metals exposure on female reproductive structure and function should deserve more attention.

9.5.1 The Effect of Heavy Metal Exposure on Reproductive Structure

The ovary plays an essential part in sexual development and female reproduction. The oocyte maturation and fertilization rates are crucial indicators of female reproductive ability. Heavy metals have a detrimental effect on both male and female gonads in adults [142, 143]. Pb exposure caused histopathological changes in oocytes including blurring the zona pellucida, shedding the granular cells and atretic follicles and ultrastructural alteration of ovaries including the rough endoplasmic reticulum swelled and the surface ribosomes disappeared in mice [144].

In addition, the Pb treatment groups significantly decreased both the fertilization rate of matured oocytes and the rate of blastocyst formation compared with the control in mice [144]. Treatment with Pb led to the gonadosomatic index of quails significantly lower than that of the control group [145]. A study of female zebrafish showed that exposure to Hg caused call cavitation, nuclear pyknosis, and loss of cytoplasmic inclusion, and thus delayed the development of oocytes [146]. To sum up, heavy metal exposures have damage on reproductive structure, as indicated by the impaired oocytes and ovaries development and structure.

9.5.2 The Effect of Heavy Metal Exposure on Sex Hormones

Sex hormones play a crucial role in regulating gonadal differentiation and development, promoting follicle development, and stimulating reproductive performance. The main sex hormones are follicle-stimulating hormone (FSH), luteinizing hormone (LH), testosterone (T), and 17β-estradiol (E_2). FSH and LH play a vital role in the expression of steroid-related genes and steroid-dependent feedback [147]. Increasing evidence demonstrate that heavy metals interfere with the synthesis, transport, and release of sex hormones. One study suggested that Cd disturbed follicle growth, differentiation, and steroidogenesis in female germ cells [7]. Besides, some epidemiological studies of women found Cd decreased and E_2/T ratio, with a significantly increase of T [148, 149]. A study about the endocrine-disrupting effect of several metals indicated the positive correlation between blood Pb levels and LH levels as well as LH/FSH ratio values, which may be related to the abnormal hypothalamic-pituitary-gonadal axis (HPG) [150]. Some similar studies also showed that Cd treatment led to a decrease in E_2 and progesterone levels in the serum of postmenopausal women, rats, and hen [142, 149, 151]. Moreover, exposure to Pb also disturbed the release of other sex hormones including FSH, and progesterone, and increased LH and T in women and rats respectively [152]. An apparent positive association with Hg and LH, T, androstenedione in the serum of women was observed [140]. Meanwhile, treatment with Pb significantly suppressed the concentration of E_2 and down-regulated the transcriptional levels of genes involved in ovarian steroidogenesis in female juvenile Japanese quails [109]. There was evidence that the concentrations of E_2 and the E_2/T ratio were reduced in female zebrafish exposed to 3 or 15 μg/L Hg^{2+} [143]. In brief, heavy metal exposure can alter sex hormone levels in female organisms and possibly the reproductive capacity to some extent.

9.5.3 The Effect of Heavy Metal Exposure on Reproductive Performance

Normal reproductive behavior is crucial to maintain the size and age structure of the population [153]. However, reproductive performance was frequently influenced by various heavy metals. Increasing evidence has emerged, indicating the reproductive toxicity induced by heavy metals in mammals [142, 143]. One study suggested that the concentration of endometrial Cd was associated with unexplained infertility in women [154].

Prenatal exposure to Pb, As, and Cr increased the risk of hypospadias [155]. Moreover, Cd decreased the survival of male and female germ cells in humans [7]. Exposure to Hg during early life stages could suppress the reproductive behavior of adult fish, such as reducing the frequency of physical contact in males and resulting in lesser spawning and fertilization [153]. Similarly, Hg exposure can cause lesser fecundity and impair subsequent reproduction in adult zebrafish [143]. Taken together, these results suggest that exposure to heavy metals impairs reproductive performance in females.

9.5.4 The Mechanisms of Heavy Metal Exposure on the Reproductive Function

Although many studies have indicated the toxicity of heavy metals on the reproductive function of humans and animals, the mechanism is unclear. According to the present published studies, here are two possible hypotheses to explain the heavy metal mechanism of exposure to reproductive function (Figure 9.1):

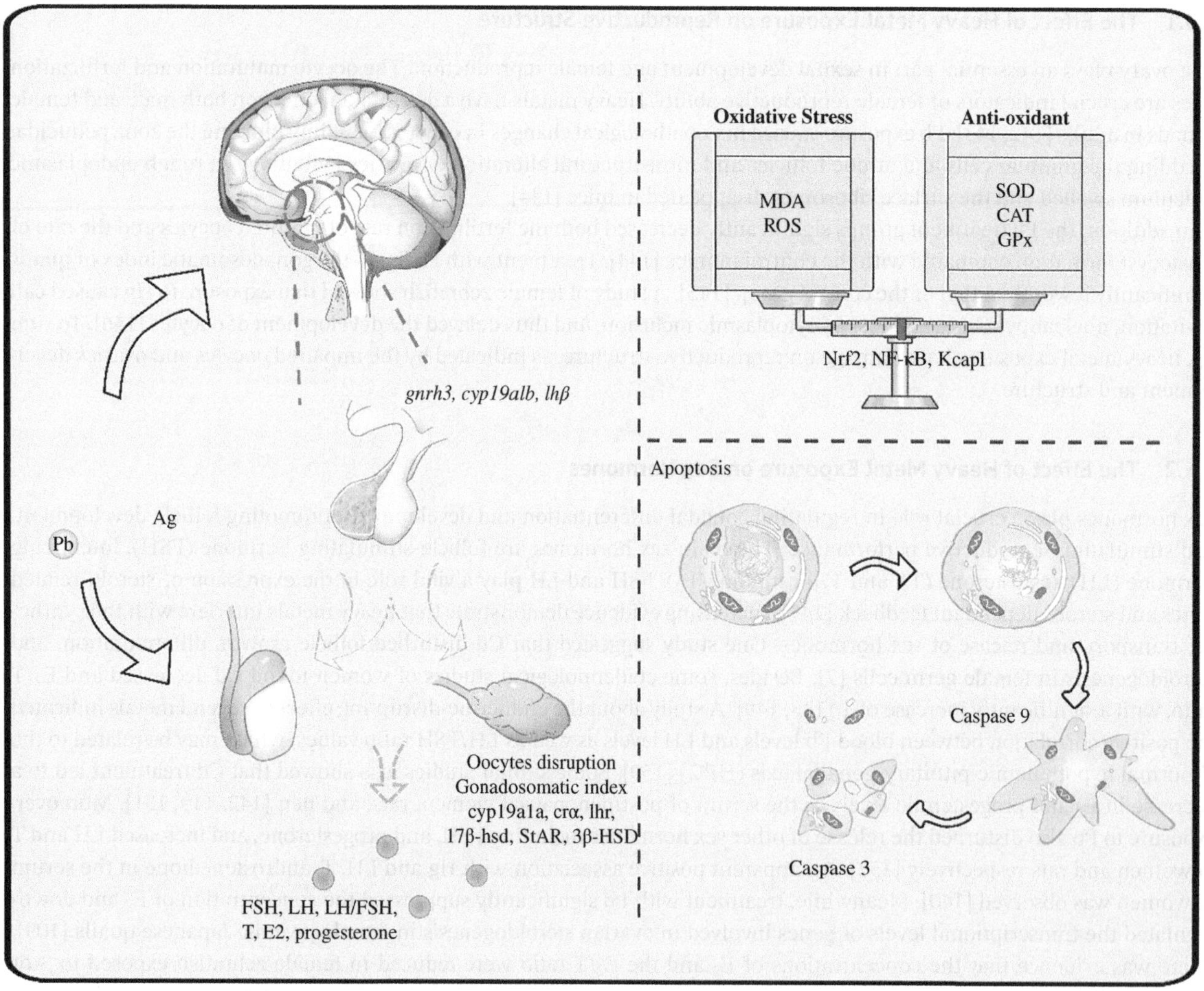

Figure 9.1 Potential toxicological effect of the metal mixture on the reproductive system.

One plausible explanation is that heavy metals (Hg, Pb, Cd, respectively) may disrupt the hypothalamic-pituitary-gonadal axis (HPG-axis) [156]. Reproduction is a complex process primarily regulated by the HPG-axis and steroid. Exposure to Pb disturbed the balance of sex hormones and gametogenesis via altering the expression of mRNA for genes along the hypothalamic-pituitary-gonadal axis, such as downregulated *star, cyp17, 3β-hsd*, and *17β-hsd* genes and upregulated *erα* and *erβ* genes [145]. Similarly, female zebrafish exposure to Hg^{2+} downregulated the expression of mRNA for *17β-hsd* and upregulated the transcriptional level of *vtg1* [146]. In adult females, exposure to Hg^{2+} downregulated *gnrh3, cyp19a1a, cyp19a1b, erα, lhβ*, and *lhr* [143] (Table 9.3). Pb was reported to have influenced the conversion of androgens in ovaries and adrenal glands and consequently on the T levels [145]. The cytochrome P450 aromatase was the only enzyme permitting the conversion of androgens, such as T or androstenedione into estrogens. The exposure of Hg may influence the estrogens synthesis by affecting the cytochrome P450 aromatase activity or the expression of *cyp19* gene [157]. Meanwhile, it is also demonstrated that exposure to Hg increased LH levels in females. GnRH-induced LH levels increased with increasing Hg concentration. These results represent that Hg contamination may disrupt GnRH input to the pituitary [158].

Another potential mechanism involves altering antioxidant defense enzyme systems and oxidative stress. ROS is formed in cells from molecular oxygen during natural metabolism, which is called endogenous source [159]. Heavy metals can increase the production of ROS, which is considered a primary contributor to heavy metal toxicity [8]. Pb and Cd-induced intracellular ROS accumulation on granulosa cells [160, 161]. A study of Hy-line white hens showed that treated with Cd increased the level of malondialdehyde (MDA), but reduced the activity of glutathione peroxidase (GPx) and the content of superoxide dismutase (SOD) in ovary and serum [142]. SOD and CAT were also reduced by Pb treatment [160]. The nuclear factor (erythroid-derived 2)-like 2 (Nrf2) and nuclear factor-kappa B (NF-κB) are the two key redox-sensitive [160]. Nrf2 is a critical regulator of cellular antioxidant response and xenobiotic metabolism [162]. NF-κB involved in the regulation of other processes, including development, cell growth, survival, and proliferation [163]. Keap1 has a negative regulatory effect on the transcriptional level of Nrf2. One report suggested that Pb activated the Nrf2/Keap1 pathway and impaired oocyte maturation and fertilization by inducing oxidative stress. Exposure to Pb increased the expression of *Nrf2, CAT, GST, HO-1, GPX, GCLC, GCLM*, and *SOD1*, and suppressed the *Keap1* in mouse oocytes [144]. In addition, exposure to Pb elevated the protein expression levels of Nrf2 and NF-κB. Pb-induced oxidative stress may be through disrupting Nrf2/NF-κB interaction [160].

Caspases are an evolutionary conserved family of cysteine proteases that play a vital role in mediating cell apoptosis [164]. The activation of caspase is mainly dependent on cytochrome c (Cyt-c) [165, 166]. In the Cyt-c-initiated apoptosis pathway, the release of Cyt-c from mitochondria was recruited and activated initiator caspases such as Casp-9, which in turn activated executioner caspases such as Casp-3 and resulted in cell death [166]. Meanwhile, Bcl-2 could inhibit the release of Cyt-c from mitochondria and prevent programmed cell death [165]. The apoptotic index of high-dose Pb exposure was all significantly higher in the ovary. Levels of Cyt-c, Casp-3, and Casp-9 mRNA were elevated, while levels of Bcl-2 mRNA were decreased [145]. The inhibited release of progesterone-induced by Hg may be associated with an increasing trend of apoptosis of porcine ovarian granulosa cells [167]. Most of the apoptotic cells in the ovary were found in the granulosa cells and interstitial cells of the ovarian region after treatment with Cd [142]. Cd increased the number of germ cell apoptosis in female mice. The activation of the apoptotic pathway is highlighted by the increased rate of cleaved caspase-3– and caspase-9–positive germ cells [7]. The apoptosis was significantly enhanced by Co, Ag addition in ovarian granulosa cells via increasing caspase-3 expression [168, 169]. Increasing apoptosis may represent programmed cell death disturbance, which may induce ovarian dysfunction and abnormal development [145].

Table 9.3 Dysregulation of genes related to the reproductive system after exposure to metal mixtures.

Genes	Species	Organ	Regulation
17β-hsd	Zebrafish	Ovary	Down
cyp19a1a, erα, lhr	Zebrafish	Ovary	Up
cyp19a	Banded Gourami	Ovary	Down
vtg1	Zebrafish	Liver	Up
gnrh3, cyp19a1b, lhβ	Zebrafish	Brain	Down

9.6 Combined Molecular Toxicity of Heavy Metal Mixtures in the Liver

Heavy metals are widely present in our immediate environmental contaminants, and a growing number of studies have shown that metal and mixed metal exposures can elaborate liver toxicity. Humans can be exposed to metals and metal mixtures through the air, water, and food. The study showed synergistic, antagonistic, or additive effects in the normal human liver cell line HL7702 by combined exposure to different metals [170].

However, the toxic effects of metal and metal mixture exposure on the liver are complex, and the mechanisms remain unclear. The liver is the main detoxification organ in mammals, plays an important role in handling metals, and is one of the first organs to be damaged by toxic substances. When absorbed, most metals and their combinations usually travel quickly via the bloodstream to the liver and build up there.

Epidemiological studies have shown that heavy metal exposure is associated with an increased risk of liver injury. In a study that quantified the relationship between urinary heavy metals and ALT (alanine aminotransferase, a biomarker of liver injury), it was shown that urinary V, Cr, Cu, As, Cd, W, and Pb showed a positive dose–response relationship with ALT in a single-metal model, respectively. In the mixed exposure model, including the seven metals mentioned above, V and Cu remained positively associated with ALT. Plasma CRP concentrations were positively correlated not only with urinary Cu and Cd but also with ALT. In addition, mediation analysis showed that CRP mediated 4.7% and 7.0% of urinary Cu and Cd-related ALT elevations [171]. Environmental Cd exposure was reported to be associated with hepatic necroinflammation, NAFLD, and NASH among 12 732 American adults [172]. In another study, Cd, Hg, and Pb were all positively correlated with liver enzymes and the NAFLD index [173]. A study of the relationship between heavy metal exposure and early liver dysfunction in rural residents aged 40–75 years in southwest China, showed that the blood levels of Fe, Cd, and Cu in the Cd-contaminated area were positively correlated with elevated ALT levels, and the blood levels of Cu were positively correlated with AST (aspartate aminotransferase, a biomarker of liver injury) levels, and the highest quartiles of Zn and Mg in the contaminated area were positively correlated with reduced ALT levels. These results suggest that industrial pollution leads to the exposure of Cd and Pb heavy metals, and the role of Fe, Cd, Cu, and Pb in Cd-contaminated areas may be the main contributor to increase the risk of liver dysfunction, while Zn in Cd-contaminated area and Mg in the uncontaminated area may be protective factors [174].

For the current studies on metal exposure-induced hepatotoxicity, the main mechanisms include hepatic metabolism, bile acid homeostasis, lipid metabolism, endocrine metabolism, and to be specific, mitochondrial function and oxidative metabolism, and ferroptosis (Figure 9.2).

As and Cu jointly interfere with ATM signaling pathways, and low levels of As and Cu exposure may promote metaplasia and carcinogenic progression, while high levels of exposure affect liver metabolism, leading to dysfunction of key organelles, such as mitochondria and lysosomes, genomic DNA damage, and impairment of cell growth and survival [175]. Sub-lethal exposures to Hg and Cd affected hepatic metabolism in gilthead sea bream by decreasing PUFA, increasing MDA, altering membrane fatty acid composition, and thus reducing total lipid levels and inducing oxidative stress [176].

Tl can accumulate in the liver and induce hepatic sinusoidal congestion and hepatocyte necrosis. In addition, both Tl (III) and Tl (I) exposure significantly affected FAs concentrations, especially unsaturated FAs. Many β-oxidation pathway-related genes were upregulated after Tl exposure, indicating an enhanced requirement for energy provision via FAs oxidation [177].

Nonalcoholic fatty liver disease (NAFLD) has gained more and more attention in toxicologic studies. Nearly half of the interactions between the 18 NAFLD-related genes were found to be co-expressed. Key molecular pathways associated with the pathogenesis of NAFLD caused by the heavy metal combination (Cd, Pb, and Hg) include activated oxidative stress, altered lipid metabolism, and increased cytokine and inflammatory responses [173]. The study showed that prolonged exposure to Pb-Cd caused oxidative stress in the liver of mice, resulting in increased SOD, LDH activity, and MDA levels. Pb-Cd also interferes with the regulation of the metabolism of linoleic acid, sphingolipids, and glycerolipids and causes disturbances in lipid metabolism. These results led to a significant increase in hepatic AST and ALT enzyme activities, hepatomegaly, and pathological liver changes in mice [178]. Cd and As can enter the body well in the liver and have deleterious effects on carbohydrate metabolic pathways such as glycolysis, glycogen, and gluconeogenesis by modifying and impairing related key enzyme activities. Disruption of hepatic glucose homeostasis plays a crucial role in the development of diabetes, and reduced liver function contributes largely to elevated blood glucose levels. The mechanism may be that these metals can potentially bring about conformational changes in these enzymes, rendering them inactive [179]. In addition, these metals promote the production of reactive oxygen species, inhibit antioxidant defense mechanisms, and produce hepatotoxicity. Another study showed that As and Sb co-exposure caused significant mitochondrial damage, autophagosomes, hepatic congestion, and nuclear sequestration in mouse liver. In addition, combined ATO and Sb exposure disrupted lipid metabolism and steatosis in liver tissue. Moreover,

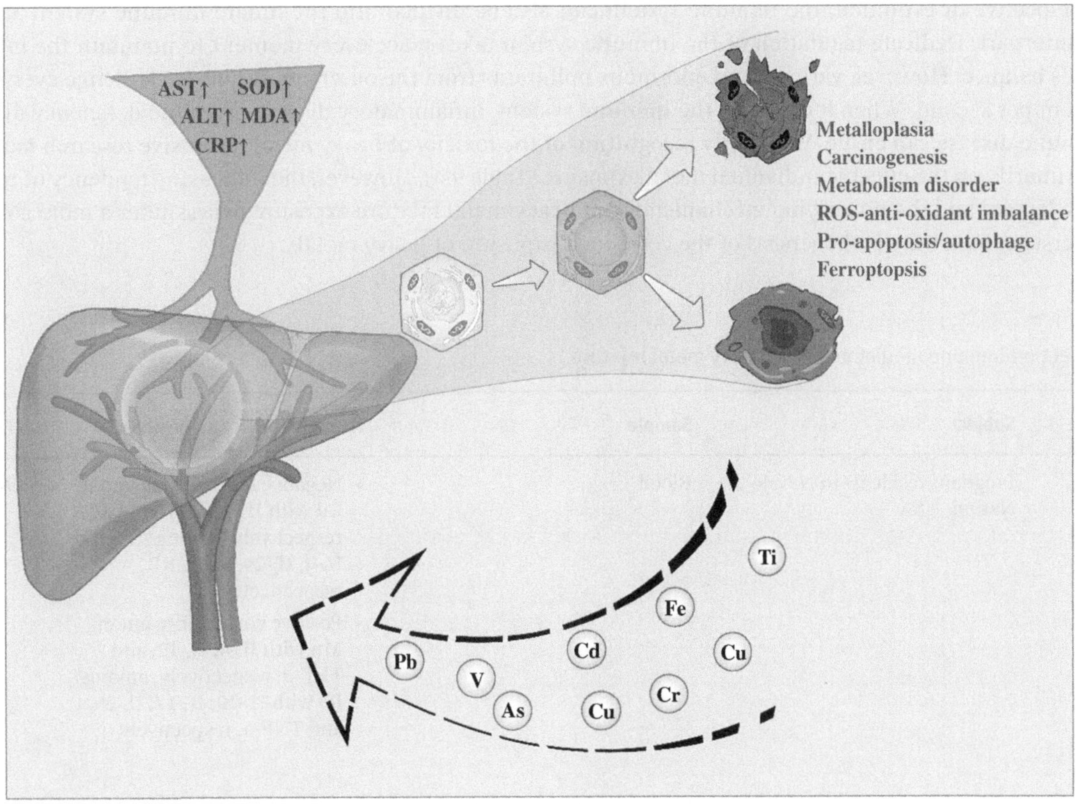

Figure 9.2 Effects of heavy metals' exposure on the liver.

the levels of the pro-apoptotic index (Caspase-3, Caspase-9, Bax, P53, Cytc) and autophagy index (LC3B, P62, PINK1, Parkin) were significantly increased in the co-exposure group. In contrast, the level of the anti-apoptotic index (Bcl-2) was significantly decreased. It indicates that co-exposure to ATO and Sb can cause abnormal hepatic energy metabolism and oxidative stress [180].

Ferroptosis caused by metal exposure is one of the latest studied mechanisms [181]. The liver plays an essential role in the storage of iron. Under pathological conditions, iron absorption from the intestine or iron uptake into hepatocytes increases, and excess iron accumulates in hepatocytes, leading to hepatocyte damage through the production of free radicals. Iron produces toxicity by catalyzing the production of ROS. ROS causes cellular damage by inducing damage to lysosomal, cytoplasmic, nuclear, and mitochondrial membranes, apoptosis by activating the caspase cascade, and hyperoxidation of fatty chains [182]. Thus, mild to moderate hepatic iron overload causes liver burden and increases the risk of developing NAFLD to nonalcoholic steatohepatitis and eventually cirrhosis [183]. Studies have shown that Cd can damage organs such as the liver, and ferroptosis may be a newly discovered regulatory cell death mode of its action. Cd exposure may cause ERS to promote ferroptosis by activating the PERK-eIF2α-ATF4-CHOP signaling pathway, which would further induce oxidative stress triggering cellular autophagy, endoplasmic reticulum stress, and ultimately hepatotoxicity [184].

Long-term exposure to metal mixtures at concentrations below environmental standards can produce toxic effects. The toxic effects and possible health hazards of metals and their mixtures in the environment-relevant concentration levels are of concern. This has become a key area of current ecotoxicology and environmental health research. To reveal the molecular mechanisms of toxicity and health hazards of metals and their mixtures, for example, studying the corresponding alteration of gene expression and pathway after exposure, are critical issues in current scientific research.

9.7 Combined Molecular Toxicity of Heavy Metal Mixtures in the Immune System

The immune system functions as a complex protection barrier of organisms, usually composed of immune surveillance, defense, and homeostasis. Different organisms possess different elements to build their immunity. Besides immune cells, there are specialized immune tissues, and even immune organs in advanced species, like mammals.

From the perspective of evolution, the immune system can also be divided into the innate immune system and the adaptive counterpart. Dedicate regulation of the immune system takes place every moment to maintain the internal environment's balance. However, various, and enormous pollutants from the outer environment challenge every creature per system per second. When it comes to the immune system, inflammatory disease, immune deficiency disease, and autoimmune disease can ensue. With early recognition of the toxicity of heavy metal, extensive research has been conducted primarily on the effect of individual metal exposure (Table 9.4). However, the increasing tendency of reports regarding the hazard and the underlying mechanism from heavy metal mixture exposure necessitates a more comprehensive understanding of the pathogenesis of the combined exposure of heavy metals.

Table 9.4 Effect on immune mediators of the heavy metal mixture.

Metal mixture	Subject	Sample	Immune mediators involved	Ref.
U, As, Cd, and Mn	Pregnant residents in Navajo Nation, USA	Blood	• Negative correlation: among Cd with IL-4 and TNF-α, respectively; among As with IL-4, IL-29, and CRP, respectively • Positive correlation: among Mn with IL-4, IL-12, and TNF-α, respectively; among Pb with IL-10, IL-17, IL-29, and TNF-α, respectively	[194]
Murine				
Cr, Mn, and Ni (soluble welding fume)	Sprague–Dawley rats	Lung Bronchoalveolar lavage fluid	• Increase in IL-6 • Decrease in TNF-α and IL-2	[222]
W, Ni, Fe, and Co	Sprague–Dawley rats (Adult male)	Lung Bronchoalveolar lavage fluid	• W, Ni, and Fe: BAL sample: increase in CINC-1, CINC-3, TNF-α, IL-1β at the high-dose group • W, Ni, and Co: increase in CINC-1, CINC-3, TNF-α, IL-1β. IL-6 at the high-dose group	[224]
Au, Zn	Sprague–Dawley rats (male)	Spinal cord	Decrease in IL-1β, TNF-α	[233]
Ag, and Ti	C57BL/6JRj mice	Jejunum	Downregulation of *Tnfα, Ifnγ, Il1β, Il17a, Il22, IL10, and Cxcl1* (with other nanoparticles [TiO_2, SiO_2])	[225]
Cu and Zn	Swiss Albino mice	Blood	Decrease in IgM and IgG	[232]
Fish				
Cd, Cr, Cu, Hg, Ni, Pb, V, Zn, and Mn	Juvenile turbots	Blood	Increase in TGF-β	[186]
Cr, Cd, and Pb	Common carp	Organs	TNF-α, IL-6, and IFN-γ were significantly expressed in the liver, kidney, and intestine with a rough time-dependent trend	[219]
Hg, and Pb	*Clarias gariepinus*	Blood Kidney	• Decrease in C3 in blood • Decrease in *il-8, il1b*, and *tgfb* in kidney	[203]

Metal mixture	Subject	Sample	Immune mediators involved	Ref.
Pb, Hg, As, and Cd	Zebrafish embryo	Embryo	• Pb with As upregulated *il1b* while Pb with Hg and Cd downregulated *il1b* • As itself or with Pb, suppressed *tnfα* • Upregulation of *ifng1* in the metal mixtures exposure groups compared to the individual exposure groups	[228]
Hg and Fe	*Nile tilapia*	Blood	A decrease in IgM could also be seen after exposure to HgCl, but it could be rescued in the context of co-exposure with Fe_3O_4 nanoparticles	[205]
Others				
Cu and Zn	Weaned piglets (Duroc × Landrace × Yorkshire)	Gut	With montmorillonite as carrier: • Decrease in IL-1β in small intestine. • Decrease in IL-6, IL-10, and TNF-α in duodenum	[234]
Cu and As	Hy-line chicken	Brain	More intense change in co-exposure group than individual exposure: • Upregulation of *IL2, IL8, IL12B,* and *IL17* in the cerebrum and cerebellum. • Downregulation of *IFNG, IL4, IL6,* and *IL10* in the cerebrum and cerebellum.	[235]
Al, Fe, As, Va, Zn, Pb, Cd, and Cu	*Ciona robusta* (Adult)	gut	Upregulation of *TNFα*.	[187]
Cd and Pb	*Bufo gargarizans*	Gut	• Upregulation of *IL1β, IL17D* • Downregulation of *IL27*	[236]
Cell or organoid				
Ag, and Ti	organoid	Gut	Upregulation of *Ccl20, Cxcl1, Il33.* (with other nanoparticles [TiO2, SiO2])	[225]
Fe, Cr, Ni, Mn, Mo, etc. (stainless steel)	Lymphocytes and monocytes/ macrophages sorted from human peripheral blood	Leukocyte	An increase in IL-1β, compared to other metal groups (Ti and its alloy), and a reverse trend were observed in the TNF-α level after 1-day intubation	[221]
As and U	CD4 T cells from the blood of adult male humans (pre-activated by CD3)	Lymphocyte	Downregulation of *IFNG* in all groups, but with higher amplitude in the mixture group	[210]
Hg, Pb and Cd	Jurkat cells (pre-stimulated by CD3)	T cell line	Decrease in IL-2 by Hg or Pb but increase in Cd, but no remarkable result in the mixture group	[208]
Au and Zn	RAW264.7	Microglia cell	Decrease in IL-1β, TNF-α	[233]

Note: The meaning of the abbreviations in uppercase: IL: interleukin; TNF: tumor necrosis factor; CINC: cytokine-induced neutrophil chemoattractant; IFN: interferon; IFNG: interferon-gamma; Cxcl: C-X-C motif chemokine ligand; Ccl: C-C Motif Chemokine Ligand in capital form.

9.7.1 Ecotoxicological Study of Immunotoxicity of Heavy Metals Mixture

The contaminated water body is a common source of real-world heavy metal mixture. Seawater rich in Cd, Cu, Pb, and Zn, in the Sørfjord, Norway, stressed *Asterias rubens* (common sea star), induced expression of cognate form of heat shock protein 70 in coelomocytes, immune cells of sea star [185]. By collecting contaminated sediments on the French coast, researchers exposed four-month juvenile turbots to a mixture of Cd, Cr, Cu, Hg, Ni, Pb, V, Zn, and Mn [186]. After a period of exposure (21 days), liver ethoxyresorufin-O-deethylase (EROD), glutathione S-transferase (GST), and catalase (CAT) were remarkably increased in the group of higher concentration of metal mixtures, indicating an imbalance of the oxidant-antioxidant system. Splenic melanomacrophage centers, an indicator of phagocytosis, also increased. Besides, an increase in the plasmatic TGF-β level was noted and collectively interpreted as immunosuppression since fundamental immunities were insufficient. Another research considered turbulence in the water body and found that adult *Ciona robusta*, a species of marine invertebrate, exposed to sediments composed of a mixture of metals (mainly Al, Fe; As, Va, Zn, Pb, Cd, Cu) and polycyclic aromatic hydrocarbons, induced inflammatory response, evidenced by the increased transcriptional level of TNF-α compared to control group while turbulence generally upregulated expression of immune-related genes [187]. Closure of crypts lumen and increased mucus production were noticed in the contaminated water group, suggesting a compromised innate immune system. Collecting mussels from different sites around the Baltic Sea, researchers showed that the phagocytic index of granulocytes was elevated and hemocytes (immune cells in invertebrates) were decreased in the mussels from the contaminated site with a generally higher concentration of heavy metals (Zn, Cu, Ni, and Pb) [188]. Effluents from a phosphate treatment plant near the Gulf of Gabes in Tunisia, with Cd, Cr, and Zn exceeding norms, impaired the capacity of production of reactive oxygen species, the phagocytosis activity, and even the viability of hemocytes isolated from *Haliotis tuberculate* [189].

Pollutant soils are another source of the heavy metal mixture. Earthworms generally had high levels of Cd, Cu, Zn, and Co, which induced an inflammatory response [190] and weakened their immunity [191]. However, mice in the contaminated sites, enriching Cd, Cu, Pb, and Zn, in the New Jersey Zinc/Mobil chemical locality reacted insignificantly to the immunization of sheep red blood cells with no significant difference in the number of splenic plaque-forming cells and proportion of splenic B cells [192]. The considerable individual variation in the research might be the clue. Dust in the air could be an inconspicuous carrier of contaminants. By intratracheally instilling with residual oil fly ash (ROFA), purified constituent metals (Ni, Fe, and V), and the mixture of purified metals, Brown Norway (BN) rats were set as the model of house dust mite (HDM) allergy and serum HDM-specific IgE and expression of pulmonary IL-5 and IL-13 were remarkably increased in the ROFA and mix group after 21 days [193]. Meanwhile, lymphocyte proliferation was also increased in the group of ROFA, Ni, V, or metal mixture, and eosinophil count in the bronchoalveolar lavage fluid was raised in the group of ROFA or Fe, compared to the HDM group (positive control). While TNF-α in bronchoalveolar lavage fluid was increased in all groups, other cytokines changed differently with exposure to the different metals, indicating distinct reactions brought by individual metals and overall additive effects in the metal mixture.

Besides, anthropogenic activities can introduce enormous heavy metal elements into the environment or concentrate initially trace metals. Improper or even absent decontamination during mineral exploitation and processing causes chronic exposure to residents for decades, such as people living in the Navajo Nation in the U.S. near obsolete uranium mines [194]. With a higher concentration of U, As, Cd, and Mn, compared to the population from The National Health and Nutrition Examination Survey (NHANES), Navajo Nation pregnant residents were recruited in the study. Analysis of the cytokines panel demonstrated a negative correlation among Cd with IL-4 and TNF-α, respectively, and among As with IL-4, IL-29, and CRP, respectively, but a positive correlation among Mn with IL-4, IL-12, and TNF-α, respectively, and among Pb with IL-10, IL-17, IL-29, and TNF-α, respectively. The comparison of cytokines between the two populations was not shown. Thus, the overall influence of metal mixture could not be inferred due to the distinctive immune-interfering effect among individual metals. The pollutant air, like fume during welding, also posed a hazard to the human body, with alteration of gene expression of the inflammatory response, oxidative stress, et al. [195]. Noticeably, downregulation of *IL8, IL1α, CXCR4, CD1D, IFNGR1, CAT, GNLY*, and *GZMA*, and upregulation of *CD97, LSP1*, and *GPX1*, might suggest depressed immune response. An epidemiological study of Pakistan lymphoma patients identified a significant elevation of Pb, Ni, Cd, Cu, and Cr in the nails of non-Hodgkin lymphoma patients and a significant elevation of Pb, Cu, Cd, and Cr in the nails of Hodgkin lymphoma patients compared with healthy volunteers, indicating potential effects of exposure to metals in the pathogenesis of tumor of the immune system [196].

9.7.2 Combined Toxicity of Heavy Metals Mixture in Immune Cells

The immune responses regulated by immune cells are often complicated. Integrative analysis of data from NHANES and the Comparative Toxicogenomics Database revealed population exposure and peripheral immune status [197]. Both Cd and Pb positively contributed to neutrophil and lymphocyte counts in the WQSR model. The overall effect of the mixture of heavy metal (mainly Cd, Pb, Hg, As) and organic pollutants favored lymphocyte-mediated response, and the possible shared pathway, the IL-6 activation pathway, was suggested through analysis. Exposing male albino Wistar rats to formulated concentrations of As, Cd, Hg, Cr, Ni, Mn, and Fe, according to the ratio of abundance of mixed metals in contaminated water body, researchers found decreases in the total leukocyte count and lymphocyte percentage in blood, and depletion of lymphoid cell population in the spleen in the high dose group after 90-day exposure [198]. The increase of leukocyte counts toward neutrophils rather than lymphocytes could also be observed in the blood of gudgeons exposed to a mixture of Cd, Cu, and Zn [199].

The immunotoxicity of Hg has been widely reported [200]. Individual Hg exposure decreased viability and proliferative response of lymphocytes at relatively higher concentrations, and such suppression could also be noticed in the combination of Hg with Cd and Pb respectively, or collectively [201–203]. La (lanthanum chloride), as a calcium channel blocker, could decrease Hg-induced lactate dehydrogenase (LDH) level in the B lymphoma cell line (A20), which could be counteracted by a calcium chelating agent [204]. Combination exposure suggested that B-cell cytotoxicity of Hg was possibly mediated through Ca^{2+} dysregulation. Antagonistic effects could also happen when Hg was mixed with certain adsorbable metal particles, such as magnetite (Fe_3O_4) [205]. After exposure to HgCl and Fe_3O_4 nanoparticles for 30 days, the residual Hg in the muscle of *Nile tilapia* significantly dropped to a level similar to that of the control group. Combined with HgCl, Fe_3O_4 nanoparticles mitigated the inflammatory response of neutrophils and lymphocytes, while reducing hepatotoxicity and nephrotoxicity resulting from single HgCl exposure. Hg alters the cytokines production of lymphocytes [206, 207]. When researchers exposed Jurkat T cells, which had been pre-stimulated by CD3, to different metals, they found that Hg or Pb (but not Cd) induced significant drops in IL-2 concentration starting from 4 hours after exposure. Furthermore, only Hg induced apoptosis at all sublethal experimental concentrations [208]. Besides, tyrosine phosphorylation was remarkable in all metal treatment groups, like that of the positive control group. However, no significant change of IL-2 by T cells was noted when treated with the mixture of metals at 1 μM concentration. Given the different effects of individual and combining metal(s) on T cells, antagonistic interaction was assumed. Pretreatment with PMA (Phorbol 12-myristate 13-acetate)/calcium ionophores increased the luciferase activity of transfected pNFATluc plasmid targeting at IL-2 in T cells exposed to Cd but diminished it in the single Hg or Pb group. Collectively, it's indicated that the potential target might be located at upstream of protein kinase C (PKC) activation or calcium mobilization since PMA activates PKC and calcium ionophores enable calcium transportation (Figure 9.3).

The advancement of total hip replacement brings internal exposure of heavy metals, like Cr, Co, and Ni, to patients, as indicated in a preliminary study [209]. The researcher calculated the total exposure amount of metals over several months or years. They then subcutaneously injected the estimated amount of metals into the hind footpad of female BALB/c mice. Following this, they assessed the proliferative response of the popliteal lymph node (PLNA). PLNA significantly proliferated in the relatively high doses of metal salt mixture ($CoCl_2$, $NiCl_2$, and $CrCl_3$), and Cr_2O_3 particles collectively with metal salts induced PLNA proliferation while individually not. Similar trends were observed in the amount of activated CD4 and CD8 T cells in the PLNA. CD4 T cells, isolated from the blood of adult male participants, were exposed to a series of concentrations of As and/or U after activation by CD3/CD28 [210]. Compared to the single U group, CD4 T cells exposed to As and As with U exhibited significant stress response, indicated by the levels of HMOX1, NQO1, GCLM, and SOD1. The transcriptome analysis showed that at the lower dose that was environmentally relevant, significant decreases in expression of the T cell activation marker genes were noted both in the mixture group of As and U and single As group, while the higher amplitudes of decrease fell into the mixture group and no significant reduction was present in the single U group. Therefore, a synergistic effect was consumed. Moreover, interferon-gamma (IFN-γ) was primarily reduced, disfavoring of Th1 phenotype and macrophage activation [211].

A total of 102 differentially expressed genes (DEGs) were shared between the mixture group and the single As group in the transcriptome analysis. DEGs of single As group at lower dose enriched mainly at the following immune-related biological processes: "*regulation of interleukin-2 production*," "*alpha-beta T cell activation*," "*Interferon-gamma production*" and "*type I interferon signaling pathway*" while DEGs of mixture group shared almost all the above-mentioned processes and enriched in: "*Leukocyte homeostasis*" and "*Regulation of cell killing*" processes. Meanwhile, mixing

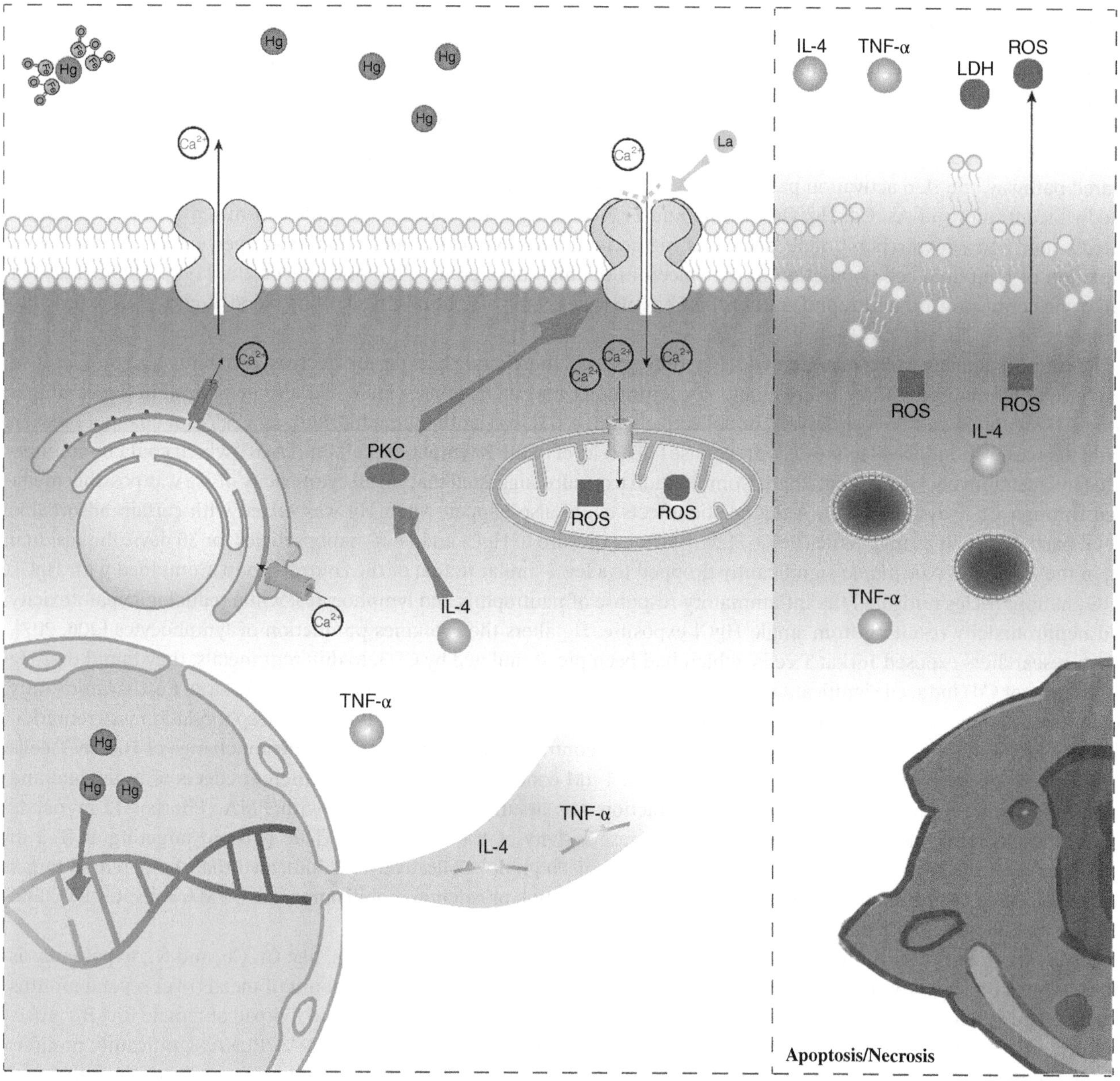

Figure 9.3 Toxicological effect of mercury on lymphocyte and interaction with other metals (Fe and La).

U with As significantly altered several metabolism pathways, including detoxification of ROS (reactive oxygen species), polyol metabolic process, phosphorylation of CD3 and TCR, and RORA-related or FOXO-related transcription, etc. The RORA is involved in the CD4 T cell differentiation [212], indicating an additive effect of U targeted at the RORA activation pathway during T cell activation. Through Reactome Analysis, the researcher concluded that adaptive immune processes altered by a mixture of As and U at environmentally relevant concentrations, included BCR/TCR signaling, co-stimulation by CD28, interaction with other lymphoid or non-lymphoid cells, RAP1 signaling and MHC I/II antigen presentation, highlighting perspectives of further research. Evaluation of an independent action model could be implemented in transcriptome analysis of such mixed contaminants to determine synergistic or antagonistic effects [213].

Other immune cells were affected by heavy metal exposure differentially. Fortier et al. found distinguished drops of human peripheral natural killer cell activity challenged by Hg at various concentrations but not by the mixture of Cd and Pb in vitro [201]. However, Pb and Cd with other industrial pollutants increased natural killer cell lytic activity at lower doses but decreased at higher doses [202]. Human monocyte sorted from peripheral blood mononucleated cells (PBMC), when under the challenge of WC (tungsten carbide) and Co, mainly upregulated the following biology processes "*nucleosome, chromatic assembly*," "*stress response*," "*glucose metabolism*," and "*apoptosis regulation*," but downregulated biology processes like "*cell-cell signaling*," "*immune response*," and "*defense response*." The upregulated genes in the apoptosis pathway are *ANGPTL4, BNIP3, EGLN3, MIF, NUPR1, SPP1*, and *SQSTM1* [214]. Besides, *HMOX1* gene was significantly expressed in the PBMC treated by WC and Co, and HIF-1α and p53 were stabilized, enabling carcinogenic potential [215]. Heavy metals, like Ni, could interfere with oxidative stress introduced by monocytes, thus compromising their anti-pathogen capacity [216]. But combining with Mn, such suppression could be relieved whether monocytes were in the inactivated or activated state. In Fortier's study, there was no remarkable change in the phagocytic activity of monocytes in Hg, Cd, or Pb groups respectively, or collectively [201]. While Ag nanoparticle-induced phenotype transition of tumor-associated macrophage from M_2 to M_1 [217], Ag nanoparticle with $AlCl_3$ at subtoxic level did not significantly interfere with phagocytic activity and M_1/M_2 polarization of THP-1 cell in vitro after 24 h exposure, denoted by trace change of expression of markers of M_1 (IL-1β, TNF, and CXCL9) and M_2 (MRC1 and DCSIGN) [218]. No significant expression of activation markers and phagocytic activity of human blood dendritic cells were also observed after exposure to a combination of Ag and Al.

9.8 Combined Toxicity of Heavy Metals Mixture in the Regulation of Immune Mediators

The immune system allocates and mobilizes different constituent cells through a sophisticated mediator system, including cytokines, chemokines, acute-phase proteins, immunoglobulins, etc. (Table 9.4). After exposure to mixtures of Cr, Cd, and Pb, TNF-α, IL-6, and IFN-γ, with antioxidant enzyme, CAT and SOD, were significantly expressed in the liver, kidney, and intestine of common carp with a rough time-dependent trend [219]. Particles of stainless steel 316 L, Ti, and its alloy (Ti_6Al_4V), the composition of implants, were supplemented into lymphocytes and monocyte/macrophages sorted from human peripheral blood to study the immune mediators panel [220]. Stainless steel, composed of Fe, Cr, Ni, Mn, and other heavy metals [221], significantly increased the production of IL-1β, compared to other metal groups, while a reverse trend was observed in the TNF-α level after 1-day intubation. The expression of other cytokines, IL-1α, IL-2, IL-4, IL-6, IL-8, and IFN- γ were not significantly changed. Stainless steel welding can cause aerosol exposure to a mixture of heavy metals. Intratracheal instillation of soluble welding fume (Cr, Mn, and Ni), and its constitutive metals individually or collectively into SD rats, bacteria colonized significantly in the lungs of the welding fume group, the mixture group, and single Cr group, indicating a predominant role of Cr [222]. Upregulation of IL-6 and downregulation of TNF-α and IL-2 indicated impairment of the phagocyte system and direct killing of cytotoxic T cells after exposure. Welding fumes exposure increased total leukocyte and monocyte counts, as well as oxidative stress in monocytes, 24hours after intratracheal exposure in rats [223]. However, the observed decrease of expression of Cxcl10, Ccl4, Cxcl2, and TNF-α in leukocytes in vitro, 24 hours after challenge by lipopolysaccharides, implied potential immunodeficiency as well. Compared to the exposure of each constitutive metal, the mixture group demonstrated a remarkable change in ROS, inflammatory cytokines, and pulmonary inflammatory cell infiltration, underlying synergistic toxicity, in a study of intratracheal exposure of W, Ni, Fe, and Co [224].

Ti and Ag are widespread nanoparticles in the atmosphere, and the mixture of them with SiO_2 suppressed the significant upregulation of *Tnfα* by individual Ti exposure, significant upregulation of *Ccl20* by individual Ag exposure, and significant upregulation of *Ccl5* and *Cxcl10* by individual Ti or Ag exposure but amplified the upregulation of *Il33* in the gut organoid model [225]. However, in the gut of male mice, the mixture group decreased expression of *Tnfα, Ifnγ, Il1β, Il17a, Il22, IL10*, and *Cxcl1* in the jejunum, but increased expression of *Tnfα* and *Il10* in the colon, while the patterns of these cytokines were different in the female mice [225].

Heavy metals, as potential teratogens, are concerns of pregnant women. Positive correlation between the concentration of heavy metals (Ni, Cu, Zn, Sn, Pb) and MCP-3 (monocyte chemotactic protein 3) in the first trimester and at term, and between the concentration of heavy metals (Ni, Zn, Sn) with IL-8 at first trimester in one Michigan study [226]. Besides, the concentration of plasmatic IL-1β was also positively associated with Mn, Ni, and Ba [227].

Exposing a mixture of Pb, Hg, As, and Cd at the concentration of the upper limit of Chinese drinking water standards to zebrafish embryos for 120 hours, disturbed the regulation of innate immunity by altering the transcriptional level of immune mediators [228]. Synergetic or antagonistic effects were observed. Pb with As upregulated interleukin 1, beta (*il1b*) gene while Pb with Hg and Cd downregulated it. Arsenic itself or with Pb, suppressed the expression of *tumor necrosis factor-alpha* (*tnfα*) gene. Expression of interferon-1γ 1 was higher in the metal mixtures exposure groups than in the individual metal exposure groups. Oxidative stress in individual metal or mixture groups was obvious, noted by decreased

Mn-SOD and Zu/Cu-SOD, and increased CAT as well as Glutathione peroxidase (GPx) in the study. It has been widely demonstrated that relationship between oxidative stress and immune response, especially when oxidative stress occurs in the immune cell [229]. Oxidative stress could be synergistically enhanced while detoxification system cytochrome P450 was suppressed when exposed to a mixture of Zn and Cu, compared to single exposure [230]. But there could be no significant change of antioxidant defense in the gudgeons challenged by a mixture of Zn, Cu, and Cd [199], and even reverse change of antioxidant activity in the *Clarias gariepinus* challenged by Hg and Pb [203], when the levels of oxidative stress were significant in both studies.

Immunoglobulin, components of the humoral immune response, could be influenced by metal exposure. Cu and Fe, heavy metals as well as trace metal elements, were added to gestating beef cattle at different concentrations in combination but no significant difference in humoral response to xenogeneic gamma-globulin was shown [231]. Oral administration of a mixture of Cu and Zn to Swiss Albino mice for 14 days perturbed the immune-biochemistry panel, as indicated by a decrease in the levels of the acute phase immunoglobulin IgM and the chronic counterpart IgG [232]. Together with hepatotoxicity, such mice tended to be immunodeficient. A decreased level of IgM could also be seen after exposure to HgCl, but it could be rescued in the context of combination with Fe_3O_4 nanoparticles [205].

9.9 Combined Toxicity of Heavy Metals Mixture in the Immune Response to the Immunogen

Organisms face diverse environmental threats, including chemicals, allergens, and pathogens. Proper immune response is vital, especially in sensitization or immunization. Metals themselves can act as immunogens. For example, Ni can induce hypersensitivity via TLR-dependent or TLR-independent pathways, i.e. adaptive or innate immune response [237]. Pre-sensitizing mice with a series of metals intraperitoneally individually and measuring the allergy after a second exposure to the same series of metals individually, researchers found that cross-reaction was identified in the Ni and Pd, and that exposure to a mixture of Ni and Pd drastically induced allergic reaction in Ni-sensitized mice compared to single Ni or Pd exposure [238]. A potentiating effect of delayed hypersensitivity reaction by mixing Ni with Co in mice could be found, indicating Ni is a potent adjuvant in metals mixture to induce immune response [239]. However, co-sensitization should be considered, as low-purity-Ni-sensitized mice also reacted noticeably with other metals of low purity [238]. Cd exposure could interfere with immune response via dysregulation of inflammatory cytokines, like TNF-α [240], and interleukins [241]. Cd with Pb at 1ppm significantly suppresses the humoral response of mice to sensitization of human serum with a significant drop of antibody titer, which could be restored by *Nigella sativa* (black Cumin) [242]. *Nigella sativa*, an antioxidant, could potentiate T cells to activate the immune system [243], suggesting oxidative stress is the underlying target of Cd-Pb toxicity. A similar result could be found after treating mice exposed to Cd and Pb with garlic, highlighting the immunomodulatory role of garlic [244]. Pretreating rats with a mixture of As, Cd, Hg, Cr, Ni, Mn, and Fe and sensitizing them with heat-extracted bacterial antigens in rats could reproduce the decline of humoral response in high-dose [198]. Furthermore, metal mixtures could weaken cellular immune responses and delayed-type hypersensitivity reactions. Inhalational exposure to engineered nanoparticles of Ag, Ti, and Si, made C57BL/6JRj mice more susceptible to developing colitis after challenging by dextran sodium sulfate [225]. Assessment of waterborne As, Mn, and plasma Pb in Bangladeshi children uncovered a negative correlation between a latent mixture of these metals and diphtheria antibody titer under a structural equation model, suggesting impaired immunization response [245].

The mechanisms by which organisms cope with pathogens are crucial to immune surveillance and clearance. Detection of pathogens initiates a corresponding immune response, involving Toll-like receptors (TLR) and other pattern recognition receptors (PRRs). Transcriptional analysis of midge *Chironomus riparius Meigen* after sublethal exposure of Cd with Cu, revealed upregulation of apoptosis-related gene, *Dronc*, and immunity-related gene, *Gp93*, ortholog to human *gp96*, chaperone for TLR [246], compared to the control group and individual metal group [247]. Single Cu exposure could increase the expression of *Dronc* compared to the Cd exposure group and increase the expression of *Gp93* compared to both the control group and Cd group. Cd potentiation of Cu was suspected to interfere with immune response. Heavy metal mixtures (Zn, Cu, Ni, Cr, Pb, and Cd) at environmentally relevant concentrations raised the white blood cell counts in *Perca fluviatilis*, which was not observed under the challenge of pathogens [248], indicating an immune-compromised reaction. The toxicity of metals in pathogens, however, sometimes aids the immune system to fight back. Ag ions are widely used for external antibacterial applications, such as silver sulfadiazine cream. [249]. Ag nanoparticles with ZnO at varying concentration ratios demonstrated diverse anti-tubercular effects on THP-1 phagocytes infected with Mycobacterium tuberculosis [250]. With a certain ratio of mixture, like 8Ag: 2ZnO [250] or 5Ag: 5ZnO [251], the mixtures exhibited potent anti-mycobacterial capacity and THP-1 cells succeeded in phagocytized the pathogen without adverse effects, suggesting a complex interaction between metals.

On the other hand, increasing attention has been laid on the interaction between intestinal microorganisms and pollutants [225, 252]. Exposing Pb with Cd reshaped the microorganism system in the digestive tract of *Bufo gargarizans*, favoring colonization of *Proteobacteria* and *Firmicutes*, and subsequent analysis revealed dysregulation of the immune system and amino acid metabolism [236]. Simultaneously, the mucosa barriers were damaged, and the level of digestive enzymes decreased drastically. In another study of larvae of *Bufo gargarizans*, Pb with Cu could disrupt the organization of intestinal mucosa and the bile acid metabolism, another essential part of intestinal immunity [253]. Intestinal dysbiosis, disturbed immune response and ensuing systemic disease are an important but previously less-attended aspect of immunotoxicity of heavy metals mixture. Despite the advancements in microbiome and other biotechnology, there is still a long way to go in accommodating the role of microbiota in the whole picture of immunity.

Collectively, there could be synergism, antagonism, or unidentified mechanisms among the interaction of different metals. The molecular toxicity of heavy metal mixture in the immune system could not be easily concluded and generalized as there are different individual and interacting effects among metals, and distinct concentration-response tendencies. Researchers have focused on the different components of the immune system, but limited specific circuits or pathways were clarified in the pathogenesis of metals mixture. The combined molecular toxicity of heavy metals mixture remains to be further elucidated.

9.10 Combined Molecular Toxicity of Heavy Metal Mixtures in the Tumorigenesis

According to statistics, cancer is the second leading cause of death after heart disease [254]. Alterations in diet, lifestyle, obesity, environment, and microbiome, and their interactions with the genome and/or genetic susceptibilities are recognized as major risk factors for cancer [255]. The incidence and mortality of cancer in the United States in 2023 were collected and estimated by the National Cancer Institute. Prostate cancer, lung, and bronchial cancer, and colorectal cancer accounted for 48% of all newly diagnosed male cases. Breast, lung, and colorectal cancers in women accounted for 52% of all new diagnoses [254].

Heavy metals are significantly associated with mortality and survival from various cancers (Table 9.5). Epidemiological studies have shown that exposure to heavy metals or compounds in the environment or workplace can cause adverse reactions, including hematological and biochemical changes [256–258]. IARC (The International Agency for Research on

Table 9.5 Exposure to heavy metals and risks of cancers.

Cancer	Heavy metals	Conclusion	Species	Potential pathway/mechanism	Ref.
Lung and Bronchus Cancer	Cr [Cr(VI)], Pb, Cd, Ni	Increase risk	–	–	[269–272]
Prostate Cancer	As, Zn, Cd, Ni, Cu, Hg	Increase risk	–	–	[276, 277]
Breast Cancer	Cu, As, Cd, Co	Increase risk	–	–	[273, 278, 279]
Esophageal Cancer	Cd	Increase risk	–	• Stimulate cancer cell stemness and Wnt/β-catenin signaling pathway • Hyper-phosphorylate PKM2, neutralized ROS, inhibited cell apoptosis	[282–284]
	Se	Lower risk	ESCC cell lines, a 4-NQO mice model	• Reduce inflammation • Induce apoptosis	[289]

(Continued)

Table 9.5 (Continued)

Cancer	Heavy metals	Conclusion	Species	Potential pathway/mechanism	Ref.
Colorectal Cancer	Cd, Pb, As	Increase risk	–	–	[286, 287]
	Cr(VI)	Increase risk	Mouse CRC model	Oxidative stress	[288]
Pancreatic cancer	Pb, Cd, As	Increase risk	–	–	[290]
	Se, Ni	Lower risk	–	–	[290]

Cancer) has criteria for classifying the levels of human carcinogenicity of metals and their compounds. Arsenic (As) and cadmium (Cd) are classified as carcinogenic in group 1. Nickel (Ni), Lead (Pb), and Cobalt (Co) are placed in group 2B as possibly carcinogenic. Chromium (Cr), Copper (Cu), Mercury (Hg), and Selenium (Se) are in the group as carcinogenicity is not classifiable. However, Manganese (Mn), Sliver (Ag), and Zinc (Zn) in group 4 as probably not carcinogenic. Some heavy metal compounds, such as inorganic compounds of nickel, arsenic, or lead, have equal or even higher carcinogenicity levels than the metals themselves [259, 260].

A study based on the US National Health and Nutrition Examination Survey (NHANES) showed that heavy metals concentration (blood [lead, cadmium, and mercury] and urine [cadmium, cobalt, lead]) were associated with increased all-cause and cancer mortality [25]. Recent studies emphasize the risks of combined heavy metal exposure, rather than exposure to a single metal, due to the chemical properties and distribution of heavy metals [261, 262].

Despite an unclear carcinogenic mechanism for heavy metals, proposed mechanisms include mitochondrial dysfunction, oxidative stress, metal-induced genotoxicity, and altered stem cell function/gene expression [263, 264]. For the completeness of the overall exposition, it will be discussed below from the perspective of tumors at different sites.

9.10.1 Lung and Bronchus Cancer

Lung and bronchus cancer incidence and mortality are increasing globally. In 2020, lung and bronchus cancer was the leading cause of cancer death in a person aged over 50 years, and the five-survival rate was 23% in all races in the United States [254]. Heavy metals and their compounds can enter the body via food and skin, but inhalation is the most dangerous route. The primary risk factors for lung cancer are air pollution and cigarettes, which include nicotine derivatives, heavy metals (including organometal compounds), and other complex compounds [265–267]. The globally rising trend in lung cancer may be due to the rising tobacco use in developing and lower- and middle-income countries [265, 268].

Some longitudinal cohort and case-control studies found that exposure to cadmium, arsenic, and hexavalent chromium [Cr (VI)] increases the risk of lung cancer [269–271]. In vitro experiments have shown that insoluble heavy metals or their compound particles stay in the lung for a long time, affect the activity of macrophages, and contact lung epithelial cells through endocytosis to cause chromosomal genetic changes [272].

9.10.2 Prostate Cancer and Breast Cancer

In 2023, the leading cause of new cancer diagnoses among American men was prostate cancer, accounting for 29% of the total [254]. Prostate cancer is classified into androgen receptor positive and negative (AR+/AR–) [273]. Breast cancer is subtyped according to molecular markers of estrogen receptors, progesterone receptors, and human epidermal growth factor 2 (ERBB2, formerly HER2) [273, 274]. Long-term heavy metal mixture exposure has adverse effects on male and female reproductive function, such as prostate cancer, menstrual disorders, endometriosis, and breast cancer [275].

Tao et al. found that heavy metal exposure is associated with changes in human sex hormones in an independent or combined manner based on NHANES data [148]. Lim et al. detected a positive association between prostate cancer risk with serum arsenic and zinc levels respectively and collectively in the Singapore prostate cancer study [276]. Chang et al. studied that serum cadmium, nickel, copper, and mercury levels were significantly higher in patients with prostate enlargement and prostate cancer [277].

Epidemiological results show that high concentrations of cadmium and cobalt in the air are associated with ER/PR negative breast cancer [278]. In vitro experiments have shown that copper, arsenic, and cadmium can induce estrogen-dependent breast cancer cell proliferation [273, 279]. It was found that high doses of arsenic and cadmium-induced cell death during acute experimental exposure, while chronic exposure triggered tumor cell proliferation and viability, increased invasiveness, and tumorigenic potential [273].

9.10.3 Esophageal Cancer

According to global population projections, the estimated age-standardized incidence and mortality rates for esophageal cancer are 6.3 and 5.6 per 100 000 population, with 957 000 new cases and 880 000 esophageal cancer deaths projected to occur by 2040. Esophageal cancer is divided into two histological subtypes, adenocarcinoma (AC) and squamous cell carcinoma (SCC). Esophageal squamous cell carcinoma (ESCC) is the most common type of esophageal cancer and the sixth leading cause of cancer death worldwide, with significant ethnic and geographic distribution [280]. Smoking, alcohol consumption, pickled food intake, and family cancer history are all risk factors [281].

It has been reported that patients with gastrointestinal cancer in southern China have significantly higher blood cadmium and lead levels than controls [282]. Cadmium chloride treatment enhanced esophageal cancer cell proliferation and invasion in vitro. Cadmium stimulated cancer cell stemness and Wnt/β-catenin signaling pathway [283]. Yang et al. found that cadmium chloride-induced hyperphosphorylation of pyruvate kinase (PKM2) in a human epithelial esophageal cell line, promoted the production of NADPH and reduced glutathione (GSH), neutralized reactive oxygen species (ROS), and ultimately inhibited cell apoptosis [284].

9.10.4 Colorectal Cancer

Colorectal cancer (CRC) is the third most common cancer in the United States. The estimated number of new colorectal cancer cases and deaths reported in the United States in 2023 is 153 020 and 52 550, respectively [254, 285].

In a cross-sectional study, the elemental composition of surgical tissues collected from patients was analyzed, and it was found that chromium and lead were higher in cancer tissues than in adjacent tissues [286]. A study from Spain found that chronic exposure to low levels of arsenic and chromium in surface soil was associated with colorectal cancer in men [287]. Cr (VI) can promote the formation of tumors in a mouse CRC model induced by 1, 2-dimethylhydrazine (DMH) through oxidative stress [288].

9.11 Combined Molecular Toxicity of Heavy Metal Mixtures in the Orofacial Clefts

Orofacial clefts (OFCs), mainly cleft lip and/or cleft palate (CLP), are among the most common structural birth defects [291]. The occurrence of orofacial clefts varies with geographic and ethnic background, and socioeconomic status, with an average rate of 1 in 700 newborns (2009) [292] or a range of 0.5–2.6 per 1000 live births (2015) [293]. Heavy metals are omnipresent in the environment, occurring in varying concentrations in air, bedrock, soil, water, and all other biological matter. Heavy metals including Arsenic (As), cadmium (Cd), lead (Pb), nickel (Ni), etc. have been an important public health concern because the environmental levels have increased due to continued worldwide anthropogenic activities [294, 295]. Epidemiological studies focus on investigating maternal exposure to heavy metals during pregnancy, given their potential role in OFC pathogenesis. It has been established that As, Cd, Pb, and Ni can cross the placental barrier caused by utero exposures to the developing fetus [296, 297].

9.11.1 Occurrence of Orofacial Clefts

Murine and human facial formation follows a similar developmental trajectory, and facial structures arise from several primordial tissues as described below [298, 299]. Facial primordia begin to form as early as the fourth week of gestation in humans or embryonic day 9.5 in mice, following the migration of cranial neural crest cells into the frontonasal prominence, paired maxillary prominences, and paired mandibular prominences [300]. By the fifth week, the medial and lateral nasal prominences outgrow rapidly on either side of the nasal pit. At the ventral junction region, these nasal prominences will subsequently fuse with the maxillary prominence to establish the upper jaw structures, including the upper lip, primary palate, and nose. Disruption of any of these early craniofaciogenic processes may result in a cleft lip with or without a cleft palate. Secondary palate formation is a multifaceted process involving a shift in growth orientation by the palatal shelves [301]. Various *WNT*-signaling genes (including *WNT3, WNT3A*, and *WNT9B*) and related genes have been associated with orofacial clefts [302].

9.11.2 Epidemiological Survey

A large number of epidemiological studies have shown that heavy metals are a high-risk factor for cleft lip and palate (Table 9.6). A case-control study in China, showed that maternal perinatal exposure to thorium might be a risk factor for offspring OFCs [303]. Higher concentrations of cadmium, lead, and nickel in newborn umbilical cords, are associated with higher rates of OFCs, according to a study in Shanxi, China [304]. A study in Bandung County, Indonesia, measured blood levels of mercury, lead, cadmium, chromium, and cobalt in infants and found that these heavy metals increased the risk of craniofacial deformities [305].

Notably, there are some negative results, a national cohort in Japan found that environmental-related exposure to Hg, Pb, Cd, and Mn was not associated with cleft lip and palate outcomes, as measured by blood composition during the first and third trimesters of pregnancy [306]. Uranium exposure might not be a risk factor for OFCs in offspring [303].

Table 9.6 Results of heavy metals and OFCs in epidemiology study.

Author	Metal	OR (95%CI)	Region	Sample	PMID
Yingnan Guo et al. [307]	U	2.08 (1.13–3.86)	Shanxi, China	Umbilical cord	31918316
Katharina Oginawati et al. [305]	Hg	8.69 (2.46–30.64)	Bandung, Indonesia	Baby's blood	35748989
	Pb	3.60 (1.13–11.48)			
	Cd	3.18 (1.02–9.86)			
	Co	0.94 (0.32–2.81)			
	Cr	2.92 (0.95–8.99)			
Wenli Ni et al. [304]	As	8.69 (2.46–30.64)	Shanxi, China	Umbilical cord	30134103
	Cd	7.22 (3.81–13.71)			
	Pb	15.32 (7.15–32.82)			
	Ni	6.79 (3.59–12.85)			
Masato Takeuch et al. [306]	Hg	0.96 (0.91–1.03)	Japan	Maternal blood	35324965
	Pb	1.01 (0.94–1.08)			
	Cd	1.00 (0.61–1.63)			
	Mn	1.00 (0.97–1.03)			
Yihui Wei et al. [303]	U	0.67 (0.39–1.16) 0.76 (0.44–1.32) 0.69 (0.40–1.20)	China	Maternal hair	31029986
	Th	2.63 (1.41–4.92) 1.98 (1.03–3.79) 2.73 (1.46–5.12)			
K.K. Hozyasz et al. [308]	Zn	2.53 (1.03–6.23)	Poland	Blood of the pregnant	19586695
	Cu	1.18 (0.44–3.15)			
Ingrid P C Krapels et al. [309]	Zn	2.0 (0.8-4.9)	Netherlands	Blood of the pregnant	15198755
	Zn	3.3 (1.3-8.0)		Baby's blood	
Ronald G. Munger et al. [310]	Zn	0.64 (0.39–1.06) 1.04 (0.67–1.60) 0.98 (0.63–1.53)	Utah, USA	Blood of the pregnant	19067407

9.11.3 Animal Data

The positive association between As, Cd, Pb, and Ni and the risk of OFCs is supported by evidence in animal models respectively. A previous animal study has shown that an organic arsenical given by gastric intubation at about 400 or 600 mg/kg/day dose on days 7–16 of gestation induced OFC phenotypes in offspring mice [311]. Hamsters injected intravenously with cadmium chloride or cadmium sulfate at 2–3 mg/kg on Day 8 of gestation gave birth to fetuses with OFC abnormalities [312]. Oral administration of Cd at 20 mg/kg dose on days 6–14 of gestation or intravenous injection of 4–12 mg/kg cadmium chloride on days 13–16 of gestation could cause OFC phenotypes in offspring rats [313, 314]. Wistar rats injected intravenously with Cd at 1.25 mg/kg on Day 10 or injected intraperitoneally on Day 12 were delivered of fetuses with OFC abnormalities [315]. Golden hamsters were given 0.05% or 0.1% lead acetate in drinking water for 4 weeks before and during pregnancy (days 1–15 of pregnancy) displayed the OFC phenotype in offspring [316]. Nickel carbonyl administered to pregnant hamsters by inhalation (0.06 mg $Ni(CO)_4$/liter/15 min) on days 4 and 5 of gestation resulted in OFCs [317]. Intraperitoneal injection of nickel chloride solution at dosages of 1.2–6.9 mg of Ni/kg on days 7–8 of gestation induced OFCs in rat fetuses [318].

9.11.4 Mechanism Research

Epidemiological and animal studies indicate an association between heavy metal exposure and neonatal cleft lip and cleft palate, but the molecular mechanisms underlying this relationship remain poorly understood.

In the section on the occurrence of OFCs, we describe that disruption of any early craniofacial processes can lead to OFCs. In terms of epigenetic inheritance, a mediating effect analysis showed that the hypermethylation of the *Wnt3a* gene was mediated in 9.32% of Pb-induced cleft lip and 15.28% of Pb-induced cleft palate [319]. In addition, lead can mimic or replace Ca^{2+} [320], which means that high levels of lead exposure may disturb normal calcium-mediated phosphorylation and destabilization of β-catenin, which is essential for the production and migration of neural crest cells (NCCs) during the early stage of craniofacial development [321]. In a study of the toxic effects of AgNP and $AgNO_3$, the expression of two typical markers, PAX6 and SIX3, was notably increased by both AgNPs and $AgNO_3$, which was similar to the effects of FGF signaling inhibition during cranial placode in vitro generation [322]. This effect may be due to the release of Ag^+, although it does not significantly affect the migration of NCCs, changes in several key genes suggest that the process of mesenchymal stem cells (MSCs) continuing to differentiate into chondrocytes and osteoblasts after transformation from NCCs is disordered. It was found that 20 ppm titanium (Ti) ions significantly decreased the viabilities of gingival epithelioid cells in cell experiments [323], suggesting heavy metals' role in nutrient systems during development and regeneration should be considered.

Alloys are widely used in oral biomaterials, so the positive function of metals on the formation and regeneration of craniofacial bone is worth exploring. It was found that increasing the content of manganese (Mn) in Mn-Zn-Ca alloy can enhance the cell differentiation, matrix generation, mineral deposition, and osteoblast gene expression of preosteoblasts, which is conducive to craniofacial construction [324].

9.11.5 Future Perspectives

The study of heavy metals and cleft lip and palate is limited by the knowledge of craniofacial bone development mechanism, so there is no set of norms. The mechanisms by which heavy metals cause orofacial clefts remain inconclusive, despite considerable epidemiological literature in this area.

Heavy metals have some common toxicities, such as oxidative stress, apoptosis, inhibition of DNA repair, and interference with nutrient transport through the umbilical cord. Excessive heavy metals may cause an increase in the production of reactive oxygen species, and result in oxidative stress, which is regarded as a common mechanism of teratogenesis [325]. Over apoptosis and necrosis may play some role in teratogenic mechanisms of As and Cd [304, 326, 327]. As, Cd, and Pb can also restrain DNA repair or altered DNA methylation respectively, which may lead to genomic instability [295, 328, 329]. Further, Cd treatment impedes gap junction assembly by interfering with the phosphorylation of certain connexins, and then prevents clonal expansion within the early embryo, which may play roles in Cd teratogenesis [326]. Pb has been shown to inhibit various enzymes of intermediate metabolism and oxidative phosphorylation, which could result in deprivation to the embryo or fetus of vital metabolic intermediates and nutrients [316].

Studies on the mechanism of craniofacial bone development continue to develop, and the key genes including *interferon regulatory factor 6* (*IRF6*), *fibroblast growth factor receptor* (*FGFR*), *transforming growth factor* (*TGF*), *bone morphogenetic proteins* (*BMP4*) and *WNT* family have been confirmed to play an important role in craniofacial development. Combining the influence of heavy metals with the mechanism of craniofacial development should be the future direction to solve the problem of how heavy metals cause cleft lip and palate.

9.12 Combined Molecular Toxicity of Heavy Metal Mixtures in the Olfactory System

Fish heavily rely on their olfactory organ since a variety of essential behaviors are mediated through this sensory system. Properly functioning olfactory organs are paramount for feeding, foraging (amino acids and nucleotides), reproduction (bile acids, steroids, and prostaglandins), and predator avoidance (injured skin releases alarm pheromones), these olfactory-mediated behaviors are essential to almost every aspect of a fish's life (Figures 9.4 and 9.5).

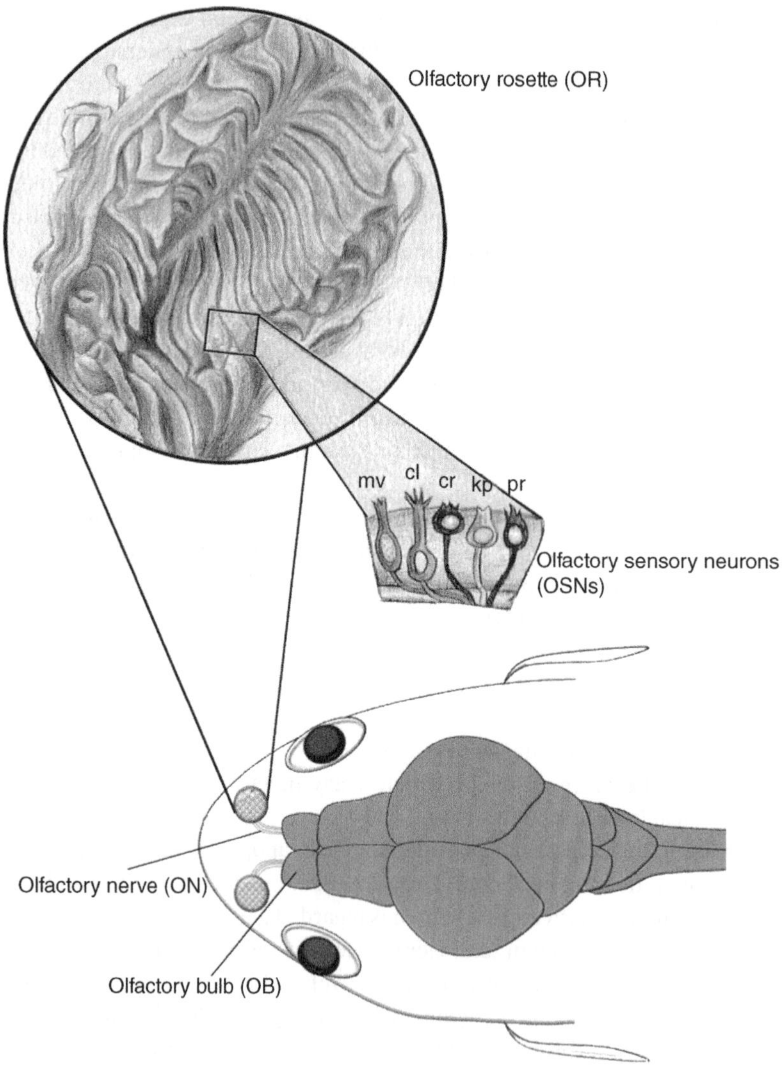

Figure 9.4 Schematic depicting the zebrafish olfactory system, showing the olfactory bulbs (OB), olfactory nerve (ON), and olfactory rosette (OR). The olfactory epithelium (OE) of OR is composed of five types of olfactory sensory neurons (OSNs, sensitized by odor or odorant molecules), including microvillous (mv-), ciliated (cl-), crypt (cr-), kappe (kp-), and pear (pr-) OSNs; OSNs extend their axons to the OB via a short olfactory nerve (ON). *Source:* The upper figure was adapted from Ref. [345]/Springer Nature/CC-BY 4.0.

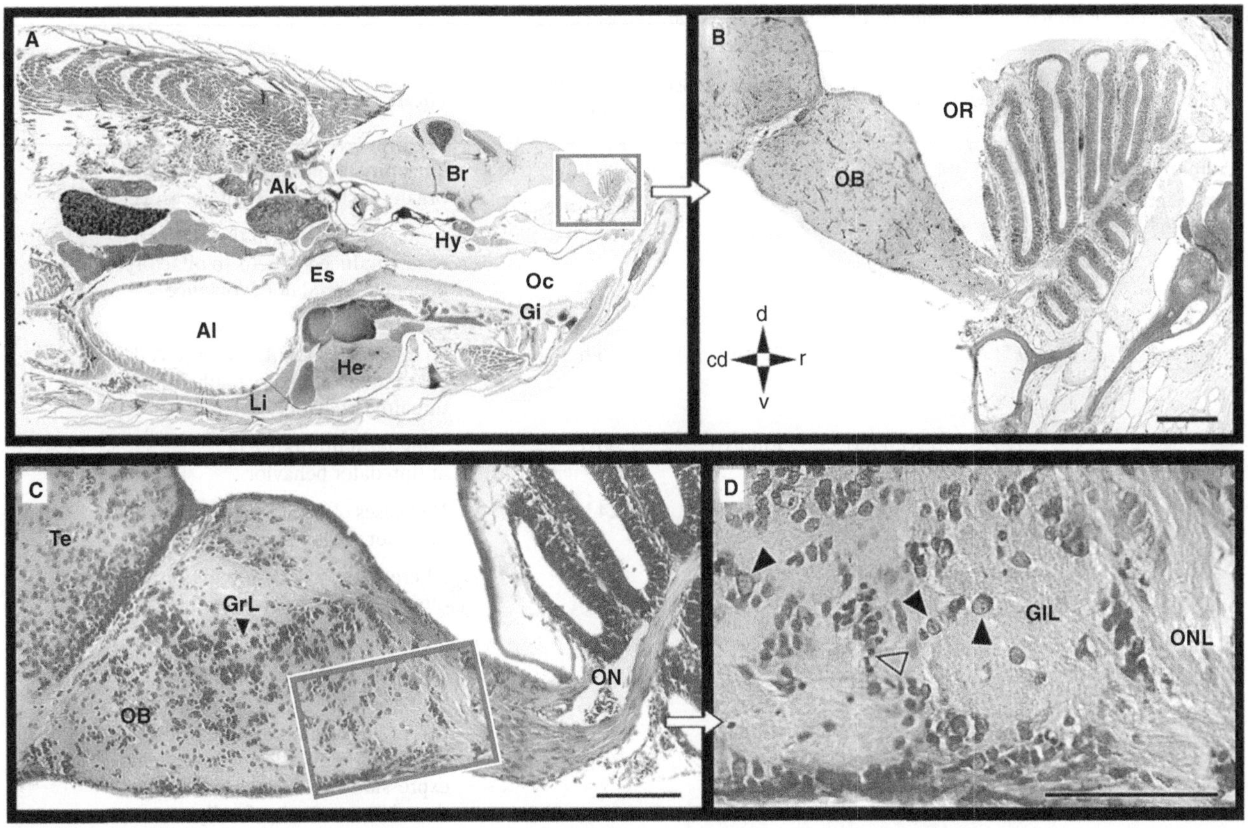

Figure 9.5 Microscopic anatomy of the zebrafish olfactory system. (A) Sagittal section of the anterior zebrafish; (B) higher magnification of the inset in (A), showing the OR and the OB; (C) sagittal section of OB; (D) higher magnification of the inset in (C), showing the olfactory nerve layer (ONL) and glomerular layer (GL). Black arrowhead, bulbous mitral cells; open arrowhead, periglomerular cells; (A,B) stained with Gallego's trichrome; (C,D) Nissl stain. Scale bars: 100 μm. *Source:* The figures were reproduced from [345]/Springer Nature / CC-BY 4.0.

However, the olfactory system is a sensitive target of metal exposure as the olfactory epithelium is almost directly in contact with the aqueous environment. Heavy metals in the water can interact with the dendritic terminations of olfactory sensory neurons, affecting their integrity and functions. Hence, the impairment or even complete cessation of olfactory perception may ensue, resulting in a considerable reduction of an individual's chance of survival and ultimately endangering the whole population. Since zebrafish (*Danio rerio*) has been one of the most popular model organisms in the study of developmental biology, the following review will focus on research in zebrafish.

9.12.1 Cadmium Exposure on the Zebrafish Olfactory System

Cadmium (Cd) is the most well-studied olfaction-impairing metal pollutant, as it can enter into the OSNs, transport through the olfactory nerve (ON), and accumulate within the olfactory bulb (OB) in the central nervous system, thus impairing the olfactory systems of fish [330] (Table 9.7). Blechinger et al. established a transgenic strain carrying a stably integrated hsp70/eGFP reporter to monitor the effect of Cd on the olfactory epithelium in whole living fish [331]. Using this model, they found 3-hour exposure to Cd ≥5 μM induced hsp70/eGFP expression specifically in the cell bodies of OSNs and in the axon termini (Figure 9.6). Moreover, 3-hour-pulse exposure to Cd resulted in olfactory cell apoptosis, and induced long-term olfactory function deficits, as estimated by predator avoidance behavioral assay, even after a 4- to 6-week depuration period [332]. Prolonged exposure duration (96 h) also generated cell death within the olfactory epithelium, impaired the cl-OSNa, and destroyed the architecture of the olfactory pit, these cellular and organizational levels contributed to the subsequent downstream effects at behavioral deficits in larvae [333]. Extend embryonic exposure for 50 days but lower Cd concentrations (87.6 nM and less) still resulted in long-term impairment of an olfactory-mediated antipredator response [334]. Adult fish reared under 5 μM Cd for 21 days had a decreased number of glomerulus and produced prominent microscopic changes in OB [335]. Interestingly, in one study, they found Zn might potentially alleviate Cd-induced olfactory dysfunction [336].

Table 9.7 List of olfactory toxic metals identified using zebrafish model.

Metallic elements	Concentration	Initial exposure stage	Exposure duration	Effects	Ref.
Cadmium chloride hemi-pentahydrate ($CdCl_2$ 2.5 H_2O)	0–125 μM	72 hpf	96 h	Olfactory tissue was one of the first organs to show eGFP fluorescence	[331]
	0.2, 5, 125 μM	80 hpf	3 h	Induced cell death in the olfactory organ; induced deficits in olfactory-dependent predator avoidance behaviors	[332]
	0.5, 1, 5 μM	72 hpf	96 h	96 h LC50 (18.8 μM), induced cell death in the olfactory epithelium; impaired olfactory function	[333]
	0.876, 8.76, 87.6 nM[a]	Embryo	50 d	Long-term hindrance of antipredator behavior	[334]
	8.76 nM–5.12 μM[a]	4 dpf	24 h	Decreases olfactory-mediated behavior	[336]
	0.044, 0.44, 4.4 μM[a]	Adult	24 h	Altered gene expression in the olfactory organ, disrupted the antipredator response	[337]
$CdCl_2$ H_2O	5 μM[a]	Adult	21 d	Produced various microscopic changes in the OB	[335]
Uranium ($UO_2(NO_3)_2$ 6H_2O)	15, 100 μg/L	Adult	3, 10 d	Damaged the anatomical structure of the OB, induced changes in gene expression encoded for members of the odorant receptor family	[338]
	250 μg/L	Adult	10 d	Severe sensory tissue alterations at the level of the olfactory	[339]
Nickel chloride hexahydrate ($NiCl_2$ 6H_2O)	0.4, 0.5, 1 mg/L	Adult (6-8 mpf)	96 h	No micro-anatomical effect of OR; decrease in thickness in the sensory area; induced differential responses of OSN populations to Ni^{2+}, cl-OSNs appear more susceptible than mv-OSNs.	[340]
Mercuric Chloride ($HgCl_2$)	2 μM	72 hpf	72 h	Mercury accumulates in the sensory cells of the OP	[341]
	0.11 μM[a]	Adult	21 d	Produced various microscopic changes in the olfactory bulb	[335]
	0.25, 0.5, 1 μM	Adult (~6 mpf)	96 h	Induced differential response of OSN populations to Hg^{2+}, mv-OSNs are more susceptible than cl-OSNs and cr-OSNs	[342]
Zinc sulfate ($ZnSO_4$)	0.2, 0.5, 1, or 5 M[b]	Adult	3 min	Rapid degeneration of the olfactory organ	[343]
	1 M[b]	Adult	3 min	Have a greater effect on cl-OSNs than mv-OSNs	
	-	Adult	-	Produced damage to olfactory epithelium, inducing overt basal cell vacuolization and intercellular edema, loss of olfaction	[344]
	7, 28, 71, 142, 313, 925, 1 560 μg/L	4 dpf	24 h	Decreases olfactory-mediated behavior	[336]

a. The concentrations were converted from μg /L to nM or μM using their molar mass.
b. Intranasal administration.

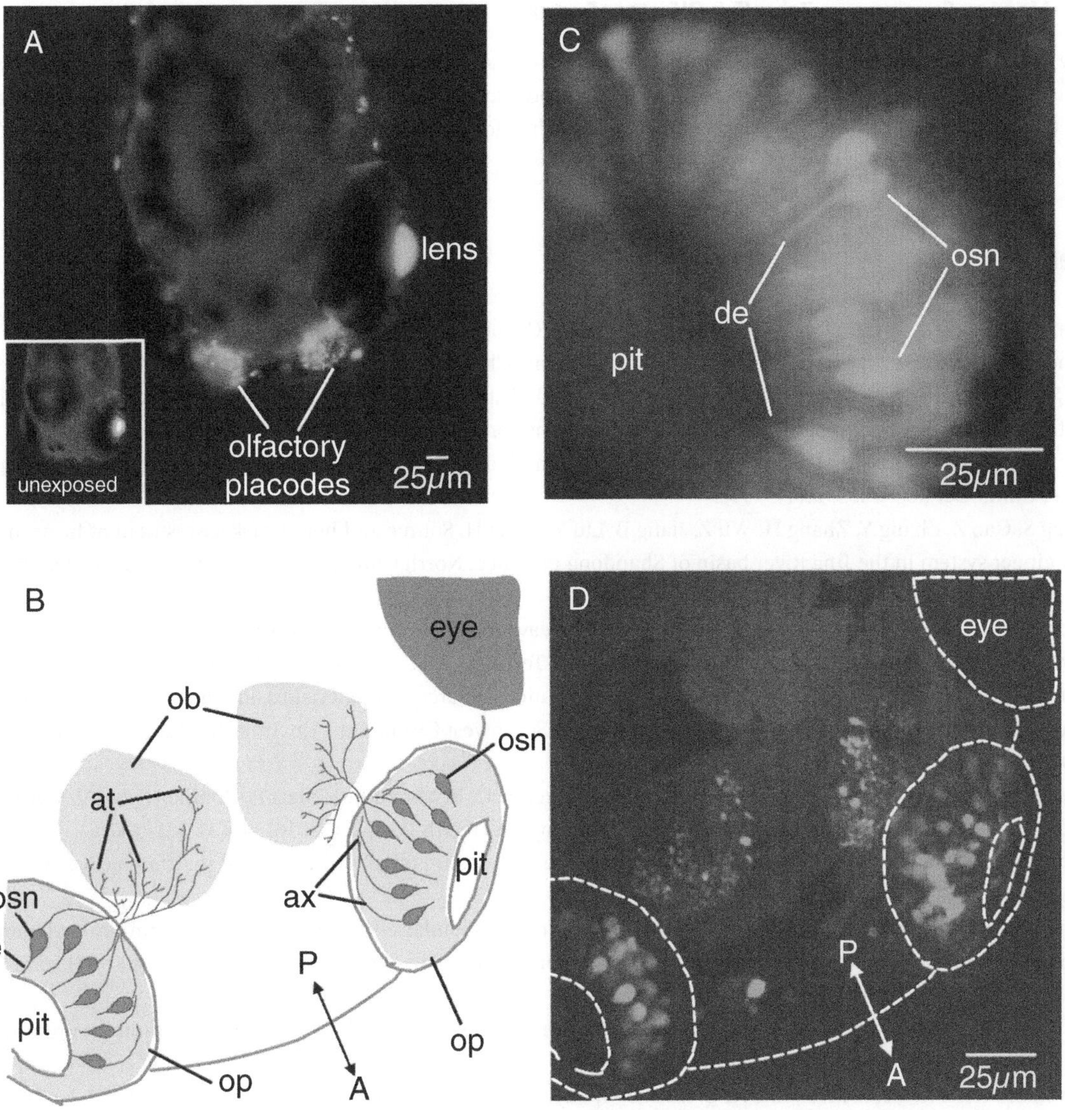

Figure 9.6 Olfactory sensory neurons express hsp70/eGFP following 3-hour exposure to 5 μM cadmium at 80 hpf. at, axon termini; ax, axons; de, dendrite; ob, olfactory bulb; op, olfactory placode; osn, olfactory sensory neuron. *Source:* The figure was reproduced with permission from Ref. [332]/with permission of Elsevier.

These studies failed to investigate the molecular response of olfactory organs, specifically OR and OB, to Cd exposure. Therefore, a recent study attempted to link molecular endpoints of olfactory organs to adverse outcomes in the whole organism, they found Cd exposure induced antioxidant response and perturbed OSN marker gene expression in the OR, and induced cellular stress in the OB, along with olfactory-mediated behavior disruption [337].

9.12.2 Uranium Exposure on the Zebrafish Olfactory System

Uranium (U) is a radioactive element that is naturally occurring in ecosystems. Depleted uranium (DU) is the by-product of the enrichment of uranium, which is a main issue for environmental protection and public health. DU contains three isotopes (^{238}U, ^{235}U, ^{234}U); all emit radiation, which can induce deleterious effects on biota via both chemical and radiological pathways. However, because it produces very little radioactivity per mass of uranium, the effects of DU on humans and animals are mainly attributed to its chemical properties [346]. DU exposure induced ultrastructure dilatations in the neuropil of the OB and changed gene expression encoded for members of the odorant receptor family (*or111-7, or102-5*, and *or107-1*) [338].

9.12.3 Mercury Exposure on Zebrafish Olfactory System

Mercury (Hg) is a well-known contaminant, which can cause high risk to aquatic species and human health [347]. MacDonald et al. used integrated high-resolution X-ray fluorescence imaging (XFI) and immunohistochemistry to assess the effects of Hg^{2+} olfactory pits (OP) on zebrafish [340]. They found Hg^{2+} accumulated in the sensory cells of OP, impairing the olfactory system.

References

1 Zhang H, Zhang F, Song J, Tan ML, Kung HT, Johnson VC. Pollutant source, ecological and human health risks assessment of heavy metals in soils from coal mining areas in Xinjiang, China. Environ Res. 2021, 202: 111702.

2 Xu X, Zhang X, Peng Y, Li R, Liu C, Luo X, Zhao Y. Spatial distribution and source apportionment of agricultural soil heavy metals in a rapidly developing area in east China. Bull Environ Contam Toxicol. 2021, 106(1): 33–39.

3 Yuan Y, Liu B, Liu H. Spatial distribution and source identification for heavy metals in surface sediments of east Dongting Lake, China. Sci Rep. 2022, 12(1): 7940.

4 Wang S, Gao Z, Zhang Y, Zhang H, Wu Z, Jiang B, Liu Y, Dong H. Source and health risk assessment of heavy metals in soil-ginger system in the Jing River basin of Shandong province, North China. Int J Environ Res Public Health. 2021, 18(13): 6749.

5 Luo P, Xu C, Kang S, Huo A, Lyu J, Zhou M, Nover D. Heavy metals in water and surface sediments of the Fenghe River Basin, China: assessment and source analysis. Water Sci Technol. 2021, 84(10-11): 3072–3090.

6 Han X, Li S, Li Z, Pang X, Bao Y, Shi J, Ning P. Concentrations, source characteristics, and health risk assessment of toxic heavy metals in PM(2.5) in a Plateau City (Kunming) in Southwest China. Int J Environ Res Public Health. 2021, 18(21): 11004.

7 Angenard G, Muczynski V, Coffigny H, Pairault C, Duquenne C, Frydman R, Habert R, Rouiller-Fabre V, Livera G. Cadmium increases human fetal germ cell apoptosis. Environ Health Perspect. 2010, 118(3): 331–337.

8 Rehman K, Fatima F, Waheed I, Akash MSH. Prevalence of exposure of heavy metals and their impact on health consequences. J Cell Biochem. 2018, 119(1): 157–184.

9 Arbi S, Bester MJ, Pretorius L, Oberholzer HM. Adverse cardiovascular effects of exposure to cadmium and mercury alone and in combination on the cardiac tissue and aorta of Sprague-Dawley rats. J Environ Sci Health A Tox Hazard Subst Environ Eng. 2021, 56(6): 609–624.

10 Siti HN, Kamisah Y, Kamsiah J. The role of oxidative stress, antioxidants and vascular inflammation in cardiovascular disease (a review). Vascul Pharmacol. 2015, 71: 40–56.

11 Steven S, Frenis K, Oelze M, Kalinovic S, Kuntic M, Bayo Jimenez MT, Vujacic-Mirski K, Helmstädter J, Kröller-Schön S, Münzel T, Daiber A. Vascular inflammation and oxidative stress: major triggers for cardiovascular disease. Oxid Med Cell Longev. 2019, 2019: 7092151.

12 Mikael LR, Paiva AMG, Gomes MM, Sousa ALL, Jardim P, Vitorino PVO, Euzébio MB, Sousa Wd, Barroso WKS. Vascular aging and arterial stiffness. Arq Bras Cardiol. 2017, 109(3): 253–258.

13 Hinderer S, Schenke-Layland K. Cardiac fibrosis - A short review of causes and therapeutic strategies. Adv Drug Deliv Rev. 2019, 146: 77–82.

14 Cocciolone AJ, Hawes JZ, Staiculescu MC, Johnson EO, Murshed M, Wagenseil JE. Elastin, arterial mechanics, and cardiovascular disease. Am J Physiol Heart Circ Physiol. 2018, 315(2): H189–H205.

15 Hechtenberg S, Beyersmann D. Inhibition of sarcoplasmic reticulum Ca(2+)-ATPase activity by cadmium, lead and mercury. Enzyme. 1991, 45(3): 109–115.

16 Patrick L. Lead toxicity part II: the role of free radical damage and the use of antioxidants in the pathology and treatment of lead toxicity. Altern Med Rev. 2006, 11(2): 114–127.

17 Chen Y, Wu F, Graziano JH, Parvez F, Liu M, Paul RR, Shaheen I, Sarwar G, Ahmed A, Islam T, Slavkovich V, Rundek T, Demmer RT, Desvarieux M, Ahsan H. Arsenic exposure from drinking water, arsenic methylation capacity, and carotid intima-media thickness in Bangladesh. Am J Epidemiol. 2013, 178(3): 372–381.

18 Osorio-Yanez C, Ayllon-Vergara JC, Aguilar-Madrid G, Arreola-Mendoza L, Hernandez-Castellanos E, Barrera-Hernandez A, De Vizcaya-Ruiz A, Del Razo LM. Carotid intima-media thickness and plasma asymmetric dimethylarginine in Mexican children exposed to inorganic arsenic. Environ Health Perspect. 2013, 121(9): 1090–1096.

19 Mateen FJ, Grau-Perez M, Pollak JS, Moon KA, Howard BV, Umans JG, Best LG, Francesconi KA, Goessler W, Crainiceanu C, Guallar E, Devereux RB, Roman MJ, Navas-Acien A. Chronic arsenic exposure and risk of carotid artery disease: The Strong Heart Study. Environ Res. 2017, 157: 127–134.

20 Sobel MH, Sanchez TR, Jones MR, Kaufman JD, Francesconi KA, Blaha MJ, Vaidya D, Shimbo D, Gossler W, Gamble MV, Genkinger JM, Navas-Acien A. Rice intake, arsenic exposure, and subclinical cardiovascular disease among US Adults in MESA. J Am Heart Assoc. 2020, 9(4): e015658.

21 Messner B, Knoflach M, Seubert A, Ritsch A, Pfaller K, Henderson B, Shen YH, Zeller I, Willeit J, Laufer G, Wick G, Kiechl S, Bernhard D. Cadmium is a novel and independent risk factor for early atherosclerosis mechanisms and in vivo relevance. Arterioscler Thromb Vasc Biol. 2009, 29(9): 1392–1398.

22 Fagerberg B, Barregard L, Sallsten G, Forsgard N, Ostling G, Persson M, Borné Y, Engström G, Hedblad B. Cadmium exposure and atherosclerotic carotid plaques--results from the Malmo diet and Cancer study. Environ Res. 2015, 136: 67–74.

23 Fagerberg B, Bergstrom G, Boren J, Barregard L. Cadmium exposure is accompanied by increased prevalence and future growth of atherosclerotic plaques in 64-year-old women. J Intern Med. 2012, 272(6): 601–610.

24 Mendy A, Gasana J, Vieira ER. Urinary heavy metals and associated medical conditions in the US adult population. Int J Environ Health Res. 2012, 22(2): 105–118.

25 Duan W, Xu C, Liu Q, Xu J, Weng Z, Zhang X, Basnet TB, Dahal M, Gu A. Levels of a mixture of heavy metals in blood and urine and all-cause, cardiovascular disease and cancer mortality: a population-based cohort study. Environ Pollut. 2020, 263(Pt A): 114630.

26 Domingo-Relloso A, Grau-Perez M, Briongos-Figuero L, Gomez-Ariza JL, Garcia-Barrera T, Duenas-Laita A, Bobb JF, Chaves FJ, Kioumourtzoglou MA, Navas-Acien A, Redon-Mas J, Martin-Escudero JC, Tellez-Plaza M. The association of urine metals and metal mixtures with cardiovascular incidence in an adult population from Spain: the Hortega Follow-Up Study. Int J Epidemiol. 2019, 48(6): 1839–1849.

27 Tyrrell J, Galloway TS, Abo-Zaid G, Melzer D, Depledge MH, Osborne NJ. High urinary tungsten concentration is associated with stroke in the National Health and Nutrition Examination Survey 1999–2010. PLoS One. 2013, 8(11): e77546.

28 Grau-Perez M, Caballero-Mateos MJ, Domingo-Relloso A, Navas-Acien A, Gomez-Ariza JL, Garcia-Barrera T, Leon-Latre M, Soriano-Gil Z, Jarauta E, Cenarro A, Moreno-Franco B, Laclaustra M, Civeira F, Casasnovas JA, Guallar E, Tellez-Plaza M. Toxic metals and subclinical atherosclerosis in carotid, femoral, and coronary vascular territories: the Aragon Workers Health Study. Arterioscler Thromb Vasc Biol. 2022, 42(1): 87–99.

29 Chen H, Lu Y, Cao Z, Ma Q, Pi H, Fang Y, Yu Z, Hu H, Zhou Z. Cadmium induces NLRP3 inflammasome-dependent pyroptosis in vascular endothelial cells. Toxicol Lett. 2016, 246: 7–16.

30 Fittipaldi S, Bimonte VM, Soricelli A, Aversa A, Lenzi A, Greco EA, Migliaccio S. Cadmium exposure alters steroid receptors and proinflammatory cytokine levels in endothelial cells in vitro: a potential mechanism of endocrine disruptor atherogenic effect. J Endocrinol Invest. 2019, 42(6): 727–739.

31 Angeli JK, Cruz Pereira CA, de Oliveira Faria T, Stefanon I, Padilha AS, Vassallo DV. Cadmium exposure induces vascular injury due to endothelial oxidative stress: the role of local angiotensin II and COX-2. Free Radic Biol Med. 2013, 65: 838–848.

32 Prozialeck WC, Edwards JR, Nebert DW, Woods JM, Barchowsky A, Atchison WD. The vascular system as a target of metal toxicity. Toxicol Sci. 2008, 102(2): 207–218.

33 Munzel T, Camici GG, Maack C, Bonetti NR, Fuster V, Kovacic JC. Impact of oxidative stress on the heart and vasculature: part 2 of a 3-part series. J Am Coll Cardiol. 2017, 70(2): 212–229.

34 Sangartit W, Kukongviriyapan U, Donpunha W, Pakdeechote P, Kukongviriyapan V, Surawattanawan P, Greenwald SE. Tetrahydrocurcumin protects against cadmium-induced hypertension, raised arterial stiffness and vascular remodeling in mice. PLoS One. 2014, 9(12): e114908.

35 Nair AR, Degheselle O, Smeets K, Van Kerkhove E, Cuypers A. Cadmium-induced pathologies: where is the oxidative balance lost (or not)? Int J Mol Sci. 2013, 14(3): 6116–6143.

36 Go YM, Orr M, Jones DP. Increased nuclear thioredoxin-1 potentiates cadmium-induced cytotoxicity. Toxicol Sci. 2013, 131(1): 84–94.

37 Dong F, Guo F, Li L, Guo L, Hou Y, Hao E, Yan S, Allen TD, Liu J. Cadmium induces vascular permeability via activation of the p38 MAPK pathway. Biochem Biophys Res Commun. 2014, 450(1): 447–452.

38 Mlynek V, Skoczynska A. [The proinflammatory activity of cadmium]. Postepy Hig Med Dosw (Online). 2005, 59: 1–8.

39 Byrne C, Divekar SD, Storchan GB, Parodi DA, Martin MB. Cadmium – a metallohormone? Toxicol Appl Pharmacol. 2009, 238(3): 266–271.

40 Cheng L, Zhang B, Zheng T, Hu J, Zhou A, Bassig BA, Xia W, Savitz DA, Buka S, Xiong C, Braun JM, Zhang Y, Zhou Y, Pan X, Wu C, Wang Y, Qian Z, Yang A, Romano ME, Shi K, Xu S, Li Y. Critical windows of prenatal exposure to cadmium and size at birth. Int J Environ Res Public Health. 2017, 14(1): 58.

41 Vaziri ND. Mechanisms of lead-induced hypertension and cardiovascular disease. Am J Physiol Heart Circ Physiol. 2008, 295(2): H454–465.

42 Farmand F, Ehdaie A, Roberts CK, Sindhu RK. Lead-induced dysregulation of superoxide dismutases, catalase, glutathione peroxidase, and guanylate cyclase. Environ Res. 2005, 98(1): 33–39.

43 Karakulak UN, Yilmaz OH, Tutkun E, Ates I, Bal C, Gunduzoz M. Evaluation of the ambulatory arterial stiffness index in lead-exposed workers. Anatol J Cardiol. 2017, 18(1): 10–14.

44 Poreba R, Gac P, Poreba M, Antonowicz-Juchniewicz J, Andrzejak R. Relationship between occupational exposure to lead and local arterial stiffness and left ventricular diastolic function in individuals with arterial hypertension. Toxicol Appl Pharmacol. 2011, 254(3): 342–348.

45 Cai R, Zheng YF, Bu JG, Zhang YY, Fu SL, Wang XG, Guo LL, Zhang JR. Effects of blood lead and cadmium levels on homocysteine level in plasma. Eur Rev Med Pharmacol Sci. 2017, 21(1): 162–166.

46 Kasperczyk S, Dobrakowski M, Kasperczyk A, Romuk E, Rykaczewska-Czerwinska M, Pawlas N, Birkner E. Effect of N-acetylcysteine administration on homocysteine level, oxidative damage to proteins, and levels of iron (Fe) and Fe-related proteins in lead-exposed workers. Toxicol Ind Health. 2016, 32(9): 1607–1618.

47 Bravo Y, Quiroz Y, Ferrebuz A, Vaziri ND, Rodriguez-Iturbe B. Mycophenolate mofetil administration reduces renal inflammation, oxidative stress, and arterial pressure in rats with lead-induced hypertension. Am J Physiol Renal Physiol. 2007, 293(2): F616–F623.

48 Navas-Acien A, Guallar E, Silbergeld EK, Rothenberg SJ. Lead exposure and cardiovascular disease--a systematic review. Environ Health Perspect. 2007, 115(3): 472–482.

49 Markovac J, Goldstein GW. Picomolar concentrations of lead stimulate brain protein kinase C. Nature. 1988, 334(6177): 71–73.

50 Fujiwara Y, Kaji T. Possible mechanism for lead inhibition of vascular endothelial cell proliferation: a lower response to basic fibroblast growth factor through inhibition of heparan sulfate synthesis. Toxicology. 1999, 133(2-3): 147–157.

51 Faria TO, Simoes MR, Vassallo DV, Forechi L, Almenara CCP, Marchezini BA, Stefanon I, Vassallo PF. Xanthine oxidase activation modulates the endothelial (Vascular) dysfunction related to HgCl(2) exposure plus myocardial infarction in rats. Cardiovasc Toxicol. 2018, 18(2): 161–174.

52 Kim SH, Sharma RP. Mercury-induced apoptosis and necrosis in murine macrophages: role of calcium-induced reactive oxygen species and p38 mitogen-activated protein kinase signaling. Toxicol Appl Pharmacol. 2004, 196(1): 47–57.

53 Cao X, Wang S, Bi R, Tian S, Huo Y, Liu J. Toxic effects of Cr(VI) on the bovine hemoglobin and human vascular endothelial cells: molecular interaction and cell damage. Chemosphere. 2019, 222: 355–363.

54 Cai Z, Zhang Y, Zhang Y, Miao X, Li S, Yang H, Ling Q, Hoffmann PR, Huang Z. Use of a mouse model and human umbilical vein endothelial cells to investigate the effect of arsenic exposure on vascular endothelial function and the associated role of calpains. Environ Health Perspect. 2019, 127(7): 77003.

55 Dorea JG. Environmental exposure to low-level lead (Pb) co-occurring with other neurotoxicants in early life and neurodevelopment of children. Environ Res. 2019, 177: 108641.

56 Eubig PA, Aguiar A, Schantz SL. Lead and PCBs as risk factors for attention deficit/hyperactivity disorder. Environ Health Perspect. 2010, 118(12): 1654–1667.

57 Parajuli RP, Fujiwara T, Umezaki M, Watanabe C. Association of cord blood levels of lead, arsenic, and zinc with neurodevelopmental indicators in newborns: a birth cohort study in Chitwan Valley, Nepal. Environ Res. 2013, 121: 45–51.

58 Baj J, Flieger W, Flieger M, Forma A, Sitarz E, Skorzynska-Dziduszko K, Grochowski C, Maciejewski R, Karakula-Juchnowicz H. Autism spectrum disorder: trace elements imbalances and the pathogenesis and severity of autistic symptoms. Neurosci Biobehav Rev. 2021, 129: 117–132.

59 Reuben A, Elliott ML, Abraham WC, Broadbent J, Houts RM, Ireland D, Knodt AR, Poulton R, Ramrakha S, Hariri AR, Hariri AR, Caspi A, Moffitt TE. Association of childhood lead exposure with mri measurements of structural brain integrity in midlife. JAMA. 2020, 324(19): 1970–1979.

60 Ijomone OM, Olung NF, Akingbade GT, Okoh COA, Aschner M. Environmental influence on neurodevelopmental disorders: potential association of heavy metal exposure and autism. J Trace Elem Med Biol. 2020, 62: 126638.

61 Wang Z, Zhao H, Xu Y, Zhao J, Song Z, Bi Y, Li Y, Lan X, Pan C, Foulkes NS, Zhang S. Early-life lead exposure induces long-term toxicity in the central nervous system: from zebrafish larvae to juveniles and adults. Sci Total Environ. 2022, 804: 150185.

62 Rocha A, Trujillo KA. Neurotoxicity of low-level lead exposure: history, mechanisms of action, and behavioral effects in humans and preclinical models. Neurotoxicology. 2019, 73: 58–80.

63 Mitra P, Sharma P. Novel direction in mechanisms underlying lead toxicity: evidence and prospective. Indian J Clin Biochem. 2019, 34(2): 121–122.

64 Li ZC, Wang LL, Zhao YS, Peng DJ, Chen J, Jiang SY, Zhao L, Aschner M, Li SJ, Jiang YM. Sodium para-aminosalicylic acid ameliorates lead-induced hippocampal neuronal apoptosis by suppressing the activation of the IP(3)R-Ca(2+)-ASK1-p38 signaling pathway. Ecotoxicol Environ Saf. 2022, 241: 113829.

65 Takeda K, Matsuzawa A, Nishitoh H, Tobiume K, Kishida S, Ninomiya-Tsuji J, Matsumoto K, Ichijo H. Involvement of ASK1 in Ca2+-induced p38 MAP kinase activation. EMBO Rep. 2004, 5(2): 161–166.

66 Zündorf G, Reiser G. Calcium dysregulation and homeostasis of neural calcium in the molecular mechanisms of neurodegenerative diseases provide multiple targets for neuroprotection. Antioxid Redox Signal. 2011, 14(7): 1275–1288.

67 Calne D, Chu N, Huang C, Lu C, Olanow W. Manganism and idiopathic parkinsonism: similarities and differences. Neurology. 1994, 44(9): 1583–1586.

68 Bouchard M, Sauvé S, Barbeau B, Legrand M, Brodeur M, Bouffard T, Limoges E, Limoges E, Bellinger DC, Mergler D. Intellectual impairment in school-age children exposed to manganese from drinking water. Environ Health Perspect. 2011, 119(1): 138–143.

69 Ijomone OM, Aluko OM, Okoh COA, Martins AC, Aschner M. Role for calcium signaling in manganese neurotoxicity. J Trace Elem Med Biol. 2019, 56: 146–155.

70 Smith M, Fernandes J, Go Y, Jones D. Redox dynamics of manganese as a mitochondrial life-death switch. Biochem Biophys Res Commun. 2017, 482(3): 388–398.

71 Xu Y, Peng T, Xiang Y, Liao G, Zou F, Meng X. Neurotoxicity and gene expression alterations in zebrafish larvae in response to manganese exposure. Sci Total Environ. 2022, 825: 153778.

72 Ma Y, Su Q, Yue C, Zou H, Zhu J, Zhao H, Song R, Liu Z. The effect of oxidative stress-induced autophagy by cadmium exposure in kidney, liver, and bone damage, and neurotoxicity. Int J Mol Sci. 2022, 23(21): 13491.

73 Wang H, Abel GM, Storm DR, Xia Z. Cadmium exposure impairs adult hippocampal neurogenesis. Toxicol Sci. 2019, 171(2): 501–514.

74 Akesson A, Barregard L, Bergdahl IA, Nordberg GF, Nordberg M, Skerfving S. Non-renal effects and the risk assessment of environmental cadmium exposure. Environ Health Perspect. 2014, 122(5): 431–438.

75 Rose C, Heywood P, Costanzo R. Olfactory impairment after chronic occupational cadmium exposure. J Occup Med. 1992, 34(6): 600–605.

76 Huang A, Zhang J, Wu K, Liu C, Huang Q, Zhang X, Lin X, Huang Y. Exposure to multiple metals and the risk of dyslexia – A case control study in Shantou, China. Environ Pollut. 2022, 307: 119518.

77 Gustin K, Tofail F, Vahter M, Kippler M. Cadmium exposure and cognitive abilities and behavior at 10 years of age: a prospective cohort study. Environ Int. 2018, 113: 259–268.

78 Kim W, Yim HS, Yoo DY, Jung HY, Kim JW, Choi JH, Yoon YS, Kim DW, Hwang IK. Dendropanax morbifera Léveille extract ameliorates cadmium-induced impairment in memory and hippocampal neurogenesis in rats. BMC Complement Altern Med. 2016, 16(1): 452.

79 Neha, Sodhi RK, Jaggi AS, Singh N. Animal models of dementia and cognitive dysfunction. Life Sci. 2014, 109(2): 73–86.

80 Wang H, Zhang L, Abel GM, Storm DR, Xia Z. Cadmium exposure impairs cognition and olfactory memory in male C57BL/6 mice. Toxicol Sci. 2018, 161(1): 87–102.

81 Song B, Xiong G, Luo H, Zuo Z, Zhou Z, Chang X. Single-cell RNA sequencing of mouse neural stem cell differentiation reveals adverse effects of cadmium on neurogenesis. Food Chem Toxicol. 2021, 148: 111936.

82 Xu Y, Zhao H, Wang Z, Gao H, Liu J, Li K, Song Z, Yuan C, Lan X, Pan C, Zhang S. Developmental exposure to environmental levels of cadmium induces neurotoxicity and activates microglia in zebrafish larvae: from the perspectives of neurobehavior and neuroimaging. Chemosphere. 2022, 291(Pt 1): 132802.

83 Wang B, Du Y. Cadmium and its neurotoxic effects. Oxid Med Cell Longev. 2013, 2013: 898034.

84 Wen S, Wang L, Zhang C, Song R, Zou H, Gu J, Liu X, Bian J, Liu Z, Yuan Y. PINK1/Parkin-mediated mitophagy modulates cadmium-induced apoptosis in rat cerebral cortical neurons. Ecotoxicol Environ Saf. 2022, 244: 114052.

85 Li J, Zheng B, He Y, Zhou Y, Chen X, Ruan S, Yang Y, Dai C, Tang L. Antimony contamination, consequences and removal techniques: a review. Ecotoxicol Environ Saf. 2018, 156: 125–134.

86 Herath I, Vithanage M, Bundschuh J. Antimony as a global dilemma: geochemistry, mobility, fate and transport. Environ Pollut. 2017, 223: 545–559.

87 Xia S, Zhu X, Yan Y, Zhang T, Chen G, Lei D, Wang G. Developmental neurotoxicity of antimony (Sb) in the early life stages of zebrafish. Ecotoxicol Environ Saf. 2021, 218: 112308.

88 Chehbani F, Gallello G, Brahim T, Ouanes S, Douki W, Gaddour N, Cervera Sanz ML. The status of chemical elements in the blood plasma of children with autism spectrum disorder in Tunisia: a case-control study. Environ Sci Pollut Res Int. 2020, 27(28): 35738–35749.

89 Kim S, Arora M, Fernandez C, Landero J, Caruso J, Chen A. Lead, mercury, and cadmium exposure and attention deficit hyperactivity disorder in children. Environ Res. 2013, 126: 105–110.

90 Farias P, Hernandez-Bonilla D, Moreno-Macias H, Montes-Lopez S, Schnaas L, Texcalac-Sangrador JL, Ríos C, Riojas-Rodriguez H. Prenatal co-exposure to manganese, mercury, and lead, and neurodevelopment in children during the first year of life. Int J Environ Res Public Health. 2022, 19(20): 13020

91 Stein CR, Wu H, Bellinger DC, Smith DR, Wolff MS, Savitz DA. Exposure to metal mixtures and neuropsychological functioning in middle childhood. Neurotoxicology. 2022, 93: 84–91.

92 Bauer JA, Devick KL, Bobb JF, Coull BA, Bellinger D, Benedetti C, Cagna G, Fedrighi C, Guazzetti S, Oppini M, Placidi D, Webster TF, White RF, Yang Q, Zoni S, Wright RO, Smith DR, Lucchini RG, Claus Henn B. Associations of a metal mixture measured in multiple biomarkers with IQ: evidence from Italian adolescents living near ferroalloy industry. Environ Health Perspect. 2020, 128(9): 97002.

93 Skogheim TS, Weyde KVF, Engel SM, Aase H, Surén P, Øie MG, Biele G, Reichborn-Kjennerud T, Caspersen IH, Hornig M, Haug LS, Villanger GD. Metal and essential element concentrations during pregnancy and associations with autism spectrum disorder and attention-deficit/hyperactivity disorder in children. Environ Int. 2021, 152: 106468.

94 Sprowles JLN, Amos-Kroohs RM, Braun AA, Sugimoto C, Vorhees CV, Williams MT. Developmental manganese, lead, and barren cage exposure have adverse long-term neurocognitive, behavioral and monoamine effects in Sprague-Dawley rats. Neurotoxicol Teratol. 2018, 67: 50–64.

95 Zhou R, Zhao J, Li D, Chen Y, Xiao Y, Fan A, Chen XT, Wang HL. Combined exposure of lead and cadmium leads to the aggravated neurotoxicity through regulating the expression of histone deacetylase 2. Chemosphere. 2020, 252: 126589.

96 Zhou F, Xie J, Zhang S, Yin G, Gao Y, Zhang Y, Bo D, Li Z, Liu S, Feng C, Fan G. Lead, cadmium, arsenic, and mercury combined exposure disrupted synaptic homeostasis through activating the Snk-SPAR pathway. Ecotoxicol Environ Saf. 2018, 163: 674–684.

97 Kim H, Lee D, Kim K. Combined exposure to metals in drinking water alters the dopamine system in mouse striatum. Int J Environ Res Public Health. 2021, 18(12): 6558.

98 Li Z, Zhou F, Yin A, Gao Y, Jiang X, Liu S, Zhang Y, Bo D, Xie J, Jia QY, Feng JG, Feng C, Fan G. [Cellular damage of low-dose combined exposure to mercury, lead and cadmium on hippocampal neurons in rats]. Zhonghua yu fang yi xue za zhi. 2018, 52(10): 976–982.

99 Carvalho CDS, Moraes Utsunomiya HS, Pasquoto-Stigliani T, Costa MJ, Fernandes MN. Biomarkers of the oxidative stress and neurotoxicity in tissues of the bullfrog, lithobates catesbeianus to assess exposure to metals. Ecotoxicol Environ Saf. 2020, 196: 110560.

100 Wang G, Fowler BA. Roles of biomarkers in evaluating interactions among mixtures of lead, cadmium and arsenic. Toxicol Appl Pharmacol. 2008, 233(1): 92–99.

101 Andrade VM, Mateus ML, Batoréu MC, Aschner M, Marreilha dos Santos AP. Lead, arsenic, and manganese metal mixture exposures: focus on biomarkers of effect. Biol Trace Elem Res. 2015, 166(1): 13–23.

102 López-Botella A, Velasco I, Acién M, Sáez-Espinosa P, Todolí-Torró JL, Sánchez-Romero R, Gómez-Torres MJ. Impact of heavy metals on human male fertility-an overview. Antioxidants (Basel). 2021, 10(9): 1473.

103 Agarwal A, Mulgund A, Hamada A, Chyatte MR. A unique view on male infertility around the globe. Reprod Biol Endocrinol. 2015, 13: 37.

104 Anyanwu BO, Orisakwe OE. Current mechanistic perspectives on male reproductive toxicity induced by heavy metals. J Environ Sci Health C Toxicol Carcinog. 2020, 38(3): 204–244.

105 Levine H, Jørgensen N, Martino-Andrade A, Mendiola J, Weksler-Derri D, Mindlis I, Pinotti R, Swan SH. Temporal trends in sperm count: a systematic review and meta-regression analysis. Hum Reprod Update. 2017, 23(6): 646–659.

106 Virtanen HE, Jørgensen N, Toppari J. Semen quality in the 21(st) century. Nat Rev Urol. 2017, 14(2): 120–130.

107 Rolland M, Le Moal J, Wagner V, Royère D, De Mouzon J. Decline in semen concentration and morphology in a sample of 26,609 men close to general population between 1989 and 2005 in France. Hum Reprod. 2013, 28(2): 462–470.

108 Revich BA. [Chemical substances in the Russian urban environment: hazard to human health and prospects for its prevention]. Vestnik Rossiiskoi Akademii Meditsinskikh Nauk. 2002, (9): 45–49.

109 Bhardwaj JK, Paliwal A, Saraf P. Effects of heavy metals on reproduction owing to infertility. J Biochem Mol Toxicol. 2021, 35(8): e22823.

110 Figà-Talamanca I, Traina ME, Urbani E. Occupational exposures to metals, solvents and pesticides: recent evidence on male reproductive effects and biological markers. Occup Med (Lond). 2001, 51(3): 174–188.

111 Bergamo P, Volpe MG, Lorenzetti S, Mantovani A, Notari T, Cocca E, Cerullo S, Di Stasio M, Cerino P, Montano L. Human semen as an early, sensitive biomarker of highly polluted living environment in healthy men: a pilot biomonitoring study on trace elements in blood and semen and their relationship with sperm quality and RedOx status. Reprod Toxicol. 2016, 66: 1–9.

112 Jenardhanan P, Panneerselvam M, Mathur PP. Effect of environmental contaminants on spermatogenesis. Semin Cell Dev Biol. 2016, 59: 126–140.

113 Yang Y, Zhang W, Wang S, Zhang H, Zhang Y. Response of male reproductive function to environmental heavy metal pollution in a free-living passerine bird, passer montanus. Sci Total Environ. 2020, 747: 141402.

114 Machado-Neves M. Effect of heavy metals on epididymal morphology and function: an integrative review. Chemosphere. 2022, 291(Pt 2): 133020.

115 Hubaux R, Becker-Santos DD, Enfield KS, Rowbotham D, Lam S, Lam WL, Martinez VD. Molecular features in arsenic-induced lung tumors. Mol Cancer. 2013, 12: 20.

116 Huang Q, Zhang J, Peng S, Tian M, Chen J, Shen H. Effects of water soluble PM2.5 extracts exposure on human lung epithelial cells (A549): a proteomic study. J Appl Toxicol. 2014, 34(6): 675–687.

117 Zubair M, Ahmad M, Qureshi ZI. Review on arsenic-induced toxicity in male reproductive system and its amelioration. Andrologia. 2017, 49(9): 12791.

118 Zeng Q, Zhou B, Feng W, Wang YX, Liu AL, Yue J, Li YF, Lu WQ. Associations of urinary metal concentrations and circulating testosterone in Chinese men. Reprod Toxicol. 2013, 41: 109–114.

119 Tian M, Wang Y-X, Wang X, Wang H, Liu L, Zhang J, Nan B, Shen H, Huang Q. Environmental doses of arsenic exposure are associated with increased reproductive-age male urinary hormone excretion and in vitro Leydig cell steroidogenesis. J Hazard Mater. 2021, 408: 124904.

120 He Y, Zou L, Luo W, Yi Z, Yang P, Yu S, Liu N, Ji J, Guo Y, Liu P, He X, Lv Z, Huang S. Heavy metal exposure, oxidative stress and semen quality: exploring associations and mediation effects in reproductive-aged men. Chemosphere. 2020, 244: 125498.

121 O'Connor D, Hou D, Ok YS, Mulder J, Duan L, Wu Q, Wang S, Tack FMG, Rinklebe J. Mercury speciation, transformation, and transportation in soils, atmospheric flux, and implications for risk management: a critical review. Environ Int. 2019, 126: 747–761.

122 Cavoura O, Brombach CC, Cortis R, Davidson CM, Gajdosechova Z, Keenan HE, Krupp EM. Mercury alkylation in freshwater sediments from Scottish canals. Chemosphere. 2017, 183: 27–35.

123 Klapstein SJ, O'Driscoll NJ. Methylmercury biogeochemistry in freshwater ecosystems: a review focusing on DOM and photodemethylation. Bull Environ Contam Toxicol. 2018, 100(1): 14–25.

124 Li S, Han B, Wu P, Yang Q, Wang X, Li J, Liao Y, Deng N, Jiang H, Zhang Z. Effect of inorganic mercury exposure on reproductive system of male mice: immunosuppression and fibrosis in testis. Environ Toxicol. 2022, 37(1): 69–78.

125 Bjørklund G, Chirumbolo S, Dadar M, Pivina L, Lindh U, Butnariu M, Aaseth J. Mercury exposure and its effects on fertility and pregnancy outcome. Basic Clin Pharmacol Toxicol. 2019, 125(4): 317–327.

126 Bellés M, Albina ML, Sánchez DJ, Corbella J, Domingo JL. Interactions in developmental toxicology: effects of concurrent exposure to lead, organic mercury, and arsenic in pregnant mice. Arch Environ Contam Toxicol. 2002, 42(1): 93–98.

127 Rao MV, Sharma PS. Protective effect of vitamin E against mercuric chloride reproductive toxicity in male mice. Reprod Toxicol. 2001, 15(6): 705–712.

128 Reis MM, Moreira AC, Sousa M, Mathur PP, Oliveira PF, Alves MG. Sertoli cell as a model in male reproductive toxicology: advantages and disadvantages. J Appl Toxicol. 2015, 35(8): 870–883.

129 Aitken RJ, Baker MA. The role of genetics and oxidative stress in the etiology of male infertility-a unifying hypothesis? Front Endocrinol (Lausanne). 2020, 11: 581838.

130 Lavranos G, Balla M, Tzortzopoulou A, Syriou V, Angelopoulou R. Investigating ROS sources in male infertility: a common end for numerous pathways. Reprod Toxicol. 2012, 34(3): 298–307.

131 Aruldhas MM, Subramanian S, Sekar P, Vengatesh G, Chandrahasan G, Govindarajulu P, Akbarsha MA. Chronic chromium exposure-induced changes in testicular histoarchitecture are associated with oxidative stress: study in a non-human primate (Macaca radiata Geoffroy). Hum Reprod. 2005, 20(10): 2801–2813.

132 El-Demerdash FM, Jebur AB, Nasr HM, Hamid HM. Modulatory effect of Turnera diffusa against testicular toxicity induced by fenitrothion and/or hexavalent chromium in rats. Environ Toxicol. 2019, 34(3): 330–339.

133 Chandra AK, Chatterjee A, Ghosh R, Sarkar M. Vitamin E-supplementation protect chromium (VI)-induced spermatogenic and steroidogenic disorders in testicular tissues of rats. Food Chem Toxicol. 2010, 48(3): 972–979.

134 Subramanian S, Rajendiran G, Sekhar P, Gowri C, Govindarajulu P, Aruldhas MM. Reproductive toxicity of chromium in adult bonnet monkeys (Macaca radiata Geoffrey). Reversible oxidative stress in the semen. Toxicol Appl Pharmacol. 2006, 215(3): 237–249.

135 Wirth JJ, Mijal RS. Adverse effects of low level heavy metal exposure on male reproductive function. Syst Biol Reprod Med. 2010, 56(2): 147–167.

136 Kumar S, Sharma A. Cadmium toxicity: effects on human reproduction and fertility. Rev Environ Health. 2019, 34(4): 327–338.

137 Tvrdá E, Kňažická Z, Lukáčová J, Schneidgenová M, Goc Z, Greń A, Szabó C, Massányi P, Lukáč N. The impact of lead and cadmium on selected motility, prooxidant and antioxidant parameters of bovine seminal plasma and spermatozoa. J Environ Sci Health A Tox Hazard Subst Environ Eng. 2013, 48(10): 1292–1300.

138 El-Demerdash FM, Yousef MI, Kedwany FS, Baghdadi HH. Cadmium-induced changes in lipid peroxidation, blood hematology, biochemical parameters and semen quality of male rats: protective role of vitamin E and beta-carotene. Food Chem Toxicol. 2004, 42(10): 1563–1571.

139 Yu X, Sidhu JS, Hong S, Faustman EM. Essential role of extracellular matrix (ECM) overlay in establishing the functional integrity of primary neonatal rat Sertoli cell/gonocyte co-cultures: an improved in vitro model for assessment of male reproductive toxicity. Toxicol Sci. 2005, 84(2): 378–393.

140 Zhang C, Xu L, Zhao Y, Wang Y. Changes in serum heavy metals in polycystic ovary syndrome and their association with endocrine, lipid-metabolism, inflammatory characteristics and pregnancy outcomes. Reprod Toxicol. 2022, 111: 20–26.

141 Farquhar CM, Bhattacharya S, Repping S, Mastenbroek S, Kamath MS, Marjoribanks J, Boivin J. Female subfertility. Nat Rev Dis Primers. 2019, 5(1): 7.

142 Yang S, Zhang Z, He J, Li J, Zhang J, Xing H, Xu S. Ovarian toxicity induced by dietary cadmium in hen. Biol Trace Elem Res. 2012, 148(1): 53–60.

143 Xie D, Chen Q, Gong S, An J, Li Y, Lian X, Liu Z, Shen Y, Giesy JP. Exposure of zebrafish to environmentally relevant concentrations of mercury during early life stages impairs subsequent reproduction in adults but can be recovered in offspring. Aquat Toxicol. 2020, 229: 105655.

144 Jiang X, Xing X, Zhang Y, Zhang C, Wu Y, Chen Y, Meng R, Jia H, Cheng Y, Zhang Y, Su J. Lead exposure activates the Nrf2/Keap1 pathway, aggravates oxidative stress, and induces reproductive damage in female mice. Ecotoxicol Environ Saf. 2021, 207: 111231.

145 He Y, Wang L, Li X, Zhao H. The effects of chronic lead exposure on the ovaries of female juvenile Japanese quails (Coturnix japonica): developmental delay, histopathological alterations, hormone release disruption and gene expression disorder. Ecotoxicol Environ Saf. 2020, 205: 111338.

146 Wang P, Du Z, Gao S, Zhang X, Giesy JP. Impairment of reproduction of adult zebrafish (Danio rerio) by binary mixtures of environmentally relevant concentrations of triclocarban and inorganic mercury. Ecotoxicol Environ Saf. 2016, 134: 124–132.

147 Martinović-Weigelt D, Wang R-L, Villeneuve DL, Bencic DC, Lazorchak J, Ankley GT. Gene expression profiling of the androgen receptor antagonists flutamide and vinclozolin in zebrafish (Danio rerio) gonads. Aquat Toxicol. 2011, 101(2): 447–458.

148 Tao C, Li Z, Fan Y, Li X, Qian H, Yu H, Xu Q, Lu C. Independent and combined associations of urinary heavy metals exposure and serum sex hormones among adults in NHANES 2013-2016. Environ Pollut. 2021, 281: 117097.

149 Ali I, Engstrom A, Vahter M, Skerfving S, Lundh T, Lidfeldt J, Samsioe G, Halldin K, Akesson A. Associations between cadmium exposure and circulating levels of sex hormones in postmenopausal women. Environ Res. 2014, 134: 265–269.

150 Liang C, Zhang Z, Cao Y, Wang J, Shen L, Jiang T, Li D, Zou W, Zong K, Liang D, Xu X, Liu Y, Tao F, Luo G, Ji D, Cao Y. Exposure to multiple toxic metals and polycystic ovary syndrome risk: endocrine disrupting effect from As, Pb and Ba. Sci Total Environ. 2022, 849: 157780.

151 Nasiadek M, Danilewicz M, Sitarek K, Swiatkowska E, Darago A, Stragierowicz J, Kilanowicz A. The effect of repeated cadmium oral exposure on the level of sex hormones, estrous cyclicity, and endometrium morphometry in female rats. Environ Sci Pollut Res Int. 2018, 25(28): 28025–28038.

152 Dumitrescu E, Cristina RT, Muselin F. Reproductive biology study of dynamics of female sexual hormones: a 12-month exposure to lead acetate rat model. Turk J Biol. 2014, 38: 581–585.

153 Chen Q, An J, Xie D, Gong S, Lian X, Liu Z, Shen Y, Li Y. Suppression and recovery of reproductive behavior induced by early life exposure to mercury in zebrafish. Comp Biochem Physiol C Toxicol Pharmacol. 2021, 239: 108876.

154 Tanrikut E, Karaer A, Celik O, Celik E, Otlu B, Yilmaz E, Ozgul O. Role of endometrial concentrations of heavy metals (cadmium, lead, mercury and arsenic) in the aetiology of unexplained infertility. Eur J Obstet Gynecol Reprod Biol. 2014, 179: 187–190.

155 White JT, Kovar E, Chambers TM, Sheth KR, Peckham-Gregory EC, O'Neill M, Langlois PH, Jorgez CJ, Lupo PJ, Seth A. Hypospadias Risk from Maternal Residential Exposure to Heavy Metal Hazardous Air Pollutants. Int J Environ Res Public Health. 2019, 16(6): 930.

156 Rana SV. Perspectives in endocrine toxicity of heavy metals--a review. Biol Trace Elem Res. 2014, 160(1): 1–14.

157 Guchhait R, Chatterjee A, Gupta S, Debnath M, Mukherjee D, Pramanick K. Molecular mechanism of mercury-induced reproductive impairments in banded gourami, Trichogaster fasciata. Gen Comp Endocrinol. 2018, 255: 40–48.

158 Tartu S, Goutte A, Bustamante P, Angelier F, Moe B, Clement-Chastel C, Bech C, Gabrielsen GW, Bustnes JO, Chastel O. To breed or not to breed: endocrine response to mercury contamination by an Arctic seabird. Biol Lett. 2013, 9(4): 20130317.

159 Abdelfattah EA, Augustyniak M, Yousef HA. Stage-, sex- and tissue-related changes in H2O2, glutathione concentration, and glutathione-dependent enzymes activity in Aiolopus thalassinus (Orthoptera: Acrididae) from heavy metal polluted areas. Ecotoxicology. 2021, 30(3): 478–491.

160 Aglan HS, Gebremedhn S, Salilew-Wondim D, Neuhof C, Tholen E, Holker M, Schellander K, Tesfaye D. Regulation of Nrf2 and NF-kappaB during lead toxicity in bovine granulosa cells. Cell Tissue Res. 2020, 380(3): 643–655.

161 Zhu M, Miao S, Zhou W, Elnesr SS, Dong X, Zou X. MAPK, AKT/FoxO3a and mTOR pathways are involved in cadmium regulating the cell cycle, proliferation and apoptosis of chicken follicular granulosa cells. Ecotoxicol Environ Saf. 2021, 214: 112091.

162 Valko M, Rhodes CJ, Moncol J, Izakovic M, Mazur M. Free radicals, metals and antioxidants in oxidative stress-induced cancer. Chem Biol Interact. 2006, 160(1): 1–40.

163 Bellezza I, Mierla AL, Minelli A. Nrf2 and NF-kappaB and their concerted modulation in cancer pathogenesis and progression. Cancers (Basel). 2010, 2(2): 483–497.

164 McIlwain DR, Berger T, Mak TW. Caspase functions in cell death and disease. Cold Spring Harb Perspect Biol. 2015, 7(4): a026716.

165 Ola MS, Nawaz M, Ahsan H. Role of Bcl-2 family proteins and caspases in the regulation of apoptosis. Mol Cell Biochem. 2011, 351(1-2): 41–58.

166 Jiang X, Wang X. Cytochrome C-mediated apoptosis. Annu Rev Biochem. 2004, 73: 87–106.

167 Roychoudhury S, Massanyi P, Slivkova J, Formicki G, Lukac N, Slamecka J, Slama P, Kolesarova A. Effect of mercury on porcine ovarian granulosa cells in vitro. J Environ Sci Health A Tox Hazard Subst Environ Eng. 2015, 50(8): 839–845.

168 Kolesarova A, Capcarova M, Sirotkin AV, Medvedova M, Kalafova A, Filipejova T, Kovacik J. In vitro assessment of molybdenum-induced secretory activity, proliferation and apoptosis of porcine ovarian granulosa cells. J Environ Sci Health A Tox Hazard Subst Environ Eng. 2011, 46(2): 170–175.

169 Kolesarova A, Capcarova M, Sirotkin AV, Medvedova M, Kovacik J. In vitro assessment of silver effect on porcine ovarian granulosa cells. J Trace Elem Med Biol. 2011, 25(3): 166–170.

170 Lin X, Gu Y, Zhou Q, Mao G, Zou B, Zhao J. Combined toxicity of heavy metal mixtures in liver cells. J Appl Toxicol. 2016, 36(9): 1163–1172.

171 Wang X, Bin W, Zhou M, Xiao L, Xu T, Yang S, Nie X, Xie L, Yu L, Mu G, Ma J, Chen W. Systemic inflammation mediates the association of heavy metal exposures with liver injury: a study in general Chinese urban adults. J Hazard Mater. 2021, 419: 126497.

172 Hyder O, Chung M, Cosgrove D, Herman JM, Li Z, Firoozmand A, Gurakar A, Koteish A, Pawlik TM. Cadmium exposure and liver disease among US adults. J Gastrointest Surg. 2013, 17(7): 1265–1273.

173 Nguyen HD, Kim MS. Cadmium, lead, and mercury mixtures interact with non-alcoholic fatty liver diseases. Environ Pollut. 2022, 309: 119780.

174 Yang Z, Li X, Tian L, Song Y, Zhang Y, Chen J, Zhang L. Heavy metals exposure is associated with early liver dysfunction among rural residents aged 40–75 years in southwest China. J Appl Toxicol. 2022, 42(6): 1044–1056.

175 Thirunavukkarasu C, Sharma Y, Tchaikovskaya T, Maslov AY, Gupta S. Transcriptional profiling reveals ataxia telangiectasia mutated pathways regulate joint copper and arsenic toxicity for hepatic metalloplasia and anti-cancer therapies. Life Sci. 2022, 305: 120787.

176 Maria B, Maria MC, Antonio B, Simona M, Rosaria A, Andrea S, Giulia M, Marianna DC, Mario S. Chemical and biochemical responses to sub-lethal doses of mercury and cadmium in gilthead seabream (Sparus aurata). Chemosphere. 2022, 307(Pt 3): 135822.

177 Li D, Yao H, Du L, Zeng X, Xiao Q. Thallium(I and III) exposure leads to liver damage and disorders of fatty acid metabolism in mice. Chemosphere. 2022, 307(Pt 1): 135618.

178 Chen H, Zhu C, Zhou X. Effects of lead and cadmium combined heavy metals on liver function and lipid metabolism in mice. Biol Trace Elem Res. 2023, 201(6): 2864–2876.

179 Sabir S, Akash MSH, Fiayyaz F, Saleem U, Mehmood MH, Rehman K. Role of cadmium and arsenic as endocrine disruptors in the metabolism of carbohydrates: inserting the association into perspectives. Biomed Pharmacother. 2019, 114: 108802.

180 Zhong G, Wan F, Wu S, Jiang X, Tang Z, Zhang X, Huang R, Hu L. Arsenic or/and antimony induced mitophagy and apoptosis associated with metabolic abnormalities and oxidative stress in the liver of mice. Sci Total Environ. 2021, 777: 146082.

181 Wang S, Liu Z, Geng J, Li L, Feng X. An overview of ferroptosis in non-alcoholic fatty liver disease. Biomed Pharmacother. 2022, 153: 113374.

182 Miyanishi K, Tanaka S, Sakamoto H, Kato J. The role of iron in hepatic inflammation and hepatocellular carcinoma. Free Radic Biol Med. 2019, 133: 200–205.

183 Fernandez M, Lokan J, Leung C, Grigg A. A critical evaluation of the role of iron overload in fatty liver disease. J Gastroenterol Hepatol. 2022, 37(10): 1873–1883.

184 He Z, Shen P, Feng L, Hao H, He Y, Fan G, Liu Z, Zhu K, Wang Y, Zhang N, Hu X, Fu Y, Wu J. Cadmium induces liver dysfunction and ferroptosis through the endoplasmic stress-ferritinophagy axis. Ecotoxicol Environ Saf. 2022, 245: 114123.

185 Matranga V, Pinsino A, Randazzo D, Giallongo A, Dubois P. Long-term environmental exposure to metals (Cu, Cd, Pb, Zn) activates the immune cell stress response in the common European sea star (Asterias rubens). Mar Environ Res. 2012, 76: 122–127.

186 Kerambrun E, Henry F, Marechal A, Sanchez W, Minier C, Filipuci I, Amara R. A multibiomarker approach in juvenile turbot, Scophthalmus maximus, exposed to contaminated sediments. Ecotoxicol Environ Saf. 2012, 80: 45–53.

187 Liberti A, Bertocci I, Pollet A, Musco L, Locascio A, Ristoratore F, Spagnuolo A, Sordino P. An indoor study of the combined effect of industrial pollution and turbulence events on the gut environment in a marine invertebrate. Mar Environ Res. 2020, 158: 104950.

188 Höher N, Köhler A, Strand J, Broeg K. Effects of various pollutant mixtures on immune responses of the blue mussel (Mytilus edulis) collected at a salinity gradient in Danish coastal waters. Mar Environ Res. 2012, 75: 35–44.

189 Ladhar-Chaabouni R, Houel T, Serpentini A, Karray S, Lebel J-M, Hamza-Chaffai A. Responses of primary cultured haemocytes derived from the marine gastropod Haliotis tuberculata to an industrial effluent exposure. Cytotechnology. 2017, 69(2): 191–200.

190 Brulle F, Cocquerelle C, Mitta G, Castric V, Douay F, Leprêtre A, Vandenbulcke F. Identification and expression profile of gene transcripts differentially expressed during metallic exposure in Eisenia fetida coelomocytes. Dev Comp Immunol. 2008, 32(12): 1441–1453.

191 Hankard PK, Svendsen C, Wright J, Wienberg C, Fishwick SK, Spurgeon DJ, Weeks JM. Biological assessment of contaminated land using earthworm biomarkers in support of chemical analysis. Sci Total Environ. 2004, 330(1-3): 9–20.

192 Biser JA, Vogel LA, Berger J, Hjelle B, Loew SS. Effects of heavy metals on immunocompetence of white-footed mice (Peromyscus leucopus). J Wildl Dis. 2004, 40(2): 173–184.

193 Lambert AL, Dong W, Selgrade MK, Gilmour MI. Enhanced allergic sensitization by residual oil fly ash particles is mediated by soluble metal constituents. Toxicol Appl Pharmacol. 2000, 165(1): 84–93.

194 Thompson González N, Ong J, Luo L, MacKenzie D. Chronic community exposure to environmental metal mixtures is associated with selected cytokines in the Navajo Birth Cohort Study (NBCS). Int J Environ Res Public Health. 2022, 19(22): 14939.

195 Wang Z, Neuburg D, Li C, Su L, Kim JY, Chen JC, Christiani DC. Global gene expression profiling in whole-blood samples from individuals exposed to metal fumes. Environ Health Perspect. 2005, 113(2): 233–241.

196 Qayyum MA, Shah MH. Disparities in trace metal levels in Hodgkin/non-Hodgkin lymphoma patients in comparison with controls. Biol Trace Elem Res. 2020, 194(1): 34–47.

197 Liu Y, Zhang Z, Han D, Zhao Y, Yan X, Cui S. Association between environmental chemicals co-exposure and peripheral blood immune-inflammatory indicators. Front Pub Health. 2022, 10: 980987.

198 Jadhav SH, Sarkar SN, Ram GC, Tripathi HC. Immunosuppressive effect of subchronic exposure to a mixture of eight heavy metals, found as groundwater contaminants in different areas of India, through drinking water in male rats. Arch Environ Contam Toxicol. 2007, 53(3): 450–458.

199 Petitjean Q, Jacquin L, Riem L, Pitout M, Perrault A, Cousseau M, Laffaille P, Jean S. Intraspecific variability of responses to combined metal contamination and immune challenge among wild fish populations. Environ Pollut. 2021, 272: 116042.

200 Kershaw JL, Hall AJ. Mercury in cetaceans: exposure, bioaccumulation and toxicity. Sci Total Environ. 2019, 694: 133683.

201 Fortier M, Omara F, Bernier J, Brousseau P, Fournier M. Effects of physiological concentrations of heavy metals both individually and in mixtures on the viability and function of peripheral blood human leukocytes in vitro. J Toxicol Environ Health A. 2008, 71(19): 1327–1337.

202 Wade MG, Foster WG, Younglai EV, McMahon A, Leingartner K, Yagminas A, Blakey D, Fournier M, Desaulniers D, Hughes CL. Effects of subchronic exposure to a complex mixture of persistent contaminants in male rats: systemic, immune, and reproductive effects. Toxicol Sci. 2002, 67(1): 131–143.

203 Mahboub HH, Shahin K, Mahmoud SM, Altohamy DE, Husseiny WA, Mansour DA, Rasheed N, Ali HS, Ismail SH, Rahman ANA. Silica nanoparticles are novel aqueous additive mitigating heavy metals toxicity and improving the health of African catfish, Clarias gariepinus. Aquat Toxicol. 2022, 249: 106238.

204 Kim SH, Sharma RP. Cytotoxicity of inorganic mercury in murine T and B lymphoma cell lines: involvement of reactive oxygen species, Ca(2+) homeostasis, and cytokine gene expression. Toxicol In Vitro. 2003, 17(4): 385–395.

205 Mahboub HH, Beheiry RR, Shahin SE, Behairy A, Khedr MHE, Ibrahim SM, Elshopakey GE, Daoush WM, Altohamy DE, Ismail TA, El-Houseiny W. Adsorptivity of mercury on magnetite nano-particles and their influences on growth, economical, hemato-biochemical, histological parameters and bioaccumulation in Nile tilapia (Oreochromis niloticus). Aquat Toxicol. 2021. 235: 105828.

206 Kennedy SN, Keogh M, Levin M, Castellini JM, Lian M, Fadely BS, Rea LD, O'Hara TM. Regional variations and relationships among cytokine profiles, white blood cell counts, and blood mercury concentrations in Steller sea lion (Eumetopias jubatus) pups. Sci Total Environ. 2021, 775: 144894.

207 Levin M, Jasperse L, Desforges JP, O'Hara T, Rea L, Castellini JM, Maniscalco JM, Fadely B, Keogh M. Methyl mercury (MeHg) in vitro exposure alters mitogen-induced lymphocyte proliferation and cytokine expression in Steller sea lion (Eumetopias jubatus) pups. Sci Total Environ. 2020, 725: 138308.

208 Colombo M, Hamelin C, Kouassi E, Fournier M, Bernier J. Differential effects of mercury, lead, and cadmium on IL-2 production by Jurkat T cells. Clin Immunol. 2004, 111(3): 311–322.

209 Tvermoes BE, Unice KM, Winans B, Kovochich M, Christian WV, Donovan E, Fung ES, Finley BL, Kimber I, Paustenbach DJ. A preliminary evaluation of immune stimulation following exposure to metal particles and ions using the mouse popliteal lymph node assay. Toxicol Appl Pharmacol. 2016, 308: 77–90.

210 Schilz JR, Dashner-Titus EJ, Luo L, Simmons KA, MacKenzie DA, Hudson LG. Co-exposure of sodium arsenite and uranyl acetate differentially alters gene expression in CD3/CD28 activated CD4+ T-cells. Toxicol Rep. 2021, 8: 1917–1929.

211 Mezouar S, Mege JL. Changing the paradigm of IFN-gamma at the interface between innate and adaptive immunity: macrophage-derived IFN-gamma. J Leukoc Biol. 2020, 108(1): 419–426.

212 Yang XO, Pappu BP, Nurieva R, Akimzhanov A, Kang HS, Chung Y, Ma L, Shah B, Panopoulos AD, Schluns KS, Watowich SS, Tian Q, Jetten AM, Dong C. T helper 17 lineage differentiation is programmed by orphan nuclear receptors ROR alpha and ROR gamma. Immunity. 2008, 28(1): 29–39.

213 Song Y, Asselman J, De Schamphelaere KAC, Salbu B, Tollefsen KE. Deciphering the combined effects of environmental stressors on gene transcription: a conceptual approach. Environ Sci Technol. 2018, 52(9): 5479–5489.

214 Lombaert N, Lison D, Van Hummelen P, Kirsch-Volders M. In vitro expression of hard metal dust (WC-Co)–responsive genes in human peripheral blood mononucleated cells. Toxicol Appl Pharmacol. 2008, 227(2): 299–312.

215 Lombaert N, Castrucci E, Decordier I, Van Hummelen P, Kirsch-Volders M, Cundari E, Lison D. Hard-metal (WC-Co) particles trigger a signaling cascade involving p38 MAPK, HIF-1alpha, HMOX1, and p53 activation in human PBMC. Arch Toxicol. 2013, 87(2): 259–268.

216 Zeromski J, Jezewska E. Functional alterations of human blood monocytes after exposure to various nickel compounds in vitro: an effect on the production of hydrogen peroxide. Immunol Lett. 1995, 45(1-2): 117–121.

217 Pal R, Chakraborty B, Nath A, Singh LM, Ali M, Rahman DS, Sengupta M. Noble metal nanoparticle-induced oxidative stress modulates tumor associated macrophages (TAMs) from an M2 to M1 phenotype: an in vitro approach. Int Immunopharmacol. 2016, 38: 332–341.

218 Roszak J, Smok-Pieniążek A, Jeżak K, Domeradzka-Gajda K, Grobelny J, Tomaszewska E, Stępnik M. Combined effect of silver nanoparticles and aluminium chloride, butylparaben or diethylphthalate on the malignancy of MDA-MB-231 breast cancer cells and tumor-specific immune responses of human macrophages and monocyte-derived dendritic cells. Toxicol In Vitro. 2020, 65: 104774.

219 Rajeshkumar S, Liu Y, Ma J, Duan HY, Li X. Effects of exposure to multiple heavy metals on biochemical and histopathological alterations in common carp, Cyprinus carpio L. Fish Shellfish Immunol. 2017, 70: 461–472.

220 Cachinho SCP, Pu F, Hunt JA. Cytokine secretion from human peripheral blood mononuclear cells cultured in vitro with metal particles. J Biomed Mater Res A. 2013, 101(4): 1201–1209.

221 Shahryari A, Omanovic S, Szpunar JA. Electrochemical formation of highly pitting resistant passive films on a biomedical grade 316LVM stainless steel surface. Mater Sci Eng C Mater Biol Appl. 2008, 28(1): 94–106.

222 Antonini JM, Roberts JR. Chromium in stainless steel welding fume suppresses lung defense responses against bacterial infection in rats. J Immunotoxicol. 2007, 4(2): 117–127.

223 Erdely A, Antonini JM, Young S-H, Kashon ML, Gu JK, Hulderman T, Salmen R, Meighan T, Roberts JR, Zeidler-Erdely PC. Oxidative stress and reduced responsiveness of challenged circulating leukocytes following pulmonary instillation of metal-rich particulate matter in rats. Part Fibre Toxicol. 2014, 11: 34.

224 Roedel EQ, Cafasso DE, Lee KWM, Pierce LM. Pulmonary toxicity after exposure to military-relevant heavy metal tungsten alloy particles. Toxicol Appl Pharmacol. 2012, 259(1): 74–86.

225 Guilloteau E, Djouina M, Caboche S, Waxin C, Deboudt K, Beury D, Hot D, Pichavant M, Dubuquoy L, Launay D, Vignal C, Choël M, Body-Malapel M. Exposure to atmospheric Ag, TiO(2), Ti and SiO(2) engineered nanoparticles modulates gut inflammatory response and microbiota in mice. Ecotoxicol Environ Saf. 2022, 236: 113442.

226 Kelley AS, Banker M, Goodrich JM, Dolinoy DC, Burant C, Domino SE, Smith YR, Song PXK, Padmanabhan V. Early pregnancy exposure to endocrine disrupting chemical mixtures are associated with inflammatory changes in maternal and neonatal circulation. Sci Rep. 2019, 9(1): 5422.

227 Aung MT, Meeker JD, Boss J, Bakulski KM, Mukherjee B, Cantonwine DE, McElrath TF, Ferguson KK. Manganese is associated with increased plasma interleukin-1β during pregnancy, within a mixtures analysis framework of urinary trace metals. Reprod Toxicol. 2020, 93: 43–53.

228 Cobbina SJ, Xu H, Zhao T, Mao G, Zhou Z, Wu X, Liu H, Zou Y, Wu X, Yang L. A multivariate assessment of innate immune-related gene expressions due to exposure to low concentration individual and mixtures of four kinds of heavy metals on zebrafish (Danio rerio) embryos. Fish Shellfish Immunol. 2015, 47(2): 1032–1042.

229 Lauridsen C. From oxidative stress to inflammation: redox balance and immune system. Poult Sci. 2019, 98(10): 4240–4246.

230 Mese Y, Tuncsoy B, Ozalp P. Effects of Cu, Zn and their mixtures on bioaccumulation and antioxidant enzyme activities in Galleria mellonella L. (Lepidoptera: Pyralidae). Ecotoxicology. 2022, 31(4): 649–656.

231 Niederman CN, Blodgett D, Eversole D, Schurig GG, Thatcher CD. Effect of copper and iron on neutrophil function and humoral immunity of gestating beef cattle. J Am Vet Med Assoc. 1994, 204(11): 1796–1800.

232 Kargin D. Changes in serum physiological and biochemical parameters of male Swiss albino mice after oral administration of metal oxide nanoparticles (ZnO, CuO, and ZnO+CuO). Biol Trace Elem Res. 2021, 199(11): 4218–4224.

233 Lin S, Li D, Zhou Z, Xu C, Mei X, Tian H. Therapy of spinal cord injury by zinc modified gold nanoclusters via immune-suppressing strategies. J Nanobiotechnology. 2021, 19(1): 281.

234 Jiao LF, Zhang QH, Wu H, Wang CC, Cao ST, Feng J, Hu CH. Influences of copper/zinc-loaded montmorillonite on growth performance, mineral retention, intestinal morphology, mucosa antioxidant capacity, and cytokine contents in weaned piglets. Biol Trace Elem Res. 2018, 185(2): 356–363.

235 Wang Y, Zhao H, Yang X, Mu M, Zong H, Luo L, Xing M. Excessive Cu(2+) deteriorates arsenite-induced apoptosis in chicken brain and resulting in immunosuppression, not in homeostasis. Chemosphere. 2020, 239: 124758.

236 Zheng R, Wang P, Cao B, Wu M, Li X, Wang H, Chai L. Intestinal response characteristic and potential microbial dysbiosis in digestive tract of Bufo gargarizans after exposure to cadmium and lead, alone or combined. Chemosphere. 2021, 271: 129511.

237 Bonefeld CM, Geisler C, Gimenéz-Arnau E, Lepoittevin J-P, Uter W, Johansen JD. Immunological, chemical and clinical aspects of exposure to mixtures of contact allergens. Contact Dermatitis. 2017, 77(3): 133–142.

238 Kinbara M, Nagai Y, Takano-Yamamoto T, Sugawara S, Endo Y. Cross-reactivity among some metals in a murine metal allergy model: metal allergy cross-reactivity in a murine model. British J Dermatol. 2011, 165(5): 1022–1029.

239 Bonefeld CM, Nielsen MM, Vennegaard MT, Johansen JD, Geisler C, Thyssen JP. Nickel acts as an adjuvant during cobalt sensitization. Exp Dermatol. 2015, 24(3): 229–231.

240 Knoell DL, Wyatt TA. The adverse impact of cadmium on immune function and lung host defense. Semin Cell Dev Biol. 2021, 115: 70–76.

241 Chen J, Pan T, Wan N, Sun Z, Zhang Z, Li S. Cadmium-induced endoplasmic reticulum stress in chicken neutrophils is alleviated by selenium. J Inorg Biochem. 2017, 170: 169–177.

242 Massadeh AM, Al-Safi SA, Momani IF, Al-Mahmoud M, Alkofahi AS. Analysis of cadmium and lead in mice organs: effect of Nigella sativa L. (Black Cumin) on the distribution and immunosuppressive effect of cadmium-lead mixture in mice. Biol Trace Elem Res. 2007, 115(2): 157–167.

243 Marefati N, Ghorani V, Shakeri F, Boskabady M, Kianian F, Rezaee R, Boskabady MH. A review of anti-inflammatory, antioxidant, and immunomodulatory effects of Allium cepa and its main constituents. Pharm Biol. 2021, 59(1): 287–302.

244 Massadeh AM, Al-Safi SA, Momani IF, Alomary AA, Jaradat QM, AlKofahi AS. Garlic (Allium sativum L.) as a potential antidote for cadmium and lead intoxication: cadmium and lead distribution and analysis in different mice organs. Biol Trace Elem Res. 2007, 120(1-3): 227–234.

245 Welch BM, Branscum A, Geldhof GJ, Ahmed SM, Hystad P, Smit E, Afroz S, Megowan M, Golam M, Sharif O, Rahman M, Quamruzzaman Q, Christiani DC, Kile ML. Evaluating the effects between metal mixtures and serum vaccine antibody concentrations in children: a prospective birth cohort study. Environ Health. 2020, 19(1): 41.

246 Anas AA, de Vos AF, Hoogendijk AJ, van Lieshout MH, van Heijst JW, Florquin S, Li Z, van ‘t Veer C, van der Poll T. Endoplasmic reticulum chaperone gp96 in macrophages is essential for protective immunity during Gram-negative pneumonia. J Pathol. 2016, 238(1): 74–84.

247 Martín-Folgar R, Martínez-Guitarte J-L. Effects of single and mixture exposure of cadmium and copper in apoptosis and immune related genes at transcriptional level on the midge Chironomus riparius Meigen (Diptera, Chironomidae). Sci Total Environ. 2019, 677: 590–598.

248 Stankevičiūtė M, Sauliutė G, Makaras T, Čapukoitienė B, Vansevičiūtė G, Markovskaja S. Biomarker responses in perch (Perca fluviatilis) under multiple stress: parasite co-infection and multicomponent metal mixture exposure. Environ Res. 2022, 207: 112170.

249 Chernousova S, Epple M. Silver as antibacterial agent: ion, nanoparticle, and metal. Angew Chem Int Ed Engl. 2013, 52(6): 1636–1653.

250 Jafari AR, Mosavi T, Mosavari N, Majid A, Movahedzade F, Tebyaniyan M, Kamalzadeh M, Dehgan M, Jafari S, Arastoo S. Mixed metal oxide nanoparticles inhibit growth of Mycobacterium tuberculosis into THP-1 cells. Int J Mycobacteriol. 2016, 5 Suppl 1: S181–S183.

251 Jafari A, Mosavari N, Movahedzadeh F, Nodooshan SJ, Safarkar R, Moro R, Kamalzadeh M, Majidpour A, Boustanshenas M, Mosavi T. Bactericidal impact of Ag, ZnO and mixed AgZnO colloidal nanoparticles on H(37)Rv Mycobacterium tuberculosis phagocytized by THP-1 cell lines. Microb Pathog. 2017, 110: 335–344.

252 Mapesa JO, Maxwell AL, Ryan EP. An exposome perspective on environmental enteric dysfunction. Environ Health Perspect. 2016, 124(8): 1121–1126.

253 Liu Y, Zhang S, Deng H, Chen A, Chai L. Lead and copper led to the dysregulation of bile acid homeostasis by impairing intestinal absorption in Bufo gargarizans larvae: an integrated metabolomics and transcriptomics approach. Sci Total Environ. 2023, 855: 159031.

254 Siegel RL, Miller KD, Wagle NS, Jemal A. Cancer statistics, 2023. CA: a Cancer J Clin. 2023, 73(1): 17–48.

255 Ugai T, Sasamoto N, Lee H-Y, Ando M, Song M, Tamimi RM, Kawachi I, Campbell PT, Giovannucci EL, Weiderpass E, Rebbeck TR, Ogino S. Is early-onset cancer an emerging global epidemic? Current evidence and future implications. Nat Rev Clin Oncol. 2022, 19(10): 656–673.

256 Lacerda LM, Garcia SC, da Silva LB, de Avila Dornelles M, Presotto AT, Lourenco ED, de Franceschi ID, Fernandes E, Wannmacher CMD, Brucker N, Sauer E, Gioda A, Machado AB, Oliveira E, Trombini TL, Feksa LR. Evaluation of

hematological, biochemical parameters and thiol enzyme activity in chrome plating workers. Environ Sci Pollut Res Int. 2019, 26(2): 1892–1901.

257 Li N, Zhai Z, Zheng Y, Lin S, Deng Y, Xiang G, Yao J, Xiang D, Wang S, Yang P, Yang S, Xu P, Wu Y, Hu J, Dai Z, Wang M. Association of 13 occupational carcinogens in patients with cancer, individually and collectively, 1990-2017. JAMA Netw Open. 2021, 4(2): e2037530.

258 Bardach AE, Ciapponi A, Soto N, Chaparro MR, Calderon M, Briatore A, Cadoppi N, Tassara R, Litter MI. Epidemiology of chronic disease related to arsenic in Argentina: a systematic review. Sci Total Environ. 2015, 538: 802–816.

259 IARC (2022) https://monographs.iarc.who.int/list-of-classifications.

260 Ma X, Yang Y, Li HL, Zheng W, Gao J, Zhang W, Yang G, Shu XO, Xiang YB. Dietary trace element intake and liver cancer risk: results from two population-based cohorts in China. Int J Cancer. 2017, 140(5): 1050–1059.

261 Anyanwu BO, Ezejiofor AN, Igweze ZN, Orisakwe OE. Heavy metal mixture exposure and effects in developing nations: an update. Toxics. 2018, 6(4): 65.

262 Wu X, Cobbina SJ, Mao G, Xu H, Zhang Z, Yang L. A review of toxicity and mechanisms of individual and mixtures of heavy metals in the environment. Environ Sci Pollut Res Int. 2016, 23(9): 8244–8259.

263 Luo Y, Ma J, Lu W. The significance of mitochondrial dysfunction in cancer. Int J Mol Sci. 2020, 21(16): 5598.

264 Caliri AW, Tommasi S, Besaratinia A. Relationships among smoking, oxidative stress, inflammation, macromolecular damage, and cancer. Mutat Res Rev Mutat Res. 2021, 787: 108365.

265 Bade BC, Dela Cruz CS. Lung cancer 2020: epidemiology, etiology, and prevention. Clin Chest Med. 2020, 41(1): 1–24.

266 Bracken-Clarke D, Kapoor D, Baird AM, Buchanan PJ, Gately K, Cuffe S, Finn SP. Vaping and lung cancer - A review of current data and recommendations. Lung Cancer. 2021, 153: 11–20.

267 Soleimani F, Dobaradaran S, De-la-Torre GE, Schmidt TC, Saeedi R. Content of toxic components of cigarette, cigarette smoke vs cigarette butts: a comprehensive systematic review. Sci Total Environ. 2022, 813: 152667.

268 Schraufnagel DE. The health effects of ultrafine particles. Exp Mol Med. 2020, 52(3): 311–317.

269 Su Z, Wei MN, Jia XH, Fan YG, Zhao FH, Zhou QH, Taylor PR, Qiao YL. Arsenic, tobacco use, and lung cancer: an occupational cohort with 27 follow-up years. Environ Res. 2022, 206: 112611.

270 Lee NW, Wang HY, Du CL, Yuan TH, Chen CY, Yu CJ, Chan CC. Air-polluted environmental heavy metal exposure increase lung cancer incidence and mortality: a population-based longitudinal cohort study. Sci Total Environ. 2022, 810: 152186.

271 Ge X, Li M, Song G-X, Zhang Z, Yin J, Ge Z, Shi Z, Liu LZ, Jiang BH, Qian X, Shen H. Chromium (VI)-induced ALDH1A1/EGF axis promotes lung cancer progression. Clin Transl Med. 2022, 12(12): e1136

272 Genchi G, Carocci A, Lauria G, Sinicropi MS, Catalano A. Nickel: human health and environmental toxicology. Int J Environ Res Public Health. 2020, 17(3): 679.

273 Zimta AA, Schitcu V, Gurzau E, Stavaru C, Manda G, Szedlacsek S, Berindan-Neagoe I. Biological and molecular modifications induced by cadmium and arsenic during breast and prostate cancer development. Environ Res. 2019, 178: 108700.

274 Waks AG, Winer EP. Breast cancer treatment: a review. JAMA. 2019, 321(3): 288–300.

275 Dutta S, Gorain B, Choudhury H, Roychoudhury S, Sengupta P. Environmental and occupational exposure of metals and female reproductive health. Environ Sci Pollut Res Int. 2022, 29(41): 62067–62092.

276 Lim JT, Tan YQ, Valeri L, Lee J, Geok PP, Chia SE, Ong CN, Seow WJ. Association between serum heavy metals and prostate cancer risk - A multiple metal analysis. Environ Int. 2019, 132: 105109.

277 Chang WH, Lee CC, Yen YH, Chen HL. Oxidative damage in patients with benign prostatic hyperplasia and prostate cancer co-exposed to phthalates and to trace elements. Environ Int. 2018, 121(Pt 2): 1179–1184.

278 Kresovich JK, Erdal S, Chen HY, Gann PH, Argos M, Rauscher GH. Metallic air pollutants and breast cancer heterogeneity. Environ Res. 2019, 177: 108639.

279 Michalczyk K, Cymbaluk-Ploska A. The role of zinc and copper in gynecological malignancies. Nutrients. 2020, 12(12): 3732.

280 Morgan E, Soerjomataram I, Rumgay H, Coleman HG, Thrift AP, Vignat J, Laversanne M, Ferlay J, Arnold M. The global landscape of esophageal squamous cell carcinoma and esophageal adenocarcinoma incidence and mortality in 2020 and projections to 2040: new estimates from GLOBOCAN 2020. Gastroenterology. 2022, 163(3):649–658.e2.

281 Zhao Y, Zhao W, Li J, Lin S, Li L, Ren Z, Lu J, Xing X, Liu X. Effect of dietary consumption on the survival of esophageal squamous cell carcinoma: a prospective cohort study. Eur J Clin Nutr. 2023, 77(1): 55–64.

282 Lin X, Peng L, Xu X, Chen Y, Zhang Y, Huo X. Connecting gastrointestinal cancer risk to cadmium and lead exposure in the Chaoshan population of Southeast China. Environ Sci Pollut Res Int. 2018, 25: 17611–17619.

283 Chen J, Zhou Z, Lin X, Liao J, Zhang Y, Xie B, Huang Y, Peng L. Environmental cadmium exposure promotes the development, progression and chemoradioresistance of esophageal squamous cell carcinoma. Front Cell Dev Biol. 2022, 10: 792933.

284 Yang J, Guo T, Liang X, Zhai Y, Cheng Y, Sun H, Cui Y, Cheng X. Cadmium inhibits apoptosis of human esophageal epithelial cells by upregulating CDK6. Ecotoxicol Environ Saf. 2020, 205: 111146.

285 Keum N, Giovannucci E. Global burden of colorectal cancer: emerging trends, risk factors and prevention strategies. Nat Rev Gastroenterol Hepatol. 2019, 16(12): 713–732.

286 Sohrabi M, Gholami A, Azar MH, Yaghoobi M, Shahi MM, Shirmardi S, Nikkhah M, Kohi Z, Salehpour D, Khoonsari MR, Hemmasi G, Zamani F, Sohrabi M, Ajdarkosh H. Trace element and heavy metal levels in colorectal cancer: comparison between cancerous and non-cancerous tissues. Biol Trace Elem Res. 2018, 183(1): 1–8.

287 Núñez O, Fernández-Navarro P, Martín-Méndez I, Bel-Lan A, Locutura JF, López-Abente G. Arsenic and chromium topsoil levels and cancer mortality in Spain. Environ Sci Pollut Res Int. 2016, 23(17): 17664–17675.

288 Zhang Z, Cao H, Song N, Zhang L, Cao Y, Tai J. Long-term hexavalent chromium exposure facilitates colorectal cancer in mice associated with changes in gut microbiota composition. Food Chem Toxicol. 2020, 138: 111237.

289 Ahsan A, Liu Z, Su R, Liu C, Liao X, Su M. Potential chemotherapeutic effect of selenium for improved canceration of esophageal cancer. Int J Mol Sci. 2022, 23(10): 5509.

290 Amaral AF, Porta M, Silverman DT, Milne RL, Kogevinas M, Rothman N, Cantor KP, Jackson BP, Pumarega JA, López T, Carrato A, Guarner L, Real FX, Malats N. Pancreatic cancer risk and levels of trace elements. Gut. 2012, 61(11): 1583–1588.

291 Kousa YA, Schutte BC. Toward an orofacial gene regulatory network. Dev Dyn. 2016, 245(3): 220–232.

292 Mossey PA, Little J, Munger RG, Dixon MJ, Shaw WC. Cleft lip and palate. Lancet. 2009, 374(9703): 1773–1785.

293 Panamonta V, Pradubwong S, Panamonta M, Chowchuen B. Global birth prevalence of orofacial clefts: a systematic review. J Med Assoc Thai. 2015, 98(Suppl 7): S11–21.

294 Garcia-Sevillano MA, Garcia-Barrera T, Gomez-Ariza JL. Application of metallomic and metabolomic approaches in exposure experiments on laboratory mice for environmental metal toxicity assessment. Metallomics. 2014, 6(2): 237–248.

295 Rani A, Kumar A, Lal A, Pant M. Cellular mechanisms of cadmium-induced toxicity: a review. Int J Environ Health Res. 2014, 24(4): 378–399.

296 Caserta D, Graziano A, Lo Monte G, Bordi G, Moscarini M. Heavy metals and placental fetal-maternal barrier: a mini-review on the major concerns. Eur Rev Med Pharmacol Sci. 2013, 17(16): 2198–2206.

297 Wang A, Holladay SD, Wolf DC, Ahmed SA, Robertson JL. Reproductive and developmental toxicity of arsenic in rodents: a review. Int J Toxicol. 2006, 25(5): 319–331.

298 Suzuki A, Sangani DR, Ansari A, Iwata J. Molecular mechanisms of midfacial developmental defects. Dev Dyn. 2016, 245(3): 276–293.

299 Szabo-Rogers HL, Smithers LE, Yakob W, Liu KJ. New directions in craniofacial morphogenesis. Dev Biol. 2010, 341(1): 84–94.

300 Cordero DR, Brugmann S, Chu YN, Bajpai R, Jame M, Helms JA. Cranial neural crest cells on the move: their roles in craniofacial development. Am J Med Genet A. 2011, 155a(2): 270–279.

301 Lough KJ, Byrd KM, Spitzer DC, Williams SE. Closing the gap: mouse models to study adhesion in secondary palatogenesis. J Dent Res. 2017, 96(11): 1210–1220.

302 Reynolds K, Kumari P, Sepulveda Rincon L, Gu R, Ji Y, Kumar S, Zhou CJ. Wnt signaling in orofacial clefts: crosstalk, pathogenesis and models. Dis Model Mech. 2019, 12(2): dmm037051.

303 Wei Y, Jin L, Li Z, Liu J, Wang L, Pi X, Yin S, Wang C, Ren A. Levels of uranium and thorium in maternal scalp hair and risk of orofacial clefts in offspring. J Environ Radioact. 2019, 204: 125–131.

304 Ni W, Yang W, Yu J, Li Z, Jin L, Liu J, Zhang Y, Wang L, Ren A. Umbilical cord concentrations of selected heavy metals and risk for orofacial clefts. Environ Sci Technol. 2018, 52(18): 10787–10795.

305 Oginawati K, Susetyo SH, Cahyani ZN, Fahimah N. Investigation of exposure to heavy metals (Hg, Pb, Cd, Co, and Cr) as the cause of congenital anomaly cases (orofacial cleft) in infants. Environ Sci Pollut Res Int. 2022, 29(54): 82169–82185.

306 Takeuchi M, Yoshida S, Kawakami C, Kawakami K, Ito S, Japan E. Children's Study G. Association of maternal heavy metal exposure during pregnancy with isolated cleft lip and palate in offspring: Japan Environment and Children's Study (JECS) cohort study. PLoS One. 2022, 17(3): e0265648.

307 Guo Y, Liu L, Ni W, Pan Y, Chen Y, Xie Q, Liu Y, Jin L, Li Z, Ren A, Wang L. Uranium concentration in umbilical cord may increase the risk for orofacial clefts. Environ Res. 2020, 182: 109103.

308 Hozyasz KK, Kaczmarczyk M, Dudzik J, Bulska E, Dudkiewicz Z, Szymanski M. Relation between the concentration of zinc in maternal whole blood and the risk of an infant being born with an orofacial cleft. Br J Oral Maxillofac Surg. 2009, 47(6): 466–469.

309 Krapels IP, Rooij IA, Wevers RA, Zielhuis GA, Spauwen PH, Brussel W, Steegers-Theunissen RP. Myo-inositol, glucose and zinc status as risk factors for non-syndromic cleft lip with or without cleft palate in offspring: a case-control study. BJOG. 2004, 111(7): 661–668.

310 Munger RG, Tamura T, Johnston KE, Feldkamp ML, Pfister R, Carey JC. Plasma zinc concentrations of mothers and the risk of oral clefts in their children in Utah. Birth Defects Res A Clin Mol Teratol. 2009, 85(2): 151–155.

311 Rogers EH, Chernoff N, Kavlock RJ. The teratogenic potential of cacodylic acid in the rat and mouse. Drug Chem Toxicol. 1981, 4(1): 49–61.

312 Gale TF, Layton WM. The susceptibility of inbred strains of hamsters to cadmium-induced embryotoxicity. Teratology. 1980, 21(2): 181–186.

313 Salvatori F, Talassi CB, Salzgeber SA, Spinosa HS, Bernardi MM. Embryotoxic and long-term effects of cadmium exposure during embryogenesis in rats. Neurotoxicol Teratol. 2004, 26(5): 673–680.

314 Ferm VH. Teratogenic effects and placental permeability of heavy metals. Curr Top Pathol. 1976, 62: 145–151.

315 Holt D, Webb M. Teratogenicity of ionic cadmium in the Wistar rat. Arch Toxicol. 1987, 59(6): 443–447.

316 Carpenter SJ. Enhanced teratogenicity of orally administered lead in hamsters fed diets deficient in calcium or iron. Toxicology. 1982, 24(3-4): 259–271.

317 Sunderman FW Jr, Shen SK, Reid MC, Allpass PR. Teratogenicity and embryotoxicity of nickel carbonyl in Syrian hamsters. Teratog Carcinog Mutagen. 1980, 1(2): 223–233.

318 Lu CC, Matsumoto N, Iijima S. Teratogenic effects of nickel chloride on embryonic mice and its transfer to embryonic mice. Teratology. 1979, 19(2): 137–142.

319 Yang W, Guo Y, Ni W, Tian T, Jin L, Liu J, Li Z, Ren A, Wang L. Hypermethylation of WNT3A gene and non-syndromic cleft lip and/or palate in association with in utero exposure to lead: a mediation analysis. Ecotoxicol Environ Saf. 2021, 208: 111415.

320 Flentke GR, Garic A, Amberger E, Hernandez M, Smith SM. Calcium-mediated repression of beta-catenin and its transcriptional signaling mediates neural crest cell death in an avian model of fetal alcohol syndrome. Birth Defects Res A Clin Mol Teratol. 2011, 91(7): 591–602.

321 da Costa MCV, Kmecick M, de Freitas PF, Ortolani-Machado CF. Lead exposure affects cephalic morphogenesis and neural crest cells in Gallus embryo. Neurotoxicol Teratol. 2021, 84: 106948.

322 Hu B, Yang RJ, Cheng ZW, Liang SJ, Liang SX, Yin NY, Faiola F. Non-cytotoxic silver nanoparticle levels perturb human embryonic stem cell-dependent specification of the cranial placode in part via FGF signaling. J Hazard Mater. 2020, 393: 122440.

323 Mine Y, Suga M, Mimura S, Minoda M, Murayama T, Nikawa H, Furue MK. Cytotoxicity assay using a human pluripotent stem cell-derived cranial neural crest cell model. In Vitro Cell Dev Biol Anim. 2020, 56(7): 505–510.

324 Prasadh S, Gupta M, Wong R. In vitro cytotoxicity and osteogenic potential of quaternary Mg-2Zn-1Ca/X-Mn alloys for craniofacial reconstruction. Sci Rep. 2022, 12(1): 8259.

325 Hansen J. Oxidative stress and signal transduction in chemical teratogenesis. Birth Defects Res A Clin Mol Teratol. 2006, 76(5): 362–362.

326 Thompson J, Bannigan J. Cadmium: toxic effects on the reproductive system and the embryo. Reprod Toxicol. 2008, 25(3): 304–315.

327 Li Y, Yu Z, Zhu H. Role of programmed cell death in mediating arsenic-induced rat embryo anomalies. Wei Sheng Yan Jiu. 1998, 27(2): 91–94.

328 Bhattacharjee P, Banerjee M, Giri AK. Role of genomic instability in arsenic-induced carcinogenicity. A review. Environ Int. 2013, 53: 29–40.

329 Pilsner JR, Hu H, Ettinger A, Sanchez BN, Wright RO, Cantonwine D, Lazarus A, Lamadrid-Figueroa H, Mercado-García A, Téllez-Rojo MM, Hernandez-Avila M. Influence of prenatal lead exposure on genomic methylation of cord blood DNA. Environ Health Perspect. 2009, 117(9): 1466–1471.

330 Gottofrey J, Tjälve H. Axonal transport of cadmium in the olfactory nerve of the pike. Pharmacol Toxicol. 1991, 69(4): 242–252.

331 Blechinger SR, Warren JT Jr, Kuwada JY, Krone PH. Developmental toxicology of cadmium in living embryos of a stable transgenic zebrafish line. Environ Health Perspect. 2002, 110(10): 1041–1046.

332 Blechinger SR, Kusch RC, Haugo K, Matz C, Chivers DP, Krone PH. Brief embryonic cadmium exposure induces a stress response and cell death in the developing olfactory system followed by long-term olfactory deficits in juvenile zebrafish. Toxicol Appl Pharmacol. 2007, 224(1): 72–80.

333 Matz CJ, Krone PH. Cell death, stress-responsive transgene activation, and deficits in the olfactory system of larval zebrafish following cadmium exposure. Environ Sci Technol. 2007, 41(14): 5143–5148.

334 Kusch RC, Krone PH, Chivers DP. Chronic exposure to low concentrations of waterborne cadmium during embryonic and larval development results in the long-term hindrance of antipredator behavior in zebrafish. Environ Toxicol Chem. 2008, 27(3): 705–710.

335 Patel UN, Patel UD, Khadayata AV, Vaja RK, Patel HB, Modi CM. Assessment of neurotoxicity following single and co-exposure of cadmium and mercury in adult zebrafish: behavior alterations, oxidative stress, gene expression, and histological impairment in brain. Wat Air and Soil Poll. 2021, 232(8): 1–18.

336 Heffern K, Tierney K, Gallagher EP. Comparative effects of cadmium, zinc, arsenic and chromium on olfactory-mediated neurobehavior and gene expression in larval zebrafish (Danio rerio). Aquat Toxicol. 2018, 201: 83–90.

337 Volz SN, Hausen J, Nachev M, Ottermanns R, Schiwy S, Hollert H. Short exposure to cadmium disrupts the olfactory system of zebrafish (Danio rerio) – Relating altered gene expression in the olfactory organ to behavioral deficits. Aquat Toxicol. 2020, 226: 105555.

338 Lerebours A, Bourdineaud JP, van der Ven K, Vandenbrouck T, Gonzalez P, Camilleri V, Floriani M, Garnier-Laplace J, Adam-Guillermin C. Sublethal effects of waterborne uranium exposures on the zebrafish brain: transcriptional responses and alterations of the olfactory bulb ultrastructure. Environ Sci Technol. 2010, 44(4): 1438–1443.

339 Faucher K, Floriani M, Gilbin R, Adam-Guillermin C. Uranium-induced sensory alterations in the zebrafish Danio rerio. Aquat Toxicol. 2012, 124–125: 94–105.

340 Lazzari M, Bettini S, Milani L, Maurizii MG, Franceschini V. Differential nickel-induced responses of olfactory sensory neuron populations in zebrafish. Aquat Toxicol. 2019, 206: 14–23.

341 MacDonald TC, Sylvain NJ, James AK, Pickering IJ, Krone PH, George GN. Effects of inorganic mercury on the olfactory pits of zebrafish larvae. Metallomics. 2016, 8(5): 514–517.

342 Lazzari M, Bettini S, Milani L, Maurizii MG, Franceschini V. Response of olfactory sensory neurons to mercury ions in zebrafish: an immunohistochemical study. Microsc Microanal. 2022, 28(1): 227–242.

343 Hentig JT, Byrd-Jacobs CA. Exposure to zinc sulfate results in differential effects on olfactory sensory neuron subtypes in adult zebrafish. Int J Mol Sci. 2016, 17(9): 1445.

344 Abreu MS, Giacomini AC, Rodriguez R, Kalueff AV, Barcellos LJ. Effects of ZnSO(4)-induced peripheral anosmia on zebrafish behavior and physiology. Behav Brain Res. 2017, 320: 275–281.

345 Villamayor PR, Arana Á J, Coppel C, Ortiz-Leal I, Torres MV, Sanchez-Quinteiro P, Sánchez L. A comprehensive structural, lectin and immunohistochemical characterization of the zebrafish olfactory system. Sci Rep. 2021, 11(1): 8865.

346 Dinocourt C, Legrand M, Dublineau I, Lestaevel P. The neurotoxicology of uranium. Toxicology. 2015, 337: 58–71.

347 Ha E, Basu N, Bose-O'Reilly S, Dórea JG, McSorley E, Sakamoto M, Chan HM. Current progress on understanding the impact of mercury on human health. Environ Res. 2017, 152: 419–433.

10

Combined Molecular Toxicity Mechanism of Pesticide Mixtures

Ali Omar Jimale[a], *Yiyun Liu*[a], *De-Sheng Pei*[a,*]

[a] *School of Public Health, Chongqing Medical University, Chongqing, China*

*Corresponding author.
E-mail: peids@cqmu.edu.cn (D.S.P.)

10.1 Introduction

10.1.1 Background

Pesticides are substances or mixtures of substances, such as insecticides and herbicides, used in the agricultural industry to prevent or destroy pests [1]. As essential tools, they are widely used in modern agriculture to control a variety of pests and pathogens and improve yields [2]. An extremely small percentage of the applied pesticides can effectively act against target organisms, and large amounts of these pesticides will eventually accumulate in soil [3]. Pesticides adversely affect soil and water biodiversity, causing environmental pollution [4, 5]. As modern agricultural practices and crop production increased worldwide, the increased demand for pesticide application resulted in pesticide resistance [6].

10.1.2 Current Situation of Pesticide Pollution

Pesticide pollution is a significant environmental concern with widespread impacts on ecosystems and human health. Pesticide runoff from agricultural fields may lead to the contamination of water bodies, such as streams, ponds, lakes, and groundwater, posing risks to plants, animals, and humans [7, 8]. Pesticide leaching into the local water sources intensifies water pollution [9, 10]. The excessive utilization of pesticides also leads to the contamination of both soil and air. Inappropriate handling and disposal of pesticides result in significant environmental pollution challenges, underscoring the necessity for effective environmental management strategies [11–13]. Furthermore, the extensive use of synthetic pesticides and the inability of agricultural workers to strictly follow pesticide handling instructions contribute to additional environmental contamination [14, 15].

The effects of pesticide pollution extend beyond environmental concerns and affect human health. Studies have revealed a noticeable spatial relationship between land-use regions and the emergence of diseases stemming from pesticide pollution, suggesting the potential health hazards associated with pesticide exposure [16, 17]. Repeated utilization of pesticides at a low rate can result in high costs and environmental pollution, stressing the importance of adopting sustainable pest management practices [18, 19]. Efforts to reduce pesticide pollution involve exploring biochar's potential to control pesticide bioavailability in soil, which may help control pesticide pollution [20]. Investigating residual pesticide concentrations in urban water networks also highlights the health threat of increasing pesticide pollutants in water resources [21].

The pesticide mixture pollution is another complex and urgent issue with significant ecological and human health implications. The pesticide mixtures have been discovered to harm various aspects of health, such as DNA damage, altered enzyme

Toxicological Assessment of Combined Chemicals in the Environment, First Edition. Edited by De-Sheng Pei and Yiyun Liu.

function, decreased survival, reduced growth rates, and changes in behavior [22, 23]. Research has revealed that mixtures of pesticides tend to have higher toxicity than individual compounds, suggesting that the cumulative effects of multiple pesticides when exposed simultaneously at low levels have substantial health consequences [24].

10.1.3 Importance of Pesticide Mixture Toxicity Mechanisms

Understanding the toxicity mechanisms of pesticide mixtures is crucial in addressing the widespread application of pesticides in agricultural areas and the consequential environmental co-occurrence, which can have harmful effects [25–27]. The complexity of their interactions, the potential for synergistic toxicity, and the challenges in determining their combined effects underline the need for comprehensive understanding [28, 29]. A deeper understanding of the toxicity mechanisms of pesticide mixtures allows the development of safer and more sustainable pesticide application practices, ensuring the protection of animal and human health.

This chapter provides a comprehensive overview of the combined molecular toxicity mechanisms of pesticide mixtures by highlighting the interactions between pesticides and the modulation of cellular signaling pathways.

10.2 Epidemiology of Pesticide Pollution

With the increasing prevalence of numerous diseases in the world, exposure to pesticides as an important environmental risk factor for many human health conditions in developed countries is designated as a significant global environmental health issue. Pesticide exposure among the general population through environmental pollution and consumption of pesticide-contaminated food and water has emerged as a major concern worldwide. The widespread application of pesticides has led to ubiquitous exposure in the Chinese general population [12]. Qi et al. [30] described that pyrethroid exposure during the first and second trimesters of pregnancy may have negative impacts on the neurodevelopmental outcomes of infants, particularly when exposure levels are high. A recent descriptive study conducted by Wang et al. [31] reported many cases of acute poisoning related to pesticides among agricultural workers. Yimaer et al. [32], in their retrospective analysis, characterized pesticide poisoning in children as a major public health issue.

10.3 Diseases Caused by Pesticides

Pesticides represent a major contributor to human health and disease. Various diseases and health issues have been linked to the use of pesticides. Evidence indicated consuming food products containing pesticides was associated with risks to human health, including nervous system dysfunction, brain injury, mental illness, and disorders of the respiratory system [33, 34]. Pesticides have also been discovered to have neurotoxin effects in humans, leading to degenerative diseases such as Alzheimer's and Parkinson's disease and birth defects [35, 36].

10.3.1 Neurodevelopmental Disorders Related to Pesticide Toxicity

Neurodevelopmental disorders are a group of conditions affecting the normal development of the nervous system, leading to difficulties in various aspects of functioning, including learning ability, behavior, social interaction, and behavior. The extensively studied neurodevelopmental disorders include attention-deficit hyperactivity disorder (ADHD), autism spectrum disorder (ASD), and impaired cognition. Liu et al. [37] presented in their review a close relationship between the etiology of ASD and exposure to pesticides during the critical developmental periods. Studies have also revealed the neurodevelopmental adverse effects of pesticide exposure in children, especially for cognitive, behavioral, and motor outcomes [38].

ADHD is a neurodevelopmental condition that arises in childhood and often lasts into adulthood. It is identified by persistent signs of inattention, hyperactivity, and impulsivity, which hinder daily activities and development. Pesticide pollution has been found as a major risk factor contributing to the development of neurodevelopmental disorders, including ADHD. Zhou et al. [39] found that high levels of Chlorpyrifos (CPF) exposure were associated with a higher risk of ADHD. ASD is another developmental disorder that affects communication and behavior prevalent in children.

10.3.2 Neurodegenerative Diseases and Pesticides

Neurodegenerative diseases represent a group of debilitating conditions that progressively damage the structure and function of the nervous system, leading to cognitive and motor dysfunction. Alzheimer's disease, Parkinson's disease, and Huntington's disease are among the most prevalent neurodegenerative disorders. While the etiology of these conditions remains multifactorial, research has suggested a potential link between neurodegenerative diseases and exposure to pesticides, raising concerns about the environmental impact on neurological health [40]. Pesticides may play a role in the development of neurodegenerative diseases through a range of complex and multilayered mechanisms. Recent studies have shown that exposure to certain pesticides can lead to oxidative stress, inflammation, mitochondrial dysfunction, and the accumulation of abnormal proteins in the brain – hallmarks of neurodegeneration [41, 42].

10.3.3 Alzheimer's Disease Caused by Pesticide Toxicity

Alzheimer's disease (AD), as a most common neurodegenerative disorder, has a profound impact on individuals and their families. The clinical manifestations of AD include memory and cognitive impairment, anxiety, confusion, and mood swings. The main neuropathological features of AD are specific neuronal degeneration, amyloid beta plaques, and neurofibrillary tangles [43]. While the causes of Alzheimer's disease are not yet fully understood, there is growing evidence to suggest that exposure to pesticides may be a contributing factor [43]. Neuronal cells in the brain communicate with each other by using neurotransmitters, which are signaling chemicals. Pesticides may contribute to Alzheimer's disease by disrupting the delicate balance of chemicals in the brain, affecting the function of neurotransmitters. Disruptions in neurotransmitter function can lead to problems with memory, learning, and cognitive function, which are all hallmark symptoms of Alzheimer's disease [43]. Besides disrupting neurotransmitter function, pesticides may also contribute to Alzheimer's disease by triggering inflammation in the brain. Chronic inflammation has been linked to a range of neurodegenerative diseases, including Alzheimer's disease[43].

10.3.4 Parkinson's Disease Caused by Pesticide Toxicity

Parkinson's disease is another common neurodegenerative disorder characterized by a constant deterioration in neuronal function, leading to brain atrophy. Pesticide exposure has been found to contribute to the development of Parkinson's disease by disrupting the nervous system functions and enhancing the overproduction of reactive oxygen species (ROS) [44].

10.3.5 Birth Defects and Pesticides

Birth defects, commonly referred to as congenital anomalies, are structural or functional abnormalities present at the time of birth. Pesticide exposure has been linked to a potential development of congenital defects. Epidemiological studies revealed an increased risk of congenital anomalies related to pesticide application [45].

10.3.6 Pesticide Exposure and Cancer Development

The widespread use of pesticides raises concerns about their potential health impacts, particularly their association with cancer development. Several studies suggest an increased risk of certain cancers, such as leukemia, lymphoma, and prostate cancer, among individuals with prolonged or high levels of pesticide exposure [46]. It is crucial to understand the essential mechanisms by which pesticides may contribute to cancer development. Some pesticides have been identified as endocrine disruptors, interfering with hormonal systems, and potentially promoting the growth of hormone-sensitive cancers [46].

10.4 The Combined Molecular Toxicity Mechanisms of Pesticide Mixtures

Pesticides are frequently used in combination to enhance their efficacy in controlling multiple pests. While the effects of individual pesticides are typically evaluated for their toxicity, the combined effects of pesticide mixtures are often disregarded. The interaction between different pesticides in a mixture can result in distinct toxicological effects that differ from those observed in individual pesticides [29, 47]. The molecular toxicity mechanism of pesticide mixtures is a complex phenomenon, influenced by the interaction between different chemicals and their toxicological characteristics when combined. One such interaction is the synergistic toxicity of pesticide mixtures, which represents the combined effect of two or more chemicals exceeding the sum of their individual effects. Table 10.1 shows the synergistic toxicities of the thiophanate-methyl and fenvalerate mixture in embryonic zebrafish. Numerous researches have emphasized the need to understand the molecular mechanisms underlying the toxicity of pesticide mixtures [48–51]. These mechanisms include oxidative stress, impairment of cytochrome P450 enzymes, disruption of cellular metabolism, and endocrine disruption [48, 51]. Figure 10.1 depicts the molecular toxicity mechanisms of pesticide mixtures.

10.4.1 Free Radicals and Oxidative Stress

Free radicals are simply referred to as molecules containing unpaired electrons [52]. Oxidative stress occurs when cells are exposed to free radicals, which are highly reactive molecules that damage cellular components, such as proteins, lipids, and DNA. As a result, cancer, heart disease, and neurodegenerative disorders are among the health problems caused by this damage [44]. Evidence has characterized oxidative stress as a key mechanism for the process of neurodegeneration in Alzheimer's and Parkinson's disease [52–54], and toxicological research has been focused on pesticide-induced oxidative stress for decades as a potential mechanism of toxicity.

Table 10.1 Synergistic toxicities of thiophanate-methyl and fenvalerate mixture to embryonic zebrafish.

Exposure time (h)	LC_{50} (95% FL)[a] mg/L		LC_{50} (95% FL)[b] mg/L		AI[c] value
	Thiophanate-methyl	Fenvalerate	Thiophanate-methyl	Fenvalerate	
24	32.6 (23.9–41.6)	35.2 (26.9–46.4)	4.2 (2.4–5.2)	4.5 (2.6–6.3)	2.9
48	26.9 (18.6–34.2)	21.1 (14.6–29.1)	2.6 (1.7–3.6)	2.0 (1.3–3.5)	4.2
72	20.7 (14.6–28.6)	7.7 (5.6–9.7)	1.6 (0.95–2.4)	0.58 (0.37–0.92)	5.6
96	12.1 (9.5–14.9)	2.8 (1.8–3.6)	0.82 (0.61–1.6)	0.18 (0.083–0.27)	6.4

a. The LC_{50} (95% fiducial limit) for thiophanate-methyl or fenvalerate individually.
b. The LC_{50} (95% fiducial limit) for thiophanate-methyl or fenvalerate in the mixture.
c. AI: additive index.
Source: The information presented in the table is taken from Wang et al. (2020)/with permission of Elsevier.

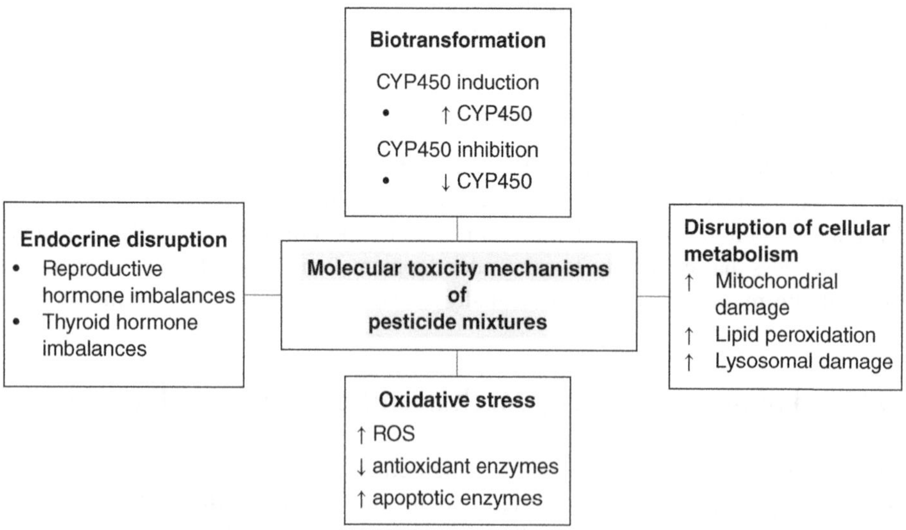

Figure 10.1 Combined molecular toxicity of pesticide mixtures.

10.4.2 Pesticide-induced Oxidative Stress

Pesticide-induced oxidative stress has been widely researched and documented in numerous fields of study. Jan et al. [55] demonstrated that pesticides cause oxidative stress in a variety of organisms, including plants, animals, and humans, leading to a range of physiological and biochemical disturbances. The mechanism of pesticide-induced oxidative stress involves the generation of reactive oxygen species (ROS) and the alteration of antioxidant defense mechanisms, which can lead to damage to biological macromolecules such as DNA, RNA, and proteins [56, 57]. Cui et al. [58] presented in their research paper that exposure to pesticides, such as insecticides and herbicides, can result in an increase in the production of free radicals, as well as lipid peroxidation and alterations in antioxidant levels. Moreover, pesticides have been found to inhibit the enzymatic and nonenzymatic antioxidant defenses, leading to oxidative stress in various biological systems [59].

Wang et al. used a mixture of thiophanate-methyl and fenvalerate and exposed to embryonic zebrafish to evaluate the underlying mixture toxicity mechanisms. The researchers' findings revealed significantly elevated malondialdehyde (MDA) levels, a key indicator of oxidative stress. This observation underscores the chemical mixture's capacity to induce overproduction of reactive oxygen species (ROS) in zebrafish, subsequently leading to extensive lipid peroxidation. The activities of antioxidant enzymes total superoxide dismutase(T-SOD) and Cu/Zn superoxide dismutase were considerably decreased. Additionally, the activities of the apoptotic enzyme (caspase 3 and caspase 9) apparently elevated in the mixture group [60]. In another study, Shukla et al. combined three pesticides, namely imidacloprid (IMD), dichlorvos (DIC), and atrazine (ATZ), to assess mixture toxicity in zebrafish. The study found a marked increase in the activities of the enzyme SOD [61]. This increased activity of the SOD enzyme implies that the enzyme contributes to the elimination of ROS induced by a mixture of pesticides. Similarly, research undertaken by Ojha et al. revealed the induction of oxidative stress by the mixture of chlorpyrifos (CPF), methyl parathion (MPT), and malathion (MLT) as evidenced by the elevated levels of MDA as well as 4-hydroxynonanal [62]. The co-exposure of the three pesticides also inhibited the activities of the antioxidant enzymes catalase (CAT), superoxide dismutase (SOD), and glutathione peroxidase (GPx) in all rat tissues [63]. The evidence has shown an exposure frequency and dose-dependent decrease in the activities of the antioxidant enzymes [62, 64].

10.4.3 Impairment of Cytochrome P450 Enzymes Due to Pesticide Toxicity

Cytochrome P450 enzymes are primarily involved in the oxidation of a wide range of substrates, contributing to the metabolism of drugs, hormones, and xenobiotics. The catalytic cycle of cytochrome P450 involves binding of the substrate, followed by electron transfer and oxygen binding, ultimately leading to the formation of a metabolite.

Pesticides inhibit cytochrome P450 enzymes through various mechanisms, including competitive binding at the enzyme's active site, interference with electron transfer, and modulation of enzyme expression [65]. Pyrethroids, as a common class of pesticides, have been shown to interfere with cytochrome P450 enzymes in rats by altering their expression levels [66]. This modulation can result in an imbalance in the metabolic pathways of endogenous compounds, leading to disruptions in homeostasis. The same study about the mixture of thiophanate-methyl and fenvalerate conducted by Wang et al. found sufficiently decreased activity of the detoxification enzyme (Cytochrome P450), which could be the underlying toxicity mechanism of the pesticide mixture [60]. However, another study published by Ojha et al. [62] presented increased activities of cytochrome P450 enzymes.

10.4.4 Endocrine Disruptions Due to Pesticide Toxicity

Endocrine-disrupting substances are chemicals that mimic the natural hormones produced by the endocrine system and bind to the receptors that are designed to receive those hormones. Endocrine disruptors interfere with the hormonal system, leading to adverse effects on the endocrine functions in humans. Endocrine disruption refers to a mechanism of toxicity that impairs the hormonal communication ability of cells, tissues, and organs, resulting in a wide range of adverse health outcomes such as reduced fertility, spontaneous abortion, male and female reproductive tract abnormalities, precocious puberty, polycystic ovary syndrome, neurobehavioral disorders, hypothyroidism, hyperthyroidism, immune dysfunction, and various cancers.The process of endocrine disruption occurs at various physiological levels, including altering hormone secretion, disrupting hormone–receptor interactions, and modifying circulating hormone metabolism by increasing or decreasing their excretion rate and/or biotransformation in the liver and other organs [67]. Jacobsen et al. performed research assessing the combined toxicity effects of five pesticides, namely procymidone, prochloraz, tebuconazole, epoxiconazole, and mancozeb. They demonstrated that the co-exposure of the chemicals induced anti-androgenic effects on the sexual differentiation of the male offspring [68].

10.4.5 Disruption of Cellular Metabolism Due to Pesticide Toxicity

Cellular metabolism is a layered process of chemical reactions in cells that allow them to perform crucial functions such as energy generation, growth, and maintenance. Disruption of the cellular metabolism processes may result in detrimental effects on cellular function and overall organism health. Chemical disruption of human cellular metabolism has become an etiological determinant of numerous disorders throughout the lifecycle, from neurodevelopmental abnormalities in utero to dementia in the elderly [69]. Pesticides are known to disrupt critical cellular mechanisms that sustain the metabolic requirements and activity of different systems, including the nervous system [70]. Falfushynska et al., in their study evaluating mixture toxicity mechanisms of chlorpyrifos and Roundup, presented increased levels of methylglyoxal, lipid peroxidation, and DNA damage. They also found evidence of lysosomal damage and decreased activities of antioxidant enzymes [71].

10.5 Summary

Pesticide pollution has been linked to a wide range of human diseases and disorders, including neurodevelopmental disorders, neurodegenerative diseases, cancers, neurotoxicity, and congenital defects. The complexity of pesticide interactions, the potential for synergistic toxicity, and the challenges in determining their combined adverse effects underline the need for a comprehensive understanding of the toxicity mechanisms of pesticide mixtures.

The molecular toxicity mechanism of pesticide pollution involves complex and layered phenomena, such as altering the body's antioxidant defensive mechanisms, increasing the accumulation of reactive oxygen species, inhibiting cytochrome P450 enzymes, disrupting cellular metabolism, and interfering with the endocrine systems.

Acknowledgments

The authors are grateful for the support from the Chongqing Technology Innovation and Application Development Sichuan-Chongqing Special Key Project (CSTB2024TIAD-CYKJCXX0017), High-level Talents Project of Chongqing Medical University (No. R4014), and Chongqing Postdoctoral Innovation Mentor Studio (X7928 D.S.P.).

Competing Interests

The authors declare no competing interests.

References

1 Tait S, Lori G, Tassinari R, La Rocca C, Maranghi F. In vitro assessment and toxicological prioritization of pesticide mixtures at concentrations derived from real exposure in occupational scenarios. Int J Environ Res Public Health. 2022, 19(9): 5202.

2 Karadurmus L, Kaya SI, Ozkan SA. Recent advances of enzyme biosensors for pesticide detection in foods. J Food Meas Charact. 2021, 15(5): 4582–4595.

3 Yao R, Yao S, Ai T, Huang J, Liu Y, Sun J. Organophosphate pesticides and pyrethroids in farmland of the pearl river delta, china: regional residue, distributions and risks. Int J Environ Res Public Health. 2023, 20(2): 1017.

4 Hvězdová M, Kosubová P, Košíková M, Scherr KE, Šimek Z, Brodský L, Šudoma M, Škulcová L, Sáňka M, Svobodová M, Krkošková L. Currently and recently used pesticides in Central European arable soils. Sci Total Environ. 2018, 613–614: 361–370.

5 Tsaboula A, Papadakis EN, Vryzas Z, Kotopoulou A, Kintzikoglou K, Papadopoulou-Mourkidou E. Environmental and human risk hierarchy of pesticides: a prioritization method, based on monitoring, hazard assessment and environmental fate. Environ Int. 2016, 91: 78–93.

6 Mei LC, Chen HM, Dong AY, Huang GY, Liu YW, Zhang X, Wang W, Hao GF, Yang GF. Pesticide Informatics Platform (PIP): an international platform for pesticide discovery, residue, and risk evaluation. J Agric Food Chem. 2022, 70(22): 6617–6623.

7 Tudi M, Daniel Ruan H, Wang L, Lyu J, Sadler R, Connell D, Chu C, Phung DT. Agriculture development, pesticide application and its impact on the environment. Int J Environ Res Public Health. 2021, 18(3): 1112.

8 Imfeld G, Meite F, Wiegert C, Guyot B, Masbou J, Payraudeau S. Do rainfall characteristics affect the export of copper, zinc and synthetic pesticides in surface runoff from headwater catchments? Sci Total Environ. 2020, 741: 140437.

9 Mottes C, Lesueur Jannoyer M, Le Bail M, Guéné M, Carles C, Malézieux E. Relationships between past and present pesticide applications and pollution at a watershed outlet: the case of a horticultural catchment in Martinique, French West Indies. Chemosphere. 2017, 184: 762–773.

10 Sharma A, Kumar V, Shahzad B, Tanveer M, Sidhu GPS, Handa N, Kohli SK, Yadav P, Bali AS, Parihar RD, Dar OI. Worldwide pesticide usage and its impacts on ecosystem. SN Appl Sci. 2019, 1(11): 1446.

11 Liu B, Fan Y, Li H, Zhao W, Luo S, Wang H, Guan B, Li Q, Yue J, Dong Z, Wang Y Control the entire journey of pesticide application on superhydrophobic plant surface by dynamic covalent trimeric surfactant coacervation. Adv Funct Mater. 2021, 31(5): 2006606.

12 Wei D, Kameya T, Urano K. Environmental management of pesticidal POPs in China: past, present and future. Environ Int. 2007, 33(7): 894–902.

13 Singh Y, Saxena MK. Insights into the recent advances in nano-bioremediation of pesticides from the contaminated soil. Front Microbiol [Internet]. 2022 [cited 2024 Jan 22];13. Available from: https://www.frontiersin.org/articles/10.3389/fmicb.2022.982611

14 Pérez-Herrera N, Vera-Avilés M, Castillo-Burguete T, Perera-Rios J, Esperón-Hernández R, Rojas-García AE, Medina-Díaz IM, Quintanilla-Vega B. Pesticide exposure index: practices among women from an agricultural community in southeast Mexico. Rev Int Contam Ambie. 2018, 34: 57–68.

15 Damalas CA, Eleftherohorinos IG. Pesticide exposure, safety issues, and risk assessment indicators. Int J Environ Res Public Health. 2011, 8(5): 1402–1419.

16 Chan DKY, Woo J, Ho SC, Pang CP, Law LK, Ng PW, Hung WT, Kwok T, Hui E, Orr K, Leung MF. Genetic and environmental risk factors for Parkinson's disease in a Chinese population. J Neurol Neurosurg Psychiatry. 1998, 65(5): 781–784.

17 Panis C, Kawassaki ACB, Crestani APJ, Pascotto CR, Bortoloti DS, Vicentini GE, Lucio LC, Ferreira MO, Prates RT, Vieira VK, Gaboardi SC. Evidence on human exposure to pesticides and the occurrence of health hazards in the Brazilian population: a systematic review. Public Health Front. 2022, 9: 787438.

18 Liu B, Wang Y, Yang F, Cui H, Wu D. Development of a chlorantraniliprole microcapsule formulation with a high loading content and controlled-release property. J Agric Food Chem. 2018, 66(26): 6561–6568.

19 Li R, Xie H, Zhang C, Sun Y, Yin H. ROS-responsive polymeric micelle for improving pesticides efficiency and intelligent release. J Agric Food Chem. 2020, 68(34): 9052–9060.

20 Bahia W. Adsorption-desorption behavior and pesticide bioavailability of biochar in soil. Sci Insights. 2022, 41(6): 725–731.

21 Aleseyyed SB, Rafati L, Heidarimoghadam R, Khodabakhshi M, Zafarmirmohammadi SA, Heidari S. Investigation of residual concentration of organochlorine, organophosphorus, and carbamate pesticides in urban drinking water networks of Hamadan Province, Iran. Avicenna J Environ Health Eng. 2021, 8(1): 9–16.

22 Tosi S, Nieh JC. Lethal and sublethal synergistic effects of a new systemic pesticide, flupyradifurone (Sivanto®), on honeybees. Proc Royal Soc Biol Sci. 2019, 286(1900): 20190433.

23 Uwizeyimana H, Wang M, Chen W, Khan K. The eco-toxic effects of pesticide and heavy metal mixtures towards earthworms in soil. Environ Toxicol Pharmacol. 2017, 55: 20–29.

24 Mai H, Cachot J, Clérandeau C, Martin C, Mazzela N, Gonzalez P, Morin B. An environmentally realistic pesticide and copper mixture impacts embryonic development and DNA integrity of the Pacific oyster, Crassostrea gigas. Environ Sci Pollut Res. 2020, 27(4): 3600–3611.

25 Backhaus T, Faust M. Predictive environmental risk assessment of chemical mixtures: a conceptual framework. Environ Sci Technol. 2012, 46(5): 2564–2573.

26 Cedergreen N. Quantifying synergy: a systematic review of mixture toxicity studies within environmental toxicology. Nazir A, editor. PLoS One. 2014, 9(5): e96580.

27 Yang M, Wang Y, Yang G, Wang Y, Liu F, Chen C. A review of cumulative risk assessment of multiple pesticide residues in food: current status, approaches and future perspectives. Trends Food Sci Technol. 2024, 144: 104340.
28 Zhu YC, Yao J, Adamczyk J, Luttrell R. Synergistic toxicity and physiological impact of imidacloprid alone and binary mixtures with seven representative pesticides on honey bee (Apis mellifera). PLoS One. 2017, 12(5): e0176837.
29 Hernández AF, Gil F, Lacasaña M. Toxicological interactions of pesticide mixtures: an update. Arch Toxicol. 2017, 91(10): 3211–3223.
30 Qi Z, Song X, Xiao X, Loo KK, Wang MC, Xu Q, Wu J, Chen S, Chen Y, Xu L, Li Y Effects of prenatal exposure to pyrethroid pesticides on neurodevelopment of 1-year- old children: a birth cohort study in China. Ecotoxicol Environ Saf. 2022, 234: 113384.
31 Wang N, Wang B, Wen J, Li X, Pan L, Zhang J, Zhu B. Types of exposure pesticide poisoning in jiangsu province, china; the epidemiologic trend between 2006 and 2018. Int J Environ Res Public Health. 2019, 16(14): 2586.
32 Yimaer A, Chen G, Zhang M, Zhou L, Fang X, Jiang W. Childhood pesticide poisoning in Zhejiang, China: a retrospective analysis from 2006 to 2015. BMC Public Health. 2017, 17(1): 602.
33 Samarghandi MR, Jaafarzadeh Haghighi Fard N, Jorfi S, Yari AR, Panahi Fard M. Pollution status of pesticide residues in food products in Iran: a mini-review within 2008-2018. Arch Hyg Sci. 2020, 9(3): 214–223.
34 Alehashem M, Peters R, Fajana HO, Eslamizad S, Hogan N, Hecker M, Siciliano SD Herbicides and pesticides synergistically interact at low concentrations in complex mixtures. Chemosphere. 2024, 353: 141431.
35 Mostafalou S, Abdollahi M. The susceptibility of humans to neurodegenerative and neurodevelopmental toxicities caused by organophosphorus pesticides. Arch Toxicol. 2023, 97(12): 3037–3060.
36 González-Alzaga B, Lacasaña M, Aguilar-Garduño C, Rodríguez-Barranco M, Ballester F, Rebagliato M, Hernández AF. A systematic review of neurodevelopmental effects of prenatal and postnatal organophosphate pesticide exposure. Toxicol Lett. 2014, 230(2): 104–121.
37 Liu F, Yuan M, Li C, Guan X, Li B. The protective function of taurine on pesticide-induced permanent neurodevelopmental toxicity in juvenile rats. FASEB J. 2021, 35(1): e21273.
38 Muñoz-Quezada MT, Lucero BA, Barr DB, Steenland K, Levy K, Ryan PB, Iglesias V, Alvarado S, Concha C, Rojas E, Vega C. Neurodevelopmental effects in children associated with exposure to organophosphate pesticides: a systematic review. Neuro Toxicol. 2013, 39: 158–168.
39 Zhou W, Deng Y, Zhang C, Dai H, Guan L, Luo X, He W, Tian J, Zhao L. Chlorpyrifos residue level and ADHD among children aged 1–6 years in rural China: a cross-sectional study. Front Pediatr [Internet]. 2022 [cited 2024 Jan 30];10. Available from: https://www.frontiersin.org/articles/10.3389/fped.2022.952559
40 Kamel F, Hoppin JA. Association of pesticide exposure with neurologic dysfunction and disease. Environ Health Perspect. 2004, 112(9): 950–958.
41 Sule RO, Condon L, Gomes AV. A common feature of pesticides: oxidative stress—the role of oxidative stress in pesticide-induced toxicity. Oxid Med Cell Longev. 2022, 2022:e5563759.
42 Kanthasamy A, Jin H, Charli A, Vellareddy A, Kanthasamy A. Environmental neurotoxicant-induced dopaminergic neurodegeneration: a potential link to impaired neuroinflammatory mechanisms. Pharmacol Ther. 2019, 197: 61–82.
43 Li Y, Fang R, Liu Z, Jiang L, Zhang J, Li H, Liu C, Li F. The association between toxic pesticide environmental exposure and Alzheimer's disease: a scientometric and visualization analysis. Chemosphere. 2021, 263: 128238.
44 Teleanu DM, Niculescu AG, Lungu II, Radu CI, Vladâcenco O, Roza E, Costăchescu B, Grumezescu AM, Teleanu RI. An overview of oxidative stress, neuroinflammation, and neurodegenerative diseases. Int J Mol Sci. 2022, 23(11): 5938.
45 Garry VF, Schreinemachers D, Harkins ME, Griffith J. Pesticide appliers, biocides, and birth defects in rural Minnesota. Environ Health Perspect. 1996, 104(4): 394–399.
46 Alavanja MCR, Ross MK, Bonner MR. Increased cancer burden among pesticide applicators and others due to pesticide exposure. CA Cancer J Clin. 2013, 63(2): 120–142.
47 Zhu W, Schmehl DR, Mullin CA, Frazier JL. Four common pesticides, their mixtures and a formulation solvent in the hive environment have high oral toxicity to honey bee larvae. PloS One. 2014, 9(1): e77547.
48 Čermak AMM, Pavičić I, Želježić D. Redox imbalance caused by pesticides: a review of OPENTOX-related research. Arch Ind Hygiene Toxicol. 2018, 69(2): 126–134.
49 Laetz CA, Baldwin DH, Collier TK, Hebert V, Stark JD, Scholz NL. The synergistic toxicity of pesticide mixtures: implications for risk assessment and the conservation of endangered pacific salmon. Environ Health Perspect. 2009, 117(3): 348–353.

50 Karami-Mohajeri S, Abdollahi M. Toxic influence of organophosphate, carbamate, and organochlorine pesticides on cellular metabolism of lipids, proteins, and carbohydrates: a systematic review. Hum Exp Toxicol. 2011, 30(9): 1119–1140.

51 Zhao G, Yang F, Li J, Xing H, Ren F, Pang G, Li YX. Toxicities of neonicotinoid-containing pesticide mixtures on nontarget organisms. Environ Toxicol Chem. 2020, 39(10): 1884–1893.

52 Abdollahi M, Ranjbar A, Shadnia S, Nikfar S, Rezaie A. Pesticides and oxidative stress: a review. Med Sci Monit. 2004, 10(6): 141–147.

53 Da B. Amyloid beta-peptide (1-42)-induced oxidative stress and neurotoxicity: implications for neurodegeneration in Alzheimer's disease brain. A review. Free Radic Res [Internet]. 2002 Dec [cited 2024 Jan 28];36(12). Available from: https://pubmed.ncbi.nlm.nih.gov/12607822/

54 Butterfield DA, Lauderback CM. Lipid peroxidation and protein oxidation in Alzheimer's disease brain: potential causes and consequences involving amyloid beta-peptide-associated free radical oxidative stress. Free Radic Biol Med. 2002, 32(11): 1050–1060.

55 Jan S, Singh R, Bhardwaj R, Ahmad P, Kapoor D. Plant growth regulators: a sustainable approach to combat pesticide toxicity. 3 Biotech. 2020, 10(11): 466.

56 Ledda C, Cannizzaro E, Cinà D, Filetti V, Vitale E, Paravizzini G, Di Naso C, Iavicoli I, Rapisarda V. Oxidative stress and DNA damage in agricultural workers after exposure to pesticides. J Occup Med Toxicol. 2021, 16(1): 1.

57 Bianchi M, Paravani EV, Acosta MG, Odetti LM, Simoniello MF, Poletta GL. Pesticide-induced alterations in zebrafish (*Danio rerio*) behavior, histology, DNA damage and mRNA expression: an integrated approach. Comp Biochem Physiol C Toxicol Pharmacol. 2024, 280: 109895.

58 Cui X, Wang C, Wang X, Li G, Liu Z, Wang H, Guo X, Xu B. Molecular mechanism of the UDP-Glucuronosyltransferase 2B20-like Gene (AccUGT2B20-like) in pesticide resistance of Apis cerana cerana. Front Genet. 2020, 11: 592595.

59 Lee KM, Park SY, Lee K, Oh SS, Ko SB. Pesticide metabolite and oxidative stress in male farmers exposed to pesticide. Ann Occup Environ Med. 2017;29: 5.

60 Wang Y, Lv L, Xu C, Wang D, Yang G, Wang X, et al. Mixture toxicity of thiophanate-methyl and fenvalerate to embryonic zebrafish (Danio rerio) and its underlying mechanism. Sci Total Environ. 2021, 756: 143754.

61 Shukla S, Jhamtani RC, Dahiya MS, Agarwal R. Oxidative injury caused by individual and combined exposure of neonicotinoid, organophosphate and herbicide in zebrafish. Toxicol Rep. 2017, 4: 240–244.

62 Ojha A, Yaduvanshi SK, Srivastava N. Effect of combined exposure of commonly used organophosphate pesticides on lipid peroxidation and antioxidant enzymes in rat tissues. Pestic Biochem Physiol. 2011, 99(2): 148–156.

63 Lv L, Jia F, Deng M, Di S, Chu T, Wang Y. Toxic mechanisms of imazalil, azoxystrobin and their mixture to hook snout carp (Opsariichthys bidens). Sci Total Environ. 2024, 926: 172022.

64 Qin LT, Lei YX, Liu M, Zeng HH, Liang YP, Mo LY. Toxic interactions at the physiological and biochemical levels of green algae under stress of mixtures of three azole fungicides. Sci Total Environ. 2024, 926: 171771.

65 Lushchak VI, Matviishyn TM, Husak VV, Storey JM, Storey KB. Pesticide toxicity: a mechanistic approach. EXCLI J. 2018, 17: 1101–1136.

66 Martínez MA, Ares I, Rodríguez JL, Martínez M, Roura-Martínez D, Castellano V, Lopez-Torres B, Martínez-Larrañaga MR, Anadón A. Pyrethroid insecticide lambda-cyhalothrin induces hepatic cytochrome P450 enzymes, oxidative stress and apoptosis in rats. Sci Total Environ. 2018, 631–632: 1371–1382.

67 Tebourbi O, Sakly M, Ben K. Molecular mechanisms of pesticide toxicity. In: M Stoytcheva (Ed.), Pesticides in the Modern World - Pests Control and Pesticides Exposure and Toxicity Assessment [Internet]. InTech; 2011 [cited 2024 Jan 29]. Available from: http://www.intechopen.com/books/pesticides-in-the-modern-world-pests-control-and-pesticides-exposure-and-toxicity-assessment/molecular-mechanisms-of-pesticide-toxicity

68 Jacobsen PR, Christiansen S, Boberg J, Nellemann C, Hass U. Combined exposure to endocrine disrupting pesticides impairs parturition, causes pup mortality and affects sexual differentiation in rats. Int J of Andrology. 2010, 33(2): 434–442.

69 Genuis SJ, Kyrillos E. The chemical disruption of human metabolism. Toxicol Mech Methods. 2017, 27(7): 477–500.

70 Rodrigues JA, Narasimhamurthy RK, Joshi MB, Dsouza HS, Mumbrekar KD. Pesticides exposure-induced changes in brain metabolome: implications in the pathogenesis of neurodegenerative disorders. Neurotox Res. 2022, 40(5): 1539–1552.

71 Falfushynska H, Khatib I, Kasianchuk N, Lushchak O, Horyn O, Sokolova IM. Toxic effects and mechanisms of common pesticides (Roundup and chlorpyrifos) and their mixtures in a zebrafish model (*Danio rerio*). Sci Total Environ. 2022, 833: 155236.

11

Combined Molecular Toxicity Mechanism of Persistent Organic Pollutant Mixtures

Ruijia Zhang[a,], Tiangang Luan[b]*

[a] *School of Environmental Science and Engineering, Hainan University, Haikou, China*
[b] *State key Laboratory of Biocontrol, School of Life Sciences, Sun Yat-sen University, Guangzhou, China*

*Corresponding author.
E-mail: zhangrj@hainanu.edu.cn (R.Z.)

11.1 Introduction

Persistent organic pollutants (POPs) are a class of organic chemical substances, possessing a particular combination of physical and chemical properties, such that they (1) persist in the environment, (2) become widely distributed throughout the environment, (3) accumulate in living organisms, and (4) are toxic to both humans and wildlife. POPs in the Stockholm Convention mainly include organochlorine pesticides (OCPs), polychlorinated biphenyls (PCBs), polybrominated diphenyl ethers (PBDEs), and per- and poly-fluoroalkylated substances (PFASs; Table 11.1). Notedly, the POPs Review Committee evaluates the proposals, and then, six compounds are considered as candidate POPs.

Table 11.1 All POPs listed in the Stockholm convention (http://chm.pops.int/TheConvention/ThePOPs/tabid/673/Default.aspx).

Annex A (Elimination)			
Aldrin ● ▲	Chlordane ● ▲	Chlordecone ●	Decabromodiphenyl ether (commercial mixture, c-decaBDE) ■
Dicofol ●	Dieldrin ● ▲	Endrin ● ▲	Heptachlor ● ▲
Hexabromobiphenyl ■	Hexabromocyclododecane (HBCDD) ■	Hexabromodiphenyl ether and Heptabromodiphenyl ether ■	Hexachlorobenzene (HCB) ● ▲
Hexachlorobutadiene ■	Alpha hexachlorocyclohexane ●	Beta hexachlorocyclohexane ●	Lindane ●
Mirex ● ▲	Pentachlorobenzene ● ■	Pentachlorophenol and its salts and esters ●	Polychlorinated biphenyls (PCBs) ■ ▲
Polychlorinated naphthalenes ■	Perfluorooctanoic acid (PFOA), its salts, and PFOA-related compounds ■	Perfluorohexane sulfonic acid (PFHxS), its salts, and PFHxS-related compounds ■	Short-chain chlorinated paraffins (SCCPs) ■
Technical endosulfan and its related isomers ●	Tetrabromodiphenyl ether and pentabromodiphenyl ether ■	Toxaphene ● ▲	

(Continued)

Toxicological Assessment of Combined Chemicals in the Environment, First Edition. Edited by De-Sheng Pei and Yiyun Liu.

Table 11.1 (Continued)

Annex B (Restriction)			
DDT ● △	Perfluorooctane sulfonic acid (PFOS), its salts, and perfluorooctane sulfonyl fluoride (PFOSF) ● ■		
Annex C (Unintentional production)			
Hexachlorobenzene (HCB)	Hexachlorobutadiene (HCBD)	Pentachlorobenzene	PCBs
Polychlorinated dibenzo-*p*-dioxins (PCDD) △	Polychlorinated dibenzofurans (PCDF) △	Polychlorinated naphthalenes	

● Pesticide; ■ Industrial chemicals; △ Initial POPs under the Stockholm Convention.

Six chemicals proposed for listing under the Convention as candidate POPs.

Dechlorane Plus	Methoxychlor	Chlorpyrifos
Chlorinated paraffins	Long-chain perfluorocarboxylic acids (LC-PFCAs)	UV-328

POPs are omnipresent on a global scale. The majority of current toxicology studies focused on the toxic effects of single compounds; while in real-world scenarios, humans are exposed to POP mixtures rather than to single compounds [1]. Besides, the potential additive, synergistic, or antagonistic effects of POP mixtures coexist with each other. Therefore, it is of importance to unveil that POP mixtures interact with adverse health effects. Nowadays, a growing body of research including epidemiological and experimental studies has explored the hazardous effects of POP mixtures [2–4]. The challenge of evaluation of toxic effects induced by POP mixtures is the preparation of POP mixtures, which are close to actual environmental exposure, so the design of the components of POP mixtures and their concentrations based on human exposure are introduced in this chapter. The toxic effects of POP mixtures mainly involved in developmental toxicity and endocrine effects as well as combined molecular toxicity mechanisms of POP mixtures are also presented.

11.2 Design of the Components of POP Mixtures and Their Concentrations

In previous studies, the compositions of POP mixtures were chosen by researchers usually according to the possibility of POP measurement as well as their prominent presence in the environmental and animal/human samples, and the concentrations of POP mixtures were prepared based on environmentally relevant concentrations derived from environmental, fluid or tissue samples and cytotoxicity testing [5–7]. Only several POPs have been concentrated by these studies so far. For example, one study selected the concentrations of POP mixtures (0.4 pM 2,3,7,8-TCDD, 3.3 pM PCB126, and 333.3 pM PCB169) based upon about 25% of maximal aryl hydrocarbon receptor (AhR)-mediated light emission induced by an AhR agonist [5]. Of course, the commercial PCB mixtures (e.g. Aroclor 1254) were also available for study, but these mixtures represent chlorine percentage by weight instead of specific compounds [8]. For instance, Aroclor 1254 is defined as 54% chlorine [9].

It is well known that various POPs exist in the human body, so it is difficult to predict the combined toxicity of POP mixtures through the reconstituted POP mixtures [10]. Thus, it is important to establish an approach for assessing the combined toxic effects of actual POP mixtures presented in animals or humans. By far, the cocktail of POPs was extracted from fish, humans, and other environmental matrices originating from contaminated areas followed by determination of combined mixture effects using in vivo and in vitro models [11–13]. One study extracted the environmental mixtures of POPs from the burbot liver and used them for zebrafish exposure [14]. Although such extracts were obtained from real specimens, a variety of substances in such extracts cannot be identified. As a result, the confounding effects of unknown compounds may bias the estimation of biological effects [15].

In the last few years, a model of an environmentally relevant mixture of POPs containing 29 different chlorinated, brominated, and perfluorinated compounds based upon a Scandinavian food and blood survey for toxicity studies was developed [15]. The POP mixture concentrations found in Scandinavian food were used to construct the levels of POP mixtures for in vivo assays (Table 11.2), whereas POP mixture levels detected in Scandinavian blood were employed for in vitro assays (Table 11.3). In addition to PFOS and PFOA, other perfluorinated compounds are not included in the Stockholm

Table 11.2 POP mixtures (29) based on human estimated daily intake (EDI) levels for in vivo assays.

Compound	Average EDI, 70 kg person (ng/d)	Daily intake (ng/kg/d)	EDI, 25 g mouse (pg/d)	EDI 5000× (ng/d)	Feed nominal 5000× (ng/g feed)[a]	Feed measured 5000× (ng/g)	Measured % of nominal 5000×
Polychlorinated biphenyl (PCB)							
PCB 28	10	0.14	3.5	18	5.8	3.1	53
PCB 52	23	0.33	8.3	41	13.8	15.0	109
PCB 101	39	0.56	14.0	70	23.3	25.4	109
PCB 118	68	0.97	24.3	121	40.4	37.2	92
PCB 138	97	1.38	34.5	173	57.5	53.8	94
PCB 153	97	1.38	34.5	173	57.5	61.4	107
PCB 180	26	0.37	9.3	46	15.4	17.4	113
Σ PCB	360	5.13	128.4	642	213.7	213.3	100
Organochlorine pesticide (OCP)							
1,1-Dichloro-2,2-bis(p-chlorophenyl) ethylene (*p,p'*-DDE)	201	2.87	71.8	359	119.6	136.0	114
Hexachlorobenzene (HCB)	84	1.20	30.0	150	50.0	37.4	75
α-Chlordane	63	0.90	22.5	113	37.5	45.0	120
Oxychlordane	21	0.30	7.5	38	12.5	9.8	78
trans-Nonachlor	21	0.30	7.5	38	12.5	14.9	119
α-Hexachlorocyclohexane (α-HCH)	36	0.52	13.0	65	21.7	21.2	98
β-Hexachlorocyclohexane (β-HCH)	29	0.42	10.5	53	17.5	22.3	127
γ-Hexachlorocyclohexane (γ-HCH; Lindane)	40	0.57	14.3	71	23.8	31.4	132
Dieldrin	126	1.80	45.0	225	75.0	70.4	94
Σ organochlorine pesticide (OCP)	621	8.88	222.1	1112	370.1	388.4	105
Σ PCB + OCP	981	14.01	350.5	1754	583.8	601.7	103

(Continued)

Table 11.2 (Continued)

Compound	Average EDI, 70 kg person (ng/d)	Daily intake (ng/kg/d)	EDI, 25 g mouse (pg/d)	EDI 5000× (ng/d)	Feed nominal 5000× (ng/g feed)[a]	Feed measured 5000× (ng/g)	Measured % of nominal 5000×
Brominated flame retardant (BFR)							
Polybrominated diphenyl ether (PBDE) 47	68	0.97	24.3	121	40.4	39.7	98
PBDE 99	13	0.19	4.8	24	7.9	8.6	109
PBDE 100	11	0.15	3.8	19	6.3	5.6	89
PBDE 153	2	0.03	0.8	4	1.3	1.5	115
PBDE 154	4	0.06	1.5	8	2.5	2.8	112
PBDE 209	105	1.50	37.5	188	62.5	64.8	104
Hexabromocyclododecane (HBCD)	21	0.30	7.5	38	12.5	9.9	79
Σ BFR	224	3.2	80.2	402	133.4	132.9	100
Polyfluoroalkyl substance (PFAS)							
Perfluorohexane sulfonic acid (PFHxS)	1.2	0.017	0.4	2	4.9	1.7	35
Perfluorooctane sulfonic acid (PFOS)	18	0.26	6.5	33	10.8	3.2	30
Perfluorooctanoic acid (PFOA)	31	0.44	11.0	55	18.3	6.0	33
Perfluorononanoic acid (PFNA)	9.5	0.14	3.5	18	5.8	2.1	36
Perfluorodecanoic acid (PFDA)	13	0.19	4.8	24	7.9	3.1	39
Perflurorundecanoic acid (PFUnDA)	6.7	0.096	2.4	12	4.0	1.6	40
Σ Perfluoroalkyl acid (PFAA)	79.4	1.14	28.6	144	51.7	17.7	34

a. Nominal concentrations of the various compounds in the feed aimed to give a daily dose of each compound equal to 5000× the EDI for a 25-g mouse consuming 3g feed/day.

Table 11.3 POP (29) mixtures based on human blood levels for in vitro assays.

Compound	Average human blood levels (ng/g) lipid[a]	Average human blood levels (ng/mL)[b]	Total mixture stock nominal concentration (mg/mL)[c]	Total mixture stock measured concentration (mg/mL)[d]	Total mixture stock measured concentration (μM)[e]	Measured % of nominal
Polychlorinated biphenyl (PCB)						
PCB 28	2.13	0.013	0.013	0.008	31.1	62
PCB 52	1.60	0.010	0.010	0.006	20.5	60
PCB 101	1.30	0.008	0.008	0.008	24.5	100
PCB 118	10.67	0.064	0.064	0.045	137.9	70
PCB 138	37.00	0.222	0.222	0.155	429.5	70
PCB 153	60.33	0.362	0.362	0.252	698.3	70
PCB 180	32.33	0.194	0.194	0.134	339.0	69
Σ PCB	145.36	0.873	0.873	0.608	1680.8	70
Organochlorine pesticide (OCP)						
1,1-Dichloro-2,2-bis(p-chlorophenyl) ethylene (*p,p'*-DDE)	83.67	0.502	0.502	0.339	1065.9	68
Hexachlorobenzene (HCB)	19.50	0.117	0.117	0.065	228.2	56
α-Chlordane	1.80	0.011	0.011	0.010	23.7	91
Oxychlordane	3.70	0.022	0.022	0.014	33.0	64
trans-Nonachlor	6.80	0.041	0.041	0.044	99.1	107
α-Hexachlorocyclohexane (α-HCH)	1.00	0.006	0.006	0.005	16.8	83
β-Hexachlorocyclohexane (β-HCH)	8.77	0.053	0.053	0.022	75.6	42
γ-Hexachlorocyclohexane (γ-HCH; Lindane)	1.00	0.006	0.006	0.005	16.8	83
Dieldrin	4.00	0.024	0.024	0.021	56.2	88
Σorganochlorine pesticide (OCP)	130.24	0.782	0.782	0.525	1615.3	67
Σ PCB + OCP	275.60	1.655	1.655	1.133	3296.1	68

(Continued)

Table 11.3 (Continued)

Compound	Average human blood levels (ng/g) lipid[a]	Average human blood levels (ng/mL)[b]	Total mixture stock nominal concentration (mg/mL)[c]	Total mixture stock measured concentration (mg/mL)[d]	Total mixture stock measured concentration (μM)[e]	Measured % of nominal
Brominated flame retardant (BFR)						
PBDE 47	1.43	0.009	0.009	0.009	17.8	100
PBDE 99	0.59	0.004	0.004	0.004	7.5	100
PBDE 100	0.36	0.002	0.002	0.002	3.8	100
PBDE 153	1.64	0.010	0.001*	0.001	2.1	100
PBDE 154	0.29	0.002	0.002	0.002	3.0	100
PBDE 209	1.81	0.011	0.011	0.009	9.4	82
Hexabromocyclododecanes (HBCDs)	4.10	0.025	0.025	0.035	54.5	140
Σ BFR	10.22	0.063	0.063	0.062	98.1	117
Polyfluoroalkyl substance (PFAS)						
Perfluorohexane sulfonic acid (PFHxS)	N/A	3.450	3.450	3.422	7809.2	99
Perfluorooctane sulfonic acid (PFOS)	N/A	29.425	29.425	22.348	41522.1	76
Perfluorooctanoic acid (PFOA)	N/A	4.523	4.523	1.743	4209.4	39
Perfluorononanoic acid (PFNA)	N/A	0.800	0.800	0.507	1092.5	63
Perfluorodecanoic acid (PFDA)	N/A	0.495	0.495	0.193	375.4	39
Perflurorundecanoic acid (PFUnDA)	N/A	0.560	0.560	0.190	336.8	34
Σ Perfluoroalkyl acid (PFAA)	N/A	39.253	39.253	28.403	55345.4	72

a. Average human blood levels of POPs based on a literature review of Scandinavian values, providing the basis for the in vitro mixtures.
b. Average human blood levels of POPs converted to ng/mL. A fat percentage of 0.6% was used. 1 mL blood was considered to have a weight of 1g.
c. Nominal concentrations of the various compounds in the total mixture stock 1 000 000× the average concentration in blood.
d. Measured concentrations of the various compounds in the total mixture stock in mg/mL.
e. Measured concentrations of the various compounds in the total mixture stock converted to μM.
* The nominal concentration of PBDE 153 included in the total mixture stock was 10 times lower than originally intended.

Convention, but these compounds were added as they were commonly found to have a long half-life of elimination in humans [15, 16]. The formulas and concentrations of 29 different POP mixtures were as close as possible to which the Scandinavian population was exposed, but it was still worth noting that different countries and regions might exhibit unique characteristic profiles of POP mixtures in the human body.

11.3 Developmental Toxicity of POP Mixtures

Fetal development is a key window in which the fetus is highly sensitive to environmental pollutants [17]. The apparent feature of POPs is their persistent presence in the human body. Therefore, the exposure of animals and humans to POPs in early life (e.g. prenatal/postnatal period) will influence the entire life cycle of animals and humans [18]. Detection of the concentrations of 22 POPs in maternal serum or placenta and human fetuses showed that more POPs were measured in fetal tissues compared with maternal serum and placenta, which indicated that fetuses were intrinsically exposed to a mixture of POPs [17]. An increasing number of studies have reported that POP mixtures induced developmental toxicity (e.g. fetal growth and brain development), as demonstrated by Developmental Origins of Health and Disease (DOHaD) hypothesis [19]. An investigation on Greenlandic population showed that the concentrations of POPs in maternal serum were inversely associated with fetal growth indices (e.g. infant's birth weight, length, and head circumference) [20]. Another epidemiological study for fetal growth found that a mixture of OCPs at low levels was negatively related to fetal growth (e.g. head circumference, abdominal circumference, and femur length), indicating that the adverse effects of POP mixtures on development could not be ignored even if the levels of POP mixtures were low [21]. As for epidemiological studies, Bayesian Kernel Machine Regression (BKMR), Bayesian Additive Regression Trees (BART), least absolute shrinkage and selection operator (LASSO), elastic net (ENET), Sparse Principal Component Analysis (SPCA), and novel Weighted Quantile Sum Regression (WQSR) models could be used to identify the most influential POPs to gain a prioritized list of POPs [22]. With regard to animal models, female mice were exposed to a human relevant mixture of 29 POPs, and then their offspring displayed the impairment of learning and memory functions and the changed expression of hippocampal genes, indicating that the POP mixtures might damage brain development [23]. In consonance with this study, a study by Christou et al. showed that exposure of zebrafish to a human relevant mixture of 29 POPs affected zebrafish's learning and memory abilities during developmental exposure [24].

Epigenetic changes are considered as one of the critical molecular mechanisms for developmental toxicity brought about by POP mixtures because most POPs do not directly attack DNA sequences [19, 25]. Epigenetic mechanisms can be classified into three types as follows [26]: (1) DNA methylation, i.e. the addition of methyl groups to the fifth position of cytosine bases; (2) histone modifications (including histone acetylation at lysine residues; histone phosphorylation at serine, threonine, and tyrosine; histone deamination; and histone ADP ribosylation at glutamate and arginine residues); and (3) noncoding RNAs (small and large noncoding RNA molecules), as shown in Figure 11.1. For example, a Korean mother–child pair study showed that the methylation level of placental long interspersed element 1 (*LINE-1*) was inversely associated with birth length in multiple-POP models [27]. 2,3,7,8-tetrachlrodibenzo-*p*-dioxin (TCDD) exposure induced cleft palate in fetal mice through high levels of global DNA methylation and the decreased methylation level of CpG site 2 in the DNA methyltransferase 3a (*Dnmt3a*) promoter region, suggesting that methylation was responsible for palate malformation in fetal mice [28]. LncRNA MEG3 could mitigate PFOS-induced placental cell growth inhibition through the lncRNA MEG3/miR-770/pentraxin 3 (*Ptx3*) axis in pregnant mice administrated with PFOS [29]. It is worth noting that these three epigenetic mechanisms are intertwined, and hence, they are able to jointly affect complex interaction networks [26]. Further studies are needed to explore not only the conversion of POP signals into epigenetic changes but also the comprehensive impacts of multiple epigenetic mechanisms on POP mixture–induced developmental toxicity instead of individual POP.

11.4 Endocrine Effects of POP Mixtures

Endocrine-disrupting chemicals (EDCs) are defined as "an exogenous substance or mixture that alters function(s) of the endocrine system and consequently causes adverse health effects in an intact organism, or its progeny, or (sub)populations" by the World Health Organization (https://apps.who.int/iris/ bitstream/handle/10665/67357/WHO_PCS_EDC_02.2.pdf?sequence=1&isAllowed=y). Nowadays, more than 1000 chemicals are considered to be EDCs. The list of EDCs can still be found in the Endocrine Disruption Exchange (TEDX) database (https://endocrinedisruption.org/), although it has been closed in 2019. Importantly, nearly all POPs (such as DDT and other OCPs, PCBs, and PFASs) are well-known EDCs [30].

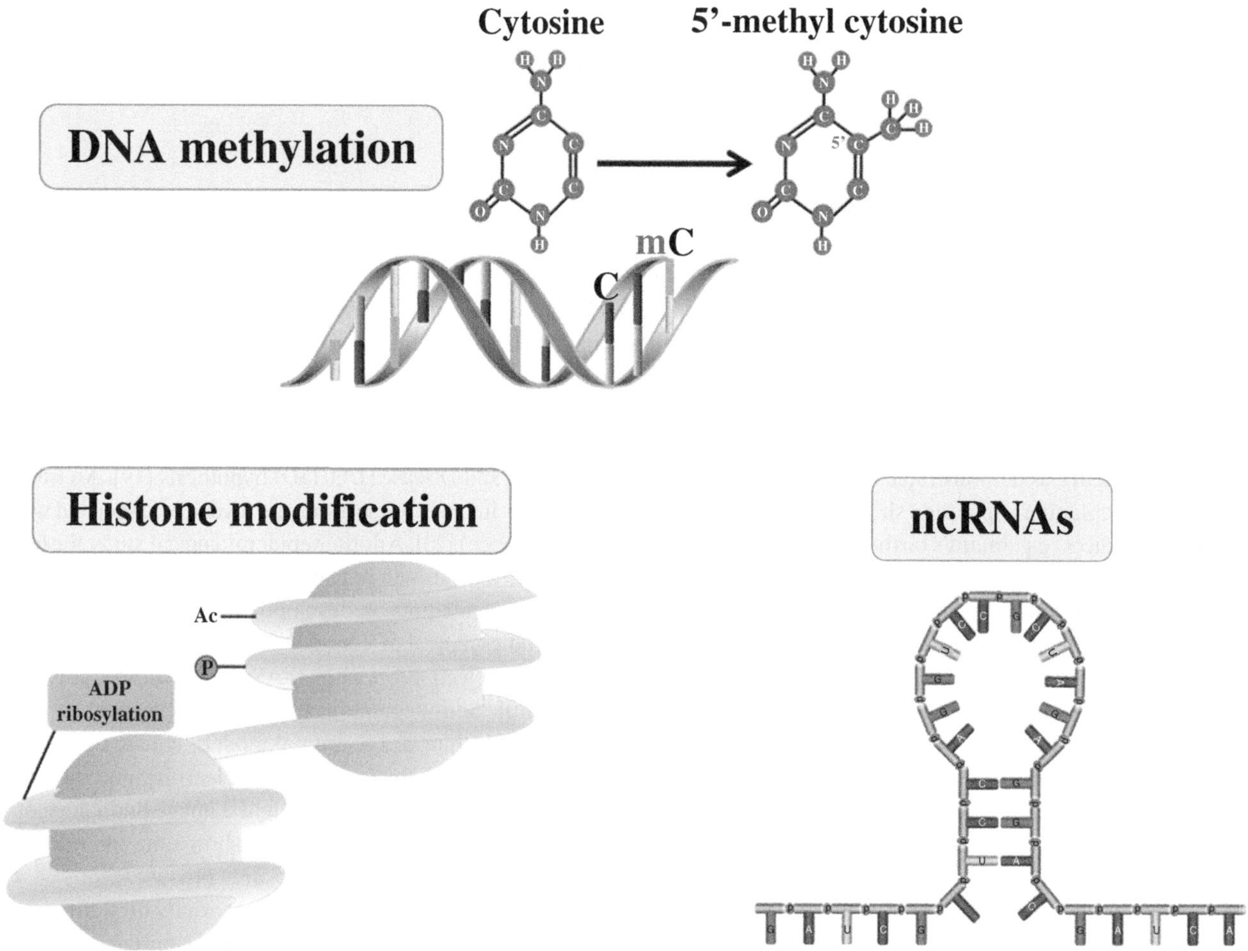

Figure 11.1 Schematics of epigenetic mechanisms in POP mixture-induced developmental toxicity. *Source:* Adapted from Patrizi et al. [26].

Human exposure to POP mixtures can bring a wide range of adverse effects. A series of diseases, such as obesity, type 2 diabetes mellitus (T2DM), cardiovascular disease, female and male reproduction, hormone-sensitive cancers, thyroid disruption, and neurodevelopment, have been reported to be associated with exposure to POP mixtures [31]. For instance, an epidemiological study has shown that an increased odds ratio of hypothyroidism (odds ratio [OR] = 98.1, 95% confidence interval [CI]: 5.51–1747) could be attributed to POP exposure [32]. A longitudinal, nested case-control study found that most POPs were positively associated with T2DM [33]. In particular, the importance of uncovering toxicity mechanisms of POPs in the endocrine system should be underscored.

Many POPs act as endocrine active by interacting with hormone receptors, including estrogen receptors (ERs), androgen receptor (AR), progesterone receptor (PR), glucocorticoid receptor (GR), constitutive androstane receptor (CAR), rodent pregnane X receptor (PXR), mineralocorticoid receptor, and thyroid hormone receptors (TRs) [34]. POPs elicited endocrine activity also through binding of AhR. The assessment of AhR agonistic and antagonistic activities of 29 POPs individually and in mixtures using three transgenic cell lines showed that 16 out of 29 individual POPs and the total POP mixtures exhibited an AhR antagonistic response, and PCB 138 and PCB 118 were identified as the two main contributors for the AhR antagonism when cells were exposed to the total POP mixtures. Moreover, this study predicted the mixed effect of POP mixtures using concentration addition or generalized concentration addition models [35]. One study reported that a combination of brominated and chlorinated mixtures antagonized AR transactivation without AR translocation, while chlorinated and perfluorinated + brominated submixtures inhibited AR translocation in vitro [36]. A total of 19 steroids in the steroid biosynthesis pathway were analyzed in H295R adrenocortical cells, which showed that steroid hormone production (e.g. glucocorticoid steroids, dehydroepiandrosterone, and testosterone) could be reduced by a mixture of brominated and perfluorinated compounds, but the increased level of aldosterone was found in cells treated with a perfluorinated mixture or a mixture of chlorinated and fluorinated compounds [37]. Additionally, additive effects on 17-hydroxyprogesterone and aldosterone levels were found in a mixture of chlorinated and fluorinated compounds compared to a single mixture of chlorinated or fluorinated compounds [37]. The trend of changes in the expression levels of genes involving *MC2R, CYP11B1, CYP11B2*, and *CYP19A1* was consistent with steroid hormone levels, whereas the alternations in gene expression were more sensitive than the changes in hormone levels [7]. Collectively, the endocrine-disrupting mechanisms of POP mixtures principally contain receptor agonism and antagonism and modulation of receptor expression and hormone metabolism (including thyroid, sex hormones, and cortisol) [38].

11.5 Molecular Toxicity Mechanisms of POPs

As mentioned above, the epigenetic mechanism and hormone pathways have been discussed, so these mechanisms herein would not be repeated. This part summarizes common molecular mechanisms, including oxidative stress, cell death–related signaling pathways, and mitochondrial pathways. In fact, multiple mechanisms are engaged in toxicity by POPs, and these mechanisms may be interrelated, such as oxidative stress and mitochondrial damage.

11.5.1 Oxidative Stress

Oxidative stress is almost regarded as a common denominator of POPs toxicity, such as neurotoxicity and hepatotoxicity. Cell culture, animal, and epidemiological studies have demonstrated the association between POPs exposure and oxidative stress [39]. Previous studies have reported that single POPs or POP mixtures (e.g. PFOA, PFOS, PCBs, and dieldrin) triggered oxidative stress via (1) inducing lipid peroxidation and protein and DNA oxidation; (2) decreasing the content of antioxidants (e.g. reduced glutathione); (3) reducing the expression levels of antioxidant genes (e.g. glutathione peroxidase [*GPX*], superoxide dismutase [*SOD*], and catalase [*CAT*]) and the activity of anti-oxidase; (4) damaging mitochondria or activating a cross talk between mitochondria and NADPH oxidase; and (5) affecting related signaling pathways, such as nuclear factor-erythroid 2-related factor 2 (NRF2), p38-mitogen-activated protein kinase (MAPK), and AhR/Cyp1a1 [40–43].

11.5.2 Cell-death-related Signaling Pathways

Cell death can be simply divided into cellular suicide (including apoptosis, necroptosis, autophagy, pyroptosis, and ferroptosis) and cellular sabotage (i.e. necrosis) [44]. Cell death types are controlled by different molecular mechanisms [45]. Enormous studies showed that POPs' exposure led to cell death in which specific molecular mechanisms were involved. These molecular mechanisms of POP-induced cell death were summarized as follows: (1) caspase activation (e.g. caspase-2/3/6/7/8/9) and Bax/bcl-2 signal pathways for apoptosis [46]; (2) the regulation of p38- and JNK-MAPK signaling cascades for apoptosis and autophagy [46, 47]; (3) the involvement of the phosphoinositide 3-kinase (PI3K) and RICTOR/protein kinase B (Akt)/mammalian target of rapamycin (mTOR) for autophagy [48]; (4) the activation of NLRP3 inflammasomes for pyroptosis [49]; and (5) GPX4-, mitochondria-, and p53-mediated ferroptosis [45].

11.5.3 Mitochondrial Pathways

Mitochondria are subcellular organelles which play an important role as a cell powerhouse, providing up to 95% of cellular energy demands in the form of adenosine triphosphate (ATP) [50]. Besides energy supply, mitochondria are responsible for oxidative stress, hormone synthesis, and apoptosis [51]. Growing evidence has highlighted that mitochondria were targeted by POPs and mitochondrial dysfunction was a hallmark of POP-induced toxicity [51–54]. Altogether, it was documented that POPs impaired mitochondria through diverse mechanisms including mitochondrial epigenetics (e.g. mitochondrial DNA [mtDNA] methylation), mitochondrial electron transfer chain (ETC), mitochondrial dynamics, mitochondrial metabolism (e.g. the production of acetyl-coA), and ion channels (e.g. Ca^{2+} channel). For example, flame retardant BDE-47 exposure disrupted methylcytosine in the mtDNA of offspring's brains [55].

11.6 Conclusions

Humans are confronted with mixtures of POPs instead of single POPs. The preparation of POP mixture levels based upon human exposure is important for toxicity assays. Another important point is to identify which POPs in the POP mixtures significantly contribute to the toxicity induced by POP mixtures. In addition, it should be noted that multiple molecular mechanisms, rather than only one molecular mechanism, regulate POP-induced toxicity.

Acknowledgments

The authors gratefully thank Dr. Xiangyu Yang of Sun Yat-sen University for language modification. The authors thank the support of Guangdong Basic and Applied Basic Research (No. 2022A1515011916), National Natural Science Foundation of China (Nos 22206207 and 22127810), and Fundamental Research Funds for the Central Universities, Sun Yat-sen University (No. 22qntd2613).

References

1 García-Villarino M, Signes-Pastor AJ, Riaño-Galán I, Rodríguez-Dehli AC, Vizcaíno E, Grimalt JO, Fernández-Somoano A, Tardón A. Serum concentrations of persistent organic pollutants mixture during pregnancy and anogenital distance in 8-year-old children from the INMA-Asturias cohort. Environ Res. 2022, 213:113607.
2 Liu X, Zhang L, Chen L, Li J, Wang J, Zhao Y, Liu L, Wu Y. Identification and prioritization of the potent components for combined exposure of multiple persistent organic pollutants associated with gestational diabetes mellitus. J Hazard Mater. 2021, 409: 124905.
3 Xie Y, Berntsen HF, Zimmer KE, Ropstad E, Verhaegen S, Connolly L. Luteolin protects against adipogenic and lipogenic potency induced by human relevant mixtures of persistent organic pollutants (POPs) in the 3T3-L1 model. Food Chem Toxicol. 2023, 173: 113608.
4 Yadav A, Verhaegen S, Filis P, Domanska D, Lyle R, Sundaram AYM, Leithaug M, Østby GC, Aleksandersen M, Berntsen HF, Zimmer KE, Fowler PA, Paulsen RE, Ropstad E. Exposure to a human relevant mixture of persistent organic pollutants or to perfluorooctane sulfonic acid alone dysregulates the developing cerebellum of chicken embryo. Environ Int. 2022, 166: 107379.
5 Bittner M, Macikova P, Giesy JP, Hilscherova K. Enhancement of AhR-mediated activity of selected pollutants and their mixtures after interaction with dissolved organic matter. Environ Int. 2011, 37(5): 960–964.
6 Gill S, Bowers WJ, Nakai JS, Yagminas A, Mueller R, Pulido O. Effects of environmentally relevant mixtures of persistent organic pollutants on the developmental neurobiology in rats. Toxicol Pathol. 2013, 41(1): 38–47.
7 van den Dungen MW, Rijk JC, Kampman E, Steegenga WT, Murk AJ. Steroid hormone related effects of marine persistent organic pollutants in human H295R adrenocortical carcinoma cells. Toxicol In Vitro. 2015, 29(4): 769–778.
8 Bowers WJ, Nakai JS, Chu I, Wade MG, Moir D, Yagminas A, Gill S, Pulido O, Meuller R. Early developmental neurotoxicity of a PCB/organochlorine mixture in rodents after gestational and lactational exposure. Toxicol Sci. 2004, 77(1): 51–62.
9 Kodavanti PR, Kannan N, Yamashita N, Derr-Yellin EC, Ward TR, Burgin DE, Tilson HA, Birnbaum LS. Differential effects of two lots of aroclor 1254: congener-specific analysis and neurochemical end points. Environ Health Perspect. 2001, 109(11): 1153–1161.
10 Bonefeld-Jørgensen EC, Ghisari M, Wielsøe M, Bjerregaard-Olesen C, Kjeldsen LS, Long M. Biomonitoring and hormone-disrupting effect biomarkers of persistent organic pollutants in vitro and ex vivo. Basic Clin Pharmacol Toxicol. 2014, 115(1): 118–128.
11 Echeveste P, Galbán-Malagón C, Dachs J, Berrojalbiz N, Agustí S. Toxicity of natural mixtures of organic pollutants in temperate and polar marine phytoplankton. Sci Total Environ. 2016, 571: 34–41.
12 Lyche JL, Grześ IM, Karlsson C, Nourizadeh-Lillabadi R, Aleström P, Ropstad E. Parental exposure to natural mixtures of persistent organic pollutants (POP) induced changes in transcription of apoptosis-related genes in offspring zebrafish embryos. J Toxicol Environ Health A. 2016, 79(13-15): 602–611.

13 Wielsøe M, Bjerregaard-Olesen C, Kern P, Bonefeld-Jørgensen EC. Receptor activities of persistent pollutant serum mixtures and breast cancer risk. Endocr Relat Cancer. 2018, 25(3): 201–215.

14 Lyche JL, Grześ IM, Karlsson C, Nourizadeh-Lillabadi R, Berg V, Kristoffersen AB, Skåre JU, Alestrøm P, Ropstad E. Parental exposure to natural mixtures of POPs reduced embryo production and altered gene transcription in zebrafish embryos. Aquat Toxicol. 2013, 126: 424–34.

15 Berntsen HF, Berg V, Thomsen C, Ropstad E, Zimmer KE. The design of an environmentally relevant mixture of persistent organic pollutants for use in in vivo and in vitro studies. J Toxicol Environ Health A. 2017, 80(16-18): 1002–1016.

16 Wang Z, DeWitt JC, Higgins CP, Cousins IT. A never-ending story of per- and Polyfluoroalkyl substances (PFASs)? Environ Sci Technol. 2017, 51(5): 2508–2518.

17 Björvang RD, Vinnars MT, Papadogiannakis N, Gidlöf S, Mamsen LS, Mucs D, Kiviranta H, Rantakokko P, Ruokojärvi P, Lindh CH, Andersen CY, Damdimopoulou P. Mixtures of persistent organic pollutants are found in vital organs of late gestation human fetuses. Chemosphere. 2021, 283: 131125.

18 Gaballah S, Swank A, Sobus JR, Howey XM, Schmid J, Catron T, McCord J, Hines E, Strynar M, Tal T. Evaluation of developmental toxicity, developmental neurotoxicity, and tissue dose in Zebrafish exposed to GenX and other PFAS. Environ Health Perspect. 2020, 128(4): 47005.

19 Trentacosta CJ, Mulligan DJ. New directions in understanding the role of environmental contaminants in child development: four themes. New Dir Child Adolesc Dev. 2020, 2020(172): 39–51.

20 Hjermitslev MH, Long M, Wielsøe M, Bonefeld-Jørgensen EC. Persistent organic pollutants in Greenlandic pregnant women and indices of foetal growth: the ACCEPT study. Sci Total Environ. 2020, 698: 134118.

21 Ouidir M, Buck Louis GM, Kanner J, Grantz KL, Zhang C, Sundaram R, Rahman ML, Lee S, Kannan K, Tekola-Ayele F, Mendola P. Association of maternal exposure to persistent organic pollutants in early pregnancy with fetal growth. JAMA Pediatr. 2020, 174(2): 149–161.

22 Vuong AM, Xie C, Jandarov R, Dietrich KN, Zhang H, Sjödin A, Calafat AM, Lanphear BP, McCandless L, Braun JM, Yolton K, Chen A. Prenatal exposure to a mixture of persistent organic pollutants (POPs) and child reading skills at school age. Int J Hyg Environ Health. 2020, 228: 113527.

23 Myhre O, Zimmer KE, Hudecova AM, Hansen KEA, Khezri A, Berntsen HF, Berg V, Lyche JL, Mandal S, Duale N, Ropstad E. Maternal exposure to a human based mixture of persistent organic pollutants (POPs) affect gene expression related to brain function in mice offspring hippocampus. Chemosphere. 2021, 276: 130123.

24 Christou M, Ropstad E, Brown S, Kamstra JH, Fraser TWK. Developmental exposure to a POPs mixture or PFOS increased body weight and reduced swimming ability but had no effect on reproduction or behavior in zebrafish adults. Aquat Toxicol. 2021, 237: 105882.

25 Tran NQV, Miyake K. Neurodevelopmental Disorders and Environmental Toxicants: Epigenetics as an Underlying Mechanism. Int J Genomics. 2017, 2017: 7526592.

26 Patrizi B, Siciliani de Cumis M. TCDD Toxicity Mediated by Epigenetic Mechanisms. Int J Mol Sci. 2018, 19(12): 4101.

27 Kim S, Cho YH, Lee I, Kim W, Won S, Ku JL, Moon HB, Park J, Kim S, Choi G, Choi K. Prenatal exposure to persistent organic pollutants and methylation of LINE-1 and imprinted genes in placenta: A CHECK cohort study. Environ Int. 2018, 119: 398–406.

28 Wang C, Yuan XG, Liu CP, Zhai SN, Zhang DW, Fu YX. Preliminary research on DNA methylation changes during murine palatogenesis induced by TCDD. J Craniomaxillofac Surg. 2017, 45(5): 678-684.

29 Li J, Quan X, Lei S, Chen G, Hong J, Huang Z, Wang Q, Song W, Yang X. LncRNA MEG3 alleviates PFOS induced placental cell growth inhibition through its derived miR-770 targeting PTX3. Environ Pollut. 2022, 293: 118542.

30 Bertram MG, Gore AC, Tyler CR, Brodin T. Endocrine-disrupting chemicals. Curr Biol. 2022, 32(13): R727-R730.

31 Gore AC. Endocrine-Disrupting Chemicals. JAMA Intern Med. 2016, 176(11): 1705–1706.

32 Dufour P, Pirard C, Petrossians P, Beckers A, Charlier C. Association between mixture of persistent organic pollutants and thyroid pathologies in a Belgian population. Environ Res. 2020, 181: 108922.

33 Charles D, Berg V, Nøst TH, Bergdahl IA, Huber S, Ayotte P, Wilsgaard T, Averina M, Sandanger T, Rylander C. Longitudinal changes in concentrations of persistent organic pollutants (1986-2016) and their associations with type 2 diabetes mellitus. Environ Res. 2022, 204(Pt B): 112129.

34 Mrema EJ, Rubino FM, Brambilla G, Moretto A, Tsatsakis AM, Colosio C. Persistent organochlorinated pesticides and mechanisms of their toxicity. Toxicology. 2013, 307: 74–88.

35 Doan TQ, Berntsen HF, Verhaegen S, Ropstad E, Connolly L, Igout A, Muller M, Scippo ML. A mixture of persistent organic pollutants relevant for human exposure inhibits the transactivation activity of the aryl hydrocarbon receptor in vitro. Environ Pollut. 2019, 254(Pt B): 113098.

36 McComb J, Mills IG, Muller M, Berntsen HF, Zimmer KE, Ropstad E, Verhaegen S, Connolly L. Human blood-based exposure levels of persistent organic pollutant (POP) mixtures antagonise androgen receptor transactivation and translocation. Environ Int. 2019, 132: 105083.

37 Ahmed KEM, Frøysa HG, Karlsen OA, Blaser N, Zimmer KE, Berntsen HF, Verhaegen S, Ropstad E, Kellmann R, Goksøyr A. Effects of defined mixtures of POPs and endocrine disruptors on the steroid metabolome of the human H295R adrenocortical cell line. Chemosphere. 2019, 218: 328–339.

38 Lisco G, Giagulli VA, Iovino M, Guastamacchia E, Pergola G, Triggiani V. Endocrine-Disrupting Chemicals: Introduction to the Theme. Endocr Metab Immune Disord Drug Targets. 2022, 22(7): 677-685.

39 Kumar J, Monica Lind P, Salihovic S, van Bavel B, Lind L, Ingelsson E. Influence of persistent organic pollutants on oxidative stress in population-based samples. Chemosphere. 2014, 114: 303–309.

40 Chen Y, Cai Y, Chen C, Li M, Lu L, Yu Z, Wang S, Fang L, Xu S. Aroclor 1254 induced inhibitory effects on osteoblast differentiation in murine MC3T3-E1 cells through oxidative stress. Front Endocrinol (Lausanne). 2022, 13: 940624.

41 Pierozan P, Kosnik M, Karlsson O. High-content analysis shows synergistic effects of low perfluorooctanoic acid (PFOS) and perfluorooctane sulfonic acid (PFOA) mixture concentrations on human breast epithelial cell carcinogenesis. Environ Int. 2023, 172: 107746.

42 Sharma H, Hirko AC, King MA, Liu B. Role of NADPH oxidase in cooperative reactive oxygen species generation in dopaminergic neurons induced by combined treatment with dieldrin and lindane. Toxicol Lett. 2018, 299: 47–55.

43 Wang QY, Zhang L, Han XY, Wang DM, Ding ML, Cheng D, He QL, Liu SZ. 2,3',4,4',5-Pentachlorobiphenyl induces mitochondria-dependent apoptosis mediated by AhR/Cyp1a1 in mouse germ cells. J Hazard Mater. 2023, 445: 130547.

44 Green DR, Llambi F. Cell death signaling. Cold Spring Harb Perspect Biol. 2015, 7(12): a006080.

45 Li J, Cao F, Yin HL, Huang ZJ, Lin ZT, Mao N, Sun B, Wang G. Ferroptosis: past, present and future. Cell Death Dis. 2020, 11(2): 88.

46 Chen H, Wang C, Li H, Ma R, Yu Z, Li L, Xiang M, Chen X, Hua X, Yu Y. A review of toxicity induced by persistent organic pollutants (POPs) and endocrine-disrupting chemicals (EDCs) in the nematode Caenorhabditis elegans. J Environ Manage. 2019, 237: 519–525.

47 Wang J, Du H, Nie Y, Wang Y, Dai H, Wang M, Wang D, Xu A. Mitochondria and MAPK cascades modulate endosulfan-induced germline apoptosis in Caenorhabditis elegans. Toxicol Res (Camb). 2017, 6(4): 412–419.

48 Liu Z, Huang Y, Jin X, Liu L, Gu H. PCB153 suppressed autophagy via PI3K/Akt/mTOR and RICTOR/Akt/mTOR signaling by the upregulation of microRNA-155 in rat primary chondrocytes. Toxicol Appl Pharmacol. 2022, 449: 116135.

49 Long Y, Liu X, Tan XZ, Jiang CX, Chen SW, Liang GN, He XM, Wu J, Chen T, Xu Y. ROS-induced NLRP3 inflammasome priming and activation mediate PCB 118- induced pyroptosis in endothelial cells. Ecotoxicol Environ Saf. 2020, 189: 109937.

50 Zhou Z, Goodrich JM, Strakovsky RS. Mitochondrial Epigenetics and Environmental Health: Making a Case for Endocrine Disrupting Chemicals. Toxicol Sci. 2020, 178(1): 16-25.

51 Duarte-Hospital C, Tête A, Brial F, Benoit L, Koual M, Tomkiewicz C, Kim MJ, Blanc EB, Coumoul X, Bortoli S. Mitochondrial Dysfunction as a Hallmark of Environmental Injury. Cells. 2021, 11(1): 110.

52 Azevedo RDS, Falcão KVG, Amaral IPG, Leite ACR, Bezerra RS. Mitochondria as targets for toxicity and metabolism research using zebrafish. Biochim Biophys Acta Gen Subj. 2020, 1864(8): 129634.

53 Meyer JN, Hartman JH, Mello DF. Mitochondrial Toxicity. Toxicol Sci. 2018, 162(1): 15–23.

54 Meyer JN, Leung MC, Rooney JP, Sendoel A, Hengartner MO, Kisby GE, Bess AS. Mitochondria as a target of environmental toxicants. Toxicol Sci. 2013, 134(1): 1–17.

55 Byun HM, Benachour N, Zalko D, Frisardi MC, Colicino E, Takser L, Baccarelli AA. Epigenetic effects of low perinatal doses of flame retardant BDE-47 on mitochondrial and nuclear genes in rat offspring. Toxicology. 2015, 328: 152–159.

12

Combined Molecular Toxicity Mechanism of Emerging Pollutant Mixtures

Xiangsheng Hong[a,*]

[a]*National Engineering Research Center of Industrial Wastewater Detoxication and Resource Recovery, Research Center for Eco-Environmental Sciences, Chinese Academy of Sciences, Beijing, China*

*Corresponding author.
E-mail: xshong@rcees.ac.cn (X.H.)

12.1 Introduction

Pharmaceuticals and personal care products (PPCPs) belong to the group of so-called emerging pollutants. As the name implies, PPCPs consist of two categories for different purposes: pharmaceuticals and personal care products (PCPs), which comprise thousands of compounds with different chemical structures. The pharmaceuticals include numerous compound classes that are used primarily to treat human and animal diseases. This category includes antibiotic, anti-inflammatory/analgesic, antipsychotic, hypolipidemic, antihypertensive, and synthetic hormone. The second category of PPCPs, known as PCPs, are used to improve the quality of daily life and include ultraviolet (UV) filters (sunscreens), biocide, fragrance, and preservatives. As emerging pollutants, numerous PPCPs have been recognized in the natural environment across the world [1, 2]. A comprehensive categorization of the representative PPCP compounds reported in the literature is presented in Table 12.1. These compounds can be continually introduced to the aquatic environment (including drinking water) as complex mixtures typically at low concentrations (ng to μg/L), but over extended periods due to increased usage followed by incomplete removal by sewage treatment plants [3, 4]. Even though many PPCPs have a short lifespan and exist in trace concentrations in water, they are often classified as "pseudo-persistent" based on the continual infusion of these compounds into the aquatic environment. The active ingredients in pharmaceuticals, as well as several personal care products, are designed to react with specific biological pathways and processes of target humans and animals and usually be worked at low therapeutic doses [5–7]. Because the primary targets for many human-use active PPCPs show high structural conservation across species, these compounds potentially jeopardize ecological and human health.

Over the past 25 years, while a large body of data are available on the occurrence, bioaccumulation, fate, and environmental and health risks of PPCPs in aquatic environments, substantial gaps exist in our knowledge of their potential deleterious effects in realistic settings, such as low concentrations and complex mixtures compared to individual compound. Current evidence shows that despite the general lack of lethality even at unrealistically high concentrations, PPCPs are capable of producing various sublethal effects on aquatic organisms and wildlife that may prey on them, even at low concentrations [10–12]. In 2022, the latest version of the watch list is contained in the revised proposal of the Water Framework Directive (WFD) published by the European Commission [13]. Several pharmaceuticals such as venlafaxine, *o*-desmethylvenlafaxine, metformin, and guanylurea have been included as priority substances for the first time in the form of a single substance or groups of substances, but without using the previous precautionary principles [14]. The presence of numerous pharmaceuticals sharing a specific mode of action (MOA) could cause significant adverse effects through additive exposures. Compared to pharmaceuticals, the effects and molecular toxicity mechanisms of PCPs remain rather deficient due to nonbiological target-designed compounds. These characteristics of PPCPs, coupled with the presence of low-level mixtures and long-term exposure, will further increase the complexity and uncertainty about their toxicological effects. Environmental scientists and government regulators are now aware that understanding the potential effects and molecular toxicity mechanisms of the PPCP mixtures in the environment on nontarget organisms has therefore become a

Toxicological Assessment of Combined Chemicals in the Environment, First Edition. Edited by De-Sheng Pei and Yiyun Liu.

Table 12.1 List of PPCPs with some representative compounds under each subcategory.

PPCP subcategory	Compound
Pharmaceuticals	
Antibiotic	Amoxicillin, Azithromycin, Erythromycin, Sulfamethazine Chloramphenicol, Metronidazole, Oxytetracycline, Ofloxacin, Ciprofloxacin, Sulfamethoxazole, Tetracycline, Trimethoprim
Anti-inflammatory/analgesic	Diclofenac acid, Indomethacine, Ketoprofen, Mefenamic acid, Naproxen, Phenazone, Paracetamol, Ibuprofen, Salicylic acid, Codeine, Tramadol
Antipsychotic	Antianxiety drugs: Alprazolam, Diazepam, Lorazepam, Midazolam, Nitrazepam, Oxazepam, Triazolam, Zolpidem Antidepressants: Amitriptyline, Clomipramine, Dosulepin, Fluoxetine, Citalopram, Sertraline, Paroxetine, Escitalopram, Venlafaxine, Duloxetine Antiepileptics: Carbamazepine, Clonazepam, Gabapentin, Levetiracetam, Oxcarbazepine, Propranolol, Metoprolol,
Hypolipidemic	Bezafibrate, Clofibric acid, Gemfibrozil, Simvastatin, Pravastatin
Antihypertensive	Atenolol, Metoprolol, Propranodol, Irbesartan, Telmisartan, Amlodipine, Diltiazem, Nifedipine
Hormone	Estrone (E1), Estradiol (E2), Ethinylestradiol (EE2)
Personal care products	
UV filters (Sunscreens)	Organic class: Benzophenone 1 (BP1), Benzophenone 2 (BP2), Benzophenone 3 (BP3), Benzophenone 4 (BP4), Benzophenone 8 (BP8), 4-Methylbenzylidene camphor (4MBC), Ethylhexyl methoxycinnamate (EHMC), Octocrylene (OC), Octyl-dimethyl para-aminobenzoic acid (OD-PABA), Butyl methoxydibenzoylmethane (BMBM), Homosalate (HMS) Inorganic class: zinc oxide (ZnO) and titanium dioxide (TiO_2)
Biocide	Terephthalylidene dicamphor sulfonic acid (TDSA), Drometrizole trisiloxane (DTS), Diethylhexyl butamido triazone (DEBT), DEET, Methylene bis-benzotriazolyl tetramethylbutylphenol (MBBT), Phenylbenzimidazole sulfonic acid (PBSA), Ethylhexyl triazone (EHT), Bis-ethylhexyloxyphenol methoxyphenyl triazine (BEMT), Benzotriazole (BZT), Triclosan (TCS), Triclocarban (TCC)
Fragrance	Acetyl hexamethyl tetralin (AHTN), Acetyl hexamethyl indan (AHDI), Hexamethylindanopyran (HHCB), Tetramethyl Acetyloctahydronaphthalenes (OTNE), Musk xylene, Musk ketone
Preservative	Phenoxyethanol (2-PE), Methyl paraben (MeP), Ethyl paraben (EtP), *N*-propyl paraben (n-PrP), I-propyl paraben (i-PrP), *N*-butyl paraben (n-BuP), *I*-butyl paraben (i-BuP)

Note: The subcategory for each pharmaceutical and PCP comprehensively refers to the MeSH pharmacological classification in PubChem database (https://meshb.nlm.nih.gov/search) and the usage class as in Molins-Delgado et al. [8] and Ziylan-Yavas et al. [9].

long-standing challenge, which was previously ranked as number 1 among the 22 scientific questions regarding PPCPs in the environment [15].

To address this issue, referring to the framework of Adverse Outcome Pathway (AOP), this chapter divides PPCP-related toxicology research literature (Web of Science [WOS] and Google Scholar [GS]) into four levels based on the integrity of the evidence chain for the molecular toxicity mechanism of PPCP mixtures (Figure 12.1). This chapter mainly synthesizes the literature in line with Level 1 and Level 2 to summarize the molecular toxicity mechanism of existing PPCP mixtures. In addition, studies on environmental hot spot PPCPs with limited data on molecular toxicity mechanisms meeting the Level 3 conditions will also be considered in this chapter.

Despite the rapid growth of the PPCP literature, limited studies on the molecular toxicity mechanism of PPCP mixtures have been conducted, which is more than an order of magnitude less than those reported for single PPCPs. Comparison of these existing toxicological data on PPCPs is significantly hampered by the large differences in research objects, experimental techniques, sample processing methods, and the rigor of the logical process of experimental design. Based on the above status quo, this chapter will discuss the following two specific topics: (1) molecular mechanisms of potential toxic effects of pharmaceutical mixtures from the perspective of adverse drug interactions (ADIs) or adverse drug reactions (ADRs) and (2) molecular mechanisms of potential toxic effects of PPCP mixtures from the perspective of environmental coexistence of multiple PPCPs.

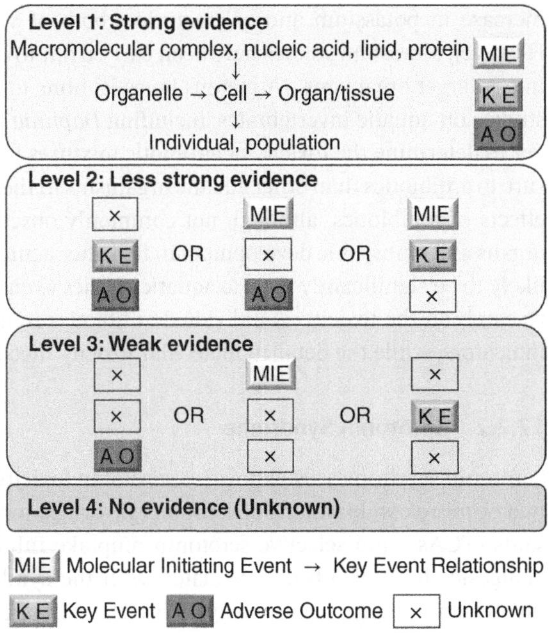

Figure 12.1 A proposed scheme for four-level classification of the literature on molecular toxicity mechanistic studies of PPCP mixtures based on the AOP framework.

12.2 ADIs or Reactions of Pharmaceutical Mixtures

The combination of drugs can enhance the efficacy of treatment and is often used as an augmentation strategy for some treatment-resistant disorders rather than the exception [16]. However, the real issue is that such co-administration or coexistence in the human body may be harmful to human health and even cause unexpected adverse effects due to adverse drug–drug interactions or/and adverse reactions rather than their positive interactions [17]. Compared to mammals, most (>90%) of the human drug targets had orthologs in zebrafish (*Danio rerio*), while only 64% were conserved in water flea (*Daphnia pulex*) and 34% in green algae (*Chlamydomonas reinhardtii*) [6]. In addition, human drug target classes associated with environmental effects data, such as enzymes (kinases and reductases), receptors, and other major drug-target ortholog s such as ion channels and transporters, are well conserved with those found in the aquatic vertebrates Western clawed frog (*Xenopus tropicalis*) and three-spined stickleback (*Gasterosteus aculeatus*) [5, 18]. ADIs or ADRs in nontarget aquatic vertebrates are therefore most likely to occur as a consequence of pharmaceutical residues in the environment [19]. Consideration of all the potential combinations of pharmaceutical mixtures is beyond the scope of our discussion in this chapter. Instead, we focus on those pharmaceuticals that have been reported to have ecological health risks and commonly coexist in the environment (e.g. Gunnarsson et al., [6], Yang et al., [20], Zhang et al., [21]), so that we can learn from the ADIs or ADRs to reveal the combined molecular toxicity mechanism of pharmaceutical mixtures. Luckily, besides the MOAs, extensive knowledge of the intracellular mechanisms of drug interactions, including the pharmacokinetic and pharmacodynamic properties, is suggested as one of the molecular initiating events (MIE) or key event (KE) for an AOP and helps to reveal their combined molecular toxicity mechanism in nontarget organisms.

12.2.1 Harmful Antibiotic Reactions and Drug Interactions

Antibiotics are specifically designed to inhibit the growth or kill the bacteria in a human or animal host instead of other microorganisms (such as archaea, viruses, protozoa, microalgae, or fungi), which distinguishes them from other microbicides or disinfectants [22]. The abuse and misuse of antibiotics are generally considered the major reasons for antibiotic resistance and can lead to the soaring increase of antibiotic-resistant bacteria, which poses a great threat to public health and has become a hot spot in the global antibiotic problem [23, 24]. Recently, a series of work by the team of Prof. Guo reported that antidepressants can also promote bacterial antibiotic resistance and spread [25–27]. Considering the subject of this chapter, the issue of antibiotic resistance will not be discussed further. Combinations of commonly used antibiotics exhibit potential ADRs and ADIs in humans [28]. For example, the combination of fluoroquinolones triggers damage to tendons, joints, muscles and nerves, and QTc prolongation, and the combination of trimethoprim–sulfamethoxazole may lead to an increase in potassium and/or creatinine, both of which are associated with significant interaction with warfarin [29]. However, almost no studies have been carried out to demonstrate whether these ADIs or ADRs of antibiotics can also occur in nontarget organisms. Numerous investigations to date have mainly focused on the lower test organisms, such as algae; studies on aquatic invertebrates, including *Daphnia magna*, *Artemia nauplii*, and *Chironomus riparius*, have been carried out to determine the toxicity of antibiotic mixtures [30–34]. It was found that fish were considered less susceptible to exposure to antibiotics than other aquatic organisms in the earlier period, such as algae and *Daphnia magna* [35, 36]. The adverse effects of antibiotics, although not commonly observed at environmental exposure levels, are known to affect multiple organs and embryonic development in fish after acute or chronic exposure [33, 37–39]. However, the antibiotic mixtures are likely to be significantly toxic to aquatic species even at concentrations below the NOECs of individual antibiotics. To date, research on the toxicity of individual antibiotics in nontarget organisms has mainly focused on alterations in the toxicity indicators, while the detailed molecular toxicity mechanisms of antibiotics are unknown, especially antibiotic mixtures.

12.2.2 Serotonin Syndrome

Serotonin syndrome (also known as serotonin toxicity) is a potentially life-threatening ADR that typically occurs following two or more combinations of serotonergic drugs, including monoamine oxidase inhibitors (MAOIs), tricyclic antidepressants (TCAs), and selective serotonin reuptake inhibitors (SSRIs) [40–42]. Apart from antidepressants, antibiotics, and analgesics have also been associated with the syndrome [43]. The mechanistic underpinnings of serotonin syndrome involve serotonergic over-activity both at the synaptic cleft of central nervous systems (CNS) receptors and peripheral nervous systems (PNS) receptors (mainly the 5-HT_{1A} and 5-HT_{2A} subtypes) [44]. The clinical symptoms of serotonin syndrome can range from barely perceptible to lethal, which are often described as neuromuscular abnormalities, autonomic hyperactivity (tachycardia with increased resting heart rate), and mental-status changes (confusion, agitation, etc.) [40, 43, 45]. ADIs leading to clinical SS in mammals have also been reproduced in different strains of rat or mice models [46], as shown by a set of abnormal behavior (e.g. forepaw treading and head weaving) and autonomic responses (e.g. tremor) [47–49].

Like other teleost fish, 5-HT is an important systemic neuromodulator in zebrafish, functionally similar to mammals with broad physiological effects in CNS and PNS [50–52]. Furthermore, the two major 5-HT receptor subtypes (5-HT_{1A} and 5-HT_{2A}) have also been identified in zebrafish [53]. An imbalance in the 5-HT levels can directly or indirectly affect behaviors such as locomotion and reproduction in zebrafish [54, 55]. The most common drug triggers of serotonin syndrome are antidepressants [56]. The work by Stewart et al. [57] provides behavioral evidence that zebrafish exposed to the combination of antidepressants exhibit surfacing behavior and hypolocomotion, which may represent previously unexplored markers of serotonin syndrome like states in zebrafish. Besides antidepressant-triggered serotonin syndrome, a few studies have discussed other combinations of serotonergic agents in aquatic animals. An example is the work by Bachour et al. [58], who first studied a binary mixture of effects of an analgesic (tramadol) and an antidepressant (citalopram) in zebrafish larvae. They showed that the binary mixture of tramadol and citalopram produces hypoactivity, which is possibly due to the ADI-induced serotonin syndrome, similar to the MOA in humans. However, the research on the combined exposure of serotonergic agents oriented to serotonin syndrome is still in its infancy, and the molecular mechanisms of its related toxic effects are needed for further studies.

12.3 Mechanistic Toxicology of PPCP Mixtures

The risks to the ecological environment and human health caused by the environmental residues of PPCPs have been recognized globally, especially the combined toxicity and molecular mechanism of PPCP mixtures, which is an area of growing concern, with various open questions relating to their adverse effects on nontarget organisms. Methodologically, the combined toxicity research of PPCPs is usually an acute toxicity test based on the rate of growth, mortality, lethality, and reproduction of cell models or in vivo research models such as bacteria, algae, and *Daphnia magna* and other major aquatic vertebrates [59, 60]. In terms of research interests, investigations have expanded to the combined effects of PPCP mixtures, as well as their MOAs, including two typical models, such as concentration addition and response addition, and four combined toxic effects, such as additive, synergistic, independent, and antagonistic effect [61]. Most of their studies, however, did not discuss the molecular mechanisms underlying the toxicity of PPCP mixtures to aquatic organisms or other model organisms. Based on the AOP framework, this chapter will systematically review the latest research progress on the combined molecular toxicity mechanism of PPCP mixtures in the following section.

12.3.1 Mixtures of Psychoactive Pharmaceuticals

Psychiatric pharmaceuticals, such as antidepressants, anxiolytics, and benzodiazepines, are among the most prescribed active substances in various environmental matrices, of which the aqueous environment is their ultimate sink [62, 63]. Over the past few decades, extensive studies have shown that they are highly persistent (mainly pseudo-persistent) and toxic to nontarget organisms. Among approaches to the characterization of toxicological effects, behavioral effects are the most frequently reported endpoints following exposure to psychoactive pharmaceuticals, including ecological effect-related behaviors such as activity, boldness, and foraging behavior [64, 65], and many other behaviors unique to different species of animals [66–70]. Besides behavioral changes, recent investigations on the toxicological effects of psychotropic pharmaceuticals on nontarget organisms have identified reproduction, malformations, and neuroendocrine or neurobehavior as the major endpoints [71–78]. Those findings highlight the fact that although each individual psychotropic pharmaceutical occurs in an aquatic environment at a low dose (low ng/L range), the total concentration when summing these levels up as a mixture was up to approximately 1 μg/L [53]. This brings exposure levels into the range that has been shown to have physiological and behavioral effects in aquatic animals in previous studies [79–81].

Several studies showed that a mixture of toxicities of psychotropic pharmaceuticals to nontarget organisms may be occurring at environmentally realistic concentrations due to their combined and synergistic effects [82–88]. Responses such as behavioral effects, histological changes, biochemical responses, and gene expression have been observed in nontarget organisms exposed to a mixture of psychotropic pharmaceuticals. However, research on the combined molecular toxicity mechanism of psychotropic pharmaceutical mixtures is limited. A binary mixture of antidepressants such as citalopram and tramadol seemingly induced antagonistic effects on zebrafish larvae [58]. An additive interaction between fluoxetine and norfluoxetine on stress-related behavior of zebrafish larvae was reported in a separate study by Zindler et al. [89]. Ternary mixture of antidepressants of fluoxetine, sertraline, and venlafaxine affected circadian rhythms of male mosquitofish (*Gambusia holbrooki*), which occurred at nominal concentrations of 1 μg/L [86]. In addition, a mixture of six psychotropic pharmaceuticals (valproate, cyamemazine, fluoxetine, sertraline, citalopram, and oxazepam) disrupted the larval locomotor behavior of Japanese medaka (*Oryzias latipes*) at concentrations of about 10–100 times higher than realistic environmental concentrations. However, the concentrations of another combination of six psychotropic pharmaceuticals (sertraline, citalopram, venlafaxine, tramadol, methamphetamine, and oxazepam) at ~1 μg/L of each compound that can alter the locomotory behavior of nontarget aquatic organisms are markedly lower than previous result by Zindler et al. [89], possibly because the differences in species as well as the composition of the mixture [87]. Two studies were dedicated to elucidating the reproductive toxicity of antidepressant mixtures at environmentally relevant concentrations to fathead minnows (*Pimephales promelas*). Schultz et al. [90] showed that 21-day exposures to antidepressant mixtures of bupropion, fluoxetine, sertraline, and venlafaxine did not affect reproductive behavior and other effects observed tended to be subtle in adult male fish. Parrott and Metcalfe [91] showed that full life cycle exposures to similar antidepressant mixtures (including bupropion, fluoxetine, sertraline, venlafaxine, and citalopram) caused no adverse effects in adult fish and F1 larvae.

Since psychotropic pharmaceuticals are intended to treat or alleviate psychiatric disorders, the environmental presence of psychotropic drugs tends to be neurotoxic to nontarget organisms. Thomas and Klaper [92] assessed the toxicity of a ternary mixture of psychotropic pharmaceuticals (fluoxetine, venlafaxine, and carbamazepine) in juvenile fathead minnows. Gene set enrichment analysis (GSEA) of brain tissue mRNA revealed that fish treated with a cocktail of psychotropic drugs elicited changes in autism-like gene expression patterns associated with idiopathic autism. Zebrafish larvae exposure to a binary mixture of norfluoxetine and venlafaxine elicited different interaction effects, such as antagonism and inverse agonism at gene expression levels, synergism in lethality, and no interaction on sensorimotor reflexes [85]. A mixture of three psychotropic pharmaceuticals (amitriptyline, nortriptyline, and clomipramine) neurotoxicity in a common carp (*Cyprinus carpio*) has also been tested. Results showed that this mixture significantly enhanced pathological changes in fish brains, which is probably because of oxidative damage of lipids. These pharmaceutical mixtures induced oxidative stress effects and were also detected in molluscans such as pond snail (*Lymnaea stagnalis*) and Asian clam (*Corbicula fluminea*) [93, 94].

12.3.2 Mixtures of UV Filters

UV filters, including organic and inorganic UV filters, are used widely in sunscreen products to protect the skin from the damage of UV radiation. For the highest UV protection, modern sunscreens use a combination of various organic and inorganic UV filters to achieve broad UV spectrum protection [95, 96]. For humans, exposure through dermal contact and inhalation pathways from UV filters containing PCPs are the two dominant external exposure pathways [97]. The photostability of some of the organic UV filters is known to cause decomposition when exposed to sunlight and generate reactive oxygen species (ROS) and toxic derivatives [59]. However, nowadays, skin penetration of organic UV filters and ROS transfer through the stratum corneum are still the subjects of controversy. Zinc oxide (ZnO) and titanium dioxide (TiO_2) nanoparticles, a size <100 nm, are the two most common inorganic UV filters in sunscreens or cosmetic preparations [98, 99]. These two compounds are commonly used as photocatalysts, can emit electrons such as free radicals and peroxides, and generate ROS, which is liable to have human health risks when exposed via inhalation [97, 100, 101]. However, compared with the organic UV filters, the inorganic UV filters ZnO and TiO_2 do not degrade under UV radiation exposure and present a limited epidermal penetration, which therefore seems to pose a low risk to human health [102–104].

In addition to the direct dermal contact and inhalation exposure, there are continuing concerns about the environmental and health impact of these UV filter cocktails or mixtures, especially considering their occurrence together in the environment [105–108]. A growing body of research focuses on the toxicity of UV filter mixtures. However, these investigations to date have mainly been performed in in vitro test systems such as bacteria, yeast, and cell and in in vitro test organisms such as invertebrate model, as well as vertebrate model fish, but usually lack combined molecular toxicity mechanism research of the UV filter mixtures [109–114]. In 2005, Heneweer et al. [111] analyzed the estrogenic effects of mixtures of four commonly used UV filters (BP-1, BP-3, 4-MBC, and OMC) and revealed the UV filter mixtures can activate the estrogen receptor in MCF-7 cells. Research from recombinant yeast assay (expressing human estrogen receptor alpha [hERα]) also showed that a mixture of four or eight UV filters displays an unexpected synergistic property and can produce estrogenic activity, even with each UV filter at its individual no-observed-effect concentration (NOEC) level [115]. However, it appears that organic UV filters in the presence of TiO_2 inorganic UV filters lead to reduced toxicity in the RTgill-W1 cell line due to their adsorption to the surface of TiO_2 [116]. These studies, however, did not further discuss the combined molecular toxicity mechanism of these UV filter mixtures. Apart from in vitro studies, several in vivo studies have reported the combined toxicity of UV filter mixtures in different fish species. For example, a ternary UV filter mixture of BP1, BP2, and 3BC displayed antagonism at lower concentrations but additivity at high concentrations, in terms of juvenile fathead minnows (*Pimephales promelas*) vitellogenesis [117]. Binary mixtures of octyl dimethyl para-aminobenzoic acid (OD-PABA) and butyl methoxydibenzoylmethane (BMBM) elicited significant changes in multi-biomarkers, such as activities of superoxide dismutase (SOD), catalase (CAT), glutathione reductase (GR), and acetylcholinesterase (AChE), which potentially causes oxidative stress and contributes to nervous system damage in juvenile crucian carp (*Carassius auratus*) [118]. Recently, Freire Sovierzoski et al. [119] assessed biochemical and genetic damage in juvenile tilapia (*Oreochromis niloticus*) simultaneously exposed to a ternary mixture of BP3, 2-ethylhexyl-*p*-methoxycinnamate (EHMC), and octocrylene (OC). Juvenile fish subchronic exposure to the ternary mixture of organic UV filter resulted in oxidative stress with increased lipid peroxidation (LPO) levels. Similar results have also been reported in the zebrafish embryo development, such as antioxidant system activation and mitochondrial dysfunction, with aqueous exposure to the same ternary UV filter mixture [120]. Notably, no significant changes in AChE activities were observed. However, a recent study on organic (benzophenone-3 [BP3]) and inorganic (TiO_2 nanoparticles) UV filter co-exposure points out that environmentally relevant concentrations of UV filter mixture can induce neurodevelopmental toxicity in zebrafish [121]. They also proposed a hypothesis that the production and accumulation of ROS is considered as the MIE, which consequently leads to apoptosis and inhibits the axonal growth in the developing zebrafish as KEs. All these effects finally resulted in AOs in the form of altered motor behaviors in zebrafish larvae.

12.3.3 Mixtures of Other PPCPS

In addition to studies of combined molecular toxicity mechanisms of PPCP mixtures of the same subcategory, combinations of different classes of PPCPs have also been studied, and the composition of the mixtures usually focuses on prioritizing selected PPCPs that are of environmental concerns and pose ecological risks [20, 122]. Several studies have already discussed the reproductive toxicity of PPCP mixtures at environmental levels in adult zebrafish. In 2013, Galus et al. [123] demonstrated that female zebrafish exposed to a mixture of acetaminophen, carbamazepine, gemfibrozil, and venlafaxine at doses ranging from 0.5 to 10 μg/L showed effects such as tissue degradation (atretic oocytes and altered ovarian histology), reduced embryo production, and increased embryo mortalities. Prolonged co-exposure to another PPCP combination (ibuprofen, carbamazepine, triclosan, erythromycin, and sulfamethoxazole) at environmental levels can cause endocrine disruption and reproductive effects in adult zebrafish [124]. The potential combined molecular toxicity of this PPCP combination was linked to activating estrogen receptors Erα and Erβ and abnormal gene expressions associated with the hypothalamic-pituitary-gonadal (HPG) axis pathway in fishes. The work by Schmitz et al. [125] evaluated the endocrine-disrupting effect of five selected pharmaceutical cocktails (carbamazepine, diclofenac, naproxen, paracetamol, and irbesartan) that co-existed in the Meuse River in Belgium on female juvenile rainbow trout. In this study, the increase in plasma levels of sex-steroid hormones and 11-ketotestosterone, as well as overexpression of several key genes related to ovarian steroidogenesis and abnormal gene expressions involved in germ cell fate, was mainly observed in fish exposed to 10–100 times the environmental concentration of the pharmaceutical cocktails.

Several in vivo studies have reported PPCP mixture's oxidative stress/damage effects on different nontarget organisms, such as crucian carp (*C. auratus*) and crayfish (*Procambarus clarkii*) [126, 127]. However, alterations in most of the selected endpoints including oxidative stress can be recovered after a week of depuration. Besides the abovementioned effects, other effects such as gut and liver toxicity as well as cardiotoxicity of PPCPs mixtures have also been reported in zebrafish [128, 129].

12.4 Conclusions and Recommendations for Future Studies

This chapter summarized previous research on the combined molecular toxicity mechanism of PPCP mixtures. PPCP mixtures can trigger different unpredictable effects on nontarget organisms and human beings. As the actual threat from the combination of PPCPs towards humans and nontarget organisms is complex, the true impact and combined molecular toxicity mechanism is largely unknown. Also, insufficient toxicology data hinder the construction of a complete AOP for the combined molecular toxicity mechanism of PPCP mixtures. Based on previous works and the latest trends in toxicology, we propose some recommendations for future investigations.

1. It is advantageous to recognize prevalent amalgamations of PPCP compounds due to the typical exposure of humans and nontarget organisms to blends of PPCPs rather than singular substances. The identification of pertinent mixtures that can be evaluated for their toxic properties is of the utmost importance. Although the potential combinations resulting from the numerous PPCPs present in the environment are seemingly limitless, the prevalent amalgamations of these PPCPs are considerably fewer in number.
2. Organoids represent a valuable tool for investigating the combined molecular toxicity mechanism of PPCPs on human health, particularly in consideration of animal welfare and ethical concerns.
3. More attention needs to be paid to the co-existence of chiral pharmaceuticals, their metabolites, and transformation products in the environment. Chiral compounds of the same pharmaceutical can exhibit significant variations in properties, and the metabolites and transformation products may retain biological activity, with some even surpassing the original compound. Nonetheless, the comprehensive evaluation of chiral pharmaceuticals, along with their metabolites and transformation products coexisting in environmental settings, remains inadequately explored in combined research endeavors.
4. Preclinical and clinical information on a pharmaceutical is used to assess its potential adverse environmental impact. Tons of information is available from mammalian studies and clinical trials. For example, the ADRs and ADIs between/among different pharmaceuticals will be beneficial to study the combined toxicity and molecular mechanism of pharmaceutical mixtures.
5. The complexity of the environmental exposure to the mixtures of PPCPs makes it extremely difficult to directly predict and reveal their combined molecular toxicity mechanism in nontarget organisms. The rise of diverse machine learning methods driven by high-quality toxicology data provides new opportunities for the combined molecular toxicity mechanisms of PPCP mixtures.

Funding

This work has been financially supported by the National Natural Science Foundation of China (grant no. 42107291).

Author Disclosure Statement

The authors have no conflicts of interest to declare.

References

1 Wilkinson JL, Boxall ABA, Kolpin DW, Leung KMY, Lai RWS, Galbán-Malagón C, Adell AD, Mondon J, Metian M, Marchant RA, Bouzas-Monroy A. Pharmaceutical pollution of the world's rivers. Proc Natl Acad Sci. 2022, 119(8): 1–10.

2 Lu S, Wang J, Wang B, Xin M, Lin C, Gu X, Lian M, Li Y. Comprehensive profiling of the distribution, risks and priority of pharmaceuticals and personal care products: a large-scale study from rivers to coastal seas. Water Res. 2023, 230: 119591.

3 Klaminder J, Brodin T, Sundelin A, Anderson NJ, Fahlman J, Jonsson M, Fick J. Long-term persistence of an anxiolytic drug (Oxazepam) in a large freshwater lake. Environ Sci Technol. 2015, 49(17): 10406–10412.

4 Guillossou R, Le Roux J, Mailler R, Vulliet E, Morlay C, Nauleau F, Gasperi J, Rocher V. Organic micropollutants in a large wastewater treatment plant: what are the benefits of an advanced treatment by activated carbon adsorption in comparison to conventional treatment? Chemosphere. 2019, 218: 1050–1060.

5 Gunnarsson L, Jauhiainen A, Kristiansson E, Nerman O, Larsson DGJ. Evolutionary conservation of human drug targets in organisms used for environmental risk assessments. Environ Sci Technol. 2008, 42(15): 5807–5813.

6 Gunnarsson L, Snape JR, Verbruggen B, Owen SF, Kristiansson E, Margiotta-Casaluci L, Österlund T, Hutchinson K, Leverett D, Marks B, Tyler CR. Pharmacology beyond the patient – the environmental risks of human drugs. Environ Int. 2019, 129: 320–332.

7 Garoche C, Boulahtouf A, Grimaldi M, Chiavarina B, Toporova L, den Broeder MJ, Legler J, Bourguet W, Balaguer P. Interspecies differences in activation of peroxisome proliferator-activated receptor γ by pharmaceutical and environmental chemicals. Environ Sci Technol. 2021, 55(24): 16489–16501.

8 Molins-Delgado D, Díaz-Cruz MS, Barceló D. Introduction: personal care products in the aquatic environment. In: Díaz-Cruz M, D Barceló (Eds.), Personal Care Products in the Aquatic Environment. The Handbook of Environmental Chemistry. Springer, 2014, pp. 1–34.

9 Ziylan-Yavas A, Santos D, Flores EMM, Ince NH. Pharmaceuticals and personal care products (PPCPs): environmental and public health risks. Environ Prog Sustain Energy. 2022, 41(4): 1–26.

10 Vera-Chang MN, St-Jacques AD, Gagné R, Martyniuk CJ, Yauk CL, Moon TW, Trudeau VL. Transgenerational hypocortisolism and behavioral disruption are induced by the antidepressant fluoxetine in male zebrafish Danio rerio. Proc Natl Acad Sci. 2018, 115(52): E12435–12442.

11 Singh A, Saidulu D, Gupta AK, Kubsad V. Occurrence and fate of antidepressants in the aquatic environment: insights into toxicological effects on the aquatic life, analytical methods, and removal techniques. J Environ Chem Eng. 2022, 10(6): 109012.

12 Adeleye AS, Xue J, Zhao Y, Taylor AA, Zenobio JE, Sun Y, Han Z, Salawu OA, Zhu Y. Abundance, fate, and effects of pharmaceuticals and personal care products in aquatic environments. J Hazard Mater. 2022, 424: 127284.

13 European Commission. Third Version of the European Surface Water. Watch List. European Commission, 2022.

14 Ellis JB. Pharmaceutical and personal care products (PPCPs) in urban receiving waters. Environ Pollut. 2006, 144(1): 184–189.

15 Rudd MA, Ankley GT, Boxall ABA, Brooks BW. International scientists' priorities for research on pharmaceutical and personal care products in the environment. Integr Environ Assess Manag. 2014, 10(4): 576–587.

16 Schatzberg AF, DeBattista C. Manual of Clinical Psychopharmacology (Eighth Edition). American Psychiatric Publishing, Inc., 2015.

17 US Food and Drug Administration. Drug Interactions: What You Should Know, US Food and Drug Administration, 2013.

18 McRobb FM, Sahagún V, Kufareva I, Abagyan R. In silico analysis of the conservation of human toxicity and endocrine disruption targets in aquatic species. Environ Sci Technol. 2014, 48(3): 1964–1972.

19 Matthee C, Brown AR, Lange A, Tyler CR. Factors determining the susceptibility of fish to effects of human pharmaceuticals. Environ Sci Technol. 2023, 57(24): 8845–8862.

20 Yang Y, Zhang X, Jiang J, Han J, Li W, Li X, Yee Leung KM, Snyder SA, Alvarez PJ. Which micropollutants in water environments deserve more attention globally? Environ Sci Technol. 2022, 56(1): 13–29.

21 Zhang H, Kato D, Ihara MO, Jürgens MD, Johnson AC, Chen J, Tanaka H, Ihara M. Biological-activity-based prioritization of antidepressants in wastewater in England and Japan. Environ Sci Technol. 2023, 57(16): 6444–6454.

22 Brandt KK, Amézquita A, Backhaus T, Boxall A, Coors A, Heberer T, Lawrence JR, Lazorchak J, Schoenfeld J, Snape JR, Zhu YG. Ecotoxicological assessment of antibiotics: a call for improved consideration of microorganisms. Environ Int. 2015, 85: 189–205.

23 McKenna M. Antibiotic resistance: the last resort. Nature. 2013, 499(7459): 394–396.

24 Willyard C. The drug-resistant bacteria that pose the greatest health threats. Nature. 2017, 543(7643): 15–15.

25 Wang Y, Yu Z, Ding P, Lu J, Mao L, Ngiam L, Yuan Z, Engelstädter J, Schembri MA, Guo J. Antidepressants can induce mutation and enhance persistence toward multiple antibiotics. Proc Natl Acad Sci. 2023, 120(5): e2208344120.

26 Lu J, Ding P, Wang Y, Guo J. Antidepressants promote the spread of extracellular antibiotic resistance genes via transformation. ISME Commun. 2022, 2(1): 63.

27 Ding P, Lu J, Wang Y, Schembri MA, Guo J. Antidepressants promote the spread of antibiotic resistance via horizontally conjugative gene transfer. Environ Microbiol. 2022, 24(11): 5261–5276.

28 Granowitz EV, Brown RB. Antibiotic adverse reactions and drug interactions. Crit Care Clin. 2008, 24(2): 421–442.

29 Agency for Healthcare Research and Quality. AHRQ Safety Program for Improving Antibiotic Use – Long-Term Care: Antibiotic Side Effects 2. Vol. 17, AHRQ Publication, 2021, p. 29.

30 Wang Y, Gao C, Qu Z, Li M. The combined toxicity of binary mixtures of antibiotics against the cyanobacterium Microcystis is dose-dependent: insight from a theoretical nonlinear combined toxicity assessment method. Environ Sci Pollut Res. 2022, 29(8): 11612–11624.

31 Yang L-H, Ying G-G, Su H-C, Stauber JL, Adams MS, Binet MT. Growth-inhibiting effects of 12 antibacterial agents and their mixtures on the freshwater microalga pseudokirchneriella subcapitata. Environ Toxicol Chem. 2008, 27(5): 1201.

32 Li G, Lijuan S, Tao Y. Growth inhibitive effect of typical antibiotics and their mixtures on Selenastrum capricornutum. J Environ Health. 2013, 30(6): 475–478.

33 Liu L, Wu W, Zhang J, Lv P, Xu L, Yan Y. Progress of research on the toxicology of antibiotic pollution in aquatic organisms. Acta Ecol Sin. 2018, 38(1): 36–41.

34 Carusso S, Juárez AB, Moretton J, Magdaleno A. Effects of three veterinary antibiotics and their binary mixtures on two green alga species. Chemosphere. 2018, 194: 821–827.

35 Crane M, Watts C, Boucard T. Chronic aquatic environmental risks from exposure to human pharmaceuticals. Sci Total Environ. 2006, 367(1): 23–41.

36 Yang C, Song G, Lim W. A review of the toxicity in fish exposed to antibiotics. Comp Biochem Physiol Part C Toxicol Pharmacol. 2020, 237: 108840.

37 Lin T, Chen Y, Chen W. Toxic effect of sulfadiazine on the growth of zebrafish embryos in the water body. J Saf Environ. 2014, 14: 324–327.

38 Laville N, Aït-Aïssa S, Gomez E, Casellas C, Porcher J. Effects of human pharmaceuticals on cytotoxicity, EROD activity and ROS production in fish hepatocytes. Toxicology. 2004, 196 (1–2): 41–55.

39 Zhang Q, Cheng J, Xin Q. Effects of tetracycline on developmental toxicity and molecular responses in zebrafish (Danio rerio) embryos. Ecotoxicology. 2015, 24(4): 707–719.

40 Martin T. Serotonin syndrome. Ann Emerg Med. 1996, 28(5): 520–526.

41 Sun-Edelstein C, Tepper SJ, Shapiro RE. Drug-induced serotonin syndrome: a review. Expert Opin Drug Saf. 2008, 7(5): 587–596.

42 Mikkelsen N, Damkier P, Pedersen SA. Serotonin syndrome - a focused review. Basic Clin Pharmacol Toxicol. 2023, 133(2): 124–129.

43 Boyer EW, Shannon M. The serotonin syndrome. N Engl J Med. 2005, 352(11): 1112–1120.

44 Beakley BD, Kaye AM, Kaye AD. Tramadol, pharmacology, side effects, and serotonin syndrome: a review. Pain Physician. 2015, 18(4): 395–400.

45 Buckley NA, Dawson AH, Isbister GK. Serotonin syndrome. BMJ. 2014, 348(feb19 6): g1626.

46 Haberzettl R, Bert B, Fink H, Fox MA. Animal models of the serotonin syndrome: a systematic review. Behav Brain Res. 2013, 256: 3328–3345.

47 Izumi T, Iwamoto N, Kitaichi Y, Kato A, Inoue T, Koyama T. Effects of co-administration of a selective serotonin reuptake inhibitor and monoamine oxidase inhibitors on 5-HT-related behavior in rats. Eur J Pharmacol. 2006, 532(3): 258–264.

48 Kennett GA, Dickinson SL, Curzon G. Central serotonergic responses and behavioural adaptation to repeated immobilisation: the effect of the corticosterone synthesis inhibitor metyrapone. Eur J Pharmacol. 1985, 119(3): 143–152.

49 Fox MA, Jensen CL, Murphy DL. Tramadol and another atypical opioid meperidine have exaggerated serotonin syndrome behavioural effects, but decreased analgesic effects, in genetically deficient serotonin transporter (SERT) mice. Int J Neuropsychopharmacol. 2009, 12(08): 1055.

50 Stewart A, Gaikwad S, Hart P, Kyzar E, Roth A, Kalueff AV. Experimental models for anxiolytic drug discovery in the era of omes and omics. Expert Opin Drug Discov. 2011, 6(7): 755–769.

51 Daubert EA, Condron BG. Serotonin: a regulator of neuronal morphology and circuitry. Trends Neurosci. 2010, 33(9): 424–434.

52 Thompson WA, Vijayan MM. Zygotic venlafaxine exposure impacts behavioral programming by disrupting brain serotonin in Zebrafish. Environ Sci Technol. 2020, 54(22): 14578–14588.

53 Gould SL, Winter MJ, Norton WHJ, Tyler CR. The potential for adverse effects in fish exposed to antidepressants in the aquatic environment. Environ Sci Technol. 2021, 55(24): 16299–16312.

54 Kawashima T. The role of the serotonergic system in motor control. Neurosci Res. 2018, 129: 32–39.

55 Prasad P, Ogawa S, Parhar IS. Role of serotonin in fish reproduction. Front Neurosci. 2015, 9: 195.

56 Francescangeli J, Karamchandani K, Powell M, Bonavia A. The serotonin syndrome: from molecular mechanisms to clinical practice. Int J Mol Sci. 2019, 20(9): 2288.

57 Stewart AM, Cachat J, Gaikwad S, Robinson KSL, Gebhardt M, Kalueff AV. Perspectives on experimental models of serotonin syndrome in zebrafish. Neurochem Int. 2013, 62(6): 893–902.

58 Bachour RL, Golovko O, Kellner M, Pohl J. Behavioral effects of citalopram, tramadol, and binary mixture in zebrafish (Danio rerio) larvae. Chemosphere. 2020, 238: 124587.

59 Huang Y, Law JCF, Lam TK, Leung KSY. Risks of organic UV filters: a review of environmental and human health concern studies. Sci Total Environ. 2021, 755: 142486.

60 Yang Y-T, Ni H-G. Predictive in silico models for aquatic toxicity of cosmetic and personal care additive mixtures. Water Res. 2023, 236: 119981.

61 Godoy AA, Kummrow F. What do we know about the ecotoxicology of pharmaceutical and personal care product mixtures? A critical review. Crit Rev Environ Sci Technol. 2017, 47(16): 1453–1496.

62 Calisto V, Esteves VI. Psychiatric pharmaceuticals in the environment. Chemosphere. 2009, 77(10): 1257–1274.

63 Cunha DL, Mendes MP, Marques M. Environmental risk assessment of psychoactive drugs in the aquatic environment. Environ Sci Pollut Res. 2019, 26(1): 78–90.

64 Brodin T, Fick J, Jonsson M, Klaminder J. Dilute concentrations of a psychiatric drug alter behavior of fish from natural populations. Science. 2013, 339(6121): 814–815.

65 Hong X, Zhao G, Zhou Y, Chen R, Li J, Zha J. Risks to aquatic environments posed by 14 pharmaceuticals as illustrated by their effects on zebrafish behaviour. Sci Total Environ. 2021, 771: 145450.

66 Valenti TW, Gould GG, Berninger JP, Connors KA, Keele NB, Prosser KN, Brooks BW. Human therapeutic plasma levels of the Selective Serotonin Reuptake Inhibitor (SSRI) sertraline decrease serotonin reuptake transporter binding and shelter-seeking behavior in adult male fathead minnows. Environ Sci Technol. 2012, 46(4): 2427–2435.

67 Saaristo M, McLennan A, Johnstone CP, Clarke BO, Wong BBM. Impacts of the antidepressant fluoxetine on the anti-predator behaviours of wild guppies (Poecilia reticulata). Aquat Toxicol. 2017, 183: 38–45.

68 Fursdon JB, Martin JM, Bertram MG, Lehtonen TK, Wong BBM. The pharmaceutical pollutant fluoxetine alters reproductive behaviour in a fish independent of predation risk. Sci Total Environ. 2019, 650: 642–652.

69 Martin JM, Saaristo M, Tan H, Bertram MG, Nagarajan-Radha V, Dowling DK, Wong BB. Field-realistic antidepressant exposure disrupts group foraging dynamics in mosquitofish. Biol Lett. 2019, 15(11): 20190615.

70 Rodrigues P, Guimarães L, Carvalho A, Oliva-Teles L. Carbamazepine, venlafaxine, tramadol, and their main metabolites: toxicological effects on zebrafish embryos and larvae. J Hazard Mater. 2023, 448: 130909.

71 Nowakowska K, Giebułtowicz J, Kamaszewski M, Adamski A, Szudrowicz H, Ostaszewska T, Solarska-Dzięciołowska U, Nałęcz-Jawecki G, Wroczyński P, Drobniewska A. Acute exposure of zebrafish (Danio rerio) larvae to environmental concentrations of selected antidepressants: bioaccumulation, physiological and histological changes. Comp Biochem Physiol Part C Toxicol Pharmacol. 2020, 229: 108670.

72 Kolpin DW, Foreman WT, Phillips PJ, Colella KM, Gray JL, Scott T-M, Furlong ET. Pharmaceutical manufacturing facility discharges can substantially increase the pharmaceutical load to U.S. wastewaters. Sci Total Environ. 2018, 636: 69–79.

73 Lopes DG, Duarte IA, Antunes M, Fonseca VF. Effects of antidepressants in the reproduction of aquatic organisms: a meta-analysis. Aquat Toxicol. 2020, 227: 105569.

74 Fraher D, Hodge JM, Collier FM, McMillan JS, Kennedy RL, Ellis M, Nicholson GC, Walder K, Dodd S, Berk M, Pasco JA. Citalopram and sertraline exposure compromises embryonic bone development. Mol Psychiatry. 2016, 21(5): 656–664.

75 Hong X, Zhang L, Zha J. Toxicity of waterborne vortioxetine, a new antidepressant, in non-target aquatic organisms: from wonder to concern drugs? Environ Pollut. 2022, 304: 119175.

76 Hong X, Chen R, Zhang L, Yan L, Li J, Zha J. Low doses and lifecycle exposure of waterborne antidepressants in zebrafish model: a survey on sperm traits, reproductive behaviours, and transcriptome responses. Sci Total Environ. 2022, 832: 155017.

77 Hong X, Chen R, Zhang L, Yan L, Xin J, Li J, Zha J. Long-term exposure to SSRI citalopram induces neurotoxic effects in zebrafish. Environ Sci Technol. 2022, 56(17): 12380–12390.

78 Venkatachalam AB, Levesque B, Achenbach JC, Pappas JJ, Ellis LD. Long and short duration exposures to the Selective Serotonin Reuptake Inhibitors (SSRIs) fluoxetine, paroxetine and sertraline at environmentally relevant concentrations lead to adverse effects on zebrafish behaviour and reproduction. Toxics. 2023, 11(2): 151.

79 Galus M, Kirischian N, Higgins S, Purdy J, Chow J, Rangaranjan S, Li H, Metcalfe C, Wilson JY. Chronic, low concentration exposure to pharmaceuticals impacts multiple organ systems in zebrafish. Aquat Toxicol. 2013, 132–133: 200–11.

80 Martin JM, Bertram MG, Saaristo M, Fursdon JB, Hannington SL, Brooks BW, Burket SR, Mole RA, Deal ND, Wong BB. Antidepressants in surface waters: fluoxetine influences mosquitofish anxiety-related behavior at environmentally relevant levels. Environ Sci Technol. 2019, 53(10): 6035–6043.

81 Wiles SC, Bertram MG, Martin JM, Tan H, Lehtonen TK, Wong BBM. Long-term pharmaceutical contamination and temperature stress disrupt fish behavior. Environ Sci Technol. 2020, 54(13): 8072–8082.

82 Henry TB, Black MC. Mixture and single-substance acute toxicity of selective serotonin reuptake inhibitors in Ceriodaphnia dubia. Environ Toxicol Chem. 2007, 26(8): 1751.

83 Henry TB, Kwon J-W, Armbrust KL, Black MC. Acute and chronic toxicity of five selective serotonin reuptake inhibitors in Ceriodaphnia dubia. Environ Toxicol Chem. 2004, 23(9): 2229.

84 Minguez L, Bureau R, Halm-Lemeille M-P. Joint effects of nine antidepressants on Raphidocelis subcapitata and Skeletonema marinoi: a matter of amine functional groups. Aquat Toxicol. 2018, 196: 117–123.

85 Rodrigues P, Cunha V, Oliva-Teles L, Ferreira M, Guimarães L. Norfluoxetine and venlafaxine in zebrafish larvae: single and combined toxicity of two pharmaceutical products relevant for risk assessment. J Hazard Mater. 2020, 400: 123171.

86 Melvin SD. Effect of antidepressants on circadian rhythms in fish: insights and implications regarding the design of behavioural toxicity tests. Aquat Toxicol. 2017, 182: 20–30.

87 Hossain MS, Kubec J, Guo W, Roje S, Ložek F, Grabicová K, Randák T, Kouba A, Buřič M. A combination of six psychoactive pharmaceuticals at environmental concentrations alter the locomotory behavior of clonal marbled crayfish. Sci Total Environ. 2021, 751: 141383.

88 Ma Y, Xu D, Li C, Wei S, Guo R, Li Y, Chen J, Liu Y. Combined toxicity and toxicity persistence of antidepressants citalopram and mirtazapine to zooplankton Daphnia magna. Environ Sci Pollut Res. 2022, 29(44): 66100–66108.

89 Zindler F, Stoll S, Baumann L, Knoll S, Huhn C, Braunbeck T. Do environmentally relevant concentrations of fluoxetine and citalopram impair stress-related behavior in zebrafish (Danio rerio) embryos? Chemosphere. 2020, 261: 127753.

90 Schultz MM, Painter MM, Bartell SE, Logue A, Furlong ET, Werner SL, Schoenfuss HL. Selective uptake and biological consequences of environmentally relevant antidepressant pharmaceutical exposures on male fathead minnows. Aquat Toxicol. 2011, 104 (1–2): 38–47.

91 Parrott JL, Metcalfe CD. Assessing the effects of environmentally relevant concentrations of antidepressant mixtures to fathead minnows exposed over a full life cycle. Sci Total Environ. 2019, 648: 1227–1236.

92 Thomas MA, Klaper RD. Psychoactive pharmaceuticals induce fish gene expression profiles associated with human idiopathic autism. Skoulakis EMC, editor. PLoS One. 2012, 7(6): e32917.

93 Gust M, Fortier M, Garric J, Fournier M, Gagné F. Effects of short-term exposure to environmentally relevant concentrations of different pharmaceutical mixtures on the immune response of the pond snail Lymnaea stagnalis. Sci Total Environ. 2013, 445–446: 210–218.

94 Bourioug M, Mazzitelli J-Y, Marty P, Budzinski H, Aleya L, Bonnafé E, Geret F. Assessment of Lemna minor (duckweed) and Corbicula fluminea (freshwater clam) as potential indicators of contaminated aquatic ecosystems: responses to presence of psychoactive drug mixtures. Environ Sci Pollut Res. 2018, 25(12): 11192–11204.

95 Villalobos-Hernández JR, Müller-Goymann CC. Sun protection enhancement of titanium dioxide crystals by the use of carnauba wax nanoparticles: the synergistic interaction between organic and inorganic sunscreens at nanoscale. Int J Pharm. 2006, 322 (1–2): 161–170.

96 Wissing SA, Müller RH. Solid lipid nanoparticles (SLN) – a novel carrier for UV blockers. Pharmazie. 2001, 56(10): 783–786.

97 Gilbert E, Pirot F, Bertholle V, Roussel L, Falson F, Padois K. Commonly used UV filter toxicity on biological functions: review of last decade studies. Int J Cosmet Sci. 2013, 35(3): 208–219.
98 Grande F, Tucci P. Titanium dioxide nanoparticles: a risk for human health? Mini-Reviews Med Chem. 2016, 16(9): 762–769.
99 Nanotechnology Task Force Report 2007. In: Administration USFD, (Ed.), 2018. Available from: https://www.fda.gov/science-research/nanotechnology-programs-fda/nanotechnology-fact-sheet.
100 Schneider SL, Lim HW. A review of inorganic UV filters zinc oxide and titanium dioxide. Photodermatol Photoimmunol Photomed. 2019, 35(6): 442–446.
101 Jesus A, Sousa E, Cruz M, Cidade H, Lobo J, Almeida I. UV Filters: challenges and prospects. Pharmaceuticals. 2022, 15(3): 263.
102 Senzui M, Tamura T, Miura K, Ikarashi Y, Watanabe Y, Fujii M. Study on penetration of titanium dioxide (TiO2) nanoparticles into intact and damaged skin in vitro. J Toxicol Sci. 2010, 35(1): 107–113.
103 Lin LL, Grice JE, Butler MK, Zvyagin AV, Becker W, Robertson TA, Soyer HP, Roberts MS, Prow TW. Time-correlated single photon counting for simultaneous monitoring of zinc oxide nanoparticles and NAD(P)H in intact and barrier-disrupted volunteer skin. Pharm Res. 2011, 28(11): 2920–2930.
104 Hanigan D, Truong L, Schoepf J, Nosaka T, Mulchandani A, Tanguay RL, Westerhoff P. Trade-offs in ecosystem impacts from nanomaterial versus organic chemical ultraviolet filters in sunscreens. Water Res. 2018, 139: 281–90.
105 Weir A, Westerhoff P, Fabricius L, Hristovski K, von Goetz N. Titanium dioxide nanoparticles in food and personal care products. Environ Sci Technol. 2012, 46(4): 2242–2250.
106 Liu Y, Gao L, Qiao L, Huang D, Lyu B, Li J, Wu Y, Zheng M. Concentrations, compound profiles, and possible sources of organic UV filters in human milk in China. Environ Sci Technol. 2022, 56(22): 15930–15940.
107 Li Z-M, Kannan K. Comprehensive survey of 14 benzophenone UV filters in sunscreen products marketed in the United States: implications for human exposure. Environ Sci Technol. 2022, 56(17): 12473–12482.
108 Yuan S, Huang J, Jiang X, Huang Y, Zhu X, Cai Z. Environmental fate and toxicity of sunscreen-derived inorganic ultraviolet filters in aquatic environments: a review. Nanomaterials. 2022, 12(4): 699.
109 Ozáez I, Morcillo G, Martínez-Guitarte J-L. The effects of binary UV filter mixtures on the midge Chironomus riparius. Sci Total Environ. 2016, 556: 154–162.
110 He T, Tsui MMP, Tan CJ, Ma CY, Yiu SKF, Wang LH, Fan TY, Lam PK, Murphy MB. Toxicological effects of two organic ultraviolet filters and a related commercial sunscreen product in adult corals. Environ Pollut. 2019, 245: 462–471.
111 Heneweer M, Muusse M, van den Berg M, Sanderson JT. Additive estrogenic effects of mixtures of frequently used UV filters on pS2-gene transcription in MCF-7 cells. Toxicol Appl Pharmacol. 2005, 208(2): 170–177.
112 Molins-Delgado D, Gago-Ferrero P, Díaz-Cruz MS, Barceló D. Single and joint ecotoxicity data estimation of organic UV filters and nanomaterials toward selected aquatic organisms. Urban groundwater risk assessment. Environ Res. 2016, 145: 126–134.
113 Park C-B, Jang J, Kim S, Kim YJ. Single- and mixture toxicity of three organic UV-filters, ethylhexyl methoxycinnamate, octocrylene, and avobenzone on Daphnia magna. Ecotoxicol Environ Saf. 2017, 137: 57–63.
114 Balázs A, Krifaton C, Orosz I, Szoboszlay S, Kovács R, Csenki Z, Urbányi B, Kriszt B. Hormonal activity, cytotoxicity and developmental toxicity of UV filters. Ecotoxicol Environ Saf. 2016, 131: 45–53.
115 Kunz PY, Fent K. Estrogenic activity of UV filter mixtures. Toxicol Appl Pharmacol. 2006, 217(1): 86–99.
116 Martin N, Wassmur B, Slomberg D, Labille J, Lammel T. Influence of TiO2 nanocomposite UV filter surface chemistry and their interactions with organic UV filters on uptake and toxicity toward cultured fish gill cells. Ecotoxicol Environ Saf. 2022, 243: 113984.
117 Kunz PY, Fent K. Estrogenic activity of ternary UV filter mixtures in fish (Pimephales promelas) — an analysis with nonlinear isobolograms. Toxicol Appl Pharmacol. 2009, 234(1): 77–88.
118 Ma B, Lu G, Liu J, Yan Z, Yang H, Pan T. Bioconcentration and multi-biomarkers of organic UV filters (BM-DBM and OD-PABA) in crucian carp. Ecotoxicol Environ Saf. 2017, 141: 178–187.
119 Freire Sovierzoski JC, Severino MA, Ribas E, Gomes MF, Rocha Martins LR, Ramsdorf WA. Biomarkers activity in Oreochromis niloticus under sub-chronic exposure to a UV filters ternary mixture. Chemosphere. 2023, 331: 138756.
120 Li AJ, Law JC-F, Chow C-H, Huang Y, Li K, Leung KS-Y. Joint effects of multiple UV filters on zebrafish embryo development. Environ Sci Technol. 2018, 52: 9460–9467.
121 Sun X, Yang Q, Jing M, Jia X, Tian L, Tao J. Environmentally relevant concentrations of organic (benzophenone-3) and inorganic (titanium dioxide nanoparticles) UV filters co-exposure induced neurodevelopmental toxicity in zebrafish. Ecotoxicol Environ Saf. 2023, 249: 114343.

122 Junaid M, Wang Y, Hamid N, Deng S, Li W-G, Pei D-S. Prioritizing selected PPCPs on the basis of environmental and toxicogenetic concerns: a toxicity estimation to confirmation approach. J Hazard Mater. 2019, 380: 120828.

123 Galus M, Jeyaranjaan J, Smith E, Li H, Metcalfe C, Wilson JY. Chronic effects of exposure to a pharmaceutical mixture and municipal wastewater in zebrafish. Aquat Toxicol. 2013, 132–133: 212–222.

124 Hamid N, Junaid M, Manzoor R, Duan J-J, Lv M, Xu N, Pei DS. Tissue distribution and endocrine disruption effects of chronic exposure to pharmaceuticals and personal care products mixture at environmentally relevant concentrations in zebrafish. Aquat Toxicol. 2022, 242: 106040.

125 Schmitz M, Beghin M, Mandiki SNM, Nott K, Gillet M, Ronkart S, Robert C, Baekelandt S, Kestemont P. Environmentally-relevant mixture of pharmaceutical drugs stimulates sex-steroid hormone production and modulates the expression of candidate genes in the ovary of juvenile female rainbow trout. Aquat Toxicol. 2018, 205: 89–99.

126 Ding J, Lu G, Li Y. Interactive effects of selected pharmaceutical mixtures on bioaccumulation and biochemical status in crucian carp (Carassius auratus). Chemosphere. 2016, 148: 21–31.

127 Trombini C, Kazakova J, Montilla-López A, Fernández-Cisnal R, Hampel M, Fernández-Torres R, Bello-López MÁ, Abril N, Blasco J. Assessment of pharmaceutical mixture (ibuprofen, ciprofloxacin and flumequine) effects to the crayfish Procambarus clarkii: a multilevel analysis (biochemical, transcriptional and proteomic approaches). Environ Res. 2021, 200: 111396.

128 Jain R, Laetz CA. Morphological and molecular analysis of the toxicity of pharmaceutical-derived aquatic contaminants (PPCPs) in zebrafish. In: Salish Sea Ecosystem Conference, Seattle, Washington, 2018, p. 236.

129 Hamid N, Junaid M, Wang Y, Pu S-Y, Jia P-P, Pei D-S. Chronic exposure to PPCPs mixture at environmentally relevant concentrations (ERCs) altered carbohydrate and lipid metabolism through gut and liver toxicity in zebrafish. Environ Pollut. 2021, 273: 116494.

13

Combined Molecular Toxicity Mechanism of Phthalate Mixtures

Yiyun Liu[a,*], *De-Sheng Pei*[a]

[a] *School of Public Health, Chongqing Medical University, Chongqing, China*

*Corresponding author.
E-mail: 191065@cqmu.edu.cn

13.1 Human Exposure of Phthalates

Phthalates (PAEs) are esters of 1,2-benzene dicarboxylic acid; the side chain of which is formed by dialkyl, alkyl, or aryl. The longer the side chain, the stronger the lipophilicity and the higher the boiling point of compounds. Short-chain PAEs are widely used as solvents for personal care products, including lotion, perfume, shampoos, hair gels, and cosmetics [1]. Long-chain PAEs are commonly used as plasticizers to enhance the durability and flexibility of plastic products. Plastic is a synthetic organic polymer with high molecular weight. Due to its lightweight, good plasticity, and nonperishable properties, it has been widely used in daily life. The global annual plastic production is about 40000 tons to meet the enormous market demand [2]. In flexible plastics such as polyvinyl chloride (PVC), PAEs can account for 80% of the final product weight [3]. Therefore, there is also a huge industrial and commercial demand for PAE compounds, including construction materials, medical devices, children's toys, food packaging, automotive manufacturing, and other fields [4].

Commonly used PAEs (Table 13.1) include di-(2-ethylhexyl) phthalate (DEHP), di-iso-nonylphthalate (DINP), di-iso-decylphthalate (DIDP), di-*n*-butyl phthalate (DBP), diethyl phthalate (DEP), dimethyl phthalate (DMP), butyl benzyl phthalate (BBP), and di-*n*-octyl phthalate (DNOP) [4, 5]. Although PAEs have strong lipophilicity, they do not covalently

Table 13.1 The use of phthalates (PAEs).

Name	Use
Diethyl phthalate (DEP)	Personal care products, cosmetics, plasticizer, detergents, aerosol sprays
Dimethyl phthalate (DMP)	Pesticide, insect repellent, solid propellant, plasticizer
Butyl benzyl phthalate (BBP)	Vinyl tiles, food conveyor belts, artificial leather, automotive trim, traffic cones
Di-*n*-butyl phthalate (DBP)	Polyvinyl chloride (PVC) plastics, latex adhesives, cosmetics, personal care products, cellulose plastics, solvent for dyes
Di(2-ethylhexyl)phthalate (DEHP)	Building products, car products, clothing, food packaging, children's products, medical devices
Di-*n*-hexyl-phthalate (DNHP)	Dip molded products, dishwasher baskets, flooring, vinyl gloves, flea collars, conveyor belts
Di-*n*-octyl-phthalate (DNOP)	Garden hoses, pool liners, flooring tiles, tarps, seam cement, bottle cap liners, conveyor belts
Di-isononyl phthalate (DINP)	Garden hoses, pool liners, flooring tiles, tarps, toys
Di-isodecyl phthalate (DIDP)	PVC plastics, covering on wires and cables, artificial leather, toys, carpet backing, pool liners
Di-isoheptyl phthalate (DIHP)	Plastics

Source: Adapted from Refs. [4, 8].

Toxicological Assessment of Combined Chemicals in the Environment, First Edition. Edited by De-Sheng Pei and Yiyun Liu.

bind with plastic polymer chains. Physical mixing alone makes it easy for PAEs to migrate from the interior of the product to the surrounding environment through wear, evaporation, and leaching, ultimately polluting water, soil, air, and dust, resulting in continuous and inevitable exposure of people to PAE mixtures [6]. In 2014, the United States Environmental Protection Agency (USEPA) listed some PAEs as priority pollutants and curbed their excessive use in personal care products. However, the use of phthalate compounds in the plastic industry has not been effectively reduced [7].

The general population may be exposed to PAEs through multiple sources of exposure, but it is believed that the main source of them is dietary exposure, followed by indoor air [9]. In food, the content of PAEs varies greatly, and packaging materials are considered relevant emission sources [10, 11]. However, other sources can also be detected: for example, Japan studied dietary exposure in duplicate food samples in 1999 and 2001 and found that DEHP had the highest concentration among all PAEs, which ranged from 10 to 4400 ng/g. Disposable PVC gloves used when preparing meals are suspected to be one of the sources of high DEHP content [12]. In indoor dust, the content of DEHP usually exceeds the content of high-volatile PAEs (such as DEP and DBP). In various studies, the median levels were as follows: DEP<10 mg/kg, DBP: 40–50 mg/kg, and DEHP: 400–700 mg/kg [13–16]. This relationship is opposite in indoor air, with a median pollution level of DEP>350–650 ng/m^3, DBP: 600–1200 ng/m^3, and DEHP: 150–450 ng/m^3 [17–19]. According to urinary metabolites, the daily exposure of the general population to DEHP is estimated to be 3–30 μg/kg [20]. But due to different living environments and lifestyles, the estimated exposure levels vary among different populations. In addition to occupational exposure, women are generally more likely to be exposed to PAEs than men, which is attributed to the more frequent use of cosmetics and toiletries containing PAEs [21]. More specifically, compared to men, the concentrations of monoethyl phthalate (MEP) and monobenzyl phthalate (MBzP) in women are significantly higher [22]. Polymer toys softened with PAEs are suspected to be a potential source of contact with PAEs through chewing activities in infants and toddlers. The estimated DINP exposure is between 5 and 44 μg/kg body weight/day [23]. For DEHP, exposure to sucking or chewing toys or other daily necessities is calculated to be up to 85 μg/kg body weight/day in children [24].

Table 13.2 Diester phthalates and their metabolites.

	Phthalates		Metabolites	
Low molecular weight (LMW)	Dimethyl phthalate	DMP	Monomethyl phthalate	MMP
	Diethyl phthalate	DEP	Monoethyl phthalate	MEP
	Di-*n*-butyl phthalate	DBP	Mono-*n*-butyl phthalate	MBP
	Di-iso-butyl phthalate	DIBP	Mono-iso-butyl phthalate	MiBP
	Dicyclohexyl phthalate	DCHP	Mono-cyclo-hexyl phthalate	MCHP
	Butyl benzyl phthalate	BBP	Monobenzyl phthalate	MBzP
	Di-*n*-pentyl phthalate	DnPP	Mono-*n*-pentyl phthalate	MnPP
High molecular weight (HMW)	Di(2-ethylhexyl)phthalate	DEHP	Mono(2-ethylhexyl)phthalate	MEHP
			Mono(2-ethyl-5-hydroxyhexyl) phthalate	MEHHP
			Mono(2-ethyl-5-oxohexyl) phthalate	MEOHP
			Mono(2-ethyl-5-carboxypentyl) phthalate	MECPP
			Mono(2-carboxy-hexyl)phthalate	MCMHP
	Di-iso-nonyl phthalate	DINP	Mono-iso-nonyl phthalates	MiNP
			Mono(hydroxy-iso-nonyl) phthalate	MHiNP
			Mono(oxo-iso-nonyl)phthalate	MOiNP
			Mono(carboxy-iso-octyl) phthalate	MCiOP
	Di-*n*-octyl phthalate	DNOP	Mono-*n*-octyl phthalate	MnOP

Source: Adapted from Refs. [31–33].

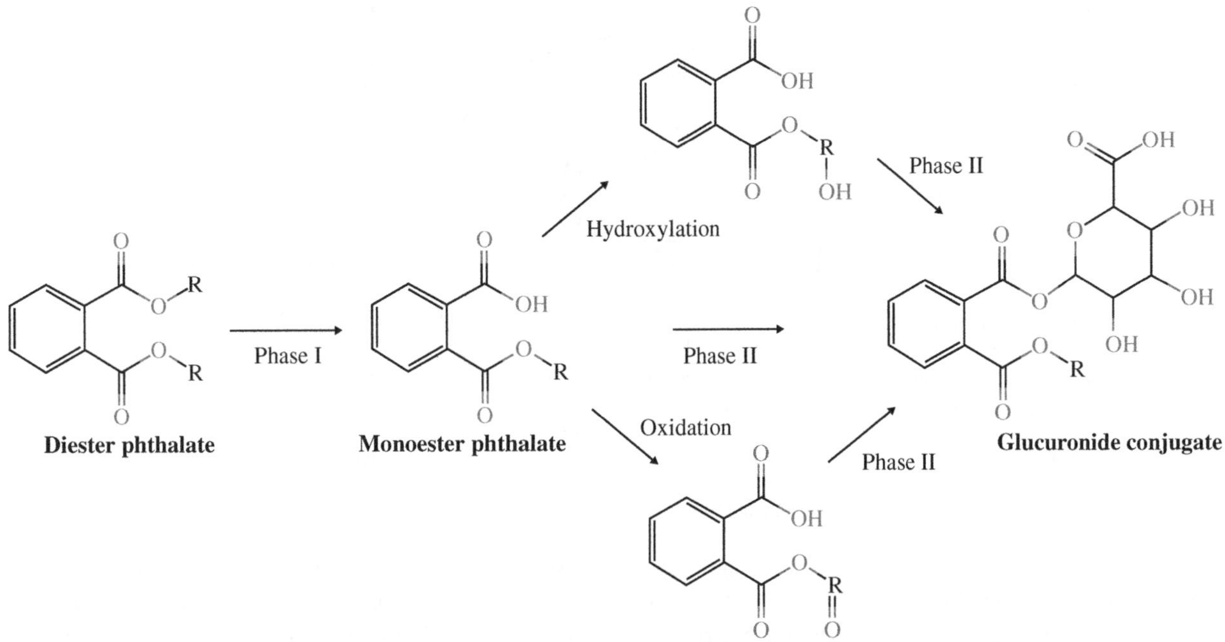

Figure 13.1 Metabolic pathways for phthalates. *Source*: Adapted from Ref. [25]/John Wiley & Sons.

After ingestion, PAEs are rapidly metabolized into various bioactive metabolites in the body (Table 13.2). Short-chain PAEs can be excreted from the body intact, while long-chain PAEs are converted into more water-soluble forms for excretion [25]. After being absorbed by the intestinal epithelium, PAEs have already been hydrolyzed into monoalkyl PAEs, and extensive continuous hydrolysis and oxidation in the body have led to the formation of various secondary metabolites [25] (Figure 13.1). For example, DEHP is rapidly hydrolyzed into mono(2-ethylhexyl) phthalate (MEHP) under the catalysis of nonspecific lipase. Then, under the catalysis of cytochrome P450 enzyme, metabolism continues to hydrolyze and oxidize side chains in multiple organs, forming dozens of metabolites [25, 26]. Although PAEs are not traditional persistent organic compounds, their exposure to the human body remains persistent due to the widespread use of plastic products. It exists for a long time in the liver, lungs, and adipose tissue, as well as in serum, urine, ovarian follicular fluid, and amniotic fluid [27–30]. The presence of PAEs in ovary and amniotic fluid means that PAE exposure is lifelong from germ cell development and fetal life. Therefore, it is necessary to further explore the health hazards and toxic effects mechanisms of PAEs.

13.2 PAEs and Diseases

13.2.1 Reproductive System Diseases

PAE exposure is associated with various diseases (Figure 13.2), but it is most closely related to reproductive diseases [34]. Around the world, PAEs are classified as endocrine-disrupting chemicals (EDCs) because they mimic the properties of human estrogen, affect the hormone balance of organisms, and alter the development and function of hormone-dependent structures in the reproductive system [35]. Data show that PAEs can lead to male sperm damage and spermatogenetic malfunction. In 105 men who visited the infertility clinic, the levels of MiNP and MEP in urine were related to the decrease in sperm concentration and sperm motility, respectively [36]. In two Polish men who went to infertility clinics, the content of PAEs such as MnBP and MEHP in semen was related to the decrease in sperm concentration, sperm motility, and sperm DNA damage [37]. In addition, male testicular damage, shortened anal genital distance, hypospadias, and cryptorchidism caused by PAEs can lead to widespread male reproductive disorders [38].

The impact of PAEs on female reproductive health is mainly caused by pregnancy dysfunction before delivery. High exposure to DEHP is associated with spontaneous abortion in 303 pregnancies in the United States and 430 pregnancies in Denmark [39, 40]. Monomethyl phthalate (MMP) and MEHP levels were significantly increased in women with pregnancy loss in a matched case-control group of 150 pairs of 22- to 35-year-old women from Shanghai, China [41]. A study involving

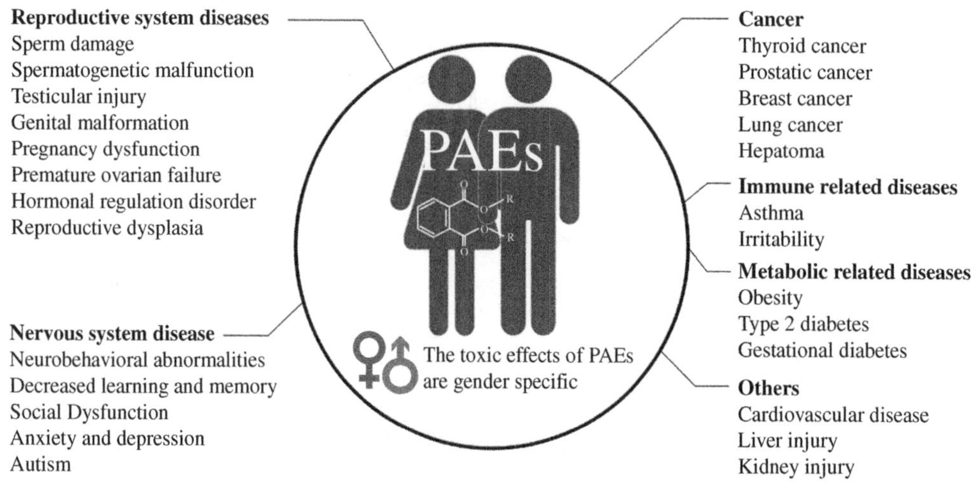

Figure 13.2 Diseases caused by phthalates (PAEs) exposure.

women aged 20–49 years from Taiwan showed a significant increase in ∑DBP (DBP and its metabolites) levels in women with recurrent miscarriage [42]. But the impact of PAEs on female reproduction also occurs in postpartum individuals, affecting ovarian function and leading to a full spectrum of reproductive-related diseases. One of these is premature ovarian failure, characterized by decreased estradiol levels, altered ovum growth patterns, and diminished ovarian reserve. The anti-estrogen activity of PAEs works by inhibiting the production of estradiol in the ovaries, leading to anovulation and premature ovarian failure [43, 44]. Among 215 women at the Massachusetts General Hospital Fertility Center in the United States, higher DEHP exposure is associated with a higher risk of diminished ovarian reserve [45].

The impact of PAEs on the reproductive system development of both adolescent males and females has produced conflicting results. For example, a Chinese case-control study reported that compared to the control group ($n = 120$), the boys ($n = 60$) at the age of 8–15 years with delayed growth and puberty had significantly higher levels of urinary MnBP and MEP, which were negatively correlated with a decrease in testosterone levels [46]. In contrast, Danish research on 84 boys aged 6–13 years showed that urinary MnBP levels are associated with pubertal precocious puberty [47]. A multiethnic longitudinal study in the United States included 1151 girls aged 6–8 years and found a positive correlation between ∑HMWP (high-molecular-weight proteins) in urine and adolescent sexual development [48]. In a study in Denmark, at the age of 7.4–9.9 years, the levels of ∑MBP and ∑DEHP in girls with precocious puberty ($n = 24$) were significantly lower than those in healthy age-matched girls ($n = 184$) [49]. These differences may be due to changes in the physiological values of different ethnic groups, metabolic processes in xenobiotic biotransformation, and exposure to different PAE mixtures [50].

13.2.2 Nervous System Disease

Prenatal exposure to PAEs is associated with poor neurodevelopment of offspring. Exposure to DEHP, DBP, DEP, BBP, and other substances during pregnancy or early life development of the mother will affect the neurodevelopment of offspring, resulting in abnormal neurobehavior, which will lead to decreased learning and memory ability, social disorder, depression, and autism spectrum disorder (ASD) [51]. In addition, the neurotoxicity of PAEs has a gender-dependent effect, which produces different toxicity in populations of different genders [52].

Chopra et al. studied the relationship between the concentration of PAE metabolites in urine with attention-deficit disorder (ADD) and learning disability (LD) in 1493 children. The probability of children suffering from ADD increases with the increase in MEHP, MEHHP, and MEOHP levels in urine, and the correlation between PAEs and ADD and LD in girls is stronger than that in boys [53]. Won et al. used child behavior checklist (CBCL) and the attention-deficit hyperactivity disorder rating scale (ARS) to evaluate neurobehavior. The social issues subscale scores were significantly correlated with MnBP only in boys. Higher levels of MEOHP and MEHHP are associated with an increase in thinking problems in girls [54]. Similarly, using CBCL assessment found that the concentrations of di-isobutyl phthalate (DIBP), DBP, and DEHP metabolites in late pregnancy urine of pregnancy cohorts in Taiwan, California, Minnesota, Missouri, and Iowa were associated with inattention, anxiety, hyperactivity disorder, aggressive behavior problems in children and even offspring during adolescence [55–57]. The ELEMENT birth cohort conducted in Mexico City used a linear regression model to estimate the impact of prenatal exposure to PAEs on the psychological and psychomotor development of children aged 24–36 months. It was found that the mental development index (MDI) was negatively correlated with the concentrations of DEHP metabolites MEHP, MEHHP, MEOHP, and MECPP in girls, but not significantly in boys [58]. Paradoxically, Kim et al. investigated the relationship between 460 pairs of prenatal DEHP and DBP exposure to MDI and psychomotor development index (PDI) in 6-month-old infants. It has been observed that there is a negative correlation between MDI and PDI in boys and MEHHP and MEOHP, while no significant linear correlation has been observed in girls [59].

ASD is a neurodevelopmental disease. The patient has insufficient social communication skills, repetitive behavior, and rigid behavior. Genetic and environmental factors are both potential causes of ASD and have a lifelong impact on patients [60, 61]. Kim et al. collected and measured five PAE metabolites in urine samples from 547 pairs of mid-pregnancy mothers and their 4-, 6-, 8-year-old children and evaluated the characteristics of autism in children using the social communication questionnaire (SCQ). The increase in DEHP metabolites, such as MEHHP and MEOHP levels, during pregnancy is associated with an increase in SCQ scores at 4 years old. Urinary MEHHP levels in children aged 4 and 8 years are associated with an increase in SCQ scores at 8 years old. The correlation between compound exposure in boys and SCQ scores is stronger, which indicates that exposure to DEHP during pregnancy is associated with autism characteristics in young children, and early exposure to DEHP in children can lead to autism characteristics in school-aged children, especially boys [62]. These findings further support the adverse association between prenatal PAE exposure and child outcomes and clearly indicate that this association is unique to gender mixing. Therefore, gender may play a regulatory role in the relationship between prenatal PAE exposure and neural development.

13.2.3 Tumor Diseases

There is limited epidemiological research linking PAEs exposure to human cancer risk. However, the carcinogenicity of PAEs in humans seems to be certain. In 1990, the carcinogenicity of PAEs was first reported in a cohort study targeting 2031 Swedish PVC processing workers. This study showed that PVC exposure significantly increased the total incidence rate of cancer and the incidence rate of respiratory cancer [63]. Due to the endocrine-disrupting effect of PAEs, it is more likely to lead to pathological changes in hormone-dependent tumors, including thyroid cancer, prostate cancer, and breast cancer.

Exposure to DEHP increases the susceptibility of patients with thyroid nodules to develop differentiated thyroid cancer [64]. MMP, MEHHP, and MEHP in human urine samples are associated with an increased risk of thyroid cancer and nodules. Gender-specific associations of MEP and MEOHP with thyroid cancer and nodules were also observed in males [65]. When analyzing variables related to prostate cancer status and prostate cancer biomarkers prostate-specific antigen, it was found that ∑DEHP (MEHP+MEHHP+MEOHP+MECPP) was associated with increased prostate cancer risk and prostate-specific antigen concentration in men aged more than 20 years. Especially high-molecular-weight phthalate metabolites, including MBzP, MEHHP, MEOHP, and MECPP, are closely related to the occurrence of prostate cancer [66]. Both studies in the United States and Taiwan have reported that the concentration of MBP and DEHP in urine is associated with an increased risk of endometriosis and leiomyoma [67, 68].

However, the epidemiological evidence of the observational study on PAEs exposure and breast cancer risk is inconsistent and shows regional and ethnic differences. DBP and DEHP are associated with an increased risk of invasive breast cancer in Indian women, and mutations of several PAEs response genes have been found in tumor tissues [69]. According to statistics of 170 women with an average age of 51 years who mainly live in Alaska, a significant positive correlation was found between MEHP, the main metabolite exposed to DEHP, and breast cancer [70]. However, an epidemiological survey in Mexico and Morgan et al.'s research on women of multiple ethnic groups in the United States found that the total amount of DEHP metabolites in urine and the concentration of each metabolite did not show any relationship with breast cancer [71, 72]. There is even a significant reverse correlation between MBP and MiBP metabolites and breast cancer risk [73].

However, compared with the epidemiological investigation, animal experiments and in vitro cell experiments have come to a more consistent conclusion that PAEs can imitate estrogen and promote the proliferation, metastasis and drug resistance of breast cancer cells [74]. In addition, for other types of tumors, it is found that oral DEHP intake can cause hepatocellular adenoma and/or cancer in mice and rats, and a significant dose-related trend of tumor incidence rate has been observed [75]. Dietary exposure to DEHP can lead to liver tumors [76], stromal cell tumors [77], and benign pancreatic tumors [78] in rats and increase the malignant phenotype of lung cancer cells in vitro [79]. Other kinds of PAEs, such as DINP and DIDP, have also been reported to increase tumor risk [80, 81].

13.2.4 Metabolic-related Diseases

Exposure to EDCs is one of the reasonable factors leading to a significant increase in the incidence of obesity and other metabolic disorders. Although most studies on EDCs focus on reproductive abnormalities, tumor diseases, and neurodevelopmental problems, more and more scientific evidence shows that they can also interfere with the metabolic process of regulating fat and insulin, thus generating the risk of obesity and diabetes.

Multiple surveys have investigated the relationship between PAEs exposure and obesity, with most results indicating a positive correlation between PAEs exposure and childhood and adult obesity. In the National Health and Nutrition Examination Survey of the United States (NHANES, 1999–2002), the levels of MBzP, MEHHP, MEOHP, and MEP in the urine of the general population are positively correlated with an increase in waist circumference (WC) and body mass index (BMI) in adult males aged 20–59 years and adolescent females [82, 83]. Urine levels of MEHP and MEP are also positively correlated with BMI and WC in Chinese men and women aged 8–15 years [84]. In the serum samples of the Swedish 70-year-old male and female cohorts, MiBP content was positively correlated with female WC, total fat mass, trunk fat mass, and subcutaneous adipose tissue, while MMP content was positively correlated with female trunk fat mass [85]. MEP, DBP, and phthalic acid in urine and serum are positively correlated with obesity (BMI value) in Korean women aged 6–14 years [86]. Dong et al. [87] also confirmed that the association between these PAEs and obesity is stronger in young women. However, there are still a few studies reporting a reverse or no association between PAEs and obesity. Evidence from South Korea also suggests that PAEs metabolites in urine samples of males and females aged 18–69 years are not associated with obesity or overweight [88], while MEHHP, MEOHP, and MnBP are not associated with BMI and WC in females aged 30–49 years in South Korea [89]. After excluding factors such as age, gender, race, smoking, and alcohol consumption, the levels of MEHP and MEHHP in urine were negatively correlated with obesity in the general population of NHANES in the United States from 2001 to 2012 [90].

Diabetes is a group of metabolic diseases, characterized by hyperglycemia caused by islet B cell dysfunction, insulin resistance, or both. Diabetes can lead to long-term complications of different organs, especially microvascular and macrovascular complications. PAEs metabolites can be detected in the urine of approximately 75% of patients in the general population of the United States [91]. Fasting blood sugar, glycated diabetes, advanced glycation end products, insulin, and homeostasis model assessment for insulin resistance (HOMA-IR) in the blood samples of 100 Saudi patients with type 2 diabetes were significantly increased, and significant concentrations of phthalate metabolites such as MEP, MEOHP, and MBP were detected in urine samples [92]. In 2007, NHANES published the first large-scale epidemiological study linking PAEs with diabetes-related results, which showed that higher concentrations of PAEs in urine were associated with increased insulin resistance [93]. There is a statistically significant strong positive correlation between increased DBP and DIBP exposure and the risk of diabetes, as well as the indicators of impaired fasting glucose and insulin resistance [94, 95]. The meta-analysis showed a positive correlation between PAEs exposure and HOMA-IR even after adjusting for multiple confounding variables, especially MMP, MnBP, MiBP, MCPP, and DEHP metabolites, which may be risk factors for diabetes mellitus [96, 97]. At the same time, exposure to DEHP and its metabolite MEHP is also associated with gestational diabetes or glucose intolerance during pregnancy [98, 99].

13.2.5 Immune-related Diseases

In the 1970s, studies showed that asthma, allergies, or some respiratory diseases may be related to PVC exposure, which led to subsequent epidemiological studies on the relationship between the large amount of PAEs plasticizers added to PVC and immune-related diseases [100, 101]. Asthma is a common chronic disease in children and adults, characterized by increased airway inflammation and mucus production, leading to airway obstruction [102]. The fundamental causes of asthma have not been fully elucidated, but genetic susceptibility and environmental factors (including allergens and chemicals such as PAEs) are considered important risk factors for the pathogenesis and progression of asthma [103]. A meta-analysis showed that the levels of MBzP in maternal urine before childbirth, as well as DEHP and BBP exposed to dust after childbirth, are significantly associated with childhood asthma [104]. The asthma status of 7765 participants in NHANES from 2007 to 2012 was evaluated using self-reported questionnaires from children and adults. Among the 10 PAEs metabolites, MBzP was also found to be positively correlated with the prevalence of self-reported asthma in children [105]. However, Hoppin et al. found that urine concentrations of MBzP metabolites were associated with self-reported asthma in adults, but not in children, using data from participants aged 6years and above in NHANES from 2005 to 2006 [106]. These inconsistencies may originate from different populations and need to be addressed through further research on the potential association between PAEs and childhood and adult asthma. In addition, PAEs may be gender-specific in the development and/or exacerbation of asthma. A prospective cohort study showed that the relationship between PAEs and asthma or wheezing symptoms may vary by gender [107]. After controlling for parental allergies and smoking among family members, 5-year-old and 8-year-old boys were more likely to develop asthma after exposure to MEHP and MBzP, while MEHP in the urine of mothers and 5-year-old children was significantly correlated with an increase of IgE in allergic children at the age of 8 years [108].

Allergen lgE refers to an immunoglobulin in the human body, which is the core of the development of allergic diseases and can be used to diagnose allergic reactions [109]. The possible association between PAEs exposure and allergic endpoints such as dermatitis, eczema, nasal conjunctivitis, and sensitization is receiving increasing attention. The symptoms of these allergic diseases are manifested in the upper and lower airways and skin, which constitute the main epithelial interface between the air and the body. Wang et al. reported that MEHP levels are positively correlated with specific IgE sensitization in boys [110]. Mutations in filaggrin (FLG) related to the occurrence and development of dermatitis are associated with elevated levels of PAEs metabolites in urine [111]. Children with high levels of PAEs in their urine are more likely to develop eczema [112]. However, Overgaard et al. [113] did not find an association between FLG loss-of-function mutations and levels of PAEs metabolites in urine, although reports suggest that dry skin and eczema are associated with increased PAEs metabolite concentrations. Hsu et al. [114] observed an association between allergic diseases and BBP in dust, but not with metabolites of PAEs in urine. Although there are differences in the results of population studies, overall, there is an association between PAEs and asthma and allergic diseases. And animal and cell experiments also support the possibility of PAEs as adjuvants, inducing respiratory and inflammatory effects in the presence of allergens and causing immune system–related diseases by altering the function of innate and adaptive immune cells [115].

13.2.6 Other Diseases

After entering the human body, PAEs are detoxified and metabolized in the liver and kidneys, and the metabolites are further transported to the whole body with the blood, leading to accumulation and impact in these three tissues and organs. Therefore, in addition to the disease types mainly explored above, PAEs exposure may also be related to cardiovascular diseases such as hypertension, atherosclerosis, coronary heart disease, and liver and kidney damage.

The high concentration of MEHP, MnBP, and MiBP in urine is positively correlated with the incidence rate of coronary heart disease [116]. The high concentrations of MBP, DINP, and DIDP in urine are positively correlated with higher systolic or blood pressure [117, 118]. There is a significant correlation between MBzP levels in pregnant women's urine samples and higher diastolic blood pressure, as well as an increased risk of pregnancy-induced hypertension [119]. Carotid-artery intima and media thickness (CIMT) is positively correlated with the prevalence of cardiovascular diseases and are important parameters for evaluating the impact of human cardiovascular system health [120]. In Taiwanese adolescents and young people, PAEs exposure is closely related to thicker CIMT [121]. In another cross-sectional study, MBzP exposure was found to be associated with the echogenicity of carotid intima-media, plaques, and CIMT [122]. Prenatal exposure to PAEs may increase the risk of liver injury, hepatitis, and liver squamous cell carcinoma in children and pose a potential risk factor for pediatric nonalcoholic fatty liver disease [123]. The urine samples of 6046 adults participating in NHANES from 2007 to 2016 showed a positive correlation between ∑DEHP and liver injury–related indicators such as serum alanine aminotransferase (ALT), γ-glutamyltransferase (GGT), and alkaline phosphatase (ALP). MBP, MCPP, and MBzP were negatively correlated with serum albumin content, indicating the potentially toxic effects of phthalate exposure on human liver [124]. The impact of PAEs on children's kidneys is not clear, but the adverse renal function parameter albumin/creatinine ratio (ACR), *β*-2 microglobulin (B2M), and *N*-acetyl-*β*-*d*-glucosaminidase (NAG) in adults are positively correlated with the levels of PAEs metabolites MBzP, MEHP, MEOHP, MEHHP, MECPP, and MCMHP, which increases the risk of kidney injury [125, 126]. After studying and analyzing PAEs from 9008 adult urine samples from NHANES from 2005 to 2016, it was found that exposure to BBP and DEHP may cause a decrease in glomerular filtration rate and an increase in ACR, thereby inducing the occurrence and development of chronic kidney disease [127].

13.3 The Toxic Molecular Mechanisms of PAEs Mixture

The mixed exposure of PAEs is not only due to their diverse sources, but also due to the comprehensive effects of various PAEs metabolites after metabolism in the organism, making the assessment of PAEs exposure and the exploration of toxic effects challenging. Currently, most research on the health hazard mechanisms of PAEs is focused on single compounds, which are conducive to an in-depth analysis of the mechanisms and can provide some ideas for studying the toxicity mechanisms of combined exposure. Therefore, this section does not make specific distinctions between individual exposure and composite exposure to PAEs.

13.3.1 The Universal Mechanisms that Affect Tissues and Cells

13.3.1.1 Oxidative Stress

Oxidative stress is defined as the imbalance between the biochemical processes of reactive oxygen species (ROS) production and removal. The normal dynamic process of ROS is crucial for many physiological functions. Under physiological conditions, ROS plays a role in cell signaling and can be easily reduced by the antioxidant defense system. However, under pathophysiological conditions, the production of ROS exceeds the buffering capacity of the antioxidant defense system, leading to cell damage and death [128]. Existing data indicate that PAEs can induce oxidative stress in the testes and epididymis of male reproductive organs and damage the spermatogenesis process by inducing oxidative stress and apoptosis in germ cells, target Sertoli cells. It also damages the function of Leydig cells by inducing ROS, thereby reducing the levels of steroid enzymes [129]. Exposure to DEHP at a dose of 750 mg/kg/d during pregnancy significantly reduced serum testosterone levels in young male Sprague Dawley (SD) rats, reduced Leydig cell size, decreased antioxidant enzyme activity, and related signals (SIRT1/PGC1α) to induce abnormal aggregation of fetal stromal cells and formation of multinucleated germ cell. Using 10 and 20 mg/kg/d Taxifolin to alter oxidative stress balance and SIRT1/PGC1α level can alleviate DEHP-induced developmental dysplasia syndrome [130]. In addition, antioxidant treatment with ellagic acid has also been reported to inhibit the number of spermatogonium, primary spermatocyte, and Sertoli cells induced by DEHP, as well as the testicular toxicity such as the reduction of sperm motility and progressive movement ability [131]. The volume of testis, the somatic cell index of seminal vesicle, and the number and vitality of epididymal sperm in DBP-exposed rats decreased significantly with the increase in abnormal sperm morphology. The levels of androgens 3β-hydroxysteroid dehydrogenase, 17β-hydroxysteroid dehydrogenase enzyme, and serum testosterone in the testes significantly decreased, while the levels of estrogen follicle-stimulating hormone (FSH) and luteinizing hormone (LH) increased. This may be closely related to oxidative stress caused by DBP, because higher oxidative stress markers (lipid peroxidation products) and lower levels of antioxidant enzymes, such as superoxide dismutase (SOD), catalase (CAT) and glutathione peroxidase (GSH-Px), were observed in the testes of male rats [132]. Zhang et al. confirmed that DBP induces oxidative stress damage in Leydig and Sertoli cells of the testes and increases the mRNA and protein levels of Annexin A5, upregulating the extracellular-signal-regulated kinase (ERK)/nuclear factor 2 (Nrf2) pathway to exert antioxidant effects [133]. For females, DEHP can induce oxidative stress in human endometrial stromal cells in vitro, increase ROS production, and reduce the activity of SOD, GSH-Px, CAT, and HO [134]. Adult female Kunming mice exposed to 0–200 mg/kg/d DINP for 14 days induced ovarian tissue apoptosis, autophagy, and oxidative stress, significantly altered the arrangement and function of granulosa cells in ovarian follicles [135].

The central nervous system is prone to oxidative damage due to its high oxygen utilization rate, relatively low concentrations of classic antioxidants and related enzymes, and high levels of easily oxidized polyunsaturated lipids [136]. Excessive levels of ROS can damage DNA, RNA, proteins, and lipids and have adverse effects on neuronal function [137, 138]. In the DEHP chemical hazard identification in vitro model constructed by the primary neuron-astrocyte co-culture system of BALB/c mice, 1nmol/L DEHP exposure can significantly increase the ROS concentration in the system, and the antioxidant vitamin C can effectively inhibit this reaction. As the protective cells of the nervous system, astrocytes initiate the proliferation mechanism under low-dose DEHP exposure to maintain neuronal homeostasis and protect neurons from damage. However, as the neurotoxic effect increases, a large area of astrocyte damage was observed at the concentration of 10 μM DEHP [139]. Luo et al. exposed female quails to DEHP of 250–1000 mg/kg/day for 45 days and observed brain histological changes through hematoxylin and eosin (H&E) staining. The antioxidant enzymes CAT, GSH-Px, GST, and SOD in the tissue showed a dose-dependent decrease, while lipid peroxide MDA significantly increased, indicating that the brain of female quails exposed to DEHP may produce oxidative stress responses through impaired antioxidant defense or increased ROS [140]. Mitochondria are the main source of intracellular ROS and are susceptible to oxidative stress. Based on Nrf2 as the main signaling pathway regulating the increase of ROS [141], Luo et al. [140] further found that after exposure to DEHP, the Nrf2 signaling pathway was activated, and mitochondrial dynamics was regulated by the content of fusion proteins MFN1, Mfn2, and Opa1 in inner mitochondrial membrane and outer membrane, thus causing brain mitochondrial dysfunction.

Allergic asthma mice exposed to DBP for a long time showed a significant increase in inflammatory cell infiltration, elevated levels of serum immunoglobulins, helper T cells, and cytokines in lung tissue, as well as typical asthma symptoms in lung histology. Blocking oxidative stress by administering melatonin (MT) can not only reduce the increase of neuropeptide calcitonin gene-related peptide level after DBP exposure but also improve asthma symptoms [142]. Exposure of mice to 15 mg/kg/d DINP or 20 mg/kg/d DIDP can also exacerbate airway wall remodeling, promote the production of IgE and IgG1, and induce AHR. And when exposed to formaldehyde and DINP together, these pathological reactions will be greatly exacerbated. Providing MT or vitamin E to block oxidative stress can alleviate pathological reactions and AHR induced by formaldehyde and DINP and inhibit NF-κB activation and thymic stromal lymphopoietin secretion. Blocking NF-κB with dehydroxymethylpolyoxyquinicin inhibits the increased expression of thymic stromal lymphopoietin and the secretion of Th2/Th17 cytokines, effectively alleviating allergic asthma-like symptoms, indicating that DINP exacerbates allergic asthma symptoms by promoting oxidative stress and activating NF-κB, and this toxic reaction is exacerbated when co-exposed to formaldehyde [143, 144]. In addition, DEHP- and DINP-induced liver/kidney lipoic damage, lipid accumulation [145, 146], and the effects of various PAEs on heart metabolism related to diabetes have been proved to be related to oxidative stress [147].

13.3.1.2 Proliferation and Apoptosis

PAEs can inhibit spermatogenesis through apoptosis. The process of spermatogenesis needs to balance pro-apoptotic and anti-apoptotic signals to maintain the optimal conditions for sperm maturation [148]. The mouse spermatocyte GC-2spd was exposed to various concentrations of DEHP (0–200 μM). As the concentration of DEHP increases, the cell survival rate gradually decreases. Exposure to DEHP at concentrations exceeding 100 μM significantly induces cell apoptosis. Compared with the control group, the mRNA level of caspase-3 significantly increased, while the mRNA level of Bcl-2 decreased. In addition, the levels of TR4, Bcl-2, and procaspase-3 proteins were significantly reduced, indicating that DEHP may induce apoptosis of GC-2spd cells through TR4/Bcl-2 pathway, causing reproductive toxicity [149]. In the study by Giammona et al., MEHP was orally administered to several rodents and strains (1 g/kg to gld mice, B6 mice, and SD rats; 2 g/kg to Fisher rats). MEHP activates external pathways and nuclear factor kappa B (NF-κB) signaling pathways of cell apoptosis [150]. The signaling pathway PI3k/Akt interacting with NF-κB is an important factor to protect germ cells from MEHP-induced apoptosis [151]. Three-week-old male SD rats were orally administered a single dose of 500 mg/kg BBP by gavage and sacrificed after treatment for 3, 12, and 24hours. Histopathology examination showed that BBP induced the destruction of Sertoli cell vimentin and actin filament, which may lead to the imbalance of physical interaction of supporting spermatid, progressive destruction, and sloughing of spermatogenic cells into the lumen and induce apoptosis of spermatid [152]. In addition, the polymorphism of apoptosis-related genes may contribute to the impact of PAEs exposure on male reproductive health, affecting sperm apoptosis and semen quality [153].

After the PAEs mixture (21% DEHP, 35% DEP, 15% DBP, 8% DIBP, 5% BBP, 15% DINP) was cultured to antral follicles from adult CD-1 mice, the sex steroid hormone was measured on the culture medium, and the atresia and gene expression of the follicle were evaluated. Compared with the control group, the PAEs mixture (100 and 500 μg/mL) began to reduce sinus follicle growth at 24 hours, reducing levels of androstenedione, testosterone, estrone, and estradiol, and inducing more oocyte fragmentation. Different dosages of PAEs mixtures have adverse effects on cell cycle regulators, antioxidant enzymes, apoptotic factors, steroidogenic enzymes, and receivers [154]. Meling et al. [155] also found that exposure to PAEs mixtures (36.7% MEP, 19.4% MEHP, 15.3% MBP, 10.2% MiBP, 10.2% MiNP, 8.2% MBzP) inhibited the growth of antral follicles from adult CD-1 mice, leading to changes in cell cycle and apoptosis-related genes. During the process of in vitro maturation of primary porcine oocytes, DINP exposure induced the activation of the continuous spindle assembly checkpoint caused by the disorganized spindle/chromosome device to prevent the oocyte from meiosis in metaphase I, weakening the meiosis ability of the primary porcine oocytes, disrupting the sperm binding dynamics and fusion protein dynamics in oocytes, and reducing the fertilization ability of cells. According to transcriptome and experimental verification, the mechanism of DINP in oocyte maturation is related to oxidative phosphorylation, apoptosis, and autophagy pathways. DINP exposure leads to mitochondrial dysfunction and increased levels of ROS, thus triggering apoptosis and protective autophagy [156]. In all MEHP treatment groups, the activity of rat ovarian granulosa cells decreased, the number of apoptotic cells increased, and caspase-3 activity and BAX expression were significantly increased. The BAX/BCL2 ratio increased in a typical dose–response relationship [157].

Salinger et al. investigated the effects of PAEs exposure on the proliferation and perinatal cell apoptosis of prenatal medial prefrontal cortex (mPFC) cells. From embryonic day 2 (E2) to the day of birth (P0), female mice consumed a mixture of PAEs (35% DEP, 21% DEHP, 15% DBP, 15% DINP, 8% DIBP, 5% BBP) at a daily dose of 0–5 mg/kg. BrdU was injected at E16 and E17 to analyze the proliferation of mPFC cells in offspring P5. It was found that the labeled cells in the PAEs exposed group decreased. After staining the mPFC tissue with triphosphate nick end labeling (TUNEL), it was found that rats treated with PAEs exhibited a higher density of apoptotic cells at P10 [158]. DEHP and microplastics have a combined toxic effect on the mouse brain, causing changes in mitochondrial function through the PI3K/AKT pathway, ultimately leading to neuronal apoptosis [159]. In the passive avoidance response test, Kunming mice exposed to DIBP exhibited learning disabilities, increased hippocampal cell apoptosis rate, and damage to hippocampal ultrastructure. The mitochondrial morphological changes caused by DIBP crossing the BBB after oral administration are important reasons for neuronal apoptosis [160]. ERK1/2 and its phosphorylation (p-ERK1/2) play a significant role in neurobehavioral responses and cognitive processes, including learning and memory [161]. Yan et al. found that DBP significantly increased oxidative damage and apoptosis of hippocampal neurons through the ERK1/2 pathway, leading to behavioral disorders and memory deficits in mice [162]. In vitro experiments have shown that the induction of neuronal apoptosis by DHEP may be related to the expression of HO-1 in the brain. Knockout of the *HO-1* gene will prevent DEHP from inducing caspase-3 expression, reducing the level of neuronal Neuro-2a apoptosis in mouse neurons [163].

P53 protein is the central sensor of cell apoptosis and plays a key and indispensable role in DEHP- and MEHP-induced hepatotoxicity. Activation of p53 mediates hepatocyte apoptosis through the internal mitochondrial pathway, inhibits anti-apoptotic Bcl-2 and Bcl xL, and induces pro-apoptotic Bax, cytochrome c, and caspases expression. The *p53* gene knocked can significantly reduce DEHP-induced cell apoptosis and liver damage [164, 165]. The expression of caspase-3 and p53 in zebrafish liver under DBP exposure also showed a dose–response relationship with the concentration of DBP [166]. However, when using immortalized normal human stem cell L02 as a model, MEHP was still able to induce cell apoptosis by specifically silencing the *p53* gene using small interfering RNA (siRNA), while upregulating Fas and FasL proteins, increasing the Bax/Bcl-2 ratio and caspase-3, 8, and 9 activity. FasL inhibitor AF-016 and caspase inhibitor *N*-benzoxycarbonyl-Val Ala Asp fluoromethylketone (Z-VAD-FMK) can prevent MEHP-induced cell apoptosis. This indicates that MEHP induces apoptosis in L02 cells, involving caspase-mediated mitochondrial signaling pathways and/or death receptor pathways, while p53 is not absolutely necessary [167].

Although we have divided the oxidative stress and apoptosis induced by PAEs into two sections to describe, the relationship between the two is actually very close. The outcome of many oxidative damages caused by PAEs is cell apoptosis, and many mechanisms of cell apoptosis are also related to oxidative stress. This may be related to the downstream chain reaction caused by the combination of PAEs with peroxisome proliferator–activated receptor (PPAR) and other nuclear receptors (NRs) after entering the organism.

13.3.1.3 NR Activation

NRs are superfamily members of the transcription factor that regulate biological processes such as development, organ homeostasis, metabolism, immune function, and reproduction. About half of human NR has been proven to be ligand-regulated transcription factors and directly respond to various endogenous hormones and metabolites [168]. The reason PAEs can interfere with steroid hormone synthesis and secretion is due to the similar chemical structure between them, which leads to PAEs being able to act as agonists or antagonists to bind to endocrine-related NRs such as estrogen receptor (ER), androgen receptor (AR), and PPAR, thereby affecting biological processes by altering downstream gene expression [169]. PAEs are known PPAR activators, and various studies have investigated the interaction between PAEs and PPAR in both computer and in vitro experiments. PAEs are known PPAR activators, and various studies have investigated the interaction between PAEs and PPAR in both computer and in vitro experiments. Multiple computer simulation docking studies have shown that the docking score between PAEs and PPAR receptors is correlated with in vitro affinity. Generally, the affinity and biological activity of PAEs monoester metabolites are stronger than their parent PAEs. The combination ability with PPAR-γ or PPAR-α increases with the increase in PAEs side chain length [170–172]. After isolating primary granulosa cells from CD-1 mice and culturing them with MNP, MiNP, MEHP, MBzP, MBP, MiBP, or MEP for 24 hours, the expression levels of known PPAR target genes *Fabp4* and *Cd36* were detected. Compared with the control group, the treatment of PAEs metabolites resulted in significant changes in the expression of Fabp4 and Cd36. Primary granulosa cell cultures were also transfected with DNA plasmids containing Luciferase expressed under the control of PPAR reaction elements. MNP, MiNP, MEHP, and MBzP caused a dose-dependent change in the expression of Luciferase, indicating that there were functional endogenous PPAR receptors in granulosa cells that responded to PAEs metabolites. By co-treating with PPAR-γ inhibitor T0070907 or PPAR-α inhibitor GW6471, the effect of PAEs metabolites on PPAR target genes is inhibited. Indicating that some PAEs metabolites may act through endogenous PPAR NRs in the ovaries, and the different structures of the compounds lead to different levels of activity [173]. In human placenta, PPAR-γ is expressed in villous cytotrophoblast and activated during its differentiation into syncytiotrophoblast. PPAR-γ mediated a U-shaped dose–response effect on trophoblast differentiation, 0.1 and 1 μM MEHP showing significantly reduced PPAR-γ activity and less villous cytotrophoblast differentiation, while doses of 10 μM have the opposite effect [174]. Extravillous trophoblast invasion is an important physiological step in early development. MEHP inhibits the expression of downstream genes *MMP-1* and *MMP-9* in human chorionic trophoblast cells HTR-8/SVneo through PPAR-γ to inhibit trophoblast invasion, which may lead to early pregnancy loss [175]. In addition, DHEP-induced neuronal apoptosis is also associated with PPAR-γ activation, leading to the expression of Trim17 protein and caspase-3 [176]. The synergistic activation of PPAR-γ and PPAR-α by MEHP may reprogram the polarization of liver macrophages, thereby affecting lipid homeostasis in mouse liver [177]. DBP can also activate the PPAR-α signaling pathway, affecting the protein expression of SREBP, FAS, and GPAT, causing hyperlipidemia and liver dysfunction in SD rats [178].

Takeuchi et al. used a highly sensitive reporter gene assay to characterize the activation effects of 22 PAEs or their metabolism on ER and AR. Dihexyl phthalate (DHP), DIHP, DEHP, and their metabolites exhibit ERα agonist and ERβ antagonistic activity. DBP and BBP have ERα agonist activity and AR antagonistic activity [179]. ER is a nuclear transcription factor involved in regulating many complex physiological processes in humans. Abnormal ER signaling can lead to the development of various diseases, such as cancer, metabolic and cardiovascular diseases, neurodegeneration, and inflammation [180]. Lv et al. evaluated the excitatory effects of DEHP, DINP, BBP, diphenyl phthalate (DPHP), and dicyclohexyl phthalate (DCHP) on ER. Compared to other compounds, DPHP enhanced ER-mediated transcriptional activity in a dose-dependent manner in reporter gene assays, increasing the expression of estrogen-responsive genes *TFF1*, *CTSD*, and *GREB1* in MCF-7 cells. The results of molecular docking also showed that the binding mode of DPHP and ER was more inclined to the agonist conformation binding than the antagonistic conformation of ER. DPHP can act as an exogenous agonist of ER, interfering with the normal function of human ER [181]. DEHP exposure induces ER expression in human endometrial stromal cells in a dose-dependent manner and affects oocyte cyst decomposition and primordial follicle formation in an ER-dependent manner [134, 182]. DBP downregulates the expression of TGF-β2 and TGF-β3 through molecular crosstalk between ER and TGF-β signaling pathways, causing ovarian granulosa cell toxicity and promoting ovarian dysfunction in female offspring [183], exposing the antral follicles of adult CD-1 mice to a PAEs mixture made of 21% DEHP, 35% DEP, 15% DBP, 8% DIBP, 5% BBP, and 15% DINP. After 24 hours, the growth of antral follicles decreased, more oocytes fragmented, and the mixture reduced levels of androstenedione, testosterone, estrone, and estradiol, which may be related to a decrease in ER receptor expression [154]. Besides reproductive system cells, ER also mediates the toxic effect of PAEs on some estrogen-regulated cells. For example, MEHP can promote the proliferation, migration, and invasion of estrogen-dependent neuroblastoma SH-SY5Y through ER and induce thyroid gland damage caused by the stress of thyroid cell Nthy-ori 3–1 endoplasmic reticulum [184, 185]. DBP promotes angiogenesis in human endothelial cells EA.hy3 through ER-dependent activation of ERK926/1, PI2K-Akt, and NO signaling pathways [185]. Christen et al. analyzed the antiandrogenic activity of PAEs mixture in human normal breast cell MDA-kb2. Isoboles and the toxic unit approach confirmed that the binary mixture BBP + DBP and DBP + DEP had a synergistic effect on the antagonism effect of AR at high concentrations [186]. During pregnancy, rats were administered antiandrogen drugs alone or in pairs at a dose of half ED50 to induce trigger hypospadias or epididymal hypoplasia. These pairs include AR antagonists: vinclozolin+procymidone, PAEs: DBP+BBP and DEHP+DBP, PAEs combined with AR antagonists: DBP+procymidone and linuron+BBP. Each chemical substance itself can induce very few deformities, but mixing any two chemicals together can cause approximately 50% of males to develop deformities. All binary combinations produce cumulative dose additive effects on androgen-dependent tissues [187]. DEHP, DBP, and BBP can significantly inhibit the expression of AR in induced pluripotent stem cells in bovine testicular cells and induce a high proportion of BAX/Bcl-2, thereby promoting cell apoptosis [188].

More and more evidence suggests that PAEs can also bind to aryl hydrocarbon receptors (AhR) to exert toxic effects. AhR is a ligand-dependent transcription factor that can be activated by ubiquitous environmental chemistry, natural compounds, and endogenous compounds [189, 190]. After activation, AhR enters the nucleus, forms a heterodimer with the AhR nuclear transporter, binds to the dioxin response element of the target gene transcriptional regulation region, and then regulates the expression of these genes. AhR signal transduction is involved in many important biological processes, such as cell differentiation, apoptosis, and interactions with other signal transduction factors [191]. DEHP can directly target AhR in human breast cancer to mediate drug resistance of tumor cells, destroy the defense process of brain neocortex cells, and thus increase their susceptibility to environmental toxins [192, 193]. DBP stimulates the activity of caspase-3 and LDH, as well as the formation of ROS in mouse neural cells in a concentration (10 nM–100 μM) and time-dependent (3–48 hours) manner, reducing the mRNA expression levels of Erα and PPAR-γ and enhancing the expression of AhR mRNA and protein. Further stimulation of caspase-3 expression and LDH activity was induced by DBP using ERα, ERβ, and PPAR-γ antagonists. After AhR silencing, DBP-induced cell apoptosis and neurotoxicity were inhibited, indicating that DBP-induced cell apoptosis and neurotoxicity were mediated by AhR rather than ER or PPAR-γ [194]. In ER-negative breast cancer cells, DBP and BBP can induce proliferation and invasion in a non-ER-dependent manner through the AhR/HDAC6/c-Myc signaling pathway [195]. BBP induces nuclear transfer of AhR protein and heterodimerization with AhR nuclear translocator (ARNT) and induces necrosis through the expression of CYP1B1 in immortalized human granulosa cell HO23 mediated by AhR [196].

13.3.2 Effects on Special Tissues and Cells

13.3.2.1 Reproductive System

Development and Function of Germ Cell In some in vitro experiments, sinus follicles were isolated from adult ovaries and exposed to solvent control or PAEs mixtures containing 35% DEP, 21% DEHP, 15% DBP, 15% DIBP, 8% DINP, and 5% BBP. The results showed that the PAEs mixture can reduce the growth of sinus follicles, increase oocyte fragmentation, and reduce the secretion of steroid hormones [154]. In addition, the environment-related PAEs mixture (as mentioned above) can damage the ovulation process in the mouse sinus follicle and reduce the progesterone regulator gene [197]. CD-1 mice were exposed to PAEs mixture (as mentioned above) before birth. Male F1 mice had smaller gonads, prostate, and seminal vesicles, accompanied by lower serum testosterone concentration, lower mRNA expression of testicular steroid genes (*STAR, CYP11*, and *CYP17*), and impaired spermatogenesis [198]. The tissue collection and fertility achievement test of female offspring mice at different ages showed that prenatal exposure to PAEs significantly increased the weight of uterus and the percentage of primordial follicles and atretic follicles of F1 offspring and reduced the content of FSH, estradiol, the percentage of anterior sinus follicles and sinus follicles, and the anogenital distance at PND 8 and 60 days. It also induced cystic ovaries at 13 months, destroyed the estrous cycle, and reduced the fertility index and the survival of F2 offspring [199, 200]. Among these mixtures, DEHP has been identified as the main contributor to toxic effects, as the ovaries are sensitive to the toxic effects of DEHP, especially prenatal exposure. The effects of acute exposure to PAEs during the developmental window can be detected throughout the entire life cycle of female offspring [201]. Prenatal DHEP exposure in CD-1 mice will increase the synaptic defect in meiosis of F1 mouse oocytes, affect follicular formation, increase ovarian cysts, reduce the anogenital distance in F2 mice and the number of follicles in F3 mice, and interfere with the secretion of testosterone, estradiol, and other sex hormones in individual generations of mice [202, 203]. PAEs also exhibit similar cross-generational toxic effects in the male reproductive system. DEHP (20–750 mg/kg/d) has been shown to cause early reproductive aging in male CD-1 mice by damaging testosterone production, reducing sperm quality, and reducing fertility [204]. CD-1 mice with scheduled pregnancy were treated with intragastric administration of vector control or DEHP (500 mg/kg). By measuring the detachment of the male penis and foreskin of PND 20, it was found that DEHP exposure delayed the onset of puberty in F3 generation male mice, increased the number of abnormal seminiferous tubules in F3 and F4 generations, and decreased sperm count [205]. In a similar study, DBP (500 mg/kg) was administered by gavage to pregnant SD rats from embryonic day 8 to day 14. It was found that DBP may lead to hypomethylation of the whole genome and decrease the content of spermatogenesis regulator follistatin like 3 by reducing betaine-homocysteine S-methyltransferase, thus reducing the number of sperm and Sertoli cells in the F1, F2, and F3 generations of male rats [206].

Exposure to 20 and 40 μg/kg/d of DEHP from lactation to PND 21 in ICR mice resulted in a significant decrease in the number of oocytes, primordial follicles, and sinus follicles in female offspring; an increase in ovarian cell apoptosis levels; and spindle abnormalities and chromosomal dislocations during oocyte maturation [207]. Exposure of CD-1 mice to 0, 20, and 40 μg/kg/d DEHP from PND 5–15 after postpartum exposure resulted in a decrease in ovarian volume, a decrease in the percentage of large sinus follicles, upregulation of apoptosis-related genes in ovarian tissue, and inhibition of proliferation-related genes, which affected the development of ovarian follicles in mice suppressed during prepuberty [208]. Prepubertal DEHP exposure leads to ferroptosis in Leydig and Sertoli cells of C57BL/6 mouse testes through hypoxia-inducible factor (HIF)-1α/heme oxygenase (HO)-1 signaling pathway [209]. DEHP (10 μM) affects the in vitro binding of sperm and ovum by producing an alternative capacity profile, weakening acrosome reactions, and increasing ROS production, posing a harmful impact on sperm fertility [210]. Intrauterine exposure to DINP will affect the expression level of steroidogenic genes in some fetal Leydig cells, the multinucleation of germ cells, and the aggregation of Leydig nuclei in testes [211].

Hypothalamic–Pituitary–Gonadal Axis Hypothalamic–pituitary–gonadal (HPG) is the foundation of normal reproductive system function and participates in the secretion of gonadotropin (GnRH), which acts on the pituitary gland and stimulates the synthesis and secretion of LH and FSH. These hormones participate in the maturation of sexual organs and control the reproductive process through steroid-generating enzymes, such as 17b-hydroxysteroid dehydrogenase (HSD17b) and cytochrome P450 aromatase (CYP19, CYP17), to produce sex steroid hormones, such as testosterone, estrone, and estradiol [212, 213]. PAEs can disrupt the reproduction of vertebrates by disrupting the HPG axis, and the type, dosage, exposure time of PAEs, as well as the species and gender of animals used in the experiment can lead to inconsistent results. Exposure of male C57/BL6 mice to PAEs mixtures (DBP, BBP, DEHP, DINP) at 0.26 mg/kg/d after birth reduces the expression of LH receptors. On the contrary, at a dose of 13 mg/kg/d, the PAEs mixture increased the expression of LH receptors and CYP17a in male C57/BL6 mice, while reducing CYP17a in female mice [214]. Dose–response studies were conducted using continuous diluents of DIBP, DPeP, DHP, DHeP, DINP, DIDP, as well as nine PAEs mixtures. Except for DIDP, all PAEs reduce the production of testosterone in male rat fetuses. Several genes involved in cholesterol transport and androgen synthesis are also downregulated in a dose–response manner through a mixture of DIBP, DPeP, DHP, DHeP, DINP, and nine PAEs mixtures, leading to reproductive tract abnormalities [215]. Long-term exposure to DBP (43.8 μg/L) impairs the reproductive function of zebrafish by reducing egg production and altering the structure of gonadal tissue. The plasma 17β-estradiol levels of female zebrafish significantly decreased in a concentration-dependent manner, while male testosterone levels significantly increased, accompanied by significant changes in gene expression of the HPG axis [216]. A Chinese rare minnow (*Gobiocypris rarus*) exposed to environmental-related concentrations of DEHP (0–117.6 μg/L) for 21 days resulted in an increase in testosterone concentration and a decrease in estradiol concentration in the plasma of female fish, accompanied by an upregulation of the HPG axis related *gene* CYP17 mRNA and a downregulation of CYP19a mRNA in the gonads. In male fish, the increase in testosterone and estradiol levels is consistent with the upregulation of CYP17 and CYP19a in the gonads [217]. Marine medaka (*Oryzias melastigma*) was exposed to DEHP (0.1 and 0.5 mg/L) from hatching to adulthood, which increased the time for females to spawn, but decreased the number of eggs. Exposure to DEHP and MEHP (0.1 and 0.5 mg/L) resulted in a decrease in the fertilization rate of oocytes produced by pairing untreated females with treated males. A significant increase in plasma 17β-estradiol and a significant decrease in testosterone/17β-estradiol ratio was observed in males, while upregulation of LDLR, STAR, CYP17a1, HSD17b, and CYP19a transcription was observed in the testes. Increased concentrations of testosterone and 17β-estradiol were observed in females, consistent with the upregulation of LDLR. The expression of brain GNRHR2, FSHβ, CYP19b, and steroid hormone receptor genes also corresponds to hormonal and reproductive changes [218]. Although DEHP, DEP, DBP, and DMP showed significant toxicity, compared to single PEAs exposure, binary mixtures, especially DMP + DBP, DMP + BBP, DMP + DEHP, DMP + DNOP, DEP + DBP, DBP + DEHP, DBP + DNOP, BBP + DEHP, BBP + DNOP, and DEHP + DNOP, showed greater toxicity to zebrafish and interference with the expression of HPG- axis-related genes *ERα, ERβ, AR, CYP19a, CYP19b*, and *VTG* [219].

13.3.2.2 Nervous System

Cerebral Cortex The cerebral cortex is the highest center of human motion and sensation. It can be divided into four regions: frontal lobe, parietal lobe, temporal lobe, and occipital lobe. The frontal lobe accounts for one-third of the brain, which participates in language and motor functions, and is also related to emotion, personality, self-consciousness. DBP exposure can damage the cerebral cortex through oxidative stress, DNA oxidation, neuroinflammation, and apoptosis protein activation and reduce the activity levels of norepinephrine, dopamine, and acetylcholinesterase in cerebral cortex nerve cells to play its role in neuronal damage [220]. The prefrontal cortex (PFC) is a control region for working memory performance and has been demonstrated as a critical liability region in schizophrenia [221–223]. SD rats exposed to DEHP showed working memory disorder and decreased expression of brain-derived neurotrophic factor (BDNF) and cannabinoid receptor 1 (CB1R) in the PFC [224]. Zhang et al. found that DEHP with a concentration of 50 mg/kg/d significantly impaired the social ability of ICR mice offspring, and the autism-like symptoms caused by DEHP may be related to the reduction of the density of dendritic spines of pyramidal neurons in the PFC and the increase of the expression of Nischarin protein in prefrontal lobe. In vitro, the inhibitory effect of DEHP on the total spinal density of PFC neurons, especially the mushroom and stubby spine populations, was reversed by knocking down the expression of Nischarin. These results indicate that the autism-like behavior of the offspring of mice after the loss of dendritic spines is caused by the expression of Nischarin induced by DEHP [225]. DEHP can also cause inflammation in the peripheral and cerebral cortex of mice, increase oxidative stress response, and reduce glutamate and glutamine in the mPFC. The mPFC is a key cortical region, which integrates information from many cortical and subcortical regions and aggregates the information into the output structure [226]. The neocortex is a conservative region of the central nervous system, which can realize complex sensory activities and higher cognitive functions [227]. Pregnancy and early postpartum are special sensitive windows susceptible to DEHP toxicity. Prenatal exposure to DEHP in mice will reduce cell proliferation and neurogenesis and increase cell death in the dorsal telencephalon and neocortex of the fetus, leading to abnormal neuronal distribution and neocortex reduction [228].

Hippocampus Exposure to DIDP, DINP, and dioctyl phthalate (DOP) leads to loose and disordered arrangement of pyramidal neurons in the hippocampal CA1 region, swelling and deformation, shortening or disappearance of apical dendrites, cytoplasmic pyknosis, and vacuolization lesions. As the concentration of PAEs increases or co-exposed with formaldehyde, neuronal damage gradually worsens [229–231]. After 14 days of exposure to BBP (0–1250 mg/kg), Kunming mice showed decreased motor activity and induced depression. Pathological changes and oxidative damage occurred in CA1 region of the hippocampus. The change in 5-HT level in the hippocampus and the decrease of CREB phosphorylation level indicate that the neurotoxicity mechanism of BBP may be caused by reducing the number of neurotransmitters and then downregulating CREB phosphorylation level through cAMP/protein kinase a (PKA)-mediated signal transduction [232]. Doublecortin/microtubule-associated protein-2 (DCX/MAP2) immunocytochemistry showed that 100 μM DBP delayed neuronal maturation by increasing the protein level of DCX, a marker of immature neurons. Further in vivo studies confirmed that the number of DCX cells in the hippocampus of DBP-treated mice significantly increased, indicating that DBP delayed neuronal maturation, which is known to be related to impaired memory retention [233]. DBP exposure increased the level of Ca^{2+} in hippocampal histiocyte and induces apoptosis of hippocampal neurons via the ERK 1/2 pathway [162].

Ahmadpour et al. orally exposed 8-week-old male C57 mice to an environmental phthalate mixture (5 μg/kg bw/d) or DEHP (5 and 50 μg/kg bw/d) for 6 weeks. It was found that blood–brain barrier (BBB) permeability increased after exposure to DEHP or phthalate mixture, influencing the endothelial accessory tight junction protein zona occludens-1 (ZO-1) and caveolae protein (Cav-1) in the medial preoptic area of the hypothalamus (mPOA) and the hippocampal CA1 and CA3 areas. In addition, the inflammation-related features, such as the activation of microglia, astrocytes, and the induced increase of nitric oxide synthase in mPOA and hippocampal CA1 and CA3 regions were enhanced. It indicates that DEHP or other phthalate mixtures in the environment can cross the BBB to interfere with the brain function of the hippocampus, and further exert other adverse effects on organisms through inflammatory reactions [234]. The pyramidal cells in the CA3 region stimulate other pyramidal cells and interneurons, and signals from the dentate gyrus are sent to the CA3 subregion and then projected to the CA1 region. Therefore, the critical characteristics of CA3 pyramidal cells may be essential for neural function [235, 236]. After MEHP (100, 300 μM) exposure, the spontaneous action potential (sAP), miniature excitatory postsynaptic currents (mEPSCs), and the electrogenic characteristics of ion channels in the hippocampal neuronal area of CA3 rats were studied. It was found that MEHP reduced the amplitude of sAP and the frequency of mEPSC, the peak current density of Ca^{2+} channels, and the peak voltage density of Na^+ and K^+ channels. It shows that MEHP can affect the excitability and synaptic plasticity of hippocampal neurons in rat CA3 by inhibiting the activity of ion channels [237]. The hippocampus is a key brain region related to learning and memory, which can edit information by regulating synaptic plasticity. The establishment of precise neural circuits depends on the normal development of dendritic length and complexity [238–241]. Pregnant Wistar rats were treated with DEHP (30, 300, and 750 mg/kg bw/d) by gavage from day 0 of pregnancy to postnatal day 21. The Golgi–Cox staining and Sholl analysis were used to measure the dendritic length and complexity of the dendritic arbors' pattern in pyramidal neurons of the hippocampus. After DEHP exposure, the total length of basal dendrites and the number of branches of male offspring were significantly reduced, and the phosphorylation of MAP2c, stathmin, and JNK1, which are important proteins for maintaining the integrity of the dendritic cytoskeleton, was significantly decreased in the hippocampus. However, the above changes were not observed in the hippocampus of female offspring [242]. Under similar exposure conditions, Dong et al. also found that in male but not female offspring, DEHP may induce changes in cytoskeleton structure through the Rac1/PAK/LIMK1/cofilin signaling pathway, reduce the density of dendritic spines and the expression of synapsin and F-actin polymerization in the hippocampus of young rats, and then affect synaptic plasticity, which leads to neurodevelopmental and learning disabilities [243].

13.3.2.3 Tumor

Tumor Proliferation Proliferation and migration are important characteristics of tumor cells and are also important reasons for the deterioration and poor prognosis of tumor diseases. Six-week-old male mice were pre-exposed to 4.0 mg/kg DEHP for 21 days before injection of B16F10 melanoma cells and then continued to be exposed to 4.0 mg/kg DEHP for 7 days. Macrophage M1 polarization markers COX-2, TNF-α, and IL-6 decreased, and the increase in M2 polarization marker IL-4 leads to the suppression of the immune ability of mice to tumors, resulting in higher tumor nodule formation and tumor volume changes compared to the control group [244]. In a tumor heterotopic inoculation model established using human prostate cancer cells LNCaP and male nude balb/c mice (5-week-old), mice exposed to DBP showed an increase in tumor volume, the number of cells in the S phase of the cell cycle while the decrease in p21 protein expression. In vitro experiments, DBP promotes the proliferation of LNCaP cells by upregulating the expression of c-myc and cyclinD1, downregulating the expression of *p21* genes, as well as the protein expression of p-smad downstream of TGF-β. The TGF-β signaling pathway is closely related to p21 and c-myc in cancer [245]. The *c-myc* gene is a cellular oncogene with the potential to modify the proliferation and apoptosis of human cancer cells [246], and its overexpression is associated with the occurrence and metastasis of cancer [247]. The ER antagonist ICI 182780 reversed these regulatory effects caused by DBP, indicating that DBP may affect the crosstalk between TGF-β and ER signals and induce the growth of LNCaP prostate cancer [248]. Similarly, 10^{-7} and 10^{-6}M BBPs significantly downregulated the expression of the main tumor suppressor genes *miRNA, miR-34a*, and *p21* in human prostate cancer LNCaP and PC-3 cells and increased the expression of cyclinD1, PCNA, and miR-34a target genes *c-myc*. Cell transfection using miR-34a mimetics and inhibitors demonstrated that BBP promotes prostate cancer cell proliferation through the miR-34a/c-myc axis [249]. The effects of a mixture of DBP, BBP, DEHP, DINP, DNOP, and DIDP at a concentration of 0.33 ppt–33.33 ppb on cell proliferation were detected on two colorectal cancer cell lines DLD1 and HT29. After treatment with PAEs with a concentration below 33.33 ppt, cell viability/proliferation increased. The mixture of PAEs induced an increase of 10.94% in HT29 proliferation at 33.33 ppt and 60.87% at 3.33 ppt, while DLD1 cells did not find this proliferation relationship at lower concentrations [250]. High concentrations of BBP (100M), DBP (10M), and DEHP (10M) mimic estrogen by increasing the ratio of Bcl-2 to Bax in MCF-7 cells, inhibiting TAM-induced apoptosis and significantly increasing tumor cell proliferation [251]. Wang et al. [252] also showed that BBP can promote the growth of breast cancer stem cells in vivo and in vitro, tumor growth, and lung metastasis in vivo by activating AhR and stimulating sphingosine kinase 1 (SPHK1)/sphingosine 1-phosphate (S1P)/sphingosine-1-phosphate receptor 3 (S1PR3) signal transduction.

Migration and Invasion Cell motility is involved in regulating the progression of many diseases, including cancer cell invasion and metastasis [253]. The ability of cancer cell migration and invasion enables them to change their position in the tissue and separate from the primary tumor, allowing tumor cells to enter lymphatic vessels and blood vessels to spread into the circulation so that cancer can colonize distant organs [254]. Lung cancer is one of the most common malignant tumors with the highest incidence rate and mortality in the world. It can be divided into small-cell lung cancer (SCLC) and non-SCLC (NSCLC). NSCLC accounts for more than 80% of all lung cancer cases [255]. Through wound healing and Transwell experiments, it was confirmed that DEHP can stimulate the migration and invasion of NSCLC A549 through NF-κB-mediated upregulation of IL-6 [256]. It is also possible to induce epithelial–mesenchymal transition (EMT) in A549 cells by increasing the expression of small extracellular vesicles and their miR-26a-5p, promoting the migration and invasion ability of NSCLC [257]. Wilms Tumor is a type of embryonic renal cell carcinoma, which is the most common malignant tumor of the urinary tract in childhood. MEHP can activate the expression, nuclear localization, and transcriptional activity of NF-κB through ER; upregulate the expression of matrix metalloproteinase-2 (MMP-2) and MMP-9; and significantly increase the migration and invasion of G401 and WiT49 cells in vitro. The use of NF-κB inhibitor BAY 11–7082 can reverse the migration and invasion of tumor cells induced by MEHP, indicating that MEHP passes through NF-κB signal stimulation of Wilms Tumor progression [258]. MEHP induced the expression of MMP-2 and c-myc in testicular embryonic carcinoma NT2/D1 cells, but did not significantly affect the expression of MMP-9. The addition of MMP-2-specific inhibitor SB-3CT inhibited MEHP-enhanced cell invasion and migration, indicating that MMP-2 plays a functional role in promoting the progression of MEHP exposure in testicular embryonic cancer. Through microarray analysis, Gap junction protein-alpha 1, vinculin, and inhibitor of DNA-binding protein-1 were significantly downregulated by MEHP treatment, while claudin-6 and beta 1-catenin expression levels were upregulated [259]; 100nM BBP can significantly trigger the migration and invasion of Hemangioma cells and induce the expression of MMPs and vimentin. The mechanism may be related to the increase of Zeb1 expression, because the targeted inhibition of Zeb1 blocks the BBP-induced cell migration. Zeb1 is a powerful transcription factor for cell migration and invasion. BBP increases the stability of Zeb1 mRNA by inhibiting miR-655 and enhances the stability of Zeb1 protein by upregulating ataxia telangiectasia mutated (ATM) [260]. BBP, DBP, DMP, and DEP can directly bind to ETS-1 by activating PXR, enhancing the activity of ETS-1, thereby inducing the expression of invasion-related ETS-1 target genes *MMP-3* and *MMP-9*, and enhancing the invasiveness of liver cancer cells in vitro and in vivo [261].

Tumor Microenvironment PAEs can affect the microenvironment of tumor cells. The microenvironment plays a vital role in the progress of any cancer. It includes stromal cells, lymphocytes, macrophages, dendritic cells, growth factors, extracellular matrix, metabolites, and cytokines. The occurrence of tumors may involve the activation of stromal cells, fibroblasts, dendritic cells, and the accumulation of cytokines [74]. Tumor-associated immune cells are markers of most solid malignant tumors, similar to macrophages, and tumor-associated dendritic cells (TADCs) are assumed to be involved in cancer progression [262]. After treating MDA-MB-231 cells with BBP, DBP, or DEHP, the conditional medium was harvested and cultured together with monocyte-derived dendritic cells, which can induce the production of inflammatory cytokine

RANTES by monocyte-derived dendritic cells, subsequently leading to MDA-MB-231 cell proliferation, migration, and invasion [263]. BBP exposure reshapes the tumor microenvironment of breast cancer by increasing CXCL1/GRO derived from breast cancer–related dendritic cells. αS100A8/A9 increases the chemotherapy resistance and angiogenesis of tumor cells to doxorubicin/cyclophosphamide [264].

13.3.2.4 Others

Lung, Liver, and Kidney Structural changes involving tissue remodeling and fibrosis are the main characteristics of many lung diseases, including asthma [265]. DINP increases airway hyperresponsiveness (AHR) in C57BL/6J mice, the number of cells in the bronchoalveolar lavage fluid, the number of inflammatory cells in the blood, and fibrosis characteristics such as mucin and collagen in lung tissue [266]. C57BL/6J mice exposed to BBP showed significant lung injury, alveolar hemorrhage, pulmonary edema, and increased neutrophil infiltration. At the same time, BBP promotes widespread neutrophil infiltration in bronchoalveolar lavage fluid and the deposition of neutral extracellular traps in lung tissue, disrupting innate immunity and leading to fibrosis of lung epithelial cells [267]; 0.05–500 mg/kg/d DEHP exposure significantly promotes carbon tetrachloride (CCl_4)-induced liver fibrosis in SD rats and inflammation, necrosis, and fibrosis in LX-2 cells in vitro, as well as increases the expression of proteins related to liver inflammation and fibrosis, including α-SMA, COL-I, COL-III, TGF-β1, P-Smad2, P-Smad3, P-p38, and P-p65. The TGF-β1/Smad and p38MAPK/NF-κB signaling pathways play a crucial role in DEHP-induced liver fibrosis [268, 269]. Zhang et al. confirmed that exposure to the highly toxic metabolite MEHP of DEHP inhibits the expression of STAT5A by causing oxidative damage to BRL-3A liver cells, thereby accelerating the expression of key fibrotic molecules MMP-2, MMP-9, TIMP2, and CTGF and promoting the occurrence of liver fibrosis [270]. Treatment of cells with low-concentration DEHP (5–25 μM) for 72 hours resulted in the downregulation of PPAR-α and PPAR-γ protein expression, inducing morphological and phenotypic changes in normal rat renal tubular epithelial cells (NRK-52E), as well as EMT markers. PPAR-γ agonist pioglitazone partially and significantly inhibits DEHP-induced EMT induction. In the C57BL/6 mouse model of folic acid–induced renal fibrosis, exposure to 50 mg/kg/d DEHP for 6 weeks also reduced the expression of PPAR-α and PPAR-β proteins; enhanced renal dysfunction, fibrosis, and mortality; and worsened renal fibrosis/nephropathy in the context of kidney disease [271]. In addition, prenatal exposure to DBP may induce oxidative stress in renal fibroblasts and tubular epithelial cells, promoting EMT in tubular epithelial cells by upregulating TGF-β1/Snail1 or RhoA/ROCK pathways, leading to renal dysplasia and fibrosis [272–274].

Adipose Tissue Adipose tissue is the main organ in the human body that maintains energy homeostasis, characterized by the presence of mature lipid storage adipocytes and preadipocytes. There are two major types of adipose tissue in mammals: brown adipose tissue and white adipose tissue, which work together to maintain the balance between fat accumulation and energy consumption [275–277]. In the mouse 3T3-L1 preadipocyte model, the adipogenic signal–induced differentiation marker PPAR-γ can drive adipocyte differentiation and lipid accumulation. The expression of target genes *aP2* and *LPL* required for PPAR-γ and its adipogenesis induced by MEHP in 3T3-L1 cells has a dose-dependent effect on adipocyte differentiation and glycerol-1-phosphate dehydrogenase (GPDH) activity [278]. And increased the expression of STAT-3 mRNA and protein, as well as P-STAT3 protein in 3T3-L1 cells, leading to lipid accumulation through the TYK-2/STAT-3 pathway [279]. The accumulation of MEHP can interfere with the energy metabolism balance of adipocytes. Adipocytes are treated by MEHP and browning agent rosiglitazone (RSG) simultaneously, and the energy metabolism activity in adipocytes is measured using the Seahorse Bioscience XF analyzer using oxygen consumption rate (OCR, an indicator of mitochondrial respiration) and extracellularly acid rate (ECAR, an indicator of glycolytic function). For mitochondrial respiration, MEHP/RSG-treated adipocytes showed significant upregulation in basal respiration, ATP production, proton leak, and nonmitochondrial respiration. For glycolysis, MEHP-/RSG-treated adipocytes showed significant upregulation in glycolysis, glycolic capacity, and nonglycolic differentiation. In addition, the expression of browning markers in adipocytes exposed to MEHP, including thermogenic genes (*Cidea, Prdm16, Ucp1*, and *Trpv1*) and adipokines (Adipor2, Apln, and Fgf21) that promote browning, increased significantly, showing brown adipose tissue cell–like characteristics [280, 281].

13.4 Conclusion

Existing studies have shown that the toxic effects of PAEs vary by gender and ethnicity, affecting reproductive, neurological, immune, metabolic, and other systems, promoting the occurrence and development of tumors and cardiovascular diseases. Our aim is to explore and summarize the effects and molecular mechanisms of PAEs mixing on human health, as human exposure to PAEs is complex and different types and proportions of compound combinations may produce different toxic effects. At present, most studies still focus on single compound exposure, especially the in-depth study of DEHP. These independent molecular mechanisms can also provide some insights and hints into the mechanism of compound toxicity. However, more exploration is still needed in the future in terms of PAEs mixture exposure.

References

1 Katsikantami I, Sifakis S, Tzatzarakis MN, Vakonaki E, Kalantzi OI, Tsatsakis AM, Rizos AK. A global assessment of phthalates burden and related links to health effects. Environ Int. 2016, 97: 212–236.

2 Annamalai J, Vasudevan N. Enhanced biodegradation of an endocrine disrupting micro-pollutant: di (2-ethylhexyl) phthalate using biogenic self-assembled monolayer of silver nanoparticles. Sci Total Environ. 2020, 719: 137115.

3 Hahladakis JN, Velis CA, Weber R, Iacovidou E, Purnell P. An overview of chemical additives present in plastics: migration, release, fate and environmental impact during their use, disposal and recycling. J Hazard Mater. 2018, 344: 179–199.

4 Heudorf U, Mersch-Sundermann V, Angerer J. Phthalates: toxicology and exposure. Int J Hyg Environ Health. 2007, 210(5): 623–634.

5 Panagiotou EM, Ojasalo V, Damdimopoulou P. Phthalates, ovarian function and fertility in adulthood. Best Pract Res Clin Endocrinol Metab. 2021, 35(5): 101552.

6 Kim DY, Sochichiu S, Kwon JH. Effects of time, temperature, and sebum layer on migration rate of plasticizers in polyvinyl chloride products. Chemosphere. 2022, 308(Pt 3): 136478.

7 Mondal T, Mondal S, Ghosh SK, Pal P, Soren T, Pandey S, Maiti TK. Phthalates - A family of plasticizers, their health risks, phytotoxic effects, and microbial bioaugmentation approaches. Environ Res. 2022, 214(Pt 3): 114059.

8 Zuccarello P, Oliveri Conti G, Cavallaro F, Copat C, Cristaldi A, Fiore M, Ferrante M. Implication of dietary phthalates in breast cancer. A systematic review. Food Chem Toxicol. 2018, 118: 667–674.

9 European Union. European Union Risk Assessment Report on Bis(2-ethylhexyl)Phthalate. Prepared by National Chemicals Inspectorate, Sweden. Draft September 2003.

10 Page BD, Lacroix GM. The occurrence of phthalate ester and di-2-ethylhexyl adipate plasticizers in Canadian packaging and food sampled in 1985–1989: a survey. Food Addit Contam. 1995, 12(1): 129–151.

11 Petersen JH, Breindahl T. Plasticizers in total diet samples, baby food and infant formulae. Food Addit Contam. 2000, 17(2): 133–141.

12 Tsumura Y, Ishimitsu S, Saito I, Sakai H, Tsuchida Y, Tonogai Y. Estimated daily intake of plasticizers in 1-week duplicate diet samples following regulation of DEHP-containing PVC gloves in Japan. Food Addit Contam. 2003, 20(4): 317–324.

13 Zhu Q, Jia J, Zhang K, Zhang H, Liao C, Jiang G. Phthalate esters in indoor dust from several regions, China and their implications for human exposure. Sci Total Environ. 2019, 652: 1187–1194.

14 Ali N, Alhakamy NA, Ismail IMI, Nazar E, Summan AS, Shah Eqani SAMA, Malarvannan G. Exposure to phthalate and organophosphate esters via indoor dust and PM10 is a cause of concern for the exposed Saudi population. Int J Environ Res Public Health. 2021, 18(4): 2125.

15 Wang Y, Yao Y, Han X, Li W, Zhu H, Wang L, Sun H, Kannan K. Organophosphate di- and tri-esters in indoor and outdoor dust from China and its implications for human exposure. Sci Total Environ. 2020, 700: 134502.

16 Guo Y, Kannan K. Comparative assessment of human exposure to phthalate esters from house dust in China and the United States. Environ Sci Technol. 2011, 45(8): 3788–3794.

17 Fromme H, Lahrz T, Piloty M, Gebhart H, Oddoy A, Rüden H. Occurrence of phthalates and musk fragrances in indoor air and dust from apartments and kindergartens in Berlin (Germany). Indoor Air. 2004, 14(3): 188–195.

18 Feng YX, Feng NX, Zeng LJ, Chen X, Xiang L, Li YW, Cai QY, Mo CH. Occurrence and human health risks of phthalates in indoor air of laboratories. Sci Total Environ. 2020, 707: 135609.

19 Liu W, Sun Y, Liu N, Hou J, Huo X, Zhao Y, Zhang Y, Deng F, Kan H, Zhao Z, Huang C, Zhao B, Zeng X, Qian H, Zheng X, Liu W, Mo J, Sun C, Su C, Zou Z, Li H, Guo J, Bu Z. Indoor exposure to phthalates and its burden of disease in China. Indoor Air. 2022, 32(4): e13030.

20 Agency for Toxic Substances and Disease Registry. Toxicological Profile for Di(2-Ethylhexyl)Phthalate (DEHP) Draft for Public Comment. E.P.A. (EPA), 2019.

21 Hannon PR, Flaws JA. The effects of phthalates on the ovary. Front Endocrinol (Lausanne). 2015, 6: 8.

22 Silva MJ, Barr DB, Reidy JA, Malek NA, Hodge CC, Caudill SP, Brock JW, Needham LL, Calafat AM. Urinary levels of seven phthalate metabolites in the U.S. population from the National Health and Nutrition Examination Survey (NHANES) 1999–2000. Environ Health Perspect. 2004, 112(3): 331–338.

23 Kavlock R, Boekelheide K, Chapin R, Cunningham M, Faustman E, Foster P, Golub M, Henderson R, Hinberg I, Little R, Seed J, Shea K, Tabacova S, Tyl R, Williams P, Zacharewski T. NTP center for the evaluation of risks to human reproduction: phthalates expert panel report on the reproductive and developmental toxicity of di(2-ethylhexyl) phthalate. Reprod Toxicol. 2002, 16(5): 529–653.

24 ATSDR. Toxicological Profile for Di(2-ethylhexyl)Phthalate. Agency for Toxic Substances and Disease, Atlanta, GA, 2002.

25 Frederiksen H, Skakkebaek NE, Andersson AM. Metabolism of phthalates in humans. Mol Nutr Food Res. 2007, 51(7): 899–911.

26 Choi K, Joo H, Campbell JL Jr, Clewell RA, Andersen ME, Clewell HJ 3rd. In vitro metabolism of di(2-ethylhexyl) phthalate (DEHP) by various tissues and cytochrome P450s of human and rat. Toxicol In Vitro. 2012, 26(2): 315–322.

27 Domínguez-Romero E, Scheringer M. A review of phthalate pharmacokinetics in human and rat: what factors drive phthalate distribution and partitioning? Drug Metab Rev. 2019, 51(3): 314–329.

28 Guo Y, Wu Q, Kannan K. Phthalate metabolites in urine from China, and implications for human exposures. Environ Int. 2011, 37(5): 893–898.

29 Li Z, Wu D, Guo Y, Mao W, Zhao N, Zhao M, Jin H. Phthalate metabolites in paired human serum and whole blood. Sci Total Environ. 2022, 824(12): 153792.

30 Silva MJ, Reidy JA, Herbert AR, Preau JL Jr, Needham LL, Calafat AM. Detection of phthalate metabolites in human amniotic fluid. Bull Environ Contam Toxicol. 2004, 72(6): 1226–1231.

31 Sree CG, Buddolla V, Lakshmi BA, Kim YJ. Phthalate toxicity mechanisms: an update. Comp Biochem Physiol C Toxicol Pharmacol. 2023, 263: 109498.

32 Koch HM, Rüther M, Schütze A, Conrad A, Pälmke C, Apel P, Brüning T, Kolossa-Gehring M. Phthalate metabolites in 24-h urine samples of the German Environmental Specimen Bank (ESB) from 1988 to 2015 and a comparison with US NHANES data from 1999 to 2012. Int J Hyg Environ Health. 2017, 220(2 Pt A): 130–141.

33 Chang WH, Herianto S, Lee CC, Hung H, Chen HL. The effects of phthalate ester exposure on human health: a review. Sci Total Environ. 2021, 786: 147371.

34 Opoku F, Flaws JA, Zelikoff JT. Reproductive effects associated with phthalate mixture exposure. Explore (NY). 2024, 20(3): 460–461.

35 Meeker JD, Ferguson KK. Urinary phthalate metabolites are associated with decreased serum testosterone in men, women, and children from NHANES 2011–2012. J Clin Endocrinol Metab. 2014, 99(11): 4346–4352.

36 Caporossi L, Alteri A, Campo G, Paci E, Tranfo G, Capanna S, Papaleo E, Pigini D, Viganò P, Papaleo B. Cross sectional study on exposure to BPA and phthalates and semen parameters in men attending a fertility center. Int J Environ Res Public Health. 2020, 17(2): 489.

37 Jurewicz J, Radwan M, Sobala W, Ligocka D, Radwan P, Bochenek M, Hawuła W, Jakubowski L, Hanke W. Human urinary phthalate metabolites level and main semen parameters, sperm chromatin structure, sperm aneuploidy and reproductive hormones. Reprod Toxicol. 2013, 42: 232–241.

38 Mesquita I, Lorigo M, Cairrao E. Update about the disrupting-effects of phthalates on the human reproductive system. Mol Reprod Dev. 2021, 88(10): 650–672.

39 Messerlian C, Wylie BJ, Mínguez-Alarcón L, Williams PL, Ford JB, Souter IC, Calafat AM, Hauser R; Earth Study Team. Urinary concentrations of phthalate metabolites and pregnancy loss among women conceiving with medically assisted reproduction. Epidemiology. 2016, 27(6): 879–888.

40 Toft G, Jönsson BA, Lindh CH, Jensen TK, Hjollund NH, Vested A, Bonde JP. Association between pregnancy loss and urinary phthalate levels around the time of conception. Environ Health Perspect. 2012, 120(3): 458–463.

41 Yi H, Gu H, Zhou T, Chen Y, Wang G, Jin Y, Yuan W, Zhao H, Zhang L. A pilot study on association between phthalate exposure and missed miscarriage. Eur Rev Med Pharmacol Sci. 2016, 20(9): 1894–1902.

42 Liao KW, Kuo PL, Huang HB, Chang JW, Chiang HC, Huang PC. Increased risk of phthalates exposure for recurrent pregnancy loss in reproductive-aged women. Environ Pollut. 2018, 241: 969–977.

43 Gallicchio L, Miller S, Greene T, Zacur H, Flaws JA. Premature ovarian failure among hairdressers. Hum Reprod. 2009, 24(10): 2636–2641.

44 Vabre P, Gatimel N, Moreau J, Gayrard V, Picard-Hagen N, Parinaud J, Leandri RD. Environmental pollutants, a possible etiology for premature ovarian insufficiency: a narrative review of animal and human data. Environ Health. 2017, 16(1): 37.

45 Messerlian C, Souter I, Gaskins AJ, Williams PL, Ford JB, Chiu YH, Calafat AM, Hauser R; Earth Study Team. Urinary phthalate metabolites and ovarian reserve among women seeking infertility care. Hum Reprod. 2016, 31(1): 75–83.

46 Xie C, Zhao Y, Gao L, Chen J, Cai D, Zhang Y. Elevated phthalates' exposure in children with constitutional delay of growth and puberty. Mol Cell Endocrinol. 2015, 407: 67–73.

47 Mouritsen A, Frederiksen H, Sørensen K, Aksglaede L, Hagen C, Skakkebaek NE, Main KM, Andersson AM, Juul A. Urinary phthalates from 168 girls and boys measured twice a year during a 5-year period: associations with adrenal androgen levels and puberty. J Clin Endocrinol Metab. 2013, 98(9): 3755–3764.

48 Wolff MS, Teitelbaum SL, Pinney SM, Windham G, Liao L, Biro F, Kushi LH, Erdmann C, Hiatt RA, Rybak ME, Calafat AM; Breast Cancer and Environment Research Centers. Investigation of relationships between urinary biomarkers of phytoestrogens, phthalates, and phenols and pubertal stages in girls. Environ Health Perspect. 2010, 118(7): 1039–1046.

49 Frederiksen H, Sørensen K, Mouritsen A, Aksglaede L, Hagen CP, Petersen JH, Skakkebaek NE, Andersson AM, Juul A. High urinary phthalate concentration associated with delayed pubarche in girls. Int J Androl. 2012, 35(3): 216–226.

50 Hlisníková H, Petrovičová I, Kolena B, Šidlovská M, Sirotkin A. Effects and mechanisms of phthalates' action on reproductive processes and reproductive health: a literature review. Int J Environ Res Public Health. 2020, 17(18): 6811.

51 Zhang Q, Chen XZ, Huang X, Wang M, Wu J. The association between prenatal exposure to phthalates and cognition and neurobehavior of children-evidence from birth cohorts. Neurotoxicology. 2019, 73: 199–212.

52 Liu Y, Guo Z, Zhu R, Gou D, Jia PP, Pei DS. An insight into sex-specific neurotoxicity and molecular mechanisms of DEHP: a critical review. Environ Pollut. 2023, 316(Pt 2): 120673.

53 Chopra V, Harley K, Lahiff M, Eskenazi B. Association between phthalates and attention deficit disorder and learning disability in U.S. children, 6–15 years. Environ Res. 2014, 128: 64–69.

54 Won EK, Kim Y, Ha M, Burm E, Kim YS, Lim H, Jung DE, Lim S, Kim SY, Kim YM, Kim HC, Lee KJ, Cheong HK, Kang HT, Son M, Sakong J, Oh GJ, Lee CG, Kim SY, Ryu JM, Kim SJ. Association of current phthalate exposure with neurobehavioral development in a national sample. Int J Hyg Environ Health. 2016, 219 (4–5): 364–371.

55 Lien YJ, Ku HY, Su PH, Chen SJ, Chen HY, Liao PC, Chen WJ, Wang SL. Prenatal exposure to phthalate esters and behavioral syndromes in children at 8 years of age: Taiwan Maternal and Infant Cohort Study. Environ Health Perspect. 2015, 123(1): 95–100.

56 Kobrosly RW, Evans S, Miodovnik A, Barrett ES, Thurston SW, Calafat AM, Swan SH. Prenatal phthalate exposures and neurobehavioral development scores in boys and girls at 6–10 years of age. Environ Health Perspect. 2014, 122(5): 521–528.

57 Hyland C, Mora AM, Kogut K, Calafat AM, Harley K, Deardorff J, Holland N, Eskenazi B, Sagiv SK. Prenatal exposure to phthalates and neurodevelopment in the CHAMACOS cohort. Environ Health Perspect. 2019, 127(10): 107010.

58 Téllez-Rojo MM, Cantoral A, Cantonwine DE, Schnaas L, Peterson K, Hu H, Meeker JD. Prenatal urinary phthalate metabolites levels and neurodevelopment in children at two and three years of age. Sci Total Environ. 2013, 461–462: 386–390.

59 Kim Y, Ha EH, Kim EJ, Park H, Ha M, Kim JH, Hong YC, Chang N, Kim BN. Prenatal exposure to phthalates and infant development at 6 months: prospective Mothers and Children's Environmental Health (MOCEH) study. Environ Health Perspect. 2011, 119(10): 1495–1500.

60 Lyall K, Croen L, Daniels J, Fallin MD, Ladd-Acosta C, Lee BK, Park BY, Snyder NW, Schendel D, Volk H, Windham GC, Newschaffer C. The changing epidemiology of autism spectrum disorders. Annu Rev Public Health. 2017, 38(1): 81–102.

61 Guo Z, Xie HQ, Zhang P, Luo Y, Xu T, Liu Y, Fu H, Xu L, Valsami-Jones E, Boksa P, Zhao B. Dioxins as potential risk factors for autism spectrum disorder. Environ Int. 2018, 121(Pt 1): 906–915.

62 Kim JI, Lee J, Lee KS, Lee YA, Shin CH, Hong YC, Kim BN, Lim YH. Association of phthalate exposure with autistic traits in children. Environ Int. 2021, 157: 106775.

63 Hagmar L, Akesson B, Nielsen J, Andersson C, Lindén K, Attewell R, Möller T. Mortality and cancer morbidity in workers exposed to low levels of vinyl chloride monomer at a polyvinyl chloride processing plant. Am J Ind Med. 1990, 17(5): 553–565.

64 Marotta V, Russo G, Gambardella C, Grasso M, La Sala D, Chiofalo MG, D'Anna R, Puzziello A, Docimo G, Masone S, Barbato F, Colao A, Faggiano A, Grumetto L. Human exposure to bisphenol AF and diethylhexylphthalate increases susceptibility to develop differentiated thyroid cancer in patients with thyroid nodules. Chemosphere. 2019, 218: 885–894.

65 Liu C, Deng YL, Zheng TZ, Yang P, Jiang XQ, Liu EN, Miao XP, Wang LQ, Jiang M, Zeng Q. Urinary biomarkers of phthalates exposure and risks of thyroid cancer and benign nodule. J Hazard Mater. 2020, 383: 121189.

66 Guo T, Meng X, Liu X, Wang J, Yan S, Zhang X, Wang M, Ren S, Huang Y. Associations of phthalates with prostate cancer among the US population. Reprod Toxicol. 2023, 116: 108337.

67 Weuve J, Hauser R, Calafat AM, Missmer SA, Wise LA. Association of exposure to phthalates with endometriosis and uterine leiomyomata: findings from NHANES, 1999–2004. Environ Health Perspect. 2010, 118(6): 825–832.

68 Huang PC, Tsai EM, Li WF, Liao PC, Chung MC, Wang YH, Wang SL. Association between phthalate exposure and glutathione S-transferase M1 polymorphism in adenomyosis, leiomyoma and endometriosis. Hum Reprod. 2010, 25(4): 986–994.

69 Mukherjee Das A, Gogia A, Garg M, Elaiyaraja A, Arambam P, Mathur S, Babu-Rajendran R, Deo SVS, Kumar L, Das BC, Janardhanan R. Urinary concentration of endocrine-disrupting phthalates and breast cancer risk in Indian women: a case-control study with a focus on mutations in phthalate-responsive genes. Cancer Epidemiol. 2022, 79: 102188.

70 Holmes AK, Koller KR, Kieszak SM, Sjodin A, Calafat AM, Sacco FD, Varner DW, Lanier AP, Rubin CH. Case-control study of breast cancer and exposure to synthetic environmental chemicals among Alaska Native women. Int J Circumpolar Health. 2014, 73(1): 25760.

71 López-Carrillo L, Hernández-Ramírez RU, Calafat AM, Torres-Sánchez L, Galván-Portillo M, Needham LL, Ruiz-Ramos R, Cebrián ME. Exposure to phthalates and breast cancer risk in northern Mexico. Environ Health Perspect. 2010, 118(4): 539–544.

72 Morgan M, Deoraj A, Felty Q, Roy D. Environmental estrogen-like endocrine disrupting chemicals and breast cancer. Mol Cell Endocrinol. 2017, 457: 89–102.

73 Fu Z, Zhao F, Chen K, Xu J, Li P, Xia D, Wu Y. Association between urinary phthalate metabolites and risk of breast cancer and uterine leiomyoma. Reprod Toxicol. 2017, 74: 134–142.

74 Mughees M, Chugh H, Wajid S. Mechanism of phthalate esters in the progression and development of breast cancer. Drug Chem Toxicol. 2022, 45(3): 1021–1025.

75 National Toxicology Program. Carcinogenesis bioassay of di(2-ethylhexyl)phthalate (CAS No.117-81-7) in F344 rats and B6C3F1 mice (feed studies). Natl Toxicol Program Tech Rep Ser. 1982, 217: 1–127.

76 Shih MS, Suk FM, Chiu WC, Lee CY, Hsu FY, Liao YJ. Long-term di-(2-ethylhexyl) phthalate exposure reduces sorafenib treatment efficacy by enhancing mesenchymal transition in hepatocellular carcinoma. Ecotoxicol Environ Saf. 2024, 273(1): 116161.

77 Voss C, Zerban H, Bannasch P, Berger MR. Lifelong exposure to di-(2-ethylhexyl)-phthalate induces tumors in liver and testes of Sprague-Dawley rats. Toxicology. 2005, 206(3): 359–371.

78 David RM, Moore MR, Finney DC, Guest D. Chronic toxicity of di(2-ethylhexyl)phthalate in mice. Toxicol Sci. 2000, 58(2): 377–385.

79 Urade R, Chou CK, Chou HL, Chen BH, Wang TN, Tsai EM, Hung CT, Wu SJ, Chiu CC. Phthalate derivative DEHP disturbs the antiproliferative effect of camptothecin in human lung cancer cells by attenuating DNA damage and activating Akt/NF-κB signaling pathway. Environ Toxicol. 2023, 38(2): 332–342.

80 Cho WS, Jeong J, Choi M, Park SN, Han BS, Son WC. 26-Week carcinogenicity study of di-isodecyl phthalate by dietary administration to CB6F1-rasH2 transgenic mice. Arch Toxicol. 2011, 85(1): 59–66.

81 European Chemicals Bureau. Bis (2-Ethylhexyl) Phthalate Risk Assessment (CAS No. 117-81-7). European Union Risk Assessment Report 2008. European Communities.

82 Christensen K, Sobus J, Phillips M, Blessinger T, Lorber M, Tan YM. Changes in epidemiologic associations with different exposure metrics: a case study of phthalate exposure associations with body mass index and waist circumference. Environ Int. 2014, 73: 66–76.

83 Hatch EE, Nelson JW, Qureshi MM, Weinberg J, Moore LL, Singer M, Webster TF. Association of urinary phthalate metabolite concentrations with body mass index and waist circumference: a cross-sectional study of NHANES data, 1999–2002. Environ Health. 2008, 7: 27.

84 Wang H, Zhou Y, Tang C, He Y, Wu J, Chen Y, Jiang Q. Urinary phthalate metabolites are associated with body mass index and waist circumference in Chinese school children. PLoS One. 2013, 8(2): e56800.

85 Lind PM, Roos V, Rönn M, Johansson L, Ahlström H, Kullberg J, Lind L. Serum concentrations of phthalate metabolites are related to abdominal fat distribution two years later in elderly women. Environ Health. 2012, 11: 21.

86 Choi J, Eom J, Kim J, Lee S, Kim Y. Association between some endocrine-disrupting chemicals and childhood obesity in biological samples of young girls: a cross-sectional study. Environ Toxicol Pharmacol. 2014, 38(1): 51–57.

87 Dong R, Zhou T, Chen J, Zhang M, Zhang H, Wu M, Li S, Zhang L, Chen B. Gender- and age-specific relationships between phthalate exposures and obesity in Shanghai adults. Arch Environ Contam Toxicol. 2017, 73(3): 431–441.

88 Lee KM, Kho Y, Kim PG, Park SH, Lee JH. Urinary levels of phthalate metabolites and associations with demographic characteristics in Korean adults. Environ Sci Pollut Res Int. 2017, 24(17): 14669–14681.

89 Hong SH, Sung YA, Hong YS, Ha E, Jeong K, Chung H, Lee H. Urinary bisphenol A is associated with insulin resistance and obesity in reproductive-aged women. Clin Endocrinol (Oxf). 2017, 86(4): 506–512.

90 Yaghjyan L, Carlsson NP, Ghita GL, Chang SH. Associations of individual characteristics and lifestyle factors with metabolism of di-2-ethylhexyl phthalate in NHANES 2001–2012. Environ Res. 2016, 149: 23–31.

91 Hauser R, Meeker JD, Park S, Silva MJ, Calafat AM. Temporal variability of urinary phthalate metabolite levels in men of reproductive age. Environ Health Perspect. 2004, 112(17): 1734–1740.

92 Al-Bazi MM, Kumosani TA, Al-Malki AL, Kannan K, Moselhy SS. Screening the incidence of diabetogensis with urinary phthalate in Saudi subjects. Environ Sci Pollut Res Int. 2022, 29(19): 28743–28748.

93 Stahlhut RW, van Wijngaarden E, Dye TD, Cook S, Swan SH. Concentrations of urinary phthalate metabolites are associated with increased waist circumference and insulin resistance in adult U.S. males. Environ Health Perspect. 2007, 115(6): 876–882.

94 Song Y, Chou EL, Baecker A, You NC, Song Y, Sun Q, Liu S. Endocrine-disrupting chemicals, risk of type 2 diabetes, and diabetes-related metabolic traits: a systematic review and meta-analysis. J Diabetes. 2016, 8(4): 516–532.

95 Sun Q, Cornelis MC, Townsend MK, Tobias DK, Eliassen AH, Franke AA, Hauser R, Hu FB. Association of urinary concentrations of bisphenol A and phthalate metabolites with risk of type 2 diabetes: a prospective investigation in the Nurses' Health Study (NHS) and NHSII cohorts. Environ Health Perspect. 2014, 122(6): 616–623.

96 Shoshtari-Yeganeh B, Zarean M, Mansourian M, Riahi R, Poursafa P, Teiri H, Rafiei N, Dehdashti B, Kelishadi R. Systematic review and meta-analysis on the association between phthalates exposure and insulin resistance. Environ Sci Pollut Res Int. 2019, 26(10): 9435–9442.

97 Zhang H, Ben Y, Han Y, Zhang Y, Li Y, Chen X. Phthalate exposure and risk of diabetes mellitus: implications from a systematic review and meta-analysis. Environ Res. 2022, 204(Pt B): 112109.

98 Liang QX, Lin Y, Fang XM, Gao YH, Li F. Association between phthalate exposure in pregnancy and gestational diabetes: a Chinese cross-sectional study. Int J Gen Med. 2022, 15: 179–189.

99 Zhang T, Wang S, Li L, Zhu A, Wang Q. Associating diethylhexyl phthalate to gestational diabetes mellitus via adverse outcome pathways using a network-based approach. Sci Total Environ. 2022, 824: 153932.

100 Sokol WN, Aelony Y, Beall GN. Meat-wrapper's asthma. A new syndrome? JAMA. 1973, 226(6): 639–641.

101 Polakoff PL, Lapp NL, Reger R. Polyvinyl chloride pyrolysis products. A potential cause for respiratory impairment. Arch Environ Health. 1975, 30(6): 269–271.

102 Hussain M, Liu G. Eosinophilic asthma: pathophysiology and therapeutic horizons. Cells. 2024, 13(5): 384.

103 Wang B, Liu F, Dong J, You M, Fu Y, Li C, Lu Y, Chen J. Maternal exposure to environmental DEHP exacerbated OVA-induced asthmatic responses in rat offspring. Sci Total Environ. 2018, 615: 253–261.

104 Li MC, Chen CH, Guo YL. Phthalate esters and childhood asthma: A systematic review and congener-specific meta-analysis. Environ Pollut. 2017, 229: 655–660.

105 Odebeatu CC, Taylor T, Fleming LE, J Osborne N. Phthalates and asthma in children and adults: US NHANES 2007–2012. Environ Sci Pollut Res Int. 2019, 26(27): 28256–28269.

106 Hoppin JA, Jaramillo R, London SJ, Bertelsen RJ, Salo PM, Sandler DP, Zeldin DC. Phthalate exposure and allergy in the U.S. population: results from NHANES 2005–2006. Environ Health Perspect. 2013, 121(10): 1129–1134.

107 Buckley JP, Quirós-Alcalá L, Teitelbaum SL, Calafat AM, Wolff MS, Engel SM. Associations of prenatal environmental phenol and phthalate biomarkers with respiratory and allergic diseases among children aged 6 and 7 years. Environ Int. 2018, 115: 79–88.

108 Ku HY, Su PH, Wen HJ, Sun HL, Wang CJ, Chen HY, Jaakkola JJ, Wang SL; TMICS Group. Prenatal and postnatal exposure to phthalate esters and asthma: a 9-year follow-up study of a Taiwanese birth cohort. PLoS One. 2015, 10(4): e0123309.

109 An SB, Yang BG, Jang G, Kim DY, Kim J, Oh SM, Oh N, Lee S, Moon JY, Kim JA, Kim JH, Song YJ, Hyun HW, Kim J, Lee K, Lee D, Kwak MJ, Kim BK, Park YK, Hong CP, Kim JH, Lim HS, Ryu MS, Jin HT, Lee SW, Chang YS, Park HS, Sung YC, Jang MH. Combined IgE neutralization and Bifidobacterium longum supplementation reduces the allergic response in models of food allergy. Nat Commun. 2022, 13(1): 5669.

110 Wang IJ, Lin CC, Lin YJ, Hsieh WS, Chen PC. Early life phthalate exposure and atopic disorders in children: a prospective birth cohort study. Environ Int. 2014, 62: 48–54.

111 Joensen UN, Jørgensen N, Meldgaard M, Frederiksen H, Andersson AM, Menné T, Johansen JD, Carlsen BC, Stender S, Szecsi PB, Skakkebæk NE, Rajpert-De Meyts E, Thyssen JP. Associations of filaggrin gene loss-of-function variants with urinary phthalate metabolites and testicular function in young Danish Men. Environ Health Perspect. 2014, 122(4): 345–350.

112 Wang IJ, Karmaus WJ. The effect of phthalate exposure and filaggrin gene variants on atopic dermatitis. Environ Res. 2015, 136: 213–218.

113 Overgaard LEK, Main KM, Frederiksen H, Stender S, Szecsi PB, Williams HC, Thyssen JP. Children with atopic dermatitis and frequent emollient use have increased urinary levels of low-molecular-weight phthalate metabolites and parabens. Allergy. 2017, 72(11): 1768–1777.

114 Hsu NY, Lee CC, Wang JY, Li YC, Chang HW, Chen CY, Bornehag CG, Wu PC, Sundell J, Su HJ. Predicted risk of childhood allergy, asthma, and reported symptoms using measured phthalate exposure in dust and urine. Indoor Air. 2012, 22(3): 186–199.

115 Bølling AK, Sripada K, Becher R, Bekö G. Phthalate exposure and allergic diseases: review of epidemiological and experimental evidence. Environ Int. 2020, 139: 105706.

116 Su TC, Hwang JJ, Sun CW, Wang SL. Urinary phthalate metabolites, coronary heart disease, and atherothrombotic markers. Ecotoxicol Environ Saf. 2019, 173: 37–44.

117 Trasande L, Attina TM. Association of exposure to di-2-ethylhexylphthalate replacements with increased blood pressure in children and adolescents. Hypertension. 2015, 66(2): 301–308.

118 Amin MM, Ebrahimpour K, Parastar S, Shoshtari-Yeganeh B, Hashemi M, Mansourian M, Poursafa P, Fallah Z, Rafiei N, Kelishadi R. Association of urinary concentrations of phthalate metabolites with cardiometabolic risk factors and obesity in children and adolescents. Chemosphere. 2018, 211: 547–556.

119 Werner EF, Braun JM, Yolton K, Khoury JC, Lanphear BP. The association between maternal urinary phthalate concentrations and blood pressure in pregnancy: the HOME study. Environ Health. 2015, 14: 75.

120 O'Leary DH, Polak JF, Kronmal RA, Manolio TA, Burke GL, Wolfson SK Jr. Carotid-artery intima and media thickness as a risk factor for myocardial infarction and stroke in older adults. Cardiovascular Health Study Collaborative Research Group. N Engl J Med. 1999, 340(1): 14–22.

121 Su TC, Hwang JS, Torng PL, Wu C, Lin CY, Sung FC. Phthalate exposure increases subclinical atherosclerosis in young population. Environ Pollut. 2019, 250: 586–593.

122 Wiberg B, Lind PM, Lind L. Serum levels of monobenzylphthalate (MBzP) is related to carotid atherosclerosis in the elderly. Environ Res. 2014, 133: 348–352.

123 Midya V, Colicino E, Conti DV, Berhane K, Garcia E, Stratakis N, Andrusaityte S, Basagaña X, Casas M, Fossati S, Gražuleviciene R, Haug LS, Heude B, Maitre L, McEachan R, Papadopoulou E, Roumeliotaki T, Philippat C, Thomsen C, Urquiza J, Vafeiadi M, Varo N, Vos MB, Wright J, McConnell R, Vrijheid M, Chatzi L, Valvi D. Association of prenatal exposure to endocrine-disrupting chemicals with liver injury in children. JAMA Netw Open. 2022, 5(7): e2220176.

124 Yu L, Yang M, Cheng M, Fan L, Wang X, Xu T, Wang B, Chen W. Associations between urinary phthalate metabolite concentrations and markers of liver injury in the US adult population. Environ Int. 2021, 155: 106608.

125 Chen J, Zhou X, Zhang H, Liu Y, Cao C, Dong R, Yuan Y, Wang M, Lu Y, Wu M, Li S, Chen B. Association between urinary concentration of phthalate metabolites and impaired renal function in Shanghai adults. Environ Pollut. 2019, 245: 149–162.

126 Kang H, Kim S, Lee G, Lee I, Lee JP, Lee J, Park H, Moon HB, Park J, Kim S, Choi G, Choi K. Urinary metabolites of dibutyl phthalate and benzophenone-3 are potential chemical risk factors of chronic kidney function markers among healthy women. Environ Int. 2019, 124: 354–360.

127 Kang H, Lee JP, Choi K. Exposure to phthalates and environmental phenols in association with chronic kidney disease (CKD) among the general US population participating in multi-cycle NHANES (2005–2016). Sci Total Environ. 2021, 791: 148343.

128 van der Pol A, van Gilst WH, Voors AA, van der Meer P. Treating oxidative stress in heart failure: past, present and future. Eur J Heart Fail. 2019, 21(4): 425–435.

129 Sedha S, Kumar S, Shukla S. Role of oxidative stress in male reproductive dysfunctions with reference to phthalate compounds. Urol J. 2015, 12(5): 2304–2316.

130 Li Q, Zhu Q, Tian F, Li J, Shi L, Yu Y, Zhu Y, Li H, Wang Y, Ge RS, Li X. In utero di-(2-ethylhexyl) phthalate-induced testicular dysgenesis syndrome in male newborn rats is rescued by taxifolin through reducing oxidative stress. Toxicol Appl Pharmacol. 2022, 456: 116262.

131 Hosseinzadeh A, Mehrzadi S, Siahpoosh A, Basir Z, Bahrami N, Goudarzi M. The ameliorative effect of ellagic acid on di-(2-ethylhexyl) phthalate-induced testicular structural alterations, oxidative stress, inflammation and sperm damages in adult mice. Reprod Biol Endocrinol. 2021, 19(1): 146.

132 Nelli G, Pamanji SR. Di-n-butyl phthalate prompts interruption of spermatogenesis, steroidogenesis, and fertility associated with increased testicular oxidative stress in adult male rats. Environ Sci Pollut Res Int. 2017, 24(22): 18563–18574.

133 Zhang L, Qin Z, Li R, Wang S, Wang W, Tang M, Zhang W. The role of ANXA5 in DBP-induced oxidative stress through ERK/Nrf2 pathway. Environ Toxicol Pharmacol. 2019, 72: 103236.

134 Cho YJ, Park SB, Han M. Di-(2-ethylhexyl)-phthalate induces oxidative stress in human endometrial stromal cells in vitro. Mol Cell Endocrinol. 2015, 407: 9–17.

135 Chen J, Yang S, Ma B, Wang J, Chen J. Di-isononyl phthalate induces apoptosis and autophagy of mouse ovarian granulosa cells via oxidative stress. Ecotoxicol Environ Saf. 2022, 242: 113898.

136 Sayre LM, Perry G, Smith MA. Oxidative stress and neurotoxicity. Chem Res Toxicol. 2008, 21(1): 172–188.

137 Spinu N, Bal-Price A, Cronin MTD, Enoch SJ, Madden JC, Worth AP. Development and analysis of an adverse outcome pathway network for human neurotoxicity. Arch Toxicol. 2019, 93(10): 2759–2772.

138 Schieber M, Chandel NS. ROS function in redox signaling and oxidative stress. Curr Biol. 2014, 24(10): 453–462.

139 Wu Y, Li K, Zuo H, Yuan Y, Sun Y, Yang X. Primary neuronal-astrocytic co-culture platform for neurotoxicity assessment of di-(2-ethylhexyl) phthalate. J Environ Sci (China). 2014, 26(5): 1145–1153.

140 Luo Y, Li XN, Zhao Y, Du ZH, Li JL. DEHP triggers cerebral mitochondrial dysfunction and oxidative stress in quail (Coturnix japonica) via modulating mitochondrial dynamics and biogenesis and activating Nrf2-mediated defense response. Chemosphere. 2019, 224: 626–633.

141 Turpaev KT. Keap1-Nrf2 signaling pathway: mechanisms of regulation and role in protection of cells against toxicity caused by xenobiotics and electrophiles. Biochemistry (Mosc). 2013, 78(2): 111–126.

142 Zhou S, Han M, Ren Y, Yang X, Duan L, Zeng Y, Li J. Dibutyl phthalate aggravated asthma-like symptoms through oxidative stress and increasing calcitonin gene-related peptide release. Ecotoxicol Environ Saf. 2020, 199: 110740.

143 Kang J, Duan J, Song J, Luo C, Liu H, Li B, Yang X, Yu W, Chen M. Exposure to a combination of formaldehyde and DINP aggravated asthma-like pathology through oxidative stress and NF-κB activation. Toxicology. 2018, 404–405: 49–58.

144 Qin W, Deng T, Cui H, Zhang Q, Liu X, Yang X, Chen M. Exposure to diisodecyl phthalate exacerbated Th2 and Th17-mediated asthma through aggravating oxidative stress and the activation of p38 MAPK. Food Chem Toxicol. 2018, 114: 78–87.

145 Gu Y, Gao M, Zhang W, Yan L, Shao F, Zhou J. Exposure to phthalates DEHP and DINP May lead to oxidative damage and lipidomic disruptions in mouse kidney. Chemosphere. 2021, 271: 129740.

146 Park CG, Sung B, Ryu CS, Kim YJ. Mono-(2-ethylhexyl) phthalate induces oxidative stress and lipid accumulation in zebrafish liver cells. Comp Biochem Physiol C Toxicol Pharmacol. 2020, 230: 108704.

147 Dong R, Chen J, Zheng J, Zhang M, Zhang H, Wu M, Li S, Chen B. The role of oxidative stress in cardiometabolic risk related to phthalate exposure in elderly diabetic patients from Shanghai. Environ Int. 2018, 121(Pt 1): 340–348.

148 Rasoulpour RJ, Boekelheide K. NF-kappaB is activated in the rat testis following exposure to mono-(2-ethylhexyl) phthalate. Biol Reprod. 2005, 72(2): 479–486.

149 Zhu L, Lu J, Tang X, Fu G, Duan P, Quan C, Zhang L, Zhang Z, Chang W, Shi Y. Di-(2-ethylhexyl) phthalate induces apoptosis of GC-2spd cells via TR4/Bcl-2 pathway. Environ Toxicol Pharmacol. 2016, 44: 18–24.

150 Giammona CJ, Sawhney P, Chandrasekaran Y, Richburg JH. Death receptor response in rodent testis after mono-(2-ethylhexyl) phthalate exposure. Toxicol Appl Pharmacol. 2002, 185(2): 119–127.

151 Rogers R, Ouellet G, Brown C, Moyer B, Rasoulpour T, Hixon M. Cross-talk between the Akt and NF-kappaB signaling pathways inhibits MEHP-induced germ cell apoptosis. Toxicol Sci. 2008, 106(2): 497–508.

152 Alam MS, Kurohmaru M. Butylbenzyl phthalate induces spermatogenic cell apoptosis in prepubertal rats. Tissue Cell. 2016, 48(1): 35–42.

153 Yang P, Gong YJ, Wang YX, Liang XX, Liu Q, Liu C, Chen YJ, Sun L, Lu WQ, Zeng Q. Effect modification by apoptosis-related gene polymorphisms on the associations of phthalate exposure with spermatozoa apoptosis and semen quality. Environ Pollut. 2017, 231(Pt 1): 694–702.

154 Zhou C, Flaws JA. Effects of an environmentally relevant phthalate mixture on cultured mouse antral follicles. Toxicol Sci. 2017, 156(1): 217–229.

155 Meling DD, Warner GR, Szumski JR, Gao L, Gonsioroski AV, Rattan S, Flaws JA. The effects of a phthalate metabolite mixture on antral follicle growth and sex steroid synthesis in mice. Toxicol Appl Pharmacol. 2020, 388: 114875.

156 Wang R, Chen J, Cui Z, Li Y, Gao Q, Miao Y, Wang H, Xiong B. Exposure to diisononyl phthalate deteriorates the quality of porcine oocytes by inducing the apoptosis. Ecotoxicol Environ Saf. 2023, 254: 114768.

157 Li N, Liu K, Yuan H, Zhu J, Yu G, Xie J, Fu S, Guo K, Ye L. The effect of mono-(2-ethylhexyl) phthalate on apoptosis of rat ovarian granulosa cells in vitro. Environ Toxicol Pharmacol. 2015, 39(2): 643–650.

158 Sellinger EP, Riesgo VR, Brinks AS, Willing J, Juraska JM. Perinatal phthalate exposure increases developmental apoptosis in the rat medial prefrontal cortex. Neurotoxicology. 2021, 87: 167–173.

159 Zhang W, Sun X, Qi X, Liu X, Zhang Y, Qiao S, Lin H. Di-(2-Ethylhexyl) phthalate and microplastics induced neuronal apoptosis through the PI3K/AKT pathway and mitochondrial dysfunction. J Agric Food Chem. 2022, 70(35): 10771–10781.

160 Ma N, Liu S, Gao P, Cao P, Xu H. [Effect of diisobutyl phthalate on learning and memory behavior and apoptosis of hippocampus cells in mice]. Wei Sheng Yan Jiu. 2013, 42(1): 57–60.

161 Johnson GL, Lapadat R. Mitogen-activated protein kinase pathways mediated by ERK, JNK, and p38 protein kinases. Science. 2002, 298(5600): 1911–1912.

162 Yan B, Sun Y, Zeng J, Chen Y, Li C, Song P, Zhang L, Yang X, Wu Y, Ma P. Combined use of vitamin E and nimodipine ameliorates dibutyl phthalate-induced memory deficit and apoptosis in mice by inhibiting the ERK 1/2 pathway. Toxicol Appl Pharmacol. 2019, 368: 1–17.

163 Aung KH, Win-Shwe TT, Kanaya M, Takano H, Tsukahara S. Involvement of hemeoxygenase-1 in di(2-ethylhexyl) phthalate (DEHP)-induced apoptosis of Neuro-2a cells. J Toxicol Sci. 2014, 39(2): 217–229.

164 Ha M, Wei L, Guan X, Li L, Liu C. P53-dependent apoptosis contributes to di-(2-ethylhexyl) phthalate-induced hepatotoxicity. Environ Pollut. 2016, 208(Pt B): 416–425.

165 Yang G, Zhou X, Wang J, Zhang W, Zheng H, Lu W, Yuan J. MEHP-induced oxidative DNA damage and apoptosis in HepG2 cells correlates with p53-mediated mitochondria-dependent signaling pathway. Food Chem Toxicol. 2012, 50(7): 2424–2431.

166 Song P, Jiang N, Zhang K, Li X, Li N, Zhang Y, Wang Q, Wang J. Ecotoxicological evaluation of zebrafish liver (Danio rerio) induced by dibutyl phthalate. J Hazard Mater. 2022, 425: 128027.

167 Yang G, Zhang W, Qin Q, Wang J, Zheng H, Xiong W, Yuan J. Mono(2-ethylhexyl) phthalate induces apoptosis in p53-silenced L02 cells via activation of both mitochondrial and death receptor pathways. Environ Toxicol. 2015, 30(10): 1178–1191.

168 Delfosse V, Maire AL, Balaguer P, Bourguet W. A structural perspective on nuclear receptors as targets of environmental compounds. Acta Pharmacol Sin. 2015, 36(1): 88–101.

169 Asai D, Tahara Y, Nakai M, Yakabe Y, Takatsuki M, Nose T, Shinmyozu T, Shimohigashi Y. Structural essentials of xenoestrogen dialkyl phthalates to bind to the estrogen receptors. Toxicol Lett. 2000, 118 (1–2): 1–8.

170 Hurst CH, Waxman DJ. Activation of PPARalpha and PPARgamma by environmental phthalate monoesters. Toxicol Sci. 2003, 74(2): 297–308.

171 MBility MT, Thompson JT, McKee RH, David RM, Butala JH, Vanden Heuvel JP, Peters JM. Activation of mouse and human peroxisome proliferator-activated receptors (PPARs) by phthalate monoesters. Toxicol Sci. 2004, 82(1): 170–182.

172 Sarath Josh MK, Pradeep S, Vijayalekshmi Amma KS, Balachandran S, Abdul Jaleel UC, Doble M, Spener F, Benjamin S. Phthalates efficiently bind to human peroxisome proliferator activated receptor and retinoid X receptor α, β, γ subtypes: an in silico approach. J Appl Toxicol. 2014, 34(7): 754–765.

173 Meling DD, De La Torre KM, Arango AS, Gonsioroski A, Deviney ARK, Neff AM, Laws MJ, Warner GR, Tajkhorshid E, Flaws JA. Phthalate monoesters act through peroxisome proliferator-activated receptors in the mouse ovary. Reprod Toxicol. 2022, 110: 113–123.

174 Gao F, Hu W, Li Y, Shen H, Hu J. Mono-2-ethylhexyl phthalate inhibits human extravillous trophoblast invasion via the PPARγ pathway. Toxicol Appl Pharmacol. 2017, 327: 23–29.

175 Shoaito H, Petit J, Chissey A, Auzeil N, Guibourdenche J, Gil S, Laprévote O, Fournier T, Degrelle SA. The role of peroxisome proliferator–activated receptor gamma (PPARγ) in mono(2-ethylhexyl) phthalate (MEHP)-mediated cytotrophoblast differentiation. Environ Health Perspect. 2019, 127(2): 27003.

176 Lin CH, Chen TJ, Chen SS, Hsiao PC, Yang RC. Activation of Trim17 by PPARγ is involved in di(2-ethylhexyl) phthalate (DEHP)-induced apoptosis on Neuro-2a cells. Toxicol Lett. 2011, 206(3): 245–251.

177 Xu M, Li Y, Wang X, Zhang Q, Wang L, Zhang X, Cui W, Han X, Ma N, Li H, Fang H, Tang S, Li J, Liu Z, Yang H, Jia X. Role of hepatocyte- and macrophage-specific PPARγ in hepatotoxicity induced by diethylhexyl phthalate in mice. Environ Health Perspect. 2022, 130(1): 17005.

178 Zhang W, Li JY, Wei XC, Wang Q, Yang JY, Hou H, Du ZW, Wu XA. Effects of dibutyl phthalate on lipid metabolism in liver and hepatocytes based on PPARα/SREBP-1c/FAS/GPAT/AMPK signal pathway. Food Chem Toxicol. 2021, 149: 112029.

179 Takeuchi S, Iida M, Kobayashi S, Jin K, Matsuda T, Kojima H. Differential effects of phthalate esters on transcriptional activities via human estrogen receptors alpha and beta, and androgen receptor. Toxicology. 2005, 210 (2–3): 223–233.

180 Chen P, Li B, Ou-Yang L. Role of estrogen receptors in health and disease. Front Endocrinol (Lausanne). 2022, 13: 839005.

181 Lv C, Wei Z, Yue B, Xia N, Huang W, Yue Y, Li Z, Li T, Zhang X, Wang Y. Characterization of diphenyl phthalate as an agonist for estrogen receptor: an in vitro and in silico study. Toxicol Mech Methods. 2022, 32(4): 280–287.

182 Mu X, Liao X, Chen X, Li Y, Wang M, Shen C, Zhang X, Wang Y, Liu X, He J. DEHP exposure impairs mouse oocyte cyst breakdown and primordial follicle assembly through estrogen receptor-dependent and independent mechanisms. J Hazard Mater. 2015, 298: 232–240.

183 Zhang J, Zhou K, Cheng R, Yang M, Shen X, Luo X, Xu L. Maternal perinatal exposure to dibutyl phthalate promotes ovarian dysfunction in adult female offspring via downregulation of TGF-β2 and TGF-β3. Reprod Sci. 2022, 29(8): 2401–2413.

184 Guo S, Zhang B, Qi W, Cao X, Xu Y, Zhang X, Li X, Zhou L, Ye L. Role of estrogen receptor alpha in MEHP-induced proliferation and invasion of SH-SY5Y cells. Toxicology. 2021, 453: 152734.

185 Xu Q, Zhou L, Ri H, Li X, Zhang X, Qi W, Ye L. Role of estrogen receptors in thyroid toxicity induced by mono (2-ethylhexyl) phthalate via endoplasmic reticulum stress: an in vitro mechanistic investigation. Environ Toxicol Pharmacol. 2022, 96: 104007.

186 Christen V, Crettaz P, Oberli-Schrämmli A, Fent K. Antiandrogenic activity of phthalate mixtures: validity of concentration addition. Toxicol Appl Pharmacol. 2012, 259(2): 169–176.

187 Rider CV, Wilson VS, Howdeshell KL, Hotchkiss AK, Furr JR, Lambright CR, Gray LE Jr. Cumulative effects of in utero administration of mixtures of "antiandrogens" on male rat reproductive development. Toxicol Pathol. 2009, 37(1): 100–113.

188 Wang SW, Wang SS, Wu DC, Lin YC, Ku CC, Wu CC, Chai CY, Lee JN, Tsai EM, Lin CL, Yang RC, Ko YC, Yu HS, Huo C, Chuu CP, Murayama Y, Nakamura Y, Hashimoto S, Matsushima K, Jin C, Eckner R, Lin CS, Saito S, Yokoyama KK. Androgen receptor-mediated apoptosis in bovine testicular induced pluripotent stem cells in response to phthalate esters. Cell Death Dis. 2013, 4(11): e907.

189 Mukai M, Kasai A, Hiramatsu N, Hayakawa K, Okamura M, Tagawa Y, Yao J, Nakamura T, Kitamura M. Blockade of the aryl hydrocarbon receptor pathway triggered by dioxin, polycyclic aromatic hydrocarbons and cigarette smoke by Phellinus linteus. Biol Pharm Bull. 2008, 31(10): 1888–1893.

190 Simones T, Shepherd DM. Consequences of AhR activation in steady-state dendritic cells. Toxicol Sci. 2011, 119(2): 293–307.

191 Zhang X, Jie D, Liu J, Zhang J, Zhang T, Zhang J, Ma E. Aryl hydrocarbon receptor regulates the expression of LmGSTd7 and is associated with chlorpyrifos susceptibility in Locusta migratoria. Pest Manag Sci. 2019, 75(11): 2916–2924.

192 Wójtowicz AK, Sitarz-Głownia AM, Szczęsna M, Szychowski KA. The action of di-(2-Ethylhexyl) phthalate (DEHP) in mouse cerebral cells involves an impairment in aryl hydrocarbon receptor (AhR) signaling. Neurotox Res. 2019, 35(1): 183–195.

193 Hsieh TH, Hsu CY, Yang PJ, Chiu CC, Liang SS, Ou-Yang F, Kan JY, Hou MF, Wang TN, Tsai EM. DEHP mediates drug resistance by directly targeting AhR in human breast cancer. Biomed Pharmacother. 2022, 145: 112400.

194 Wójtowicz AK, Szychowski KA, Wnuk A, Kajta M. Dibutyl phthalate (DBP)-induced apoptosis and neurotoxicity are mediated via the aryl hydrocarbon receptor (AhR) but not by estrogen receptor alpha (ERα), estrogen receptor beta (ERβ), or peroxisome proliferator-activated receptor gamma (PPARγ) in mouse cortical neurons. Neurotox Res. 2017, 31(1): 77–89.

195 Hsieh TH, Tsai CF, Hsu CY, Kuo PL, Lee JN, Chai CY, Wang SC, Tsai EM. Phthalates induce proliferation and invasiveness of estrogen receptor-negative breast cancer through the AhR/HDAC6/c-Myc signaling pathway. FASEB J. 2012, 26(2): 778–787.

196 Chen HS, Chiang PH, Wang YC, Kao MC, Shieh TH, Tsai CF, Tsai EM. Benzyl butyl phthalate induces necrosis by AhR mediation of CYP1B1 expression in human granulosa cells. Reprod Toxicol. 2012, 33(1): 67–75.

197 Land KL, Lane ME, Fugate AC, Hannon PR. Ovulation is inhibited by an environmentally relevant phthalate mixture in mouse antral follicles in vitro. Toxicol Sci. 2021, 179(2): 195–205.

198 Barakat R, Seymore T, Lin PP, Park CJ, Ko CJ. Prenatal exposure to an environmentally relevant phthalate mixture disrupts testicular steroidogenesis in adult male mice. Environ Res. 2019, 172: 194–201.

199 Zhou C, Gao L, Flaws JA. Prenatal exposure to an environmentally relevant phthalate mixture disrupts reproduction in F1 female mice. Toxicol Appl Pharmacol. 2017, 318: 49–57.

200 Gill S, Brehm E, Leon K, Chiu J, Meling DD, Flaws JA. Prenatal exposure to an environmentally relevant phthalate mixture alters ovarian steroidogenesis and folliculogenesis in the F1 generation of adult female mice. Reprod Toxicol. 2021, 106: 25–31.

201 Fletcher EJ, Santacruz-Márquez R, Mourikes VE, Neff AM, Laws MJ, Flaws JA. Effects of phthalate mixtures on ovarian folliculogenesis and steroidogenesis. Toxics. 2022, 10(5): 251.

202 Mirihagalle S, You T, Suh L, Patel C, Gao L, Rattan S, Qiao H. Prenatal exposure to di-(2-ethylhexyl) phthalate and high-fat diet synergistically disrupts mouse fetal oogenesis and affects folliculogenesis†. Biol Reprod. 2019, 100(6): 1561–1570.

203 Brehm E, Rattan S, Gao L, Flaws JA. Prenatal exposure to di(2-Ethylhexyl) phthalate causes long-term transgenerational effects on female reproduction in mice. Endocrinology. 2018, 159(2): 795–809.

204 Barakat R, Lin PP, Rattan S, Brehm E, Canisso IF, Abosalum ME, Flaws JA, Hess R, Ko C. Prenatal exposure to DEHP induces premature reproductive senescence in male mice. Toxicol Sci. 2017, 156(1): 96–108.

205 Doyle TJ, Bowman JL, Windell VL, McLean DJ, Kim KH. Transgenerational effects of di-(2-ethylhexyl) phthalate on testicular germ cell associations and spermatogonial stem cells in mice. Biol Reprod. 2013, 88(5): 112.

206 Yuan B, Wu W, Chen M, Gu H, Tang Q, Guo D, Chen T, Chen Y, Lu C, Song L, Xia Y, Chen D, Rehan VK, Sha J, Wang X. From the cover: metabolomics reveals a role of betaine in prenatal DBP exposure-induced epigenetic transgenerational failure of spermatogenesis in rats. Toxicol Sci. 2017, 158(2): 356–366.

207 Liu JC, Xing CH, Xu Y, Pan ZN, Zhang HL, Zhang Y, Sun SC. DEHP exposure to lactating mice affects ovarian hormone production and antral follicle development of offspring. J Hazard Mater. 2021, 416: 125862.

208 Li L, Liu JC, Lai FN, Liu HQ, Zhang XF, Dyce PW, Shen W, Chen H. Di (2-ethylhexyl) phthalate exposure impairs growth of antral follicle in mice. PLoS One. 2016, 11(2): e0148350.

209 Wu Y, Wang J, Zhao T, Chen J, Kang L, Wei Y, Han L, Shen L, Long C, Wu S, Wei G. Di-(2-ethylhexyl) phthalate exposure leads to ferroptosis via the HIF-1α/HO-1 signaling pathway in mouse testes. J Hazard Mater. 2022, 426: 127807.

210 Khasin LG, Della Rosa J, Petersen N, Moeller J, Kriegsfeld LJ, Lishko PV. The impact of di-2-ethylhexyl phthalate on sperm fertility. Front Cell Dev Biol. 2020, 8: 426.

211 Li L, Bu T, Su H, Chen Z, Liang Y, Zhang G, Zhu D, Shan Y, Xu R, Hu Y, Li J, Hu G, Lian Q, Ge RS. In utero exposure to diisononyl phthalate caused testicular dysgenesis of rat fetal testis. Toxicol Lett. 2015, 232(2): 466–474.

212 Acevedo-Rodriguez A, Kauffman AS, Cherrington BD, Borges CS, Roepke TA, Laconi M. Emerging insights into hypothalamic-pituitary-gonadal axis regulation and interaction with stress signalling. J Neuroendocrinol. 2018, 30(10): e12590.

213 Ma YB, Jia PP, Junaid M, Yang L, Lu CJ, Pei DS. Reproductive effects linked to DNA methylation in male zebrafish chronically exposed to environmentally relevant concentrations of di-(2-ethylhexyl) phthalate. Environ Pollut. 2018, 237: 1050–1061.

214 Repouskou A, Panagiotidou E, Panagopoulou L, Bisting PL, Tuck AR, Sjödin MOD, Lindberg J, Bozas E, Rüegg J, Gennings C, Bornehag CG, Damdimopoulou P, Stamatakis A, Kitraki E. Gestational exposure to an epidemiologically defined mixture of phthalates leads to gonadal dysfunction in mouse offspring of both sexes. Sci Rep. 2019, 9(1): 6424.

215 Hannas BR, Lambright CS, Furr J, Evans N, Foster PM, Gray EL, Wilson VS. Genomic biomarkers of phthalate-induced male reproductive developmental toxicity: a targeted RT-PCR array approach for defining relative potency. Toxicol Sci. 2012, 125(2): 544–557.

216 Hu J, Jiang K, Tang X, Liu H, Zhang H, Yang X, Nie X, Luo H. Chronic exposure to di-n-butyl phthalate causes reproductive toxicity in zebrafish. J Appl Toxicol. 2020, 40(12): 1694–1703.

217 Wang X, Yang Y, Zhang L, Ma Y, Han J, Yang L, Zhou B. Endocrine disruption by di-(2-ethylhexyl)-phthalate in Chinese rare minnow (Gobiocypris rarus). Environ Toxicol Chem. 2013, 32(8): 1846–1854.

218 Ye T, Kang M, Huang Q, Fang C, Chen Y, Shen H, Dong S. Exposure to DEHP and MEHP from hatching to adulthood causes reproductive dysfunction and endocrine disruption in marine medaka (Oryzias melastigma). Aquat Toxicol. 2014, 146: 115–126.

219 Hamid N, Junaid M, Manzoor R, Jia PP, Pei DS. Prioritizing phthalate esters (PAEs) using experimental in vitro/vivo toxicity assays and computational in silico approaches. J Hazard Mater. 2020, 398: 122851.

220 Kassab RB, Lokman MS, Essawy EA. Neurochemical alterations following the exposure to di-n-butyl phthalate in rats. Metab Brain Dis. 2019, 34(1): 235–244.

221 Molnár Z, Clowry GJ, Šestan N, Alzu'bi A, Bakken T, Hevner RF, Hüppi PS, Kostović I, Rakic P, Anton ES, Edwards D, Garcez P, Hoerder-Suabedissen A, Kriegstein A. New insights into the development of the human cerebral cortex. J Anat. 2019, 235(3): 432–451.

222 Lett TA, Voineskos AN, Kennedy JL, Levine B, Daskalakis ZJ. Treating working memory deficits in schizophrenia: a review of the neurobiology. Biol Psychiatry. 2014, 75(5): 361–370.

223 Chayer C, Freedman M. Frontal lobe functions. Curr Neurol Neurosci Rep. 2001, 1(6): 547–552.

224 Wang DC, Lin HT, Lee YJ, Yu HF, Wu SR, Qamar MU. Recovery of BDNF and CB1R in the prefrontal cortex underlying improvement of working memory in prenatal DEHP-exposed male rats after aerobic exercise. Int J Mol Sci. 2020, 21(11): 3867.

225 Zhang X, Huang J, Zheng G, Liang J, Hu B, Lou Z, Li A, Ding Y. Prenatal exposure to di (2-ethylhexyl) phthalate causes autism-like behavior through inducing Nischarin expression in the mouse offspring. Biochem Biophys Res Commun. 2021, 585: 29–35.

226 Xu P, Chen A, Li Y, Xing X, Lu H. Medial prefrontal cortex in neurological diseases. Physiol Genomics. 2019, 51(9): 432–442.

227 Watanabe M, Buth JE, Vishlaghi N, de la Torre-Ubieta L, Taxidis J, Khakh BS, Coppola G, Pearson CA, Yamauchi K, Gong D, Dai X, Damoiseaux R, Aliyari R, Liebscher S, Schenke-Layland K, Caneda C, Huang EJ, Zhang Y, Cheng G, Geschwind DH, Golshani P, Sun R, Novitch BG. Self-organized cerebral organoids with human-specific features predict effective drugs to combat Zika virus infection. Cell Rep. 2017, 21(2): 517–532.

228 Komada M, Gendai Y, Kagawa N, Nagao T. Prenatal exposure to di(2-ethylhexyl) phthalate impairs development of the mouse neocortex. Toxicol Lett. 2016, 259: 69–79.

229 Ge S, Yan B, Huang J, Chen Y, Chen M, Yang X, Wu Y, Shen D, Ma P. Diisodecyl phthalate aggravates the formaldehyde-exposure-induced learning and memory impairment in mice. Food Chem Toxicol. 2019, 126: 152–161.

230 Peng L. Mice brain tissue injury induced by diisononyl phthalate exposure and the protective application of vitamin e. J Biochem Mol Toxicol. 2015, 29(7): 311–320.

231 Feng W, Wu X, Mao G, Zhao T, Wang W, Chen Y, Zhang M, Yang L, Wu X. Neurological effects of subchronic exposure to dioctyl phthalate (DOP), lead, and arsenic, individual and mixtures, in immature mice. Environ Sci Pollut Res Int. 2020, 27(9): 9247–9260.

232 Min A, Liu F, Yang X, Chen M. Benzyl butyl phthalate exposure impairs learning and memory and attenuates neurotransmission and CREB phosphorylation in mice. Food Chem Toxicol. 2014, 71: 81–89.

233 Lee S, Lee W, Yang S, Suh YJ, Hong DG, Chang SC, Kim HS, Lee J. Di-n-butyl phthalate disrupts neuron maturation in primary rat embryo neurons and male C57BL/6 mice. J Toxicol Environ Health A. 2022, 85(2): 56–70.

234 Ahmadpour D, Mhaouty-Kodja S, Grange-Messent V. Disruption of the blood-brain barrier and its close environment following adult exposure to low doses of di(2-ethylhexyl)phthalate alone or in an environmental phthalate mixture in male mice. Chemosphere. 2021, 282: 131013.

235 Spencer W, Kandel ER. Hippocampal neuron responses to selective activation of recurrent collaterals of hippocampofugal axons. Exp. Neurol. 1961, 4: 149–161.

236 Manns JR, Eichenbaum H. Time and treason to the trisynaptic teachings: theoretical comment on Kesner et al. (2005). Behav Neurosci. 2005, 119(4): 1140–1143.

237 Lu Y, Wang H, Yang J, Jiang W, Xin H, Luo Y, Loya-López S, Gu H, Ran D. Distinct role of mono-2-ethylhexyl phthalate in neuronal transmission in rat CA3 hippocampal neurons: involvement of ion channels. Molecules. 2022, 27(10): 3082.

238 Hering H, Sheng M. Dendritic spines: structure, dynamics and regulation. Nat Rev Neurosci. 2001, 2(12): 880–888.

239 Schlichting ML, Zeithamova D, Preston AR. CA1 subfield contributions to memory integration and inference. Hippocampus. 2014, 24(10): 1248–1260.

240 Park S, Lee JM, Kim JW, Cheong JH, Yun HJ, Hong YC, Kim Y, Han DH, Yoo HJ, Shin MS, Cho SC, Kim BN. Association between phthalates and externalizing behaviors and cortical thickness in children with attention deficit hyperactivity disorder. Psychol Med. 2015, 45(8): 1601–1612.

241 Häusser M, Spruston N, Stuart GJ. Diversity and dynamics of dendritic signaling. Science. 2000, 290(5492): 739–744.

242 You M, Dong J, Fu Y, Cong Z, Fu H, Wei L, Wang Y, Wang Y, Chen J. Exposure to di-(2-ethylhexyl) phthalate during perinatal period gender-specifically impairs the dendritic growth of pyramidal neurons in rat offspring. Front Neurosci. 2018, 12: 444.

243 Dong J, Fu H, Fu Y, You M, Li X, Wang C, Leng K, Wang Y, Chen J. Maternal exposure to di-(2-ethylhexyl) phthalate impairs hippocampal synaptic plasticity in male offspring: involvement of damage to dendritic spine development. ACS Chem Neurosci. 2021, 12(2): 311–322.

244 Lee JW, Park S, Han HK, Gye MC, Moon EY. Di-(2-ethylhexyl) phthalate enhances melanoma tumor growth via differential effect on M1-and M2-polarized macrophages in mouse model. Environ Pollut. 2018, 233: 833–843.

245 Danielpour D. Functions and regulation of transforming growth factor-beta (TGF-beta) in the prostate. Eur J Cancer. 2005, 41(6): 846–857.

246 Miller DM, Thomas SD, Islam A, Muench D, Sedoris K. C-Myc and cancer metabolism. Clin Cancer Res. 2012, 18(20): 5546–5553.

247 Bouchalova K, Cizkova M, Cwiertka K, Trojanec R, Hajduch M. Triple negative breast cancer—current status and prospective targeted treatment based on HER1 (EGFR), TOP2A and C-MYC gene assessment. Biomed Pap Med Fac Univ Palacky Olomouc Czech Repub. 2009, 153(1): 13–17.

248 Lee HR, Hwang KA, Choi KC. The estrogen receptor signaling pathway activated by phthalates is linked with transforming growth factor-β in the progression of LNCaP prostate cancer models. Int J Oncol. 2014, 45(2): 595–602.

249 Zhu M, Wu J, Ma X, Huang C, Wu R, Zhu W, Li X, Liang Z, Deng F, Zhu J, Xie W, Yang X, Jiang Y, Wang S, Geng S, Xie C, Zhong C. Butyl benzyl phthalate promotes prostate cancer cell proliferation through miR-34a downregulation. Toxicol In Vitro. 2019, 54: 82–88.

250 Yurdakok Dikmen B, Alpay M, Kismali G, Filazi A, Kuzukiran O, Sireli UT. In vitro effects of phthalate mixtures on colorectal adenocarcinoma cell lines. J Environ Pathol Toxicol Oncol. 2015, 34(2): 115–123.

251 Kim IY, Han SY, Moon A. Phthalates inhibit tamoxifen-induced apoptosis in MCF-7 human breast cancer cells. J Toxicol Environ Health A. 2004, 67 (23–24): 2025–2035.

252 Wang YC, Tsai CF, Chuang HL, Chang YC, Chen HS, Lee JN, Tsai EM. Benzyl butyl phthalate promotes breast cancer stem cell expansion via SPHK1/S1P/S1PR3 signaling. Oncotarget. 2016, 7(20): 29563–29576.

253 Yamaguchi H, Condeelis J. Regulation of the actin cytoskeleton in cancer cell migration and invasion. Biochim Biophys Acta. 2007, 1773(5): 642–652.

254 Polacheck WJ, Zervantonakis IK, Kamm RD. Tumor cell migration in complex microenvironments. Cell Mol Life Sci. 2013, 70(8): 1335–1356.

255 DeSantis CE, Lin CC, Mariotto AB, Siegel RL, Stein KD, Kramer JL, Alteri R, Robbins AS, Jemal A. Cancer treatment and survivorship statistics, 2014. CA Cancer J Clin. 2014, 64(4): 252–271.

256 Wang Y, Zhao M, Liu J, Ni J, Jiao Y, Bai C. Up regulation of IL-6 is involved in di (2-ethylhexyl) phthalate (DEHP) induced migration and invasion of non small cell lung cancer (NSCLC) cells. Biomed Pharmacother. 2017, 89: 1037–1044.

257 Qin Y, Zhang J, Avellán-Llaguno RD, Zhang X, Huang Q. DEHP-elicited small extracellular vesicles miR-26a-5p promoted metastasis in nearby normal A549 cells. Environ Pollut. 2021, 272: 116005.

258 Wang Z, Shao M, Liu Y. Promotion of Wilms' tumor cells migration and invasion by mono-2-ethyhexyl phthalate (MEHP) via activation of NF-κB signals. Chem Biol Interact. 2017, 270: 1–8.

259 Yao PL, Lin YC, Richburg JH. Mono-(2-ethylhexyl) phthalate (MEHP) promotes invasion and migration of human testicular embryonal carcinoma cells. Biol Reprod. 2012, 86(5): 160, 1–10.

260 Cui S, Wang L, Zhao H, Lu F, Wang W, Yuan Z. Benzyl butyl phthalate (BBP) triggers the migration and invasion of hemangioma cells via upregulation of Zeb1. Toxicol In Vitro. 2019, 60: 323–329.

261 Du Y, Shi X, Ma W, Wen P, Yu P, Wang X, Fang P, Chen A, Gao Z, Cui K. Phthalates promote the invasion of hepatocellular carcinoma cells by enhancing the interaction between Pregnane X receptor and E26 transformation specific sequence 1. Pharmacol Res. 2021, 169: 105648.

262 Kan JY, Wu DC, Yu FJ, Wu CY, Ho YW, Chiu YJ, Jian SF, Hung JY, Wang JY, Kuo PL. Chemokine (C-C Motif) ligand 5 is involved in tumor-associated dendritic cell-mediated colon cancer progression through non-coding RNA MALAT-1. J Cell Physiol. 2015, 230(8): 1883–1894.

263 Hsu YL, Hsieh CJ, Tsai EM, Hung JY, Chang WA, Hou MF, Kuo PL. Didymin reverses phthalate ester-associated breast cancer aggravation in the breast cancer tumor microenvironment. Oncol Lett. 2016, 11(2): 1035–1042.

264 Hsu YL, Hung JY, Tsai EM, Wu CY, Ho YW, Jian SF, Yen MC, Chang WA, Hou MF, Kuo PL. Benzyl butyl phthalate increases the chemoresistance to doxorubicin/cyclophosphamide by increasing breast cancer-associated dendritic cell-derived CXCL1/GROα and S100A8/A9. Oncol Rep. 2015, 34(6): 2889–2900.

265 Liu G, Philp AM, Corte T, Travis MA, Schilter H, Hansbro NG, Burns CJ, Eapen MS, Sohal SS, Burgess JK, Hansbro PM. Therapeutic targets in lung tissue remodelling and fibrosis. Pharmacol Ther. 2021, 225: 107839.

266 Hwang YH, Lee Y, Paik MJ, Yee ST. Inhibitions of HMGB1 and TLR4 alleviate DINP-induced asthma in mice. Toxicol Res (Camb). 2019, 8(5): 621–629.

267 Wang W, Liu Z, Zhang Y, Wang L, Meng D, Li X, Zhang J, Wu Y, Zhou X, Liu G. Benzyl butyl phthalate (BBP) induces lung injury and fibrosis through neutrophil extracellular traps. Environ Pollut. 2022, 309: 119743.

268 Zhao ZB, Ji K, Shen XY, Zhang WW, Wang R, Xu WP, Wei W. Di(2-ethylhexyl) phthalate promotes hepatic fibrosis by regulation of oxidative stress and inflammation responses in rats. Environ Toxicol Pharmacol. 2019, 68: 109–119.

269 Lee CY, Suk FM, Twu YC, Liao YJ. Long-term exposure to low-dose di-(2-ethylhexyl) phthalate impairs cholesterol metabolism in hepatic stellate cells and exacerbates liver librosis. Int J Environ Res Public Health. 2020, 17(11): 3802.

270 Zhang Y, Hui J, Xu Y, Ma Y, Sun Z, Zhang M, Nie L, Ye L. MEHP promotes liver fibrosis by down-regulating STAT5A in BRL-3A hepatocytes. Chemosphere. 2022, 295: 133925.

271 Wu CT, Wang CC, Huang LC, Liu SH, Chiang CK. Plasticizer di-(2-ethylhexyl)phthalate induces epithelial-to-mesenchymal transition and renal fibrosis in vitro and in vivo. Toxicol Sci. 2018, 164(1): 363–374.

272 Zhu YP, Chen L, Wang XJ, Jiang QH, Bei XY, Sun WL, Xia SJ, Jiang JT. Maternal exposure to di-n-butyl phthalate (DBP) induces renal fibrosis in adult rat offspring. Oncotarget. 2017, 8(19): 31101–31111.

273 Zhao S, Jiang JT, Li D, Zhu YP, Xia SJ, Han BM. Maternal exposure to di-n-butyl phthalate promotes Snail1-mediated epithelial-mesenchymal transition of renal tubular epithelial cells via upregulation of TGF-β1 during renal fibrosis in rat offspring. Ecotoxicol Environ Saf. 2019, 169: 266–272.

274 Ye Q, Zhao S, Zhang Y, Su YM, Chen M, Zhao J, Jia GZ, Han BM, Jiang JT. Activation of the RhoA/ROCK pathway contributes to renal fibrosis in offspring rats induced by maternal exposure to di-n-butyl phthalate. Toxicology. 2020, 443: 152573.

275 Halberg N, Wernstedt-Asterholm I, Scherer PE. The adipocyte as an endocrine cell. Endocrinol Metab Clin North Am. 2008, 37(3): 753–768.

276 Lo KA, Sun L. Turning WAT into BAT: a review on regulators controlling the browning of white adipocytes. Biosci Rep. 2013, 33(5): e00065.

277 Armani A, Mammi C, Marzolla V, Calanchini M, Antelmi A, Rosano GM, Fabbri A, Caprio M. Cellular models for understanding adipogenesis, adipose dysfunction, and obesity. J Cell Biochem. 2010, 110(3): 564–572.

278 Hao C, Cheng X, Xia H, Ma X. The endocrine disruptor mono-(2-ethylhexyl) phthalate promotes adipocyte differentiation and induces obesity in mice. Biosci Rep. 2012, 32(6): 619–629.

279 Qi W, Zhou L, Zhao T, Ding S, Xu Q, Han X, Zhao Y, Song X, Zhao T, Zhang X, Ye L. Effect of the TYK-2/STAT-3 pathway on lipid accumulation induced by mono-2-ethylhexyl phthalate. Mol Cell Endocrinol. 2019, 484: 52–58.

280 Hsu JW, Nien CY, Yeh SC, Tsai FY, Chen HW, Lee TS, Chen SL, Kao YH, Tsou TC. Phthalate exposure causes browning-like effects on adipocytes in vitro and in vivo. Food Chem Toxicol. 2020, 142: 111487.

281 Chiang HC, Kuo YT, Shen CC, Lin YH, Wang SL, Tsou TC. Mono(2-ethylhexyl)phthalate accumulation disturbs energy metabolism of fat cells. Arch Toxicol. 2016, 90(3): 589–601.

14

Combined Molecular Toxicity Mechanism of Microplastics Mixtures

Ya Liu[a], *Yin Liu*[a,*]

[a] *School of Environment, Hangzhou Institute for Advanced Study, University of Chinese Academy of Sciences, Hangzhou, China*

*Corresponding author.
E-mail: liuyin19860406@163.com (Y.L.)

14.1 Introduction

Microplastics (MPs) were first introduced by Richard C. Thompson in 2004 [1], which are defined as plastic particles or fragments less than 5mm in diameter, formed by photolysis, mechanical friction, and microbial decomposition of large waste plastics [2]. MPs can be classified according to the source (primary and secondary MPs) [3], polymer types [polyethylene (PE), polypropylene (PP), polyvinyl chloride (PVC), and so on], and morphology (granular, fibrous, film, foam, and so on) [4]. Due to their small size, high mobility, and large surface area, MPs have been reported to exist in the environment as "Trojan horses" carrying other environmental substances such as heavy metals, persistent organic pollutants (POPs), pathogens, engineered nanomaterials (ENMs), and other contaminants, which will be facilitated by their dissemination [5]. Also, the mixture could be ingested by organisms, and the adsorbed contaminants can be released into the complex biological environment and bioaccumulated in the food chain, which will lead to ecological environmental destruction and adverse health outcomes in organisms and humans. Consequently, the focus of research has shifted from MPs' ubiquitous occurrence to interactive toxicity, including analyses of toxic effects on various species of organisms and organs. Based on recent studies on the interaction of MPs with several representative environmental pollutants, this chapter provides a comprehensive review of the processes and mechanisms of their interaction and systematically describes the biotoxic effects of representative hazardous mixtures (Figure 14.1) [3, 6–8]

14.2 Heavy Metals

The term "heavy metals" refers to metal elements with significant biological toxicity, such as Hg, Cd, Pb, Cr, and metalloid As. Heavy metals enter the environment through natural or anthropogenic activities and accumulate in nature and living organisms through physical, chemical, and biological actions [9]. Once exceeding the risk threshold, they can accumulate in tissues and lead to adverse effects. The heavy metals accounted for a large proportion of the inorganic pollutants adsorbed by MPs [10]. The interaction of MPs with heavy metals can change their environmental behavior, bioavailability, and potential toxicity, leading to ecological risks.

14.2.1 Sources of Heavy Metals

Heavy metals found in MPs mainly come from two sources, such as manufacturing processes and environmental sorption. (1) Nondestructive X-ray fluorescence techniques have demonstrated the presence of heavy metal additives in polymer matrices [11]. Instead of pure forms, MPs are combined with monomeric ingredients or additives to enhance the properties of the material [12]. Heavy metals added during production are relatively stable and have little tendency to migrate, as most

Toxicological Assessment of Combined Chemicals in the Environment, First Edition. Edited by De-Sheng Pei and Yiyun Liu.

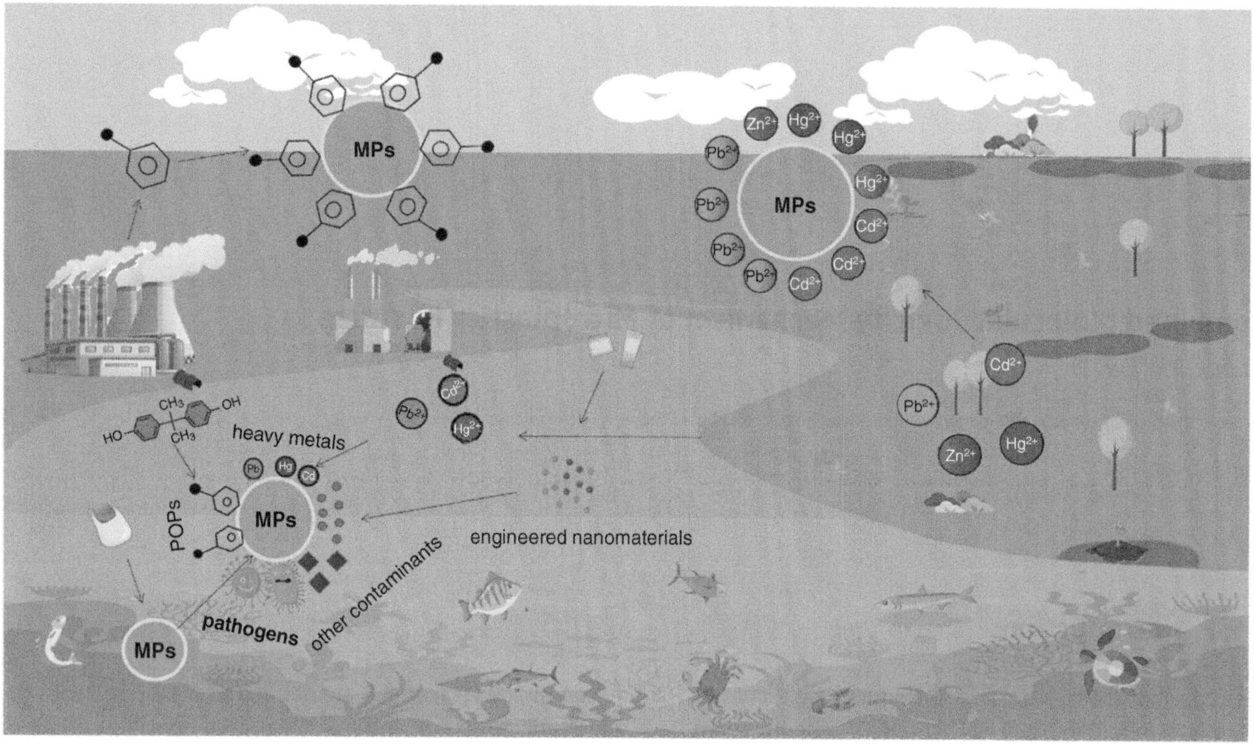

Figure 14.1 Environmental complex exposure of microplastic/nanoplastic and other contaminants.

metal compounds are added to plastic in the liquid phase, making them difficult to release. However, when they are fragmented into smaller sizes by external forces, there is the potential for heavy metals to migrate continuously along the concentration gradient to the surface of the MPs [10]. (2) MPs have a strong affinity to heavy metals and can quickly gather heavy metals from their surroundings [13]. Using energy-dispersive X-ray (EDX) spectroscopy, Bradeny et al. [14] found that metals transported by MPs are not from the plastic itself but from the environment, which indicated that the metals on MPs are accumulated from surface deposits.

14.2.2 Interaction Mechanisms Between MPs and Heavy Metals

MPs have large specific surface areas, aging properties, and surface functional groups [15]. They can act as carriers of various heavy metals, which are transported and transferred to the environment. The interaction of MPs with heavy metals was first reported in 2010 [16]. The adsorption of heavy metals on MPs can be attributed to physical adsorption, including electrostatic interactions of metal ions with charged or polar regions. After extensive weathering, wear, and photooxidation in the natural environment, MPs are negatively charged and have a variety of oxygen-containing groups on their surfaces. As a result, they rely on electrostatic attraction to adsorb metal cations from the environment. Pb was adsorbed on aged MPs mainly by electrostatic attraction and complexation with –COOH and –OH [17]. And the adsorption of Cu and Zn by PVC and PS is mainly through the surface complexation of metal ions with MPs. In addition, the adsorption and accumulation of biofilm and organic matter substances on the surface of MPs changed the surface structure of MPs. Biofilms are rich in functional groups, which in turn facilitate the adsorption of heavy metals by MPs through complexation. Organic matter attached to the surface of MPs usually increases its surface charge, roughness, porosity, hydrophilicity, etc., thus increasing the adsorption of heavy metals by MPs [17].

14.2.3 Toxicity Mechanisms of MPs Combined with Heavy Metals

Based on the strong interaction between MPs and heavy metals, their mixtures could be a potential source of heavy metals-enriched MPs that pose a high risk to organisms [13]. Micro- and macro-algae act as major producers of food chains/webs in the marine environment, which supply food for a range of zooplankton, crustaceans, mollusks, and macrofauna. This phytoplankton is responsible for nearly half of the photosynthesis on earth, and their habitat is the ocean where plastic pollution accumulates. Laboratory experiments have shown that Ag-containing MPs have a significant effect on Duckweed root development [18]. In the rhizosphere of wheat, polyethylene terephthalate (PET) acts as a carrier of heavy metal ions such as Pb, Zn, and Cd [14]. In addition, after these plants are consumed, heavy metals enter the food chain/web and pose threats to a variety of organisms and humans. The coexistence of MPs and heavy metals may lead to complex toxic effects on marine, freshwater, and terrestrial organisms, such as synergistic, antagonistic, or potentiating effects, according to the current in vivo experimental evidence [2]. The heavy metals containing MPs can be ingested by a variety of organisms through direct feeding, filter feeding, and food chain transfer [19]. The heavy metals can enter the organism through the gastrointestinal tract and gradually accumulate in the organism. Because of the low pH conditions in their gastrointestinal tract, the accumulation of heavy metals by MPs increases the likelihood of bioaccumulation, which has been documented in both aquatic and terrestrial environments. A significant increase in bioaccumulation and toxicity of Ni in the presence of MPs has been reported in *Daphnia magna* (*D. magna*) [20]. By co-exposing MPs and Cd to the earthworm *Eisenia foetida* (*E. foetida*), the decreased growth rate and increased mortality were observed, the study of the mechanism indicated that exposure to MPs may induce oxidative damage, and the presence of Cd accelerates the adverse effects of MPs. The accumulation of Cd in earthworms also increased from 9.7% to 161.3% compared with Cd alone [21]. Oral exposure is a typical route for MPs to enter organisms and humans, and the damage to the intestinal barrier has also been well-studied for the co-exposure of heavy metals containing MPs. In a tractable *Drosophila melanogaster* (*D. melanogaster*) model, Zhang et al. found that co-exposure to MPs and Cd led to extensive particle size depended on gut damage early in life and an enhancement of Cd-induced inhibition of locomotor-behavioral function in adult flies. And a dramatically enhanced Cd-induced epigenetic gene silencing via position-effect variegation in somatic tissues was detected by co-exposure with MPs [22]. It has also been reported that exposure to MPs combined with Hg or Cd inhibits acetylcholinesterase (AchE), which reduces cell survival in European seabass and carp [12]. According to Barboza et al., MPs have an impact on the bioconcentration of Hg in the gills of *Dicentrarchus labrax* and their bioaccumulation in the liver. It has been established that MPs and Hg have additive toxicological interactions in the gills and additive or synergistic interactions in the liver [7]. Through sex-specific behavioral changes, redox status, and regulation of detoxification-related genes in *D. magna*, Lee et al. showed additive or synergistic effects of MPs with Zn. This study also shows that Zn can be transported by MPs for *D. magna* in the aquatic environment, lowering the biological concentration of Zn and increasing toxicity. Finally, it was proven that heavy metal Zn and MPs work in concert [6]. When heavy metals were absorbed on the surface of MPs, their concentrations were significantly decreased in the exposure medium. Changes in the heavy metals' bioavailability caused by MPs may lead to antagonistic effects in organisms. Kim et al. investigated the toxic effects of Ni-adsorbed two types of MPs (PS and PS-COOH) on *D. magna* and found that the toxicity of *D. magna* exposed to Ni in combination with PS was lower than Ni in combination with PS-COOH. Calculation of mixture toxicity using a TU model based on these results revealed that PS MPs had a slight antagonistic effect on Ni toxicity. This could be attributed to the fact that Ni is not a substance that adsorbs, much unlike a hydrophobic substance [20]. The experimental results in another study by Khan et al. where Ag was first adsorbed by MPs and then exposed to zebrafish show a significant reduction in Ag uptake by zebrafish. The authors attributed this to the fact that PE MPs with a density of 0.96 g/cm^3 (10–106 μm) tend to float on the water surface and the bioavailability of Ag to zebrafish in the adsorbed state is reduced [23]. Toxicological interactions of MPs and Cr(VI) were observed in early juveniles of common goby, *Pomatoschistus microps*. The results indicated that the presence of MPs slightly decreased the mortality and the negative effects on predatory performance induced by Cr alone tested on the fish [24]. Based on the antagonistic effects of heavy metals with MPs, hypotheses were put forward to elucidate the behavior of MPs–heavy metals' mixture. (1) Heavy metals in the environment can be absorbed by MPs, which reduces their concentration in the biological exposure medium [25]. (2) Heavy metals may bind to MPs but are not bioavailable and are removed from the organism with MPs.

14.3 Persistent Organic Pollutants

POPs are natural organic substances with long-term residual, bioaccumulative, and highly toxic properties; simultaneously, they can be transported over long distances through various environmental media. To date, both the epidemiological investigation and experimental studies have pointed out their endocrine-disrupting potential, carcinogenicity, and other serious effects on organisms and human health. For example, the cross-placental transfer of polyfluoroalkyl substances (PFASs) has been verified in several studies, which may pose multiple health threats to pregnant women and cause adverse birth outcomes for fetuses [26].

14.3.1 Sources of POPs

Most POPs are man-made compounds and include pesticides and industrial chemicals; a few other POPs are unintentional by-products of industrial processes. Halogenation of (F, Cl, or Br) aliphatic or aromatic rings is the main characteristic of POPs. These structural features render POPs low water solubility, high lipid solubility, and high resistance to chemical, biological, and photolytic degradation. In the list of Stockholm Convention POPs, aldrin, chlordane, DDT, dieldrin, endrin, heptachlor, hexachlorobenzene, mirex, and toxaphene are organochlorine pesticides (OCPs); polychlorinated biphenyls (PCBs) are industrial chemicals; and polychlorinated dibenzodioxin (PCDDs) or polychlorinated dibenzofurans (PCDFs) are by-products in incomplete combustion process [27].

14.3.2 Interaction Mechanisms Between MPs and POPs

It has well been accepted that MPs have a strong sorption capacity of POPs, and the adsorption of these organic contaminants in MPs is reported ~100 folds compared to sediment and soil [28]. These behaviors can be attributed to the hydrophobic properties of both MPs and POPs. MPs in the environment are often loaded with a variety of organic pollutants, including pharmaceuticals and personal care products (PPCPs), endocrine disruptors (EDCs), polycyclic aromatic hydrocarbons (PAHs), PCBs, and polybrominated diphenyl ethers (PBDEs) [29]. Due to the variety of polymer types, particle sizes, surface roughness, and aging of the MPs themselves, as well as the various physicochemical properties of POPs in the aqueous and terrestrial environment, the mechanism of sorption between MPs and POPs is complex. Partitioning and surface adsorption are the two most important mechanisms at work when it comes to the adsorption of organic contaminants on MPs.

14.3.3 Toxic Effects of MPs and Organic Pollutants

The hydrophobicity of MPs and their large specific surface area leads to the direct adsorption of organic pollutants from the environment, while organic additives [phthalates, nonylphenols, bisphenol A (BPA)] are added during the production of MPs, and there are also residues of harmful raw materials from MPs. As a result, organic contaminants are both carried and produced by MPs [30]. MPs that combined with organic pollutants can worsen environmental contamination and have a detrimental effect on biological health [31]. Numerous studies have shown that the combined toxic effects of MPs and POPs can mainly be divided into synergistic and antagonistic effects. The synergistic effects of MPs and POPs on aquatic species have become one of the most debated issues in the field of aquatic ecotoxicology because they frequently coexist in the environment [32]. Recent research has demonstrated that the presence of MPs plays a key role in the bioaccumulation and toxicity of POPs. Chen et al. documented that co-exposure of MPs and BPA in adult zebrafish increased the BPA uptake by 2.2 and 2.6 fold in the head and viscera compared with BPA alone treatment after 3 days of exposure. Moreover, for co-exposure treatment, enhanced neurotoxic effects in both the central neural system (CNS) and dopaminergic system were observed [33]. The initial absorption of POPs in the intestinal tract is a key factor for their circulation and accumulation in vivo. Cormier et al. employed artificial intestinal fluid to investigate the desorption behavior of PFOS pre-adsorbed on PE MPs. The results indicated that 70%–80% of PFOS was released into artificial intestinal fluid, which may lead to potential biological toxicity [34]. Furthermore, several studies have shown that MPs and POPs have a synergistic effect on the intestine of test organisms. Di-(2-ethylhexyl) phthalate (DEHP) can increase the elasticity and toughness of plastics and is widely used in the plastics industry. Deng et al. exposure to DEHP-contaminated MPs in mice for 30 days resulted in a significant increase in intestinal permeability and enhanced intestinal inflammation compared to individual MPs and DEHP exposure. Moreover, gut microbiota analysis indicated that the mixture exposure also caused alternations in gut microbiota composition, especially the energy metabolism and immune function–related bacteria. The mechanism study attributed these phenomena to increased DEHP accumulation, changed exposure pathway, and shared toxic mechanisms [35]. The interaction between MPs and POPs promotes the formation of complexes, reduces the bioavailability of POPs in the environment, and ultimately decreases the toxicity of POPs to living organisms. Dibutyl phthalate (DBP) used as a plasticizer for PVC is easily released into the environment during processing and use, causing adverse effects on the organism such as cancer, development deformity, reproductive toxicity, and gene mutation. Li et al. observed more than 20%

reduction in the bioavailability of DBP in the presence of PS MPs (20mg/L). While the volume, morphological complexity, and chlorophyll fluorescence intensity of microalgal cells were disturbed by both DBP and PS MPs. The antagonistic effect of PS MPs could be partially attributed to the combination of hetero- and homo-aggregation and the reduced bioavailability of DBP [36]. Zhu et al. also confirmed the antagonistic effects of triclosan (TSC)-adsorbed MPs on growth inhibition and oxidative stress in microalgae *Skeletonema costatum*. The joint toxicity of TSC with PE, PS, and PVC were all antagonism, and the inhibition effect order was consistent with the adsorption capacity of TSC [37]. Roxithromycin (ROX) is widely used for treating respiratory tract, urinary, and soft tissue infections and has been detected in aquatic environments worldwide, at concentrations ranging from ng/L to μg/L [38]. Zhang et al. investigated the combined effects of PS MPs and ROX on freshwater red tilapia. The results of the study showed that when the two were combined, hepatic malondialdehyde (MDA) levels were significantly reduced, superoxide dismutase (SOD) activity was increased, and AchE activity was inhibited, indicating that the presence of MPs helped to attenuate oxidative damage in the liver and reduce ROX neurotoxicity to the hepatobiliary system [39].

14.4 Pathogens

Pathogens are mainly referred to some pathogenic and harmful microorganisms, and they were first observed attached to plastic debris by Maso et al. [40]. In 2013, Zettler et al. [41] first described the "plastisphere," highlighting the potential for marine MPs to house distinct communities of microbes on their surface. The pathogens are called "hitchhikers" once they are adsorbed on the surface of MPs, by which they can transmit in the water and soil environment [42]. The adsorbed pathogens acting as the congruent body of MPs can be harmful to the environment and human health because of the trophic transmission of pathogens down the food chain.

14.4.1 Sources of Pathogens

The pathogen is a parasitic organism and its natural host includes animals, plants, and humans. A relatively large number of pathogens are found in solid and liquid organic waste (noroviruses) groups. For example, more than 120 different types of viruses may be excreted in feces, including members of the enteroviruses, rotavirus, enteric adenoviruses, and human caliciviruses (noroviruses) groups. The ubiquitous occurrence of MPs and pathogens contained in the same place can enter coastal water through rivers, causing a wider range of ecological pollution.

14.4.2 Interaction Mechanisms Between MPs and Pathogens

The primary of plastic particles shows a smooth, hydrophobic, and non-net charge surface, but once they get into the natural environment, including aquatic and terricolous, organic materials, biomolecules, nutrients, bacteria, and other hydrophobic contaminants quickly adsorb to the surface of plastic particles and create a unique eco-corona [43]. Based on this progress, MPs can act as a unique microhabitat and facilitate the formation of biofilm, which was observed as a complex structure group of single or multiple species of microorganisms attached to a solid surface and encapsulated by an extracellular polysaccharide matrix [42]. Vibrios, in particular, have been found in high abundance within MPs' biofilm, particularly in summer [44]. Living within a biofilm is highly beneficial and can cause microbes to become more infectious than when free-living [45]. Therefore, it is urgently necessary to pay attention to the biological and ecological toxicity of pathogens containing MPs.

14.4.3 Toxic Effects of MPs and Pathogens

After being ingested, the pathogens can be released into organisms and cause toxicity. The transfer and accumulation of MPs along with the food chain is the key factor that threatens wildlife, food safety, and human health. Fabra et al. found that the oxygen consumption and respiration rate of oysters exposed to *Escherichia coli*–coated MPs increased significantly over time, while virgin MPs did not produce any measurable significant physiological responses. These phenomena can be attributed to the immune system response of oysters to the presence of *E. coli*. The results also suggested that MPs did not bioaccumulate in oyster tissues, but the *E.coli* assimilated by the ingestion of coated MPs can transfer to high trophic levels [46]. In another study, *E. coli* was also found released from the surface of MPs to the intestinal tissue after being ingested by coral [47]. The influence of biofilm on zooplankton ingestion rates of MPs in the aquatic environment has also been studied. Lucy et al. found that the intake of biofilm-coated MPs by *D. magna* was higher compared with virgin MPs, and the difference may come from nutrient intake by zooplankton [48]. The gut microbiota in the host can also be affected by the ingested pathogen-coated MPs. Yan et al. found that the microflora on the surface of MPs can significantly affect the microbial environment in the gut of shrimp with the function of diversity and enrichment. These results indicated that except for the toxicity mentioned above, the effects on intestinal microbiota induced by MPs were possible because of the biofilm on their surface, causing notable impacts on aquatic animals [49]. However, the widespread biofilm-coated MPs can lead to a significant safety risk to the organisms and human health.

14.5 Engineered Nanomaterials

Accompanied by the application of ENMs, the ecological and health risks caused by their widespread distribution have been raised for several years. Most different from other contaminants, ENMs have a large specific surface area and show quantum effects. These properties are similar to MPs, which attracted more interest and attention to the environmental health risks caused by their co-exposure.

14.5.1 Sources of ENMs

ENMs are artificially manufactured microscopic materials that have at least one dimension of nanometer size (1–100 nm). ENMs have been widely used in many fields, such as nanocatalysts in industrial production, food additives in daily life, personal care products (PCPs), and nanomedicine. Whether directly or indirectly released, ENMs have been widely detected in soil, water, atmosphere, and other media. This increases the opportunity for photooxidation with other contaminants, such as MPs. MPs have also been found to act as vectors of ENMs such as ZnO NPs, TiO_2 NPs, and Au NPs and to aggregate with Ag NPs, in turn modifying the fate, migration, bioavailability, and toxicity of these metal (oxide) NPs [50].

14.5.2 Interaction Mechanisms Between MPs and ENMs

The interaction between MPs and NPs includes electrostatic interaction, π–π interactions, and pore-filling.[5] The electrostatic interaction plays a key role in the aggravation of PS MPs aggregation in the presence of CeO_2 NPs. These interactions can be evidenced by the pH-dependent charge changes causing significant differences in PS MPs diameter increase and colloidal stability [51]. Li et al. investigate the removal effect of AgNPs by employing PE MPs, PP MPs, and PS MPs in aquatic environments. Determined by the π–π interactions, the AgNPs are significantly captured on PS MPs surface in the form of Ag^0 rather than Ag^+ [52]. Nevertheless, fully understanding the interaction mechanisms between MPs and ENMs is beneficial to predict the behavior and toxic effects of the complex mixture.

14.5.3 Toxic Effects of MPs and ENMs

The ability of MPs to act as vectors for ENMs and promote their accumulation and toxicity has been demonstrated in animals and plants. Rotifer plays an important role in the energy flow of the marine ecosystems. Sun et al. showed that co-exposed ZnO NPs and MPs significantly prolonged rotifer maturation time and decreased total offspring. Accompanied by the rising concentration, more serious joint deleterious effects on survival, development, and reproduction have been detected [53]. Microalgae are primary producers that contribute to the growth and sustainability of freshwater ecosystems [54]. Das et al. found that the mixture of MPs and TiO_2 NP significantly increased the oxidative stress parameters like total ROS (superoxide and hydroxyl radical) in comparison with their individual counterparts. The lipid peroxidation and the antioxidant enzyme activities were correlated well with the reactive species generation results in the algal cells [55]. The chemical transformation of nanomaterials can also be influenced by the presence of MPs. Tong et al. demonstrated that PS MPs remarkably promoted Ag^+ release, altered the Ag^+:Ag^0 ratio, and presented vehicle effects on Ag^+ toxicity to *D. magna*. The ion-related toxicity of Ag NPs was significantly increased by PS MPs [56]. In contrast with these results, Zhang et al. found that Ag NPs can heteroaggregate with PS MPs through electrostatic interactions and co-precipitate from the water phase. This heteroaggregation can stabilize Ag NPs by inhibiting Ag NPs surface photooxidation and suppressing Ag^+ release and significantly suppressing acute toxicity of Ag^+ to *E. coli*, *Selenastrum capricornutum*, *D. magna*, and zebrafish [57]. Au NPs are one of the most widely used ENMs mainly due to their biocompatibility and unique optical and electronic properties [58]. Pacheco et al. indicated that the mix of Au NPs (5 nm) and MPs (1–5 μm) caused higher toxicity in *D. magna* than mixture components alone. Based on mortality, the antagonism effect of mixture components was converted to a synergism effect from low to high concentrations [59]. Investigating the different kinds of ENMs on the mixture toxicity with MPs would be a great opportunity to fully assess their exposure risks in the environment.

14.6 Other Contaminants

Except for the environmental contaminations mentioned above, there are still different kinds of pollutants that can interact with MPs in the environment, such as organic pesticides [60], pharmaceutical preparations [61], plastic additives [62], and flame retardants [63]. The co-existence of these pollutions can be classified as endogenously contained and exogenously adsorbed. The interaction mechanisms between the MPs and other contaminations can also be divided into electrostatic interaction, hydrophobic force, π–π interactions, H-bonding, and others. The composition of these mixtures is a highly dynamic process, influenced by temperature, time, salinity, light, and other factors. In conclusion, the interfacial effects of MPs make them an important site for the adsorption–desorption of pollutants, which requires urgent attention to their potential biosafety risks.

Biological uptake and accumulation assisted by the adsorption of pollutants on the surface of MPs can greatly improve the bioavailability of contaminants. The results of toxicity studies indicated that enhanced biosafety risks were detected when the subject contaminants are combined [64], but we should also emphasize that higher sorption to MPs does not imply higher toxicity. When chemicals in the mixture do not interact with each other, they can cause toxic effects independently. Horton et al. [65] indicated that the deltamethrin toxicity in the *D. magna* was not changed by the presence of PS MPs.

14.7 Conclusion and Prospect

Concerns about MPs' exposure risks have risen in recent years. The wide distribution of MPs has caused several negative ecological and toxic effects, especially in the organisms located at the bottom of the food chain. In another aspect, the hydrophobic surface and high specific surface area of MPs facilitate competitive adsorption of heavy metals, POPs, pathogens, ENMs, and other contaminants, which may increase the concerns about biosafety risks caused by the mixture. In this chapter, we focused on the sources of various pollutants and the mechanisms by which they interact with MPs. The complex toxic effects and mechanisms in various organisms have also been reviewed. Therefore, we suggest that future MPs toxicity studies should include a more complex mixture and consider deeper investigations in more cell and animal models.

References

1 Thompson RC, Olsen Y, Mitchell RP, Davis A, Rowland SJ, John AWG, McGonigle D, Russell AE. Lost at sea: where is all the plastic? Science. 2004, 304(5672): 838–838.
2 Khalid N, Aqeel M, Noman A, Khan SM, Akhter N. Interactions and effects of microplastics with heavy metals in aquatic and terrestrial environments. Environ Pollut. 2021, 290: 118104.
3 Fang S, Yu W, Li C, Liu Y, Qiu J, Kong F. Adsorption behavior of three triazole fungicides on polystyrene microplastics. Sci Total Environ. 2019, 691: 1119–1126.
4 Zhao M, Huang L, Arulmani SRB, Yan J, Wu L, Wu T, Zhang H, Xiao T. Adsorption of different pollutants by using microplastic with different influencing factors and mechanisms in wastewater: a review. Nanomaterials. 2022, 12(13): 2256.
5 Hu L, Zhao Y, Xu H. Trojan horse in the intestine: a review on the biotoxicity of microplastics combined environmental contaminants. J Hazard Mater. 2022, 439: 129652.
6 Lee Y, Yoon D-S, Lee YH, Kwak JI, An Y-J, Lee J-S, Park JC. Combined exposure to microplastics and zinc produces sex-specific responses in the water flea Daphnia magna. J Hazard Mater. 2021, 420: 126652.

7 Antao Barboza LG, Vieira LR, Branco V, Carvalho C, Guilherminol L. Microplastics increase mercury bioconcentration in gills and bioaccumulation in the liver, and cause oxidative stress and damage in Dicentrarchus labrax juveniles. Scientific Reports 2018, 8: 15655.

8 Mammo FK, Amoah ID, Gani KM, Pillay L, Ratha SK, Bux F, Kumari S. Microplastics in the environment: Interactions with microbes and chemical contaminants. Sci Total Environ. 2020, 743: 140518.

9 Li W, Lo H-S, Wong H-M, Zhou M, Wong C-Y, Tam NF-Y, Cheung S-G. Heavy metals contamination of sedimentary microplastics in Hong Kong. Marine Pollut Bull. 2020, 153: 110977.

10 Cao Y, Zhao M, Ma X, Song Y, Zuo S, Li H, Deng W. A critical review on the interactions of microplastics with heavy metals: mechanism and their combined effect on organisms and humans. Sci Total Environ. 2021, 788: 147620.

11 Turner A, Holmes L, Thompson RC, Fisher AS. Metals and marine microplastics: adsorption from the environment versus addition during manufacture, exemplified with lead. Water Res. 2020, 173: 115577.

12 Naqash D. Interaction of freshwater microplastics with biota and heavy metals: a review. Environ Chem Lett. 2020, 18(6): 1813–1824.

13 Brennecke D, Duarte B, Paiva F, Cacador I, Canning-Clode J. Microplastics as vector for heavy metal contamination from the marine environment. Estuar Coast Shelf Sci. 2016, 178: 189–195.

14 Abbasi S, Moore F, Keshavarzi B, Hopke PK, Naidu R, Rahman MM, Oleszczuk P, Karimi J. PET-microplastics as a vector for heavy metals in a simulated plant rhizosphere zone. Sci Total Environ. 2020, 744: 140984.

15 Zhou Y, Liu X, Wang J. Characterization of microplastics and the association of heavy metals with microplastics in suburban soil of central China. Sci Total Environ. 2019, 694: 133798.

16 Ashton K, Holmes L, Turner A. Association of metals with plastic production pellets in the marine environment. Marine Pollution Bulletin 2010, 60 (11): 2050–2055.

17 Tang S, Lin L, Wang X, Feng A, Yu A. Pb(II) uptake onto nylon microplastics: interaction mechanism and adsorption performance. Journal of Hazardous Materials 2020, 386: 133798.

18 Kalčíková G, Skalar T, Marolt G, Jemec Kokalj A. An environmental concentration of aged microplastics with adsorbed silver significantly affects aquatic organisms. Water Res. 2020, 175: 115644.

19 Jinhui S, Sudong X, Yan N, Xia P, Jiahao Q, Yongjian X. Effects of microplastics and attached heavy metals on growth, immunity, and heavy metal accumulation in the yellow seahorse, Hippocampus kuda Bleeker. Marine Pollut Bull. 2019, 149: 110510.

20 Kim D, Chae Y, An Y-J. Mixture toxicity of nickel and microplastics with different functional groups on daphnia magna. Environ Sci Technol. 2017, 51(21): 12852–12858.

21 Zhou Y, Liu X, Wang J. Ecotoxicological effects of microplastics and cadmium on the earthworm Eisenia foetida. J Hazard Mater. 2020, 392: 122273.

22 Zhang Y, Wolosker MB, Zhao Y, Ren H, Lemos B. Exposure to microplastics cause gut damage, locomotor dysfunction, epigenetic silencing, and aggravate cadmium (Cd) toxicity in Drosophila. Sci Total Environ. 2020, 744: 140979.

23 Oliveira M, Ribeiro A, Hylland K, Guilhermino L. Single and combined effects of microplastics and pyrene on juveniles (0 + group) of the common goby Pomatoschistus microps (Teleostei, Gobiidae). Ecol Indic. 2013, 34: 641–647.

24 Luís LG, Ferreira P, Fonte E, Oliveira M, Guilhermino L. Does the presence of microplastics influence the acute toxicity of chromium(VI) to early juveniles of the common goby (Pomatoschistus microps)? A study with juveniles from two wild estuarine populations. Aquat Toxicol. 2015, 164: 163–174.

25 Bhagat J, Nishimura N, Shimada Y. Toxicological interactions of microplastics/nanoplastics and environmental contaminants: current knowledge and future perspectives. J Hazard Mater. 2021, 405: 123913.

26 Ma D, Lu Y, Liang Y, Ruan T, Li J, Zhao C, Wang Y, Jiang G. A critical review on transplacental transfer of per- and polyfluoroalkyl substances: prenatal exposure levels, characteristics, and mechanisms. Environ Sci Technol. 2022, 56(10): 6014–6026.

27 Guo Y, Kannan K. Chapter 1 - analytical methods for the measurement of legacy and emerging persistent organic pollutants in complex sample matrices. In: Zeng EY (Ed.), Comprehensive Analytical Chemistry. Vol. 67. Elsevier, 2015, pp. 1–56.

28 Richard H, Carpenter EJ, Komada T, Palmer PT, Rochman CM. Biofilm facilitates metal accumulation onto microplastics in estuarine waters. Sci Total Environ. 2019, 683: 600–608.

29 Fisner M, Majer A, Taniguchi S, Bícego M, Turra A, Gorman D. Colour spectrum and resin-type determine the concentration and composition of Polycyclic Aromatic Hydrocarbons (PAHs) in plastic pellets. Marine Pollut Bull. 2017, 122(1): 323–330.

30 Mato Y, Isobe T, Takada H, Kanehiro H, Ohtake C, Kaminuma T. Plastic resin pellets as a transport medium for toxic chemicals in the marine environment. Environ Sci Technol. 2001, 35(2): 318–324.

31 Fu L, Li J, Wang G, Luan Y, Dai W. Adsorption behavior of organic pollutants on microplastics. Ecotoxicol Environ Saf. 2021, 217: 112207.

32 Tang Y, Rong J, Guan X, Zha S, Shi W, Han Y, Du X, Wu F, Huang W, Liu G. Immunotoxicity of microplastics and two persistent organic pollutants alone or in combination to a bivalve species. Environ Pollut. 2020, 258: 113845.

33 Chen Q, Yin D, Jia Y, Schiwy S, Legradi J, Yang S, Hollert H. Enhanced uptake of BPA in the presence of nanoplastics can lead to neurotoxic effects in adult zebrafish. Sci Total Environ. 2017, 609: 1312–1321.

34 Cormier B, Borchet F, Kärrman A, Szot M, Yeung LWY, Keiter SH. Sorption and desorption kinetics of PFOS to pristine microplastic. Environ Sci Pollut Res. 2022, 29(3): 4497–4507.

35 Deng Y, Yan Z, Shen R, Wang M, Huang Y, Ren H, Zhang Y, Lemos B. Microplastics release phthalate esters and cause aggravated adverse effects in the mouse gut. Environ Int. 2020, 143: 105916.

36 Li Z, Yi X, Zhou H, Chi T, Li W, Yang K. Combined effect of polystyrene microplastics and dibutyl phthalate on the microalgae Chlorella pyrenoidosa. Environ Pollut. 2020, 257: 113604.

37 Zhu Z-L, Wang S-C, Zhao F-F, Wang S-G, Liu F-F, Liu G-Z. Joint toxicity of microplastics with triclosan to marine microalgae Skeletonema costatum. Environ Pollut. 2019, 246: 509–517.

38 Liu J, Lu G, Xie Z, Zhang Z, Li S, Yan Z. Occurrence, bioaccumulation and risk assessment of lipophilic pharmaceutically active compounds in the downstream rivers of sewage treatment plants. Sci Total Environ. 2015, 511: 54–62.

39 Zhang S, Ding J, Razanajatovo RM, Jiang H, Zou H, Zhu W. Interactive effects of polystyrene microplastics and roxithromycin on bioaccumulation and biochemical status in the freshwater fish red tilapia (Oreochromis niloticus). Sci Total Environ. 2019, 648: 1431–1439.

40 Masó M, Garcés E, Pagès F, Camp J. Drifting plastic debris as a potential vector for dispersing Harmful Algal Bloom (HAB) species. Sci Mar. 2003, 67(1): 107–111.

41 Zettler ER, Mincer TJ, Amaral-Zettler LA. Life in the "plastisphere": microbial communities on plastic marine debris. Environ Sci Technol. 2013, 47(13): 7137–7146.

42 Bowley J, Baker-Austin C, Porter A, Hartnell R, Lewis C. Oceanic hitchhikers – assessing pathogen risks from marine microplastic. Trends Microbiol. 2021, 29(2): 107–116.

43 Zhang P, Liu Y, Zhang L, Xu M, Gao L, Zhao B. The interaction of micro/nano plastics and the environment: effects of ecological corona on the toxicity to aquatic organisms. Ecotoxicol Environ Saf. 2022, 243: 113997.

44 Frère L, Maignien L, Chalopin M, Huvet A, Rinnert E, Morrison H, Kerninon S, Cassone A-L, Lambert C, Reveillaud J, Paul-Pont I. Microplastic bacterial communities in the Bay of Brest: influence of polymer type and size. Environ Pollut. 2018, 242: 614–625.

45 Lyons MM, Ward JE, Gaff H, Hicks RE, Drake JM, Dobbs FC. Theory of island biogeography on a microscopic scale: organic aggregates as islands for aquatic pathogens. Aquat Microb Ecol. 2010, 60(1): 1–13.

46 Fabra M, Williams L, Watts JEM, Hale MS, Couceiro F, Preston J. The plastic Trojan horse: biofilms increase microplastic uptake in marine filter feeders impacting microbial transfer and organism health. Sci Total Environ. 2021, 797: 149217.

47 Rotjan RD, Sharp KH, Gauthier AE, Yelton R, Lopez EMB, Carilli J, Kagan JC, Urban-Rich J. Patterns, dynamics and consequences of microplastic ingestion by the temperate coral, Astrangia poculata. Proc Royal Soc B: Biol Sci. 2019, 286(1905): 20190726.

48 Polhill L, de Bruijn R, Amaral-Zettler L, Praetorius A, van Wezel A. Daphnia magna's favorite snack: biofouled plastics. Environ Toxicol Chem. 2022, 41(8): 1977–1981,

49 Yan M, Li W, Chen X, He Y, Zhang X, Gong H. A preliminary study of the association between colonization of microorganism on microplastics and intestinal microbiota in shrimp under natural conditions. J Hazard Mater. 2021, 408: 124882.

50 Fries E, Dekiff JH, Willmeyer J, Nuelle M-T, Ebert M, Remy D. Identification of polymer types and additives in marine microplastic particles using pyrolysis-GC/MS and scanning electron microscopy. Environ Sci Process Impact. 2013, 15(10): 1949–1956. doi:10.1039/C3EM00214D.

51 Li X, He E, Xia B, Van Gestel CAM, Peijnenburg WJGM, Cao X, Qiu H. Impact of CeO2 nanoparticles on the aggregation kinetics and stability of polystyrene nanoplastics: importance of surface functionalization and solution chemistry. Water Res. 2020, 186: 116324.

52 Li P, Zou X, Wang X, Su M, Chen C, Sun X, Zhang H. A preliminary study of the interactions between microplastics and citrate-coated silver nanoparticles in aquatic environments. J Hazard Mater. 2020, 385: 121601.

53 Sun Y, Qian Y, Geng S, Wang P, Zhang L, Yang Z. Joint effects of microplastics and ZnO nanoparticles on the life history parameters of rotifers and the ability of rotifers to eliminate harmful phaeocystis. Chemosphere. 2023, 310: 136939.

54 Khoshnamvand M, Hanachi P, Ashtiani S, Walker TR. Toxic effects of polystyrene nanoplastics on microalgae Chlorella vulgaris: changes in biomass, photosynthetic pigments and morphology. Chemosphere. 2021, 280: 130725.

55 Das S, Thiagarajan V, Chandrasekaran N, Ravindran B, Mukherjee A. Nanoplastics enhance the toxic effects of titanium dioxide nanoparticle in freshwater algae Scenedesmus obliquus. Comp Biochem Physiol Part C Toxicol Pharmacol. 2022, 256: 109305.

56 Tong L, Duan P, Tian X, Huang J, Ji J, Chen Z, Yang J, Yu H, Zhang W. Polystyrene microplastics sunlight-induce oxidative dissolution, chemical transformation and toxicity enhancement of silver nanoparticles. Sci Total Environ. 2022, 827: 154180.

57 Zhang W, Song K, Ding R, Han H, Yao L, Ji M, Chen Z, Yu H, Wu C, Fang T. Role of polystyrene microplastics in sunlight-mediated transformation of silver in aquatic environments: mechanisms, kinetics and toxicity. J Hazard Mater. 2021, 419: 126429.

58 Dedeh A, Ciutat A, Treguer-Delapierre M, Bourdineaud J-P. Impact of gold nanoparticles on zebrafish exposed to a spiked sediment. Nanotoxicology. 2015, 9(1): 71–80.

59 Pacheco A, Martins A, Guilhermino L. Toxicological interactions induced by chronic exposure to gold nanoparticles and microplastics mixtures in Daphnia magna. Sci Total Environ. 2018, 628–629: 474–483.

60 Zocchi M, Sommaruga R. Microplastics modify the toxicity of glyphosate on Daphnia magna. Sci Total Environ. 2019, 697: 134194.

61 Grgić I, Cetinić KA, Karačić Z, Previšić A, Rožman M. Fate and effects of microplastics in combination with pharmaceuticals and endocrine disruptors in freshwaters: Insights from a microcosm experiment. Sci Total Environ. 2023, 859: 160387. doi:https://doi.org/10.1016/j.scitotenv.2022.160387; Wagstaff A, Lawton LA, Petrie B. Polyamide microplastics in wastewater as vectors of cationic pharmaceutical drugs. Chemosphere. 2022, 288: 132578.

62 Hasegawa T, Mizukawa K, Yeo BG, Sekioka T, Takada H, Nakaoka M. The significance of trophic transfer of microplastics in the accumulation of plastic additives in fish: an experimental study using brominated flame retardants and UV stabilizers. Mar Pollut Bull. 2022, 185: 114343.

63 Cheng H, Luo H, Hu Y, Tao S. Release kinetics as a key linkage between the occurrence of flame retardants in microplastics and their risk to the environment and ecosystem: a critical review. Water Res. 2020, 185: 116253.

64 Vieira Y, Lima EC, Foletto EL, Dotto GL. Microplastics physicochemical properties, specific adsorption modeling and their interaction with pharmaceuticals and other emerging contaminants. Sci Total Environ. 2021, 753: 141981.

65 Horton AA, Vijver MG, Lahive E, Spurgeon DJ, Svendsen C, Heutink R, van Bodegom PM, Baas J. Acute toxicity of organic pesticides to Daphnia magna is unchanged by co-exposure to polystyrene microplastics. Ecotoxicol Environ Saf. 2018, 166: 26–34.

15

Combined Molecular Toxicity Mechanism of Flame Retardant Mixtures

Xinyan Li[a, b, c], *Tiangang Luan*[b, c, d,*]

[a] *School of Biomedical and Pharmaceutical Sciences, Guangdong University of Technology, Guangzhou 510006, China*
[b] *Guangdong Provincial Laboratory of Chemistry and Fine Chemical Engineering Jieyang Center, Jieyang 515200, China*
[c] *Smart Medical Innovation Technology Center, Guangdong University of Technology, Guangzhou 510006, China*
[d] *School of Environmental and Chemical Engineering, Wuyi University, Jiangmen 529020, China*

*Corresponding author.
E-mail: cesltg@mail.sysu.edu.cn

15.1 Introduction

Flame retardants (FRs) are chemical additives in products preventing combustion and fire spread. FRs have been used in various industries, such as electronics [1], building materials [2], plastics, textiles [3], and furniture [4]. Generally, FRs are classified as organic or inorganic. Although most inorganic FRs are typically considered low in toxicity and have been rarely detected in the environment [5, 6], the superior flame retardance and processing performance of organic FRs make them more widely applicable [1, 7]. As a consequence, organic FRs were detected frequently in water, soil, air [8–10], indoor dust [11], and in biological samples from marine animals [12] and humans, including serum [13], urine, and breast milk [8], posing a great threat to the public health, which have generated concern and attention [6, 14]. The most widely used organic FRs are brominated flame retardants (BFRs) and organophosphorus flame retardants (OPFRs) [1, 8, 15, 16]. Starting from the late 1990s, BFRs gradually withdrew from the market because of their persistence, bioaccumulation, long-range atmospheric transport, and various toxicity evidence, including cancer induction, endocrine disruption [17], fatty liver disease [18], and reproductive toxicity [1, 4, 8–10, 19]. In particular, polybrominated diphenyl ethers (PBDEs) and polychlorinated biphenyl (PCBs) have been prohibited in the global market. However, owing to the continuous release during recycling processes or from residual products [20], BFRs have been widely detected in the environment. To substitute the BFRs, OPFRs have emerged as the dominant FRs over the last decade, owing to their lower cost, less bioaccumulation, and reduced persistence [1, 3, 19, 21]. Correspondingly, OPFRs, like their predecessors, have been detected in various environmental media and human samples [16, 22].

Considering the growing evidence of FR exposure in humans, requests for further investigation and comprehensive understanding have been put forward. It is important to recognize that current toxicity studies of FRs mainly focus on individual exposure, whereas the toxic effects in natural environments are an outcome of complex interactions among various pollutants [3, 16, 19]. Combined toxicity occurs when pollutants interact with each other, resulting in harmful effects on living organisms or ecosystems and can be categorized as synergistic, antagonistic, or additive [23–25]. In evaluating the combined toxic effects of chemical mixtures, a common approach is to compare the observed toxicity to the predicted toxicity, which is calculated by adding the individual toxicities of each chemical in the mixture. Synergism, additivity, and antagonism are observed when the mixture's toxicity is greater than, equal to, or less than the predicted toxicity, respectively. Considering the ability of FRs to bind with or absorb other environmental pollutants, the potential toxicity suggests a greater or concealed threat posed by FR mixtures. For example, co-exposure to BDE-47 and BDE-209 was found to enhance the immunotoxicity on *Daphnia magna* [26], while commercial FR FM550, containing both BFRs and OPFRs, was found to aggravate the neurodevelopmental toxicity [27, 28], endocrine disruption [29–31], and growth inhibition [32] in mammalian models. In addition, studies also discussed the

Toxicological Assessment of Combined Chemicals in the Environment, First Edition. Edited by De-Sheng Pei and Yiyun Liu.

combined toxicity between FRs and other pollutants, e.g., microplastics, graphene oxide, and nano-TiO_2. For example, the combination of nano-TiO_2 and tert-butyl hydroperoxide (TBHP) was found to antagonize lipid metabolism disorders in zebrafish embryos [33], while co-exposure to MPS and BDE-47 increased oxidative stress in mussels [34]. These findings highlight the importance of investigating the combined toxicity of FR mixtures and elucidating the underlying mechanisms.

The objective of this chapter is to provide a comprehensive overview of the combined toxicity and underlying mechanisms of organic flame-retardant mixtures in biological systems, including neurotoxicity, DNA damage, oxidative stress, endocrine disruption, reproductive toxicity, and growth and developmental inhibition. The detailed information on the FRs discussed in this chapter are provided in Table 15.1. Given the complexity of organic flame-retardant mixtures, we address their synergistic and antagonistic effects, as well as their interactions with other environmental pollutants, highlighting the current limitations of research and proposing avenues for future development.

15.2 Synergistic Effects of OFR Co-exposure on Oxidative Stress and DNA Damage

Exposure to OFR mixtures may induce oxidative stress, which is characterized by the excessive production of reactive oxygen species (ROS), malondialdehyde (MDA), and lipid peroxidation (LPO), that can attack various intracellular and extracellular biological molecules [44]. This disruption of the cellular redox balance can lead to a range of physiological and pathological reactions, including DNA damage [45], cell cycle dysregulation [46], and interference with immune, endocrine, and neurodevelopmental responses, posing significant risks to the body. Table 15.2 summarizes the detailed information on oxidative stress and the corresponding mechanisms induced by co-exposure to OFRs.

15.2.1 Oxidative Stress

Exposure to organic FRs can increase the generation of intracellular ROS, which may attack biological molecules such as proteins [60], lipids [61], and nucleic acids [62], leading to oxidative damage and cell dysfunction. Kling et al. found that exposure of tetrabromobisphenol A (TBBPA) and hexabromocyclododecane (HBCD) to zebrafish liver (ZFL) cells upregulated the expression of isocitrate dehydrogenase 2 (IDH2) [55], a nicotinamide adenine dinucleotide phosphate (NADP)-dependent enzyme, consuming NADP and converting into NADPH during catalysis. Then NADPH oxidase (NOX) catalyzes the oxidation of NADPH to $NADP^+$, releasing ROS in the process. Therefore, excessive NOX activity is considered as a warning sign of oxidative stress. Additionally, Albano et al. observed exposure to a mixture of $CdCl_2$/BDEs activated NOX-4 in human lung adenocarcinoma cells, resulting in the inhibition of Nrf2 and TFAM proteins. Nrf2 is an oxidative stress–sensitive transcription factor for antioxidant defense, while TFAM is responsible for maintaining the integrity of the mitochondrial respiratory chain and balance between antioxidant and oxidative processes. This may represent an indirect mechanism by which co-exposure enhances oxidative stress by increasing NOX activity and inhibiting cellular antioxidant defense. Similarly, DE-71 (a mixture of BDE-47, BDE-99, and BDE-153) [63] was found to activate NOX and increase ROS production in neutrophils by activating tyrosine kinase (TK), phosphoinositide 3-kinase (PI3k), protein kinase C (PKC), and phospholipase C (PLC) [52]. TK is a crucial intracellular signaling molecule involved in various cellular physiological and pathological processes, and studies have shown that its activation may lead to an increase in NOX. PI3K is involved in regulating intracellular signaling transduction and metabolic pathways and can increase the expression of NOX by activating the Akt/mTOR signaling pathway. The activation of PLC can lead to the breakdown of PI into inositol-1,4,5-triphosphate (IP3) and diacylglycerol (DAG) in the cell. IP3 promotes the release of intracellular calcium ions, which activate enzyme molecules such as PKC and CaMKII, while DAG can directly activate PKC. Furthermore, the literature suggests that PKC can activate NOX directly or through transcription factors such as nuclear factor-κB (NF-κB) and AP-1. For example, Giordano et al. [53] reported that exposure to DE-71 significantly reduced intracellular GSH levels and increased ROS and MDA levels in primary mouse astrocytes, possibly due to PKC translocation, a critical step for PKC activation.

In addition to ROS overproduction, co-exposure to OFRs can also damage the antioxidant defense system by decreasing the activity and expression of antioxidant enzymes. Enhanced synergistic oxidative stress of FR mixture has been observed in aquatic organisms and mammals. For instance, BDE-209 and DBDPE were found to suppress the activity of superoxide dismutase (SOD), catalase (CAT), and glutathione peroxidase (GPx), inducing higher oxidative stress in freshwater fish [51], while long-term treatment with BDE-47 and BDE-209 inhibited the activity of glutathione S-transferase (GST) in large flea [26]. Furthermore, exposure to other pollutants such as heavy metals and microparticles/nanoparticles can also promote oxidative stress. Co-exposure to graphene and TPHP was found to increase ROS and 8-hydroxy-deoxyguanosine (8-OHDG) in zebrafish and downregulate the level of gstp1 protein, leading to a significant decrease in GST enzyme

Table 15.1 Name, structure, properties, and bioconcentration factors of organic flame retardants (OFRs) discussed in this chapter.

Classification	Compounds	Abbreviation	Structure	*S* (mg/L)	Molecular mass	Log K_{ow}	BCF	Ref.
PBDEs	Tetrabromodiphenyl ether to hexabromobiphenyl ether	BDE-47, BDE-99, BDE-100, BDE-153, BDE-154, BDE-183 (Take BDE-47 as an example)	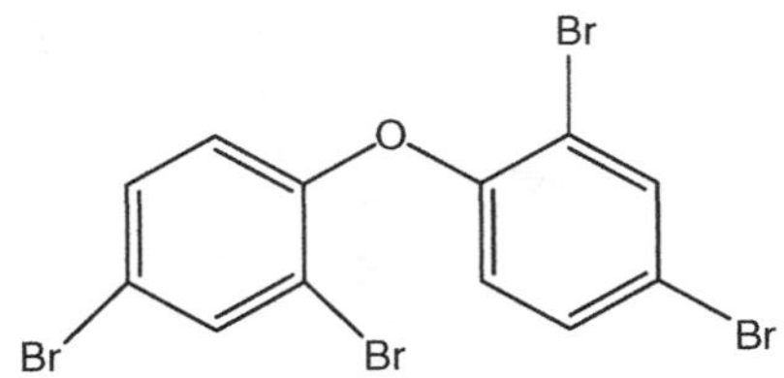	5×10^{-5}	485.8	5.9–6.2	1×10^{5} to 3×10^{5}	[35]
	Decabromodiphenyl ether	BDE-209	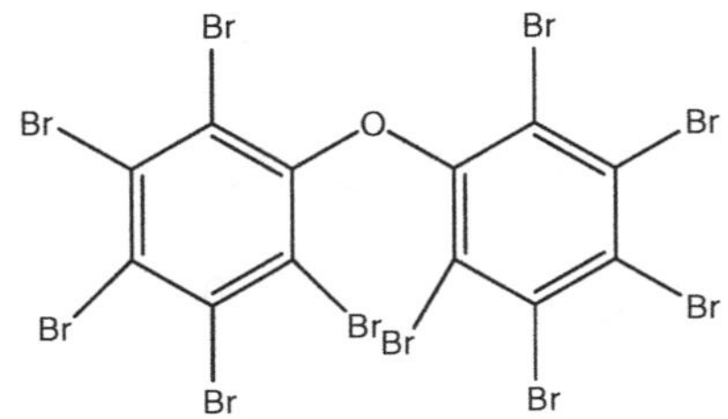	2×10^{-11}	959.2	9.4	—	[35]
PCBs	Polychlorinated biphenyl	PCB-153 (Take PCB-153 as an example)	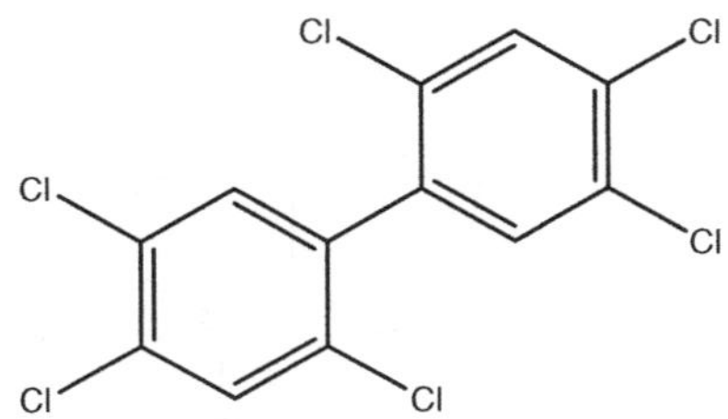	1.6×10^{-11}	257.5	6.5	1×10^{5} -1×10^{6}	[36]
TBBPA	Tetrabromobisphenol A	TBBPA	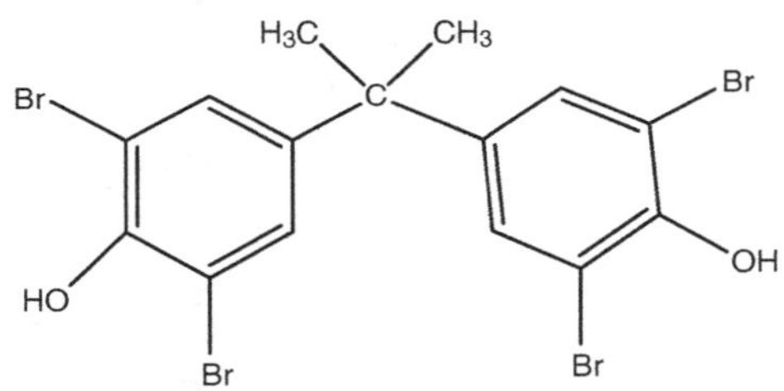	0.5–5	543.9	4.5-5.3	30–485	[37]

(Continued)

Table 15.1 (Continued)

Classification	Compounds	Abbreviation	Structure	S (mg/L)	Molecular mass	Log K_{ow}	BCF	Ref.
Monoaromatic hydrocarbon NBFRs	Bis(2-ethylhexyl)-2,3,4,5-tetrabromophthalate	TBPH	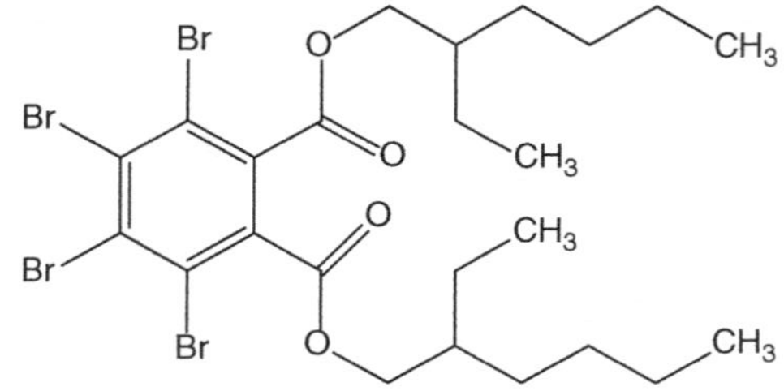	1×10^{-4}	706.1	11.9	—	[38]
	2-Ethylhexyl 2,3,4,5-tetrabromobenzoate	TBB	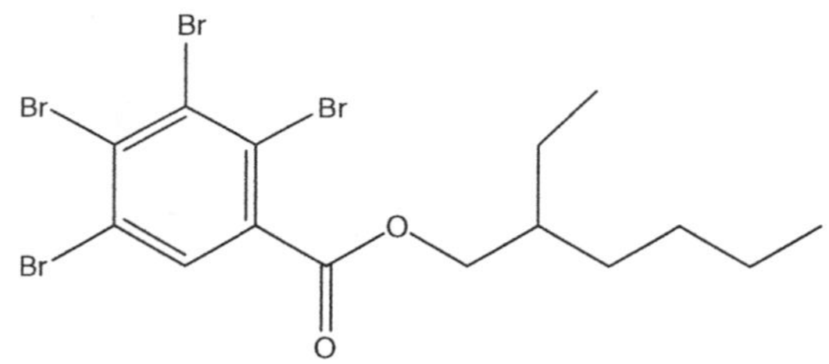	1×10^{-4}	543.6	7.2	16–2000	[39]
Polyaromatic hydrocarbons NBFRs	Decabromodiphenyl ethane	DBDPE	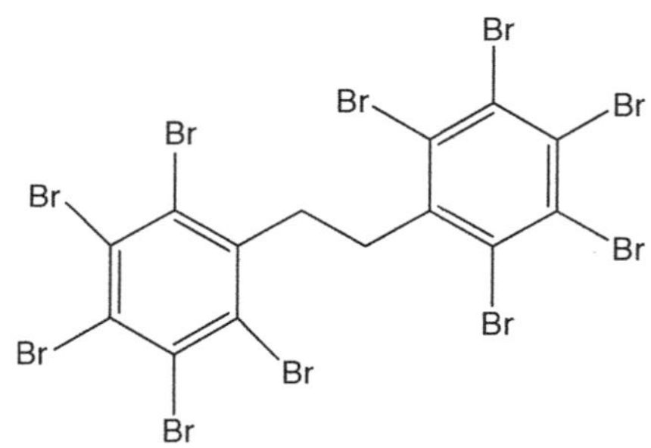	0.01–0.1	971.2	13.6	—	[38]
	Hexabromocyclododecane	HBCD	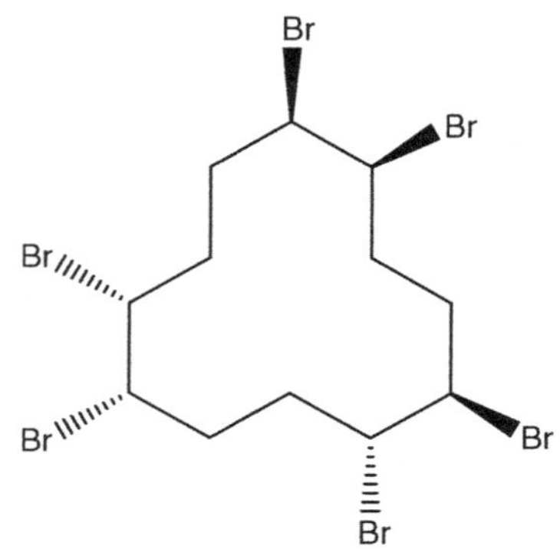	0.5	641.7	5.9	6×10^3 -9×10^3	[37]

Chloro OPFRs	Tris(2-chloroethyl) phosphate	TCEP	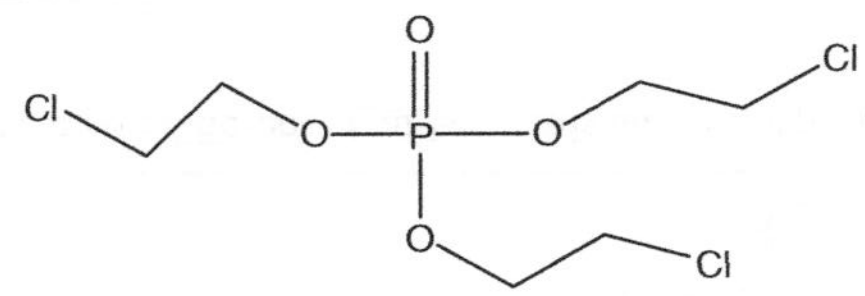	10–20	286.7	1.4	1.37	[4]
	Tris(1,3-dichloroisopropyl) phosphate	TDCPP	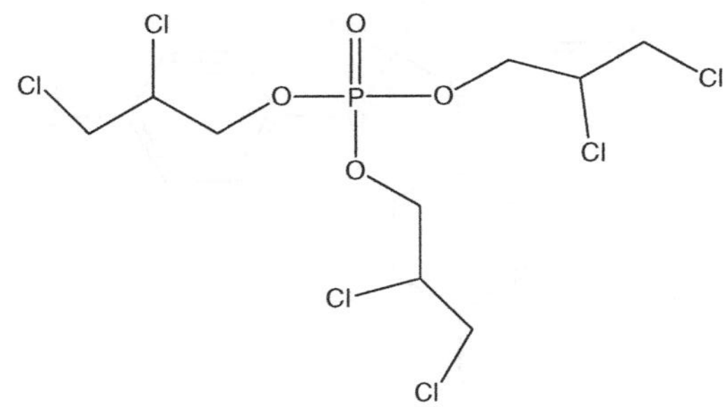	1.5	430.9	3.8	13.5	[4]
Alkyl OPFRs	Tripropyl phosphate	TPHP	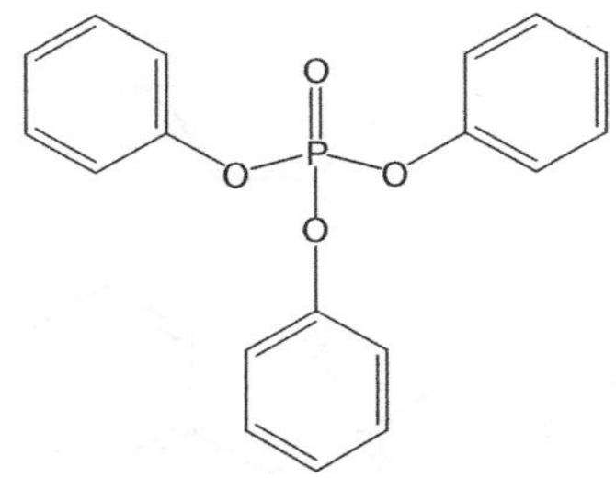	0.029	326.3	2.7	63.1	[4]
	Cresyl diphenyl phosphate	CDP	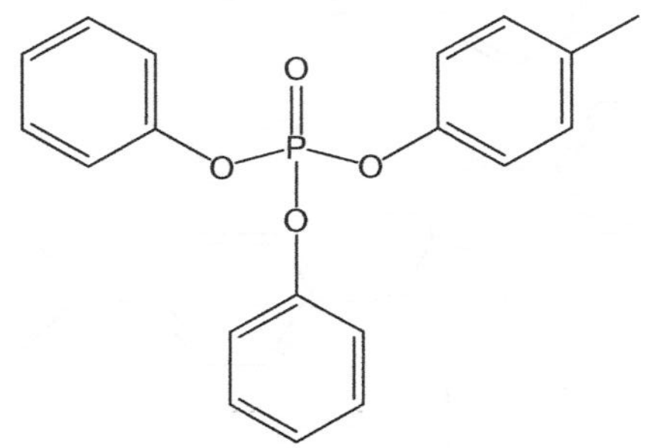	0.234	340.3	5.3	364	[40]

(Continued)

Table 15.1 (Continued)

Classification	Compounds	Abbreviation	Structure	S (mg/L)	Molecular mass	Log K_{ow}	BCF	Ref.
	Phenol, isopropylated, phosphate	IPPP	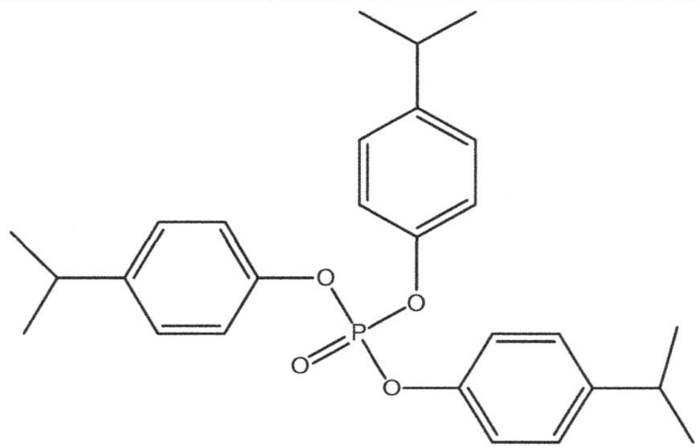	0.4	452.5	0.1	—	[41]
	2-Ethylhexyldiphenyl phosphate	EHDPP	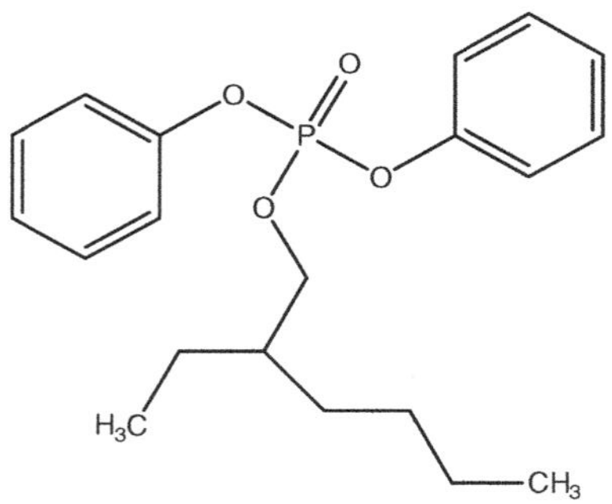	—	362.4	5.7	269.2	[42]
	Tricresyl phosphate	TCP	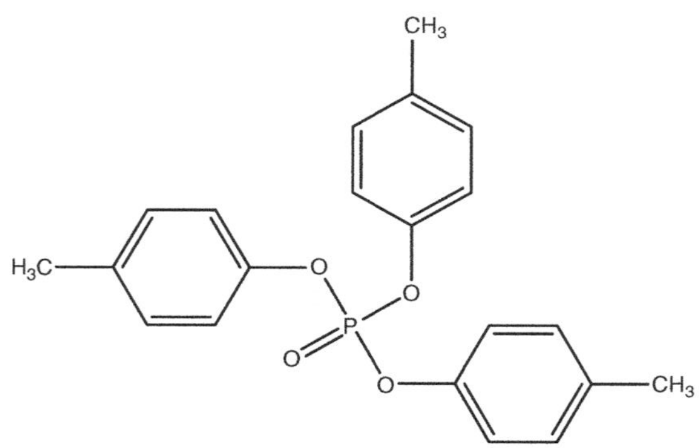	Insoluble	368.4	5.1	—	[43]

S, solubility; BCF, bioconcentration factors, the specific data depending on different organisms, this table only provides reference information; —, no data.

Table 15.2 Synergistic effects of organic flame retardant (OFR) co-exposure on oxidative stress and DNA damage.

Type of FRs	Dose and time	Organisms	Phenotypes	Mechanisms	Ref.
BDE-47	1 mg/L	*Epinephelus moara*	Increasing MDA; promoting CAT activity; inhibiting GSH and SOD activity	Downregulating *SLC3A2*	[47]
BDE-209	1 mg/L				
MP	750 mg/L				
BDE-47	2, 4 μg/L	Large flea	Inhibiting GST activity	Downregulating *GSTP1*	[26]
BDE-209	400 μg/L				
BDE-47	100 nM	Human lung adenocarcinoma cells	Increasing ROS	Downregulating Nrf2 and TFAM; Increasing NOX-4	[48]
BDE-209	100 nM				
$CdCl_2$	100 nM				
BDE-47	39 mg/L	THP-1 cell	Increasing ROS;	Downregulating glutathione biosynthesis; Upregulating nucleotide metabolism	[49]
HBCD	34 mg/L				
TBBPA	22 mg/L				
Cu^{2+}	5 mg/L				
Cd^{2+}	5 mg/L				
Hg^{2+}	5 mg/L				
BDE-99	200 μg/mL	Rat	Increasing LPO, H_2O_2, NO; Decreasing SOD, GPx, and GSH activity	Downregulating *SOD1, SOD2*; Downregulating *GSTP1*	[50]
MeHg	20 μg/mL				
BDE-209	10, 50, 100 mg/kg	Crucian	Decreasing GSH; Increasing MDA; Inhibiting SOD, CAT, and GPx activity	Downregulating *SOD1, SOD2*; Downregulating *GSTP1*; Downregulating *CAT*	[51]
DBDPE	10, 50, 100 mg/kg				
DBDPE	25 mg/L	Human neuroblastoma SK-N-SH cells	Increasing ROS; Inhibiting SOD and GSH activity; Causing mitochondrial fusion and fission imbalance	Inhibiting Nrf2 signaling pathway; Downregulating the expression of downstream target genes *HO-1, NQO1, Keap1*, and *GCLC*	[45]
ZnO NPs	50 mg/L				
DE-71	1–12 μM	Neutrophils	Increasing ROS; Activating NADPH oxidase	Activating tyrosine kinases, PI3-kinase, PKC, and PLC signaling pathway	[52]
DE-71	25 μM	Primary mouse astrocyte	Increasing ROS and MDA; Significantly decreasing GSH; Altering cytoplasmic and mitochondrial calcium homeostasis	Increasing translocation of PKC	[53]

(Continued)

Table 15.2 (Continued)

Type of FRs	Dose and time	Organisms	Phenotypes	Mechanisms	Ref.
DE-71	0.36, 4.64, 51.4 μg/μL	Common terns American kestrels	Increasing 8-OHDG; Leading to DNA strand breaks	Increasing ROS	[54]
TBBPA	0.5, 5, 187 μM	ZFL cell	Increasing NADPH oxidase; Increasing ROS	Upregulating expression of IDH2	[55]
HBCD	0.5, 5, 220 μM				
TBBPA	0.54-81.6 μg/mL	Rainbow trout gill cells	Increasing NADPH oxidase; Increasing ROS; Decreasing the percentage of tail DNA	Interfering mitochondrial electron transport chain	[56]
PSNPs	200 μg/mL				
TBBPA	16 $\mu g/m^3$	Mice	Increasing MDA; Promoting SOD and CAT activity; Downregulating GPx expression; Inducting LDH	Downregulating GPx2 and SOD	[57]
Cd	8 $\mu g/m^3$				
TBPH	744.7 ± 85.97 μg/g	Pimephales promelas	Increasing ROS	Activating the PPAR signaling pathway	[58]
TBB	1658 ± 198.9 μg/g				
TPHP	5, 20, 50 μg/L	Zebrafish larvae	Increasing ROS, 8OHDG; Inhibiting GST activity	Downregulating expression of *Gstp1* Activating mTOR protein	[59]
Graphene	10, 100 μg/L				

hpf, Hour postfertilization; dpf, days postfertilization; ROS, reactive oxygen species; GSH, glutathione; GPx, glutathione peroxidase; GR, glutathione reductase; ACP, acid phosphatase; LDH, lactate dehydrogenase; SOD, superoxide dismutase; MDA, malondialdehyde; GST, glutathione S-transferase; CAT, catalase; LPO, lipid peroxidation; 8OHDG, 8-hydroxy-deoxyguanosine; HO-1, heme oxygenase-1; NQO1, NAD(P)H quinone oxidoreductase 1; Keap1, Kelch-like ECH-associated protein 1; GCLC, glutathione cysteine ligase regulatory subunit; IDH2, isocitrate dehydrogenase 2.

activity, demonstrating the disruption of antioxidant system. The researchers also found that co-exposure activated the mTOR pathway, which may be a mechanism of oxidative stress–induced DNA damage in zebrafish. mTOR is a protein kinase involved in cell atrophy, apoptosis, and cell death, and its activation can lead to p53 protein phosphorylation, transcriptional activation of pro-apoptotic proteins (e.g., Bax family), and mitochondrial cell death pathway activation. In addition, Li et al. reported that co-exposure to microplastics (MP) and BDEs downregulated the expression of *SLC3A2*, a cysteine/glutamate antiporter gene, contributing to GSH antioxidant system inhibition in *Epinephelus moara* [47]. Similarly, DBDPE and nanoparticles of zinc oxide (ZnO-NPs) downregulated Nrf2 signaling pathway and downstream antioxidant genes. Nrf2 could translocate to the nucleus upon exposure to oxidative stress and trigger the production of antioxidant enzymes to restore redox balance [45]. Therefore, downregulation of Nrf2 and its downstream antioxidant genes can inhibit the antioxidant system. Similar results were observed in co-exposure with BDE-99 and methylmercury [50], TBBPA, and Cd [57], as well as BDE-47, HBCD, and TBBPA with Cu^{2+}, Cd^{2+}, and Hg^{2+} [49]. These findings underscore the potential risks of oxidative stress enhancement induced by combined treatment of FR mixtures.

15.2.2 Mitochondria Damage and DNA Damage

Mitochondria are crucial organelles in cells, mainly responsible for energy metabolism. The abundance of oxidases and mitochondrial DNA molecules makes mitochondria one of the primary targets for oxidative damage [64–66]. Giordano et al. found that the BFR mixture DE-71 disrupted mitochondrial calcium homeostasis, causing an imbalance in redox reactions and ion concentrations within mitochondria, thereby inducing oxidative stress [53]. Additionally, ZnO-NPs were reported to synergistically enhance the imbalance of mitochondrial fusion and fission caused by DBDPE, resulting in abnormal mitochondrial morphology [45]. Notably, the excessive generation of ROS has been confirmed as a major cause for DNA damage, including DNA strand breakage [49] and base damage [54]. For example, treating rainbow trout gill cells with TBBPA and polystyrene nanoplastics (PSNPs) mixtures resulted in cellular redox imbalance and subsequent DNA damage [56]. Moreover, Zhang et al. reported that the combination of graphene and TPHP activated mTOR and apoptosis-promoting signaling pathways, resulting in excessive ROS and subsequent DNA damage [59]. Similarly, co-exposure to TBPH and TBB synergistically activated the peroxisome proliferator activated receptor (PPAR) pathway through oxidative stress, leading to significant DNA strand breakage in liver cells of *Pimephales promelas* [58]. The DNA damage caused by OFRs may lead to various adverse effects, including apoptosis, mutations, and carcinogenesis, which necessitate further exploration of molecular mechanisms.

15.3 Synergistic Effects of OFR Co-exposure on Endocrine Disruption and Reproduction Toxicity

Studies have indicated that prolonged exposure to organic FRs may damage the endocrine system of organisms. Some OFRs such as TPHP, TDCPP, BDE-47, and BDE-209 have been classified as endocrine- disrupting chemicals (EDCs). These EDCs can impair the normal function of the endocrine system by mimicking or interfering with hormone activity [67–69]. In this section, we will, respectively, discuss the effects of OFRs on thyroid hormones and sex hormones, as well as the corresponding pathways involved.

15.3.1 Thyroid Hormone Disturbance

Organic flame-retardant mixtures have been found to interfere with the synthesis and secretion of hormones. FM 550, a mixture of BFRs (TBHP, TBB) and OPFRs (TPHP, IPPP), was observed to decrease the thyroid hormone level in the serum of rats [31]. Similarly, Stoker et al. [70] reported that DE-71 reduced thyroid hormones and luteinizing hormones in rats by activating hepatic CYP1A1, CYP2B, and T4-UDGPT, which was also supported by the results that DE-71 decreased total triiodothyronine (T3) levels by upregulating CYP1A1 in mink [71]. Apart from the activation of the CYP family, some studies have found that combined OFRs affected the synthesis and metabolism of thyroid hormones via hypothalamic–pituitary–thyroid (HPT) axis. The HPT axis stimulates the hypothalamus to secrete thyrotropin-releasing hormone (TRH), which in turn stimulates the anterior pituitary gland to release thyroid-stimulating hormone (TSH), encouraging the release of thyroid hormones (T3 and T4) [72, 73]. Zhou et al. found that exposing zebrafish larvae to BDE-209 and Nano-TiO_2 upregulated genes (*tshβ, tg, dio2*) in the HPT axis, leading to a significant increase in total thyroid hormones (T3 and T4) [74]. Specifically, the *tshβ* gene, responsible for TSH coding, is located in TRH neurons of hypothalamus and is positively correlated with thyroid hormone production. Located in the thyroid gland, the *tg* and *dio2* have different functions: the former encodes thyroglobulin (TG), which is in charge of coupling iodinated tyrosine and tyrosine to produce T3 and T4, while the latter encodes the endogenous thyroid hormone metabolism enzyme DIO2, converting T4 into the more active T3 [74]. However, a recent study by Zhang et al. reported opposite findings that co-exposure to PCBs and MPs significantly decreased T4 levels in Japanese halibut by suppressing the expression of key genes involved in HPT axis, such as *tshβ, typo*, and *crh* [75]. The corticotropin-releasing hormones (CRH) secreted from the hypothalamus share the same mechanism as the HPT axis to release thyroid hormones, and thyroid peroxidase (TPO) is a crucial enzyme for thyroid hormone synthesis [76]. The additional MPs aggravated not only the reduction of these genes but also the damage of fish gill tissue. Furthermore, the inhibition of oxygen consumption and energy metabolism induced by gill injury may also be blamed for the decreased synthesis of thyroid hormones.

Studies have also indicated that OFRs could alter the responsiveness to hormones by acting on hormone receptors. In the case of transthyretin (TTR), the principal thyroid hormone–binding protein in fish species, it is susceptible to environmental pollutants since it can be competitively bound or have its transport capacity inhibited by pollutants [77, 78]. Zhao et al. discovered that compared to a single exposure to BDE-47, the presence of perchlorate (PER) aggravated the downregulation of TTR as well as the inhibition of thyroglobulin (TG) in zebrafish larvae, resulting in reduced thyroid hormone synthesis [79]. Arkoosh et al. found that a mixture of BDE-47 and BDE-99 competed with T3 and T4, binding to TTR in salmon, also leading to a decrease of thyroid hormone in plasma [80].

15.3.2 Sex Hormone Disturbance and Reproductive Toxicity

Apart from thyroid hormone disorders, OFR mixtures have also been reported to disrupt sex hormones, including estrogen and androgen. As a result, the development and function of the reproductive system may exhibit varying degrees of abnormality, such as abnormal development of the testes and ovaries, genital malformations, infertility, and precocious puberty [81]. Lefevre et al. found that combined exposure to DE-71, DE-79, and BDE-209 in female rats resulted in elevated testosterone levels, increasing the potential risk of sex reversal [82]. They postulated that this may be due to the inhibited expression of 17α-hydroxylase (*Cyp17a1*), a rate-limiting enzyme for androgen synthesis. Additionally, significant downregulation of *Insl3* has been incriminated as another factor in increased testosterone levels in serum, since INSL3 can increase luteinizing hormone (LH) synthesis in the female placenta, further increasing testosterone production.

Some studies have shown that OFR mixtures could affect the synthesis and metabolism of sex hormones via the hypothalamus–pituitary–gonadal (HPG) axis, which plays an important role in regulating the development and function of the reproductive system. First, hypothalamus produces gonadotropin-releasing hormone (GnRH), which then stimulates the anterior pituitary to release follicle-stimulating hormone (FSH), luteinizing hormone (LH), and prolactin. Next, FSH and LH act on the testes in males and the ovaries in females to promote the synthesis and release of sex hormones [83]. Stoker et al. reported that DE-71 obviously increased progesterone, androstenedione, and testosterone levels in Wistar rats [70], which was attributed to *lhb* gene from the HPG encoding luteinizing hormone (LH). Once LH binds to its receptors in corpus luteum cells of the ovary, the synthesis and secretion of progesterone are triggered [84].

It is interesting to notice that in aquatic organisms, the endocrine-disrupting effects of combined exposure on female fish appear to be less obvious than those on male fish, probably because female fish have higher E2 content and are less susceptible to exogenous estrogen [85–87]. Saunders et al. found that combined exposure to TBHP and TBB significantly decreased FSH and LH in the plasma of Japanese medaka, which may be due to the downregulation of mRNA expression of genes in the HPG axis, including *GnRHs* (*cGnRH II, sGnRH*) and their receptors *GnRHRs* (*GnRH RI/II/III*), as well as the subsequent gonadotropin subunits, *GTHα* and *LH-β* [88]. Similarly, He et al. discovered that MPs or nanoplastics (NPs) synergistically enhanced reproductive toxicity compared to treatment with TPHP alone in zebrafish, resulting in enlarged gonads and reduced mature sperm and oocytes [89]. It was suggested that particles make OFRs easier to enter into the gonads, and the co-exposure induced over-activation of genes *vtg* and *CYP19A1*, both of which are involved in the HPG axis [90, 91]. CYP19A1 is an enzyme converting the male hormone testosterone into the female hormone estradiol, which is known as the aromatization. Therefore, the over-activation of *CYP19A1* resulted in the excessive estradiol (E2) and vitellogenin (vtg) in male fish, which may also be the reason for the significant decrease in the amount of spawning, fertilization rate, and hatching rate of zebrafish in the combined exposure group.

In some cases, OFRs may act as an alternative to hormones, binding to professional receptors, such as the estrogen receptor (ER) and androgen receptors (AR). Ren et al. [92] have demonstrated that the combined exposure of nanoscale TiO_2 and TDCPP results in a synergistic effect, leading to a significant reduction in the levels of the sex hormones estradiol (E2) and testosterone (T), compared to exposure to TDCPP alone. This effect may be attributed to the adsorption of TDCPP onto the surface of nanoparticles and forming aggregates with TiO_2. These aggregates, when attached to the skin of zebrafish, can promote the uptake of TDCPP through the skin and facilitate its entry into internal organs. The study has also revealed that TDCPP upregulates the expression of *vtg1* gene in the liver of zebrafish, leading to an increase in plasma VTG levels, particularly in female fish. VTG is considered as a biomarker of estrogen endocrine disruption, and the increased VTG levels resulted in a decrease in the number of eggs. Similar results have also been observed in mammals. Female rats treated with $CdCl_2$ and TBBPA were reported to increase uterine weight, decrease ovarian weight, change ovarian morphology, and inhibit follicular development, which are related to the combination of cadmium and ER [93]. Additionally, DE-71 or BFRs have been reported to cause an increase in anogenital distance and preputial separation in male rat offspring, while inhibiting mammary gland development in female offspring [94]. Research suggests that BFRs exhibit anti-androgenic and anti-estrogenic effects, leading to delayed preputial separation (PPS) in male offspring. PPS refers to the delayed or abnormal development of reproductive organs such as the penis, scrotum, and perineum during reproductive development [95]. This effect may be due to DE-71 inhibiting the binding of AR to testosterone, thereby reducing its response to circulating androgens or inhibiting the function of endogenous androgens.

Besides receptors of sex hormones, OFRs have also been observed to disturb other receptor pathways such as the farnesoid X receptor (FXR)/retinoid X receptor (RXR), and liver X receptor (LXR)/RXR, causing endocrine disruptions in organisms [30]. Recently, it has been discovered that FM550 can activate the retinoic acid receptor (RAR) and form heterodimers with RXR, leading to differential expression in the aryl hydrocarbon receptor (AHR). This may be the reason why male fetuses exhibit disrupted AHR signaling and glutamate degradation [96, 97]. The AHR signaling pathway is a sensitive indicator of environmental pollutants and endocrine disruptors [30].This suggests that exposure to FM550 during pregnancy may affect the development of the brain and placenta through the RXR-/RAR-mediated signaling pathway, indicating that FM550 may interfere with the normal functioning of the endocrine system. RXR is a nuclear receptor (NR) that forms heterodimers with other important NRs involved in endocrine and metabolic regulation as well as fetal development, especially the transport and metabolism of cholesterol and triglyceride [98]. Therefore, abnormal lipid metabolism can be an emerging clue to endocrine disruption (Table 15.3).

15.4 Synergistic Effects of OFR Co-exposure on Neurotoxicity

Co-exposure to OFRs has also been reported to induce neurodevelopmental toxicity in organisms, leading to impairments in motor skills, habits, learning, social abilities, anxiety, and neurodevelopmental disorders that compromise normal growth and development.

15.4.1 Neuronal Signal Transduction

Combined exposure to OFRs can disrupt the function of neurotransmitter receptors and ion channels on neuronal membranes, leading to a disturbance in communication between neurons and resulting in neurotoxicity. This mechanism of

Table 15.3 Endocrine disruption and reproduction toxicity of combined exposure to organic flame retardants (OFRs).

Type of FRs	Dose and time	Organisms	Phenotypes	Mechanisms	Ref.
BDE-47	0.7–690 ng/g	Salmon	Decreasing circulating thyroid hormone in plasma	Significantly reducing activity of DI; Inhibiting T3 and T4 binding to TTR	[80]
BDE-99	0.7–690 ng/g				
BDE-47	10 μg/L	Zebrafish embryos	Decreasing T4 levels	Significantly upregulating mRNA expression and protein level of TG; Significantly down-regulating TTR expression	[79]
PER	35 μg/L				
TBPH	1422 ± 156.4 μg/g	Japanese medaka	Decreasing the concentration of FSH and LH	Significantly downregulating the mRNA expression of GnRHs, GnRHRs, and gonadotropin subunits (*GTHα* and *LH-β*)	[88]
TBB	1474 ± 265.9 μg/g				
BDE-209	0.08, 0.38 mg/L	Zebrafish larvae	Significantly increasing T3 and T4	Upregulating several gene transcriptions (*tshβ, tg, dio2*) in the HPT axis	[74]
Nano-TiO_2	0.1 mg/L				
PCBs	0.5 μg/L	Juvenile Japanese halibut	Decreasing levels of L-thyroxine and L-tri-iodothyronine; Inducing thyroid follicular colloid defects; Severely damaging the thyroid tissue and gill morphology	Significantly downregulating expression of *TSHβ, TPO, ID1* and *ID2*;	[75]

(Continued)

Table 15.3 (Continued)

Type of FRs	Dose and time	Organisms	Phenotypes	Mechanisms	Ref.
MPs	2, 20, 200 μg/L				
TBBPA	50 mg/kg	Female rats	Increasing uterus weight; Decreasing ovarian weight; Altering the morphological structure of the ovaries; Inhibiting follicular development	Promoting binding of cadmium to the estrogen receptor	[93]
$CdCl_2$	1, 10, 100 mg/L				
DE-71	0.1, 0.5, 25 mg/L	Pasture mink	Significantly reducing total T3	Upregulating CYP1A1; Causing direct toxic effects on the thyroid gland; Altering hepatic phase II enzymes and deiodinases	[71]
DE-71	60 mg/kg	Rat	Significantly increasing luteinizing hormone; Nonsignificantly increasing androstenedione and testosterone; Significantly decreasing thyroid hormones;	Upregulating hepatic CYP1A1, CYP2B and T4-UDGPT activities; Activating *lhb* gene	[70]
DE-71	1.7, 10.2, 30.6 mg/kg	Long-Evans rats	Decreasing androgen; Increasing anogenital distance and causing foreskin separation;	Interfering the signaling of AR	[94]
DE-71	0.06, 20, 60 mg/kg	Female rat	Significantly decreasing androgen	Downregulating expression of *Insl3*; Downregulating expression of *Cyp17a1*	[82]
DE-79	0.06, 20, 60 mg/kg				
BDE-209	0.06, 20, 60 mg/kg				
TPHP	0.7, 1, 1.2, 1.5 mg/L	Adult zebrafish	Increasing secretion of E2 and vtg	Activating *vtg* gene and *CYP19A1* gene	[89]
MPS	2 mg/L				
NPS	2 mg/L				
TDCPP	5, 20, 80 μg/L	Zebrafish	Decreasing secretion of Testosterone, E2; Increasing plasma VTG levels	Significantly upregulating the expression of *vtg*1 in liver	[92]
Nano-TiO_2	0.1 mg/L				
FM 550	300, 1000 μg/day	Wistar rat	Degrading glutamate; Decreasing cholesterol metabolism	Interrupting AHR signal; Disrupting the RXR pathway; Interrupting the FXR/RXR, LXR/RXR pathways;	[30]
FM 550	100, 1000 μg/day	Rat	Increasing levels of serum thyroxine	Reducing hepatic activity of carboxylesterase	[31]

GD, gestation day; PND, postnatal day; T3, tri-iodothyronine; T4, thyroxine; TTR, thyroxine transporter; TG, thyroglobulin; CRH, corticotropin-releasing Hormone; TSHβ, thyroid-stimulating hormone beta subunit; HPT, hypothalamic–pituitary–thyroid; RAR, retinoic acid receptors; Atp5mc1, ATP synthase membrane subunit c locus 1; AR, androgen receptor; E2, estradiol; FSH, follicle-stimulating hormone; LH, luteinizing hormone; VTG, vitellogenin; Cyp17a1, 17α-hydroxylase, a rate-limiting enzyme for androgen synthesis; AHR, aryl hydrocarbon receptor.

combined toxicity is prominently manifested in mammalian experimental models, where multiple studies have demonstrated that OFRs affect enzymes involved in the degradation and metabolism of acetylcholine in synaptic neurons, altering neurotransmitter levels and disturbing neuronal function [99, 100].

It has been observed that BDE-47 and BDE-209 decreased the activity of acetylcholinesterase (Che) in large fleas, leading to overstimulation of cholinergic neurons and disordered behavior such as tremors and unstable swimming [26]. Similarly, co-exposure of DBDPE and microplastics (polylactic acid, PLA-MPs; polyethylene, PE; polypropylene, PP) has been shown to exhibit strong synergistic toxicity towards earthworms by significantly increasing acetylcholinesterase (AChE) levels responsible for the degradation of the neurotransmitter acetylcholine (ACh). The excessive elevation of AChE content can lead to a reduction in ACh levels, which may cause a decrease in cholinergic neurotransmission efficiency in earthworms, resulting in cognitive impairment and memory loss. Specifically, PLA-MPs exhibit a stronger synergistic toxicity, as co-exposure to PLA-MPs caused the most severe vacuolization, partial hyperplasia, and epidermal damage in skin cells, which may exacerbate the neurotoxic effects of DBDPE. Additionally, co-exposure was found to significantly upregulate the levels of the neurotransmitter glutamate (Glu) and Ca^{2+}. Excessive Glu can cause excitatory neurotoxicity and even cell death, while too little Glu can hinder the transmission of nerve signals and feedback. Ca^{2+} is a fundamental regulatory factor for cell survival or death and is maintained at a relatively stable concentration within a narrow range. Excessive Glu in co-exposure induced the overactivation of Glu receptors and *N*-methyl-D-aspartate receptors (NMDAr), causing an influx of excessive Ca^{2+} and overload, triggering cell death and disrupting neurons and their structures [101].

Neuronal transport proteins are also a susceptible target for combined OFRs. In mice, DE-71 can suppress the dopamine neuronal transporter, resulting in reduced dopamine release and potentially leading to neurological disorders such as Parkinson's disease and depression [102–104]. Besides, Bradner et al. discovered that DE-71 reduced the activity of vesicular monoamine transporter 2 (VMAT2), which affects proteins involved in mediating GABA and glutamate neurotransmission [105].

Studies have shown that co-exposure to OFRs affects the connections and signal transmission between neurons by inhibiting the growth and maintenance of neuronal axons and synapses. For instance, in adult zebrafish, TCP and CDP co-exposure downregulated the expression of neuronal genes such as *α1-tubulin, elavl3, shha*, and *gap43* [106], which play important roles in microtubule assembly and stability, growth and maintenance of neuronal axons, and formation of synapses [107, 108]. Additionally, co-exposure of nanoscale TiO_2 and TPHP in zebrafish larvae was found to decrease the activity of Elavl3 and neurogenin-1 (ngn1), which may hinder the formation of synapses, inhibit neuronal growth, and affect the connections and function of neurons [109]. Furthermore, exposure to DE-71 in zebrafish larvae was observed to decrease the expression of myelin basic protein, synaptotagmin IIa, and genes involved in serotonin synthesis (TPH2, TPH1, TRHR) and neurotransmission (SETA/B, HTR8Aa/B), leading to decreased activity and content of related proteins and neurotransmitters [110]. Reduction of synaptic protein IIa and myelin basic protein expression could lead to slower nerve conduction velocity, while decrease in serotonin synthesis could affect the normal function of the nervous system, including mood, sleep, appetite, and memory regulation [111, 112]. In addition, DE-71 was also found to disrupt calcium ion homeostasis leading to inhibition of neuronal growth, cone migration, synaptic signal transmission, and microtubule assembly in rats [113].

Interestingly, OFRs have been found to induce neurotoxicity through endocrine signaling pathways. For instance, Gillera et al. reported that exposure to FM550 was found to induce abnormal social behavior in field voles, including anxiety, attachment, and memory, which was linked to the interruption of the oxytocin (OT)/arginine vasopressin (AVP) signaling pathway [27]. OT and AVP are hormones produced by the posterior pituitary gland that play a role in social bonding, affiliation, and other prosocial behaviors in animals. They interact with the midbrain dopamine system and modulate the activity of the hypothalamic–pituitary–adrenal (HPA) axis and ER in a gender-dependent manner. OT and AVP are also closely related to physiological and psychological processes such as neuronal development and maturation, social memory, and emotional regulation. Abnormal levels of these hormones have been associated with the development of neuropsychiatric disorders such as autism, depression, and anxiety [114, 115]. Therefore, disruption of OT/AVP signaling pathway may reduce the content of these hormones or hinder their interaction with other systems, which could be a possible cause of abnormal social behavior in voles.

15.4.2 Oxidative Damage of Neurons

Co-exposure of OFRs can induce neurotoxicity by regulating the cellular oxidation state. In mice, the BFR mixture DE-71 induced apoptosis of neurons and astrocytes due to excessive production of ROS [53]. Similarly, co-exposure to DBDPE and ZnO-NPs induced oxidative stress in neuronal cells, leading to neuronal apoptosis and neurotoxicity. This was manifested by a significant increase in ROS production, as well as a significant decrease in the activities of SOD and GSH. Impaired mitochondrial function was identified as a deep-seated cause of neuronal apoptosis, as evidenced by the observed reduction in mitochondrial membrane potential. This reduction was likely due to excessive ROS production, which altered mitochondrial permeability and led to the release of cytochrome C into the cytoplasm, thereby activating the caspase cascade and executing cell death. Additionally, the authors observed LDH release in neuronal cells, confirming that damage to the cell membrane was one of the mechanisms underlying neurotoxicity. However, the exact cause of this membrane damage remains unclear.

Furthermore, co-exposure to BDE-99 and low-dose methylmercury (MeHg) resulted in enhanced oxidative stress, as indicated by the decrease in antioxidant enzyme activity and the increase in LPO and H_2O_2 in the cerebellum of rats. Importantly, no changes in neurobehavior or oxidative stress were observed in the groups exposed to low-dose MeHg or BDE-99 alone during the adolescent period of rats. The experimental data indicated a synergistic effect of co-exposure on neurotoxicity, which may be due to the high content of polyunsaturated fatty acids (PUFAs) in brain tissue, which are highly susceptible to oxidation. Oxidation of PUFAs may lead to damage to neuronal cell membranes and inflammatory responses, affecting the normal function of neurons and even leading to neuronal death, thereby affecting the overall function of the nervous system. The cerebellum was found to be the most sensitive region to co-exposure, as MeHg targets the cerebellum, while BDE-99 was more likely to cause toxicity in the hippocampus. The neurotoxicity was specifically manifested in delayed negative geotaxis reflex and impaired motor coordination in rats. Similarly, compared to PCB153 and BDE-99, MeHg was also discovered to enhance the affinity for mAChRs in the cerebellum of rats and mice, respectively, impairing the toxic spontaneous movement behavior and habituation ability [50, 116]. Excessive production or insufficient clearance of ROS can result in oxidative stress, causing damage and death of neurons and ultimately leading to neurotoxicity. Therefore, reducing ROS production or enhancing cellular antioxidant capacity may be beneficial in preventing and treating neurotoxicity-related diseases (Table 15.4).

15.5 Synergistic Effects of OFR Co-exposure on Immunotoxicity

Co-exposure to OFRs has been reported to induce immunotoxicity, impairing immune function and attenuating the antiviral and/or antibacterial efficacy, which can increase susceptibility to pathogens. The underlying molecular mechanisms are multifaceted and include interference with immune cell function, metabolism, and immune regulation mechanisms. Aluru et al. found that the combination of TCPP and EHDPP elevated the levels of cholesterol and triglycerides in plasma levels, promoted lipid droplet accumulation in the liver of cod fish, and triggered inflammation, which can damage immune cells. The combined toxicity also altered the expression of extracellular matrix (ECM) proteins and disrupted the expression of PPARα subtype b gene [117]. PPARα is a key regulatory factor in the inflammatory response, and its disruption can lead to an increase in pro-inflammatory cytokine production.

Co-exposure to BDCIPP, TPHP, and DPHP can induce CpG hypomethylation in zebrafish, affecting the differentiation and function of T and B cells. Specifically, CpG hypomethylation can result in abnormal Th1/Th2/Th17 cell differentiation in T cells [118–120], thereby affecting the generation and regulation of immune responses. Additionally, CpG hypomethylation can also affect B cell development and function, such as the localization in lymph nodes and antibody production [121], and ultimately disrupting immune response generation and regulation in zebrafish [122, 123]. Furthermore, Cd was found to synergistically enhance the activation of genes encoding phase I detoxification enzymes (*CYP1A1, CYP1A2, CYP2B10*, and *CYP3A11*) and inhibit genes encoding phase II detoxification enzymes (*GST, Nqo1, Hmox1*, and *UGTs*) induced by TBBPA in male mice, causing elevated hepatic enzyme levels in serum, local necrosis, and inflammation in the liver, which may be responsible for the increasing susceptibility to inflammatory diseases [57].

15.6 Synergistic Effects of OFR Co-exposure on Growth, Development, and Organ

15.6.1 Damage to Growth and Development

Many studies have found that exposure to OFRs could negatively impact biological development, including height, weight, and various systems as well as the development of organs. Severe damage could even lead to death of cells even organisms. Zhang et al. reported that exposing zebrafish larvae to a combination of triphenyl phosphate (TPHP) and graphene downregulated genes and proteins involved in cytoskeleton construction, cartilage, and retinal development.

Table 15.4 Study on neurotoxicity caused by combined exposure to organic flame retardants (OFRs).

Type of FRs	Dose and time	Organisms	Phenotypes	Mechanisms	Ref.
BDE-47	0.5, 1, 2, 4, 8 μg/L	Large flea	Reducing serum ChE activity; Causing tremors and unstable swimming	Leading to overstimulation of cholinergic neurons	[26]
BDE-209	25, 50, 100, 200, 400 μg/L				
BDE-99	20 μg/mL	Rat	Delay appearance of negative geotaxis reflexes; Impairing motor coordination	Affecting muscarinic cholinergic receptors in cerebellum	[50]
MeHg	0.2 mg/kg				
DE-71	30 mg/kg	Mice	Cell death in dopaminergic cell lines; Decreasing dopamine	Inhibiting VMAT2	[102]
DE-71	30 mg/kg	Mice	Reducing primary cultured neuronal and neurite growth in frontal cortical detachment	Altering mediating GABA and glutamate neurotransmission; Including GAD67, vGAT, vGlut, and GABA(A)2α receptor subunits	[105]
DE-71	0–100 μg/L	Zebrafish larvae	Decreasing serotonin neuronal; Decreasing systemic serotonin levels	Significantly decreasing protein content (myelin basic protein, synaptophycin IIa); Reducing the transcription of genes involved in serotonergic synthesis (*TPH2, TPH1, TRHR*) and neurotransmission (*SETA/B, HTR8Aa/B*)	[110]
DE-71	25 μM	Primary mouse astrocytes	Apoptosis of neurons and astrocytes in different brain regions (cerebellum, hippocampus, cerebral cortex)	Decreasing levels of intracellular GSH	[53]
DE-71	30.6 mg/kg/day	Pregnant Long-Evans rats	Inhibiting neuronal growth, cone migration, synaptic signaling, and microtubule assembly	Affecting calcium signaling; inducing oxidative stress	[113]
PCB153	0.51 mg/kg	Mice	Leading to spontaneous interruption, lack of habits; Decreasing cognitive function	Affecting muscarinic cholinergic receptors in the cerebellum	[116]
MeHg	4, 40 mg/kg				
DBDPE	25 mg/L	Human neuroblastoma SK-N-SH cells	Damaging cell membranes	Increasing release of LDH	[45]
ZnO NPs	50 mg/L				
DBDPE	10 mg/kg	Earthworm	Leading to cognitive impairment and memory loss	Increasing AchE; Significantly activating NMDAr; Significantly upregulating Glu	[101]
PLA-MPs	10–1000 μm				
PP	10–1000 μm				

(Continued)

Table 15.4 (Continued)

Type of FRs	Dose and time	Organisms	Phenotypes	Mechanisms	Ref.
PE	10–1000 μm				
TCP	5 μL/L	Adult zebrafish	Decreasing swimming intensity; Affecting neuronal cell division and the formation of neural circuits; Causing defects in mitochondrial activity and neuronal electrical activity of the cell; Causing axonal damage to nerve cells	The toxic effect on neuronal genes (*α1-tubulin, elavl3, shha,* and *gap43*); Affecting acetylcholine esterase activity in earthworms	[106]
CDP	25 μL/L				
TPHP	72, 144 μg/L	Zebrafish larvae	Abnormalities in spontaneous movement and swimming behavior; Inhibiting secondary motor neurons axon growth; Reducing content of the neurotransmitter serotonin	Decreasing expression of *ngn1* and *ELAVL3* proteins	[109]
Nano-TiO_2	100 μg/L				
FM 550	500, 1000, 2000 μg/day	Prairie voles	Sex-specific effects on aspects of sociability, anxiety, attachment, and memory; Increasing ASD risk	Disrupting OT/AVP pathways	[27]

GD, gestation day; PND, postnatal day; ASD, autism spectrum disorder; NDDs, neurodevelopmental disorders; ChE, cholinesterase; AVP, arginine vasopressin; GABA, gamma-aminobutyric acid; NMDAr, the Glu receptor *N*-methyl-D-aspartic acid receptor; Glu, Glutamate; AChE, acetylcholinesterase.

The downregulation of genes encoding important structural and functional proteins like collagen, matrilin, and keratin can cause significant damage to biological bones, muscles, and cell cytoskeleton. Furthermore, downregulation of genes such as *cdk5, gads,* and *dctn1a*, which are involved in retinal development, can inhibit retinal neuron differentiation, synaptogenesis, and function [59]. *Cdk5* plays an important role in the migration, differentiation, and synaptogenesis of retinal neurons, while *gmds* participates in sugar and lipid metabolism pathways such as PPP (pentose phosphate pathway) and TCA (tricarboxylic acid cycle). *Dctn1a* participates in the transport and synaptogenesis of retinal neurons. Therefore, significant downregulation of these genes can negatively impact retinal development and function. Similarly, Blake et al. found that exposure to DE-71 activated the osteopontin gene promoter in rats, which is closely associated with skeletal health. Osteopontin is an extracellular calcium-binding matrix protein that plays a crucial role in multiple physiological processes, such as bone remodeling, inflammation, cell adhesion and migration, and prevention of cell apoptosis [124]. However, overactivation of the osteopontin promoter can lead to excessive expression of osteopontin, which may cause malignant transformation of cells, promote proliferation and invasion of cancer cells, and ultimately contribute to tumor occurrence and development [125, 126]. Furthermore, excessive expression of osteopontin may also lead to the development of diseases such as osteoporosis [127].

In some cases, co-exposure could also damage the cell membrane structure, leading to the loss of cell integrity and inhibiting organism's growth and reproduction. Chen et al. [128] discovered that the addition of Cd enhanced the disruptive effects on cellular membrane integrity and permeability of earthworms induced by TBBPA, significantly reducing the lethal dose. Furthermore, Aluru et al. investigated the joint exposure of Atlantic cod to TCPP and EDHPP and found that genes involved in cholesterol synthesis were downregulated. Cholesterol is an important component of cell membranes and a crucial mediator for maintaining membrane fluidity and permeability. Changes in membrane composition can affect the transport of nutrients, osmolytes, and other important molecules, as well as the ability to adapt to temperature changes. Therefore, the researchers

Table 15.5 Effects of combined exposure to OFRs on biological growth and development.

Type of FRs	Dose and time	Organisms	Phenotypes	Mechanisms	Ref.
DE-71	60 μg/kg	Rat	Increasing thyroid gland weight; Increasing ovarian osteopontin	Activating the osteopontin gene promoter	[124]
DE-71	0.06, 20, 60 mg/kg	Adult female rats	Decreasing ossification of the sixth sternebra; Causing significant liver and kidney enlargement	Down-regulating *Runx2*; Causing *HOXD13* mutations or deletions	[129]
DE-79	0.06, 20, 60 mg/kg				
BDE-209	0.06, 20, 60 mg/kg				
PCBs	3, 6 mg/kg	Female Rat and cubs	Impairing cochlear function	Binding to RyR and stabilizing the receptor Increasing intracellular calcium concentration	[136]
PBDEs	5.7, 11.4 mg/kg				
BDE-47	10 μg/L	Zebrafish embryos	Reducing body weight; Increasing the rate of malformation	Significantly upregulating the expression of TH synthesis-related genes (*NIS* and *Nkx2.1a*); Significantly downregulating the expression of HPT axis–regulated genes (*CRH* and *TSH beta*); Significantly upregulating the expression of *TG* genes and proteins in the co-exposed group; Significantly downregulating the expression of TTR	[79]
PER	3.5 μg/L				
PCBs	2, 20, 200 μg/L	Juvenile Japanese halibut	Reducing body length and weight; Inducing thyroid follicular colloid defects; Impairing the structure and function of the intestine	Inducing endocrine disturbance by downregulating expression of TSHβ, TPO, ID1, and ID2	[75]
MPs	10000 μg/L				
TBBPA	10, 50, 100, 500 mg/kg	Earthworm	Aggravating the deterioration of skin cells	Damaging the integrity and permeability of biofilms	[128]
Cd	1 mg/kg				
TPHP	5, 20, 50 μg/L	Zebrafish larvae	Causing axial spinal curvature and pericardial and yolk edema; Causing defects of cartilage and bone development; Significantly inhibiting rate of mortality; Significantly inhibiting heartbeat	Downregulating genes and proteins of cytoskeletal construction and cartilage (*col1a1a, col2a1a, col5a3b, col6a1, matn1, matn4, krt18a.1, krt8, smyhc2, cthrc1a, tnni2b.2, myl12.1*, and *mylz3*) Downregulating genes and proteins of retina development (*cdk5, gads*, and *dctn1a*)	[59]
Graphene	10, 100 μg/L				

CM, cerebral white matter; GD, gestation day; RyR, ryanodine receptor; HPT, hypothalamic–pituitary–thyroid axis; *Runx2*, runt-related transcription factor 2; *HOXD13*, homeobox D13.

suggested that the disruption of membrane integrity may be one of the possible reasons for the decreased reproductive, growth, and behavioral performance, as well as the reduced ability to adapt to temperature changes in Atlantic cod [117].

15.6.2 Damage to Organ

Studies have shown that combined exposure can impair organ performance through direct action on the tissues or binding to related receptors, then inhibiting normal organ physiology and development. As an illustration, Wang et al. found that MPs can aggravate intestinal dysfunction in juvenile Japanese flounder induced by PCBs, resulting in decrease in body length and weight [75]. In addition, concurrent exposure to DE-71, DE-79, and BDE-209 has been shown to reduce ossification of the sternum in Sprague–Dawley rat fetuses, as well as digit anomalies such as soft tissue syndactyly or distal phalangeal dislocation in the low-dose co-exposure group [129]. Studies have suggested that OFRs can inhibit osteoblast differentiation and skeletal development by inducing downregulation of *Runx2* gene expression [130, 131], resulting in decreased fetal sternum ossification. Besides *Runx2*, mutations or deletions in the HOX genes are another major cause of various finger and toenail malformations [132, 133]. Furthermore, Poon et al. reported that the mixture of PCBs and PBDEs could stabilize the high-affinity conformation of the ryanodine receptor (RyR), increasing intracellular calcium ion concentration and finally cause cochlear dysfunction in female rats and cubs. RyR is a transmembrane protein responsible for calcium ion release in muscle cells, and its stable high-affinity conformation can maintain the channel in an open state, leading to an increase in intracellular calcium ion concentration. Excessively high calcium ion concentrations can cause abnormal movement of the outer hair cells in the cochlea, which can affect the transmission and processing of auditory signals. Therefore, abnormal movement of outer hair cells may be a cause of cochlear defects [134, 135] (Table 15.5).

15.7 Antagonism Effects of OFR Co-exposure on Growth, Development, and Organ

In all the aforementioned descriptions, we have focused on the synergistic effects caused by OFRs. However, according to the existing literature, it has also been reported that OFRs can exhibit antagonistic effects on other environmental pollutants, such as nano-TiO_2, graphene, and MPs. This provides new insights for acquiring a comprehensive understanding of the toxic effects of these pollutants in real environment. The following is a discussion on the antagonism effects of OFRs and their underlying mechanisms.

In the previous discussion, we explored the inhibitory effect of co-exposure of TPHP and graphene oxide (GO) on antioxidant oxidase activity in zebrafish, demonstrating that co-exposure of TPHP and graphene induces stronger oxidative stress [59]. Interestingly, Meng et al. found that TPHP could inhibit the surface activity of graphene and downregulated the expression of genes related to stress responses (*MyD88a*), cytoskeleton (*MHC1, PMyo*, and *TMyo*), as well as reproduction (*CP450* and *HSD*) in the co-exposure group compared to graphene alone, thereby alleviating the developmental inhibition, oxidative stress, and reproductive toxicity of graphene on mussels [137]. Following that, they further elucidated that graphene could alleviate TPHP-induced oxidative stress by upregulating *NF-κB, Bcl-2*, and *Ras* genes [138]. The varying experimental outcomes on the toxicity of graphene and oxidized graphene may be attributed to their distinct chemical properties and biological interactions. GO contains numerous oxygen functional groups on surface, which can react with biomolecules and induce cellular toxicity and inflammatory responses. The surface properties make it susceptible to adsorb biomolecules and impact biological functions [139]. In contrast, graphene lacks functional groups and exhibits relatively stable chemical properties. The toxicity of graphene is primarily due to its nanoscale size and high surface area, causing mechanical stimulation and cell membrane damage [140]. Researchers suggest that TPHP reduced the surface sharpness of graphene, thereby reducing its toxicity. These findings highlight the importance of understanding the chemical and biological properties of nanomaterials in evaluating their potential toxicity.

In a co-exposure experiment with TDCPP and GO in zebrafish, TDCPP was found to downregulate the expression of proteins responsible for skeletal, cartilage, and muscle development. Deficiencies in these proteins can lead to growth defects in zebrafish. TDCPP also downregulated MRC (mitochondrial electron transport and mitochondrial respiratory chain complex IV), while GO increased the expression levels of proteins associated with complex I and IV subunits, such as *ndufa10, ndufa8, sdha*, and *cox5aa*. These findings suggest that GO can counteract the inhibitory effect of TDCPP on MRC by regulating cellular energy metabolism [141]. Furthermore, NADH dehydrogenase plays a crucial role in intracellular redox reactions by converting NADH to NAD+ and releasing electrons to alleviate changes in the intracellular NADH/NAD+ ratio induced by oxidative stress. Studies have also shown that upregulating NADH dehydrogenase can increase the levels of reduced glutathione, thereby mitigating oxidative stress–induced cellular damage [142, 143]. The comparative analysis of combined exposure experiments of GO with TPHP (hereafter referred to as TPG) and GO with TDCPP (hereafter referred to as TDG) revealed that exposure to GO alone can induce oxidative stress by significantly elevating cellular ROS levels. However, in TPG, the sole exposure of TPHP did not significantly induce oxidative stress, whereas in TDG, the exposure to TDCPP increased the ROS production, with ROS levels twice as high as those in the control group. Similarly, in the presence of PS, TPHP induced lower levels of oxidative stress and developmental toxicity in marine microalgae, revealing an antagonistic effect [144]. Furthermore, regarding the co-exposure of DBDPE and ZnO-NPs, an antagonistic effect was observed at low concentrations of ZnO-NPs, as DBDPE adsorbed free Zn^{2+}. However, at higher concentrations of ZnO-NPs, as more Zn^{2+} was released and DBDPE was adsorbed onto the surface of ZnO-NPs, the toxicity shifted from antagonistic to synergistic, thereby increasing the exposure time of SK-N-SH cells to DBDPE and enhancing its bioavailability and toxicity [45]. Therefore, it is plausible that the synergistic and antagonistic effects of combined exposure may be attributed to the chemical properties of OFRs or their binding affinity with other pollutants [64, 141].

Additionally, Ma et al. found that TCEP could be adsorbed onto MPs, forming large-sized complexes that cannot penetrate the epidermal surface of nematodes at an observable rate, therefore alleviating the TCEP-induced oxidative stress responses [145]. Agglomeration will occur among co-exposed pollutants, enlarging the size of exposure and reducing the possibility of substances' intake into the organism. As an illustration, the co-exposure of nano TiO_2 and TBPH led to the formation of agglomeration, blocking the entry of TBPH into zebrafish, due to the restricted size of chorion pores. This may explain the alleviation of lipid metabolism disorder caused by TBPH in zebrafish [33].

15.8 Summary and Perspectives

The natural environment is a dynamic and intricate system containing a variety of pollutants, and the ecological consequences are actually the comprehensive results of long-term interactions between multiple pollutants and complex ecosystems. Conducting comprehensive and systematic research on the health risks of pollutants is imperative from a management perspective and is in line with the development of environmental science research. However, the study on combined toxicity is still in the early stage, with complex and diverse content, and no established laws or regulations have been developed to guide research in this area. In this chapter, we extensively discussed the combined toxicity and molecule mechanisms of FR mixtures, as well as the combination with other pollutants.

Unlike drugs, which have specific targets and clear modes of action, FRs are originally designed to serve industrial purposes without a defined biological target. Consequently, exposure to FRs, especially the co-exposure, may result in the coexistence of multiple toxicities, which is actually consistent with our investigation results. According to the comprehensive summary, it is found that oxidative stress and endocrine disruption are the most frequently reported effects of combined toxicity of FR mixtures, followed by neurotoxicity and inhibition of growth and development. The different toxic endpoints induced by FR mixtures can share the common mechanisms. For example, hormone receptors also play a role in modulation of behavior and neural perception. And increased ROS, the marker of oxidative stress, has been commonly accepted as the initial event for a variety of damage, including but not limited to DNA damage and neuronal apoptosis. In addition, the research interest is currently moving away from examining the toxic effects of organic pollutants to investigating the combined toxic effects of organic pollutants and materials, such as graphene, nanoparticles, and microplastics. Notably, the synergistic effects have been more commonly reported, while antagonistic effects were almost exclusively observed in the co-exposure of FRs and other materials, especially graphene oxide. There are still significant knowledge gaps in our understanding of the interactions and co-metabolism between two or more FRs within the body. Instead, explanations based on the physical properties of materials such as particles appear to be more accessible.

In conclusion, further investigations into the molecular mechanisms of FR mixtures and the development of AOP frameworks are necessary, which is of great significance for pollution control and ecological protection.

Acknowledgments

The authors thanked the support of National Natural Science Foundation of China (No.22306033, No.22127810, and No.22076027); the Smart Medical Innovation Technology Center, GDUT (ZYZX24-013); and the Open Research Project of the State Key Laboratory of Industrial Control Technology, China (ICT2024B28).

References

1 Chupeau Z, Bonvallot N, Mercier F, Le Bot B, Chevrier C, Glorennec P. Organophosphorus flame retardants: a global review of indoor contamination and human exposure in Europe and epidemiological evidence. Int J Environ Res Public Health. 2020, 17(18): 6713.

2 Bika SH, Adeniji AO, Okoh AI, Okoh OO. Spatiotemporal distribution and analysis of organophosphate flame retardants in the environmental systems: a review. Molecules. 2022, 27(2): 573.

3 Yao C, Yang H, Li Y. A review on organophosphate flame retardants in the environment: occurrence, accumulation, metabolism and toxicity. Sci Total Environ. 2021, 795: 148837.

4 Wei G-L, Li D-Q, Zhuo M-N, Liao Y-S, Xie Z-Y, Guo T-L, Li J-J, Zhang S-Y, Liang Z-Q. Organophosphorus flame retardants and plasticizers: sources, occurrence, toxicity and human exposure. Environ Pollut. 2015, 196: 29–46.

5 Elbasuney S. Novel multi-component flame retardant system based on nanoscopic aluminium-trihydroxide (ATH). Powder Technol. 2017, 305: 538–545.

6 van der Veen I, de Boer J. Phosphorus flame retardants: properties, production, environmental occurrence, toxicity and analysis. Chemosphere. 2012, 88(10): 1119–1153.

7 Alaee M, Arias P, Sjodin A, Bergman A. An overview of commercially used brominated flame retardants, their applications, their use patterns in different countries/regions and possible modes of release. Environ Int. 2003, 29(6): 683–689.

8 Feiteiro J, Mariana M, Cairrao E. Health toxicity effects of brominated flame retardants: from environmental to human exposure. Environ Pollut. 2021, 285: 117475.

9 Hales BF, Robaire B. Effects of brominated and organophosphate ester flame retardants on male reproduction. Andrology. 2020, 8(4): 915–923.

10 Klincic D, Dvorscak M, Jagic K, Mendas G, Romanic SH. Levels and distribution of polybrominated diphenyl ethers in humans and environmental compartments: a comprehensive review of the last five years of research. Environ Sci Pollut Res Int. 2020, 27(6): 5744–5758.

11 Kacew S, Blais MS, Hayes AW, Droege W, Osimitz TG. Benefit versus risk associated with the use of brominated flame retardants. Curr Opin Toxicol. 2020, 22: 19–24.

12 Anisuzzaman S, Whalen MM. Tetrabromobisphenol A and hexabromocyclododecane alter secretion of IL-1 from human immune cells. J Immunotoxicol. 2016, 13(3): 403–416.

13 Bjermo H, Aune M, Cantillana T, Glynn A, Lind PM, Ridefelt P, Darnerud PO. Serum levels of brominated flame retardants (BFRs: PBDE, HBCD) and influence of dietary factors in a population-based study on Swedish adults. Chemosphere. 2017, 167: 485–491.

14 Stapleton HM, Sharma S, Getzinger G, Ferguson PL, Gabriel M, Webster TF, Blum A. Novel and high volume use flame retardants in US couches reflective of the 2005 PentaBDE phase out. Environ Sci Technol. 2012, 46(24): 13432–13439.

15 Dong J, Li G, Gao J, Zhang H, Bi S, Liu S, Liao C, Jiang G. Catalytic degradation of brominated flame retardants in the environment: new techniques and research highlights. Sci Total Environ. 2022, 848: 157695.

16 Du J, Li H, Xu S, Zhou Q, Jin M, Tang J. A review of organophosphorus flame retardants (OPFRs): occurrence, bioaccumulation, toxicity, and organism exposure. Environ Sci Pollut Res Int. 2019, 26(22): 22126–22136.

17 Hoffman K, Sosa JA, Stapleton HM. Do flame retardant chemicals increase the risk for thyroid dysregulation and cancer? Curr Opin Oncol. 2017, 29(1): 7–13.

18 Negi CK, Khan S, Dirven H, Bajard L, Blaha L. Flame retardants-mediated interferon signaling in the pathogenesis of nonalcoholic fatty liver disease. Int J Mol Sci. 2021, 22(8): 4282.

19 Yang J, Zhao Y, Li M, Du M, Li X, Li Y. A review of a class of emerging contaminants: the classification, distribution, intensity of consumption, synthesis routes, environmental effects and expectation of pollution abatement to organophosphate flame retardants (OPFRs). Int J Mol Sci. 2019, 20(12): 2874.

20 Blum A, Behl M, Birnbaum LS, Diamond ML, Phillips A, Singla V, Sipes NS, Stapleton HM, Venier M. Organophosphate ester flame retardants: are they a regrettable substitution for polybrominated diphenyl ethers? Environ Sci Technol Lett. 2019, 6(11): 638–649.

21 Zhao J-Y, Zhan Z-X, Lu M-J, Tao F-B, Wu D, Gao H. A systematic scoping review of epidemiological studies on the association between organophosphate flame retardants and neurotoxicity. Ecotoxicol Environ Saf. 2022, 243: 113973.

22 Yang Y, Chen P, Ma S, Lu S, Yu Y, An T. A critical review of human internal exposure and the health risks of organophosphate ester flame retardants and their metabolites. Crit Rev Environ Sci Technol. 2022, 52(9): 1528–1560.

23 Wang F, Wang X, Song N. Polyethylene microplastics increase cadmium uptake in lettuce (Lactuca sativa L.) by altering the soil microenvironment. Sci Total Environ. 2021, 784: 147133.

24 Deng R, Lin D, Zhu L, Majumdar S, White JC, Gardea-Torresdey JL, Xing B. Nanoparticle interactions with co-existing contaminants: joint toxicity, bioaccumulation and risk. Nanotoxicology. 2017, 11(5): 591–612.

25 Zhu Z-l, Wang S-C, Zhao F-F, Wang S-G, Liu F-F, Liu G-Z. Joint toxicity of microplastics with triclosan to marine microalgae Skeletonema costatum. Environ Pollut. 2019, 246: 509–517.

26 Xiong Q, Shi Y, Lu Y, Pan K, Dakhil MA, Zhang L, Xiao Y. Sublethal or not? Responses of multiple biomarkers in Daphnia magna to single and joint effects of BDE-47 and BDE-209. Ecotoxicol Environ Saf. 2018, 164: 164–171.

27 Gillera SEA, Marinello WP, Horman BM, Phillips AL, Ruis MT, Stapleton HM, Reif DM, Patisaul HB. Sex-specific effects of perinatal FireMaster (R) 550 (FM 550) exposure on socioemotional behavior in prairie voles. Neurotoxicol Teratol. 2020, 79: 106840.

28 Gillera SEA, Marinello WP, Cao KT, Horman BM, Stapleton HM, Patisaul HB. Sex-specific disruption of the prairie vole hypothalamus by developmental exposure to a flame retardant mixture. Endocrinology. 2021, 162(8): bqab100.

29 Rock KD, St Armour G, Horman B, Phillips A, Ruis M, Stewart AK, Jima D, Muddiman DC, Stapleton HM, Patisaul HB. Effects of prenatal exposure to a mixture of organophosphate flame retardants on placental gene expression and serotonergic innervation in the fetal rat brain. Toxicol Sci. 2020, 176(1): 203–223.

30 Rock KD, Horman B, Phillips AL, McRitchie SL, Watson S, Deese-Spruill J, Jima D, Sumner S, Stapleton HM, Patisaul HB. EDC IMPACT: Molecular effects of developmental FM 550 exposure in Wistar rat placenta and fetal forebrain. Endocr Connect. 2018, 7(2): 305–324.

31 Patisaul HB, Roberts SC, Mabrey N, McCaffrey KA, Gear RB, Braun J, Belcher SM, Stapleton HM. Accumulation and endocrine disrupting effects of the flame retardant mixture FireMaster (R) 550 in rats: an exploratory assessment. J Biochem Mol Toxicol. 2013, 27(2): 124–136.

32 Tung EWY, Ahmed S, Peshdary V, Atlas E. FireMaster (R) 550 and its components isopropylated triphenyl phosphate and triphenyl phosphate enhance adipogenesis and transcriptional activity of peroxisome proliferator activated receptor (PPAR gamma) on the adipocyte protein 2 (aP2) promoter. PLOS One. 2017, 12(4): e0175855.

33 Zhou Y, Lei L, Chen P, Guo W, Guo Y, Yang L, Han J, Hu B, Zhou B. Effects of nano-TiO2 on the bioavailability and toxicity of bis (2-ethylhexyl)-2,3,4,5-tetrabromophthalate (TBPH) in developing zebrafish. Chemosphere. 2022, 295: 133862.

34 Gu H, Wei S, Hu M, Wei H, Wang X, Shang Y, Li'ang L, Shi H, Wang Y. Microplastics aggravate the adverse effects of BDE-47 on physiological and defense performance in mussels. J Hazard Mater. 2020, 398: 122909.

35 Waaijers SL, Parsons JR. Biodegradation of brominated and organophosphorus flame retardants. Curr Opin Biotechnol. 2016, 38: 14–23.

36 Krivokapic M. Assessment of PCBs and OCPs in anchovy (Engraulis encrasicolus) and sardine (Sardina pilchardus) from the Adriatic Sea, Bay of Herceg Novi (alongside Kumbor Marine Channel). Acta Adriatica. 2020, 61(1): 27–38.

37 Hardy ML. A comparison of the fish bioconcentration factors for brominated flame retardants with their nonbrominated analogues. Environ Toxicol Chem. 2004, 23(3): 656–661.

38 Martinez G, Niu J, Takser L, Bellenger J-P, Zhu J. A review on the analytical procedures of halogenated flame retardants by gas chromatography coupled with single quadrupole mass spectrometry and their levels in human samples. Environ Pollut. 2021, 285: 117476.

39 Lee H-J, Jung J-H, Kwon J-H. Evaluation of the bioaccumulation potential of selected alternative brominated flame retardants in marine fish using in vitro metabolic transformation rates. Sci Total Environ. 2019, 653: 1333–1342.

40 Yan Z, Feng C, Jin X, Liu D, Hong Y, Qiao Y, Bai Y, Moon H-B, Qadeer A, Wu F. In vitro metabolic kinetics of cresyl diphenyl phosphate (CDP) in liver microsomes of crucian carp (Carassius carassius). Environ Pollut. 2021, 274: 116586.

41 Page-Lariviere F, Chiu S, Jones SP, Farhat A, Crump D, O'Brien JM. Prioritization of 10 organic flame retardants using an avian hepatocyte toxicogenomic assay. Environ Toxicol Chem. 2018, 37(12): 3134–3144.

42 Li J, Zhao L, Letcher RJ, Zhang Y, Jian K, Zhang J, Su G. A review on organophosphate Ester (OPE) flame retardants and plasticizers in foodstuffs: levels, distribution, human dietary exposure, and future directions. Environ Int. 2019, 127: 35–51.

43 Bruchajzer E, Frydrych B, Szymaniska JA. Organophosphorus flame retardants - toxicity and influence on human health. Med Pr. 2015, 66(2): 235–264.

44 Lushchak VI. Free radicals, reactive oxygen species, oxidative stress and its classification. Chem Biol Interact. 2014, 224: 164–175.

45 Dong L, Wang S, Zhang L, Liu D, You H. DBDPE and ZnO NPs synergistically induce neurotoxicity of SK-N-SH cells and activate mitochondrial apoptosis signaling pathway and Nrf2-mediated antioxidant pathway. J Hazard Mater. 2023, 441: 129872.

46 Ruiz CE, Manuguerra S, Curcuraci E, Santulli A, Messina CM. Carbamazepine, cadmium chloride and polybrominated diphenyl ether-47, synergistically modulate the expression of antioxidants and cell cycle biomarkers, in the marine fish cell line SAF-1. Mar Environ Res. 2020, 154: 104844.

47 Li H, Li Y, Maryam B, Ji Z, Sun J, Liu X. Polybrominated diphenyl ethers as hitchhikers on microplastics: sorption behaviors and combined toxicities to Epinephelus moara. Aquat Toxicol. 2022, 252: 106317.

48 Albano GD, Bonanno A, Montalbano AM, Di Sano C, Anzalone G, Gagliardo R, Ruggieri S, Profita M. Cadmium and cadmium/BDE (47 or 209) exposure affect mitochondrial function, DNA damage/repair mechanisms and barrier integrity in airway epithelial cells. Atmosphere. 2022, 13(2): 201.

49 Wang B, Wang H, Han D, Chen J, Yin Y. Studying the mixture effects of brominated flame retardants and metal ions by comet assay. Environ Pollut. 2020, 267: 115677.

50 Zhao W, Cheng J, Gu J, Liu Y, Fujimura M, Wang W. Assessment of neurotoxic effects and brain region distribution in rat offspring prenatally co-exposed to low doses of BDE-99 and methylmercury. Chemosphere. 2014, 112: 170–176.

51 Feng M, Li Y, Qu R, Wang L, Wang Z. Oxidative stress biomarkers in freshwater fish Carassius auratus exposed to decabromodiphenyl ether and ethane, or their mixture. Ecotoxicology. 2013, 22(7): 1101–1110.

52 Reistad T, Mariussen E. A commercial mixture of the brominated flame retardant pentabrominated diphenyl ether (DE-71) induces respiratory burst in human neutrophil granulocytes in vitro. Toxicol Sci. 2005, 87(1): 57–65.

53 Giordano G, Kavanagh TJ, Costa LG. Neurotoxicity of a polybrominated diphenyl ether mixture (DE-71) in mouse neurons and astrocytes is modulated by intracellular glutathione levels. Toxicol Appl Pharmacol. 2008, 232(2): 161–168.

54 Rattner BA, Lazarus RS, Heinz GH, Karouna-Renier NK,Schultz SL, Hale RC. Comparative embryotoxicity of a pentabrominated diphenyl ether mixture to common terns (Sterna hirundo) and American kestrels (Falco sparverius). Chemosphere. 2013, 93(2): 441–447.

55 Kling P, Forlin L. Proteomic studies in zebrafish liver cells exposed to the brominated flame retardants HBCD and TBBPA. Ecotoxicol Environ Saf. 2009, 72 (7): 1985-1993.

56 Soto-Bielicka P, Tejeda I, Peropadre A, Hazen MJ, Fernandez Freire P. Detrimental effects of individual versus combined exposure to tetrabromobisphenol A and polystyrene nanoplastics in fish cell lines. Environ toxicol pharmacol. 2023, 98: 104072–104072.

57 Chen Y, Hu Y, Liu S, Zheng H, Wu X, Huang Z, Li H, Peng B, Long J, Pan B, Huang C, Dong Q. Whole-body aerosol exposure of cadmium chloride (CdCI2) and tetrabromobisphenol A (TBBPA) induced hepatic changes in CD-1 male mice. J Hazard Mater. 2016, 318: 109–116.

58 Bearr JS, Stapleton HM, Mitchelmore CL. Accumulation and DNA damage in fathead minnows (Pimephales promelas) exposed to 2 brominated flame-retardant mixtures, firemaster® 550 and firemaster® BZ-54. Environ Toxicol Chem. 2010, 29(3): 722–729.

59 Zhang X, Zhou Q, Li X, Zou W, Hu X. Integrating omics and traditional analyses to profile the synergistic toxicity of graphene oxide and triphenyl phosphate. Environ Pollut. 2020, 263: 114473.

60 Li B, Zhang T, Tang M. Toxicity mechanism of nanomaterials: focus on endoplasmic reticulum stress. Sci Total Environ. 2022, 834: 155417.

61 Petan T, Jarc E, Jusovic M. Lipid droplets in cancer: guardians of fat in a stressful world. Molecules. 2018, 23(8): 1941.

62 Vaishampayan A, Grohmann E. Antimicrobials functioning through ROS-mediated mechanisms: current insights. Microorganisms. 2022, 10(1): 61.

63 Sanders JM, Burka LT, Smith CS, Black W, James R, Cunningham ML. Differential expression of CYP1A, 2B, and 3A genes in the F344 rat following exposure to a polybrominated diphenyl ether mixture or individual components. Toxicol Sci. 2005, 88(1): 127–133.

64 Burgos-Moron E, Abad-Jimenez Z, Martinez de Maranon A, Iannantuoni F, Escribano-Lopez I, Lopez-Domenech S, Salom C, Jover A, Mora V, Roldan I, Sola E, Rocha M, Victor VM. Relationship between oxidative stress, ER stress, and inflammation in type 2 diabetes: the battle continues. J Clin Med. 2019, 8(9): 1385.

65 Sinha K, Das J, Pal PB, Sil PC. Oxidative stress: the mitochondria-dependent and mitochondria-independent pathways of apoptosis. Arch Toxicol. 2013, 87(7): 1157–1180.

66 Bhatti JS, Bhatti GK, Reddy PH. Mitochondrial dysfunction and oxidative stress in metabolic disorders - a step towards mitochondria based therapeutic strategies. Biochim Biophys Acta Mol Basis Dis. 2017, 1863(5): 1066–1077.

67 Padhye LP, Yao H, Kung'u FT, Huang C-H. Year-long evaluation on the occurrence and fate of pharmaceuticals, personal care products, and endocrine disrupting chemicals in an urban drinking water treatment plant. Water Res. 2014, 51: 266–276.

68 Kumar M, Sarma DK, Shubham S, Kumawat M, Verma V, Prakash A, Tiwari R. Environmental endocrine-disrupting chemical exposure: role in non-communicable diseases. Front Public Health. 2020, 8: 553850.

69 Yilmaz B, Terekeci H, Sandal S, Kelestimur F. Endocrine disrupting chemicals: exposure, effects on human health, mechanism of action, models for testing and strategies for prevention. Rev Endocr Metab Disord. 2020, 21(1): 127–147.

70 Stoker TE, Cooper RL, Lambright CS, Wilson VS, Furr J, Gray LE. In vivo and in vitro anti-androgenic effects of DE-71, a commercial polybrominated diphenyl ether (PBDE) mixture. Toxicol Appl Pharmacol. 2005, 207(1): 78–88.

71 Zhang S, Bursian SJ, Martin PA, Chan HM, Tomy G, Palace VP, Mayne GJ, Martin JW. Reproductive and developmental toxicity of a pentabrominated diphenyl ether mixture, DE-71, to ranch mink (Mustela vison) and hazard assessment for wild mink in the Great Lakes region. Toxicol Sci. 2009, 110(1): 107–116.

72 Duntas LH. New insights into the hypothalamic-pituitary-thyroid axis. Acta Endocrinol (Buchar). 2016, 12(2): 125–129.

73 Jing L, Zhang Q. Intrathyroidal feedforward and feedback network regulating thyroid hormone synthesis and secretion. Front Endocrinol (Lausanne). 2022, 13: 992883.

74 Wang Q, Chen Q, Zhou P, Li W, Wang J, Huang C, Wang X, Lin K, Zhou B. Bioconcentration and metabolism of BDE-209 in the presence of titanium dioxide nanoparticles and impact on the thyroid endocrine system and neuronal development in zebrafish larvae. Nanotoxicology. 2014, 8: 196–207.

75 Wang J, Li X, Li P, Li L, Zhao L, Ru S, Zhang D. Porous microplastics enhance polychlorinated biphenyls-induced thyroid disruption in juvenile Japanese flounder (Paralichthys olivaceus). Mar Pollut Bull. 2022, 174: 113289.

76 Godlewska M, Gawel D, Buckle AM, Banga JP. Thyroid peroxidase revisited – what's new? Horm Metab Res. 2019, 51(12): 765–769.

77 Yamauchi K. Evolution of thyroid hormone distributor proteins in fish. Gen Comp Endocrinol. 2021, 305: 113735.

78 Mimoto MS, Refetoff S. Clinical recognition and evaluation of patients with inherited serum thyroid hormone-binding protein mutations. J Endocrinol Invest. 2020, 43(1): 31–41.

79 Zhao X, Wang S, Li D, You H, Ren X. Effects of perchlorate on BDE-47-induced alteration thyroid hormone and gene expression of in the hypothalamus-pituitary-thyroid axis in zebrafish larvae. Environ Toxicol Pharmacol. 2013, 36(3): 1176–1185.

80 Arkoosh MR, Van Gaest AL, Strickland SA, Hutchinson GP, Krupkin AB, Dietrich JP. Alteration of thyroid hormone concentrations in juvenile Chinook salmon (Oncorhynchus tshawytscha) exposed to polybrominated diphenyl ethers, BDE-47 and BDE-99. Chemosphere. 2017, 171: 1–8.

81 Amir S, Shah STA, Mamoulakis C, Docea AO, Kalantzi O-I, Zachariou A, Calina D, Carvalho F, Sofikitis N, Makrigiannakis A, Tsatsakis A. Endocrine disruptors acting on estrogen and androgen pathways cause reproductive disorders through multiple mechanisms: a review. Int J Environ Res Public Health. 2021, 18(4): 1464.

82 Lefevre PLC, Berger RG, Ernest SR, Gaertner DW, Rawn DFK, Wade MG, Robaire B, Hales BF. Exposure of female rats to an environmentally relevant mixture of brominated flame retardants targets the ovary, affecting folliculogenesis and steroidogenesis. Biol Reprod. 2016, 94(1): 9.

83 Kanda S. Evolution of the regulatory mechanisms for the hypothalamic-pituitary-gonadal axis in vertebrates hypothesis from a comparative view. Gen Comp Endocrinol. 2019, 284: 113075.

84 Przygrodzka E, Plewes MR, Davis JS. Luteinizing hormone regulation of inter-organelle communication and fate of the corpus luteum. Int J Mol Sci. 2021, 22(18): 9972.

85 Davis KB, McEntire M. Influence of reproductive status, sex hormones and temperature on plasma IGF-I concentrations in sunshine bass (Morone chrysops x Morone saxatilis). Comp Biochem Physiol A Mol Integr Physiol. 2011, 158(1): 13–16.

86 Bergeron JM, Gahr M, Horan K, Wibbels T, Crews D. Cloning and in situ hybridization analysis of estrogen receptor in the developing gonad of the red-eared slider turtle, a species with temperature-dependent sex determination. Dev Growth Differ. 1998, 40(2): 243–254.

87 Iorga A, Cunningham CM, Moazeni S, Ruffenach G, Umar S, Eghbali M. The protective role of estrogen and estrogen receptors in cardiovascular disease and the controversial use of estrogen therapy. Biol Sex Differ. 2017, 8: 33.

88 Saunders DMV, Podaima M, Codling G, Giesy JP, Wiseman S. A mixture of the novel brominated flame retardants TBPH and TBB affects fecundity and transcript profiles of the HPGL-axis in Japanese medaka. Aquat Toxicol. 2015, 158: 14–21.

89 He J, Yang X, Liu H. Enhanced toxicity of triphenyl phosphate to zebrafish in the presence of micro- and nano-plastics. Sci Total Environ. 2021, 756: 143986.

90 Ying S, Qin J, Dai Z, An H, Zhu H, Chen R, Yang X, Wu W, Shi Z. Effects of LPS on the secretion of gonadotrophin hormones and expression of genes in the hypothalamus-pituitary-ovary (HPG) axis in laying Yangzhou geese. Animals. 2020, 10(12): 2259.

91 Zhu L, Li W, Zha J, Wang M, Yuan L, Wang Z. Butachlor causes disruption of HPG and HPT axes in adult female rare minnow (Gobiocypris rarus). Chem Biol Interact. 2014, 221: 119–126.

92 Ren X, Zhao X, Duan X, Fang Z. Enhanced bio-concentration of tris(1,3-dichloro-2-propyl) phosphate in the presence of nano-TiO2 can lead to adverse reproductive outcomes in zebrafish. Environ Pollut. 2018, 233: 612–622.

93 Yu Y, Ma R, Yu L, Cai Z, Li H, Zuo Y, Wang Z, Li H. Combined effects of cadmium and tetrabromobisphenol a (TBBPA) on development, antioxidant enzymes activity and thyroid hormones in female rats. Chem Biol Interact. 2018, 289: 23–31.

94 Kodavanti PRS, Coburn CG, Moser VC, MacPhail RC, Fenton SE, Stoker TE, Rayner JL, Kannan K, Birnbaum LS. Developmental exposure to a commercial PBDE mixture, DE-71: neurobehavioral, hormonal, and reproductive effects. Toxicol Sci. 2010, 116(1): 297–312.

95 Rayner JL, Enoch RR, Wolf DC, Fenton SE. Atrazine-induced reproductive tract alterations after transplacental and/or lactational exposure in male Long-Evans rats. Toxicol Appl Pharmacol. 2007, 218(3): 238–248.

96 Hayashida Y, Kawamura T, Hori-e R, Yamashita I. Retionic acid and its receptors are required for expression of aryl hydrocarbon receptor mRNA and embryonic development of blood vessel and bone in the medaka fish, Oryzias latipes. Zoolog Sci. 2004, 21(5): 541–551.

97 Janosek J, Hilscherova K, Blaha L, Holoubek I. Environmental xenobiotics and nuclear receptors - interactions, effects and in vitro assessment. Toxicol In Vitro. 2006, 20(1): 18–37.

98 Kakuta H. Challenge of creating single-agents for the treatment of type 1 and 2 diabetes by targeting retinoid x receptor. Yakugaku Zasshi-J Pharmaceut Soc Japan. 2011, 131(6): 917–923.

99 Prado MAM, Reis RAM, Prado VF, de Mello MC, Gomez MV, de Mello FG. Regulation of acetylcholine synthesis and storage. Neurochem Int. 2002, 41(5): 291–299.

100 Ferreira-Vieira TH, Guimaraes IM, Silva FR, Ribeiro FM. Alzheimer's disease: targeting the cholinergic system. Curr Neuropharmacol. 2016, 14(1): 101–115.

101 Han Y, Fu M, Wu J, Zhou S, Qiao Z, Peng C, Zhang W, Liu F, Ye C, Yang J. Polylactic acid microplastics induce higher biotoxicity of decabromodiphenyl ethane on earthworms (Eisenia fetida) compared to polyethylene and polypropylene microplastics. Sci Total Environ. 2023, 862: 160909.

102 Bradner JM, Suragh TA, Wilson WW, Lazo CR, Stout KA, Kim HM, Wang MZ, Walker DI, Pennell KD, Richardson JR, Miller GW, Caudle WM. Exposure to the polybrominated diphenyl ether mixture DE-71 damages the nigrostriatal dopamine system: role of dopamine handling in neurotoxicity. Exp Neurol. 2013, 241: 138–147.

103 Sackner-Bernstein J. Estimates of intracellular dopamine in Parkinson's disease: a systematic review and meta-analysis. J Parkinsons Dis. 2021, 11(3): 1011–1018.

104 Pan X, Kaminga AC, Wen SW, Wu X, Acheampong K, Liu A. Dopamine and dopamine receptors in Alzheimer's disease: a systematic review and network meta-analysis. Front Aging Neurosci. 2019, 11: 175.

105 Bradner JM, Suragh TA, Caudle WM. Alterations to the circuitry of the frontal cortex following exposure to the polybrominated diphenyl ether mixture, DE-71. Toxicology. 2013, 312: 48–55.

106 Ren Z, Poopal R-K, Ramesh M. Synthetic organic chemicals (flame retardants and pesticides) with neurotoxic potential induced behavioral impairment on zebrafish (Danio rerio): a non-invasive approach for neurotoxicology. Environ Sci Pollut Res Int. 2021, 28(28): 37534–37546.

107 Atkins M, Nicol X, Fassier C. Microtubule remodelling as a driving force of axon guidance and pruning. Semin Cell Dev Biol. 2023, 140: 35–53.

108 Udvadia AJ. 3.6 kb Genomic sequence from Takifugu capable of promoting axon growth-associated gene expression in developing and regenerating zebrafish neurons. Gene Expr Patterns. 2008, 8(6): 382–388.

109 Fan B, Dai L, Liu C, Sun Q, Yu L. Nano-TiO2 aggravates bioaccumulation and developmental neurotoxicity of triphenyl phosphate in zebrafish larvae. Chemosphere. 2022, 287: 132161.

110 Wang X, Yang L, Wang Q, Guo Y, Li N, Ma M, Zhou B. The neurotoxicity of DE-71: effects on neural development and impairment of serotonergic signaling in zebrafish larvae. J Appl Toxicol. 2016, 36(12): 1605–1613.

111 Shulman Y, Stavsky A, Fedorova T, Mikulincer D, Atias M, Radinsky I, Kahn J, Slutsky I, Gitler D. ATP binding to synaspsin IIa regulates usage and clustering of vesicles in terminals of hippocampal neurons. J Neurosci. 2015, 35(3): 985–998.

112 van Hooft JA, Vijverberg HPM. 5-HT3 receptors and neurotransmitter release in the CNS: a nerve ending story? Trends Neurosci. 2000, 23(12): 605–610.

113 Kodavanti PRS, Royland JE, Osorio C, Winnik WM, Ortiz P, Lei L, Ramabhadran R, Alzate O. Developmental exposure to a commercial PBDE mixture: effects on protein networks in the cerebellum and hippocampus of rats. Environ Health Perspect. 2015, 123(5): 428–436.

114 Le Melledo JM, Bradwejn J, Koszycki D, Bellavance F, Bichet D. Arginine-vasopressin and oxytocin response to cholecystokinin-tetrapeptide. Peptides. 2001, 22(8): 1349–1357.

115 Morris JF, Pow DV, Mohr E, Richter D. Production of hybrid oxytocin/vasopressin precursors and accumulation of oxytocin precursors in the rough endoplasmic reticulum of rat magnocellular neurons. Adv Exp Med Biol. 1995, 395: 49–58.

116 Fischer C, Fredriksson A, Eriksson P. Neonatal co-exposure to low doses of an ortho-PCB (PCB 153) and methyl mercury exacerbate defective developmental neurobehavior in mice. Toxicology. 2008, 244(2-3): 157–165.

117 Aluru N, Hallanger IG, McMonagle H, Harju M. Hepatic gene expression profiling of Atlantic cod (Gadus morhua) liver after exposure to organophosphate flame retardants revealed altered cholesterol biosynthesis and lipid metabolism. Environ Toxicol Chem. 2021, 40(6): 1639–1648.

118 Chang H, Lin Y, Lei K, Wang F, Zhang Q-Q, Zhang Q-Y. Role of hypomethylation of suppressor of cytokine signaling in T helper 17 cell/regulatory T cell imbalance in children with Henoch-Schonlein purpura. Zhongguo Dang Dai Er Ke Za Zhi. 2019, 21(1): 38–44.

119 Rose S, Lichtenheld M, Foote MR, Adkins B. Murine neonatal CD4(+) cells are poised for rapid Th2 effector-like function. J Immunol. 2007, 178(5): 2667–2678.

120 Janson PCJ, Marits P, Thorn M, Ohlsson R, Winqvist O. CpG methylation of the IFNG gene as a mechanism to induce immunosupression in tumor-infiltrating lymphocytes. J Immunol. 2008, 181(4): 2878–2886.

121 Martin-Subero JI, Oakes CC. Charting the dynamic epigenome during B-cell development. Semin Cancer Biol. 2018, 51: 139–148.

122 Dasguta S, Vliet SMF, Cheng, V, Mitchell CA, Kirkwood J, Vollaro A, Hur M, Mehdizadeh C, Volz DC. Complex interplay among nuclear receptor ligands, cytosine methylation, and the metabolome in driving tris(1,3-dichloro-2-propyl) phosphate-induced epiboly defects in zebrafish. Environ Sci Technol. 2019, 53(17): 10497–10505.

123 Klinman DM. Immunotherapeutic uses of CpG oligodeoxynucleotides. Nat Rev Immunol. 2004, 4(4): 248–257.

124 Blake CA, McCoy GL, Hui YY, LaVoie HA. Perinatal exposure to low-dose DE-71 increases serum thyroid hormones and gonadal osteopontin gene expression. Exp Biol Med. 2011, 236(4): 445–455.

125 Hu Z, Lin DM, Yuan JS, Xiao T, Zhang HS, Sun WY, Han NJ, Ma Y, Di XB, Gao MX, Ma JF, Zhang JH, Cheng SJ, Gao YN. Overexpression of osteopontin is associated with more aggressive phenotypes in human non-small cell lung cancer. Clin Cancer Res. 2005, 11(13): 4646–4652.

126 Ahmed M, Behera R, Chakraborty G, Jain S, Kumar V, Sharma P, Bulbule A, Kale S, Kumar S, Mishra R, Raja R, Saraswati S, Kaur R, Soundararajan G, Kumar D, Thorat D, Sanyal M, Ramdasi A, Ghosh P, Kundu GC. Osteopontin: a potentially important therapeutic target in cancer. Exp Opin Therap Target. 2011, 15(9): 1113–1126.

127 Vancea A, Serban O, Fodor D. Relationship between osteopontin and bone mineral density. Acta Endocrinol (Buchar). 2021, 17(4): 509–516.

128 Chen X, Gu X, Zhao X, Ma X, Pan Y, Wang X, Ji R. Species-dependent toxicity, accumulation, and subcellular partitioning of cadmium in combination with tetrabromobisphenol A in earthworms. Chemosphere. 2018, 210: 1042–1050.

129 Berger RG, Lefevre PLC, Ernest SR, Wade MG, Ma Y-Q, Rawn DFK, Gaertner DW, Robaire B, Hales BF. Exposure to an environmentally relevant mixture of brominated flame retardants affects fetal development in Sprague-Dawley rats. Toxicology 2014, 320: 56–66.

130 Yan H, Hales BF. Effects of organophosphate ester flame retardants on endochondral ossification in ex vivo murine limb bud cultures. Toxicol Sci. 2019, 168(2): 420–429.

131 Kakutani H, Yuzuriha T, Akiyama E, Nakao T, Ohta S. Complex toxicity as disruption of adipocyte or osteoblast differentiation in human mesenchymal stem cells under the mixed condition of TBBPA and TCDD. Toxicol Rep. 2018, 5: 737–743.

132 Haggard DE, Das SR, Tanguay RL. Comparative toxicogenomic responses to the flame retardant mITP in developing zebrafish. Chem Res Toxicol. 2017, 30(2): 508–515.

133 Bastida MF, Perez-Gomez R, Trofka A, Zhu J, Rada-Iglesias A, Sheth R, Stadler HS, Mackem S, Ros MA. The formation of the thumb requires direct modulation of Gli3 transcription by Hoxa13. Proc Natl Acad Sci USA. 2020, 117(2): 1090–1096.

134 He W, Burwood G, Fridberger A, Nuttall AL, Ren T. An outer hair cell-powered global hydromechanical mechanism for cochlear amplification. Hear Res. 2022, 423: 108407.

135 Patuzzi R. Ion flow in cochlear hair cells and the regulation of hearing sensitivity. Hear Res. 2011, 280(1-2): 3–20.

136 Poon E, Powers BE, McAlonan RM, Ferguson DC, Schantz SL. Effects of developmental exposure to polychlorinated biphenyls and/or polybrominated diphenyl ethers on cochlear function. Toxicol Sci. 2011, 124(1): 161–168.

137 Meng X, Li F, Wang X, Liu J, Ji C, Wu H. Combinatorial immune and stress response, cytoskeleton and signal transduction effects of graphene and triphenyl phosphate (TPP) in mussel Mytilus galloprovincialis. J Hazard Mater. 2019, 378: 120778.

138 Meng X, Li F, Wang X, Liu J, Ji C, Wu H. Toxicological effects of graphene on mussel Mytilus galloprovincialis hemocytes after individual and combined exposure with triphenyl phosphate. Mar Pollut Bull. 2020, 151: 110838.

139 Chen L, Li J, Chen Z, Gu Z, Yan L, Zhao F, Zhang A. toxicological evaluation of graphene-family nanomaterials. J Nanosci Nanotechnol. 2020, 20(4): 1993–2006.

140 Yang K, Li Y, Tan X, Peng R, Liu Z. Behavior and toxicity of graphene and its functionalized derivatives in biological systems. Small. 2013, 9(9-10): 1492–1503.

141 Zou W, Zhang X, Ouyang S, Hu X, Zhou Q. Graphene oxide nanosheets mitigate the developmental toxicity of TDCIPP in zebrafish via activating the mitochondrial respiratory chain and energy metabolism. Sci Total Environ. 2020, 727: 138486.

142 Wu J, Jin Z, Zheng H, Yan L-J. Sources and implications of NADH/NAD(+) redox imbalance in diabetes and its complications. Diabetes Metab Syndr Obes. 2016, 9: 145–153.

143 Yan L-J. Redox imbalance stress in diabetes mellitus: role of the polyol pathway. Animal Model Exp Med. 2018, 1(1): 7–13.

144 Wang S-C, Gao Z-Y, Liu F-F, Chen S-G, Liu G-Z. Effects of polystyrene and triphenyl phosphate on growth, photosynthesis and oxidative stress of Chaetoceros meülleri. Sci Total Environ. 2021, 797: 149180.

145 Ma W, Sun Z, Zhang X, Tang X, Zhang X. Alleviation of tris(2-chloroethyl) phosphate toxicity on the marine rotifer Brachionus plicatilis by polystyrene microplastics: features and molecular evidence. Int J Mol Sci. 2022, 23(9): 4934.

16

Adverse Outcome Pathways (AOPs) of Combined Pollutant Mixtures and Their Toxicogenetic Endpoints

Hossam El Din H. Abdelhafez[a,*], *Mohamed Bedair M. Ahmed*[b], *Magdy Moheb El-Dein Saad*[b]

[a] *Mammalian and Aquatic Toxicology Department, Central Agricultural Pesticides Laboratory, Agricultural Research Center, Dokki, Giza, Egypt*
[b] *Department of Food Toxicology and Contaminants, Institute of Food Industries & Nutrition, National Research Centre, Dokki, Cairo, Egypt*

*Corresponding author.
E-mail: hossam4003@live.com

16.1 Introduction

Chemicals that are known to be hazardous under a variety of categories, such as metals, organic solvents, and usage classes, include cosmetics, medicines, endocrine disruptors (EDs), and industrial chemicals. About 218 chemicals, including 102 pesticides, 41 organic solvents, 48 other organic substances, and 27 metals or inorganic compounds, are categorized as extremely toxic [1]. Moreover, pollutants in the environment rarely occur as individual substances; instead, they exist in complex mixtures that might have synergism or antagonism. While individual levels of chemicals may be below the lowest observed adverse effect concentrations (LOAEC) or detection limits, the mixture as a whole may still cause negative impacts [2]. Furthermore, because of synergistic effects, it is generally known that "mixture effects" might be higher than those caused by the most potent single chemical in a mixture [3]. Models, exposure, toxicity, risk, and impacts are used to assess the mixture of chemicals' effects on the environment, and the combination of these models should provide reliable evidence for decisions regarding regulatory risk management [4]. Consequently, biological effect screening is required for a more reliable risk assessment of the status of mixture pollution in the environment. For example, scientific reports support associations between exposure to chemicals and endocrine disruptive effects, leading to disorders in humans and wildlife and inducing a feeling of threat to communities. Chemical exposure may be a key factor in the high cancer incidence, and the constant finding of novel substances; assessing the carcinogenic potential of compounds is critical in risk assessment [5], Additionally, it is necessary for building decision-making, assessing, and visual tools that provide for the adverse impacts of outcomes.

Currently, a vast variety of chemical pollutants have the potential to cause cancer, which is assessed using genotoxicity procedures. Genotoxicity investigations are frequently conducted using a (tiered) test approach, in which a positive result calls for more testing and an in vitro positive result requires in vivo confirmation [6]. Testing for cytotoxicity and mutagenicity is of the utmost significance and relies on well-established in vitro systems using cell lines from both bacterial and mammalian sources. Model rodents may then be used for in vivo testing. New approach techniques, either alone or in conjunction with conventional in vivo tests, aim to improve, reduce, or ultimately replace animal testing. These tests may enable the achievement of both protection and prediction goals [7]. Genomic, or "Omics" technologies provide an accurate analysis of biological events down to the level of miRNAs, proteins, and metabolites, utilizing genetic technology in toxicology is termed toxicogenomics [8]. Toxicogenomics is the use of microarray technology as an element of omics technologies in toxicology to determine molecular gene expression patterns (also known as molecular biomarkers) that can be used to predict toxicity, as well as the identification of molecular pathways that lead to the development of toxicological effects. Now, scientists can simultaneously assess the reaction of thousands of genes and their products from a single sample due to major developments in OMIC technologies [9]. Automated analytical techniques (bioinformatics), computational tools,

Toxicological Assessment of Combined Chemicals in the Environment, First Edition. Edited by De-Sheng Pei and Yiyun Liu.

and systems biology approaches are being created to organize and analyze the data [10]. These improved molecular insights can deepen our understanding of the mechanisms causing adverse outcomes (AOs), thereby facilitating our transition toward more predictive and informative screening methods [11].

Considering the growing human exposure to chemical pollutants in the environment and the limitations of traditional toxicity tests, there is a need today for advanced risk assessment techniques [12]. The field of toxicology faced a paradigm shift at the beginning of the twenty-first century, moving from animal-based (in vivo) methods to new approach methodologies (NAMs), which mostly rely on molecule- and cell-based (in vitro), and computational (in silico) procedures. Concurrently, next-generation risk assessment (NGRA) is a method for minimizing the use of animal testing that raises many ethical, health-related costs and efficacy [13]. A clear, evidence-based mechanistic knowledge framework that would simultaneously connect molecular perturbations to AOs relevant to human health is essential for increasing regulatory acceptance of information from NAMs and risk assessment studies [14]. To assist in establishing these bridges between NAMs and NGRA, adverse outcome pathways (AOPs) seem to be a useful tool. According to the International OECD Environmental, Health and Safety (EHS) Programme, AOPs have become a worldwide framework [15]. The use of AOPs, which links pollutant exposure to physiological responses and subsequently to their impacts on ecosystems and populations, is one of the main challenges for environmental risk assessment (ERA) [16]. The consequences of "real-life" chemical mixtures present a challenge to risk experts. AOP is a crucial tool in this situation for examining the impacts of mixtures to prioritize remediation efforts by bringing attention to the chemicals and AOPs that actually have an impact [17].

Pesticides and endocrine-disrupting chemicals (EDS) are periodically exposed to the population and animals in the environment. According to recent findings from several workers, the combination of EDS and pesticides may have a more severe impact on human health than each of them individually. Few studies have reported the synergistic effects of heavy metals and plasticizers on reproductive toxicity, genotoxicity, and endocrine hormones as biological events *in vivo* or epidemiological studies [18]. To reduce their negative environmental impacts, further research is needed to assess the relationships between pesticide, heavy metal, and phthalate concentrations in biological systems. European pesticide regulations require extensive animal testing, including standard toxicity studies, and specialized assessments of endocrine disruption

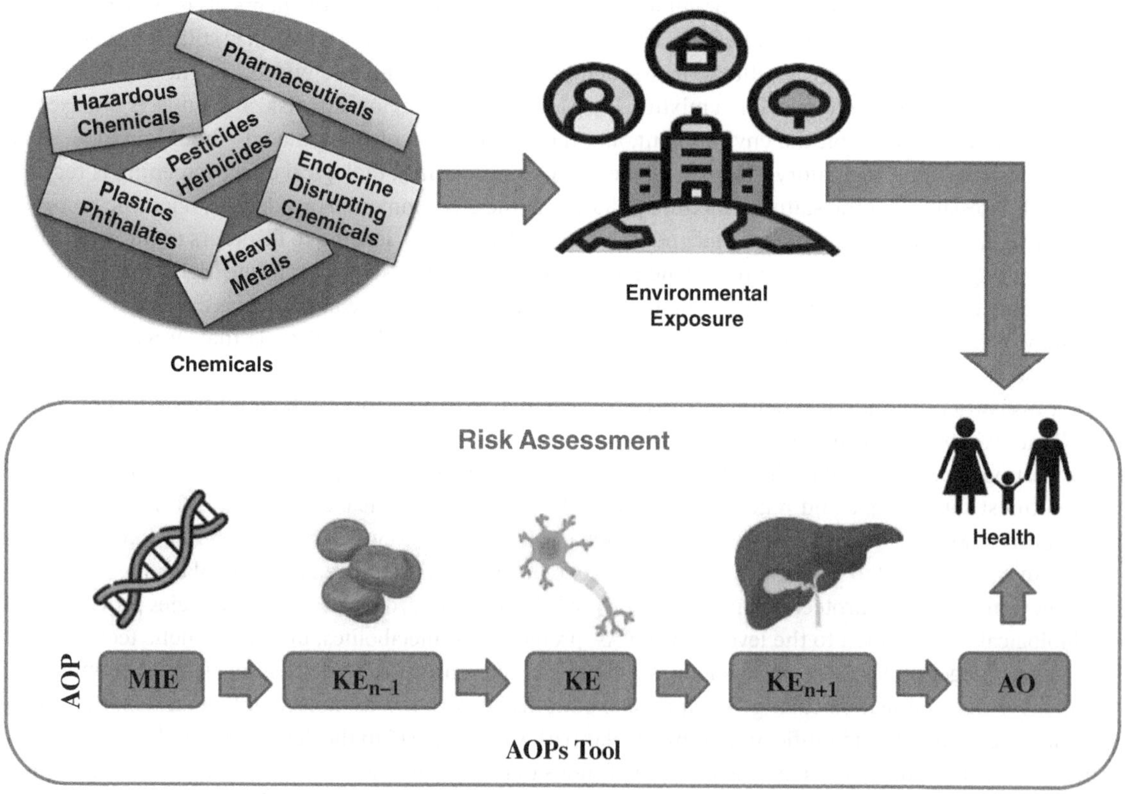

Figure 16.1 Graphical diagram showing the chapter's overview.

and developmental neurotoxicity. Additionally, the law mandates the use of human observational evidence and in vitro evidence in risk assessment, if appropriate and available. The AOP framework serves as a practical tool for "building" these many evidence streams together [19].

This chapter introduced the concept of screening AOP at various biological levels, enabling a link between environmental, molecular, physiological, and ecological responses (as shown in Figure 16.1). It also provided information on omics applications in AOP and case studies for AOP by toxicogenetics endpoints using two different chemical types (insecticides and EDS). The article also covered challenges and prospective developments.

16.2 Originated AOP and Approaches of Research in Public Databases

A paradigm shift in toxicology testing was advocated in a report published by the National Research Council (NRC) in 2007 titled "Toxicity testing in the twenty-first century: a vision and a strategy." The NRC report promoted the use of a testing platform based on in vitro methods rather than in vivo, and that takes a pathway approach to study perturbations of biological systems and important toxicity pathways using a combination of bioinformatics analysis and a wide range of high-throughput in vitro tests, preferably with tissues and cells. This NRC's call to action gave origin to the AOP framework [20]. In this research, multiple databases were utilized for the development of the AOP framework for assessing chemical safety. The concept draws into novel approaches for predicting toxicity in the twenty-first century until reducing cost, time, and animal models. Based on the analysis, it is possible to connect the definitions of AOPs in Table 16.1. They have been published in the scientific literature with their constituent parts (the MIE, the key events [KEs], and the AO) in a variety of ways, depending on the available information. As a result, linkages, plausible or hypothetical [16], can be established and predicted by analyzing associated scientific literature in the ScienceDirect, Web of Science, PubMed, and Google Scholar databases that are published regarding the pollutants that had been addressed previously, modeling techniques, and software tools in Table 16.2.

The core of a framework for toxicological knowledge is built up of fully developed AOPs, which are anticipated to lead experimental testing methodologies and support risk assessments and regulatory decision-making. Although there are now more than 100 AOPs being developed, only a small number of them may be considered complete or substantially complete. Even incomplete AOPs, however, may help in chemical classification and "read-across" (predicting a molecule's toxicity based on structural similarities to other molecules). An integrated approach to testing and assessment (IATA) and the creation of integrated testing strategies (ITS) could be based on AOPs that are more comprehensive and have linkages that incorporate quantitative information [21, 22]. The *Environmental and Molecular Mutagenesis* (EMM) and the Environmental Mutagenesis and Genomics Society (EMGS) have been encouraging the development of AOPs in the field of genetic toxicology ever since the Organization for Economic Co-operation and Development (OECD) AOP Programme was established [23]. In recent years, several large-scale initiatives have been launched to develop, collect, organize, and leverage high-dimensional data derived from advanced biological testing methods, particularly those incorporating omics technologies. These ambitious programs transcend the scope of individual studies and smaller-scale consortia, aiming to

Table 16.1 Definitions of some elements and terms relating to AOP.

Adverse outcome pathway (AOP)	An AOP is a chain of events that starts with exposure to a single or mixture of chemical substances and final adverse toxic effects on the individual or population
Intermediate event	Biological events that lie "between the molecular initiating event and the apical outcome"
Key event (KE)	An observable biological event or its marker that is crucial to the mode of action that determines the outcome (i.e. necessary but not always sufficient in and of itself); important occurrences are measurable and repeatable
Key event relationship (KER)	One key event is connected to another by a scientific relationship that establishes a directed relationship between them (designating one as upstream and the other as downstream) and makes it easier to extrapolate or infer the state of the downstream key event from the state of the upstream key event
Mode of action (MoA)	A set of crucial events that can be explained biologically and lead to an impact
Initiating event (MIE)	The initial event of chemical–biological interaction within the organism that starts the pathway

Table 16.2 AOP networks (computation tools) to analyze and understand different pathways.

Tool	Description	URL
AOP-Wiki	A free platform that serves as a single storage place for developed or in-development AOPs, allowing for the sharing of KEs and collaboration among AOP developers	https://aopwiki.org/
AOP-helpFinder	An application for doing a comprehensive review of the literature from the PubMed database to aid in the construction of AOPs. By using text mining tools in conjunction with graph theory, the automatic discovery of documented links between stressors and critical events is possible	http://aophelpfinder.u-paris-sciences.fr/
Biovista Vizit	A graphical exploratory tool that allows scholars to see, study, and create interactive networks of AOPs and KEs of interest	https://aopkb.biovista.com/
Wiki Kaptis	Improves AOP knowledge, visualization, and interrogating using AOP network functionality	https://wikikaptis.lhasacloud.org/
AOPXplorer	A computational tool that allows for the automated visual illustration of AOPs and the networks that connect them	http://datasciburgoon.github.io/aopxplorer

revolutionize toxicological assessments. Prominent among these are the Environmental Protection Agency's (EPA) ToxCast program; the Toxicology in the Twenty-first Century Consortium (Tox21C); and an unprecedented tri-agency collaboration involving the EPA, National Institutes of Health (NIH), and Food and Drug Administration (FDA).

16.3 AOP Development Program

An AOP is a crucial framework, a global paradigm, used to assess the biological events that explain how molecular perturbations, starting at the lower levels of biological organization, lead to higher-order toxicological effects in populations that are caused by environmental pollutants until reducing our chemical tests in animal models to safeguard human and environmental genetic health [24]. AOP is a conceptual structure and practical technique for organizing and evaluating toxicology data that demonstrate how molecular perturbations brought on by stresses from the environment might pose risks to individuals or populations [25]. For regulatory toxicological scientists, the greatest challenge is evaluating an increasing variety of chemicals while at the same time decreasing the utilization of animal models, funding, and time for research [26]. In the same context, building blocks for the development of AOP are discussed, such as causal relationships (known as key event relationships, or KERs) between quantifiable/observable biological changes that are crucial (known as KEs) to the progression from a molecular initiating event (MIE) to an AO, based on a wide range of mechanistically based data in risk assessment that is considered crucial for regulatory decision-making [16]. Likewise, the idea of identifying a series of measurable biological processes is to visualize a toxicological response that improves the weight of the evidence for toxicological effects [27, 28] and that the systematic and transparent assembly of data based on AOP descriptions is a consequence of extrapolating initial or intermediate assessments of biological disruption. It is also necessary to deeply clarify the KEs and KERs that connect an MIE and AO to enable the use of a variety of mechanistically based data in risk assessment as shown in Figure 16.2.

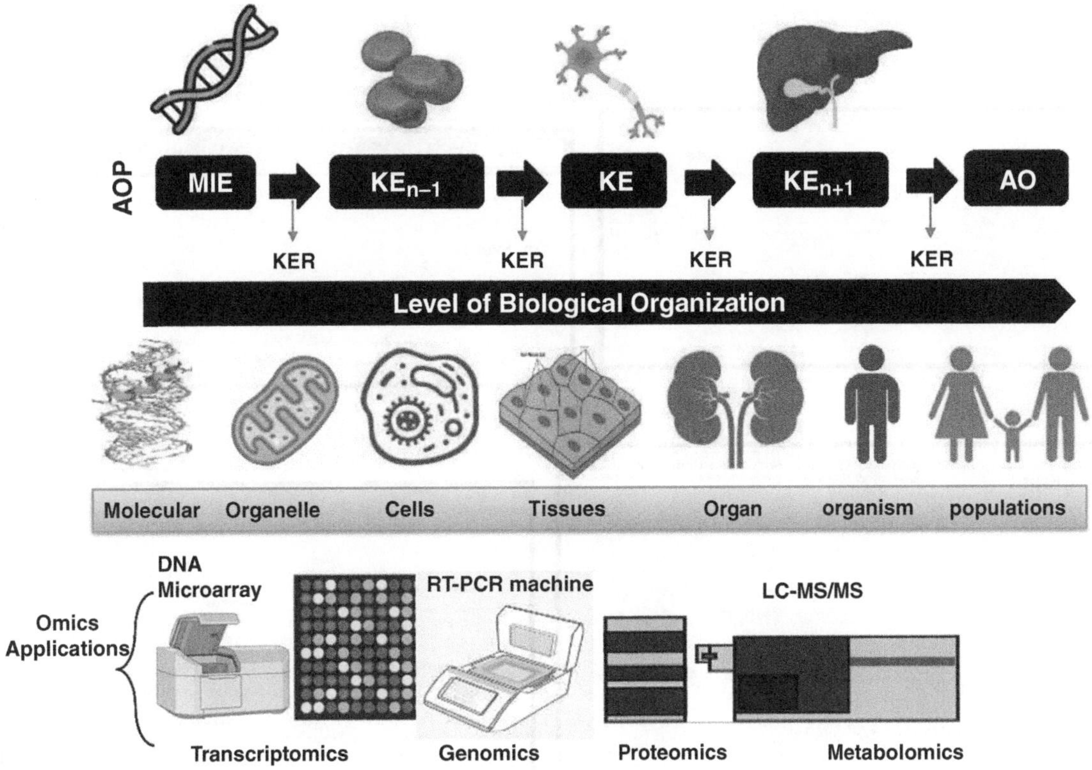

Figure 16.2 AOP building blocks, biological organization levels, and omics applications.

It became evident early in the AOP development program that producing static documents was not an effective method for advancing or disseminating AOP expertise. An AOP-Wiki was developed as a user-friendly, open-source interface as a result, supporting both collaborative AOP development and knowledge exchange about AOP (www.aopwiki.org, accessed 9 October 2014) [17]. Table 16.2 presents the comprehension of how many pathways of action interact can be improved by learning how to analyze the relationship between multiple chemical exposures and detrimental health effects utilizing AOP networks (computational tools). Figure 16.3 shows an example of using the AOP knowledgebase to create an AO from AOP-WiKi by inserting MIE interaction chemicals as biological events and ID belonging to KE from AOP-Kaptis.

16.3.1 Strategies, Principles, and Best Practices for AOP Development

Villeneuve et al. (2014) have outlined six strategies for the development of AOPs. The first strategy involves the developer starting with an apical AO of interest and then progressively delving down to lower levels of biological organization to identify a specific MIE. This approach is referred to as top-down AOP development. However, this strategy has its limitations as it lacks predictive tests and is not well-understood [29]. Second, the developer starts with a well-defined MIE and begins linking to effects at higher levels of biological organization. Bottom-up AOP development, separate from antagonistic effects, and to connect binding to risks, additional tests, and measures are needed [30]. Third is the middle-out AOP development, where the developer starts with an observable phenotype or biological measurement (i.e. a KE) that is neither thought to be directly perturbed by exogenous chemicals or stressors nor considered to have regulatory relevance in and of itself [31]. Fourth, the developer starts with a well-defined sequence of biological events linking an MIE to an AO for a single, well-studied chemical and then assembles evidence supporting generalization of that "motif of failure" for other chemicals/stressors that cause the same type of perturbations; this addressed as AOP development from case-study. Fifth, the developer begins with an AOP that has been thoroughly examined in a specific animal model or group of species. Research efforts are focused on identifying which KEs and KERs from established AOPs are conserved across different

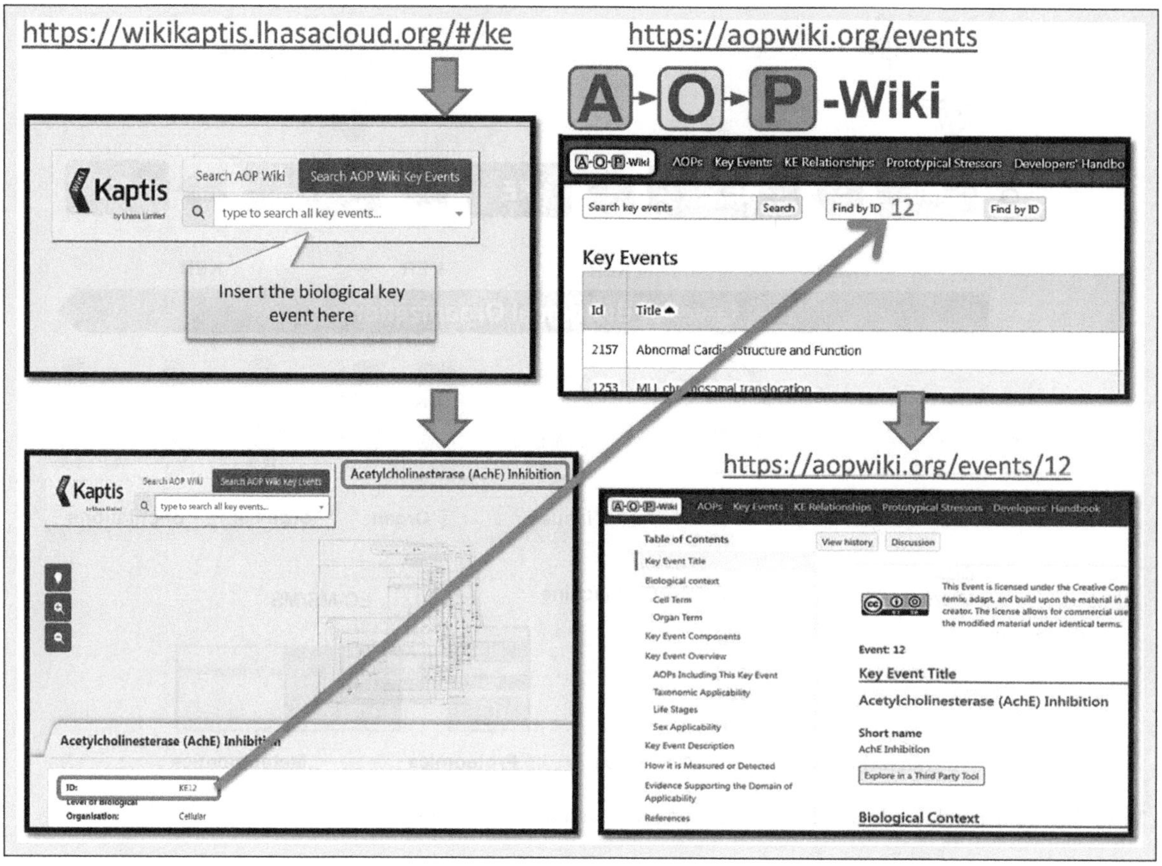

Figure 16.3 A simple example of using AOP knowledgebase to create an AO from AOP-Kaptis and AOP-Wiki.

organisms or taxonomic groups. When certain KEs or KERs are not conserved, scientists work to develop alternative, species-specific KEs and KERs. This process, known as "AOP development by analogy," allows researchers to adapt existing AOPs to new biological contexts, facilitating cross-species extrapolation in toxicology. Similarly, the analogy between AOPs developed for fish and those that may occur in birds has been considered [32]. Sixth, in AOP development from a data-mining strategy, the developers use high content and/or high throughput data sets (such as "omics," ToxCast screening data), as well as other sorts of automated literature and data-base mining tools, to extrapolate (usually statistically) correlations between KEs. This strategy is most often used in the early stages of AOP development.

Villeneuve et al. (2014) have outlined five core principles for the development of AOPs which are as follows. First, AOPs are not chemical-specific. Second, AOPs are modular and made up of reusable parts. In particular, KEs are observable changes in biological state that are necessary but not always sufficient for progression from an MIE to an AO [28, 33]. In terms of best practices, it is preferable to identify at least one KE at each level of biological organization (molecular, cellular, tissue, organ, organ system, and individual). This includes (1) a description of the biological state being observed; and (2) a description of the types of methods that can be used to observe or measure. Third, one AOP can be viewed as a practical unit for development and assessment. Fourth, a set of AOPs that share KEs and KERs is referred to as an "AOP network," and a network will serve as the functional unit of prediction. Fifth, AOPs are live documents that can change over time with updates in existing knowledge and developed an AOP in three operationally defined stages.

Putative AOP: Development of the building of a hypothesized collection of KEs and KERs based mostly on biological plausibility and/or statistical inference. Assembled partial AOPs with insufficient connectivity between the MIE and AO due to known gaps and uncertainties.

Qualitative formal AOP development: KEs assembled with descriptions of how the KEs can be quantified, and KERs assembled with direct proof in addition to plausibility or statistical inference. Formal in the knowledge that the information in the descriptions is assembled under internationally harmonized OECD guidance.

Quantitative AOP development: Assembly of KEs based on descriptions of how the KEs can be measured and the precision and accuracy with which measurements are made, as well as KERs supported by a quantitative understanding of what magnitude and/or duration of change in the upstream KE, is required to elicit some magnitude of change in the downstream KE.

16.3.2 The Applications of Omics Analysis and Their Contribution to the Development of AOPs

On the basic toxicology side, a lot of information is collected via histopathology, toxicokinetics xenobiotic inhibition enzymes (XME assays), functional reproductive assays, and information about animal experiments. However, the growing toxicogenomics and "omics" technologies produce much more extraordinary and complex data [34]; "omics," which includes genomics, proteomics, and metabolomics, provides the analysis of toxicant-induced changes in macromolecule networks that involve perturbations in gene, protein, and metabolite space [35]. To determine potency estimations, it can also be used quantitatively. Implementing omics also helps develop alternatives to animal testing. However, more research at the gene expression level examining dose–response relationships for a wide range of chemicals is needed to identify whether important changes in gene expression occur in a dose-dependent way and how these alterations should be interpreted in terms of adversity. Toxicogenomics must also be globally standardized in many areas before it can be used in risk assessment [6].

The two main processes in the present method for assessing hazards are: (1) defining the hazard of cause for concern, and (2) quantifying the risk by concentration–response measurements to obtain a toxicity threshold or value of reference based on an apical outcome [36, 37]. One major challenge for regulation toxicologists in hazard assessment is the growing demand for managing a huge number of synthetic substances in the environment [38]. In this context, researchers have focused on leveraging high-content datasets to elucidate the precise mechanisms by which chemicals interfere with biological processes [39]. Within the AOP framework, omics technologies are increasingly being recognized as invaluable tools [40]. Through the many techniques and models, such as rats, zebrafish (*Danio rerio*) [20, 41, 42], and *Caenorhabditis elegans* [43]. Omics datasets can be used to verify and support existing evidence of KEs across many taxonomic groups and levels of biological organization. Omics data allow readouts of changes in gene expression through a particular biological pathway, which can be utilized to choose measurable and relevant biomarkers [44, 45]. Although the fact that omics have been around for more than 10 years, regulated applications have been slow to accept the omics data. The toxicology community is debating how to use omics data in a legal context and how to employ AOPs to improve decision-making [46].

In this section, the omics are divided into four main categories: genomics, transcriptomics, proteomics, and metabolomics. First, genomics stands out as a pivotal tool for investigating the effects of a wide array of substances on biological systems. This field enables researchers to comprehensively assess how pharmaceuticals, industrial pollutants, and various environmental stressors alter gene expression patterns, potentially disrupting cellular functions [47]. The sequencing of genomics technologies began with conventional capillary electrophoresis (CE) sequencing, also known as "Sanger" or "first-generation" sequencing methods, which were replaced by next-generation sequencing (NGS) to enable massively parallel sequencing of billions of DNA molecules simultaneously. Ultrafast DNA sequencing is the third generation of DNA sequencing. In another area, technology like oligonucleotide microarrays have enabled great advances and given rise to entirely new disciplines like epigenetics and hybrid technologies, such as chromatin immunoprecipitation coupled to DNA microarray (ChIP-chip) or sequencing, and are another form of technology. At the forefront of genomic innovation, CRISPR/Cas9 technology has emerged as a groundbreaking tool, revolutionizing the field of toxicology. This precise gene-editing system enables toxicologists to directly modify specific gene sequences, providing unprecedented insights into how genetic alterations influence gene function and, consequently, an organism's response to toxicants [48, 49]. Applications of genomics in AOP that highlight mechanism-of-action research and discoveries will focus research on developing disease-specific therapies. These investigations will provide new perspectives on the toxicity mechanisms and KEs and suggest strategies for minimizing, avoiding, or bypassing the natural cell responses to reduce AOs. Second, transcriptomics shows that, although there are 21 000 genes in the human genome, only a small portion of those genes are active. To understand toxicological mechanisms or mode-of-action through evaluating the transcript, it is necessary to analyze gene expression in response to environmental contaminants. The transcriptome consists of about 100 000 MiRNA molecules, including microRNAs (miRNAs) and long noncoding RNAs (lncRNAs). Oligonucleotide microarrays, NGS methods, or real-time polymerase chain reaction (RT-PCR) are used in transcriptomics investigations; however, RT-PCR is less common than microarrays or sequencing. Microarrays and RNA-seq provide quantitative measurements of mRNA expression and capture dynamic gene expression patterns for thousands of genes after exposure to pollutants. In contrast, gene sequencing provides a static measurement of DNA. Microarrays can be used to connect gene changes to chemically induced phenotypes, enabling predictive toxicity and understanding of mode-of-action [50]. The three most prominent uses of transcriptomics in AOP can be summarized as grouping/classifying chemicals, elucidating mechanisms of action, and identifying biomarkers. RNA-seq shows promise due to its higher sensitivity, better accuracy, and greater dynamic range. Proteomics, on the other hand, poses more complex and challenging measurements compared to the previous two types because it is highly dynamic, varies from cell to cell, and involves interactions with other molecules. Many proteomics studies are carried out using liquid chromatography/mass spectrometry (LC/MS) due to its accuracy, speed, selectivity, sensitivity, and throughput [51]. Due to its inherent limitation, which prevents the detection of low-abundance proteins, selected reaction monitoring (SRM) or multiple reaction monitoring (MRM) is employed. SRM/MRM enables the detection of low-abundance proteins and is inherently multiplexed, allowing for the detection of tens to hundreds of proteins, and exhibits enhanced reproducibility, enabling multiple laboratories to generate consistent results. Proteomics finds application in AOP through human hazard assessment of chemicals, aiding in the identification of toxicant protein targets to elucidate mechanisms of action (MOA). Another application involves the development and validation of biomarkers.

Metabolomics, the study of metabolites, encompasses amino acids, lipids, organic acids, and sugars that serve as intermediates or end products of metabolism. It involves characterizing specific metabolites based on their concentration or other parameters or identifying all metabolites present in each cell or organism at a specific time (global metabolite profiling). Mass spectrometry-based metabolomics workflows facilitate targeted or untargeted metabolomics experiments. Targeted metabolomics workflows measure the relative abundances and concentrations of preselected metabolites, typically associated with specific metabolic pathways. Researchers can link metabolites to biological processes through an interpretation phase. The metabolome sensitively reflects even minor alterations in DNA or proteins. It also portrays an organism's response to environmental changes. Metabolomics contributes to informed decision-making by aiding in biomarker discovery and drug safety assessment [46].

16.4 Pollutants

Complex mixtures may have negative long-term consequences on ecosystems and pollutants frequently affecting an organism's behavior, gene expression, and physiological function, which can affect its bioenergetics, endocrine, and neuroendocrine functioning, as well as its immunity; even at concentrations below their no-observable-adverse-effect concentrations [52]. Moreover, chemical interactions within complex mixtures can produce undesirable effects. A plausible MOA can be postulated and used to inform quantitative adverse outcome pathways (qAOPs) for establishing safety criteria and assessments for environmental agents. Long-term bioassays in animals are conducted to understand the molecular basis of cancer generated by AOs [28, 53].

16.4.1 Case Study I: Summary of AOP Linking Mode of Action in Pesticides Mixture

Due to the large number of pesticide mixtures in the environment [54], robust predictive models capable of estimating the toxicity of mixtures with acceptable accuracy are necessary [55, 56]. The researchers provide a strategy for AOP development that considers toxicity data from a wide range of organophosphate and carbamate insecticides that work by inhibiting AChE as MOA. Furthermore, much of the existing material on AChE inhibition is the MIE that connects KEs to reach acute mortality in populations as AO. It can be discovered in various tissues, including nerves and muscles, central and peripheral tissues, and motor and sensory fibers, but it has the greatest distribution in the blood, brain, and muscles. The neurotransmitter acetylcholine (ACh) is hydrolyzed by it [57]. The interaction of the chemical with the anionic site of the enzyme blocks the site for ACh, resulting in a build-up of ACh at synapses (KE2) and unregulated excitation (KE3) at neuromuscular junctions, preganglionic neurotransmitters, and postganglionic nerve-endings in the autonomous nervous system and neurotransmitters in the brain and CNS [58] as shown in Figure 16.4.

Recently, low (16 g/L) and high (32 g/L) concentrations of fipronil and low (42 mg/L) and high (64 mg/L) concentrations of imidacloprid were confirmed to affect ATPase-linked membrane transport and oxidase reductase activity as KES by RNA sequencing and microarray in a *Daphnia magna* model [59]. Moreover, Reinwald et al. investigated the use of insecticide

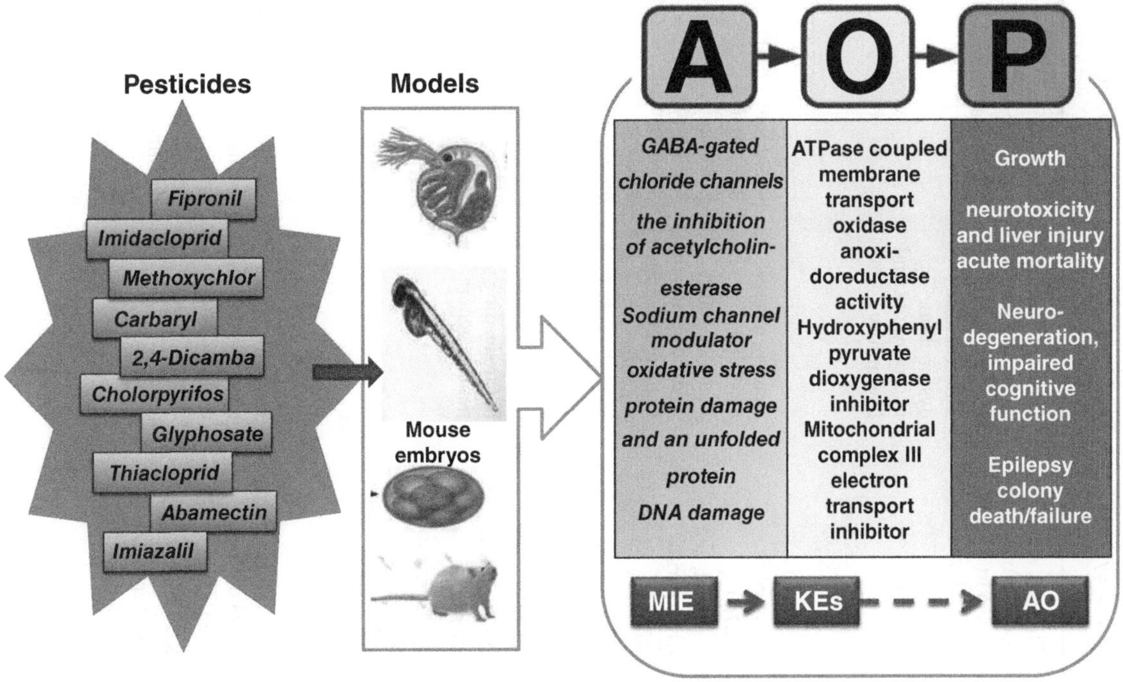

Figure 16.4 Case studies on some pesticides employing adverse pathway frameworks using some pesticides and intuition multiple models (rat, mouse embryonic cells, Zebrafish embryos, and Daphnia magna) with following three blocks of AOP (Molecular Initiation Events MIE; Biological Key Events KEs; and AO adverse outcome).

mixtures such as (Abamectin, Carbaryl, Chlorpyrifos, Fipronil, AImidacloprid, and Methoxychlor) in Zebrafish embryos model-induced multiple biological KEs such as glutamate-gated chloride channel modulator AChE inhibitor, AChE inhibitor, GABA-gated chloride channel antagonist, nACh competitive modulator, and sodium channel modulator by qPCR and library sequence in Bioinformatics analysis tool endpoint which leads to AOs such as acute mortality, neurodegeneration, and impaired cognitive function [11]. Besides, 2,4-D, dicamba, and glyphosate appeared separately and in combination. The effects gradually appeared, first in 2,4-D, which was found to induce oxidative stress and an unfolded protein response as AO, then in dicamba, which slightly increased oxidative stress, and finally in glyphosate, which only had a minor impact on WT mouse embryonic stem cells. These effects were caused by oxidative stress (*Srxn1* and *Blvrb*), protein damage, and an unfolded protein response (*Ddit3*), the activation of a DNA damage response (*Bscl2* and *Rtkn*), and p53-mediated cellular stress (Btg2) as KEs by cytotoxicity testing/by flow cytometry (Luminex) and ToxTracker assay system [60]. Similar findings were made by Sprenger et al. who discovered that different doses of imazalil, thiacloprid, and clothianidin exposure to adult female rats/HepaRG cells caused DNA damage as MIE, which led to KEs involving P13K-like kinase activation ATM and ATR kinases MRN Sensor complex, and upregulation transcription regulator (*TERF2*). Mismatch repair less upregulation RFC5 enzyme DNA synthesis less by Comet assay, γH2AX assay (real-time RT-PCR), and RNA sequencing [7]. Finally, for example, putative AOP, a review of animal data from (OECD) repeats dosage studies on Wistar and Sprague Dawley rats conducted between 1974 and 2016. Exposed to pesticides (18 compounds), including herbicide, fungicide, insecticide, and acaricide, to investigate links between toxicity findings and carcinogenicity studies through (28 and 90 days). These compounds appeared disordered molecule-by-molecule such as hydroxyphenylpyruvate dioxygenase inhibitor, succinate dehydrogenase inhibitor, acetylcholinesterase inhibitor, and mitochondrial complex II electron transport inhibitor by cytotoxicity endpoint tools [61] (Table 16.3).

16.4.2 Case Study II: Summary of AOP Linking Mixture Endocrine Disruption Pathways

EDS have raised concerns regarding their effects on sexual and reproductive health. People are exposed to these compounds on a huge scope when they leak into the environment from a wide range of consumer products, including electronic gadgets, building materials, cosmetics, food packaging, and medical devices [62]. Exposure to EDC mixtures, whether short-term or long-term, can lead to their significant accumulation in organisms. This can cause various adverse effects, including disruption

Table 16.3 Examples of AOP in mixtures of pesticides by Wiki Kaptis and AO by AOP-Wiki.

Chemical	Model (in vivo/ in vitro)	Doses	Description	MoA (mode of action)	KEs (ID by Wiki Kaptis)	AO by AOP-Wiki	Tool endpoint	Ref.
Fipronil imidacloprid	Daphnia magna	Concentrations for fipronil were LE 16 μg/L and HE 32 μg/L, imidacloprid was LE 42 mg/L, and HE 64 mg/L	Fipronil interfered with molecular initiation involved in ATPase-coupled transmembrane transport and transcription regulation, imidacloprid affected oxidoreductase activity	Fipronil GABA-gated chloride channels Imidacloprid the inhibition of acetylcholinesterase	Fipronil: ATPase coupled membrane transport KE1446 Imidacloprid oxidoreductase activity KE887	Growth inhibition by Decreased Na-K ATPase Mitochondria dysfunction, neurotoxicity, and liver injury	RNA sequence and Microarray	[59]
Abamectin, Carbaryl, Chlorpyrifos, Fipronil, Imidacloprid, and Methoxychlor	Zebrafish embryos	Sub-lethal concentrations of pesticides were divided into low, medium, and high doses	Identified biological processes that cause harm as suitable for MoA, which was connected to the three main phases of cardiac muscle, oxygen transport hypoxic stress neural development, and plasticity	Abamectin: Glutamate gated chloride channel modulator Carbaryl, and Chlorpyrifos AChE inhibitor AChE inhibitor Fipronil GABA-gated chloride channel antagonist Imidacloprid nAChE competitive modulator Methoxychlor Sodium-channel modulator	Abamectin: KE760 Carbaryl, and Chlorpyrifos KE12 Fipronil KE1359 Imidacloprid KE559 Methoxychlor KE1602	Leading to acute mortality Acute mortality Neurodegeneration, impaired cognitive function Epilepsy Colony death/ failure leading to neurodegeneration	Real-time RT-PCR and library sequencing Bioinformatic data processing	[11]
2,4-D, dicamba and glyphosate alone and their mixture	Wild-type mouse embryonic stem	With a maximum concentration of 5 mM, cells were exposed to 20 different concentrations of the test chemicals	Beginning with 2,4-D, which was found to be a potent inducer of oxidative stress and an unfolded protein response, the effects develop gradually. Next comes dicamba, which modestly increases oxidative stress, and glyphosate, which plays a minimal influence	Oxidative stress (*Srxn1* and *Blvrb*), protein damage and an unfolded protein response (*Ddit3*), the activation of a DNA damage response (*Bscl2* and *Rtkn*) and p53-mediated cellular stress (Btg2)	*Srxn1*, *Blvrb*, and *Ddit3* unknown DNA damage KE97 *Bscl2* and *Rtkn* Unknown p53 KE1923 Btg2 Unknown	Heritable mutations, cancer 1, cancer 2, and reduced sperm count Impaired lung function	Cytotoxicity testing/by flow cytometry (Luminex) and ToxTracker assay system	[60]

Thiacloprid Imazalil, and clothianidin	Female Wistar rats/HepaRG cells	288 mg/kg b.w. of CTD IMZ 93 mg/kg b.w. or THI 108 mg/kg b.w.		DNA damage by clothianidin followed by activation of P13K, ATM, and upregulation transcription regulator (*TERF2*), Mismatch repair less upregulation RFC5 enzyme, and DNA synthesis less upregulation PCNA and POLQ enzymes	DNA damage KE97	Heritable mutations, cancer 1, cancer 2, and reduced sperm count	Comet assay / γH2AX assay / (real-time RT-PCR) RNA sequencing Illumina NovaSeq6000 platform	[7]
Pesticides (18 compounds) Herbicide, Fungicide Insecticide, and Acaricide	Wistar and Sprague Dawley rats	Animal data were reviewed from (OECD) which used repeat doses conducted between 1974 and 2016	This study was to investigate links between toxicity findings and carcinogenicity studies from short-term rat studies (28 and 90 days) of agrochemical compounds (18 compounds)	Hydroxyphenyl pyruvate dioxygenase inhibitor, Succinate dehydrogenase inhibitor, Acetylcholinesterase inhibitor, and Mitochondrial complex III electron transport inhibitor	Hydroxyphenyl pyruvate dioxygenase inhibitor KE724 Succinate dehydrogenase inhibitor As AOP 474 Acetylcholinesterase inhibitor KE12 Mitochondrial complex III electron transport inhibitor KE1653	Lead to cancer	MLA; HPRT MGM; C'Abs, MN UDS; BM MN; PB MN	[61]

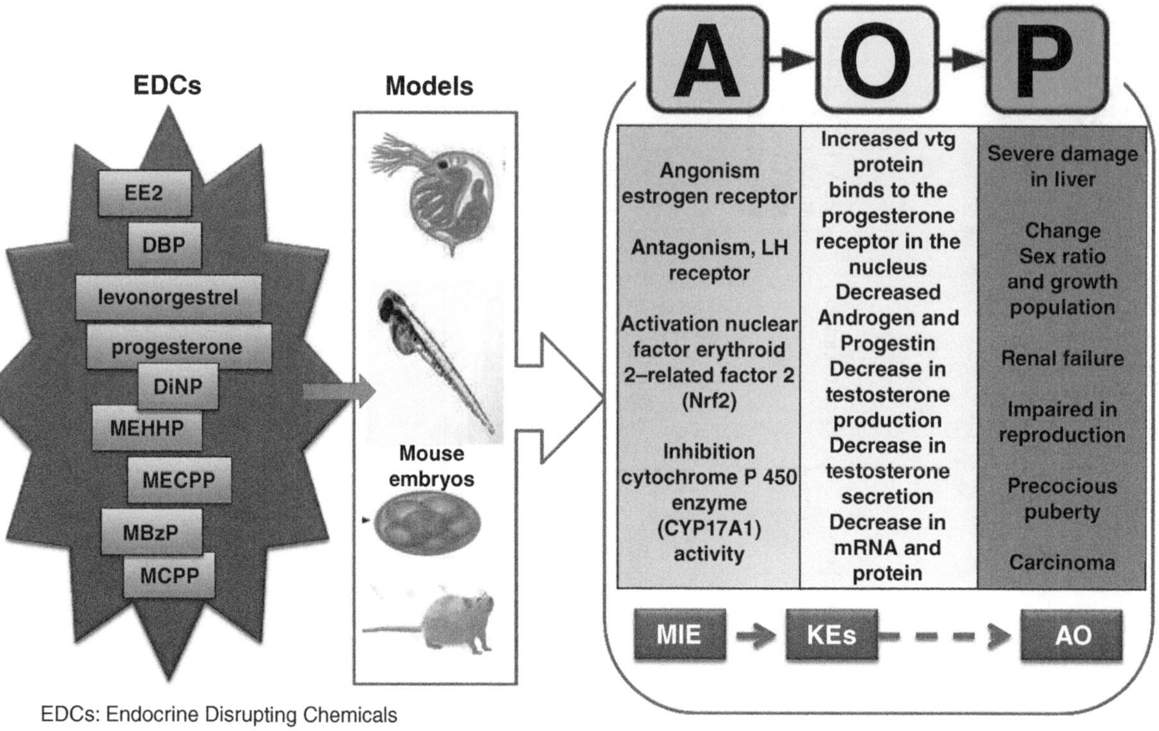

Figure 16.5 Case studies on endocrine-disrupting chemicals in adverse pathway frameworks employing multiple models (rat, mouse embryonic cells, Zebrafish embryos, and Daphnia magna) with three blocks of AOP (Molecular Initiation Events MIE; Biological Key Events KEs; and AO adverse outcome).

of androgen metabolism, reduced fertility, and other serious abnormalities [63]. Here, a substance needs to meet all three requirements to be classified as an EDC: It must (1) cause a detrimental effect in an intact organism or its offspring; (2) do so by altering the function(s) of the endocrine system; and (3) the adverse effect must be a logical result of the endocrine mode-of-action [64]. Early exposure to EDCs can have immediate effects on the establishment of the various gonad cell types or the development of the reproductive system and later effects may be found on hormonal homeostasis, gamete production, gamete quality, and somatic cell differentiation. These effects depend on the type of EDCs, the dose, and the time of exposure [65, 66]. To obtain a better understanding of the molecular processes by which EDCs influence KEs and have long-term impacts on reproduction, one should use basic studies, mostly in fish and rodents, which investigated sex determination, gonadal differentiation, and the mechanisms underlying the events during the embryo and early life stages that are crucial to the development of the reproductive system and germ cell lineage [62]. It is still challenging to predict in vivo outcomes in EDCs because alternative methods data that include complex toxicokinetic parameters (absorption, distribution, metabolism, and excretion; ADME), as well as sensitive windows of exposure, species-to-species differences, sex differences, quantitative understanding of dose–response relationships, and so on as shown in Figure 16.5.

Following the prior literature, Chen et al. discovered that 17-ethinyl estradiol (EE2) and dibutyl phthalate (DBP) as typical estrogen and phthalate ester had negative effects on zebrafish development and their gonad, liver, and gill morphology, which changed the sex ratio and caused decrease in population growth, at doses of (5 ng/L and 20 ng/L), and 0.1 mg/L and 0.5 mg/L respectively [67]. In addition, Svensson et al. (2015) observed that exposure to levonorgestrel caused endometrial cancer and impaired reproduction in 100% of male zebrafish and the developing effect occurred very quickly in gonadal genes *amh, CYP11B*, and *CYP19a1a*'s transcript levels [68]. Besides, when female rats were exposed to different doses at 10, 100, 500, and 1000 mg/kg by oral gavage from gestational day (GD) 12 to 21, Diisononyl phthalate caused decreasing in testosterone production and testosterone secretion with hyperplasia as well as downregulation mRNA and protein level snsl3 and 3hsd as molecular disorders which lead to a decrease in population growth rate adenomas carcinoma as AO [69]. Previous research suggested that some phthalate mixtures altered the expression of genes involved in trans-membrane transport, metabolism, differentiation, chemotaxis, and cell adhesion, including (*CCDC163*), (*DCAKD*), (*IFI6*), (*PFAS*), and (*SPRY4-AS1*). These changes may have adverse effects on reproduction and cause a variety of morbidities, including infertility [70] (Table 16.4).

Table 16.4 Examples of AOP in mixtures of EDs by Wiki Kaptis and AO by AOP-Wiki.

Chemicals	Species	Doses	Describing	KEs ID by Wiki Kaptis	AO Wiki-AOP	Tool endpoints	Ref.
Using 17α-ethinyl estradiol (EE2) and dibutyl phthalate (DBP) as typical estrogen and phthalate esters	Zebrafish (*Danio rerio*)	EE2 (5 ng/L and 20 ng/L) and DBP (0.1 mg/L and 0.5 mg/L) either individually or in mixture	The combined long-term negative effects of EE2 and DBP on the development of zebrafish, however, influence on morphology of gonad, liver, and gill generated by exposure to the mixture of EE2 and DBP	Increased vtg protein at 45d KE 418	Severe damage to liver Change sex ratio and growth population ER agonism leads to reduced survival due to renal failure	Histological examination and sex ratio and growth	[67]
Levonorgestrel and progesterone	Zebrafish (*Danio rerio*)	5.5, 79, and 834 ng/L 3.7, 77, and 1122 ng/L	Exposure to levonorgestrel resulted in 100% of males, whereas progesterone did not affect the sex ratio. The gonadal genes *amh, CYP11B*, and *CYP19a1a*'s transcript levels suggested that levonorgestrel's developing effect happened extremely quickly. Levonorgestrel and progesterone both triggered precocious puberty in fish	Binds to the progesterone receptor in the nucleus and decreased androgen and progestin KE 1064	Impaired in reproduction Precocious puberty leading to endometrial carcinoma in mammals	Reverse transcription and real-time qPCR	[68]
Diisononyl phthalate (DiNP)	Dawley rats	10, 100, 500, and 1000 mg/kg by oral gavage from gestational day (GD) 12 to 21	Low-dose DINP increases fetal Leydig cell size and increases fetal Leydig cell aggregation DINP inhibits fetal Leydig cell-specific protein expressions	Decrease in testosterone production KE1612 Decrease in testosterone secretion LC hyperplasia KE740 Decrease in mRNA and protein levels nsl3 and 3βhsd	Decrease population growth rate Adenomas carcinoma	Quantitative real-time PCR	[69]
Phthalate mixture (MEHHP), (MECPP), (MBzP), (MCPP), (MEP), and (MBP)	Primary ovarian stromal cell and granulosa cell culture	Studied phthalates at 1× or 100× concentrations.	Altered estradiol levels, differentially expressed genes (DEGs) in human cell models The five universally shared DEGs were coiled-coil domain-containing protein 163 (CCDC163), dephospho-CoA kinase domain containing (DCAKD), interferon alpha inducible protein(IFI6), phosphoribosylformylglycinamidine synthase (PFAS), and SPRY4 antisense RNA 1 (SPRY4-AS1)	Altered estradiol levels KE966	Aromatase inhibition female rats	CytoTox-Glo™ cytotoxicity assay and RNA-sequencing and data analysis	[70]

16.5 Challenges and Future Perspectives

AOPs are frequently challenged these days for being overly simplistic and unable to accurately represent intricate toxicological processes. AOPs are portrayed as independent, linear occurrences, but the truth is probably considerably more complicated because parallel cascades and pathway crossings may be involved. It is crucial that while using AOPs, the larger toxicological picture is not lost. AOPs are to be viewed as open, adaptable structures that should be continuously improved by incorporating both old and new data [15]. On this subject, every KE in an AOP should be necessary, meaning that without the event, the further downstream events illustrated in the AOP cannot advance. Additionally, if the KE does occur, it suggests a chance that the disturbance will have other biological effects (i.e. the next KE in the chain) [28, 71]. In addition, a threshold value (chemical concentration and time of exposure) that causes KE downstream to an extent that overcomes adaptive adjustments and compensatory mechanisms is largely unknown. As a result, such as the available AOPs for adult and developmental neurotoxicity are now mostly qualitative rather than quantitative [72]. Therefore, when developing an AOP, it is important to consider both more general events like hormonal impacts and adaptive responses and classical kinetic factors like absorption, distribution, metabolism, and excretion until periodic improvement activities between KEs and AOs are made [13].

The weight of evidence of data that are provided to justify an AOP is a further hurdle that must be overcome soon [73]. Basically, anyone can suggest an AOP, but not all AOPs have enough evidence to support them. To encourage the growth of the global AOP niche, the OECD has established a so-called AOP knowledge storage and is currently running an AOP development program with up to 30 AOP projects in the works which summarized in complying with the overall complexity of toxicological processes, quantification and inclusion of dose–response relationships, implementation of exposure data and toxicokinetic data, and establishment of a transparent and objective evaluation system. Such studies are predicted to produce robust and reliable AOP tools that can be used soon for a range of purposes related to toxicology and risk assessment.

Despite scenarios focusing on single compounds, they can be used for prospective assessments of mixtures. These data can be evaluated using recently formed cumulative assessment groups based on chemicals with comparable effects or MoA [74]. However, comprehensive approaches are required since identifying concerns from combined exposure to various substances under separate legislation makes implementing risk management measures challenging [75]. Nonetheless, the toxicology of mixtures is a difficult project, particularly for environmental toxicologists and ecotoxicologists, who deal with chemical mixtures, many of which contain unknowns. Despite the costs and demanding computations, the systems toxicology framework, of which "omics" is a key component, aims to extract adverse pathways for complex mixtures [76]. The implementation of omics approaches becomes complicated by a key challenge, which needs to analyze very large data sets, advanced molecular and analytical techniques, highly specialized training, and sophisticated bioinformatics tools are required. Furthermore, researchers must have a good basic understanding of biology as well as the sensitivity of the approaches that may lead to the detection of KEs [74]. A shortage of standardization and validation throughout the omics (particularly proteomics and metabolomics) is a problem that must be overcome [46]. Finally, there could be a multitude of biological "pathways" that are influenced by chemicals. AOP development has been voluntary, with pathways chosen primarily based on specific expertise and interest. To build AOP, the priority of AOP development would be determined, for example, by regulatory or human or environmental health priorities, as well as the pathways most frequently implicated in understanding the MIEs [77]. The success of the improved systems biology approach to toxicology was dependent on additional information such as medicine and disease research, as well as additional expertise stakeholders such as additional regulatory sectors (e.g. pharmaceutical) and regions (e.g. other non-OECD countries such as Brazil, China, and Russia). Regulators must be involved at multiple levels, including pathway and assessment plan formulation, effective ways to employ these new approaches within the context of existing rules or by altering regulations as needed, and communicating this information to the regulated community. Communication with interested public stakeholders is crucial for addressing concerns early on and publicly communicating aims, methods, and findings [78].

16.6 Conclusion

The main ecotoxicological categorized pollutants of pesticides, endocrine disruptive agents, and heavy metals which involved unlimited numbers of individual chemicals could be easily applied to such a recommended approach. AOPs represent an acceptable, plausible framework to assess the risk of such hazards and/or their mixtures as common in the environment to the human population, avoiding many disadvantages including high cost and consequences of experiments on laboratory animals. Omics could be employed as a more specific tool to differentiate between mode and MOA. Following names and arrows and utilizing the available literature and data, risk assessment could be easily predicted. This attempt needs to be encouraged for many reasons; besides it is extended to involve additional variables and continual improvement.

References

1 Grandjean P, Landrigan PJ. Neurobehavioural effects of developmental toxicity. Lancet Neurol. 2014, 13: 330–338.

2 Silva E, Rajapakse N, Kortenkamp A. Something from "nothing" – eight weak estrogenic chemicals combined at concentrations below NOECs produce significant mixture effects. Environ Sci Technol. 2002, 36: 1751–1756.

3 Pistollato F, de Gyves EM, Carpi D, Bopp SK, Nunes C, Worth A, Bal-Price A. Assessment of developmental neurotoxicity induced by chemical mixtures using an adverse outcome pathway concept. Environ Health. 2020, 19: 1–26.

4 McCarty L, Borgert C, Posthuma L. The regulatory challenge of chemicals in the environment: toxicity testing, risk assessment, and decision-making models. Regul Toxicol Pharmacol. 2018, 99: 289–295.

5 Bodar C, Berthault F, De Bruijn J, Van Leeuwen C, Pronk M, Vermeire T. Evaluation of EU risk assessments existing chemicals (EC Regulation 793/93). Chemosphere. 2003, 53: 1039–1047.

6 Hernandez L, Mennes W, Van Benthem J, Luijten M. Toxicogenomics in Risk Assesment. RIVM, 2011.

7 Sprenger H, Kreuzer K, Alarcan J, Herrmann K, Buchmüller J, Marx-Stoelting P, Braeuning A. Use of transcriptomics in hazard identification and next generation risk assessment: a case study with clothianidin. Food Chem Toxicol. 2022, 166: 113212.

8 Nuwaysir EF, Bittner M, Trent J, Barrett JC, Afshari CA. Microarrays and toxicology: the advent of toxicogenomics. Mol Carcinog. 1999, 24: 153–159.

9 Cote I, Andersen ME, Ankley GT, Barone S, Birnbaum LS, Boekelheide K, Bois FY, Burgoon LD, Chiu WA, Crawford-Brown D. The next generation of risk assessment multi-year study—highlights of findings, applications to risk assessment, and future directions. Environ Health Perspect. 2016, 124: 1671–1682.

10 Attene-Ramos MS, Huang R, Michael S, Witt KL, Richard A, Tice RR, Simeonov A, Austin CP, Xia M. Profiling of the Tox21 chemical collection for mitochondrial function to identify compounds that acutely decrease mitochondrial membrane potential. Environ Health Perspect. 2015, 123: 49–56.

11 Reinwald H, Alvincz J, Salinas G, Schäfers C, Hollert H, Eilebrecht S. Toxicogenomic profiling after sublethal exposure to nerve-and muscle-targeting insecticides reveals cardiac and neuronal developmental effects in zebrafish embryos. Chemosphere. 2022, 291: 132746.

12 Lu H, Yang D, Shi Y, Chen K, Li P, Huang S, Cui D, Feng Y, Wang T, Yang J. Toxicogenomics scoring system: TGSS, a novel integrated risk assessment model for chemical carcinogenicity prediction. Ecotoxicol Environ Saf. 2023, 250: 114466.

13 Andersen ME, Krewski D. The vision of toxicity testing in the 21st century: moving from discussion to action. Toxicol Sci. 2010, 117: 17–24.

14 Krewski D, Andersen ME, Tyshenko MG, Krishnan K, Hartung T, Boekelheide K, Wambaugh JF, Jones D, Whelan M, Thomas R. Toxicity testing in the 21st century: progress in the past decade and future perspectives. Arch Toxicol. 2020, 94: 1–58.

15 Vinken M. The adverse outcome pathway concept: a pragmatic tool in toxicology. Toxicology. 2013, 312: 158–165.

16 Ankley GT, Bennett RS, Erickson RJ, Hoff DJ, Hornung MW, Johnson RD, Mount DR, Nichols JW, Russom CL, Schmieder PK. Adverse outcome pathways: a conceptual framework to support ecotoxicology research and risk assessment. Environ Toxicol Chem. 2010, 29: 730–741.

17 Villeneuve DL, Crump D, Garcia-Reyero N, Hecker M, Hutchinson TH, LaLone CA, Landesmann B, Lettieri T, Munn S, Nepelska M. Adverse outcome pathway (AOP) development I: strategies and principles. Toxicol Sci. 2014, 142: 312–320.

18 Pant N, Kumar G, Upadhyay A, Patel D, Gupta Y, Chaturvedi P. Reproductive toxicity of lead, cadmium, and phthalate exposure in men. Environ Sci Pollut Res Int. 2014, 21: 11066–11074.

19 Bennekou S. A new path for pesticide assessment: using the AOP framework as a tool in risk assessment. Toxicol Lett. 2021, 350: S27–S27.

20 Garcia-Reyero N, Murphy CA. A Systems Biology Approach to Advancing Adverse Outcome Pathways for Risk Assessment. Springer, 2018.

21 No P. Users' Handbook Supplement to the Guidance Document for Developing and Assessing Adverse Outcome Pathways. OECD Publishing, 2018.

22 Tollefsen KE, Scholz S, Cronin MT, Edwards SW, de Knecht J, Crofton K, Garcia-Reyero N, Hartung T, Worth A, Patlewicz G. Applying adverse outcome pathways (AOPs) to support integrated approaches to testing and assessment (IATA). Regul Toxicol Pharmacol. 2014, 70: 629–640.

23 Delrue N, Sachana M, Sakuratani Y, Gourmelon A, Leinala E, Diderich R. The adverse outcome pathway concept: a basis for developing regulatory decision-making tools. Altern Lab Anim. 2016, 44: 417–429.

24 O'Brien JM, Yauk CL. Introducing AOP reports: collaborative review and publication of adverse outcome pathways. Environ Mol Mutagen. 2022, 63: 116–117.

25 Vinken M, Knapen D, Vergauwen L, Hengstler JG, Angrish M, Whelan M. Adverse outcome pathways: a concise introduction for toxicologists. Arch Toxicol. 2017, 91: 3697–3707.

26 Krewski D, Acosta Jr D, Andersen M, Anderson H, Bailar III JC, Boekelheide K, Brent R, Charnley G, Cheung VG, Green Jr S. Toxicity testing in the 21st century: a vision and a strategy. J Toxicol Environ Health B Crit Rev. 2010, 13: 51–138.

27 Boobis AR, Cohen SM, Dellarco V, McGregor D, Meek M, Vickers C, Willcocks D, Farland W. IPCS framework for analyzing the relevance of a cancer mode of action for humans. Crit Rev Toxicol. 2006, 36: 781–792.

28 Boobis AR, Doe JE, Heinrich-Hirsch B, Meek M, Munn S, Ruchirawat M, Schlatter J, Seed J, Vickers C. IPCS framework for analyzing the relevance of a noncancer mode of action for humans. Crit Rev Toxicol. 2008, 38: 87–96.

29 Kimber I, Dearman RJ, Basketter DA, Boverhof DR. Chemical respiratory allergy: reverse engineering an adverse outcome pathway. Toxicology. 2014, 318: 32–39.

30 Schmieder PK, Tapper MA, Denny JS, Kolanczyk RC, Sheedy BR, Henry TR, Veith GD. Use of trout liver slices to enhance mechanistic interpretation of estrogen receptor binding for cost-effective prioritization of chemicals within large inventories. Environ Sci Technol. 2004, 38: 6333–6342.

31 Villeneuve D, Volz DC, Embry MR, Ankley GT, Belanger SE, Léonard M, Schirmer K, Tanguay R, Truong L, Wehmas L. Investigating alternatives to the fish early-life stage test: a strategy for discovering and annotating adverse outcome pathways for early fish development. Environ Toxicol Chem. 2014a, 33: 158–169.

32 LaLone CA, Villeneuve DL, Burgoon LD, Russom CL, Helgen HW, Berninger JP, Tietge JE, Severson MN, Cavallin JE, Ankley GT. Molecular target sequence similarity as a basis for species extrapolation to assess the ecological risk of chemicals with known modes of action. Aquat Toxicol. 2013, 144: 141–154.

33 OECD (2013). Guidance Document on Developing and Assessing Adverse Outcome Pathways. Organisation for Economic Co-operation and Development, Paris, France, p. 45.

34 Bao W, Schmid JE, Goetz AK, Ren H, Dix DJ. A database for tracking toxicogenomic samples and procedures. Reprod Toxicol. 2005, 19: 411–419.

35 Boverhof DR, Zacharewski TR. Toxicogenomics in risk assessment: applications and needs. Toxicol Sci. 2006, 89: 352–360.

36 Forum U.S.E.P.A.R.A. Guidelines for Ecological Risk Assessment, Risk Assessment Forum. US Environmental Protection Agency, 1998.

37 SE D, Schöpf TJ, Renner N. Regulation (EC) No. 1907/2006 of the European Parliament and of the Council of 18 December 2006 concerning the Registration, Evaluation, Authorisation and Restriction of Chemicals (REACH), 2019.

38 Dorato MA, Engelhardt JA. The no-observed-adverse-effect-level in drug safety evaluations: use, issues, and definition (s). Regul Toxicol Pharmacol. 2005, 42: 265–274.

39 Sturla SJ, Boobis AR, FitzGerald RE, Hoeng J, Kavlock RJ, Schirmer K, Whelan M, Wilks MF, Peitsch MC. Systems toxicology: from basic research to risk assessment. Chem Res Toxicol. 2014, 27: 314–329.

40 Van Aggelen G, Ankley GT, Baldwin WS, Bearden DW, Benson WH, Chipman JK, Collette TW, Craft JA, Denslow ND, Embry MR. Integrating omic technologies into aquatic ecological risk assessment and environmental monitoring: hurdles, achievements, and future outlook. Environ Health Perspect. 2010, 118: 1–5.

41 Moreira JD, Knorr L, Ganzella M, Thomazi AP, de Souza CG, de Souza DG, Pitta CF, e Souza TM, Wofchuk S, Elisabetsky E. Omega-3 fatty acids deprivation affects ontogeny of glutamatergic synapses in rats: relevance for behavior alterations. Neurochem Int. 2010, 56: 753–759.

42 Tilton FA, Bammler TK, Gallagher EP. Swimming impairment and acetylcholinesterase inhibition in zebrafish exposed to copper or chlorpyrifos separately, or as mixtures. Comp Biochem Physiol C Toxicol Pharmacol. 2011, 153: 9–16.

43 Viñuela A, Snoek LB, Riksen JA, Kammenga JE. Genome-wide gene expression regulation as a function of genotype and age in C. elegans. Genome res. 2010, 20: 929–937.

44 Gatzidou ET, Zira AN, Theocharis SE. Toxicogenomics: a pivotal piece in the puzzle of toxicological research. J Appl Toxicol. 2007, 27: 302–309.

45 Nimse SB, Sonawane MD, Song K-S, Kim T. Biomarker detection technologies and future directions. Analyst. 2016, 141: 740–755.

46 McBride MT. The application of omics data to the development of AOPs. In: Garcia-Reyero N, Murphy C (Eds.), A Systems Biology Approach to Advancing Adverse Outcome Pathways for Risk Assessment. Springer, 2018, pp. 177–198.

47 Collins FS, Green ED, Guttmacher AE, Guyer MS, US National Human Genome Research Institute. A vision for the future of genomics research. Nature. 2003, 422: 835–847.

48 Konermann S, Brigham MD, Trevino AE, Joung J, Abudayyeh OO, Barcena C, Hsu PD, Habib N, Gootenberg JS, Nishimasu H. Genome-scale transcriptional activation by an engineered CRISPR-Cas9 complex. Nature. 2015, 517: 583–588.

49 Swiech L, Heidenreich M, Banerjee A, Habib N, Li Y, Trombetta J, Sur M, Zhang F. In vivo interrogation of gene function in the mammalian brain using CRISPR-Cas9. Nat Biotechnol. 2015, 33: 102–106.

50 Cui Y, Paules RS. Use of transcriptomics in understanding mechanisms of drug-induced toxicity. Pharmacogenomics. 2010, 11: 573–585.

51 Chen G, Pramanik BN. Application of LC/MS to proteomics studies: current status and future prospects. Drug Discov Today. 2009, 14: 465–471.

52 Monosson E. Chemical mixtures: considering the evolution of toxicology and chemical assessment. Environ Health Perspect. 2005, 113: 383–390.

53 Sonich-Mullin C, Fielder R, Wiltse J, Baetcke K, Dempsey J, Fenner-Crisp P, Grant D, Hartley M, Knaap A, Kroese D. IPCS conceptual framework for evaluating a mode of action for chemical carcinogenesis. Regul Toxicol Pharmacol. 2001, 34: 146–152.

54 Gilliom RJ, Barbash JE, Crawford CG, Hamilton PA, Martin JD, Nakagaki N, Nowell LH, Scott JC, Stackelberg PE, Thelin GP. Pesticides in the Nation's Streams and Ground Water, 1992–2001. US Geological Survey, 2006.

55 Lydy M, Belden J, Wheelock C, Hammock B, Denton D. Challenges in regulating pesticide mixtures. Ecol Soc. 2004, 9: 1.

56 Vighi M, Altenburger R, Arrhenius Å, Backhaus T, Bödeker W, Blanck H, Consolaro F, Faust M, Finizio A, Froehner K. Water quality objectives for mixtures of toxic chemicals: Problems and perspectives. Ecotoxicol Environ Saf. 2003, 54: 139–150.

57 Quinn DM. Acetylcholinesterase: enzyme structure, reaction dynamics, and virtual transition states. Chem Rev. 1987, 87: 955–979.

58 Hodges G, Gutsell S, Taylor N, Brockmeier E, Butler E, Rendal C, Colbourne J. Invertebrate model species in AOP development. In: Garcia-Reyero N, C Murphy (Eds.), A Systems Biology Approach to Advancing Adverse Outcome Pathways for Risk Assessment, Springer, 2018, pp. 75–106.

59 Pfaff J, Reinwald H, Ayobahan SU, Alvincz J, Göckener B, Shomroni O, Salinas G, Düring R-A, Schäfers C, Eilebrecht S. Toxicogenomic differentiation of functional responses to fipronil and imidacloprid in Daphnia magna. Aquat Toxicol. 2021, 238: 105927.

60 Mesnage R, Brandsma I, Moelijker N, Zhang G, Antoniou MN. Genotoxicity evaluation of 2, 4-D, dicamba and glyphosate alone or in combination with cell reporter assays for DNA damage, oxidative stress and unfolded protein response. Food Chem Toxicol. 2021, 157: 112601.

61 Mistry P, McInnes EF, Beevers C, Wolf D, Currie RA, Salimraj R, Parsons P. An evaluation of carcinogenicity predictors from short-term and sub chronic repeat-dose studies of agrochemicals in rats: opportunities to refine and reduce animal use. Toxicol Lett. 2021, 351: 18–27.

62 Delbès G, Blázquez M, Fernandino JI, Grigorova P, Hales B, Metcalfe C, Navarro-Martin L, Parent L, Robaire B, Rwigemera A. Effects of endocrine disrupting chemicals on gonad development: mechanistic insights from fish and mammals. Environ Res. 2022, 204: 112040.

63 Kabir ER, Rahman MS, Rahman I. A review on endocrine disruptors and their possible impacts on human health. Environ Toxicol Pharmacol. 2015, 40: 241–258.

64 Solecki R, Kortenkamp A, Bergman Å, Chahoud I, Degen GH, Dietrich D, Greim H, Håkansson H, Hass U, Husoy T. Scientific principles for the identification of endocrine-disrupting chemicals: a consensus statement. Arch Toxicol. 2017, 91: 1001–1006.

65 Johansson HKL, Svingen T, Fowler PA, Vinggaard AM, Boberg J. Environmental influences on ovarian dysgenesis—developmental windows sensitive to chemical exposures. Nat Rev Endocrinol. 2017, 13: 400–414.

66 Wohlfahrt-Veje C, Main KM, Skakkebæk NE. Testicular dysgenesis syndrome: foetal origin of adult reproductive problems. Clin Endocrinol. 2009, 71: 459–465.

67 Chen P, Li S, Liu L, Xu N. Long-term effects of binary mixtures of 17α-ethinyl estradiol and dibutyl phthalate in a partial life-cycle test with zebrafish (Danio rerio). Environ Toxicol Chem. 2015, 34: 518–526.

68 Svensson J, Mustafa A, Fick J, Schmitz M, Brunström B. Developmental exposure to progestins causes male bias and precocious puberty in zebrafish (Danio rerio). Aquatic Toxicol. 2016, 177: 316–323.

69 Li L, Bu T, Su H, Chen Z, Liang Y, Zhang G, Zhu D, Shan Y, Xu R, Hu Y. In utero exposure to diisononyl phthalate caused testicular dysgenesis of rat fetal testis. Toxicol Lett. 2015, 232: 466–474.

70 Tarvainen I, Soto DA, Laws MJ, Björvang RD, Damdimopoulos A, Roos K, Li T, Kramer S, Li Z, Lavogina D. Identification of phthalate mixture exposure targets in the human and mouse ovary in vitro. Reprod Toxicol. 2023, 119: 108393.

71 Meek M, Boobis A, Cote I, Dellarco V, Fotakis G, Munn S, Seed J, Vickers C. New developments in the evolution and application of the WHO/IPCS framework on mode of action/species concordance analysis. J Appl Toxicol. 2014, 34: 1–18.

72 Bal-Price A, Crofton KM, Leist M, Allen S, Arand M, Buetler T, Delrue N, FitzGerald RE, Hartung T, Heinonen T. International STakeholder NETwork (ISTNET): creating a developmental neurotoxicity (DNT) testing road map for regulatory purposes. Arch Toxicol. 2015, 89: 269–287.

73 Meek M, Palermo CM, Bachman AN, North CM, Jeffrey Lewis R. Mode of action human relevance (species concordance) framework: evolution of the Bradford Hill considerations and comparative analysis of weight of evidence. J Appl Toxicol. 2014b, 34: 595–606.

74 Committee ES. Guidance on the structure and content of EFSA's scientific opinions and statements. EFSA Journal. 2014, 12: 3808.

75 Kienzler A, Bopp SK, van der Linden S, Berggren E, Worth A. Regulatory assessment of chemical mixtures: requirements, current approaches and future perspectives. Regul Toxicol Pharmacol. 2016, 80: 321–334.

76 Martins C, Dreij K, Costa PM. The state-of-the art of environmental toxicogenomics: challenges and perspectives of "omics" approaches directed to toxicant mixtures. Int J Environ Res Public Health. 2019, 16: 4718.

77 Allen TE, Goodman JM, Gutsell S, Russell PJ. Defining molecular initiating events in the adverse outcome pathway framework for risk assessment. Chem Res Toxicol. 2014, 27: 2100–2112.

78 Garcia-Reyero N, Escalon L, Prats E, Faria M, Soares AM, Raldúa D. Targeted gene expression in zebrafish exposed to chlorpyrifos-oxon confirms phenotype-specific mechanisms leading to adverse outcomes. Bull Environ Contam Toxicol. 2016, 96: 707–713.

17

Mathematical Model for Combined Toxicity Prediction

Fasiha Javaid[a,b], De-Sheng Pei[a,]*

[a] *School of Public Health, Chongqing Medical University, Chongqing, China*
[b] *Quaid-i-Azam University, Islamabad, Pakistan*

*Corresponding author.
E-mail: peids@cqmu.edu.cn (D.S.P)

17.1 Significance of Predicting Combined Toxicity

One of the most significant endeavors in the modern environmental and public health realms is predicting combination toxicity. The assessment of combined toxicity in chemical mixtures is a critical aspect of environmental risk management [1]. This chapter will provide a comprehensive overview of the significance of predicting combined toxicity in the context of environmental risk assessment. The convergence of numerous chemical substances provides a multidimensional issue in an era characterized by expanding chemical production and widespread environmental contamination [2]. With the increasing number of chemicals in use [3] and the limited information available on their toxicity, the assessment of chemical mixtures remains a challenging task [1]. The assessment of combined toxicity is essential for understanding the potential risks posed by chemical mixtures to both human health and the environment [4]. In recent decades, uncontrollable environmental pollution has emerged as a major health issue worldwide [5]. Human beings and other living organisms are constantly exposed to a wide range of chemicals, and the synergistic effect of chemicals with each other [6] is a serious health concern [7]. Therefore, combined toxicity evaluation is crucial for a more accurate risk assessment of environmental pollutants [8]. The implementation of alternative novel strategies for chemical toxicity assessment is of more attraction to researchers and regulatory bodies [9]. We can assess the possible dangers and decide on exposure levels and risk management techniques by predicting the combined toxicity [10]. It is essential to comprehend the combined effects of these substances for several significant reasons [11].

Real-world exposure situations: It is unusual for human environments and natural ecosystems to come into contact with single chemical entities; instead, complex combinations are encountered through a variety of channels, including food, water, and air [12]. Accurately assessing dangers posed to natural and human systems depends on predicting the cumulative toxicity of these complicated matrix structures.

Cumulative health effects: The interactions of chemicals within mixtures can increase or decrease the toxicities of individual components, resulting in antagonistic or synergistic effects [13]. To determine the actual risk that chemical combinations entail and develop efficient risk mitigation plans, it is essential to anticipate these cumulative effects.

Environmental impact assessment: Chemical pollutants are processed into complex interactions and complicated alterations that produce new chemicals/compounds that may have different toxicity characteristics, compared to a single chemical, when they are added to environmental matrices. To assess these mixtures' ecological effects on ecosystems, biodiversity, and important environmental processes, it is imperative to project their combined toxicity [14].

Toxicological Assessment of Combined Chemicals in the Environment, First Edition. Edited by De-Sheng Pei and Yiyun Liu.

Regulatory compliance and policy development: Concerning the need to protect both human health and the environment, regulatory agencies are pushing for risk assessment frameworks to include an assessment of combined toxicity. Combining projections of combined toxicity makes it less challenging to create effective rules and policies for managing chemicals and reducing pollution [15].

Resource allocation and prioritization: Prioritizing substances of concern becomes critical when resources are limited. Policymakers can identify high-risk chemical mixtures and allocate resources towards their monitoring and regulatory control by using predictive models for combined toxicity [16].

Public health protection: In addition to environmental pollution, there are other ways that people are exposed to chemical mixtures, such as through consumer goods and work environments. It is necessary to predict combined toxicity to reduce the possible health concerns that may arise from exposure to these substances [17].

Advancement of toxicology and risk assessment science: Toxicological and risk assessment developments are driven by predictive models for combined toxicity, which makc usc of multidisciplinary knowledge from chemistry, biology, pharmacology, and computer sciences [18]. These models promote the development of more accurate and reliable risk assessment processes (Figure 17.1a).

17.1.1 The Challenges in Assessing Combined Toxicity

There are several challenges in assessing the combined toxicity of chemical mixtures. These include data gaps, the need to address interactions between chemicals, and the complexity of dose–response relationships and modes of action [20]. Additionally, there is a lack of standardized methods for assessing combined toxicity, which can lead to inconsistencies in predictions and recommendations [21].

In conclusion, the assessment of combined toxicity in chemical mixtures is a complex and challenging task. The significance of predicting combined toxicity lies in its potential to inform risk management decisions and protect human health and the environment [22]. By understanding the challenges and approaches in assessing combined toxicity, we can develop more effective strategies for managing chemical risks [23].

Overview of the goals and organization of the chapter: This chapter embarks on a multi-faceted investigation of combined toxicity in chemical combinations. We begin by establishing its critical role in environmental risk assessment and hazard evaluation. This thing makes a ground for a deeper dive into the complexities and potential outcomes of various chemical interactions within mixtures. The chapter then aims to enlighten the multifaceted nature of combined toxicity. We will delve into its definition, practical applications, and the diverse types of interactions that can occur. Through a systematic exploration of additive, antagonistic, and synergistic effects, readers will gain a comprehensive understanding of the complicated dynamics at play when chemicals combine. This newfound knowledge will equip them to grasp the potential impacts of these interactions on human health and the environment.

By intertwining together these crucial elements, the chapter aspires to equip the readers with a comprehensive understanding of combined toxicity, its complexities, and its significance in environmental hazard evaluation. This knowledge will ultimately empower them to advocate for effective risk management strategies in the face of ever-present chemical mixtures in our environment.

Moreover, the chapter further indulged in the importance of hazard evaluation for combined chemicals, highlighting the challenges imposed by the intricacy of chemical combinations in real-world exposure circumstances. It emphasizes how important it is to employ all-encompassing strategies, including cumulative risk assessment, to effectively manage these challenges. It also focuses on regulatory factors that influence how integrated toxicity assessment and risk management are practiced (Figure 17.1).

17.2 Fundamental Concepts of Combined Toxicity

It is very important to describe the proper and precise definition of the scope of this complicated issue while delving into the concept of combined toxicity. The combined or interacting effects of various chemical substances when they are present simultaneously in an environment, organism, or system are referred to as combined toxicity, also known as mixture

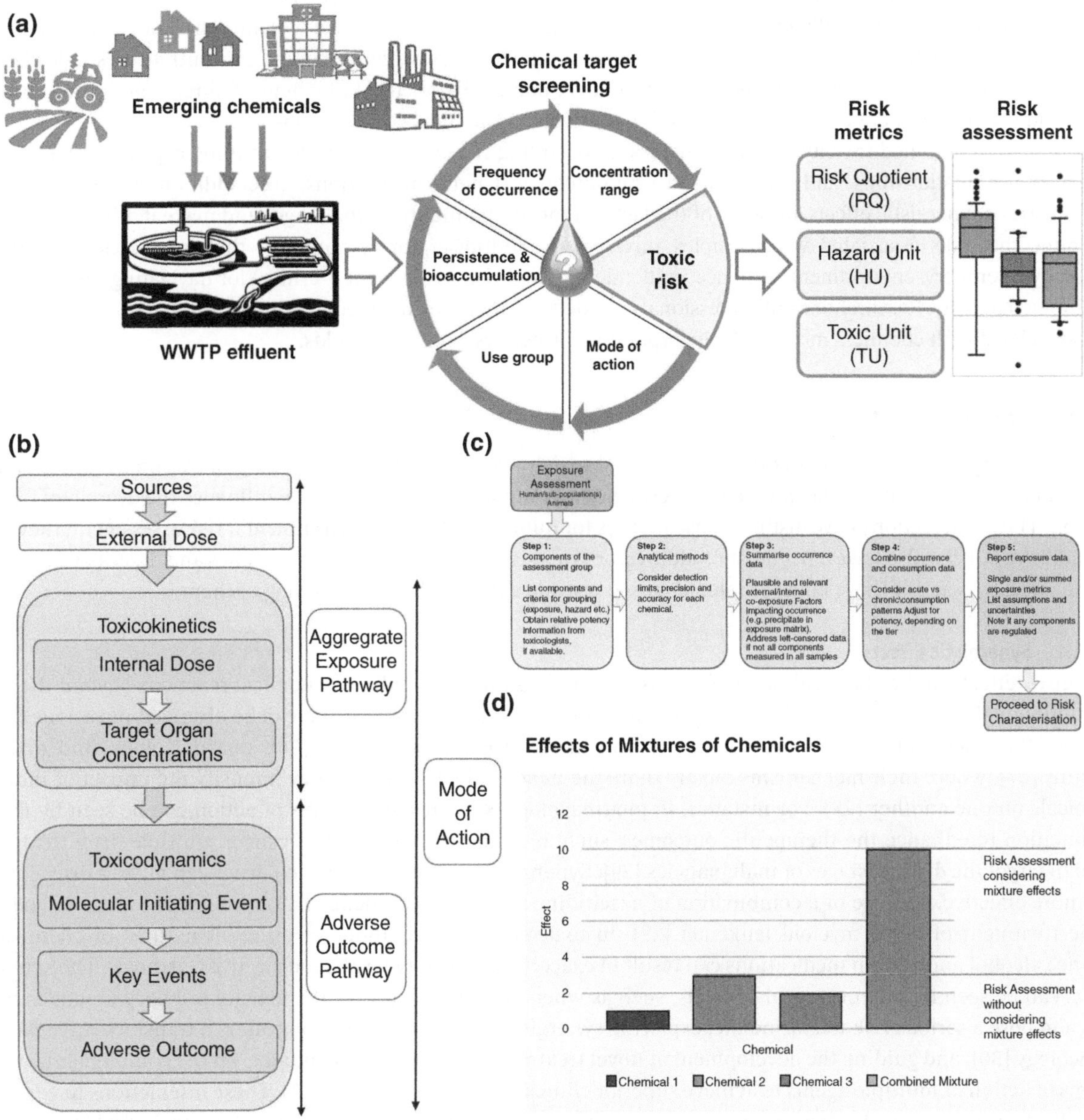

Figure 17.1 (a) Graphical representation of risk-based assessment approach for chemical mixtures from wastewater treatment plant effluents [19]/Elsevier/CC BY 4.0; (b) mode of action, adverse outcome pathway, and aggregate exposure pathway [15]/MDPI/CC BY 4.0; (c) stepwise approach for exposure to multiple chemicals using the component-based approach [15]/MDPI/CC BY 4.0; (d) the hypothetical example presented below shows a mixture of chemicals (Chemical 1, Chemical 2, and Chemical 3) with a combined effect that is greater than the effect of its most toxic component (Chemical 2).

toxicity or cocktail effect. Fundamentally, combined toxicity refers to the concept that the combined presence of several substances may have consequences that differ or exceed those predicted from the toxicity of any one component taken alone. Numerous mechanisms, such as additive, antagonistic, and synergistic interactions among the substances involved, give genesis to this phenomenon.

The range of combined toxicity encompasses various sectors, including consumer goods, pharmaceuticals, occupational health, and the environment [24]. It shows up in environmental contexts as the intricate interactions between pollutants in the air, water, soil, and food, endangering both human populations and ecosystems. Workers may come into contact with chemical combinations in the workplace that intensify health risks beyond what would be expected from exposure to each material alone. Comparably, consumer goods and pharmaceutical products frequently include several components whose combined effects may be partially known or sufficiently evaluated. The importance of comprehending combined toxicity stems from its ability to avoid traditional risk assessment methodologies, which generally concentrate on single chemicals alone. Conventional risk assessment methods can undervalue or distort the full extent of threats presented to human health and the environment by ignoring the combined effects of chemical combinations.

Additionally, combined toxicity poses distinct issues regarding the challenges of determining exposure, characterizing dose–response relationships, and projecting adverse outcomes. Chemical reactions, dose addition, and the presence of antagonistic or synergistic effects are some of the factors that compound these challenges and make it difficult to precisely determine the risks associated with complex mixtures. A multidisciplinary approach that integrates concepts from toxicology, chemistry, environmental science, epidemiology, and risk assessment is critical for navigating the complexities of combined toxicity. Researchers and professionals can develop more resilient techniques for assessing and mitigating the risks associated with chemical mixtures by incorporating principles from other fields.

17.2.1 Types of Interactions

Comprehending the intricacies of combined chemical toxicity demands an understanding of the processes of chemical interactions within mixtures. The interactions between the compounds in a mixture can influence the overall toxicological reaction. These interactions have distinct consequences for human health and environmental risk assessment; they are not only additive but can also have antagonistic, additive, or synergistic effects. This section outlines and describes the mechanics and effects of each of these kinds of interactions to provide a thorough understanding of them.

17.2.1.1 Synergistic Effects

Synergistic effects in the chemicals are defined as the amplification of combined chemical reactions beyond the sum of their individual effects (Figure 17.1d); this phenomenon is of great importance and can be observed in various fields of science [25]. These interactions are commonly seen in environmental science [26], pharmacology, and toxicology domains [27], where their mechanisms mostly share the biological pathways or they intensify the impact of individual chemicals on one another [28]. For instance, in pharmacology, similar mechanisms of action can be seen by the drug combination to enhance the therapeutic outcomes, such as in the management of cancer multiple drug treatment is opted to target the distinct stages of malignancies [29]. Synergetic interactions between drugs can make a drug combination more effective. The use of a combination of azacitidine with a noncovalent FLT3 inhibitor shows its effectiveness in the treatment of acute myeloid leukemia [29]. In toxicology, combinations of different natures of chemicals for instance alcohol and certain medications can result in exacerbated toxicity levels, such as liver cirrhosis. The synergistic effects can be seen in environmental contexts, such as when pesticides increase the toxicity of heavy metals in ecosystems and pose a serious risk to marine life, exploring synergism in drug combinations offers a transformative paradigm in oncology [30], and guiding the development of novel treatment regimens that capitalize on the complementary mechanisms of action of multiple agents to achieve superior clinical responses (Figure 17.2). These interactions have a significant effect on human and other organisms' health, necessitating work on risk assessment, and combined chemical exposures rather than evaluating individual effects of chemicals. This strategy will give a way for both regulatory agencies and researchers to ensure precise assessments of chemical safety and environmental impact and provide a comprehensive understanding of synergistic interactions and their long-term consequences [31].

17.2.1.2 Antagonistic Effects

Antagonistic effects, in which the combined effect of a chemical is less than the overall effect of an individual one [33], this phenomenon is prevalent in many scientific fields (Figure 17.3b). For the evaluation of combined chemical risks, these interactions are crucial through diverse mechanisms including the chemical inactivation process, competitive inhibition phenomena, metabolic interference, and functional antagonism. For instance, in pharmacology antagonistic interactions of drugs can adversely affect therapeutic procedures by competitive inhibition or metabolic processes [34]. In the toxicology the chemicals could lessen the toxicity of other chemicals; it can be observed by the way chelating agents reduce the toxicity of heavy metals by sequestering the metals [35]. Understanding antagonistic interactions is necessary to restore environmental health where microorganisms counteract the effects of pollutants through decomposition reactions. However, because these interactions may underestimate the entire toxicity of chemicals, there is a dire need for in-depth investigations and risk assessment [31].

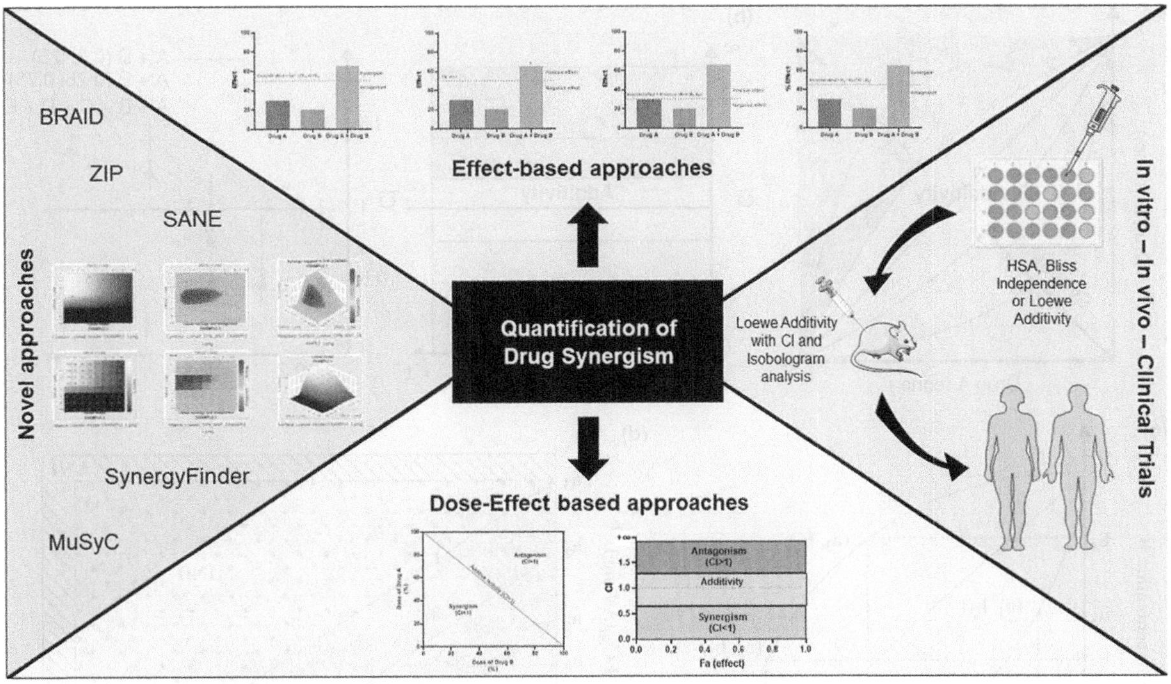

Figure 17.2 Graphical representation of evaluation of synergism in drug combinations and reference models for future orientations in oncology [32]/with permission of Elsevier.

Furthermore, on the therapeutic side, the identification of antagonistic medication interactions is indispensable for assuring patient safety and increasing the treatment procedures. Similarly, in environmental science, the assessment of antagonistic relationships facilitates the development of efficient pollution management plans for ecosystem protection. Therefore, it is critical to comprehend antagonistic effects across the disciplines of science to fully guide risk assessment and management approaches [38]. The antagonistic chemical interaction can be seen in the summary of articles listed in Table 17.1.

17.2.1.3 Additive Effects

Additive effect interaction is also one of the most imperative features of chemical interactions in mixtures, it occurs when the overall sum of the individual effects of a chemical is equal to the effect when chemicals are combined [6]. The main feature of additive effects is that it happens when chemicals interact in such a way that the separate toxicological features of chemicals remain intact. This interaction of chemicals is opposite to antagonistic effects of chemicals, in which the combined effects of chemicals decrease [39], and additive effects are also opposite as compared to synergistic effects, where the combined interaction of chemicals intensifies beyond the total sum of individual effects (Figure 17.3a).

This effect is applicable in pharmacology when two drugs with the same mechanism of action are processed at the same time, it yields a combined effect, which is equal to the sum of the effect of each drug [40]. In toxicological studies, a similar scenario can be observed when two different drugs act independently or collectively to produce an increased level of toxicity. Additive effects should be acknowledged and taken into consideration for the account of risk management strategies and evaluation of the combined effect of chemical exposure on the environment and human health. Researchers and regulatory organizations need to integrate additive effects for understanding risk assessment models to ensure problem identification, analyses, and efficient management of chemical risks [41]. The additive effect can be seen from the articles listed in Table 17.1.

According to Ismael Rodea-Palomares et al., understanding the effects of exposure to chemical mixtures is one of the common goals of ecotoxicology and pharmacology, given that additivity models can predict the result of any chemical combination using only individual chemical information, determining the extent to which they may be applied in risk assessment-oriented ecotoxicology has received a lot of attention in the last two decades [1]. Loewe additivity, which gave rise to the concentration addition (CA) additivity model, is the gold standard for additivity in ecotoxicology. Over the last two decades, researchers in pharmacology have also been drawn to the hunt for interactions or departures from additivity (synergism and antagonism), which has led to the development and use of the combination index (CI) theorem. Loewe's

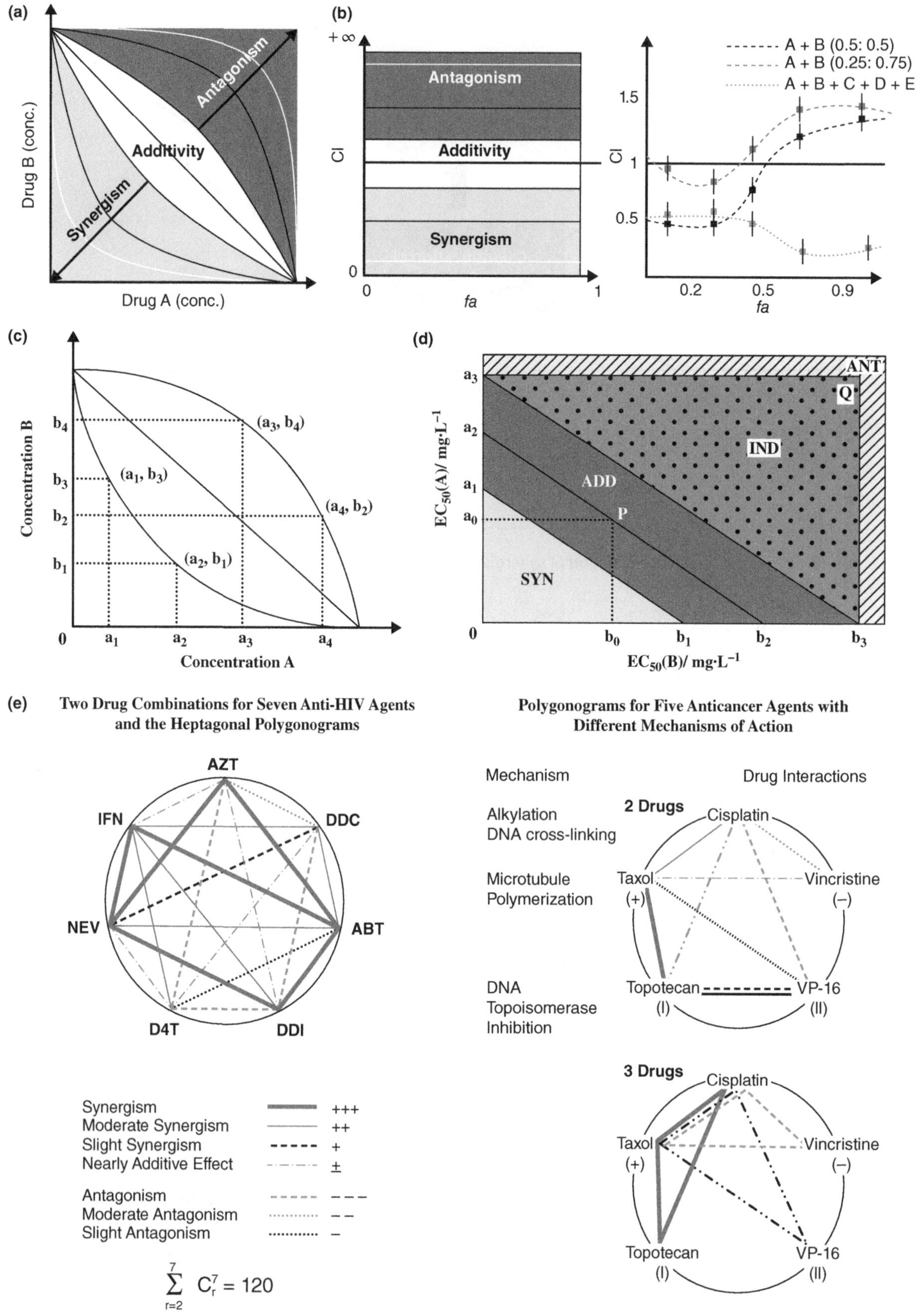
(a)
Drug B (conc.)
Drug A (conc.)
Antagonism
Additivity
Synergism
(b)
+ ∞
CI
0
Antagonism
Additivity
Synergism
0
fa
1
A + B (0.5: 0.5)
A + B (0.25: 0.75)
A + B + C + D + E
1.5
1
0.5
0.2
0.5
0.9
fa
(c)
Concentration B
Concentration A
b_4
b_3
b_2
b_1
0
a_1
a_2
a_3
a_4
(a_3, b_4)
(a_1, b_3)
(a_4, b_2)
(a_2, b_1)
(d)
$EC_{50}(A)/ mg \cdot L^{-1}$
$EC_{50}(B)/ mg \cdot L^{-1}$
a_3
a_2
a_1
a_0
0
b_0
b_1
b_2
b_3
ANT
Q
IND
ADD
P
SYN
(e)
Two Drug Combinations for Seven Anti-HIV Agents and the Heptagonal Polygonograms
AZT
IFN
DDC
NEV
ABT
D4T
DDI
Synergism +++
Moderate Synergism ++
Slight Synergism +
Nearly Additive Effect ±
Antagonism − − −
Moderate Antagonism − −
Slight Antagonism −
$\sum_{r=2}^{7} C_r^7 = 120$
Polygonograms for Five Anticancer Agents with Different Mechanisms of Action
Mechanism
Drug Interactions
Alkylation
DNA cross-linking
Microtubule
Polymerization
DNA
Topoisomerase
Inhibition
2 Drugs
Cisplatin
Taxol
(+)
Vincristine
(−)
Topotecan
(I)
VP-16
(II)
3 Drugs
Cisplatin
Taxol
(+)
Vincristine
(−)
Topotecan
(I)
VP-16
(II)

Figure 17.3 (a) Schematic isobologram illustrates combined effects of Drug A and Drug B at a specific effect level (fa). The additivity line represents the theoretical additive effect, with uncertainty shown by the white region. Synergism occurs when combined doses (A + B) achieve the desired effect (fa) below theoretical summation, plotted below the additivity line. Antagonism is indicated by points above additivity line, surpassing theoretical summation [1]; (b) the fa-CI plot serves as a schematic representation of combined drug effects, focusing on fractional effect levels (*x*-axis). The additivity line, marked by CI = 1, denotes theoretical additive effect, with uncertainty depicted by the white area. Synergism is indicated by CI < 1, while CI > 1 denotes antagonism; (c) the combined effect of components A + B at any mixture ratio in the Cartesian plane should theoretically align on a straight line. Upward concave lines indicate antagonism, while downward concave lines suggest synergy; (d) the Isobologram method shows varied joint effects of joint toxicity due to different toxic ratios [36]; (e) sample illustration of polygonograms; a, heptagonal polysomnogram for seven anti-HIV agents using two-drug combinations in vitro (AZT, D4T, DDC, and DDI – all nucleoside reverse transcriptase inhibitors; NEV, a non-nucleoside reverse transcriptase inhibitor; INF, interferon-2a; and ABT-538, a protease inhibitor). *Source:* Adapted from Refs. [1, 37].

Table 17.1 Studies related to chemical interactions for toxicity assessment.

Article	Summary	Main findings	Methodology	Types of interactions
Risk Assessment of Combined Exposure to Multiple Chemicals at the European Food Safety Authority: Principles, Guidance Documents, Applications and Future Challenges [15]	The default assumption for combined toxicity of the chemicals is applied unless evidence for response addition or interactions is available	The primary results include the availability of standardized frameworks for evaluating the risks associated with chemicals and substances, real-time case studies done for the demonstration of the application of the guidance, and further suggestions for future research in the fields of human and other organisms' health	The study evaluates combined chemical exposure as a whole, focusing on general assumptions, such as dose addition, standardized methods, and hierarchical analysis according to the availability of data. Additionally, it emphasizes database construction, integration of data, and high throughput data for hazard identification	The default assumption for combined toxicity of the chemicals is applied unless evidence for response addition or interactions is available
Toxicity Weighting for Human Biomonitoring Mixture Risk Assessment: A Proof of Concept [17]	The combined toxicity of DCA varied with the type and concentration of TDNMs	The finding of the study summarizes the hazardous quotients weighted by related health-based SOPs, by designing a biomonitoring hazard index. This is validated for the identification of seven communities within the data, with N-acetyl-S-(2-carbamoyl-ethyl) cysteine (AAMA) being the only biomarker by an SOP value specified in the highest hazard community	The methodology used in the study encompassed several processes, including data preparation, network analysis utilizing of graphical lasso, handling of incomplete data, and implementing the hazard index approach assuming additive effects of mixture components with a default mode of action	Additive effects of mixture components with a common mode of action
An Insight into the Combined Toxicity of 3,4-Dichloroaniline with Two-Dimensional Nanomaterials: From Classical Mixture Theory to Structure–Activity Relationship [45]	The combined toxicity of DCA varied with the type and concentration of TDNMs	Two-dimensional nanomaterials (TDNMs) and DCA together show additive, antagonistic, and synergistic effects by evaluating toxicity. The comparison of the conventional mixture model and the prediction model which included both the Freundlich adsorption coefficient (KF) and the adsorption energy (Ea) provided a more accurate estimate of the combined chemical toxicity	The methodology involved assessing the toxicity of TDNMs with DCA on freshwater microalgae, using classical mixture theory, structure–activity relationships, adsorption experiments, molecular simulations, and predictive modeling	Additive effect Antagonistic effect Synergistic effect

(Continued)

Table 17.1 (Continued)

Article	Summary	Main findings	Methodology	Types of interactions
Toxicity Characterization of Environment-Related Pollutants Using a Biospectroscopy-Bioreporter-Coupling Approach: Potential for Real-World Toxicity Determination and Source Apportionment of Multiple Pollutants [46]	There is a possibility to estimate and allocate the toxicities for multiple pollutant combinations easily using the spectroscopy-bioreporter-coupling technique	The research effectively demonstrated the integration of Raman spectroscopy and a toxicity bioreporter, exhibited the potential of Raman spectral alterations to differentiate biological indicators for multiple toxicity pathways, and also validated the high performance in determining and allocating the toxicities of specific toxins	A toxicity bioreporter and Raman spectroscopy were used for the investigation of pollutant toxicity, and validating their combination. To determine the toxicity, a support vector machine model was also employed parallel, and biomarkers for several toxicity pathways were identified in the Raman spectrum through alterations	Additive effects
Toxicity of Tetracycline and Metronidazole in *Chlorella pyrenoidosa* [47]	The toxicity of TC to *C. pyrenoidosa* was higher than that of MTZ.	The 96-hour effective concentration (EC50) of TC and MTZ was 8.72 mg/L and 45.125 mg/L, respectively. The toxic effect of TC and MTZ was enhanced by the combination of a 1:1 ratio, resulting in a synergistic toxicity effect	Different concentration gradients for TC and MTZ in the EC50 experiments were done by using a standard curve to measure algal cell density. The experiments were conducted over 7 d	Synergistic after the combined action of a 1:1 toxicity ratio
Ecological Harm and Economic Damages of Chemical Contamination to Linked Aquatic-Terrestrial Food Webs: A Study-Design Tool for Practitioners [48]	Contamination of aquatic ecosystems can have cascading effects on terrestrial consumers by altering the availability and quality of aquatic insect prey	AT2 tool was developed and used for generating contamination effect studies for linked aquatic-terrestrial food webs which are the main findings in the study, as well as demonstrate the need to consider ecological endpoints and financial modeling while designing the research	The methodological approach involves the innovation of an aquatic-terrestrial assessment tool (AT2) with ecological and economic decision trees to help in the designing of contaminant effect studies for linked aquatic-terrestrial food webs, generated for regulatory claims and research purposes	Additive effects
The influence of soil organic matter content on the toxicity of pesticides to soil invertebrates: a review [49]	The study determined that the grouping of chemicals based exclusively on their lipophilicity may not be an appropriate approach in the assessment of risk	The main findings are that the relationship between toxicity and soil organic matter content is chemical dependent, while soil organic matter influences diverse toxicity endpoints in different ways	The methodology in the study involved the collection of soil invertebrate toxicity data from studies on organic chemicals in soils with different organic matter content, through the literature search and data on toxicity	Additive or synergistic effects

additivity serves as the foundation for CI, which is concentrated on identifying and measuring synergism and antagonistic relationships [1]. Even though additive models show a very high predictive capacity in risk assessment of chemical mixtures, there are still issues since in some experimental settings, additive models might exhibit unanticipated synergism or antagonism which can be seen in Figure 17.3.

17.3 Relevance of Combined Toxicity in Environmental Hazard Evaluation

Understanding the intricate interactions of chemical mixtures and their implications for ecosystem health and human well-being requires an understanding of combined toxicity in the context of environmental hazard assessment. The significance of considering combined toxicity in the assessment of environmental hazards is discussed in this section, along with its applicability in other fields and implications for risk assessment and management strategies. Environmental health assessment covers combined toxicity, which encompasses additive, antagonistic, and synergistic interactions between chemical substances. In their natural environments, organisms are often subjected to intricate mixtures of pollutants that originate from diverse sources, such as industrial processes, farming methods, and urbanization. The cumulative toxicity of these mixtures may be greater than the sum of the individual poisons, increasing the risk to biodiversity and the integrity of ecosystems. This phenomenon is known as synergistic effects [42].

In assessing environmental hazards, combined toxicity is essential since it affects human health in addition to the environment. The importance of combined toxicity in environmental hazard evaluation goes beyond ecological considerations and includes human health implications [42]. Humans, particularly those living in cities or near industrial facilities, may be more exposed to chemical combinations found in the air, water, soil, and food. Synergistic interactions between pollutants can have negative health consequences ranging from respiratory problems to carcinogenic effects, emphasizing the necessity of taking combined toxicity into account when assessing environmental dangers. Antagonistic interactions within chemical mixtures deserve attention in environmental hazard assessment because they can reduce the hazardous effects of some pollutants. Understanding these interactions is critical for accurately assessing the total risk presented by chemical mixtures and developing effective risk management plans. For example, the presence of natural antioxidants in environmental matrices may offset the oxidative stress caused by contaminants, reducing their negative impact on ecosystem health. Additive interactions, while less complex than synergistic or antagonistic interactions, add to the overall toxicity of chemical mixtures and demand thorough risk assessments. Regulatory agencies and researchers must include knowledge of additive effects into risk assessment frameworks to ensure comprehensive evaluations of environmental hazards and effective chemical pollution control [43].

Integrating integrated toxicity assessment into environmental hazard evaluation frameworks necessitates interdisciplinary collaboration and the creation of novel approaches [6]. Researchers, politicians, and regulatory agencies must collaborate to deepen our understanding of chemical interactions, improve risk assessment procedures, and create strong regulatory frameworks that handle the problems posed by complex chemical mixtures. To summarize, understanding the importance of combined toxicity in environmental hazard evaluation is critical for enhancing our ability to identify and minimize the dangers caused by chemical pollutants. By taking a comprehensive strategy that takes into account the interacting impacts of chemical mixtures, we may better conserve ecosystems, protect human health, and assure long-term environmental management for future generations [44].

17.3.1 Significance of Hazard Evaluation for Combined Chemicals

According to recent estimates from the World Health Organization, approximately 23% of global human mortality is attributable to environmental risk factors, which encompass air and water pollution, as well as chemical exposures [50]. Many chemical pollutants in the environment are byproducts of the biological, physical, or chemical decomposition of various products. Accordingly, numerous chemicals with identical compositions exhibit varying degrees of environmental harm. To minimize the potential adverse impacts of chemical pollutants on the environment and human health, thorough risk assessments should be conducted for all existing and newly synthesized chemicals.

It is essential to assess the hazards of combined chemical mixtures in the environment for several key reasons. Firstly, these mixtures can interact synergistically, intensifying their toxicity beyond the sum of individual chemicals. Such interactions must be thoroughly investigated to accurately evaluate hazardous risks to human health and environmental safety. Secondly, it is crucial to assess the cumulative impacts of multiple chemicals to which organisms and humans are simultaneously exposed from various sources to determine the overall population risk. Finally, evaluating the impact of complex transformations of chemical mixtures that occur in the environment is necessary to determine their distribution, persistence, and toxicity in the surrounding environment.

Regulatory compliance requires hazard assessments for chemical combinations to establish safety limits and guidelines. Assessing chemical mixtures provides a more realistic risk evaluation than examining individual chemical effects because mixtures better represent real-world exposures. Moreover, chemical mixtures impact entire ecosystems, not just individual organisms, highlighting the importance of assessing chemical hazards to protect ecosystem health.

In conclusion, innovative methodologies are needed to address the challenges of evaluating complex chemical mixtures, including emerging pollutants and their mixtures. Hazard evaluation for chemical mixtures is indispensable for protecting human health, safeguarding ecosystems, providing an accurate understanding of the hazards associated with environmental contaminants, and ensuring regulatory compliance.

17.4 Mathematical Models for Combined Toxicity Prediction

17.4.1 Role of Mathematical Models in Predicting Combined Effects

Many different mathematical models are employed to predict the combined toxicity of chemical mixtures in the area of risk assessment and environmental toxicology. Mathematical models play a pivotal role in predicting the combined effects of chemical mixtures, offering valuable tools for understanding complex interactions and quantifying the overall toxicity of combined exposures. These models integrate data from various sources, including toxicity studies, exposure assessments, and chemical properties, to provide insights into the additive, synergistic, or antagonistic effects of mixtures. The following points elucidate the significant role of mathematical models in predicting combined effects:

Quantifying combined toxicity: Mathematical models enable researchers to quantify the combined toxicity of chemical mixtures by integrating dose–response relationships, exposure data, and chemical properties [51]. By applying mathematical principles, such as CA and response addition, these models can predict the overall toxicity of mixtures under different exposure scenarios. This quantitative approach allows for the estimation of risk levels and the identification of critical exposure thresholds for regulatory decision-making.

Understanding of interaction mechanisms: Mathematical models facilitate the understanding of interaction mechanisms between chemical components within mixtures. By simulating various interaction scenarios and analyzing dose–response curves, these models can elucidate the underlying mechanisms driving additive, synergistic, or antagonistic effects. This mechanistic insight is essential for identifying key pathways of toxicity and designing targeted interventions to mitigate adverse effects.

Prediction of complex exposure scenarios: Real-world exposure scenarios often involve complex mixtures of chemicals with varying concentrations and exposure routes. Mathematical models enable researchers to predict the combined effects of these mixtures under different environmental conditions and exposure scenarios. By integrating data on chemical fate and transport, exposure pathways, and biological responses, these models can simulate realistic exposure scenarios and assess the cumulative risks posed to human health and the environment.

Enhancing risk assessment frameworks: Mathematical models contribute to the development of comprehensive risk assessment frameworks that consider the combined effects of chemical mixtures. These models provide quantitative tools for integrating data from multiple sources, evaluating exposure-response relationships, and estimating population-level risks. By incorporating mathematical models into risk assessment frameworks, policymakers can make informed decisions regarding chemical management and regulatory priorities.

In conclusion, mathematical models play a critical role in predicting the combined effects of chemical mixtures, offering valuable tools for quantifying toxicity, understanding interaction mechanisms, predicting exposure scenarios, and enhancing risk assessment frameworks. By leveraging mathematical principles and computational methods, researchers can advance our understanding of combined toxicity and develop evidence-based strategies for protecting human health and the environment from the adverse effects of chemical exposures. The mathematical models for risk assessment of environmental pollutants are summarized in the Figure 17.4.

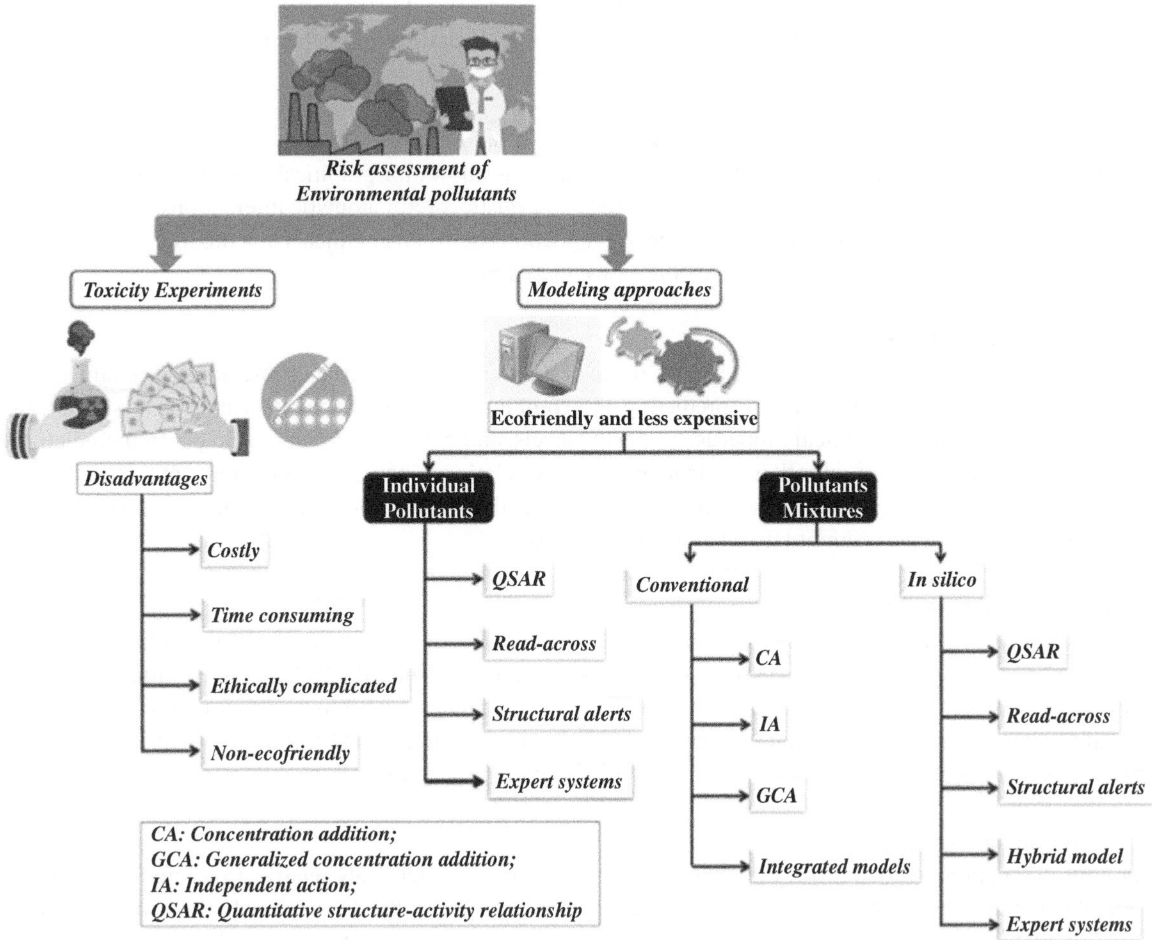

Figure 17.4 Recent advances in modeling the toxicity of environmental pollutants for risk assessment. *Source*: Adapted from Ref. [9].

17.4.2 Classification of Mathematical Models used for Combined Toxicity Prediction

17.4.2.1 Concentration Addition

Concentration addition (CA) assumes that the toxicity of a mixture is solely determined by the combined concentrations of its components, regardless of their specific modes of action or interactions. Under CA, the combined effect of a mixture is calculated by summing the toxicities of each component at their respective concentrations, assuming the additivity of effects. CA is applicable when chemicals in a mixture act independently or additively on the same biological target or pathway. It is commonly used for mixtures of chemicals with similar toxicological properties or mechanisms of action.

17.4.2.2 Response Addition

Response addition (RA) focuses on the combined biological responses induced by multiple chemicals in a mixture, considering their potency or efficacy rather than their concentrations. RA assumes that the combined effect of a mixture is determined by summing the individual responses elicited by each component, accounting for their relative potency or activity [43]. RA is applicable when chemicals in a mixture have diverse modes of action but produce similar biological responses. It is commonly used for assessing the combined effects of chemicals with different toxicological properties or targets.

17.4.2.3 Independent Action

Independent action (IA) assumes that the effects of chemicals in a mixture act independently or multiplicatively, with no interactions between components. Under IA, the combined effect of a mixture is calculated by multiplying the individual probabilities or risks of each component, assuming no interference or modulation of toxicity. IA is applicable when chemicals in a mixture act independently and their effects are not influenced by the presence of other components. It is commonly used for mixtures of chemicals with diverse toxicological properties or mechanisms.

17.4.2.4 Isobole Analysis

Isobole analysis involves plotting dose–response curves for individual chemicals and their combinations to assess the nature of interactions (e.g. synergism, antagonism). Isobole Analysis allows for graphical visualization of the combined effects of chemicals, facilitating the identification of additive, synergistic, or antagonistic interactions. Isobole Analysis is applicable for characterizing the nature and magnitude of interactions between chemicals in a mixture, providing insights into the combined toxicity beyond simple additive or response-based models.

17.4.2.5 Mixtures Toxicity Models

Mixtures toxicity models (MTMs) encompass a diverse range of mathematical approaches, including mechanistic, statistical, and machine learning-based models, for predicting the combined toxicity of chemical mixtures. MTMs integrate various data sources, including chemical properties, exposure levels, and biological endpoints, to develop predictive models of combined toxicity based on quantitative structure–activity relationships (QSARs) or other modeling techniques. MTMs are applicable across a wide range of scenarios and contexts, allowing for flexible and comprehensive assessment of combined toxicity in diverse environmental and biological systems.

17.4.2.6 Fractional Factorial Design

Fractional factorial design (FFD) involves systematically selecting a subset of components from a mixture to evaluate their combined toxicity, reducing the number of experimental trials required while maintaining statistical validity. FFD employs experimental designs that allow for the examination of main effects and interactions between components, facilitating efficient exploration of the combined toxicity landscape. FFD is suitable for evaluating the combined effects of mixtures with numerous components, offering a cost-effective and resource-efficient approach to toxicity testing and analysis.

17.4.2.7 Toxic Unit Approach

The toxic unit (TU) approach quantifies the toxic potency of individual chemicals relative to a reference substance or threshold, allowing for the calculation of total toxic units in a mixture. Under the TU approach, the toxicity of each component is expressed as a fraction of its concentration relative to a reference concentration or threshold, with total toxic units representing the sum of individual contributions. The TU approach is applicable for comparing the toxicity of different chemicals in a mixture and assessing their combined effects relative to established toxicity thresholds or benchmarks.

17.4.2.8 Physiologically Based Pharmacokinetic Modeling

Physiologically based pharmacokinetic (PBPK) modeling integrates physiological, pharmacokinetic, and toxicokinetic parameters to simulate the absorption, distribution, metabolism, and excretion of chemicals in biological systems. PBPK models incorporate physiological and biochemical processes to predict the internal concentrations of chemicals and their metabolites, enabling the assessment of combined toxicity based on internal dosimetry. PBPK modeling is applicable for predicting the combined effects of chemicals with complex pharmacokinetic profiles and metabolism, providing mechanistic insights into the disposition and toxicity of chemical mixtures.

17.4.2.9 Toxicological Interaction Models

Toxicological interaction models encompass a range of mathematical formulations for quantifying the nature and magnitude of interactions between chemicals in a mixture. These models may include dose–response models for characterizing synergistic, antagonistic, or additive interactions, as well as mechanistic models based on biochemical or physiological pathways. Toxicological interaction models are applicable for elucidating the mechanisms underlying combined toxicity and predicting the outcomes of chemical interactions in biological systems.

These mathematical models provide valuable frameworks and tools for predicting and assessing the combined toxicity of chemical mixtures, aiding in the understanding of complex interactions and guiding informed decision-making for environmental and human health protection.

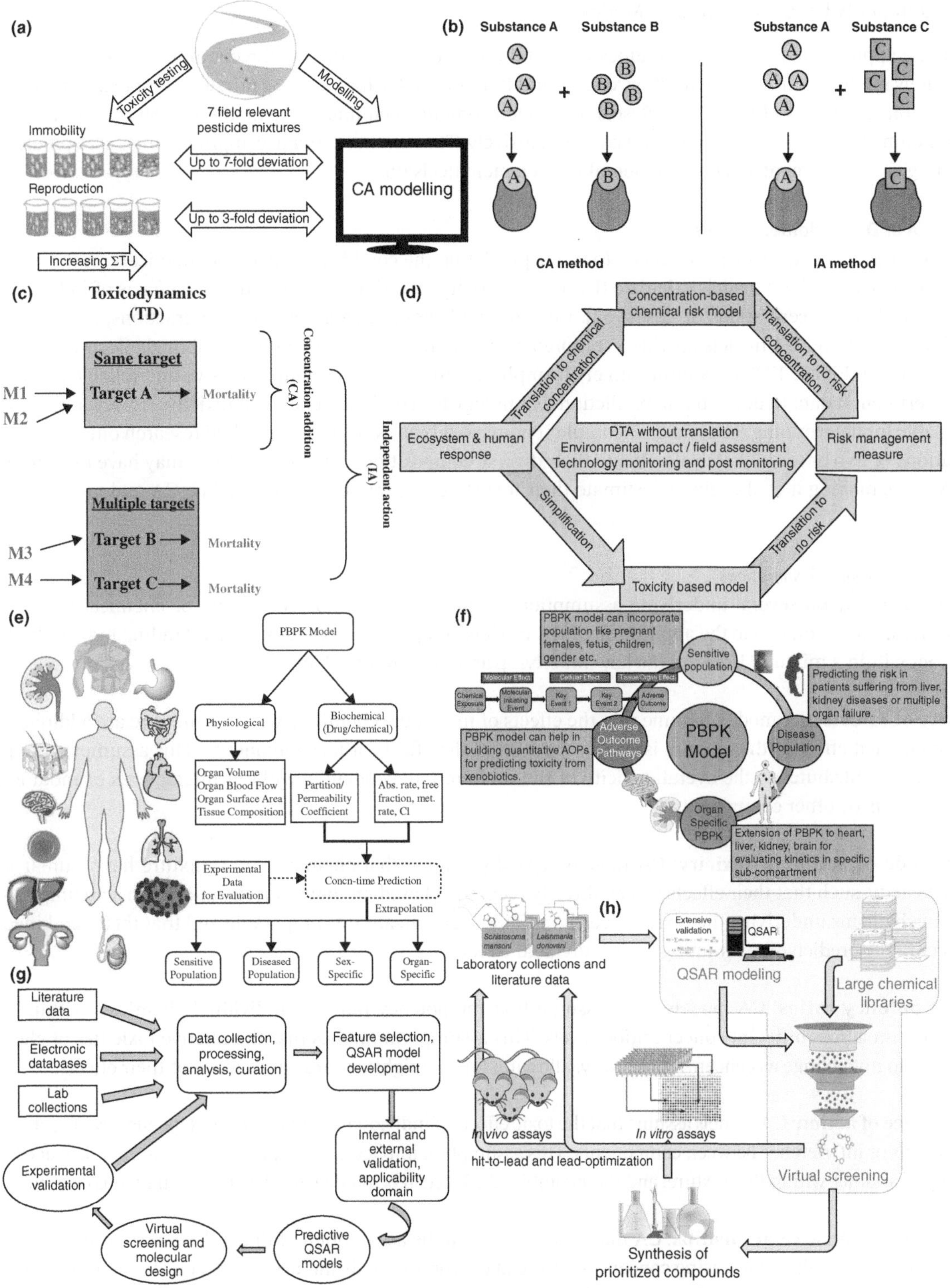

Figure 17.5 (a) Graphical representation of assessing the effects of field-relevant pesticide mixtures for their compliance with the CA model [54]/with permission of Elsevier; (b) the diverse toxic mode of actions of classical joint prediction model of CA and independent action [36]/Frontiers Media S.A/CC BY 4.0; (c) graphical representation of CA and independent action model [35]; (d) flow chart of direct toxicity assessment, methods, evaluation, and interpretation [55]; (e) conceptual diagram for the PBPK model representing physiological and biochemical parameters and extrapolation of the model for human health risk assessment [56]/MDPI/CC BY 4.0; (f) applications of PBPK models in different fields (sensitive populations, diseased populations, organ-specific, and adverse outcomes) for therapeutics and toxicological risk assessment [56]/MDPI/CC BY 4.0; (g) flowchart of QSAR-based model in drug design and discovery; (h) QSAR-based virtual screening workflow [57]/Frontiers Media S.A/CC BY 4.0.

17.4.3 Commonly Used Mathematical Models

Selecting the appropriate modeling techniques is crucial for accurately predicting the combined effects of chemical mixtures. Various methods exist, each with its strengths and limitations. Factors, such as mixture complexity, data availability, and regulations guide modeling choices. Researchers often combine techniques like mathematical models, in vitro assays, and simulations to assess combined toxicity comprehensively. This multidisciplinary approach increases prediction reliability, aiding risk assessment and environmental management decisions.

17.4.3.1 CA or Dose Addition Models

CA or dose addition (DA) models are crucial tools for predicting the combined toxicity of chemicals on the environment and living organisms. These models calculate the overall toxicity by estimating the combined effects of individual chemicals based on their concentrations or doses [52]. Considering factors, such as chemical interactions, bioavailability, and mode of action, CA and DA models provide a comprehensive approach to assessing the combined effects of chemical mixtures, as shown in Figure 17.5a–c. Sunmi Kim et al. emphasize the need for a comprehensive and reliable database of scientific experimental data to develop risk prediction technology for combined chemical toxicity. However, there is limited data available for determining the toxicity of complex chemical mixtures, as most published research only provides data on combinations of two or three chemicals. Due to their diverse composition, complex mixtures may have multiple modes of action (MoAs), making it challenging to estimate their toxicity using traditional models like CA and independent action (IA) [53].

17.4.3.1.1 Underlying Assumptions

CA/DA models rely on several underlying assumptions to predict the combined toxicity of chemical mixtures. These assumptions are fundamental to the application of CA models and provide a basis for understanding how multiple chemicals interact within a mixture. Here are the key underlying assumptions of CA models:

Additivity of effects: CA models assume that the effects of individual chemicals within a mixture are additive, meaning that the combined effect of the mixture is simply the sum of the effects of its components. This assumption implies that each chemical contributes to the overall toxicity of the mixture independently, without interactions or modifications of their toxic effects by other components.

Similar mode of action or toxicity: CA models typically assume that chemicals in a mixture have similar modes of action or toxicity, such that their effects can be directly combined based on their concentrations. This assumption implies that the mechanisms underlying the toxic effects of individual chemicals are comparable and that their combined toxicity can be accurately predicted by summing their individual effects.

Constant potency ratios: CA models often assume that the potency ratios of individual chemicals within a mixture remain constant across different concentration levels. This assumption implies that the relative toxicities of chemicals in the mixture do not change as concentrations vary, allowing for a straightforward summation of their effects.

Independence of action: CA models assume that the toxic effects of individual chemicals act independently of each other, with no interactions or interference between components. This assumption implies that the presence of one chemical does not affect the toxicity of another within the mixture, and their combined effect can be predicted solely based on their concentrations.

Linear dose–response relationships: CA models often rely on the assumption of linear dose–response relationships for individual chemicals, where increasing concentrations lead to proportionally increasing effects. This assumption simplifies the prediction of combined toxicity by allowing for direct summation of effects based on concentration levels.

Overall, these underlying assumptions of CA or DA models provide a framework for predicting the combined toxicity of chemical mixtures based on the concentrations and individual toxicities of their components. While these assumptions simplify the modeling process, it's essential to recognize their limitations and consider the potential for deviations from additivity in real-world scenarios.

17.4.3.1.2 Mathematical Formulation

The mathematical formulation of CA or DA models involves a straightforward approach based on the principle of additivity of effects. The basic formulation of CA or DA models involves summing the effects of individual chemicals within a mixture to predict the combined toxicity. Here's the general mathematical formulation:

$$E_{\mathrm{m}} = \sum_{i=1}^{n} c_i \times E_i \tag{17.1}$$

where

E_{m} is the combined effect of the mixture;
c_i is the concentration of the ith chemical in the mixture;
E_i is the effect of the ith chemical.

The formulation assumes that the effects of individual chemicals are additive, meaning that the combined effect of the mixture is simply the sum of the effects of its individual components. This formulation relies on the assumption that the potency ratios of individual chemicals remain constant across different concentration levels. Chemical concentrations are often represented in mass or volume per unit volume of the mixture/compound (mg/L), whereas reaction can be estimated using a variety of endpoints (for example, morality or inhibition).

17.4.3.1.3 Strengths and Limitations

CA or DA models have several strengths and limitations, which are important to consider when applying these models in toxicological assessments.

Strengths

Simplicity: CA models offer a straightforward and intuitive approach to predicting the combined toxicity of chemical mixtures. They are easy to understand and implement, making them accessible to researchers and regulators.

Widely accepted: CA models are widely accepted in the field of toxicology and environmental risk assessment. They are commonly used by regulatory agencies for assessing the risks associated with chemical mixtures in various environmental and human health contexts [58].

Conservative estimation: CA models tend to provide conservative estimates of combined toxicity by assuming additivity of effects. This conservative approach ensures that potential risks associated with chemical mixtures are adequately considered, even if interactions between components are present.

Applicability across chemical classes: CA models can be applied to mixtures containing chemicals with diverse modes of action and toxicological properties. They do not require detailed knowledge of the mechanisms underlying interactions, making them applicable across a wide range of chemical classes.

Limitations

Additivity assumption: The primary limitation of CA models is their reliance on the assumption of additivity of effects. In reality, interactions between chemicals in mixtures may result in synergistic or antagonistic effects that deviate from additivity [54].

Lack of mechanistic insight: CA models do not provide mechanistic insights into the interactions between chemicals or the underlying biological pathways. They treat all chemicals as equally contributing to the combined toxicity without considering differences in mode of action or potency.

Nonlinear dose–response relationships: While CA models typically predict a linear correlation between dose and response for individual substances, this isn't always the case. The presence of nonlinear associations can result in either an overestimate or underestimate of the cumulative toxicity.

Limited predictive accuracy: While CA models are useful for screening purposes and initial risk assessment, they may have limited predictive accuracy in complex real-world scenarios. Factors, such as mixture composition, exposure routes, and biological variability can influence the actual combined toxicity of mixtures.

Data requirements: CA models require data on the individual toxicities and concentrations of chemicals within the mixture. Obtaining accurate and comprehensive data for all components can be challenging, particularly for complex mixtures or chemicals with limited toxicity data.

While CA or DA models are easy-to-use and provide a reasonable estimate of combined toxicity, besides this they also have their limitations, such as a lack of mechanistic insight, a dependence on additivity predictions, and the possibility of wrongly predicting nonlinear dose–response relationships. While applying the CA models in toxicological investigation studies the strengths and limitations should be considered important.

17.4.3.1.4 Application

CA or DA models find application across diverse fields, aiding in the understanding and prediction of the behavior and impacts of chemical substances in various contexts. Here's how they are applied:

Environmental fate and transport

Partitioning: CA or DA models predict the distribution of chemicals among different environmental compartments, such as air, water, soil, and sediment. By accounting for solubility, volatility, and sorption, these models facilitate estimating chemical fate and transport in natural systems.

Bioaccumulation potential: These models assess the bioaccumulation potential of chemicals within living organisms. By forecasting the partitioning of chemicals between environmental media and biota, CA or DA models assist in identifying substances that may accumulate in food chains, thereby posing risks to higher trophic levels.

Toxicokinetics and pharmacokinetics: Absorption, Distribution, Metabolism, and Excretion (ADME): CA or DA models simulate the ADME processes of chemicals within biological systems. They predict the absorption, distribution, metabolism, and excretion of chemicals in organisms, contributing to the design of pharmaceuticals and enhancing understanding of their pharmacokinetic profiles.

Dose–response relationships: These models elucidate the relationship between chemical concentration or dose and biological response. By integrating toxicokinetic data with toxicity endpoints, CA or DA models facilitate the prediction of dose–response relationships, thereby informing assessments of chemical safety and efficacy.

Chemical risk assessment: Exposure Assessment: CA or DA models estimate human and ecological exposures to chemical substances. They consider factors, such as environmental concentrations, exposure pathways, and receptor characteristics to quantify exposure levels and evaluate potential risks to health and the environment.

Risk characterization: These models integrate exposure and toxicity data to characterize the risks associated with chemical exposures. They predict adverse health effects, assess the likelihood of harm, and provide insights to guide risk management decisions aimed at safeguarding public health and environmental quality.

Regulatory compliance: CA or DA models support regulatory agencies in the registration and evaluation of chemical substances. They furnish predictive tools for assessing environmental fate, toxicological properties, and potential risks of chemicals, thereby facilitating compliance with regulatory requirements and guidelines [59].

Chemical classification and labeling: These models assist in the classification and labeling of chemicals based on their physicochemical properties and hazard profiles. They aid in the identification of substances posing risks to human health or the environment, enabling appropriate labeling and effective risk communication.

Environmental Management and Remediation

Contaminant remediation: CA or DA models guide the remediation of contaminated sites by predicting the behavior of chemicals in soil and groundwater. They optimize remediation strategies, such as soil washing, bioremediation, and chemical oxidation by simulating the transport and fate of contaminants.

Environmental monitoring: These models support environmental monitoring endeavors by predicting the spatial and temporal distribution of chemicals in environmental media. They assist in designing monitoring programs, selecting sampling locations, and interpreting monitoring data to track changes in environmental quality over time.

In summary, CA or DA models play a pivotal role in comprehending and forecasting the behavior and effects of chemical substances across various environmental, biological, and regulatory domains. They furnish invaluable tools for assessing chemical safety, managing environmental risks, and facilitating informed decision-making to safeguard human health and the environment [60].

Case study Related to CA/Dose–Response Model

Spurgeon et al. evaluated mixture toxicity due to organic chemicals in English surface and groundwaters using monitoring data and the CA model. Semiquantitative GC-MS and LC-MS measurements were combined with chronic species sensitivity distribution (SSD) HC50 values to calculate substance-specific hazard quotients and mixture effects. Analysis showed increased hazard from multiple substances compared to single chemicals. These findings provide important information for future policy on managing mixture risks in water resources and demonstrate the utility of the CA model for assessing mixture toxicity and guiding risk management strategies [61].

*GC-MS: Gas chromatography-mass spectrometry

*LC-MS: Liquid chromatography Tandem mass spectrometry

Title	Proportional contributions to organic chemical mixture effects in groundwater and surface water.
Background	This study assessed the toxicity of organic chemicals in surface water and groundwater in England using semiquantitative GC-MS and LC-MS measurements from national monitoring programs.
Experimental Setup	Samples were taken from groundwater and surface H_2O, from two separate national monitoring programs. Samples were taken for various reasons including site assessment and compliance monitoring
Data Collection	Hazard values were collected from published chronic species sensitivity distribution (SSD) HC50 values. Samples were analyzed using liquid-liquid extraction for GC-MS measurements and Waters Oasis HLB SPE cartridges for LC-MS measurements.
Application of Mathematical Model	Individual substance hazard quotients and mixture effects were calculated using a concentration addition model, which involved comparing measured concentrations of chemicals to determine their relative toxicity and contribution to the overall mixture hazard.
Validation and Interpretation	Analysis revealed that the presence of multiple substances increased hazard compared to any single chemical. The difference in hazard attributed to the most important chemical was large.
Outcome	This study identified the most significant substances contributing to mixture hazard and assessed their relative contributions using the concentration addition model.
Implications	These findings offer critical information to help guide future policies for effectively managing mixture risks in groundwater and surface waters. The concentration addition model can be a practical tool for evaluating mixture toxicity.

17.4.3.2 Risk Assessment Models

Risk assessment models are essential tools for evaluating the potential risks associated with chemical mixtures. These models integrate data on exposure levels, toxicological properties, and dose–response relationships to quantify the overall risk posed by combined chemical exposures [62]. Risk assessment models employ mathematical algorithms and statistical analyses to estimate the likelihood and magnitude of adverse effects on human health and the environment. By considering factors, such as chemical interactions, exposure pathways, and population demographics, risk assessment models provide valuable insights into the complex dynamics of combined toxicity, guiding regulatory decision-making and risk management strategies (Figure 17.5d).

17.4.3.2.1 Underlying Assumptions

Exposure homogeneity: Risk assessment models often assume that the exposure to the chemical is homogeneous within the population being studied, meaning that all individuals are exposed to the same concentration or dose of the chemical [63].

Population homogeneity: Risk assessment models may assume homogeneity within the exposed population, meaning that individuals have similar susceptibilities and responses to chemical exposure.

No synergistic effects: Risk assessment models typically assume that there are no synergistic effects among different chemicals or environmental factors, meaning that the combined effect of multiple exposures is simply the sum of their individual effects [64].

Conservative estimation: Risk assessment models often use conservative estimation techniques, assuming higher than predicted risk to ensure public health and safety.

Threshold effects: Some risk assessment models assume threshold effects, implying no adverse health effects below a certain exposure level and no response below a certain dose.

Constant exposure duration: Risk assessment models can assume a constant exposure duration (n), implying that the length of exposure to the chemical is uniform across all individuals and scenarios under consideration.

Predictive validity: Risk assessment models assume that the data and parameters used in the model accurately represent real-world conditions and that the model's predictions are valid for the intended population or scenario [65].

Linear dose–response: Some risk assessment models may assume a linear dose–response relationship, where the magnitude of the response is directly proportional to the dose or concentration of the chemical.

No genetic variability: In certain cases, risk assessment models may assume no genetic variability within the population being studied, meaning that all individuals have identical genetic susceptibilities to the effects of chemical exposure [66].

Temporal stability: Risk assessment models may assume temporal stability in exposure and response relationships, meaning that the effects of exposure to the chemical remain consistent over time.

These assumptions underlie risk assessment modeling, informing data interpretation and predictions of potential health risks from chemical exposures [67]. However, recognizing the assumptions' limitations and uncertainties is crucial when applying model results to real-world situations.

17.4.3.2.2 Mathematical Formulation

Risk assessment models use a mathematical approach to estimate the potential risks of exposure to hazards. Different models and techniques are available, but many risk assessment models share a common goal: to estimate the likelihood and seriousness of harmful effects from exposure to a hazard. Here is a general mathematical formula for risk assessment models:

Exposure Assessment

$$\text{Exposure Concentration (EC): EC} = C \times T \tag{17.2}$$

where C is the concentration of the hazard in the environment (e.g. air, water, soil) and T is the duration of exposure.

Exposure Dose

$$\text{ED} = \text{EC} \times \text{BW} \tag{17.3}$$

where BW is the body weight of the exposed individual or population.

Dose–Response Assessment

$$\text{Dose} - \text{response curve: Response} = f(\text{Dose}) \tag{17.4}$$

where Dose represents the exposure dose or concentration, and f is the mathematical function describing the relationship between dose and response (e.g. linear, sigmoidal).

Toxicity Assessment

$$\text{Hazard Quotient (HQ): HQ} = \frac{\text{ED}}{\text{RfD}} \tag{17.5}$$

where RfD is the reference dose, representing the level of exposure at which no adverse effects are expected to occur.

Cancer Risk

$$\text{Risk} = \text{CSF} \times \text{ED} \tag{17.6}$$

where CSF is the cancer slope factor, representing the incremental cancer risk associated with a unit increase in ED exposure dose.

Risk Characterization

$$\text{Overall Risk (OR): OR} = \sum_{i=1}^{n} \text{HR}_i \tag{17.7}$$

where HR_i is the hazard risk associated with each hazard or exposure pathway.

Total Risk

$$\text{TR} = \sum_{j=1}^{m} \text{OR}_j \tag{17.8}$$

where OR_j is the overall risk associated with each scenario or exposure pathway.

Probabilistic Risk Assessment

Probability density function (PDF): Characterizing the uncertainty in model inputs using probability distributions.

Monte Carlo simulation: Generating multiple simulations by sampling from input parameter distributions to estimate the probability distribution of model outcomes.

These mathematical formulations provide a framework for quantifying and evaluating the risks associated with exposure to hazards, helping decision-makers to prioritize the interventions and allocate resources to mitigate risks effectively. It's essential to customize these formulations based on the specific hazard, exposure scenario, and available data to ensure accurate risk assessment outcomes.

17.4.3.2.3 Strengths and Limitations

Risk assessment models play a vital role in predicting the combined chemical toxicity by integrating data on exposure levels, toxicity of individual chemicals, and potential interactions between them. Here are some strengths and limitations of risk assessment models in predicting combined chemical toxicity:

Strengths

Quantitative estimation: Risk assessment models provide a quantitative estimation of the combined toxicity of multiple chemicals, allowing for the comparison of risks across different exposure scenarios and chemicals.

Systematic framework: These models offer a systematic framework for integrating data from various sources, such as toxicity studies, exposure assessments, and dose–response relationships, to evaluate the overall risk associated with chemical mixtures.

Regulatory guidance: Risk assessment models are often used to inform regulatory decisions and set exposure limits or guidelines for chemical mixtures, helping to protect public health and the environment.

Risk prioritization: They aid in prioritizing chemicals or exposure scenarios for further investigation or risk management actions based on the magnitude of their combined toxicity (24).

Transparency and communication: Risk assessment models provide a transparent and structured approach to evaluating combined chemical toxicity, facilitating communication of risks to stakeholders, policymakers, and the public.

Limitations

Additivity assumption: Numerous risk assessment models are predicated on the assumption of additive effects. However, this assumption may not consistently align with real-world scenarios. In these contexts, chemicals could potentially interact in a synergistic or antagonistic manner.

Data limitations: Availability and quality of data on exposure levels, toxicity of individual chemicals, and interactions between chemicals can vary, leading to uncertainties in risk assessments.

Complexity: Predicting combined chemical toxicity involves dealing with complex interactions between chemicals, exposure pathways, and biological systems, which may exceed the capabilities of current risk assessment models.

Cumulative exposures: Risk assessment models may not adequately account for cumulative exposures to multiple chemicals across different sources and pathways, leading to underestimation of risks.

Variability and uncertainty: They often involve inherent variability and uncertainty due to factors, such as variability in individual susceptibility, differences in exposure scenarios, and limitations in data and modeling assumptions.

Extrapolation: Extrapolating findings from laboratory studies or single exposure scenarios to real-world situations can introduce uncertainties and may not capture the full range of potential risks.

Overall, while risk assessment models provide valuable tools for predicting combined chemical toxicity, it's essential to recognize their limitations and uncertainties and use them as part of a comprehensive risk management strategy that considers multiple lines of evidence and approaches.

17.4.3.2.4 Application

Risk assessment (RA) models are essential tools used to evaluate the potential risks associated with exposure to chemical substances in various contexts. Here's how they are applied:

Chemical Safety Evaluation

Hazard identification: Risk assessment models assess the inherent hazards of chemical substances by analyzing their physicochemical properties, toxicity data, and mode of action. They identify potential adverse effects on human health and the environment, such as carcinogenicity, genotoxicity, and ecotoxicity.

Exposure assessment: Risk assessment models estimate the extent of human and environmental exposures to chemical substances. They consider exposure pathways, routes, durations, and receptor characteristics to quantify exposure levels and evaluate potential risks to exposed populations and ecosystems [66].

Regulatory Compliance

Regulatory submissions: Risk assessment models support regulatory submissions by providing predictive tools for assessing chemical safety and compliance with regulatory requirements. They generate data on hazard identification, exposure assessment, and risk characterization to demonstrate the safety of chemical substances and products.

Risk management: Risk assessment models inform risk management decisions by identifying risk factors, evaluating control measures, and prioritizing interventions to mitigate potential risks. They assist regulatory agencies in setting exposure limits, establishing safety thresholds, and implementing risk reduction strategies to protect public health and environmental quality.

Product Development and Formulation

Product safety assessment: Risk assessment models guide the safety assessment of consumer products, pharmaceuticals, and industrial chemicals during the development and formulation stages. They predict potential hazards, exposures, and risks associated with product ingredients, formulations, and manufacturing processes.

Alternative assessment: Risk assessment models facilitate the evaluation of alternative chemicals, materials, and technologies to identify safer and more sustainable alternatives. They compare the hazards, exposures, and risks of different options to support informed decision-making and product innovation.

Occupational Health and Safety

Workplace risk assessment: Risk assessment models assess occupational exposures to chemical substances in workplaces to protect workers' health and safety. They quantify exposure levels, evaluate health risks, and recommend control measures to minimize workplace hazards and prevent occupational illnesses and injuries.

Exposure monitoring: Risk assessment models support exposure monitoring programs by predicting potential exposures to hazardous chemicals in different occupational settings. They assist in the design of sampling strategies, selection of monitoring methods, and interpretation of monitoring data to ensure compliance with occupational exposure limits and regulations.

Environmental Management and Remediation

Environmental risk assessment: Risk assessment models evaluate the risks posed by chemical pollutants to environmental receptors, such as air, water, soil, and biota. They predict the fate, transport, and effects of contaminants in ecosystems, guiding pollution control measures and environmental management strategies.

Contaminated site remediation: Risk assessment models support the remediation of contaminated sites by assessing the risks associated with exposure to hazardous substances. They optimize remediation technologies, assess cleanup goals, and monitor remediation progress to restore environmental quality and protect human health.

In summary, risk assessment models play a crucial role in assessing and managing the risks associated with chemical substances in diverse settings [68]. They provide valuable tools for regulatory compliance, product safety evaluation, occupational health and safety, and environmental management, supporting informed decision-making and risk reduction efforts to safeguard public health and environmental quality.

Case study related to Risk Assessment Model
Fan and Xu investigated health risks in coal mines using a combination of deep learning algorithms and mathematical models. They collected and analyzed gas samples from four working areas in a Shanxi Province coal mine for harmful substances. Using mathematical evaluation models, they calculated the risk index for each area, showing correlations between coal cutting activities, induced airflow velocities, and concentrations of harmful substances. The deep learning mathematical model proved effective in health risk assessment, highlighting the importance of employing such models for safety management in the mining industry, especially in reducing exposure to toxic chemicals [69].

(Continued)

(Continued)

Title	Health risks of occupational exposure to toxic chemicals in coal mine workplaces based on risk assessment mathematical model based on deep learning
Background	As China's energy demands surge, coal mining has become more prevalent, necessitating an evaluation of the health risks associated with occupational exposure to toxic chemicals in coal mines. This study aims to do so by combining deep learning algorithms and mathematical models.
Experimental Setup	Four areas within a coal mine in Shanxi Province were selected for a study. Gas samples were collected from the work environment to analyze them for harmful substances.
Data Collection	Employee lung conditions were evaluated using mathematical models to analyze gas samples and single-factor evaluation scales.
Application of Mathematical Model	The risk assessment model, leveraging deep learning algorithms and mathematical models, calculated a risk index for each working area.
Validation and Interpretation	Correlations exist between coal cutting operations, induced airflow velocities, and concentrations of hazardous substances in work environments, as revealed by experimental findings.
Outcome	This study concluded that the risk assessment model effectively evaluated health risks associated with occupational exposure to toxic chemicals in coal mines.
Implications	The findings underscore the importance of employing risk assessment models to ensure worker safety management in the mining industry, particularly in mitigating exposure to hazardous substances.

17.4.3.3 Interaction-based Models

Interaction-based models are mathematical frameworks used to quantify the interactions between chemicals in a mixture and predict their combined toxicity [1]. These models consider various interactions, including synergism, antagonism, and additivity, to provide insights into the combined effects of chemical mixtures. Interaction-based models assume that the combined effect of a mixture is influenced by the interactions between its components. These interactions can alter the overall toxicity of the mixture, leading to effects greater or less than expected based on the individual toxicities of the components.

17.4.3.3.1 Underlying Assumptions

Interaction-based models for predicting mixture toxicity rely on certain assumptions to understand how chemicals interact within a mixture. Here are the key assumptions underlying these models:

Additivity of effects: Interaction-based models typically assume that the effects of individual chemicals in a mixture are additive: the combined effect is the sum of the effects of each component. This assumption implies that each chemical contributes to toxicity independently, i.e. without interactions or modifications of their effects by other components.

Similar mode of action or toxicity: Interaction-based models typically assume that chemicals in a mixture have similar modes of action or toxicity, such that their effects can be directly combined based on their concentrations. This assumption implies that the mechanisms underlying the toxic effects of individual chemicals are comparable and that their combined toxicity can be accurately predicted by summing their individual effects.

Constant potency ratios: These models often assume that the potency ratios of individual chemicals within a mixture remain constant across different concentration levels. This assumption implies that the relative toxicities of chemicals in the mixture do not change as concentrations vary, allowing for a straightforward summation of their effects.

Independence of action: Interaction-based models assume that the toxic effects of individual chemicals act independently of each other, with no interactions or interference between components. This assumption implies that the presence of one chemical does not affect the toxicity of another within the mixture, and their combined effect can be predicted solely based on their individual concentrations.

Linear dose–response relationships: These models often rely on the assumption of linear dose–response relationships for individual chemicals, where increasing concentrations lead to proportionally increasing effects. This assumption simplifies the prediction of combined toxicity by allowing for direct summation of effects based on concentration levels.

These underlying assumptions provide a framework for predicting the combined toxicity of chemical mixtures within interaction-based models. While these assumptions simplify the modeling process, it's essential to recognize their limitations and consider the potential for deviations from additivity or interactions in real-world scenarios.

17.4.3.3.2 Mathematical Formulation

Interaction-based models in the prediction of combined chemical toxicity aim to quantify the combined effect of multiple chemicals considering potential interactions between them. These models typically incorporate mathematical formulations that account for the additive, synergistic, or antagonistic effects of chemicals within a mixture. Here's an overview of the mathematical formulations commonly used in interaction-based models:

Additive Model In the additive model, the combined effect of a mixture is calculated as the sum of the individual effects of its components:

$$E_{\text{comb}} = E_{\text{A}} + E_{\text{B}} + E_{\text{C}} + \ldots \ldots E_n \tag{17.9}$$

where E_{comb} is the combined effect of the mixture, and E_{A}, E_{B}, ... are the individual effects of chemicals A, B, etc.

Synergistic Model/Hewlett Equation The Hewlett equation is a mathematical model used in the context of Interaction-Based Models to predict the combined toxicity of chemical mixtures, specifically considering synergistic interactions. It provides a quantitative framework for estimating the combined effect of two or more chemicals based on their toxicities and a synergy factor that accounts for their interaction.

The Hewlett equation is expressed as:

$$E_{\text{comb}} = E_{\text{A}} + E_{\text{B}} + (E_{\text{A}} \times E_{\text{B}} \times S) \tag{17.10}$$

where E_{comb} is the combined effect of the mixture, E_{A} and E_{B} are the individual effects of chemicals A and B, respectively. S is the synergy factor, representing the extent of synergistic interaction between chemicals A and B. In the Hewlett equation, the first two terms (E_{A} and E_{B}) represent the individual effects of chemicals A and B, assuming no interaction between them. The third term (captures the additional effect resulting from their interaction, where ($E_{\text{A}} \times E_{\text{B}} \times S$) captures the additional effect resulting from their interaction, where S is typically a value greater than 1 for synergistic interactions.

Antagonistic Model In the antagonistic model, the combined effect is less than the sum of the individual effects, indicating antagonistic interactions:

$$E_{\text{comb}} = E_{\text{A}} + E_{\text{B}} - (E_{\text{A}} \times E_{\text{B}} \times A) \tag{17.11}$$

where A_{A} is the antagonism factor, representing the extent of antagonistic interaction between chemicals A and B.

Isobolographic analysis: Isobolographic analysis is a graphical method used to assess the nature and extent of interaction between chemicals within a mixture [70]. It involves plotting dose–response curves for individual chemicals and comparing them to the predicted combined effect based on additivity, synergy, or antagonism.

Response Surface Methodology (RSM): Response surface methodology is a statistical technique used to model and analyze the effects of multiple factors, including chemical concentrations, on a response variable (e.g. toxicity) [71]. It allows for the characterization of complex interactions between chemicals and the prediction of combined toxicity based on experimental data.

Generalized Concentration Addition (GCA): GCA is a widely used concept in toxicology and ecotoxicology for predicting the combined toxicity of chemical mixtures. It assumes that the individual chemicals in a mixture act independently and their toxic effects can be predicted based on their concentrations alone. Here's how GCA can be incorporated into an interaction-based mathematical model for combined chemical toxicity:

Mathematically, the GCA model for predicting combined toxicity TC_{mix} can be expressed as:

$$TC_{\text{mix}} = \sum_{i=1}^{n} TC_i \times f_i \tag{17.12}$$

where TC_{mix} is the predicted combined toxicity of the mixture, TC_i is the toxicity of chemical i, f_i is the fraction or proportion of chemical i in the mixture.

17.4.3.3.3 Strengths and Limitations

Interaction-based models offer a valuable framework for predicting the combined toxicity of chemical mixtures by considering potential interactions between individual chemicals. While these models have strengths, they also face limitations. Let's explore both:

Strengths

Capturing synergistic effects: Interaction-based models can account for synergistic interactions between chemicals, where the combined effect is greater than the sum of individual effects. This allows for a more accurate estimation of the combined toxicity of chemical mixtures, especially in scenarios where synergistic effects may lead to heightened risks.

Addressing antagonistic interactions: These models can also capture antagonistic interactions between chemicals, where the combined effect is less than the sum of individual effects. By quantifying antagonism, they provide insights into scenarios where the presence of one chemical mitigates the toxic effects of another, potentially reducing overall toxicity.

Enhancing predictive accuracy: Interaction-based models improve the predictive accuracy of combined toxicity assessments by accounting for complex interactions between chemicals. This enables better-informed decision-making in risk assessment and management, leading to more effective strategies for protecting human health and the environment.

Informing regulatory decision-making: By providing quantitative estimates of combined toxicity that consider interactions between chemicals, these models contribute to regulatory decision-making processes. They help regulators set exposure limits and guidelines that adequately account for the combined effects of chemical mixtures, ensuring the protection of public health and the environment.

Limitations

Data requirements: Interaction-based models often require extensive data on the individual toxicities of chemicals and their interactions, which may not always be available. A lack of comprehensive data can limit the applicability and accuracy of these models, especially for complex mixtures with numerous interacting components.

Complexity: Modeling interactions between chemicals can be complex and challenging, particularly when considering multiple chemicals with diverse modes of action and varying levels of toxicity. The complexity of these interactions may lead to uncertainties in model predictions and require simplifying assumptions that could affect the accuracy of results.

Assumption of additivity: Some interaction-based models assume additivity of effects, which may not always hold in real-world scenarios. Deviations from additivity, such as synergistic or antagonistic interactions, may not be adequately captured, potentially leading to underestimation or overestimation of combined toxicity.

Interpretation and validation: Interpreting and validating the results of interaction-based models can be challenging due to the inherent complexity of chemical interactions and the limited availability of experimental data for validation [72]. Ensuring the reliability and robustness of model predictions requires scrutiny and validation against empirical data.

In summary, while interaction-based models offer significant advantages in predicting combined chemical toxicity, they also face challenges related to data availability, model complexity, and validation. Addressing these limitations through improved data collection, model refinement, and validation efforts can enhance the utility and reliability of these models in informing risk assessment and management decisions.

17.4.3.3.4 Applications

Risk assessment: Interaction-based models contribute to the assessment of combined chemical toxicity in environmental, occupational, and consumer settings. By considering interactions between chemicals, these models provide a more accurate estimation of overall risk, enabling regulators and policymakers to set appropriate exposure limits and guidelines to protect public health.

Environmental monitoring: These models help environmental scientists and regulators evaluate the combined effects of chemical pollutants in air, water, soil, and sediments. By predicting the combined toxicity of chemical mixtures, they support efforts to monitor and mitigate environmental contamination and assess the effectiveness of pollution control measures.

Occupational health and safety: In occupational settings where workers may be exposed to multiple chemicals simultaneously, interaction-based models aid in assessing the combined health risks associated with exposure [18]. By accounting for synergistic or antagonistic interactions, these models inform workplace safety measures and help minimize occupational hazards.

Pharmaceutical development: Interaction-based models are used in pharmaceutical research and development to evaluate potential interactions between drugs or drug combinations. By predicting the combined pharmacological effects and potential adverse reactions of drug mixtures, these models support the development of safer and more effective medication regimens.

Toxicological research: These models contribute to toxicological research by elucidating the mechanisms underlying interactions between chemicals and their effects on biological systems. By providing insights into the complex interactions between chemicals, they enhance our understanding of toxicological processes and support the development of predictive models for hazard identification and risk assessment.

Chemical safety testing: Interaction-based models are used in chemical safety testing to assess the combined toxicity of chemical mixtures and prioritize chemicals for further testing or regulatory action. By predicting the combined effects of chemicals, these models inform risk-based approaches to chemical safety assessment and help identify chemicals of concern.

Public health policy: Interaction-based models inform public health policies aimed at reducing exposure to harmful chemical mixtures and protecting vulnerable populations. By quantifying the combined toxicity of chemical mixtures, these models support evidence-based decision-making and facilitate the development of targeted interventions to mitigate health risks.

Overall, interaction-based models play a critical role in various fields, including risk assessment, environmental monitoring, occupational health, pharmaceutical development, toxicological research, chemical safety testing, and public health policy. By considering interactions between chemicals, these models provide valuable insights into the complex nature of combined chemical toxicity and support efforts to safeguard human health and the environment.

Case Studies Related to Interaction-based Models

Teuschler et al. evaluated Prop-RA by statistically analyzing data on binary combinations of four regulated trihalomethanes (THMs), which were the subject of a toxicology study designed to evaluate Prop-RA. The experimental design provided hepatotoxicity data on both single THMs and binary combinations. Prop-RA predicts the contribution of each component to mixture toxicity based on its fraction in the mixture and its response at the total mixture dose. The analysis found statistically significant departures from the Prop-RA prediction for some evaluations, suggesting that Prop-RA works best with chemicals whose effective dose ranges overlap. Overall, the study demonstrates the application of Prop-RA for mixture risk estimation and toxicological interaction assessment in cases where the toxic modes of action may be unknown. [72].

*(Prop-RA): Proportional Response Addition

(Continued)

(Continued)

Title	Evaluation of a Proportional Response Addition Approach to Mixture Risk Assessment and Predictive Toxicology Using Data on Four Trihalomethanes from the U.S. EPA's Multiple-Purpose Design Study
Background	The study evaluates the Proportional Response Addition (Prop-RA) model's ability to predict responses from chemical mixture exposure. The study focuses on four regulated trihalomethanes (THMs) in a toxicology study designed for this purpose.
Experimental Setup	Female CD-1 mice were exposed to varying doses of individual THMs and binary combinations using a common set of doses for all components and mixtures, providing hepatotoxicity data.
Data Collection	Data on relative liver weight, serum enzymes, mortality, and organ weights were collected to assess hepatotoxicity.
Application of Mathematical Model	Prop-RA predicts mixture toxicity based on the fractions of each component in the mixture and their responses at the total mixture dose.
Validation and Interpretation	The primary analysis involved 160 evaluations, with statistically significant departures from Prop-RA predictions found in seven cases.
Outcome	Prop-RA demonstrated effectiveness, especially with chemicals exhibiting overlapping effective dose ranges. It accommodates mixtures affecting the same organ/system.
Implications	The study underscores Prop-RA's utility in mixture risk estimation and toxicological interaction assessment, offering insights into chemical mixture toxicity prediction and its implications for regulatory decisions.

17.4.3.4 Physiologically Based Pharmacokinetic/Pharmacodynamic Models

Physiologically based pharmacokinetic/pharmacodynamic (PBPK/PD) models are sophisticated mathematical tools used to simulate the absorption, distribution, metabolism, and excretion (ADME) of chemicals in the body, as well as their pharmacological effects [56]. These models incorporate physiological parameters, such as blood flow, tissue volumes, enzyme kinetics, and receptor binding to predict the concentration-time profiles of chemicals in different organs and tissues [73]. By integrating pharmacokinetic and pharmacodynamic data, PBPK/PD models can elucidate the relationship between chemical exposure and biological response, helping to assess the risks associated with combined chemical exposures. PBPK/PD models are particularly valuable for extrapolating data across species, estimating internal doses, and identifying sensitive populations, making them essential tools for regulatory risk assessment and decision-making.

17.4.3.4.1 Underlying Assumptions

Mass balance: PBPK/PD models adhere to the principle of mass balance, ensuring that the total amount of the chemical or drug in the body remains constant over time. This assumption is fundamental for tracking the distribution and elimination of the compound within the body [74].

First-order kinetics: PBPK/PD models often assume that the processes governing absorption, distribution, metabolism, and excretion follow first-order kinetics. This assumption means that the rate of these processes is proportional to the concentration of the compound, simplifying the mathematical representation of these processes.

Passive diffusion: PBPK/PD models typically assume that the movement of chemicals or drugs across biological membranes occurs via passive diffusion. This assumption simplifies the modeling of absorption and distribution processes, although active transport mechanisms can also be incorporated into more complex models.

Constant physiology: PBPK/PD models often assume that physiological parameters, such as organ volumes, blood flows, and metabolic rates, remain constant over the simulation period. While this assumption facilitates model development and interpretation, it may not accurately reflect changes in physiology under different conditions (e.g. age, disease states).

Instantaneous mixing: PBPK/PD models assume instantaneous mixing of the compound within each tissue compartment, allowing for rapid equilibration of concentrations. This assumption simplifies the modeling of tissue distribution kinetics but may not accurately represent the actual dynamics of compound distribution within tissues.

Linear pharmacokinetics: PBPK/PD models often operate under the assumption of linear pharmacokinetics, where the rates of absorption, distribution, metabolism, and excretion are proportional to the dose or concentration of the compound.

This assumption, while simplifying model development, may not be universally applicable to all compounds, particularly at elevated doses [75].

These foundational assumptions serve as the bedrock for the development and application of PBPK/PD models, steering the mathematical depiction of intricate physiological and pharmacological processes. While these assumptions streamline model development and interpretation, it's crucial to rigorously assess their validity and bear in mind their limitations when employing PBPK/PD models in research, risk assessment, and regulatory decision-making.

17.4.3.4.2 Mathematical Formulation

PBPK/PD models strive to delineate the absorption, distribution, metabolism, and excretion (ADME) of drugs within the body, in addition to their pharmacodynamic effects. These models generally encompass sets of ordinary differential equations (ODEs) that represent various physiological compartments and processes. Although the specific equations can fluctuate based on the complexity of the model and the drug under study, the following are some general components and equations frequently employed in PBPK/PD models:

Pharmacokinetic Equations

Absorption

First-order absorption:

$$\frac{dA}{dt} = K_a \times \mathrm{F} \times \mathrm{Dose} - K_a \times A \tag{17.13}$$

where A is the amount of drug in the absorption compartment, F is the fraction of drug absorbed, Dose is the drug dose, and k_a is the absorption rate constant.

Distribution

$$\frac{dC_i}{dt} = \kappa_{\mathrm{in},i} \times \sum_{j=1}^{N} Q_{ji} \times (C_j - C_i) - k_{\mathrm{out},i} \times C_i \tag{17.14}$$

where C_i is the drug concentration in compartment i, Q_{ji} is the flow rate from compartment j to compartment i, and $k_{\mathrm{in},\,i}$ and $k_{\mathrm{out},\,i}$ are the rate constants for drug entering and leaving compartment i, respectively.

Metabolism

First-order elimination:

$$\frac{dC}{dt} = -k_{\mathrm{elim}} \times C \tag{17.15}$$

where C is the drug concentration and k_{elim} is the elimination rate constant.

Excretion

Renal excretion:

$$\frac{dC}{dt} = k_{\mathrm{renal}} \times C \tag{17.16}$$

where k_{renal} is the renal clearance rate constant.

Pharmacodynamic Equations

Drug-Receptor Interaction

E_{max} model:

$$E = E_{\mathrm{max}} \times \frac{C}{C + EC_{50}} \tag{17.17}$$

where E is the effect (e.g. response or biomarker), E_{max} is the maximum effect, C is the drug concentration, and EC_{50} is the concentration at which half of the maximum effect is observed.

Physiological Parameters

Volume of distribution (V_{d}):

$$V_d = \frac{\mathrm{Dose}}{C_0} \tag{17.18}$$

where C_0 is the initial drug concentration.

Clearance (CL)

$$\mathrm{CL} = \frac{\mathrm{Dose}}{\mathrm{AUC}} \tag{17.19}$$

where AUC is the area under the plasma concentration-time curve.

These equations can be combined and extended to represent more complex physiological systems and drug behaviors in PBPK/PD modeling. Additionally, PBPK models often incorporate anatomical and physiological data specific to the species being studied, such as organ volumes, blood flows, tissue partition coefficients, and enzyme kinetic parameters.

17.4.3.4.3 Strengths and Limitations

PBPK/PD modeling is a powerful approach for predicting the combined toxicity of chemical mixtures. However, it also has its strengths and limitations:

Strengths

Mechanistic understanding: PBPK/PD models incorporate detailed physiological and biochemical processes, allowing for a mechanistic understanding of how chemicals interact with biological systems. This enables more accurate predictions of toxicokinetic (absorption, distribution, metabolism, and excretion) and toxicodynamic (biological response) processes.

Interindividual variability: PBPK/PD models can account for interindividual variability in biological parameters, such as metabolic rates, organ volumes, and receptor sensitivities. This allows for personalized predictions of toxicity, considering factors including age, sex, genetic polymorphisms, and comorbidities.

Dose–response relationships: PBPK/PD models can elucidate dose–response relationships by integrating exposure data with biological response data [76]. This helps in understanding how changes in exposure levels translate into changes in biological effects and enables the derivation of dose–response curves for different endpoints.

Extrapolation across species: PBPK/PD models can extrapolate toxicity data across different species, facilitating the translation of preclinical data to human risk assessment. This is particularly useful for predicting human toxicity based on animal studies and for estimating safe exposure levels in humans.

Mixture toxicity prediction: PBPK/PD models can assess the combined toxicity of chemical mixtures by considering the pharmacokinetic interactions between individual compounds and their effects on biological pathways. This allows for the evaluation of synergistic, additive, or antagonistic effects of mixtures.

Limitations

Data requirements: PBPK/PD models require extensive data on chemical properties, physiological parameters, and biological mechanisms. Obtaining such data can be challenging, particularly for novel chemicals or complex mixtures where limited experimental data are available.

Complexity: PBPK/PD models are often complex, involving numerous differential equations and parameters. This complexity can make model development, calibration, and validation labor-intensive and may require specialized expertise in pharmacokinetics, pharmacodynamics, and mathematical modeling.

Assumption sensitivity: The accuracy of PBPK/PD predictions can be sensitive to the validity of model assumptions and the quality of input data. Assumptions regarding tissue partition coefficients, metabolic pathways, and receptor interactions can influence model predictions and may require careful validation.

Uncertainty: PBPK/PD models inherently involve uncertainty due to variability in biological processes, parameter estimation, and model structure. Sensitivity analyses and uncertainty quantification techniques are necessary to assess the robustness of model predictions and identify sources of uncertainty.

Validation: PBPK/PD models require rigorous validation against experimental data to ensure their accuracy and reliability. However, obtaining comprehensive validation datasets for complex toxicological endpoints or combined chemical exposures can be challenging.

In summary, PBPK/PD modeling offers a mechanistic framework for predicting combined chemical toxicity, but its application requires careful consideration of data availability, model complexity, assumptions, and uncertainty [77]. Despite its limitations, PBPK/PD modeling is a valuable tool for risk assessment and decision-making in toxicology and chemical safety.

17.4.3.4.4 Applications

Dose scaling and extrapolation: PBPK/PD models can be used to scale doses across different species, routes of exposure, and exposure durations. This allows for the extrapolation of toxicity data from animal studies to humans and from high-dose to low-dose exposures, aiding in the assessment of combined chemical toxicity across various scenarios.

Metabolism and clearance prediction: PBPK models incorporate information on the metabolism and clearance pathways of individual chemicals, allowing for the prediction of metabolite formation and elimination rates. This is particularly useful for assessing the combined toxicity of chemicals with shared metabolic pathways and potential interactions that affect their metabolism.

Organ-specific toxicity assessment: PBPK/PD models can simulate the distribution of chemicals to different organs and tissues based on their physicochemical properties and physiological parameters. By integrating toxicodynamic models with organ-specific concentration data, PBPK/PD models can predict the toxic effects of combined chemical exposures on specific target organs or systems.

Synergistic and antagonistic interactions: PBPK/PD models can evaluate the combined toxicity of chemicals by assessing potential interactions, such as synergistic or antagonistic effects. By simulating the pharmacokinetic interactions between chemicals and their effects on relevant biological pathways, PBPK/PD models can predict whether combined exposures result in enhanced or reduced toxicity compared to individual exposures.

Route of exposure consideration: PBPK/PD models can simulate different routes of exposure (e.g. inhalation, oral, dermal) and account for differences in absorption, distribution, metabolism, and excretion kinetics. This allows for the prediction of combined toxicity resulting from exposures via multiple routes and helps identify the most relevant exposure scenarios for risk assessment.

Chemical mixtures evaluation: PBPK/PD models can assess the combined toxicity of chemical mixtures by considering the pharmacokinetic interactions between individual compounds. By incorporating data on mixture composition, exposure levels, and toxicity endpoints, PBPK/PD models can predict the overall toxicity of complex mixtures and identify potential synergies or antagonisms among mixture components.

Risk assessment and decision support: PBPK/PD models provide quantitative predictions of combined chemical toxicity, which can be used for risk assessment and decision-making in regulatory contexts. By integrating exposure data with toxicity predictions, PBPK/PD models help prioritize chemicals for further testing, establish safe exposure levels, and inform risk management strategies [77].

In summary, PBPK/PD models play a crucial role in predicting combined chemical toxicity by integrating pharmacokinetic and pharmacodynamic information, assessing potential interactions, and providing mechanistic insights into the toxicological effects of chemical mixtures.

Case Study Related to PBPK/PD Model

Schuhmacher et al. studied the integration of a dynamic signal transduction pathway into the tissue dosimetry model, which they claim represents a significant advancement in computational toxicology. By coupling existing system biological models with PBPK/PD models, the framework of an integrated PBPK/PD coupled mechanistic system pathway model, termed system toxicology, was developed. This approach enables the description of both the kinetics of chemicals and biomolecules, providing insights into the dynamic and steady-state behaviors of molecular pathways under perturbed conditions. The objective of the article is to illustrate this system toxicology-based approach by developing

(Continued)

(Continued)

an integrated PBPK/PD coupled miRNA-BDNF pathway model and demonstrating its application in a case study of PFOS-mediated neurotoxicity. Therefore, the discussion includes the utilization and application of PBPK/PD models in assessing chemical toxicity [73].

Title	Developing integrated PBPK/PD coupled mechanistic pathway model (miRNA-BDNF): An approach towards system toxicology
Background	By integrating dynamic signal transduction pathways into tissue dosimetry models, a novel approach in computational toxicology was adopted. This integrated approach, called system toxicology, combines existing system biological models with (PBPK/PD) models to elucidate the kinetics of both chemicals and biomolecules.
Experimental Setup	This study develops an integrated PBPK/PD model coupled to a miRNA-BDNF pathway to investigate PFOS-mediated neurotoxicity. The dynamic signaling involves miRNA-mediated BDNF regulation, which is critical for neuronal cell processes.
Data Collection	PFOS-mediated neurotoxicity data is incorporated into the developed model, encompassing PFOS kinetics and the miRNA-BDNF pathway.
Application of Mathematical Model	The integrated PBPK/PD model estimates biomolecule concentrations and neuronal survival using mass balance equations and ordinary differential equations.
Validation and Interpretation	Model predictions are validated against experimental data. Any deviations are interpreted to refine the model and enhance accuracy.
Outcome	Insights into PFOS neurotoxicity are gained by elucidating the dynamics of the miRNA-BDNF pathway and its impact on neuronal cell survival.
Implications	Integrating PBPK/PD models with dynamic signal transduction pathways provides a holistic approach for chemical toxicity assessment. This enhances our understanding of toxicity mechanisms and facilitates the development of more precise risk assessment strategies for safeguarding environmental and human health.

17.4.3.5 Quantitative Structure–Activity Relationship Models

Quantitative structure–activity relationship (QSAR) models are computational tools used to predict the biological activity or toxicity of chemicals based on their chemical structure and physicochemical properties [78]. These models quantify the relationship between chemical structure and biological activity through mathematical equations, allowing for the prediction of toxicological endpoints, such as acute toxicity, carcinogenicity, and mutagenicity [79]. QSAR models utilize molecular descriptors, which represent the structural and electronic features of chemicals, to generate predictive models using statistical or machine learning algorithms [80]. By analyzing large datasets of chemical structures and corresponding biological activities, QSAR models can identify structural features associated with specific toxicological outcomes, providing valuable insights into the mechanisms of action and potential hazards of chemicals [81]. QSAR models are widely used in chemical risk assessment, virtual screening of chemical libraries, and prioritization of chemicals for further testing, making them valuable tools for predicting the combined effects of chemical mixtures [82].

17.4.3.5.1 Underlying Assumptions

The underlying assumptions of QSAR models in the prediction of combined chemical toxicity may vary depending on the specific context and application. However, some common underlying assumptions include:

Additivity of effects: QSAR models often assume that the effects of individual chemicals in a mixture are additive, meaning that the combined toxicity of the mixture can be predicted by summing the toxicities of its components. This presumption might not be consistently accurate, particularly when chemical interactions lead to either synergistic or antagonistic outcomes [83].

Independence of chemicals: QSAR models may assume that the toxicological properties of individual chemicals are independent of each other. In other words, the presence of one chemical in a mixture does not affect the toxicity of other chemicals in the same mixture [83]. However, this assumption may not be valid in cases where chemicals interact with each other, leading to changes in their toxicological properties.

Linearity of relationships: QSAR models typically assume linear relationships between chemical descriptors (such as molecular structure or physicochemical properties) and toxicity. While linear models are often used due to their simplicity and interpretability, they may not capture nonlinear relationships that exist in complex biological systems [84].

Homogeneity of mixtures: QSAR models may assume that the chemicals in a mixture are homogeneous and well-mixed, meaning that the concentrations of individual components remain constant throughout the exposure period [85]. However, in real-world scenarios, the distribution and interactions of chemicals within mixtures can vary spatially and temporally, leading to heterogeneous exposures and potentially nonuniform toxicological effects [86].

These assumptions simplify the modeling process and facilitate the development of predictive QSAR models for combined chemical toxicity. Nonetheless, it's crucial to acknowledge that these suppositions may not consistently align with real-world scenarios. The unique properties of the chemicals and mixtures under investigation should be meticulously evaluated. Furthermore, strides are being made to create advanced QSAR models capable of accommodating interactions, nonlinearity, and diversity in predicting mixture toxicity [87].

17.4.3.5.2 Mathematical Formulation

QSAR models for predicting combined chemical toxicity aim to estimate the toxicological effects of mixtures of chemical compounds based on their molecular structures. The formulation of such models involves extending traditional QSAR approaches to handle the complexities of mixture toxicity. Here's an outline of the mathematical formulation and equations for QSAR models in the context of predicting combined chemical toxicity:

Descriptors for mixtures: In QSAR models for mixture toxicity prediction, descriptors are extended to capture the characteristics of chemical mixtures. These descriptors may include information about the individual compounds in the mixture, their concentrations, physicochemical properties, interactions, and any other relevant features.

Combined toxicity model: The goal is to develop a mathematical model that relates the descriptors of chemical mixtures to their combined toxicological effects. This model should account for potential interactions between the components of the mixture, such as synergistic or antagonistic effects.

Linear model: A linear model for predicting the combined toxicity of chemical mixtures may be formulated as:

$$Y = \beta_0 + \beta_1 X_1 + \beta_2 X_2 + \dots + \beta_\eta X_\eta + \epsilon \tag{17.20}$$

where Y is the combined toxicity or effect of the mixture, $X_1, X_2, \dots, X_n$ are the descriptors representing the mixture, β_0, β_1, $\beta_2, \dots, \beta_n$ are the regression coefficients, ϵ is the error term.

Nonlinear model: Nonlinear relationships between mixture descriptors and combined toxicity can be captured using functions, such as sigmoidal functions, exponential functions, or machine learning algorithms like neural networks.

Model training: Similar to traditional QSAR models, the combined toxicity model is trained using a dataset containing descriptors and corresponding combined toxicity values for a variety of chemical mixtures. Model parameters are adjusted during training to minimize the difference between predicted and observed combined toxicity values.

Model evaluation and validation: The performance of the QSAR model for combined toxicity prediction is evaluated using appropriate metrics, such as R_2 (coefficient of determination), RMSE (root mean squared error), or other statistical measures. The model should also be validated using independent datasets to ensure its robustness and generalization capability.

Application: Once validated, the QSAR model can be used to predict the combined toxicity of new chemical mixtures based on their molecular characteristics. These predictions can help prioritize testing efforts, assess potential risks, and guide regulatory decision-making in fields, such as environmental toxicology, drug combination therapy, and chemical safety assessment.

17.4.3.5.3 Strengths and Limitations

When considering the strengths and limitations of QSAR models in the prediction of combined chemical toxicity, the following points can be highlighted:

Strengths

Mechanistic insight: QSAR models provide mechanistic insight into the relationship between chemical structure and toxicity. By identifying structural features associated with toxic effects, QSAR models can help elucidate underlying biological mechanisms and guide further experimental studies.

High-throughput screening (HTS): QSAR models offer a cost-effective and efficient means of screening large chemical libraries for potential toxicity. By predicting toxicity based on chemical structure alone, QSAR models enable rapid prioritization of chemicals for further testing, thereby streamlining the drug discovery and chemical safety assessment process.

Interpretability: QSAR models are often interpretable, allowing researchers to identify key structural features that contribute to toxicity. This information can be used to design safer chemicals with reduced toxicity or to modify existing chemicals to improve their safety profile.

Versatility: QSAR models can be applied to a wide range of chemical classes and toxicity endpoints, including acute and chronic toxicity, carcinogenicity, mutagenicity, and ecotoxicity. This versatility makes QSAR models valuable tools for assessing the toxicity of diverse chemical substances.

Limitations

Data availability and quality: QSAR models rely on high-quality data for model development and validation. However, data on combined chemical toxicity may be limited or of variable quality, particularly for complex mixtures or less-studied chemicals. This can hinder the accuracy and reliability of QSAR predictions.

Assumption of additivity: Numerous QSAR models operate under the presumption that the toxic impacts of individual substances in a combination are cumulative. Yet, this assumption may not always be valid, given that chemical interactions can result in synergistic or antagonistic effects, which QSAR models may not precisely forecast.

Complexity of biological systems: QSAR models oversimplify the complexity of biological systems by focusing solely on chemical structure-toxicity relationships. In reality, toxicity is influenced by a multitude of factors, including metabolism, bioavailability, target specificity, and environmental conditions, which may not be adequately captured by QSAR models.

Domain of applicability: QSAR models have a limited domain of applicability and may not perform well outside the chemical space and toxicity endpoints for which they were developed. Extrapolating QSAR predictions to new chemicals or endpoints requires careful consideration of model uncertainty and validation in relevant contexts.

QSAR models offer valuable insights into combined chemical toxicity prediction, they also have limitations related to data availability, model assumptions, and the complexity of biological systems. Addressing these limitations requires ongoing efforts to improve data quality, model robustness, and the integration of QSAR with other predictive approaches.

17.4.3.5.4 Application

QSAR models have several applications in the prediction of combined chemical toxicity:

Mixture toxicity prediction: QSAR models can predict the combined toxicity of chemical mixtures by integrating information on the chemical structures of individual components and their toxicological properties. By considering the additive or interactive effects of mixture components, QSAR models provide insights into the overall toxicity of complex mixtures.

Identification of synergistic or antagonistic interactions: QSAR models can assess the potential for synergistic or antagonistic interactions between chemicals in mixtures. By analyzing the structural features of mixture components and their effects on biological targets or pathways, QSAR models can predict whether combined exposures result in enhanced or reduced toxicity compared to individual exposures [88].

Prioritization of testing: QSAR models can help prioritize chemicals or mixtures for further testing based on their predicted toxicity. By screening large chemical libraries or environmental samples, QSAR models identify compounds or mixtures with the highest likelihood of adverse effects, enabling more efficient allocation of resources for toxicity testing and risk assessment.

Structure-based design of safer chemicals: QSAR models can guide the design of safer chemicals with reduced toxicity [89]. By identifying structural features associated with toxic effects, QSAR models inform the modification of chemical structures to mitigate toxicity while maintaining desired properties. This approach supports the development of environmentally friendly chemicals and pharmaceuticals.

Risk assessment and regulatory decision-making: QSAR models contribute to risk assessment and regulatory decision-making by providing quantitative predictions of combined chemical toxicity. By integrating QSAR predictions with exposure data and toxicological endpoints, regulators can evaluate the potential risks associated with chemical mixtures and establish appropriate safety thresholds and regulatory limits [90].

Environmental monitoring and pollution control: QSAR models assist in environmental monitoring and pollution control efforts by predicting the toxicity of chemical mixtures in environmental samples. By analyzing the composition of complex environmental matrices and predicting their toxicological effects, QSAR models help identify sources of contamination and prioritize remediation efforts.

Overall, QSAR models play a crucial role in predicting combined chemical toxicity by leveraging computational approaches to analyze chemical structures, assess toxicity mechanisms, and make quantitative predictions of mixture toxicity. These applications support various domains, including drug discovery, chemical safety assessment, environmental monitoring, and regulatory decision-making.

Case Study Related to QSAR models

Lopes et al. studied to evaluates the toxicity and mutagenicity of a synthetic effluent containing the Azo dye Amido Black 10B (AB10B) before and after treatment with Advanced Oxidation Processes (AOP). Various bioassays and in silico analyses were used to assess the effects of the effluent and its degradation products. The study found that AOP treatment reduced the mutagenicity of the effluent, with residual hydrogen peroxide and degradation byproducts potentially contributing to toxicity. In silico (QSAR) models were employed to predict the mutagenic and carcinogenic potential of AB10B and its byproducts, providing valuable insights for effluent treatment strategies and risk assessment [91].

Title	Evaluation of toxicity and mutagenicity of a synthetic effluent containing azo dye after Advanced Oxidation Process treatment
Background	Assessment of chemical toxicity and prediction using mathematical models, specifically in the context of evaluating the mutagenicity and carcinogenicity of Amido Black 10B (AB10B) and its degradation products.
Experimental Setup	Preparation of a synthetic effluent containing AB10B and subjecting it to Advanced Oxidation Process (AOP) treatment.
Data Collection	Acute toxicity, chromosomal mutations, and gene mutations were assessed before (Gross Synthetic Effluent, GSE) and after (Treated Synthetic Effluent, TSE) AOP treatment.
Application of Mathematical Model	(QSAR) model implemented in the VEGA suite to predict carcinogenicity and mutagenicity of AB10B and its byproducts.
Validation and Interpretation	The in-silico models predicted potential mutagenic and carcinogenic effects of AB10B and its degradation products.
Outcome	AOP treatment effectively reduced the mutagenicity of the effluent containing AB10B. The observed effects may be attributed to residual hydrogen peroxide from AOP treatment and byproducts generated during azo bond cleavage.
Implications	Effluent treatment using AOP shows promise in mitigating the mutagenicity of azo dye-containing effluents.

17.4.3.6 Machine Learning/Artificial Intelligence-based Models

Machine learning and artificial intelligence (AI) techniques have gained prominence in predicting combined toxicity due to their ability to analyze large and complex datasets efficiently. These models utilize algorithms that enable computers to learn patterns and relationships from data without being explicitly programmed. In the context of predicting combined toxicity, machine learning and AI-based models can integrate diverse sources of information, including chemical structures, biological activities, environmental factors, and exposure data, to generate predictive models with high accuracy. Several computational toxicology algorithms based on machine learning and deep learning techniques have recently been developed and put into practice to predict the toxicity of chemicals and mixtures. One benefit of computational toxicology is its ability to rapidly evaluate vast datasets and efficiently determine the toxicity of combined chemicals, for different organisms even at minimal exposure levels. Therefore, computational toxicology techniques are anticipated to emerge as the primary technique to evaluate and predict the risk assessment of combined chemical toxicity [92].

17.4.3.6.1 Underlying Assumptions

Machine learning/AI-based models for the prediction of combined chemical toxicity rely on several underlying assumptions, including:

Independence of features: These models operate under the presumption that the input attributes characterizing chemicals, such as structural descriptors and physicochemical properties, are mutually exclusive. This assumption streamlines the modeling procedure, but it may not always be accurate, particularly when these features are interrelated or influence each other.

Additivity or linearity of effects: Similar to QSAR models, machine learning/AI-based models may assume that the effects of individual chemicals in a mixture are additive or follow linear relationships [93]. This assumption facilitates the prediction of combined toxicity by summing the contributions of individual chemicals. However, it may not accurately capture nonlinear interactions or synergistic/antagonistic effects between mixture components.

Homogeneity of data: Machine learning models assume that the data used for training and testing are homogeneous and representative of the target population or domain. In the context of combined chemical toxicity prediction, this assumption implies that the training data adequately cover the range of chemical combinations, concentrations, and toxicity endpoints of interest.

Generalization across data distributions: Machine learning models assume that patterns learned from training data generalize well to unseen data drawn from similar distributions. In the context of combined chemical toxicity prediction, this assumption implies that the model can accurately predict toxicity for new chemical mixtures that were not present in the training data [94].

Relevance of features: Machine learning models assume that the input features used for prediction are relevant to the target task (i.e. combined chemical toxicity prediction). While feature selection and dimensionality reduction techniques can help mitigate this assumption, it's essential to ensure that the selected features adequately capture the relevant characteristics of chemical mixtures.

Data quality and label accuracy: Machine learning models assume that the training data are high quality, with accurate and reliable labels (e.g., toxicity measurements). However, data quality issues, such as measurement errors, missing values, or mislabeled instances can impact model performance and interpretation.

Stationarity of relationships: Machine learning models assume that the relationships between input features and the target variable (combined toxicity) remain stationary over time and across different datasets. This assumption, while facilitating the streamlining of model training and deployment, may not always be applicable in environments that are constantly changing or evolving.

Overall, while machine learning/AI-based models offer powerful tools for predicting combined chemical toxicity, it's essential to be mindful of these underlying assumptions and their potential implications for model performance and interpretation. Careful validation and sensitivity analysis can help assess the validity of these assumptions and ensure the robustness of model predictions.

17.4.3.6.2 Mathematical Formulation

In the context of predicting combined chemical toxicity using machine learning or AI-based models, the formulation involves extending traditional ML/AI approaches to handle the complexities of mixture toxicity. Here's a general outline of the mathematical formulation:

Data representation: Similar to traditional machine learning, the data is represented as feature vectors describing the characteristics of chemical mixtures. These features may include descriptors of individual compounds, their concentrations, physicochemical properties, interactions, and other relevant information.

Target variable: The target variable is the combined toxicity or effect of the chemical mixture. This variable can be represented numerically, such as IC50 values or toxicity scores, indicating the level of toxicity exhibited by the mixture.

Model selection: Various machine learning algorithms can be applied, including regression, classification, or ensemble methods, depending on the nature of the toxicity prediction task. The choice of model depends on factors, such as the complexity of interactions between mixture components and the availability of labeled data.

Model Formulation

Regression model: If the toxicity prediction is a continuous variable, a regression model can be used. For example, a linear regression model may be formulated as:

$$Y = \beta_0 + \beta_1 X_1 + \beta_2 X_2 + \dots + \beta_\eta X_\eta + \epsilon \tag{17.21}$$

where Y is the predicted combined toxicity, $X_1, X_2, \dots, X_n$ are the descriptors of the chemical mixture, $\beta_0, \beta_1, \beta_2, \dots, \beta_n$ are the regression coefficients, and ϵ is the error term.

Classification model: If the toxicity prediction is a categorical variable (e.g. toxic or nontoxic), classification algorithms, such as logistic regression, decision trees, random forests, or support vector machines can be employed.

Model training: The model is trained using a dataset containing features of chemical mixtures and corresponding combined toxicity values. During training, the model parameters are adjusted to minimize the discrepancy between predicted and observed toxicity values, typically using optimization algorithms, such as gradient descent.

Model evaluation and validation: The performance of the model is evaluated using metrics appropriate for regression or classification tasks, such as R^2 (for regression) or accuracy, precision, recall, and F1-score (for classification). Cross-validation techniques may be employed to assess the model's generalization ability.

Hyperparameter tuning: Hyperparameters of the machine learning algorithm, such as regularization strength, learning rate, tree depth, etc., are tuned using techniques, such as grid search or randomized search to optimize model performance.

Prediction and application: Once trained and validated, the model can be used to predict the combined toxicity of new chemical mixtures based on their features. These predictions can aid in risk assessment, prioritization of testing efforts, and regulatory decision-making in fields, such as environmental toxicology, drug combination therapy, and chemical safety assessment.

To summarize, the mathematical formulation of machine learning or AI-based models for predicting combined chemical toxicity involves representing data, selecting appropriate models, formulating predictive equations, training and evaluating the models, tuning hyperparameters, and applying the models for toxicity prediction.

17.4.3.6.3 Strengths and Limitations

Strengths

Nonlinear relationships: ML models can capture complex, nonlinear relationships between chemical properties and toxicity, which may not be easily discernible using traditional approaches.

Data-driven approach: ML models learn from data, allowing them to adapt and generalize to diverse chemical mixtures and toxicity endpoints. They can incorporate a wide range of input features, including chemical structures, physicochemical properties, and biological activities.

Predictive performance: Well-trained ML models can achieve high predictive accuracy, effectively predicting the combined toxicity of chemical mixtures across different exposure scenarios and biological systems [95].

Feature importance analysis: ML models can provide insights into the importance of different chemical features or descriptors in predicting combined toxicity. This information can help identify key determinants of toxicity and guide further experimental investigations [96].

Scalability: ML-based approaches can handle large datasets with numerous chemicals and toxicity endpoints, making them suitable for HTS and risk assessment applications.

Limitations

Data availability and quality: ML models require large, high-quality datasets for training, which may be lacking for certain chemical mixtures or toxicity endpoints. Biased or incomplete datasets can lead to model inaccuracies and limited generalizability.

Interpretability: Some ML algorithms, especially deep learning models, are often referred to as "black box" models due to their complex, opaque internal workings [97]. This lack of interpretability hinders understanding of the underlying mechanisms driving toxicity predictions.

Overfitting: ML models may overfit the training data, capturing noise or spurious correlations that do not generalize well to unseen data. Regularization techniques and cross-validation can help mitigate overfitting, but careful model evaluation and validation are necessary.

Domain expertise: ML models may not capture domain-specific knowledge or mechanistic insights into chemical toxicity. Integrating expert knowledge into model development and interpretation can enhance model performance and credibility.

Uncertainty estimation: ML models typically provide point predictions of toxicity, but they may not always quantify uncertainty or provide confidence intervals around predictions. Probabilistic models and uncertainty quantification techniques can address this limitation by estimating prediction uncertainty.

Overall, while ML-based models offer promising opportunities for predicting combined chemical toxicity, careful consideration of data quality, model interpretability, and uncertainty estimation is essential for their effective application in risk assessment and decision-making. Collaboration between data scientists, toxicologists, and domain experts is crucial for developing robust and interpretable ML models for predicting combined chemical toxicity.

17.4.3.6.4 Applications

Machine learning (ML) and AI-based models are increasingly applied in toxicity prediction to assess the potential health risks associated with exposure to chemical substances. Here's how these models are applied in toxicity prediction:

Prediction of mixture toxicity: Machine learning models can predict the combined toxicity of chemical mixtures by integrating information on the chemical structures, physicochemical properties, and biological activities of individual components. These models provide quantitative predictions of mixture toxicity, facilitating risk assessment and regulatory decision-making.

Identification of synergistic or antagonistic interactions: Machine learning models can identify synergistic or antagonistic interactions between chemicals in mixtures by analyzing their structural features and biological activities. By

Figure 17.6 (a) Machine learning construction outline [99]; (b) MRA Toolbox v. 1.0: a web-based toolbox for predicting mixture toxicity of chemical substances in chemical products [92]; (c) Artificial intelligence-based toxicity prediction of environmental chemicals: future directions for chemical management applications environmental science technology [100]/American Chemical Society.

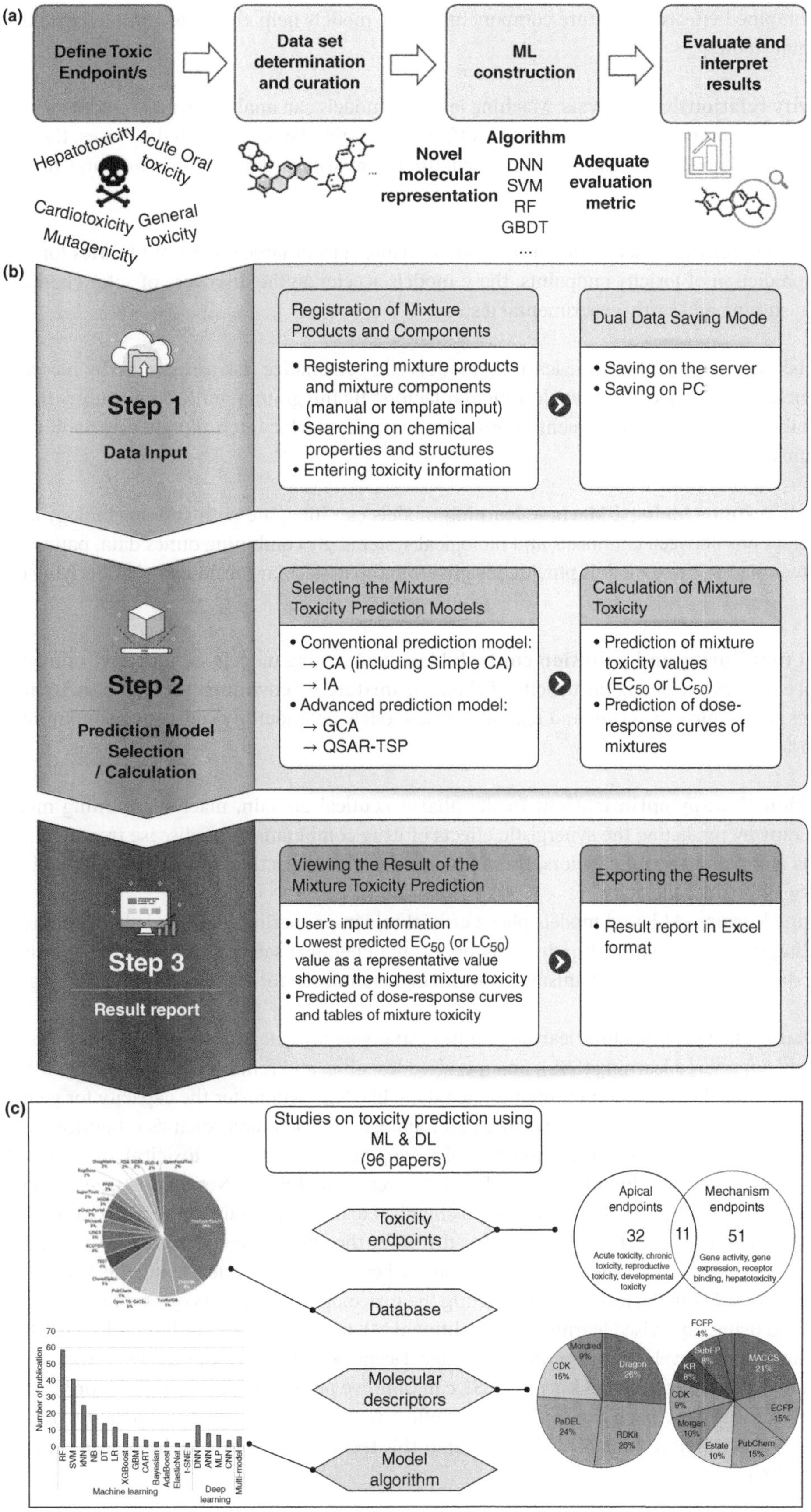
(a)
Define Toxic Endpoint/s
Data set determination and curation
ML construction
Evaluate and interpret results
Hepatotoxicity
Acute Oral toxicity
Cardiotoxicity
General toxicity
Mutagenicity
Novel molecular representation
Algorithm
DNN
SVM
RF
GBDT
...
Adequate evaluation metric
(b)
Step 1
Data Input
Registration of Mixture Products and Components
• Registering mixture products and mixture components (manual or template input)
• Searching on chemical properties and structures
• Entering toxicity information
Dual Data Saving Mode
• Saving on the server
• Saving on PC
Step 2
Prediction Model Selection / Calculation
Selecting the Mixture Toxicity Prediction Models
• Conventional prediction model:
→ CA (including Simple CA)
→ IA
• Advanced prediction model:
→ GCA
→ QSAR-TSP
Calculation of Mixture Toxicity
• Prediction of mixture toxicity values (EC_{50} or LC_{50})
• Prediction of dose-response curves of mixtures
Step 3
Result report
Viewing the Result of the Mixture Toxicity Prediction
• User's input information
• Lowest predicted EC_{50} (or LC_{50}) value as a representative value showing the highest mixture toxicity
• Predicted of dose-response curves and tables of mixture toxicity
Exporting the Results
• Result report in Excel format
(c)
Studies on toxicity prediction using ML & DL (96 papers)
Toxicity endpoints
Database
Molecular descriptors
Model algorithm
Apical endpoints
Mechanism endpoints
32
11
51
Acute toxicity, chronic toxicity, reproductive toxicity, developmental toxicity
Gene activity, gene expression, receptor binding, hepatotoxicity
Number of publication
RF
SVM
kNN
NB
DT
LR
XGBoost
GBM
CART
Bayesian
AdaBoost
ElasticNet
t-SNE
DNN
ANN
MLP
CNN
Multi-model
Machine learning
Deep learning
Mordred 9%
CDK 15%
Dragon 26%
PaDEL 24%
RDKit 26%
FCFP 4%
SubFP 8%
KR 8%
MACCS 21%
CDK 9%
ECFP 15%
Morgan 10%
Estate 10%
PubChem 15%

predicting the combined effects of mixture components, these models help elucidate complex toxicity mechanisms and guide experimental studies.

Structure-activity relationship analysis: Machine learning models can analyze structure-activity relationship (SAR) to identify key structural features associated with toxic effects. By mining large chemical databases, these models help prioritize chemicals for further testing based on their structural similarity to known toxicants and their predicted toxicity profiles.

High-throughput screening: Machine learning models enable HTS of large chemical libraries for potential toxicity. By automating the prediction of toxicity endpoints, these models accelerate the discovery of safer chemicals and reduce the need for time-consuming and costly experimental testing.

Personalized risk assessment: Machine learning models can personalize risk assessment by incorporating individual-level data on genetic, physiological, and environmental factors. By integrating multi-omic data with toxicity predictions, these models enable tailored risk assessment approaches that account for interindividual variability in susceptibility to chemical exposures.

Integration with systems biology: Machine learning models can integrate with systems biology approaches to model the complex interactions between chemicals and biological systems. By combining omics data, pathway analysis, and network modeling techniques, these models provide insights into the molecular mechanisms underlying combined chemical toxicity.

Environmental monitoring and pollution control: Machine learning models support environmental monitoring and pollution control efforts by predicting the toxicity of chemical mixtures in environmental samples. By analyzing spatial and temporal patterns of chemical exposure and toxicity, these models help identify sources of contamination and prioritize remediation efforts.

Drug combination therapy optimization: In the pharmaceutical domain, machine learning models optimize drug combination therapy by predicting the synergistic effects of drug combinations on disease treatment. By analyzing drug-drug interactions and patient-specific factors, these models guide the selection and dosing of combination therapies for improved efficacy and reduced toxicity.

Overall, machine learning/AI-based models play a crucial role in predicting combined chemical toxicity across various domains, including environmental health, drug discovery, and chemical safety assessment. These models offer valuable tools for understanding toxicity mechanisms, prioritizing chemicals for testing, and informing risk management strategies.

According to Jiarui Chen et al. machine learning training strategies are categorized into various types, such as supervised learning (SL), semi-supervised learning (SSL), unsupervised learning, and reinforcement learning. It has been researched that methods that opt for the utilization of unannotated data, like SSL, will foster the capacity for generalization capacity of prediction models. For semi-supervised learning methodologies labeled data, such as (chemicals with known toxicity information) and unlabeled data for instance (chemical having no information of toxicity) are required. The researchers labeled the data, which is used in this study as the Tox21 dataset from MoleculeNet. They suggested a Graph Convolution Neural Network (GCN) for the prediction of combined chemical toxicity, and trained the network using the Mean Teacher (MT) SSL algorithm (Figure 17.6b). This approach was driven by the effectiveness of semi-supervised learning (SSL) algorithms in the previous experiments. The results show that the best SSL-GCN models using the Tox21 data acquire an average ROC-AUC score of 0.757 in the test set for predicting the toxicological endpoints of chemicals, which is 6% better than GCN models trained using supervised learning and traditional ML methodologies. Additionally, SSL-GCN models produce better results than models developed using the integrated DeepChem ML techniques. This work demonstrates that by using unannotated data to inform model learning, SSL can improve prediction accuracy. A 1:1 or 4:1 ratio of unannotated to annotated data is ideal. This work shows that SSL is effective in predicting chemical toxicity; by using the enormous chemical databases that are now available, it is likely that this technique will also be helpful in other chemical property prediction studies [98].

Table 17.2 Studies related to the assessment of chemical toxicity (using mathematical models for prediction and assessment).

Article	Summary	Model	Chemical interaction	Ref.
Predicting Chemical Toxicity Effects Based on Chemical-Chemical Interactions	This study established an order to predict toxic effects based on chemical-chemical interactions, assuming compounds with similar interactions share toxicity profiles. Through Jackknife testing on 10 groups, the classifier achieved a high accuracy of approximately 79.50%, outperforming random guessing (25.43%). This method shows promise for screening out highly toxic drugs in drug discovery.	Machine Learning/ AI-Based Model	Chemical–chemical interactions	[102]
Ten years of Research on Synergisms and Antagonisms in Chemical Mixtures: A Systematic Review and Quantitative Reappraisal of Mixture Studies	This review assessed mixture effects in ecotoxicology and human/mammalian toxicology, finding few claims of synergistic effects stood up to scrutiny upon quantitative reappraisal. Notably, certain pesticide combinations showed potential for synergistic interactions at environmentally relevant doses.	Dose Addition Model	Synergistic effects	[103]
Machine Learning Toxicity Prediction: Latest Advances by Toxicity End Point	The paper discusses the use of machine learning models in predicting the toxicity of small molecules, emphasizing their significance in early drug discovery stages. It examines recent advancements, challenges, and various toxic endpoints predicted in silico, along with molecular representations, algorithms, and evaluation metrics used in toxicity prediction studies.	Machine Learning/ AI-Based Model	–	[99]
Formalization of the Task of Creating a Mathematical Model of Combined Wastewater Treatment Processes	The article discovers the formalization of mathematical modeling for treatment of wastewater, focusing on the effects of introducing oxidants (H_2O_2) and toxicants (sulfur compounds) on activated sludge. It proposes a decision support system and pH measurements, aiming to prevent activated sludge death.	Multiple Linear Regression Model	Synergistic interactions	[104]
Improving Sediment Toxicity Testing for Very Hydrophobic Chemicals: Part 1-Spiking, Equilibrating, and Exposure Quantification	This study examines challenges in sediment toxicity testing with very hydrophobic organic chemicals (VHOCs), proposing methods to mitigate experimental artifacts. It evaluates spiking techniques, sediment equilibration, and passive sampling for VHOC exposure quantification, highlighting the importance of intensive mixing and prolonged equilibration for accurate toxicity assessments.	–	Physical and chemical interactions	[105]
Quantitative Adverse outcome Pathway (qAOP) Models for Toxicity Prediction	This is the review article measuring the quantitative adverse outcome pathway (qAOP) models, aiming to standardize their development and evaluation for regulatory applications. It analyzes existing qAOP definitions, identifies common features, and evaluates published qAOP models and associated software tools.	Computational, Machine Learning-based Models	–	[106]
Estimating the Combined Toxicity by a Two-Step Prediction Model on the Complicated Chemical Mixtures from Wastewater Treatment Plant Effluents	The study investigates the toxicity of chemical mixtures from wastewater treatment plant effluents using a two-step prediction (TSP) model with Daphnia magna. It employs the CA and IA models, finding that while CA predictions align with observed results, IA predictions do not. The TSP model provides a statistically significant and accurate prediction of mixture toxicity, suggesting its validity for diverse chemical mixtures.	Concentration Addition (CA) Independent Action (IA) Models	Additive effects	[107]

(Continued)

Table 17.2 (Continued)

Article	Summary	Model	Chemical interaction	Ref.
In Silico Prediction of Toxicity and Its Applications for Chemicals at Work	The methodology involved a review of in silico prediction of chemical toxicity, utilizing recent reports after 2010 and data from before 2010, with a focus on determining credibility and confidence levels of in silico predictions.	Quantitative Structure–Activity Relationships (QSAR) Computational Modeling	Additive effects	[18]
Mathematical Modeling Approaches for Assessing the Joint Toxicity of Chemical Mixtures Based on Luminescent Bacteria	The article discusses the complex interactions between chemicals in mixtures, highlighting various methods like isobologram analysis and response surface methodology to assess their effects. It emphasizes the importance of understanding these interactions for accurate risk assessment and environmental management.	Quantitative Structure–Activity Relationship (QSAR)	Additive effects Independent effects	[36]
Development of In Silico Methods to Aid Chemical Risk Assessment	The research develops computational tools, including QSAR models and PBK mixture models, for chemical risk assessment, focusing on bisphenols and PFAS. Findings suggest that chemicals in human blood may affect thyroid hormone pathways, while zebrafish studies did not show significant mixture effects on ADME. Competitive plasma protein binding may influence ADME properties of bisphenols and PFAS at certain concentration levels.	Quantitative Structure–Activity Relationship (QSAR) Model Physiologically Based Kinetic (PBK) Mixture Model	Independent effect	[108]
Prediction of Aquatic Toxicity of Chemical Mixtures by the QSAR Approach Using 2D Structural Descriptors	The article develops QSAR models using partial least squares regression to predict aquatic toxicity of chemical mixtures, considering structural descriptors and endpoints for marine bacteria and freshwater algae. The models, compliant with OECD guidelines, prove robust and highly predictive, offering a mechanistic interpretation for untested mixtures' toxicity within their domain of applicability.	Quantitative Structure–Activity Relationship (QSAR) Model	–	[109]
Computational MODELING Approaches for Developing a Synergistic Effect Prediction Model of Estrogen Agonistic Activity	The study focuses on predicting the synergistic effects of estrogenic activity in chemical mixtures using computational models. By employing deep learning algorithms and chemical-protein network descriptors, a high-performance binary classification model is developed, aiding in the screening of mixture toxicity during product development.	Machine Learning-Based Model	Synergistic effects	[25]
Non-Target and Suspected-Target Screening for Potentially Hazardous Chemicals in Food Contact Materials: Investigation of Paper Straws	A nontarget screening method was developed to assess hazardous chemicals in paper food contact materials (FCMs) using high-resolution mass spectrometry, identifying 74 suspect compounds, with 40 detected confidently. A tentative toxicity assessment revealed two compounds as mutagenic and three as carcinogenic. Sampling included 17 commercially available straws, analyzed via LC-HRMS, with data processed using Compound Discoverer 2.1 for compound identification.	Quantitative Structure–Activity Relationship (QSAR) Model	–	[110]
Predicting Mixture Effects over Time with Toxicokinetic-Toxicodynamic Models (GUTS): Assumptions, Experimental Testing, and Predictive Power	The General Unified Threshold model for Survival (GUTS) offers a framework for assessing mixture toxicity considering temporal effects, integrating toxicokinetic-toxicodynamic (TKTD) models. GUTS-derived models, including the GUTS reduced (GUTS-RED) models, account for both independent action and concentration addition of mixture components, demonstrating predictive power in binary mixture studies with *Enchytraeus crypticus*.	Quantitative Structure–Activity Relationship (QSAR) Toxicodynamic Models	–	[71]

Article	Summary	Model	Chemical interaction	Ref.
Toxicity Predictions for Mycotoxins: A Combined in Silico Approach on Enniatin-Like Cluster	Climate change may exacerbate mycotoxin contamination in crops, posing risks to human health via direct consumption and through animal products. Using in silico tools, 12 mycotoxins were identified with potential carcinogenic activity, highlighting a link to lung cancer through gene pathway analysis and data mining.	Quantitative Structure–Activity Relationship (QSAR) Model	–	[111]
Prioritization of mycotoxins Based on Mutagenicity and Carcinogenicity Evaluation Using Combined In Silico QSAR Methods	To address the challenge of evaluating the toxicity of numerous mycotoxins, new approach methodologies (NAMs) were employed, particularly quantitative structure–activity relationship (QSAR) models. Initially, 904 mycotoxins were compiled, with selected mycotoxins used to optimize QSAR tools for predicting mutagenicity and carcinogenicity, leading to the identification of 95 mycotoxins with both mutagenic and carcinogenic potential for prioritized evaluation.	Quantitative Structure–Activity Relationship (QSAR) Model	–	[82]
Evaluation of time-Dependent Toxicity and Combined Effects for a Series of Mono-Halogenated Acetonitrile-Containing Binary Mixtures	The study evaluated the mixture and time-dependent toxicity (TDT) of mono-halogenated acetonitrile combinations using *Aliivibrio fischeri*. Concentration-response curves were determined for individual chemicals and mixtures at different time points, with TDT values calculated accordingly. Results showed variations in mixture toxicity, exhibiting dose-additive effects while others demonstrated less toxicity than predicted. TDT values were found to be useful in predicting chemical mixture toxicity over time.	Dose Addition Model	Additive effect	[112]
Predicting joint Toxicity of Chemicals by Incorporating a Weighted Descriptor into a Mixture Model: Cases for Binary Antibiotics and Binary Nanoparticles	The study assessed the toxicity of advanced two-dimensional nanomaterials (TDNMs) combined with an organic chemical (3,4-dichloroaniline, DCA) to freshwater microalgae. It found varied toxicity levels influenced by TDNM type, concentration, and algal species, with interactions exhibiting additive, antagonistic, and synergistic effects. The predictive model integrating the Freundlich adsorption coefficient (KF) and adsorption energy (Ea) demonstrated higher accuracy in predicting combined toxicity compared to classical mixture theory, offering insights for assessing ecotoxicological risks of nanomaterials in mixed pollution scenarios.	Concentration Addition (XGCA) Model	Additive effect	[113]
Reproductive Toxicity of Combined Effects of Endocrine Disruptors on Human reproduction	The review discusses the impact of endocrine-disrupting chemicals (EDCs) on human reproduction, emphasizing combined toxicity effects and transgenerational epigenetic consequences. It highlights exposure sources, such as food, air, water, soil, and occupational settings, detailing how EDCs affect female reproductive functions with varying evidence on the extent of their impact.	Concentration Addition Model	Synergistic effect Antagonistic effect	[13]
Machine Learning-Driven QSAR Models for Predicting the Mixture Toxicity of Nanoparticles	This study combines experimental data and ML methods to predict the toxicity of seven metallic-engineered nanoparticles (ENPs) for *Escherichia coli* in binary combinations. Machine learning models, particularly neural networks with specific molecular descriptors, outperform traditional component-based models, offering a promising approach for ecological risk assessment of ENP mixtures.	Quantitative Structure–Activity Relationship (QSAR) Model	–	[114]

(Continued)

Table 17.2 (Continued)

Article	Summary	Model	Chemical interaction	Ref.
Critical Features Identification for Chemical Chronic Toxicity Based on Mechanistic Forecast Models	This study develops machine learning models using rotifer population abundance as endpoints to predict chronic aquatic toxicity of chemicals. The best model, incorporating 1D and 2D molecular descriptors with random trees, achieved high accuracy and identified key factors and substructures influencing toxicity. Insights from the study enhance understanding of toxicity mechanisms and offer guidance for designing environmentally safer alternatives.	Machine Learning Models	–	[115]

Case Study Related to ML/AI-based Models

Hou et al. studied the limitation of USEtox, a widely used impact assessment method, which lacks ecotoxicity characterization factors for many chemicals because of insufficient laboratory experiment data. To overcome this limitation, the researchers developed machine learning models to estimate ecotoxicity hazardous concentrations 50% (HC50) based on the physical-chemical properties of chemicals and their mode of action classification. The model was validated using 10 test sets, demonstrating its predictive performance. The results indicated that the random forest model performed the best, outperforming traditional QSAR and linear regression models. The validated model was then used to estimate missing ecotoxicity characterization factors for 552 chemicals in USEtox. Therefore, machine learning is employed to assess chemical toxicity and fill data gaps in a widely used impact assessment method [101].

*USEtox: a scientific consensus model for characterizing human toxicological/ ecotoxicological impacts of chemical emissions in life cycle assessment.

Title	Estimate ecotoxicity characterization factors for chemicals in life cycle assessment using machine learning models
Background	In life cycle assessment (LCA), characterization factors are essential for converting the amount of chemicals and pollutants in a product's life cycle to standardized impact categories, like ecotoxicity.
Experimental Setup	Researchers developed machine learning models to estimate ecotoxicity hazardous concentrations (HC50) in USEtox using physical-chemical properties from EPA's CompTox Chemical Dashboard and mode of action classification.
Data Collection	Data included physical-chemical properties of chemicals from CompTox Dashboard and mode of action classification using ToxTree software. The dataset comprised 2307 observations with HC50 as the response variable and 14 predictor variables.
Application of Mathematical Model	Six machine learning models (KNN, SVM, NN, RF, AdaBoost, GBM) were considered for regression analysis. Cross-validation was used to select the best parameters for each model.
Validation and Interpretation	The average RMSE on test sets was 0.761, and the average R2 was 0.630, indicating that 63% of HC50 variability in USEtox could be explained by the predicted HC50.
Outcome	The validated random forest model was used to estimate missing ecotoxicity characterization factors for 552 chemicals in USEtox.
Implications	The machine learning-based approach offers a promising solution to address data gaps in impact assessment methods like USEtox, improving the accuracy and completeness of ecotoxicity characterization factors.

These models can capture nonlinear relationships and interactions between chemicals, allowing for the identification of complex synergistic or antagonistic effects within chemical mixtures. Machine learning algorithms, such as random forests, support vector machines, neural networks, and deep learning architectures are commonly used to develop predictive models for combined toxicity. By leveraging advanced computational techniques, machine learning and AI-based models can provide valuable insights into the cumulative effects of chemical mixtures on human health and the environment, facilitating risk assessment, hazard identification, and decision-making processes.

Each of these mathematical modeling techniques offers unique advantages and capabilities for predicting combined toxicity and assessing the risks associated with chemical exposures. By integrating these models into comprehensive risk assessment frameworks, researchers and regulators can gain valuable insights into the potential hazards posed by chemical mixtures and develop effective strategies for protecting human health and the environment (Table 17.2).

17.5 Evaluation of Model and Selection Criteria

When evaluating and selecting appropriate mathematical models for the prediction of combined chemical toxicity, several criteria and considerations should be taken into account:

Data availability: The availability of high-quality, comprehensive data on combined chemical toxicity is essential for model development and evaluation. This includes data on chemical structures, toxicity endpoints, exposure conditions, and potential interactions between mixture components.

Impact: Limited or biased data can lead to model overfitting, poor generalization, or biased predictions. Therefore, models should be evaluated based on their performance with available data and validated using external datasets to assess their robustness and generalization ability.

Model complexity: Model complexity should be commensurate with the available data and the complexity of the underlying biological mechanisms involved in combined chemical toxicity. Overly complex models may lead to overfitting, making them difficult to interpret and generalize.

Impact: Complex models may achieve high accuracy on the training data but perform poorly on unseen data. Simpler models are preferred if they can achieve comparable predictive performance, as they are easier to interpret, validate, and deploy in practice.

Accuracy: Model accuracy is a critical criterion for evaluating predictive performance. Accuracy metrics, such as mean squared error, root mean squared error, or coefficient of determination (R^2), provide quantitative measures of model performance.

Impact: Models with higher accuracy provide more reliable predictions of combined chemical toxicity, which is crucial for risk assessment, decision-making, and regulatory applications. However, accuracy should be balanced with other factors, such as interpretability and generalization ability.

Interpretability: Interpretability refers to the ease with which the model's predictions can be understood and explained. Simple, transparent models with interpretable coefficients or feature importance rankings enhance the model's explanatory power.

Impact: Interpretable models provide insights into the underlying relationships between chemical descriptors and toxicity, facilitating risk assessment, hypothesis generation, and decision-making. Additionally, interpretability enhances the model's trustworthiness and acceptance by stakeholders, including regulators and domain experts.

When evaluating and selecting mathematical models for predicting combined chemical toxicity, it is essential to consider factors, such as data availability, model complexity, accuracy, and interpretability. Balancing these factors ensures that the selected model achieves high predictive performance while remaining interpretable, robust, and applicable to real-world scenarios.

17.6 Future Directions and Research Needs

17.6.1 Emerging Trends and Future Research Directions

Multi-omics integration: Integrating multi-omics data (genomics, transcriptomics, proteomics, metabolomics) with computational approaches can provide comprehensive insights into the mechanisms underlying combined chemical toxicity. Future research may focus on developing integrative models that leverage multi-omics data to predict toxicity endpoints more accurately [116].

Systems toxicology: Systems toxicology approaches, which combine computational modeling with experimental data, can elucidate complex interactions between chemicals and biological systems. Future research may explore systems-level modeling techniques to capture the dynamic responses of biological pathways to chemical exposures and identify key determinants of combined toxicity [101].

Machine learning advancements: Advancements in machine learning algorithms, such as deep learning and ensemble methods, offer opportunities to improve the accuracy and robustness of predictive models for combined toxicity. Future research may investigate novel machine learning architectures and feature selection techniques tailored to the challenges of mixture toxicity prediction [117].

Dynamic modeling: Dynamic modeling approaches, such as pharmacokinetic-pharmacodynamic (PK-PD) modeling and agent-based modeling, can capture the temporal dynamics of chemical exposures and biological responses. Future research may focus on developing dynamic models that account for the time-dependent interactions between mixture components and their effects on biological pathways.

Exposure science: Integrating exposure science principles with computational modeling can enhance the accuracy of combined toxicity predictions by accounting for variations in exposure routes, concentrations, and durations. Future research may explore advanced exposure modeling techniques, such as PBPK modeling, to better characterize human exposure profiles and their implications for toxicity assessment.

17.6.2 Potential Integration of Computational Approaches

Computational approaches: Computational approaches, such as QSAR modeling, molecular docking, and molecular dynamics simulations, can provide insights into the structure–activity relationships of chemical mixtures and their interactions with biological targets. Integrating these computational approaches with experimental data enables more accurate predictions of combined toxicity [118].

Big data analytics: Big data analytics techniques, including data mining, machine learning, and network analysis, can leverage large-scale chemical and biological datasets to identify patterns, correlations, and predictive models of combined toxicity [119]. By analyzing diverse datasets, big data analytics can uncover hidden relationships and facilitate the discovery of novel toxicity mechanisms [120].

HTS techniques: HTS techniques, such as HTS assays and omics technologies, generate large volumes of data on chemical toxicity and biological responses. Integrating HTS data with computational models allows for rapid and cost-effective screening of chemical mixtures and prioritization of chemicals for further testing [121].

In silico toxicology platforms: In silico toxicology platforms, which integrate computational models, databases, and visualization tools, provide user-friendly interfaces for exploring chemical toxicity data and predicting combined toxicity [122]. These platforms enable researchers to access and analyze diverse datasets, perform virtual screening experiments, and generate actionable insights for risk assessment and decision-making.

In summary, the future of combined toxicity prediction lies in the integration of computational approaches, big data analytics, and HTS techniques. By leveraging advances in these fields, researchers can develop more accurate, comprehensive, and mechanistically informed models of combined toxicity, leading to improved risk assessment and management strategies.

17.7 Conclusion

Mathematical models play a crucial role in combined toxicity prediction, providing quantitative frameworks for hazard evaluation and risk assessment. Key points emphasizing their importance include:

Quantitative assessment: Mathematical models allow for the quantitative prediction of combined toxicity, enabling the estimation of potential health risks associated with exposure to mixtures of chemicals. By integrating data on chemical properties, exposure levels, and biological responses, these models provide objective measures of hazard and risk.

Mechanistic understanding: Mathematical models provide mechanistic insights into the underlying processes driving combined toxicity, elucidating the interactions between mixture components and their effects on biological systems. By modeling dose–response relationships and toxicity pathways [123], these models help identify key determinants of toxicity and inform hypothesis-driven research.

Predictive accuracy: Mathematical models strive to accurately predict combined toxicity across a range of conditions and chemical mixtures. By leveraging computational approaches, machine learning algorithms, and QSAR, these models achieve high levels of predictive accuracy, enhancing the reliability of risk assessments and regulatory decisions.

Data integration: Mathematical models integrate diverse datasets, including chemical properties, exposure data, toxicological endpoints, and mechanistic insights, into a unified framework. By synthesizing information from multiple sources, these models provide a holistic view of combined toxicity and enable the interpretation of complex interactions between mixture components [124].

Risk management: Mathematical models support risk management efforts by identifying high-risk chemical mixtures, prioritizing chemicals for further testing, and guiding the development of mitigation strategies. By quantifying potential health risks and uncertainties [125], these models help regulators, policymakers, and stakeholders make informed decisions to protect human health and the environment.

In summary, mathematical models for combined toxicity prediction are indispensable tools in hazard evaluation and risk assessment. They provide quantitative, mechanistic, and predictive frameworks for understanding the complex interactions between chemical mixtures and biological systems, informing regulatory decisions, and safeguarding public health and environmental quality. By advancing the development and application of mathematical models, researchers and policymakers can address emerging challenges in chemical safety and ensure the effective management of chemical risks in the modern world.

Acknowledgment

The authors are grateful for the support from the Chongqing Technology Innovation and Application Development Sichuan-Chongqing Special Key Project (CSTB2024TIAD-CYKJCXX0017), High-level Talents Project of Chongqing Medical University (No. R4014), and Chongqing Postdoctoral Innovation Mentor Studio (X7928 D.S.P.).

Competing Interests

The authors declare no competing interests.

References

1 Rodea-Palomares I, González-Pleiter M, Martín-Betancor K, Rosal R, Fernández-Piñas F. Additivity and interactions in ecotoxicity of pollutant mixtures: some patterns, conclusions, and open questions. Toxics. 2015, 3(4): 342–369.

2 Subramanian V, Peijnenburg WJGM, Vijver MG, Blanco CF, Cucurachi S, Guinée JB. Approaches to implement safe by design in early product design through combining risk assessment and Life Cycle Assessment. Chemosphere. 2023, 311 (Pt 1): 137080.

3 Feron VJ, Groten JP. Toxicological evaluation of chemical mixtures. Food Chem Toxicol. 2002, 40(6): 825–839.

4 Nawaz A, Šotek PE, Molnárová M. Reciprocal effects of metal mixtures on phytoplankton. Phycology. 2024, 4(1): 117–138.

5 Escher BI, Altenburger R, Blüher M, Colbourne JK, Ebinghaus R, Fantke P, Hein M, Köck W, Kümmerer KA, Leipold S, Li X, Scheringer M, Scholz S, Schloter M, Schweizer PJ, Tal T, Tetko I, Traidl-Hoffmann C, Wick LY, Fenner K. Modernizing persistence–bioaccumulation–toxicity (PBT) assessment with high throughput animal-free methods. Arch Toxicol. 2023, 97(5): 1267–1283.

6 Kar S, Leszczynski J. Exploration of computational approaches to predict the toxicity of chemical mixtures. Toxics. 2019, 7(1): 15.

7 Cheng Y, Ding J, Davila Arenas CE, Brinkmann M, Ji X. A brief review on the assessment of potential joint effects of complex mixtures of contaminants in the environment. Environ Sci: Adv Royal Soc Chem. 2024, 3: 661–675.

8 San Diego Kim J, Kim S, Schaumann GE. Reliable predictive computational toxicology methods for mixture toxicity: toward the development of innovative integrated models for environmental risk assessment. Rev Environ Sci Biotechnol. 2013, 12: 235–256.

9 Binet MT, Golding LA, Adams MS, Robertson T, Elsdon TS. Advantages of model averaging of species sensitivity distributions used for regulating produced water discharges. Integr Environ Assess Manag. 2024, 20(2): 498–517.

10 Pérez Santín E, Rodríguez Solana R, González García M, García Suárez MDM, Blanco Díaz GD, Cima Cabal MD, Moreno Rojas JM, López Sánchez JI. Toxicity prediction based on artificial intelligence: a multidisciplinary overview. Wiley Interdiscip Rev Comput Mol Sci. 2021, 11: e1516. John Wiley and Sons Inc.
11 Braun G, Escher BI. Prioritization of mixtures of neurotoxic chemicals for biomonitoring using high-throughput toxicokinetics and mixture toxicity modeling. Environ Int. 2023, 171: 107680.
12 Rider CV, Simmons JE. Chemical mixtures and combined chemical and nonchemical stressors: exposure, toxicity, analysis, and risk. Springer International Publishing, 2018, pp. 1–556.
13 Dutta S, Sengupta P, Bagchi S, Chhikara BS, Pavlík A, Sláma P, Roychoudhury S. Reproductive toxicity of combined effects of endocrine disruptors on human reproduction. Front Cell Dev Biol. 2023, 11: 1162015.
14 Sigmund G, Ågerstrand M, Antonelli A, Backhaus T, Brodin T, Diamond ML, Erdelen WR, Evers DC, Hofmann T, Hueffer T, Lai A. Addressing chemical pollution in biodiversity research. Glob Chang Biol. 2023, 29(12): 3240–3255.
15 Cattaneo I, Kalian AD, Di Nicola MR, Dujardin B, Levorato S, Mohimont L, Nathanail AV, Carnessechi E, Astuto MC, Tarazona JV, Kass GEN, Liem AKD, Robinson T, Manini P, Hogstrand C, Price PS, Dorne JLCM. Risk assessment of combined exposure to multiple chemicals at the European food safety authority: principles, guidance documents, applications and future challenges. Toxins (Basel). 2023, 15(1): 40.
16 Singh AK, Bilal M, Iqbal HMN, Raj A. Trends in predictive biodegradation for sustainable mitigation of environmental pollutants: recent progress and future outlook. Sci Total Environ. 2021, 770: 144561.
17 Loh MM, Schmidt P, Christopher de Vries Y, Vogel N, Kolossa-Gehring M, Vlaanderen J, Lebret E, Luijten M. Toxicity weighting for human biomonitoring mixture risk assessment: a proof of concept. Toxics. 2023, 11(5): 408.
18 Rim KT. In silico prediction of toxicity and its applications for chemicals at work. Toxicol Environ Health Sci. 2020, 12: 191–202. Korean Society of Environmental Risk Assessment and Health Science.
19 Finckh S, Beckers LM, Busch W, Carmona E, Dulio V, Kramer L, Krauss M, Posthuma L, Schulze T, Slootweg J, Von der Ohe PC, Brack W. A risk based assessment approach for chemical mixtures from wastewater treatment plant effluents. Environ Int. 2022, 164: 107234.
20 Savitz DA, Hattersley AM. Evaluating chemical mixtures in epidemiological studies to inform regulatory decisions. Environ Health Perspect. 2023, 131(4): 45001.
21 Lasch A, Lichtenstein D, Marx-Stoelting P, Braeuning A, Alarcan J. Mixture effects of chemicals: the difficulty to choose appropriate mathematical models for appropriate conclusions. Environ Pollut. 2020, 260: 113953.
22 Knudsen TB, Keller DA, Sander M, Carney EW, Doerrer NG, Eaton DL, Fitzpatrick SC, Hastings KL, Mendrick DL, Tice RR, Watkins PB. FutureTox II: in vitro data and in silico models for predictive toxicology. Toxicol Sci. 2015, 143(2): 256–267.
23 Marx-Stoelting P, Rivière G, Luijten M, Aiello-Holden K, Bandow N, Baken K, Cañas A, Castaño A, Denys S, Fillol C, Herzler M, Iavicoli I, Karakitsios S, Klanova J, Kolossa-Gehring M, Koutsodimou A, Vicente JL, Lynch I, Namorado S, Norager S, Pittman A, Rotter S, Sarigiannis D, Silva MJ, Theunis J, Tralau T, Uhl M, van Klaveren J, Wendt-Rasch L, Westerholm E, Rousselle C, Sanders P. A walk in the PARC: developing and implementing 21st century chemical risk assessment in Europe. Arch Toxicol. 2023, 97: 893–908. Springer Science and Business Media Deutschland GmbH.
24 Nowak PM, Wietecha-Posłuszny R, Płotka-Wasylka J, Tobiszewski M. How to evaluate methods used in chemical laboratories in terms of the total chemical risk? A ChlorTox Scale. Green Anal Chem. 2023, 5: 100056.
25 Seo M, Choi J, Park J, Yu WJ, Kim S. Computational modeling approaches for developing a synergistic effect prediction model of estrogen agonistic activity. Chemosphere. 2024, 349: 140926.
26 Alehashem M, Peters R, Fajana HO, Eslamizad S, Hogan N, Hecker M, Siciliano SD. Herbicides and pesticides synergistically interact at low concentrations in complex mixtures. Chemosphere. 2024, 353: 141431.
27 He Y, Wang X, Zhang C, Sun J, Xu J, Li D. Near-infrared light-mediated cyclodextrin metal-organic frameworks for synergistic antibacterial and anti-biofilm therapies. Small. 2023, 19(35): e2300199.
28 Meng F, Wang Y, Cao Q. Synergistic enhancement of redox pairs and functional groups for the removal of phenolic organic pollutants by activated PMS using silica-composited biochar: mechanism and environmental toxicity assessment. Chemosphere. 2023, 337: 139441.
29 Yang J, Friedman R. Synergy and antagonism between azacitidine and FLT3 inhibitors. Comput Biol Med. 2024, 169: 107889.
30 Wang Y, Sharma A, Ge F, Chen P, Yang Y, Liu H, Liu H, Zhao C, Mittal L, Asthana S, Schmidt-Wolf IGH. Non-oncology drug (meticrane) shows anti-cancer ability in synergy with epigenetic inhibitors and appears to be involved passively in targeting cancer cells. Front Oncol. 2023, 13: 1157366.
31 Vandenberg LN, Rayasam SDG, Axelrad DA, Bennett DH, Brown P, Carignan CC, Chartres N, Diamond ML, Joglekar R, Shamasunder B, Shrader-Frechette K, Subra WA, Zarker K, Woodruff TJ. Addressing systemic problems with exposure assessments to protect the public's health. Environ Health. 2023, 21(Suppl 1): 121.

32 Duarte D, Vale N. Evaluation of synergism in drug combinations and reference models for future orientations in oncology. Curr Res Pharmacol Drug Discov. 2022, 3: 100110.

33 Xia Q, Raoji P, Yang C, Xiong W, Peng X, Xia J, Wang W, Hai M. A review on the chemical properties, plant sources, anti-shock effects, pharmacokinetics, toxicity, and clinical applications of anisodamine. Chem Biodivers. 2024, 28: e202301477.

34 Montañés MT, García-Gabaldón M, Giner-Sanz JJ, Mora-Gómez J, Pérez-Herranz V. Effect of the anode material, applied current and reactor configuration on the atenolol toxicity during an electrooxidation process. Heliyon. 2024, 10(5): e27266.

35 Gao Y, Feng J, Kang L, Xu X, Zhu L. Concentration addition and independent action model: which is better in predicting the toxicity for metal mixtures on zebrafish larvae. Sci Total Environ. 2018, 610-611: 442–450.

36 Wang D, Wang S, Bai L, Nasir MS, Li S, Yan W. Mathematical modeling approaches for assessing the joint toxicity of chemical mixtures based on luminescent bacteria: a systematic review. Front Microbiol. 2020, 11: 1651.

37 Chou TC. Theoretical basis, experimental design, and computerized simulation of synergism and antagonism in drug combination studies. Pharmacol Rev. 2006, 58(3): 621–681.

38 Zeng J-P, Zhang J, Hong J-H, Zhao Y-F, Zhang J, Zhang Y, Huang XH, Xie FZ. Predicting the occurrence of antagonism within ternary guanidine mixture pollutants based on the concentration ratio of components. Sci Total Environ. 2024, 913: 169380.

39 Guo L, Tan J, Kaya S, Leng S, Li Q, Zhang F. Multidimensional insights into the corrosion inhibition of 3,3-dithiodipropionic acid on Q235 steel in H2SO4 medium: a combined experimental and in silico investigation. J Colloid Interface Sci. 2020, 570: 116–124.

40 González-Trujano ME, Ventura-Martínez R, Silveira D, Déciga-Campos M. Editorial: pharmacological interaction between drugs and medicinal plants, Volume II. Front Pharmacol. 2024, 15: 1372366.

41 Sarkhandia S, Sharma G, Mahajan R, Koundal S, Kumar M, Chadha P, Saini HS, Kaur S. Synergistic and additive interactions of Shewanella sp., Pseudomonas sp. and Thauera sp. with chlorantraniliprole and emamectin benzoate for controlling Spodoptera litura (Fabricius). Sci Rep. 2023, 13(1): 14648.

42 Chen C, Wang Y, Qian Y, Zhao X, Wang Q. The synergistic toxicity of the multiple chemical mixtures: implications for risk assessment in the terrestrial environment. Environ Int. 2015, 77: 95–105.

43 Bil W, Govarts E, Zeilmaker MJ, Woutersen M, Bessems J, Ma Y, Thomsen C, Haug LS, Lignell S, Gyllenhammar I, Palkovicova Murinova L, Fabelova L, Tratnik JS, Kosjek T, Gabriel C, Sarigiannis D, Pedraza-Diaz S, Esteban-López M, Castaño A, Rambaud L, Riou M, Franken C, Colles A, Vogel N, Kolossa-Gehring M, Halldorsson TI, Uhl M, Schoeters G, Santonen T, Vinggaard AM. Approaches to mixture risk assessment of PFASs in the European population based on human hazard and biomonitoring data. Int J Hyg Environ Health. 2023, 247: 114071.

44 Cavasotto CN, Scardino V. Machine learning toxicity prediction: latest advances by toxicity end point. ACS Omega. 2022, 7(51): 47536–47546.

45 Wang Z, Yu L. An insight into the combined toxicity of 3,4-dichloroaniline with two-dimensional nanomaterials: from classical mixture theory to structure-activity relationship. Int J Mol Sci. 2023, 24(4): 3723.

46 Jin N, Yang K, Li J, Song Y, Ding A, Sun Y, Li G, Zhang D, Li G, Zhang D. Toxicity characterization of environment-related pollutants using a biospectroscopy–bioreporter-coupling approach: potential for real-world toxicity determination and source apportionment of multiple pollutants. Anal Chem. 2023, 95(5): 4291–4300.

47 Li J, Wang Y, Fan Z, Tang P, Wu M, Xiao H, Zeng Z. Toxicity of tetracycline and metronidazole in chlorella pyrenoidosa. Int J Environ Res Public Health. 2023, 20(4): 3623.

48 Kraus JM, Skrabis K, Ciparis S, Isanhart J, Kenney A, Hinck JE. Ecological harm and economic damages of chemical contamination to linked aquatic-terrestrial food webs: a study-design tool for practitioners. Environ Toxicol Chem. 2023, 42(9): 2029–2039.

49 van Hall BG, Sweeney C, Bottoms M, van Gestel CAM. Influence of soil organic matter content on the toxicity of pesticides to soil invertebrates: a review. Integr Environ Assess Manag. 2023, 19: 1457–1472. John Wiley and Sons Inc.

50 Gajewicz-Skretna A, Wyrzykowska E, Gromelski M. Quantitative multi-species toxicity modeling: does a multi-species, machine learning model provide better performance than a single-species model for the evaluation of acute aquatic toxicity by organic pollutants?. Sci Total Environ. 2023, 861: 160590.

51 Fishkis O, Strassemeyer J, Pöllinger F, Roß CA, Koch HJ. Toxicological risk assessment of mechanical-chemical vs. chemical weed control techniques in sugar beet in Germany using SYNOPS-GIS. Front Agron. 2023, 5.

52 Hamid N, Junaid M, Pei DS. Combined toxicity of endocrine-disrupting chemicals: a review. Ecotoxicol Environ Saf. 2021, 215: 112136.

53 Kim S, Kang K, Kim H, Seo M. In vitro toxicity screening of fifty complex mixtures in hepg2 cells. Toxics. 2024, 12(2): 126.

54 Schell T, Goedkoop W, Zubrod JP, Feckler A, Lüderwald S, Schulz R, Bundschuh M. Assessing the effects of field-relevant pesticide mixtures for their compliance with the concentration addition model–an experimental approach with Daphnia magna. Sci Total Environ. 2018, 644: 342–349.

55 Gruiz K, Fekete-Kertész I, Kunglné-Nagy Z, Hajdu C, Feigl V, Vaszita E, Molnár M. Direct toxicity assessment—Methods, evaluation, interpretation. Sci Total Environ. 2016, 563–564: 803–812.

56 Deepika D, Kumar V. The role of "physiologically based pharmacokinetic model (pbpk)" new approach methodology (nam) in pharmaceuticals and environmental chemical risk assessment. Int J Environ Res Public Health. 2023, 20(4): 3473.

57 Neves BJ, Braga RC, Melo-Filho CC, Moreira-Filho JT, Muratov EN, Andrade CH. QSAR-based virtual screening: advances and applications in drug discovery. Front Pharmacol. 2018, 9: 1275.

58 Zilber D, Messier K. Reflected generalized concentration addition and Bayesian hierarchical models to improve chemical mixture prediction. PLoS One. 2024, 19(3): e0298687.

59 Escher B, Braun G, Zarfl C. Exploring the concepts of concentration addition and independent action using a linear low-effect mixture model. Environ Toxicol Chem. 2020, 39(12): 2552–2559.

60 Gray LE, Conley JM, Bursian SJ. Dose addition models accurately predict the subacute effects of a mixture of perfluorooctane sulfonate and perfluorooctanoic acid on Japanese quail (Coturnix japonica) chick mortality. Environ Toxicol Chem. 2024, 43(1): 97–104.

61 Spurgeon D, Wilkinson H, Civil W, Hutt L, Armenise E, Kieboom N, Sims K, Besien T. Proportional contributions to organic chemical mixture effects in groundwater and surface water. Water Res. 2022, 220: 118641.

62 Lee S, Ok SY, Moon HB, Seo SC, Ra JS. Developing a novel read-across concept for ecotoxicological risk assessment of phosphate chemicals: a case study. Toxics. 2024, 12(1): 96.

63 Spilsbury F, Kisielius V, Bester K, Backhaus T. Ecotoxicological mixture risk assessment of 35 pharmaceuticals in wastewater effluents following post-treatment with ozone and/or granulated activated carbon. Sci Total Environ. 2024, 906: 167440.

64 Morovati R, Badeenezhad A, Najafi M, Azhdarpoor A. Investigating the correlation between chemical parameters, risk assessment, and sensitivity analysis of fluoride and nitrate in regional groundwater sources using Monte Carlo. Environ Geochem Health. 2023, 46(1): 5.

65 Ouabane M, Zaki K, Tabti K, Alaqarbeh M, Sbai A, Sekkate C, Bouachrine M, Lakhlifi T. Molecular toxicity of nitrobenzene derivatives to tetrahymena pyriformis based on SMILES descriptors using Monte Carlo, docking, and MD simulations. Comput Biol Med. 2024, 169: 107880.

66 Sigurnjak Bureš M, Cvetnić M, Miloloža M, Kučić Grgić D, Markić M, Kušić H, Bolanča T, Rogošić M, Ukić S. Modeling the toxicity of pollutants mixtures for risk assessment: a review. Environ Chem Lett. 2021, 19: 1629–1655. Springer Science and Business Media Deutschland GmbH.

67 Doménech E, Martorell S. Review of the terminology, approaches, and formulations used in the guidelines on quantitative risk assessment of chemical hazards in food. Foods. 2024, 13(5): 714.

68 Sample BE, Johnson MS, Hull RN, Kapustka L, Landis WG, Murphy CA, Sorensen M, Mann G, Gust KA, Mayfield DB, Ludwigs JD, Munns WR Jr. Key challenges and developments in wildlife ecological risk assessment: problem formulation. Integr Environ Assess Manag. 2024, 20(3): 658–673.

69 Fan Z, Xu F. Health risks of occupational exposure to toxic chemicals in coal mine workplaces based on risk assessment mathematical model based on deep learning. Environ Technol Innov. 2021, 22: 101500.

70 Bart S, Jager T, Robinson A, Lahive E, Spurgeon DJ, Ashauer R. Predicting mixture effects over time with toxicokinetic-toxicodynamic models (guts): assumptions, experimental testing, and predictive power. Environ Sci Technol. 2021, 55(4): 2430–2439.

71 Lei H, Wang J, Sun Y, Wu Z, Wang X, Wang Y, Wang X. Thermally activated persulfate (TAP)-enhanced tris(2-chloroethyl) phosphate removal in real-world waters based on a response-surface approach as well as toxicological evaluation on its degradation products. Ecotoxicol Environ Saf. 2024, 270: 115924.

72 Teuschler LK, Hertzberg RC, McDonald A, Sey YM, Simmons JE. Evaluation of a proportional response addition approach to mixture risk assessment and predictive toxicology using data on four trihalomethanes from the U.S. EPA's multiple-purpose design study. Toxics. 2024, 12(4): 240.

73 Sharma RP, Schuhmacher M, Kumar V. Developing integrated PBPK/PD coupled mechanistic pathway model (miRNA-BDNF): an approach towards system toxicology. Toxicol Lett. 2017, 280: 79–91.

74 Wu Y, Sinclair G, Narasimhan RA, Pecquet A. Physiologically based kinetic (PBK) modeling of propiconazole using a machine learning 2 based read-across approach for interspecies extrapolation [Internet]. 2024. Available from: https://ssrn.com/abstract=4729300.

75 Simon L. Estimation of volatile organic compound exposure concentrations and time to reach a specific dermal absorption using physiologically based pharmacokinetic modeling. J Occup Environ Hyg. 2024, 21(1): 1–12.

76 Zhu J, Zhou S, Wang L, Zhao Y, Wang J, Zhao T, Li T, Shao F. Characterization of pediatric rectal absorption, drug disposition, and sedation level for midazolam gel using physiologically based pharmacokinetic/pharmacodynamic modeling. Mol Pharm. 2024, 21(5): 2187–2197.

77 Cuquerella-Gilabert M, Merino-Sanjuán M, García-Arieta A, Mangas-Sanjuán V, Reig-López J. Regulatory requirements and applications of physiologically based pharmacokinetic models. In: Talevi A, Quiroga PA (Eds.), ADME Processes in Pharmaceutical Sciences. Springer, Cham, 2024, pp. 193–219.

78 Lin X, Li X, Lin X. A review on applications of computational methods in drug screening and design. Molecules. 2020, 25(6): 1375.

79 Tropsha A, Isayev O, Varnek A, Schneider G, Cherkasov A. Integrating QSAR modelling and deep learning in drug discovery: the emergence of deep QSAR. Nat Rev Drug Discov. 2024, 23(2): 141–155.

80 Wang MWH, Goodman JM, Allen TEH. Machine learning in predictive toxicology: Recent applications and future directions for classification models. Chem Res Toxicol. 2021, 34: 217–239.

81 Lemée P, Fessard V, Habauzit D. Prioritization of mycotoxins based on mutagenicity and carcinogenicity evaluation using combined in silico QSAR methods. Environ Pollut. 2023, 323: 121284.

82 Saavedra LM, Garro Martinez JC, Duchowicz PR. Advances of the QSAR approach as an alternative strategy in the environmental risk assessment. In: QSAR in Safety Evaluation and Risk Assessment. Academic Press, 2023, pp. 117–137.

83 Khan PM, Jillella GK, Roy K. Recent advancements in QSAR and machine learning approaches for risk assessment of organic chemicals. In: QSAR in Safety Evaluation and Risk Assessment. Academic Press, 2023, pp. 167–185.

84 Puzyn T, Gajewicz A, Leszczynska D, Leszczynski J. Nanomaterials – the next great challenge for QSAR modelers. In: Challenges and Advances in Computational Chemistry and Physics. Springer, 2010, pp. 383–409.

85 Mansouri K, Lowe CN, Moreira-Filho JT, Charest N, Martin T, Tkachenko V, Judson R, Conway M, Kleinstreuer NC, Williams AJ. Free and open-source QSAR-ready workflow for automated standardization of chemical structures in support of QSAR modeling. J Cheminform. 2024, 16(1): 19.

86 Altenburger R, Nendza M, Schüürmann G. Mixture toxicity and its modeling by quantitative structure-activity relationships. Environ Toxicol Chem. 2003, 22: 1900–1915.

87 Tropsha A. Development of QSAR models as reliable computational tools for regulatory assessment of chemicals for acute toxicity. In: QSAR in Safety Evaluation and Risk Assessment. Academic Press, 2023, pp. 13–24.

88 Kim SY, Tariq S, Heo SK, Yoo CK. Interpretable attention-based multi-encoder transformer based QSPR model for assessing toxicity and environmental impact of chemicals. Chemosphere. 2024, 350: 141086.

89 Xu T, Ngan DK, Huang R. Application of QSAR models based on machine learning methods in chemical risk assessment and drug discovery. In: QSAR in Safety Evaluation and Risk Assessment. Academic Press, 2023, pp. 245–258.

90 Shi W, Guo J, Bao T. QSAR tools for toxicity prediction in risk assessment—Comparative analysis. In: QSAR in Safety Evaluation and Risk Assessment. Academic Press, 2023, pp. 203–218.

91 Alderete BL, da Silva J, Godoi R, da Silva FR, Taffarel SR, da Silva LP, Garcia AL, Júnior HM, de Amorim HL, Picada JN. Evaluation of toxicity and mutagenicity of a synthetic effluent containing azo dye after advanced oxidation process treatment. Chemosphere. 2021, 263: 128291.

92 Kim J, Seo M, Choi J, Na M. MRA Toolbox v. 1.0: a web-based toolbox for predicting mixture toxicity of chemical substances in chemical products. Sci Rep. 2022, 12(1): 8880.

93 Abd El-Hafeez T, Shams MY, Elshaier YAMM, Farghaly HM, Hassanien AE. Harnessing machine learning to find synergistic combinations for FDA-approved cancer drugs. Sci Rep. 2024, 14(1): 2428.

94 Nowak D, Huczyński A, Bachorz RA, Hoffmann M. Machine learning application for medicinal chemistry: colchicine case, new structures, and anticancer activity prediction. Pharmaceuticals (Basel). 2024, 17(2): 173.

95 Moukheiber L, Mangione W, Moukheiber M, Maleki S, Falls Z, Gao M, Samudrala R. Identifying protein features and pathways responsible for toxicity using machine learning and tox21: implications for predictive toxicology. Molecules. 2022, 27(9): 3021.

96 Yang Y, Zhong J, Shen S, Huang J, Hong Y, Qu X, Chen Q, Niu B. Application and progress of machine learning in pesticide hazard and risk assessment. Med Chem (Los Angeles). 2023, 20(1): 2–16.

97 Hassija V, Chamola V, Mahapatra A, Singal A, Goel D, Huang K, Scardapane S, Spinelli I, Mahmud M, Hussain A. Interpreting black-box models: a review on explainable artificial intelligence. Cogn Comput. 2024, 16: 45–74. Springer.

98 Chen J, Si YW, Un CW, Siu SWI. Chemical toxicity prediction based on semi-supervised learning and graph convolutional neural network. J Cheminform. 2021, 13(1): 93.

99 Cavasotto CN, Scardino V. Machine learning toxicity prediction: latest advances by toxicity end point. Am Chem Soc. 2022, 7: 47536–47546.

100 Jeong J, Choi J. Artificial intelligence-based toxicity prediction of environmental chemicals: future directions for chemical management applications. Environ Sci Technol. 2022, 56(12): 7532–7543.

101 Hou P, Jolliet O, Zhu J, Xu M. Estimate ecotoxicity characterization factors for chemicals in life cycle assessment using machine learning models. Environ Int. 2020, 135: 105393.

102 Chen L, Lu J, Zhang J, Feng KR, Zheng MY, Cai YD. Predicting chemical toxicity effects based on chemical-chemical interactions. PLoS One. 2013, 8(2): e56517.

103 Martin O, Scholze M, Ermler S, McPhie J, Bopp SK, Kienzler A, Parissis N, Kortenkamp A. Ten years of research on synergisms and antagonisms in chemical mixtures: a systematic review and quantitative reappraisal of mixture studies. Environ Int. 2021, 146: 106206.

104 Alekseevsky DG, Chernysh YeYu, Shtepa VN. Formalization of the task of creating a mathematical model of combined wastewater treatment processes. J Eng Sci. 2021, 8(2): H1–H7.

105 Jonker MTO, Diepens NJ. Improving sediment toxicity testing for very hydrophobic chemicals: Part 1-Spiking, equilibrating, and exposure quantification. Environ Toxicol Chem. 2024, 43: 1717–1727.

106 Spinu N, Cronin MTD, Enoch SJ, Madden JC, Worth AP. Quantitative adverse outcome pathway (qAOP) models for toxicity prediction. Arch Toxicol. 2020, 94: 1497–510. Springer.

107 Ra JS, Lee BC, Chang NI, Kim SD. Estimating the combined toxicity by two-step prediction model on the complicated chemical mixtures from wastewater treatment plant effluents. Environ Toxicol Chem. 2006, 25(8): 2107–2113.

108 Golosovskaia E. Development of *in silico* methods to aid chemical risk assessment: focusing on kinetic interactions in mixtures. Umeå University, 2024.

109 Chatterjee M, Roy K. Prediction of aquatic toxicity of chemical mixtures by the QSAR approach using 2D structural descriptors. J Hazard Mater. 2021, 408: 124936.

110 Rusko J, Perkons I, Rasinger JD, Bartkevics V. Non-target and suspected-target screening for potentially hazardous chemicals in food contact materials: investigation of paper straws. Food Addit Contam Part A Chem Anal Control Expo Risk Assess. 2020, 37(4): 649–664.

111 Habauzit D, Lemée P, Botana LM, Fessard V. Toxicity predictions for mycotoxins: a combined in silico approach on enniatin-like cluster. Expo Health [Internet]. 2023, 15(2): 315–331.

112 Dawson DA, Guinn D, Pöch G. Evaluation of time-dependent toxicity and combined effects for a series of mono-halogenated acetonitrile-containing binary mixtures. Toxicol Rep. 2016, 3: 572–583.

113 Wang Z, Zhang F, Wang DG. Predicting joint toxicity of chemicals by incorporating a weighted descriptor into a mixture model: cases for binary antibiotics and binary nanoparticles. Ecotoxicol Environ Saf. 2022, 236: 113472.

114 Zhang F, Wang Z, Peijnenburg WJGM, Vijver MG. Machine learning-driven QSAR models for predicting the mixture toxicity of nanoparticles. Environ Int. 2023, 177: 108025.

115 Wang X, Li F, Chen J, Teng Y, Ji C, Wu H. Critical features identification for chemical chronic toxicity based on mechanistic forecast models. Environ Pollut. 2022, 307: 119584.

116 PavelA. Knowledge Graphs, Network Models and Health Data Science Approaches for Toxicology and Pharmacology. 2024 April 12.

117 Kamkar M, Leonard KC, Ferrer I, Loo SCJ, Biddinger EJ, Brady D, Carrier DJ, Gathergood N, Han H, Hermans I, Hii KK. Artificial intelligence (AI) for sustainable resource management and chemical processes. ACS Sustain Chem Eng. 2024, 12(8): 2924–2926.

118 Park M, Lee S, Lee H, Denna MCFJ, Jang J, Oh D, Bae MS, Jang KS, Park K. New health index derived from oxidative potential and cell toxicity of fine particulate matter to assess its potential health effect. Heliyon. 2024, 10(3): e25310.

119 Setiya A, Jani V, Sonavane U, Joshi R. MolToxPred: small molecule toxicity prediction using machine learning approach. RSC Adv. 2024, 14(6): 4201–4220.

120 Singh AV, Varma M, Rai M, Pratap Singh S, Bansod G, Laux P, Luch A. Advancing Predictive Risk Assessment of Chemicals via Integrating Machine Learning, Computational Modeling, and Chemical/Nano-Quantitative Structure-Activity Relationship Approaches. Advanced Intelligent Systems. John Wiley and Sons Inc, 2024.

121 Collins SP, Mailloux B, Kulkarni S, Gagné M, Long AS, Barton-Maclaren TS. Development and application of consensus in silico models for advancing high-throughput toxicological predictions. Front Pharmacol. 2024, 15: 1307905.

122 Uesawa Y. Efficiency of pharmaceutical toxicity prediction in computational toxicology. Toxicol Res. 2024, 40(1): 1–9.

123 Kumari M, Kumar A. Identification of component-based approach for prediction of joint chemical mixture toxicity risk assessment with respect to human health: a critical review. Food Chem Toxicol. 2020, 143: 111458.

124 Smith TP, Clegg T, Ransome E, Martin-Lilley T, Rosindell J, Woodward G, Pawar S, Bell T. High-throughput characterization of bacterial responses to complex mixtures of chemical pollutants. Nat Microbiol. 2024, 9(4): 938–948.

125 Merecz-Sadowska A, Isca VMS, Sitarek P, Kowalczyk T, Małecka M, Zajdel K, Zielińska-Bliźniewska H, Jęcek M, Rijo P, Zajdel R. Exploring the anticancer potential of semisynthetic derivatives of 7α-acetoxy-6β-hydroxyroyleanone from plectranthus sp.: an in silico approach. Int J Mol Sci. 2024, 25(8): 4529.

18

Novel Quantitative Structure–Activity Relationship Tox21 Techniques for Combined Toxicity Prediction

Na Li[a,*]

[a] *Key Laboratory of Drinking Water Science and Technology, Research Center for Eco-Environmental Sciences, Chinese Academy of Sciences, Beijing, China*

*Corresponding author.
E-mail: nali@rcees.ac.cn

18.1 Introduction

There is an increasing concern about the toxicity evaluation and risk assessments of chemical mixtures, because human activities have caused continuous discharge and accumulation of various pollutants in the environment, leading to humans and animals being exposed to mixtures of various environmental pollution. The necessity of mixtures' toxicity assessment has got more and more attention by regulatory authorities and toxicologist, and chemical risk assessment has been shifting from single chemicals to mixtures: the General Unified Threshold model of Survival (GUTS) framework has extended for mixture toxicity assessment (https://cefic-lri.org/toolbox/guts-general-unified-threshold-model-of-survival); Agency for Toxic Substances and Disease Registry (ATSDR) has developed a direction for chemical mixtures, which offers more weighting on physiologically based pharmacokinetic (PBPK) and physiologically based pharmacodynamic (PBPD) modeling (https://www.atsdr.cdc.gov/mixtures/index.html); Agencies of National Institute of Environmental Health Sciences (NIEHS), National Institute for Occupational Safety and Health (NIOSH), and National Toxicology Program (NTP) have made substantial progress in learning how exposures to various mixtures affect human health (https://www.niehs.nih.gov/health/topics/science/chemical-mixtures/index.cfm); recently, the European Union's (EU's) "Chemicals Strategy for Sustainability Towards a Toxic-Free Environment" is to take account of the effects of chemical mixtures (https://environment.ec.europa.eu/strategy/chemicals-strategy_en).

However, until now, most of the toxicity evaluation and risk assessment of environmental pollution are related to single compounds. Characterizing the risk of chemical mixtures is particularly challenging: (1) there are both concentration diversity and mixture ratio diversity of chemical mixtures in environmental media; (2) the dose–response of toxicity effect and exposure parameters are required for all the chemicals in the mixture; (3) the interaction between chemicals, e.g. synergism or antagonism effect, and the integrating data across chemicals were required; and (4) most of the environmental pollution lack traditional chemical hazard information as animal test is time-consuming and expensive; therefore, it is impossible to test the toxicity of chemical mixture for potential harmful effects on a large scale. As such, it is necessary to develop effective methods for mixture toxicity and risk evaluation.

18.2 Computational Approaches and the Policy

Fortunately, the computational approaches of quantitative structure–activity/property/toxicity relationship (QSAR/QSPR/QSTR) have proven good practical potentials for predicting the activities or properties of industrial chemicals to overcome the lack of data, which have been used widely in drug design and the toxicity data gap filling of environmental

Toxicological Assessment of Combined Chemicals in the Environment, First Edition. Edited by De-Sheng Pei and Yiyun Liu.

pollutions by making use of structural information. In recent years, computational toxicology has been gradually applied to chemical mixture toxicity prediction, which is a new discipline based on the technology of biology, chemistry, physics, and computational science. Computational toxicology is also commonly referred to as "in silico toxicology," and the computational toxicology models include models developed based on toxicological data sets, molecular modeling to study intermolecular interactions, prediction models using structure–effect relationships, and cross-referenced toxicity prediction models.

The US EPA has specially set up the National Computational Toxicology Research Center, and the Registration, Evaluation, Authorization, and Restriction of Chemicals (REACH) regulations implemented by the EU also require the use of "integrated/intelligent testing strategy" with computational simulation technology as the core to obtain environmental safety information of chemicals [1]. The International Council for Harmonisation of Technical Requirements for Pharmaceuticals for Human Use (ICH), US Food and Drug Administration (FDA), Japan Pharmaceutical and Medical Device Corporation Pharmaceuticals and Medical Devices Agency (PMDA), and drug regulatory agencies such as European Medicines Agency (EMA) and National Medical Products Administration (NMPA) allow the use of computational methods to predict the potential mutagenicity of drug impurities [2–4]. In addition, the Toxic Substances Control Act (TSCA) integrated statistical predictive modeling and expert review as part of the overall toxicity assessment (https://www.epa.gov/tsca-screening-tools/about-using-predictive-models-and-tools-assess-chemicals-under-tsca). Recently, a cooperative assessment project participated by the International Chemical Council Association (ICCA) and the Organization for Economic Co-operation and Development (OECD) has proved that the hazard levels of 261 chemicals prediction by cross-reference method and computational toxicology model has reduced the usage of 108 500–142 500 rodents and the cost of $54–$67.5 million [5]. OECD has also developed a QSAR software named QSAR Toolbox, which has been widely used by governments, the chemical industry, and other stakeholders in filling gaps in (eco)toxicity data, making QSAR technology readily accessible and transparent and helping reduce animal tests.

18.3 The Methods for QSAR Models

The commonly used QSAR/QSPR/QSTR method includes multiple linear regression (MLR), partial least squares (PLS), and genetic function approximation (GFA), as well as machine-learning methods such as artificial neural network (ANN), support vector machine (SVM), and random forest (RF). MLR, which involves more than one explanatory variable, can model the linear relationship between the explanatory (independent) variables and response (dependent) variables. PLS regression is very useful when the predictors are highly collinear, can reduce the predictors to a smaller set of uncorrelated components, and performs least squares regression on the components. GFA, which creates multiple models by evolving random initial models using a genetic algorithm, can replace regression analysis and generate statistical models, including not only linear polynomials but also nonlinear models containing splines [6]. The machine-learning methods such as ANNs are information processing networks simulating the nervous system of the human brain [7, 8], as well as radial basis function (RBF) ANN is a kind of typical feedforward neural network, which possesses distinct advantages including universal approximation abilities, no local minimum problem, and a faster learning algorithm over other ANNs [9, 10].

18.4 Tox21 and ToxCast

Combined in vitro high-content screening (HCS) or high-throughput screening (HTS) generated in Tox21 and ToxCast, computational prediction models, and chemo-informatic methods, the new approach methodologies (NAMs) for toxicology can fill the data gap in hazard assessment and provide mechanistic chemical hazard information (https://www.epa.gov/sciencematters/epa-releases-updated-new-approach-methodologies-nams-work-plan). Tox21 is an agreement on "high-throughput screening, toxicity pathway profiling, and biological interpretation of findings" reached by the US EPA's National Center for Computational Toxicology, the National Human Genome Research Institute/National Institutes of Health Chemical Genomics Center, the NIEHS/National Toxicology Program, and the FDA during 2008–2010 (https://www.epa.gov/chemical-research/toxicology-testing-21st-century-tox21).

ToxCast is complemented by the Tox21 project being jointly carried out by the US NIH Chemical Genomics Center (NCGC), the US National Toxicology Program (NTP), and the EPA to screen chemicals against a large battery of in vitro HTS assays (https://cfpub.epa.gov/si/si_public_record_report.cfm?Lab=NCCT&dirEntryId=218296). Tox21 and ToxCast have generated dose–response information for >10K chemicals to build toxicity prediction models linking in vitro activity to in vivo adversity, prioritize data-poor chemicals for more in-depth testing, and better understand the mechanistic basis of chemical toxicity including developmental and reproductive toxicity, cancer, and endocrine signaling [11, 12].

Based on these increasing testing data of toxic target and biology pathways, the Adverse Outcome Pathway (AOP) and Aggregate Exposure Pathway (AEP) frameworks which link environmental chemical sources to an adverse outcome through a cascade of pathway-specific biological key events that occur at multiple levels of biological organization can help to address the adsorption, distribution, metabolism, and excretion of a chemical and its metabolites and then connected the external dose and internal concentration. The internal concentrations are referred to as the target site exposures, which are related to the molecular initiating event (MIE) [13]. The common events, MIEs, as well as adverse outcomes of multiple AOPs can be combined to form AOP networks, which can quantify the combined effects of chemical mixtures from MIE to adverse outcomes [14]. The development of AOPs aids in hypothesis generation related to the potential effects of exposure to environmentally relevant chemicals and the molecular targets that they may impact.

Based on the promotion of policy and the development of environmental toxicology studies, Kar and Leszczynski [15] have reviewed the computational approaches to predict the toxicity of chemical mixtures, elaborated on the importance of toxicity evaluation and prediction of chemical mixtures, and demonstrated how QSAR models were built and validated (Figure 18.1).

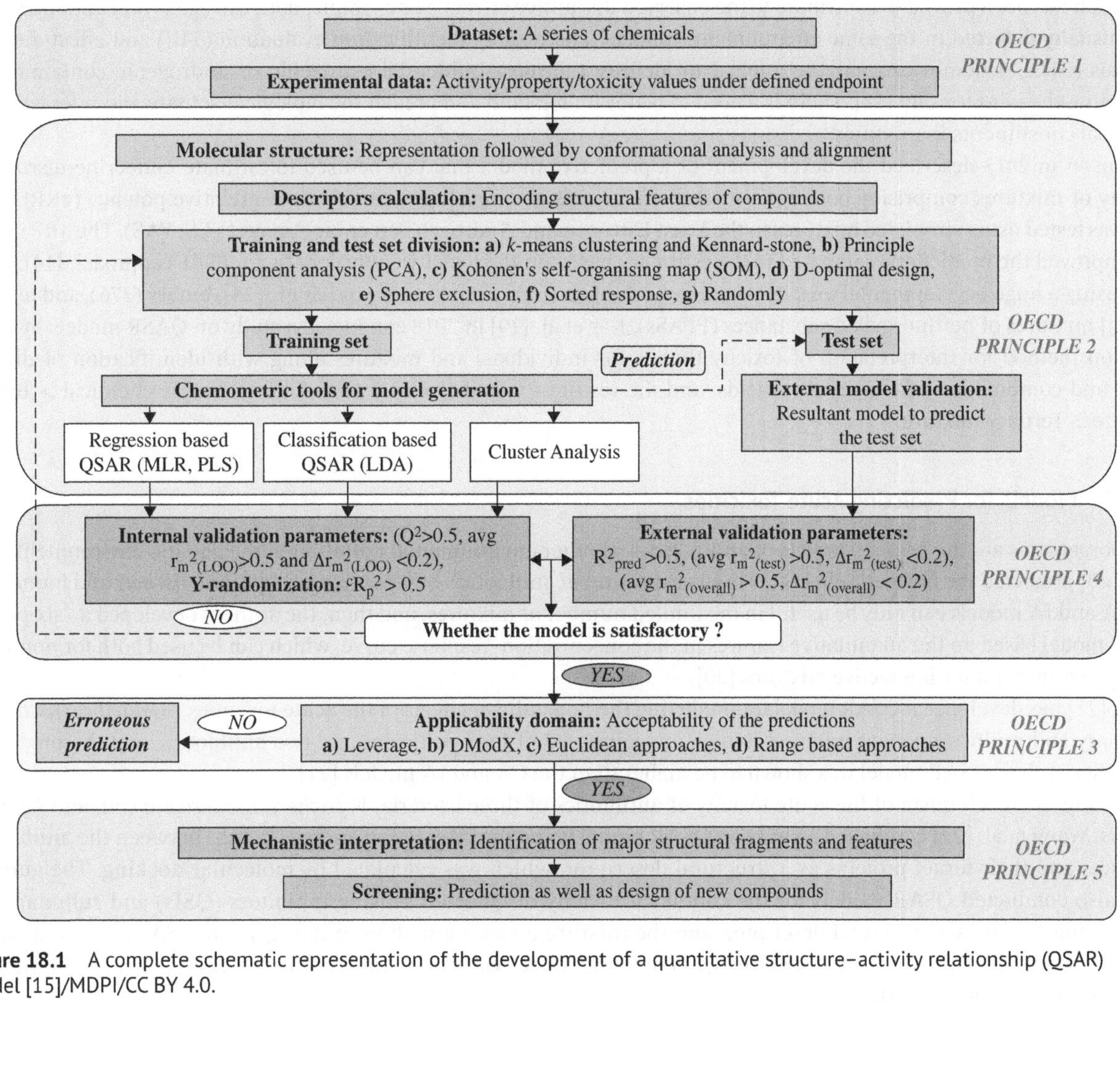

Figure 18.1 A complete schematic representation of the development of a quantitative structure–activity relationship (QSAR) model [15]/MDPI/CC BY 4.0.

18.5 The Cases of QSAR Studies for Prediction of the Toxicity of Chemical Mixtures

18.5.1 Comparison of Traditional Models of Joint Toxicity

Traditional methodologies for evaluating the effects of chemical mixtures were the concentration addition (CA) and independent action (IA) models, which have been used to describe the chemical mixtures having similar and dissimilar modes of action (MOAs). In previous studies, the joint effect of the majority of chemical mixtures has usually been predicted using the reference of CA. However, this approach has been increasingly challenged and replaced.

Based upon the assumption that the polycyclic aromatic hydrocarbons (PAHs) impact the growth rate of the crustacean *Daphnia magna* by a common MOA, four typical PAHs (pyrene, phenanthrene, fluoranthene, and naphthalene) were selected to construct the mixture QSAR models using the experimental parameters of the individual chemicals. However, the mixtures experiments revealed that the combined toxicity of the PAHs better conforms to IA model as opposed to the CA models [16]. In another work, a multiple linear regression model called ICIM (integrated CA with IA based on the multiple linear regression), was proposed for predicting the toxicity of a noninteractive mixture. The results showed that ICIM had a strong predictive power for the mixture toxicities and its prediction is better than CA and IA models [17].

18.5.2 Models for Predicting Endocrine-disrupting Activities

Environmental endocrine-disrupting chemicals (EDCs) are widespread in the environment, not only including the chemicals that have nuclear receptor agonist activity but also including the chemicals that have nuclear receptor antagonist activity, which has been proved to contribute to the observed disruptive effects. For example, both estrogens and antiandrogens were usually detected in the same environment samples. The toxicity identification evaluation (TIE) and effect-directed analysis (EDA) schemes can only be applied to identify the most influential estrogenic or androgenic contaminants, neglecting the joint toxicity [18]. There is a need to better understand and predict the biological activity and interaction of chemical constituents in mixtures.

Johnson in 2013 described the development of a predictive model that can be used to estimate endocrine-disrupting activity of mixtures comprising both agonists and antagonists and developed an alternative relative potency (aRP) value that was tested using simulated mixtures in the Yeast Estrogen and Androgen Screening Assays (YES, YAS). The aRP model has improved the predictions compared to the estimates based on the toxicity equivalent factor (TEF) approach [14].

By using a huge developmental toxicity dataset on zebrafish (*Danio rerio*) embryos [single (24), binary (276), and tertiary (2024)] mixtures of perfluoroalkyl substances (PFASs), Kar et al. [19] in 2018 conducted a study on QASR models by combination method for interpretation of toxicity threats for individuals and mixtures along with identification of diverse range and combination of toxicity thresholds, and the result of toxicity trend was as follows: Single chemical > Binary mixture > Tertiary mixture.

18.5.3 Models for Predicting Acute Toxicities

Microorganisms are the primary targets of antibiotics and other environmental pollution in the aquatic environment. Zou et al. [20] predicted the hormetic effects (24 hours exposure) of antibiotic's binary mixtures to *Vibrio fischeri* and found that the CA and IA models can only be useful in the limited number of mixtures, and then, the authors developed a "six-point" QSAR model based on the quantitative features in the concentration–response curve, which can be used both for noninteractive mixtures and for interactive mixtures [20].

Qin [21] has developed a QSAR model for predicting the nonadditive effects of the acute toxicities to *Aliivibrio fischeri* of 45 binary and multicomponent mixtures, which are composed of four pesticides and two antibiotics, and the predictive capability of their QSAR model was shown to be higher than the CA and IA models [21].

To study the mechanism of the acute toxicity of antibiotics of three bacteria, *V. fischeri*, *Escherichia coli*, and *Bacillus subtilis*, Wang et al. [22] conducted a study on QSAR model using the interaction energy (Ebind) between the antibiotics mixtures and their target proteins as a structural descriptor, which was calculated by molecular docking. The authors have also conducted QSAR models for the combined toxicity of Quorum sensing inhibitors (QSIs) and sulfonamides (SAs) using Ebind as a structural descriptor, and the mixture toxicity test showed that QSIs and SAs presented either additive or antagonistic joint effects and the QSIs in the mixtures were more involved in the chemical–protein interaction than the antibiotics [23].

Within the CORAL software, Carnesecchi et al. [24] have conducted three different QSAR models to predict the acute contact mixtures potency for honey bees (*Apis mellifera*), including two regression models and one classification-based model, using the test data retrieved from EFSA databases including binary mixtures ($n = 123$) (LD50-mix) and corresponding Toxic Unit values.

The pseudo-persistence of cosmetics and personal care additives (CPCAs) can cause long-term harmful effects on aquatic organisms. The test and prediction of the mixture toxicity of CPCAs are essential for ecological risk assessment because these chemicals generally exist in the aquatic environment as mixtures. An acute toxicity test of a typical CPCA mixture of the aquatic *D. magna* CA and IA models was used to study the combined toxicity which was tested previously, and a total of 18 QSAR models were built based on three dataset partitioning methods, two descriptor screening methods (genetic algorithm and stepwise multiple linear regression), and three regression methods. The result showed that the QSAR model can be used to predict the aquatic toxicity of binary CPCA mixtures with arbitrary proportions in the application domain, and the author appealed that in the risk management of CPCAs, joint exposure to other substances in the aquatic environment should be avoided as much as possible [25].

18.5.4 Models Based on Machine-learning Method

Using the composite descriptors according to the fractional concentrations of the mixture components, the mixture toxicity prediction model of nonpolar narcotic chemicals was conducted by forward stepwise multilinear regression (MLR) and radial basis function neural networks (RBFNNs).The result showed that the MLR-based QSAR model was simple, reliable, and credible, and fast toxicity of nonpolar narcotic chemicals, as well as the RBFNN model, gave even improved results for the prediction mixture [26].

Based on the backpropagation neural network (BPNN), which is one of the most maturely studied neural network algorithms because of its good self-learning, self-adapting, and generalization abilities and robustness, a new approach (BNNmix) was developed for predicting the toxicities of seven-component mixtures (consisting of one heavy metal, two pesticides, two ionic liquids, and two substituted phenols) on *Caenorhabditis elegans*. Using the BNNmix approach, the author found that the CA and IA models could not accurately predict the toxicities of these mixtures [27].

18.5.5 Models Based on Biomolecular Interaction Networks

Using information from protein–protein and chemical–protein interaction networks, Kim et al. [28] developed a synergistic toxicity prediction (STP) model by machine-learning method to qualitatively predict the joint toxicity of binary mixtures to the marine luminescent bacterium (*V. fischeri*). The STP model could classify chemical mixtures to have synergistic or nonsynergistic toxicity with an area under the receiver operating characteristic (ROC) curve (AUC) up to 0.73, and the result revealed that the numbers of shared targets and their neighborhood were the most important features for classifying chemicals into synergistic and nonsynergistic groups.

18.5.6 Toxicokinetic–Toxicodynamic Models

The classic mixture effects prediction models were mostly based on dose–response curves at a single time point and could not identify a mixture interaction where exposure varies in time. Using toxicokinetic–toxicodynamic modeling, Bart et al. [29] designed a dedicated mechanistic interaction module coupled with the GUTS mixture model to identify chemical interactions and predict the effect of synergistic and antagonistic mixtures, including two species (*Enchytraeus crypticus* and *Mamestra brassicae*) and six chemicals of prochloraz, imidacloprid, cypermethrin, azoxystrobin, chlorothalonil, and chlorpyrifos. The results showed an accurate simulation of synergistic and antagonistic effects accounting for the entire time of exposure. With a case study on neurotoxicity, Braun et al. [30] proposed a high-throughput approach using experimental toxicokinetic data to predict the mixture effects, which could correctly identify and prioritize known chemicals and identify new potential neurotoxicants to which a virtual population is exposed.

18.5.7 Models Based on High-throughput Transcriptomics Approach

In recent years, high-throughput transcriptomics (HTT) has been used in investigating genome-scale perturbation of chemical exposure in a dose-dependent manner, which can also be used as a high-throughput and cost-effective approach for complex chemical mixture safety assessment. Xia et al. [31] conducted a newly developed human-reduced transcriptomics (RHT) approach to assess biological pathway-based profiles of single chemicals and chemical mixtures, which were based on a single chemical-based prediction model. In their study, three typical pollutants, including 2,3,7,8-tetrachlorodibenzo-*p*-dioxin (TCDD), triclosan (TCS), and 5-chloro-6-hydroxy-2,2′,4,4′-tetrabromodiphenyl ether (5-Cl-6-OH-BDE-47), were used as model toxicant, and their dose-dependent RHT was used to qualitatively and quantitatively differentiate pathway-based patterns. Transcriptomics of individual chemicals were used to predict the transcriptional potency of mixtures by CA and IA models, and the result showed that some multiple biological processes such as hormone signaling and DNA damage can be predicted using CA models for mixtures.

Recently, a zebrafish embryo toxicity bioassay and a concentration-dependent reduced zebrafish transcriptome (CRZT) approach have been conducted to test the toxicities and transcriptomic effects of two artificial chemical mixtures including 12 component chemicals, which could identify the main bioactive compounds and provide a powerful tool for assessment of biological pathways perturbed by chemicals mixtures [32].

18.5.8 Models Based on Geospatial Modeling Approach

To integrate human exposures with hazard data from in vitro high-throughput (HTS) assays, Eccles et al. [33] have demonstrated a workflow for evaluating the geospatial risk of chemical mixtures that perturb molecular targets in human health adverse outcomes (AOs) based on real-world exposures (Figure 18.2). Using the response data of environmental chemicals for molecular targets that cause early events in AEP-AOP networks from curated HTS (cHTS) assays generated in Tox21 and ToxCast, the joint action of 41 modeled environmental chemicals from air exposures from the US-wide National Air Toxics Assessment were assessed to identify counties where exposure to the local chemical mixture may perturb a common biological target. The result demonstrated that NAMs can be used to predict early-stage biological perturbations of environmental chemical mixtures that can lead to AOs, and the author emphasized that the geospatially explicit nature of exposure to environmental chemical mixtures was important and should be considered in future mixtures' research.

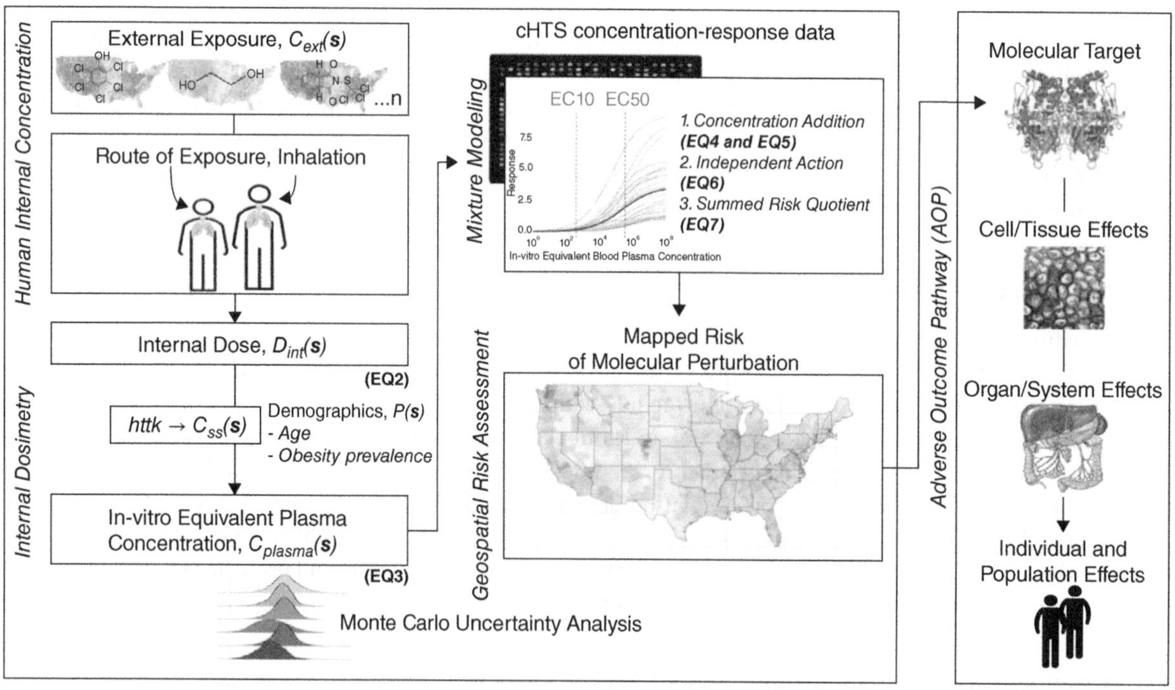

Figure 18.2 Schematic overview of the methods used to quantify the geospatial risk of molecular perturbation [33]/with permission of Elsevier.

18.6 Future Avenues of Chemical Mixture Toxicity Prediction Research

To characterize the human health risk and ecological risk, the risk assessment requires information on the potential adverse effect, the degree of response at differing exposure levels, and the dose of chemical mixtures; however, due to the limited understanding of the toxicity effect and toxicity mechanism of most chemicals, it is impossible and impractical to test the toxicity of chemical mixtures on every possible combination. Therefore, chemical mixture toxicity prediction research is useful and important.

Until now, it has been acknowledged that available QSAR models for the prediction of chemical mixture toxicity are insufficient. First, to increase the uptake of QSARs for mixtures, more fundamental knowledge is required about mixture toxicology. Most of the data from TOX21, ToxCast, and the publication are about the toxicity of individual compounds, and there is a huge gap in the difficulty of toxicity test between individual chemicals and chemical mixtures. Second, the evaluation of the toxicity of chemical mixtures needs to be at a system level. Although in vitro HCS and HTS methods offer many advantages for toxicity evaluation, care must be taken to understand that many combination toxicity effects are dependent on the biological environment, and the in vitro toxicity effects might not lead to the same biological effects in vivo. In addition, even in the same biological setting, combinations as characterized by different routes of administration, the toxicokinetics–toxicodynamics (TK-TD) of the individual chemicals, and their interaction with each other were different. Fortunately, omics advances in animal toxicology and ecotoxicology allow the assessment of the broadest coverage of disrupted biological pathways by chemical mixtures. Using omics and other high-throughput technologies will help us understand the biological effects of chemical mixtures. Third, individuals are actually exposed to chemical mixtures from environmental sources and vary over time and space. However, traditional chemical health risk assessments and ecological risk assessments are based on animal studies of the potential effects caused by environmentally relevant chemicals and the molecular targets that they may impact. The localized chemical exposure data integrated via molecular targets can provide an early warning.

In general, the prediction of the toxicity of chemical mixtures needs to be integrated for a comprehensive systems-level analysis to identify their toxicity effect by integrating chemical bioactivity data including chemical bioactivity, targets and pathways, gene expression, and protein interactions, thereby finding out and measuring the early events in their AEP-AOP network from combined exposure to multiple environmental chemicals, as well as the localized chemical exposure data. Such studies could have a significant impact on the understanding of the toxicity effect and MOA for chemical mixtures and will help provide a comprehensive and solid foundation on which future chemical mixtures prediction and analyses could be performed.

References

1 Bopp SK, Kienzler A, Richarz AN, van der Linden SC, Paini A, Parissis N, Worth AP. Regulatory assessment and risk management of chemical mixtures: challenges and ways forward. Crit Rev Toxicol. 2019, 49: 174–189.
2 ICH. Guidance on genotoxicity testing and data interpretation for pharmaceuticals intended for human use. In: International Conference on Harmonisation of Technical Requirements for Registration of Pharmaceuticals for Human Use, 2008.
3 EPA. TSCA New Chemicals Program NCP Chemical Categories. United States Environmental Protection Agency, 2010.
4 ICH. ICH M7 Assessment and Control of DNA Reactive (Mutagenic Impurities in Pharmaceuticals to Limit Potential Carcinogenic Risk. European Medicines Agency, 2017.
5 Stanton K, Kruszewski FH. Quantifying the benefits of using read-across and in silico techniques to fulfill hazard data requirements for chemical categories. Regul Toxicol Pharmacol. 2016, 81: 250–259.
6 Rogers D, Hopfinger AJ. Application of genetic function approximation to quantitative structure-activity relationships and quantitative structure-property relationships. J Chem Inf Compu Sci. 1994, 34: 854–866.
7 Bagheri M, Mirbagheri SA, Ehteshami M, Bagheri Z. Modeling of a sequencing batch reactor treating municipal wastewater using multi-layer perceptron and radial basis function artificial neural networks. Pro Saf Environ Protect. 2015, 93: 111–123.
8 Iliyas SA, Elshafei M, Habib MA, Adeniran AA. RBF neural network inferential sensor for process emission monitoring. Control Eng Pract. 2013, 21: 962–970.
9 Jin L, Bai P. QSPR study on normal boiling point of acyclic oxygen containing organic compounds by radial basis function artificial neural network. Chemom Intell Lab Syst. 2016, 157: 127–132.
10 Zhao Z, Lou Y, Chen Y, Lin H, Li R, Yu G. Prediction of interfacial interactions related with membrane fouling in a membrane bioreactor based on radial basis function artificial neural network (ANN). Bioresource Technol. 2019, 282: 262–268.

11 Judson RS, Houck KA, Kavlock RJ, Knudsen TB, Martin MT, Mortensen HM, Reif DM, Rotroff DM, Shah I, Richard AM, Dix DJ. In vitro screening of environmental chemicals for targeted testing prioritization: the ToxCast project. Environ Health Perspect. 2010, 118: 485–492.

12 Tice RR, Austin CP, Kavlock RJ, Bucher JR. Improving the human hazard characterization of chemicals: a Tox21 update. Environ Health Perspect. 2013, 121: 756–765.

13 Ankley GT, Bennett RS, Erickson RJ, Hoff DJ, Hornung MW, Johnson RD, Mount DR, Nichols JW, Russom CL, Schmieder PK, Serrrano JA, Tietge JE, Villeneuve DL. Adverse outcome pathways: a conceptual framework to support ecotoxicology research and risk assessment. Environ Toxicol Chem. 2010, 29: 730–741.

14 Knapen D, Angrish MM, Fortin MC, Katsiadaki I, Leonard M, Margiotta-Casaluci L, Munn S, O'Brien JM, Pollesch N, Smith LC, Zhang X, Villeneuve DL. Adverse outcome pathway networks I: development and applications. Environ Toxicol Chem. 2018, 37: 1723–1733.

15 Kar S, Leszczynski J. Exploration of computational approaches to predict the toxicity of chemical mixtures. Toxics. 2019, 7: 15.

16 Olmstead AW, Leblanc GA. Joint action of polycyclic aromatic hydrocarbons: predictive modeling of sublethal toxicity. Aquat Toxicol. 2005, 75: 253–262.

17 Qin LT, Liu SS, Zhang J, Xiao QF. A novel model integrated concentration addition with independent action for the prediction of toxicity of multi-component mixture. Toxicology. 2011, 280: 164–172.

18 Burgess RM, Ho KT, Brack W, Lamoree M. Effects-directed analysis (EDA) and toxicity identification evaluation (TIE): complementary but different approaches for diagnosing causes of environmental toxicity. Environ Toxicol Chem. 2013, 32: 1935–1945.

19 Kar S, Ghosh S, Leszczynski J. Single or mixture halogenated chemicals? Risk assessment and developmental toxicity prediction on zebrafish embryos based on weighted descriptors approach. Chemosphere. 2018, 210: 588–596.

20 Zou X, Lin Z, Deng Z, Yin D. Novel approach to predicting hormetic effects of antibiotic mixtures on Vibrio fischeri. Chemosphere. 2013, 90: 2070–2076.

21 Qin LT, Chen YH, Zhang X, Mo LY, Zeng HH, Liang YP. QSAR prediction of additive and non-additive mixture toxicities of antibiotics and pesticide. Chemosphere. 2018, 198: 122–129.

22 Wang D, Shi J, Xiong Y, Hu J, Lin Z, Qiu Y, Cheng J. A QSAR-based mechanistic study on the combined toxicity of antibiotics and quorum sensing inhibitors against Escherichia coli. J Hazard Mater. 2018, 341: 438–447.

23 Wang D, Wu X, Lin Z, Ding Y. A comparative study on the binary and ternary mixture toxicity of antibiotics towards three bacteria based on QSAR investigation. Environ Res. 2018, 162: 127–34.

24 Carnesecchi E, Toropov AA, Toropova AP, Kramer N, Svendsen C, Dorne JL, Benfenati E. Predicting acute contact toxicity of organic binary mixtures in honey bees (A. mellifera) through innovative QSAR models. Sci Total Environ. 2020, 704: 135302.

25 Yang YT, Ni HG. Predictive in silico models for aquatic toxicity of cosmetic and personal care additive mixtures. Water Res. 2023, 236: 119981.

26 Luan F, Xu X, Liu H, Cordeiro MN. Prediction of the baseline toxicity of non-polar narcotic chemical mixtures by QSAR approach. Chemosphere. 2013, 90: 1980–1986.

27 Wang P, Xia P, Wang Z, Zhang X. Evidence-based assessment on environmental mixture using a concentration-dependent transcriptomics approach. Environ Pollut. 2020, 265: 114839.

28 Kim J, Fischer M, Helms V. Prediction of synergistic toxicity of binary mixtures to vibrio fischeri based on biomolecular interaction networks. Chem Res Toxicol. 2018, 31: 1138–1150.

29 Bart S, Short S, Jager T, Eagles EJ, Robinson A, Badder C, et al. How to analyse and account for interactions in mixture toxicity with toxicokinetic-toxicodynamic models. Sci Total Environ. 2022, 843: 157048.

30 Braun G, Escher BI. Prioritization of mixtures of neurotoxic chemicals for biomonitoring using high-throughput toxicokinetics and mixture toxicity modeling. Environ Int. 2023, 171: 107680.

31 Xia P, Zhang H, Peng Y, Shi W, Zhang X. Pathway-based assessment of single chemicals and mixtures by a high-throughput transcriptomics approach. Environ Int. 2020, 136: 105455.

32 Wang ZJ, Liu SS, Feng L, Xu YQ. BNNmix: a new approach for predicting the mixture toxicity of multiple components based on the back-propagation neural network. Sci Total Environ. 2020, 738: 140317.

33 Eccles KM, Karmaus AL, Kleinstreuer NC, Parham F, Rider CV, Wambaugh JF, Messier KP. A geospatial modeling approach to quantifying the risk of exposure to environmental chemical mixtures via a common molecular target. Sci Total Environ. 2023, 855: 158905.

19

Challenges and Prospects in the Application of Experimental, Analytical, and Predictive Models in Combined Toxicity Assessment

Naima Hamid[a], Qiang Yue[b], Jun Wang[c], Muhammad Junaid[b,c,]*

[a] *Faculty of Science and Marine Environment, Universiti Malaysia Terengganu, 21030 Kuala Nerus, Terengganu, Malaysia*
[b] *Guangdong Provincial Key Laboratory of Utilization and Conservation of Food and Medicinal Resources in Northern Region, Shaoguan University, Shaoguan 512005, China*
[c] *Joint Laboratory of Guangdong Province and Hong Kong Region on Marine Bioresource Conservation and Exploitation, College of Marine Sciences, South China Agricultural University, Guangzhou 510641, China*

*Corresponding author.
E-mail: junaid@scau.edu.cn, junaidyaha@gmail.com (MJ.)

19.1 Introduction

Combining toxicological assessments is a practical way to evaluate chemical interactions and associated health risks [1]. Moreover, regulatory authorities have recognized mixture effects as a significant concern in environmental risk assessment [2]. The existence of multicomponent mixtures as parent compounds, metabolites, and conversion products is ubiquitous in the environment [3, 4]. Assessing the toxicity of chemical mixtures has posed a challenge for the scientific community in recent decades [5–8]. Fortunately, substantial advancements have been made in examining the effects of individual chemicals, enabling us to gain a deeper understanding of their possible combined toxicity [5, 9].

There are numerous chemical pollutants in the environment, including dioxins, flame retardants, detergents, antibiotics, anti-depressants, contraceptive pills, micro(nano)plastics, phthalates (PAEs), polychlorinated biphenyls (PBCs), polyfluoroalkyl substances (PFASs), polycyclic aromatic hydrocarbons (PAHs), personal care products (PCPs), and nonsteroidal anti-inflammatory drugs (NSAIDs), that cause various health effects [9–11]. Moreover, it is crucial to identify various sources of emerging pollutants to implement control measures [12]. There are various potential sources of emerging pollutants including manufacturing industries, agricultural activities and household processes, and distribution of raw materials [13]. In addition, domestic wastewater, hospital effluents, pharmaceutical, and cosmetic industries are sources of pharmaceuticals and personal care products (PPCPs) [14]. The extensive production and use of plastics lead to the emission of microplastics (size >1 μm to <5 mm) and nanoplastics (size: 1nm to 1000nm) from various materials including rubber, plastic industry, nitrocellulose, and ethyl cellulose [15].

Scientists can now better understand the combined toxicity of individual chemicals through novel analytical and omics approaches, which facilitate understanding the association between chemical mixtures and biological responses [16]. In general, chemical mixtures interact with biomolecules such as enzymes and genes, disturb their expression levels, and mimic their entire metabolism mechanism [14]. It is imperative to take into account the diverse effects chemical mixtures can have on tissues, organs, and various biological systems [9, 17]. Previously, chemical toxicity studies have focused on short-term effects using bioassay systems [14]. Thus, there is still a need for in-depth research identifying the chronic effects of mixtures, as they may induce additional adverse effects via distinct toxicological pathways.

For a comprehensive toxicological assessment, it is crucial to combine dose-based effect screening through in vivo/in vitro bioassays and in silico mathematical modeling. Furthermore, toxic chemical mixtures must be prioritized for safer chemical research. In addition, various parameters, including high precision, rapidity, economy, and the ability to predict possible toxicity scenarios, should be taken into account when choosing the experimental design. Previously, it was reported that

Toxicological Assessment of Combined Chemicals in the Environment, First Edition. Edited by De-Sheng Pei and Yiyun Liu.

factorial design is considered the most effective method used to formulate chemical mixtures to identify responses based on individual and multiple combinations [13]. Similarly, the exposure dose selection should be environmentally relevant to foresee the realistic impacts of combined exposure. Considering the above discussion, this chapter aims to elucidate the combined chemical effects on various in vivo and in vitro animal models to identify the challenges and future perspectives of applying experimental, analytical, and predictive models used in combined toxicology. A summary of the combined toxicological effects of legacy and emerging contaminants in in vivo and in vitro models is presented in Table 19.1.

19.2 Experimental Models

19.2.1 In Vivo Models

There are various animal models used for in vivo mixture toxicity assessment, i.e., zebrafish *(Danio rerio)*, marine medaka (*Oryzias melastigma*), *Caenorhabditis elegans, Daphnia magna, Carassius auratus, Mus musculus*, honey bees (*apis mellifera*), wild bees (*bombus spp.*), and solitary bees (*Osmia spp.*) [17, 41, 42]. In addition, ecosystem relevance, sensitivity, standardization, reproducibility, and animal ethics should be carefully considered for mixture toxicity experiments [13]. Bioaccumulation of microplastics was reported in *D. magna* and zebrafish in previous studies, showing that India's coastline needed rapid attention to cope with plastic pollution. In addition, microplastics may absorb toxins and act as endocrine disruptors, increasing the ecological danger to aquatic life [43]. A study demonstrated that co-exposure to microplastics and heavy metals in female marine medaka experienced significant reproductive toxicity [44]. Similarly, the single and joint effects of polyfluoroalkyl substances (PFAS) demonstrated higher thyroid levels quantified in maternal plasma samples from Boston, USA [45]. In summary, the combined effects of PFAS affected normal thyroid function, impaired normal growth, and caused serious reproductive toxicity [45].

Rats exposed to the six PAE mixtures displayed increased estrogenic activity in comparison to groups exposed to individual chemical treatment [46]. Pollock et al. [26] investigated that quinary mixtures also affected the 17b-estradiol (E2) function crucial for regulating estrogen metabolism. Similarly, binary mixtures of antibiotics induced reproductive abnormalities in offspring [33]. In addition, male zebrafish exposed to anti-inflammatory antibiotics exhibited elevated *vtg* genetic levels and caused hepatotoxicity [27]. Significant decreases in oxidative stress, particularly SOD, GSH-Px, CAT, and MDA enzymes, were observed in adult male rats treated with 250 mg/kg DBP + BaP for 12weeks [19]. Increased antioxidant responses such as SOD, CAT, GSH, and MDA were observed in *Carassius auratus* fish after DBP + DEHP + Cu mixture exposure (500 μg/L) [18]. Similarly, binary DEP + DEHP mixtures increased the expression of genes related to lipid metabolism and reproduction in *C. elegans* [20]. Moreover, PPCPs and phthalate mixtures perturb the androgen metabolism in adult rats [23].

The co-exposure of mycotoxins and fungicides on zebrafish embryos revealed significant alteration in oxidative stress and the HPT axis pathway [47]. Similarly, binary mixtures of antibiotics also showed significant reproductive toxicity in four generations of *Daphnia magna* [33]. The combined co-exposure of pesticides myclobutanil (MYC) and thiamethoxam (THI) showed altered genetic expression involved in apoptosis, immunity, antioxidation, and endocrine system in zebrafish [48]. Likewise, the binary mixtures of pesticides tested on honey bees (*apis mellifera*), wild bees (*bombus spp.*), and solitary bees (*Osmia spp.*) species revealed compromised cytpchrome P450 (CYP) expression levels [49]. In zebrafish, exposure to binary mixtures of sulfonamides (SAs) antibiotics caused high developmental toxicity and significantly disturbed the detoxification pathway in comparison to individual SAs [50]. The co-exposure of cadmium and aflatoxin B_1 (AFB_1) in Kunming mice revealed the severely affected organs were the blood, kidney, and liver [51]. Furthermore, the chronic exposure to PPCP chemicals caused gut and liver toxicity in adult zebrafish [52]. Long-term exposure to PPCP mixtures even at low concentrations also affected the normal metabolic functions in both female and male zebrafish. Similarly, the combined exposure of BPA and polystyrene nanoplastics on *Lctalurus punctatus* significantly elevated the MDA, SOD, and CAT levels in the liver and intestine [53]. In addition, the joint exposure of MNPs and various organic and inorganic pollutants impairs normal growth, oxidative stress, and reproduction in aquatic organisms [54].

19.2.2 In Vitro Models

Previously, numerous studies have been conducted regarding the in vitro toxicity potential of chemical mixtures [55–57]. For instance, the joint in vitro cytotoxicity was evaluated after co-exposure of PbO and CuO nanoparticles using human

Table 19.1 Combined toxicological effects of the legacy and emerging contaminants in in vivo and in vitro models.

Compound name	Mixture	Mixture concentration/ratios	Animal model	Observation	Ref.
Phthalates, metal	DBP, DEHP, Cu	Cu 50 μg/L, 100 μg/L + DBP 50 μg/L, 500 μg/L; Cu 50 μg/L, 100 μg/L, + DEHP 50 μg/L, 500 μg/L	*Carassius auratus*	Antioxidant response, SOD, CAT, MDA, GSH	[18]
Phthalates, PAH	DBP, BaP	250 mg/kg DBP, 5 mg/kg BaP or 250 mg/kg DBP + 5 mg/kg BaP	Adult male rats	SOD, GSH-Px, CAT, and MDA	[19]
Phthalates	DEHP, DEP	DEHP 1, 10 μM + DEP 1, 10 μM	*Caenorhabditis elegans*	Gene expression, lipid staining, fecundity assay	[20]
Pharmaceuticals (oral contraceptives)	EE2, NOR	EE2 (5.88 μg/L) + NOR (94.12 μg/L)	*Daphnia magna*	Reproductive effects	[21]
Phthalates, PCBs	DEHP, Aroclor 1254	DEHP (15 mg/kg bw/day), Aroclor 1254 (7.5 mg/kg bw/day)	Adult rats	Metabolic responses	[22]
Phthalates, GMS	DEHP, DEP, DMP, BBP, DEP, DNOP, Glycerin monostearate (GMS)	DMP:DEP:DBP:BBP:DEHP:DNOP: GMS = 1:1:1:1:1:1. The exposure levels were 1.6-, 16-, 160 mg/kg/day	Adult rats	Reproductive system, explore the mechanisms	[23]
Synthetic Estrogen EE2, ZM	Ethynylestradiol (EE2) Antiestrogen ZM	50 ng EE2 with 100 ng ZM/L	*Pimephales promelas* gonads	Individual treatment–related expressions increase/decrease	[24]
Plasticizers	DEHP, DBP, ATBC	50 μg/L (DEP, DEHP, ATBC)	Zebrafish (*Danio rerio*)	Reproductive toxicity	[25]
Phthalate, parabens	PP, DEHP, BP, TRI, TBBPA	0.1 mg PP, 0.1 mg DEHP, 0.1 mg BP, 0.1 mg triclosan, 0.1 mg TBBPA,	Adult rats	Influences of multiple EDCs, using ERCs	[26]
Antibiotics	CBZ, F, PHO, SMX, TMP,	CBZ (178.0 ng/L), FA (70.3 ng/L), PHO (3.18 ng/L), SMX (53.3 ng/L), TMP (15.7 ng/L)	Zebrafish (*Danio rerio*)	Liver cytohistological parameters, in zebrafish	[27]
Phthalates	BBP, DBP, DEHP, DIDP, DINP, DNOP	1.5, 0.06, and 0.01ppm for 1:1 mixture	*Oryzias melastigma*, *Danio rerio*	Developmental toxicity and EDR estrogenic of phthalates	[28]
Phthalates, metal	DBP, DEHP, Cu	DBP and Cu were 0.25, 0.5, 0.625, 0.75, 0.875 and a (a was 390 mg/L for DBP) and 7 mg/L for Cu(II)	*Daphnia magna*, photobacterium phosphoreum	Combined effects of Cu and PAEs	[29]
Plasticizers	BPA, DBP	BPA 285.4 mg/L + DBP 285.4 mg/L	Male rats	Combined subchronic toxicity	[30]
Antibiotics	CBZ, F, PHO, SMX, TMP	CBZ (178.0 ng/L), FA (70.3 ng/L), PHO (3.18 ng/L), SMX (53.3 ng/L) TMP (15.7 ng/L)	Zebrafish (*Danio rerio*)	The toxicity potential of mixtures	[31]

(Continued)

Table 19.1 (Continued)

Compound name	Mixture	Mixture concentration/ratios	Animal model	Observation	Ref.
Phthalates, metal	DBP, As, MBP	As 10 mg/L + DBP 50 mg/kg, As 10 mg/L, MBP 50 mg/kg, DBP 50 mg/kg, MBP 50 mg/kg, As 10 mg/L, DBP 50 mg/kg, MBP 50 mg/kg	Rats	Spatial cognition associated with hippocampal apoptosis	[32]
Plasticizers	BPA, DBP	10 mg/L BPA + 50 mg/L DBP (500mL each)	Zebrafish embryos *(Danio rerio)*	Molecular action at proteome level	[15]
Antibiotics	CBZ, DIC, EE2, MET	CBZ (0.50 μg/L), DIC (0.36 μg/L), EE2 (0.10 μg/L), MET (1.20 μg/L)	*Daphnia magna*	The influence of pharmaceutical mixtures over six generations	[33]
Phthalates, metal	Pb, DBP	Pb (0,0.01 mg/L,1 mg/L), DBP (0,0.005 mg/L, 0.5 mg/L)	Zebrafish embryos (*Danio rerio*)	Neurological and apoptosis-related gene expressions	[32]
Phthalates	DBP, DEHP	DBP (1.0 g/kg), DEHP (1.7 g/kg)	Female rats	Cytochrome P540 enzyme activity	[34]
Antiandrogen	Cyproterone, 17β-trenbolone	TB + CAL (500 ng/L TB + 20 μg/L CA) and TB + CAH (500 ng/L TB + 200 μg/L CA	Pimephales promelas	H-NMR metabolic profile of male urine samples	[35]
Phthalates	DBP, DEHP	DEHP (1.6 g/kg), DBP (0.9 g/kg)	Male rats	Spermatogenesis of male mice	[36]
Phthalates	DBP, DEHP	DBP (1.0 g/kg), DEHP (1.7 g/kg)	Female rats	SOD, GSH, and GSH-Px in the serum	[34]
Phthalates, antibiotics	CYP, DEHP	DEHP (500 mg/kg) + CYP (80 mg/kg)	Male rats	Combined toxic effects on adult male rats	[37]
Antibiotics	Caf, Ibu, Cbz, Tmx	Low mix: 0.05 μM of Caf, Ibu, Cbz and 0.003 μM of Tmx, and high mix: 5 μM of Caf, Ibu, Cbz and 0.3 μM of Tmx	Zebrafish	Gene expression and enzymatic activity	[38]
Plasticizers	BPA, DEHP, DBP	BPA (1%, 50 mg/kg/day), DEHP (0.025%, 750 mg/kg/day), and DBP (0.8%, 66 mg/kg/day)	Rats	To examine the mixture effects on epigenetic transgenerational inheritance of adult rats	[39]
Plasticizers+personal care products	BPA, BP, DBP, DEHP, 4-MBC, OMC, DDE, epoxiconazole, linuron, prochloraz, procymidone, vinclozolin, epoxiconazole, mancozeb, prochloraz, procymidone, tebuconazole	BPA (0.005), BP (0.26), DBP (0.030), DEHP (0.043), 4-MBC (0.19), OMC (0.34), DDE (0.003), epoxiconazole (0.025), linuron (0.002), prochloraz (0.031), procymidone (0.044), vinclozolin (0.026), epoxiconazole (0.09), mancozeb (0.06), prochloraz (0.18), procymidone (0.35), and tebuconazole (0.32)	H295R cell-line	Investigated the pros and cons of the CA, IA, and GCA models and measured effects of single chemicals and mixtures on steroid synthesis in H295R cells.	[40]

DBP: di-*n*-butyl phthalate; DEHP: di-2-ethylhexyl phthalate; PAH: polycyclic aromatic hydrocarbon; BaP: benzo[a]pyrene; DEP: diethyl phthalate; NOR: norethindrone; BBP: butyl benzyl phthalate; DNOP: di-*n*-octyl phthalate; ATBC: acetyl tributyl citrate; CBZ: carbamazepine; PHO: propanolol; SMX: sulfamethoxazole; TMP: trimethoprim; 4-MBC: 4-methylbenzylidene camphor; OMC: octylmethoxycinnamate; SOD: superoxide dismutase; CAT: catalase; GCA: generalized concentration addition; IA: independent action.

fibroblasts cell line. There was a significant decrease in cellular proliferation, ATP content, and cellular dehydrogenase activity [56]. It is important to mention that the mixtures of BPA and its derivatives had significant effects on androgen metabolism in HeLa9903 cells and human breast MCF-7 cells during in vitro testing [58]. The joint binary exposure of methyl mercury with copper, nickel, and cadmium on HepG2 cells showed prominent cytotoxicity with the modification of intracellular proteins [57]. Similarly, the co-exposure of mycotoxin and cadmium in human L-02 liver cells revealed synergistic cytotoxicity [59]. POP mixtures, primarily, perfluorinated alkylated acids (PFAA), disturb the translocation and transactivation of the androgen receptor in vitro [55]. The potential embryonic toxicity was determined after co-exposure to BPA and genistein (GEN) via an embryonic stem cell test model. It was found that both chemicals showed an additive effect and affected the embryonic development that led to the birth defect and behavioral alterations [60]. Similarly, the co-exposure of anthraquinones and their derivatives exhibited significant hepatotoxicity in two human liver cell lines HepG2/C3A and HuH-7 [61]. The PAE binary mixtures also revealed synergistic toxicity on the inhibition of *Acinetobacter sp. Tox2* in vitro [62].

19.3 Analytical Methods

There are various analytical techniques available to quantify combined chemicals exposed to various species. These methods include gas chromatography–mass spectrometry (GC-MS), liquid chromatography–mass spectrometry (LC-MS), flow cytometry, inductively coupled plasma mass spectrometry (ICP-MS), Fourier transform infrared spectroscopy (FTIR), and Raman spectroscopy. However, each technique is used for specific chemical detection. Similarly, for the emerging pollutants detection, there are various analytical techniques available, which are discussed in Figure 19.1. Furthermore, toxicologists used numerous approaches to reveal the joint toxicological effects of legacy and emerging pollutants, such as transcriptomics, proteomics, and metabolomics via nuclear magnetic resonance spectroscopy (NMR) or mass spectroscopy.

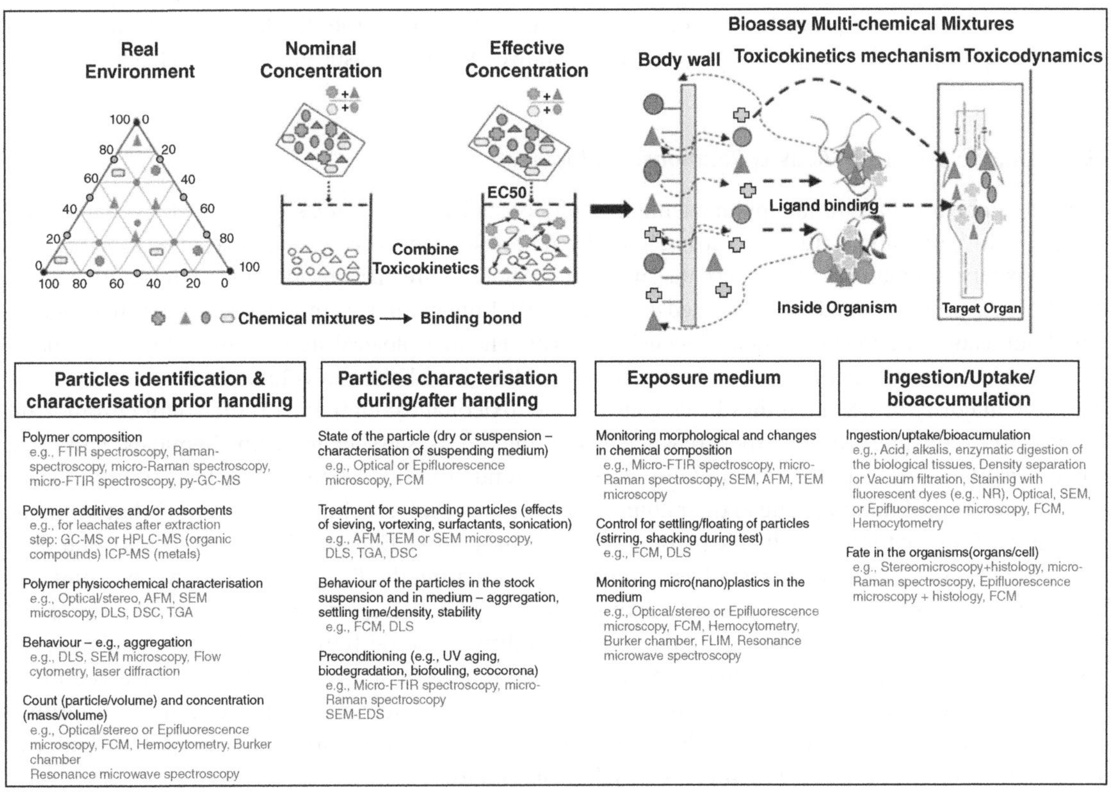

Figure 19.1 Analytical techniques applied in the quantification of joint emerging pollutants ecotoxicity studies. *Source:* Adapted from Refs. [13, 63].

19.4 Mathematical Models

19.4.1 Dose–Response Predictive Models (CA, IA, and CI)

There are several modeling techniques available to investigate the chemical interactions in samples. The modeling approach for chemical interactions in samples must be chosen with careful consideration of various factors. These factors include exposure models, input parameters, and the volume and patterns of chemical mixtures. Concentration addition (CA), independent action (IA), and combination index (CI) are the dose–response predictive models used to determine chemical mixture toxicity. The CA model is efficient when dealing with compounds that have similar mechanisms, provided they react at the same target site or have a similar mode of action (MoA) [16]. This concept was introduced by Loewe in 1926 [16]. Using this equation, the CA model can be calculated as $\frac{cA}{ECxA}+\frac{cB}{ECxB}=1$ [5, 18]. Contrarily, in 1939, Bliss proposed the IA model hypothesis for chemical mixtures with dissimilar MoA or where the chemical mixtures react with multiple target sites [62]. The equation used for the IA model is: $C_{AB} = C_A + C_B - C_{AB}$ [40]. In the twentieth century, these models were known as the "classic models" because they provided a quantitative method for comprehending the nature of interactions among mixtures. In addition, these models are still widely used in mixture toxicity studies [18, 62, 64]. It has been noted through previous research that the CA model is able to provide more precise estimations for combined effects only if the tested compounds share similar molecular mechanisms or MoA [5, 65, 66]. In contrast, a study reported that the CA and IA models have low efficacy in dealing with mixtures with high potency [40]. In short, choosing the right model for predicting chemical mixture effects is crucial for accurate and applicable results.

Chou's CI model, presented in 1976, exhibits a clear relationship between dose–effect mechanisms [67]. The simulation of synergism and antagonism at different effect levels and concentrations of chemicals in a mixture is automated by CompuSyn Software [41, 50]. The PAE binary mixtures showed antagonistic and additive effects on in vitro *Acinetobacter sp. Tox2* using CI and IA models [62]. Conversely, the sulfonamide binary mixtures revealed synergistic effects simulated by CI and IA models on the in vitro *Acinetobacter sp. Tox2* [50]. The IA model used to assess the toxicity of a 17-ethinylestradiol (EE2) and salicylic acid (SA) mixture on *Mytilus galloprovincialis* mussels revealed greater synergistic effects [68]. Likewise, the BPA+DBP mixtures have been shown to have additive or synergistic effects on male rats, increasing genetic levels of thyroid and progesterone hormone receptors (PRs) [30]. From the literature, it can be inferred that most of the mixtures displayed synergistic toxicity behavior followed by antagonistic and additive effects.

19.4.2 Quantitative Structure–Activity Relationship Models

There is scope for applying high-throughput modeling techniques such as QSARs to assess the toxicity of a wider range of mixtures, both additive and nonadditive. Moreover, it has been prevalent across various industrial sectors for simulating a broad range of toxicity endpoints and bridging data gaps [69]. Predictions are made by identifying the relationship between measurable chemical properties and the concerned endpoints [70]. Konemann [71] reported that mixture toxicity could be predicted without using experimental toxicity data using QSARs. The toxicological studies used a variety of species; however, in QSAR models, the majority focused on bacterial and cell line–based bioassays (Table 19.2).

Previously, the majority of QSAR studies investigated acute toxicological endpoints rather than chronic effects [13, 54]. It is important to consider chronic exposure when evaluating the risks associated with chemicals. This is especially crucial when organisms are exposed to pollutants over an extended period of time. Limited research exists on the chronic effects of chemical mixtures due to their complexity and cost [86]. It is interesting to note that utilizing molecular docking as a QSAR tool is an invaluable method for accurately predicting a multitude of toxicity end points [70]. Moreover, it promotes the use of mechanism-based models by simulating the E_{binding} values between chemical (ligand) interactions and receptors [87]. Specifically with mixture descriptors, which are derived by summing the E_{binding} values pertaining to the individual components using the integral additive method [69]. A molecular docking-based QSAR model was applied by Hamid et al. [87], where the chronic reproductive toxicity of PPCPs was discussed in zebrafish. The docking technique was also used to determine the binary mixture toxicological effects using quorum sensing inhibitors (QSIs) [73]. In conclusion, QSAR-based techniques are important to simulate mixture toxicity trends. However, available literature is limited to drawing conclusions (Figure 19.2).

Table 19.2 An overview of the major QSAR characteristics used in joint toxicity studies.

Chemicals	Mixture type	Taxa/model system	Molecular descriptor	Modeling technique	Ref.
Priority pollutants	Binary	Cell-line	Molecular structure	Regression analysis	[72]
Pharmaceuticals and biocides	Binary, ternary, and quaternary	Bacteria	Molecular structure	Regression analysis	[12]
Pharmaceuticals	Binary	Zebrafish	Integral additive	Molecular docking	[52]
Pharmaceuticals	Binary	Bacteria	Integral additive	Molecular docking	[73]
Industrial	Binary	Bacteria and algae	Integral additive	Regression analysis	[74]
Biocides	Supra-quinary	Bacteria	Structural similarity	Machine learning, CA, and IA	[75]
Pharmaceuticals	Binary	Bacteria	Integral additive	Machine learning	[76]
Priority pollutants	Binary	Zebrafish	Integral additive	Molecular docking	[62]
Industrial	Binary	Bacteria	Single-variable component	Regression analysis	[77]
Pharmaceuticals	Binary	Bacteria	Distribution coefficient	Regression analysis	[78]
Industrial	Binary	Bacteria	Integral nonadditive	Regression analysis	[79]
Biocides	Supra-quinary	Algae	Structural similarity	CA and IA	[80]
Industrial	Binary, ternary, and quaternary	Amphibian	Integral additive	Regression analysis	[81]
Industrial	Binary	Bacteria	Integral nonadditive	Regression analysis	[82]
Industrial	Binary	Bacteria	Quantum chemical	Regression analysis	[83]
Priority pollutants	Binary	Not stated	Molecular docking and physicochemical	Assumed regression	[84]
Biocides	Binary, ternary, quaternary, and quinary	Embryos	Physicochemical	Regression analysis	[85]

QSAR: quantitative structure–activity relationship; CA: concentration addition; IA: independent action.

19.5 Challenges and Future Prospects Associated with Combined Toxicity Assessment

Adding mixture toxicity to regulatory frameworks is a highly intricate and challenging task that demands considerable expertise. However, there are some cons, gaps, and uncertainties to consider when addressing the mixture effects. Choosing mixture components is a primary concern for experimental design. When designing experiments, it is essential to consider factors such as cost, time, and accuracy to accurately represent the environment being studied. At the same time, the exposure levels should be environmentally relevant. Previously, even at low exposure levels, significant toxicological effects could be observed [88]. Research on mixture transformation products is limited, including metabolites from photoaging and biodegradation as well as parent compounds.

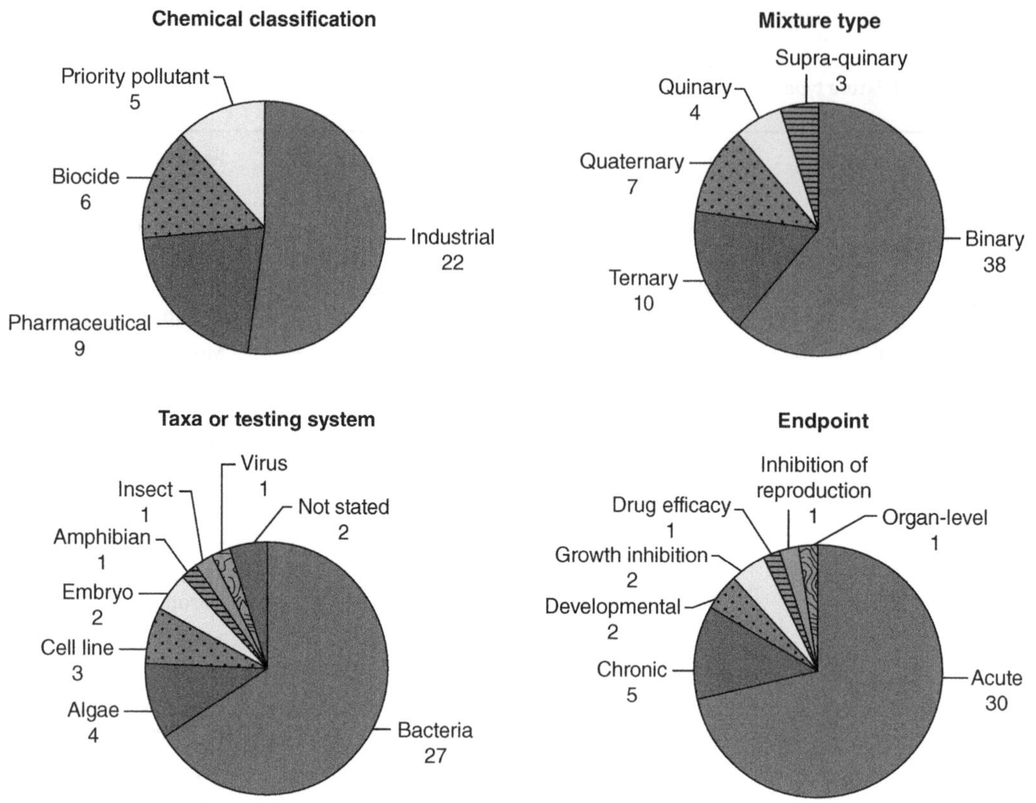

Figure 19.2 An overview of the previously reported quantitative structure–activity relationship (QSAR) studies. *Source:* Reproduced with permission from Belfield et al. [69]/with permission of Elsevier.

It is important to note that the molecular MoA can vary for different mixtures, resulting in a wide range of effects. While acute effects have been thoroughly studied, the chronic effects of mixture toxicity have yet to be fully explored. Developing predictive methods to quantify interspecies toxicogenetic differences is crucial to advancing the risk assessment process. In order to comprehend metabolic pathways and biomarkers, a combination of knowledge from various fields such as computational biology, genetics, molecular biology, toxicology, and metabolomics is required. There is limited literature available on the transgenerational mixture effects of EDCs, so more research is needed in this area. Finally, it is important to carefully evaluate the limitations of predictive models such as CI, IA, and CI in order to accurately determine toxicological trends, despite their significance.

In silico techniques also play a crucial role when it comes to assessing the toxicity of mixtures. Nonetheless, it is essential to acknowledge the various challenges and gaps that currently exist in this field. QSAR model–based molecular docking is a novel approach that has been widely used in various studies [13, 54]. In order to make advancements in this field, it is crucial to enhance our understanding of co-exposure scenarios, component interactions, and grouping criteria. In order to ensure safe exposure to mixtures, it is necessary to implement strategic transdisciplinary initiatives and utilize in silico approaches for accurate safety assessment. It is imperative to acknowledge that the present QSAR workflows for scrutinizing mixtures are insufficient.

Addressing concerns related to combined toxicity may prove to be a daunting task, but it is a crucial responsibility that demands utmost attention. While machine learning has become a popular method for in silico prediction, it is important to recognize that it has its limitations. Developing reliable models often requires a significant amount of data. Interpreting results from QSAR workflows analyzing mixtures can be challenging due to overfitting and the black box nature of machine learning. These factors can limit the ability to derive mechanistic knowledge.Experts are discovering innovative methods to reveal the harmful genetic effects of emerging pollutants. It is crucial for regulatory bodies to address the toxicological concerns related to various chemicals in a thorough manner.

19.6 Conclusion

This chapter was aimed at determining the joint chemical effects on various species both in vivo and in vitro to elucidate the current challenges, problems, and future prospects of applying experimental, analytical, and predictive models used in mixture toxicology. Further investigation is necessary to comprehend the impact of chemicals on reproduction, especially in nonmammalian creatures, in order to safeguard the environment and its inhabitants. Fixed-ratio designs are effective for mixture studies, minimizing experiments while maximizing the effect concentration range. The majority of the legacy and emerging chemical mixtures affected the androgen, estrogen, and thyroid receptors, leading to oxidative stress, reproductive deformities, a disturbed estrogen cycle, and hepatotoxicity. In addition to adverse effects on hormones and the environment, the combination of chemicals can also lead to multigenerational epigenetic effects. Studies have shown that DNA methylation has reduced progenies in both the F3 and F4 generations. Mathematical models, like CA and IA models, are widely used to predict and determine the joint toxic effects of the mixtures in vitro. Different species were studied to determine toxicity, although the models that relied on QSAR mostly used bioassays based on bacteria and cell lines. Finally, more extensive studies are necessary to incorporate innovative machine learning–based in silico methods to reveal the negative health impacts of the chemical mixtures.

Acknowledgments

This study was funded by Fund for International Young Scientists, National Natural Science Foundation of China (42150410389).

Declaration of Competing Interest

The authors declare that they have no known competing financial interests or personal relationships that could have appeared to influence the work reported in this paper.

References

1 Yang RSH. 1.24-Toxicology and risk assessment of chemical mixtures and multiple stressors. In: McQueen CA (Ed.), Comprehensive Toxicology (Third Edition). Elsevier, Oxford, 2018, pp. 489–518.
2 Altenburger R, Scholz S, Schmitt-Jansen M, Busch W, Escher BI. Mixture toxicity revisited from a toxicogenomic perspective. Environ Sci Technol. 2012, 46: 2508–2522.
3 Cleuvers M. Mixture toxicity of the anti-inflammatory drugs diclofenac, ibuprofen, naproxen, and acetylsalicylic acid. Ecotoxicol Environ Saf. 2004, 59: 309–315.
4 Liu L, Wu W, Zhang J, Lv P, Xu L, Yan Y. Progress of research on the toxicology of antibiotic pollution in aquatic organisms. Acta Ecologica Sinica. 2018, 38: 36–41.
5 Cedergreen N. Quantifying synergy: a systematic review of mixture toxicity studies within environmental toxicology. PLoS One. 2014, 9: e96580.
6 Brosche S, Backhaus T. Toxicity of five protein synthesis inhibiting antibiotics and their mixture to limnic bacterial communities. Aquat Toxicol. 2010, 99: 457–465.
7 Backhaus T, Scholze M, Grimme LH. The single substance and mixture toxicity of quinolones to the bioluminescent bacterium Vibrio fischeri. Aquat Toxicol. 2000, 49: 49–61.
8 Ding K, Lu L, Wang J, Wang J, Zhou M, Zheng C, Liu J, Zhang C, Zhuang S. In vitro and in silico investigations of the binary-mixture toxicity of phthalate esters and cadmium (II) to Vibrio qinghaiensis sp.-Q67. Sci Total Environ. 2017, 580: 1078–1084.
9 Pal A, Gin KY-H, Lin AY-C, Reinhard M. Impacts of emerging organic contaminants on freshwater resources: review of recent occurrences, sources, fate and effects. Sci Total Environ. 2010, 408: 6062–6069.

10 Christensen AM, Ingerslev F, Baun A. Ecotoxicity of mixtures of antibiotics used in aquacultures. Environ Toxicol Chem. 2006, 25: 2208–2215.

11 Caliman FA, Gavrilescu M. Pharmaceuticals, personal care products and endocrine disrupting agents in the environment – a review. CLEAN Soil Air Water. 2009, 37: 277–303.

12 Qin L-T, Chen Y-H, Zhang X, Mo L-Y, Zeng H-H, Liang Y-P. QSAR prediction of additive and non-additive mixture toxicities of antibiotics and pesticide. Chemosphere. 2018, 198: 122–129.

13 Hamid N, Junaid M, Pei D-S. Combined toxicity of endocrine-disrupting chemicals: a review. Ecotoxicol Environ Saf. 2021, 215: 112136.

14 Escher BI, Fenner K. Recent advances in environmental risk assessment of transformation products. Environ Sci Technol. 2011, 45: 3835–3847.

15 Dong X, Qiu X, Meng S, Xu H, Wu X, Yang M. Proteomic profile and toxicity pathway analysis in zebrafish embryos exposed to bisphenol A and di-n-butyl phthalate at environmentally relevant levels. Chemosphere. 2018, 193: 313–320.

16 Cedergreen N, Christensen AM, Kamper A, Kudsk P, Mathiassen SK, Streibig JC, Sørensen H. A review of independent action compared to concentration addition as reference models for mixtures of compounds with different molecular target site. Environ Toxicol Chem. 2008, 27: 1621–1632.

17 Vasquez MI, Lambrianides A, Schneider M, Kümmerer K, Fatta-Kassinos D. Environmental side effects of pharmaceutical cocktails: what we know and what we should know. J Hazard Mater. 2014, 279: 169–189.

18 Huang B, Feng M, Li D, Yang Y. Antagonistic joint toxicity assessment of two current-use phthalates with waterborne copper in liver of Carassius auratus using biochemical biomarkers. Ecotoxicol Environ Saf. 2015, 116: 107–112.

19 Zhang Z, Hu Y, Zhao L, Li J, Bai H, Zhu D, Hu J. Estrogen agonist/antagonist properties of dibenzyl phthalate (DBzP) based on in vitro and in vivo assays. Toxicol Lett. 2011, 207: 7–11.

20 Pradhan A, Olsson PE, Jass J. Di(2-ethylhexyl) phthalate and diethyl phthalate disrupt lipid metabolism, reduce fecundity and shortens lifespan of Caenorhabditis elegans. Chemosphere. 2018, 190: 375–382.

21 Goto T, Hiromi J. Toxicity of 17α-ethynylestradiol and norethindrone, constituents of an oral contraceptive pill to the swimming and reproduction of cladoceran Daphnia magna, with special reference to their synergetic effect. Mar Pollut Bull. 2003, 47: 139–142.

22 Zhang J, Yan L, Tian M, Huang Q, Peng S, Dong S, Shen H. The metabonomics of combined dietary exposure to phthalates and polychlorinated biphenyls in mice. J Pharm Biomed Anal. 2012, 66: 287–297.

23 Gao H-T, Xu R, Cao W-X, Di Q-N, Li R-X, Lu L, Xu Q, Yu S-Q. Combined effects of simultaneous exposure to six phthalates and emulsifier glycerol monosterate on male reproductive system in rats. Toxicol Appl Pharmacol. 2018, 341: 87–97.

24 Garcia-Reyero N, Kroll KJ, Liu L, Orlando EF, Watanabe KH, Sepulveda MS, Villeneuve DL, Perkins EJ, Ankley GT, Denslow ND. Gene expression responses in male fathead minnows exposed to binary mixtures of an estrogen and antiestrogen. BMC Genomics. 2009, 10: 308.

25 Gao H-T, Xu R, Cao W-X, Qian L-L, Wang M, Lu L, Xu Q, Yu S-Q. Effects of six priority controlled phthalate esters with long-term low-dose integrated exposure on male reproductive toxicity in rats. Food Chem Toxicol. 2017, 101: 94–104.

26 Pollock T, Weaver RE, Ghasemi R, deCatanzaro D. A mixture of five endocrine-disrupting chemicals modulates concentrations of bisphenol A and estradiol in mice. Chemosphere. 2018, 193: 321–328.

27 Madureira TV, Rocha MJ, Cruzeiro C, Rodrigues I, Monteiro RAF, Rocha E. The toxicity potential of pharmaceuticals found in the Douro River estuary (Portugal): evaluation of impacts on fish liver, by histopathology, stereology, vitellogenin and CYP1A immunohistochemistry, after sub-acute exposures of the zebrafish model. Environ Toxicol Pharmacol. 2012, 34: 34–45.

28 Chen X, Xu S, Tan T, Lee ST, Cheng SH, Lee FW, Xu SJ, Ho KC. Toxicity and estrogenic endocrine disrupting activity of phthalates and their mixtures. Int J Environ Res Public Health. 2014, 11: 3156–3168.

29 Huang B, Li D, Yang Y. Joint toxicity of two phthalates with waterborne copper to daphnia magna and photobacterium phosphoreum. Bull Environ Contam Toxicol. 2016, 97: 380–386.

30 Zhang WZ, Yong L, Jia XD, Li N, Fan YX. Combined subchronic toxicity of bisphenol A and dibutyl phthalate on male rats. Biomed Environ Sci. 2013, 26: 63–69.

31 Madureira TV, Cruzeiro C, Rocha MJ, Rocha E. The toxicity potential of pharmaceuticals found in the Douro River estuary (Portugal)—Experimental assessment using a zebrafish embryo test. Environ Toxicol Pharmacol. 2011, 32: 212–217.

32 Mao G, Zhou Z, Chen Y, Wang W, Wu X, Feng W, Cobbina SJ, Huang J, Zhang Z, Xu H, Yang L, Wu X. Neurological toxicity of individual and mixtures of low dose arsenic, mono and di (n-butyl) phthalates on sub-chronic exposure to mice. Biol Trace Elem Res. 2016, 170: 183–193.

33 Dietrich S, Ploessl F, Bracher F, Laforsch C. Single and combined toxicity of pharmaceuticals at environmentally relevant concentrations in Daphnia magna – a multigenerational study. Chemosphere. 2010, 79: 60–66.

34 Ling L, Tian X, Song Q. Combined toxic effects of di-n-butyl phthalate (DBP) and di-2-ethylhexyl phthalate (DEHP) on reproductive ability in female rats. J Environ Health. 2010, 27: 857–860.

35 Collette TW, Teng Q, Jensen KM, Kahl MD, Makynen EA, Durhan EJ, Villeneuve DL, Martinovic-Weigelt D, Ankley GT, Ekman DR. Impacts of an anti-androgen and an androgen/anti-androgen mixture on the metabolite profile of male fathead minnow urine. Environ Sci Technol. 2010, 44: 6881–6886.

36 Hao Y, Zheng G, Li Q, Xu H, Zhang Y, Yan L. Combined toxic effects of DBP and DEHP on sperm in male mice. In: W Du (Eds.), Informatics and Management Science. Springer, London, 2013, pp. 729–734.

37 Li X, Cai D. [Single and combined toxic effects of di-2-ethylhexyl phthalate and cypermethrin on fertility and development in the the prepubertal male rats]. Wei sheng yan jiu. 2012, 41: 710–716.

38 Aguirre-Martínez GV, Reinardy HC, Martín-Díaz ML, Henry TB. Response of gene expression in zebrafish exposed to pharmaceutical mixtures: implications for environmental risk. Ecotoxicol Environ Saf. 2017, 142: 471–479.

39 Manikkam M, Tracey R, Guerrero-Bosagna C, Skinner MK. Plastics derived endocrine disruptors (BPA, DEHP and DBP) induce epigenetic transgenerational inheritance of obesity, reproductive disease and sperm epimutations. PLoS One. 2013, 8: e55387.

40 Hadrup N, Taxvig C, Pedersen M, Nellemann C, Hass U, Vinggaard AM. Concentration addition, independent action and generalized concentration addition models for mixture effect prediction of sex hormone synthesis in vitro. PLoS One. 2013, 8: e70490.

41 Chen C, Wang Y, Qian Y, Zhao X, Wang Q. The synergistic toxicity of the multiple chemical mixtures: implications for risk assessment in the terrestrial environment. Environ Int. 2015, 77: 95–105.

42 Saud S, Yang A, Jiang Z, Ning D, Fahad S. New insights in to the environmental behavior and ecological toxicity of microplastics. J Hazard Mater. 2023, 10: 100298.

43 Pandey D, Singh A, Ramanathan A, Kumar M. The combined exposure of microplastics and toxic contaminants in the floodplains of north India: a review. J Environ Manage. 2021, 279: 111557.

44 Yan W, Hamid N, Deng S, Jia P-P, Pei D-S. Individual and combined toxicogenetic effects of microplastics and heavy metals (Cd, Pb, and Zn) perturb gut microbiota homeostasis and gonadal development in marine medaka (Oryzias melastigma). J Hazard Mater. 2020, 397: 122795.

45 Preston EV, Webster TF, Claus Henn B, McClean MD, Gennings C, Oken E, Rifas-Shiman SL, Pearce EN, Calafat AM, Fleisch AF, Sagiv SK. Prenatal exposure to per- and polyfluoroalkyl substances and maternal and neonatal thyroid function in the Project Viva Cohort: a mixtures approach. Environ Int. 2020, 139: 105728.

46 Rider CV, Furr J, Wilson VS, Gray Jr LE. A mixture of seven antiandrogens induces reproductive malformations in rats. Int J Androl. 2008, 31: 249–262.

47 Wang R, Liu N, Huang N, Shu F, Lou Y, Zhang S, Zhu H, Chen C. Combined toxicity assessment of a naturally occurring toxin and a triazole fungicide on different biological processes through toxicogenomic data mining with mixtures. Pestic Biochem Physiol. 2023, 193: 105440.

48 Shen W, Yang G, Guo Q, Lv L, Liu L, Wang X, Lou B, Wang Q, Wang Y. Combined toxicity assessment of myclobutanil and thiamethoxam to zebrafish embryos employing multi-endpoints. Environ Pollut. 2021, 269: 116116.

49 Carnesecchi E, Svendsen C, Lasagni S, Grech A, Quignot N, Amzal B, Toma C, Tosi S, Rortais A, Cortinas-Abrahantes J, Capri E, Kramer N, Benfenati E, Spurgeon D, Guillot G, Dorne JLCM. Investigating combined toxicity of binary mixtures in bees: meta-analysis of laboratory tests, modelling, mechanistic basis and implications for risk assessment. Environ Int. 2019, 133: 105256.

50 Hamid N, Junaid M, Pei D-S. Individual and combined mechanistic toxicity of sulfonamides and their implications for ecological risk assessment in the Three Gorges Reservoir Area (TGRA), China. J Hazard Mater. 2020, 382: 121106.

51 Zhao Q, Yang Z-S, Cao S-J, Chang Y-F, Cao Y-Q, Li J-B, Yao Z-X, Wen Y-P, Huang X-B, Wu R, Yan Q-G, Huang Y, Ma X-P, Han X-F, Wu Y. Acute oral toxicity test and assessment of combined toxicity of cadmium and aflatoxin B1 in kunming mice. Food Chem Toxicol. 2019, 131: 110577.

52 Hamid N, Junaid M, Wang Y, Pu S-Y, Jia P-P, Pei D-S. Chronic exposure to PPCPs mixture at environmentally relevant concentrations (ERCs) altered carbohydrate and lipid metabolism through gut and liver toxicity in zebrafish. Environ Pollut. 2021, 273: 116494.

53 Zheng Q, Liao H, Junaid M, Li Z, Liu S, Gao D, Kong C, Chen X, Pan T, Zheng Y. Combined exposure to polystyrene nanoplastics and bisphenol a induces hepato-and intestinal-toxicity and disturbs gut microbiota in channel catfish (Lctalurus Punctatus). Sci Total Environ. 891: 164319. Available at SSRN 4427917.

54 Junaid M, Liu S, Chen G, Liao H, Wang J. Transgenerational impacts of micro(nano)plastics in the aquatic and terrestrial environment. J Hazard Mater. 2023, 443: 130274.

55 McComb J, Mills IG, Muller M, Berntsen HF, Zimmer KE, Ropstad E, Verhaegen S, Connolly L. Human blood-based exposure levels of persistent organic pollutant (POP) mixtures antagonise androgen receptor transactivation and translocation. Environ Int. 2019, 132: 105083.

56 Bushueva T, Minigalieva I, Panov V, Kuznetsova A, Naumova A, Shur V, Shishkina E, Gurvich V, Privalova L, Katsnelson B. More data on in vitro assessment of comparative and combined toxicity of metal oxide nanoparticles. Food Chem Toxicol. 2019, 133: 110753.

57 Akiyama M, Shinkai Y, Yamakawa H, Kim Y-G, Kumagai Y. Potentiation of methylmercury toxicity by combined metal exposure: In vitro and in vivo models of a restricted metal exposome. Chemosphere. 2022, 299: 134374.

58 Park C, Song H, Choi J, Sim S, Kojima H, Park J, Iida M, Lee Y. The mixture effects of bisphenol derivatives on estrogen receptor and androgen receptor. Environ Pollut. 2020, 260: 114036.

59 Liu X, Yin S, Zhao C, Fan L, Hu H. Glycyrol alleviates the combined toxicity of fumonisin B1 and cadmium in vitro and in vivo. Toxicon. 2021, 200: 165–172.

60 Kong D, Xing L, Liu R, Jiang J, Wang W, Shang L, Wei X, Hao W. Individual and combined developmental toxicity assessment of bisphenol A and genistein using the embryonic stem cell test in vitro. Food Chem Toxicol. 2013, 60: 497–505.

61 Liu Y, Mapa MST, Sprando RL. Liver toxicity of anthraquinones: a combined in vitro cytotoxicity and in silico reverse dosimetry evaluation. Food Chem Toxicol. 2020, 140: 111313.

62 Hamid N, Junaid M, Manzoor R, Jia P-P, Pei D-S. Prioritizing phthalate esters (PAEs) using experimental in vitro/vivo toxicity assays and computational in silico approaches. J Hazard Mater. 2020, 398: 122851.

63 Silva ALP, Silva SAM, Duarte A, Barceló D, Rocha-Santos T. Analytical methodologies used for screening micro(nano) plastics in (eco)toxicity tests. Green Anal Chem. 2022, 3: 100037.

64 González-Pleiter M, Gonzalo S, Rodea-Palomares I, Leganés F, Rosal R, Boltes K, Marco E, Fernández-Piñas F. Toxicity of five antibiotics and their mixtures towards photosynthetic aquatic organisms: implications for environmental risk assessment. Water Res. 2013, 47: 2050–2064.

65 Fang S, Wang D, Zhang X, Long X, Qin M, Lin Z, Liu Y. Similarities and differences in combined toxicity of sulfonamides and other antibiotics towards bacteria for environmental risk assessment. Environ Monit Assess. 2016, 188: 429.

66 Magos L. RSH Yang (ed.). Toxicology of chemical mixtures case studies, mechanisms, and novel approaches. Academic Press, San Diego, 1994;720 pp., $175.0. J Appl Toxicol. 1995, 15: 243–243.

67 De Liguoro M, Di Leva V, Gallina G, Faccio E, Pinto G, Pollio A. Evaluation of the aquatic toxicity of two veterinary sulfonamides using five test organisms. Chemosphere. 2010, 81: 788–793.

68 Cunha M, Silva MG, De Marchi L, Morgado RG, Esteves VI, Meucci V, Battaglia F, Mvm Soares A, Pretti C, Freitas R. Toxic effects of a mixture of pharmaceuticals in Mytilus galloprovincialis: the case of 17α-ethinylestradiol and salicylic acid. Environ Pollut. 2023, 324: 121070.

69 Belfield SJ, Firman JW, Enoch SJ, Madden JC, Erik Tollefsen K, Cronin MTD. A review of quantitative structure-activity relationship modelling approaches to predict the toxicity of mixtures. Comput Toxicol. 2023, 25: 100251.

70 Leedale J, Sharkey KJ, Colley HE, Norton ÁM, Peeney D, Mason CL, Sathish JG, Murdoch C, Sharma P, Webb SD. A combined in vitro/in silico approach to identifying off-target receptor toxicity. iScience. 2018, 4: 84–96.

71 Könemann H. Fish toxicity tests with mixtures of more than two chemicals: a proposal for a quantitative approach and experimental results. Toxicol. 1981, 19: 229–238.

72 Hoover G, Kar S, Guffey S, Leszczynski J, Sepúlveda MS. In vitro and in silico modeling of perfluoroalkyl substances mixture toxicity in an amphibian fibroblast cell line. Chemosphere. 2019, 233: 25–33.

73 Wang D, Shi J, Xiong Y, Hu J, Lin Z, Qiu Y, Cheng J. A QSAR-based mechanistic study on the combined toxicity of antibiotics and quorum sensing inhibitors against Escherichia coli. J Hazard Mater. 2018, 341: 438–447.

74 Chang C, Ou Y, Liu T-C, Lu S-Y, Wang M-K. A quantitative structure–activity relationship approach for assessing toxicity of mixture of organic compounds. SAR QSAR Environ Res. 2016, 27: 441–453.

75 Kim J, Kim S, Schaumann G. Development of QSAR-based two-stage prediction model for estimating mixture toxicity. SAR QSAR Environ Res. 2013, 24: 841–861.

76 Zou X, Zhou X, Lin Z, Deng Z, Yin D. A docking-based receptor library of antibiotics and its novel application in predicting chronic mixture toxicity for environmental risk assessment. Environ Monit Assess. 2013, 185: 4513–4527.

77 Su L, Zhang X, Yuan X, Zhao Y, Zhang D, Qin W. Evaluation of joint toxicity of nitroaromatic compounds and copper to Photobacterium phosphoreum and QSAR analysis. J Hazard Mater. 2012, 241: 450–455.

78 Zou X, Lin Z, Deng Z, Yin D, Zhang Y. The joint effects of sulfonamides and their potentiator on photobacterium phosphoreum: differences between the acute and chronic mixture toxicity mechanisms. Chemosphere. 2012, 86: 30–35.

79 Zhang L, Zhou P-J, Yang F, Wang Z-D. Computer-based QSARs for predicting mixture toxicity of benzene and its derivatives. Chemosphere. 2007, 67: 396–401.

80 Mwense M, Wang X, Buontempo FV, Horan N, Young A, Osborn D. QSAR approach for mixture toxicity prediction using independent latent descriptors and fuzzy membership functions. SAR QSAR Environ Res. 2006, 17: 53–73.

81 Huang H, Wang X, Shao Y, Chen D, Dai X, Wang L. QSAR for prediction of joint toxicity of substituted phenols to tadpoles (Rana japonica). Bull Environ Contam Toxicol. 2003, 71: 1124–1130.

82 Lin Z, Zhong P, Yin K, Wang L, Yu H. Quantification of joint effect for hydrogen bond and development of QSARs for predicting mixture toxicity. Chemosphere. 2003, 52: 1199–1208.

83 Yuan X, Lu G, Zhao J. QSAR study on the joint toxicity of 2, 4-dinitrotoluene with aromatic compounds to Vibrio fischeri. J Environ Sci Health A Tox Hazard Subst Environ Eng. 2002, 37: 573–578.

84 Wang T, Zhou X, Wang D, Yin D, Lin Z. Using molecular docking between organic chemicals and lipid membrane to revise the well known octanol–water partition coefficient of the mixture. Environ Toxicol Pharmacol. 2012, 34: 59–66.

85 Wang H, Li Y, Huang H, Xu X, Wang Y. Toxicity evaluation of single and mixed antifouling biocides using the Strongylocentrotus intermedius sea urchin embryo test. Environ Toxicol Chem. 2011, 30: 692–703.

86 Junaid M, Siddiqui JA, Liu S, Lan R, Abbas Z, Chen G, Wang J. Adverse multigeneration combined impacts of micro (nano) plastics and emerging pollutants in the aquatic environment. Sci Total Environ. 2023, 882: 163679.

87 Hamid N, Junaid M, Manzoor R, Duan J-J, Lv M, Xu N, Pei D-S. Tissue distribution and endocrine disruption effects of chronic exposure to pharmaceuticals and personal care products mixture at environmentally relevant concentrations in zebrafish. Aquat Toxicol. 2022, 242: 106040.

88 You R, Sun H, Yu Y, Lin Z, Qin M, Liu Y. Time-dependent hormesis of chemical mixtures: a case study on sulfa antibiotics and a quorum-sensing inhibitor of Vibrio fischeri. Environ Toxicol Pharmacol. 2016, 41: 45–53.

20

Future Research Perspectives of Combined Toxicology

Tariq Mehmood[a,*], *Anam Ashraf*[b], *Kousar Parveen*[c], *Muhammad Azher Hassan*[d], *Muhammad Fahad Sardar*[e], *Muhammad Faheem*[f], *Licheng Peng*[e], *Shakeel Ahmad*[g], *Tauseef Ahmad*[e], *Predrag Ilić*[h]

[a] *Helmholtz Centre for Environmental Research - UFZ, Department of Environmental Engineering, Permoserstr. 15, D-04318 Leipzig, Germany*
[b] *Landscape Lab, Xishuangbanna Tropical Botanical Garden, Chinese Academy of Sciences, Menglun, Mengla, Yunnan, 666303, China*
[c] *The Department of Environmental Sciences, The Women University Multan, 66000, Pakistan*
[d] *Tianjin Key Lab of Indoor Air Environmental Quality Control, School of Environmental Science and Engineering, Tianjin University, Tianjin 300072, China*
[e] *Key Laboratory of Ecological Prewarning, Protection and Restoration of Bohai Sea, Ministry of Natural Resources, School of Life Sciences, Shandong University, Qingdao, 266237, China*
[f] *Department of Civil Infrastructure and Environmental Engineering, Khalifa University of Science and Technology, 127788 Abu Dhabi, United Arab Emirates*
[g] *School of Environment, Tsinghua University, Beijing 100084, China*
[h] *PSRI Institute for Protection and Ecology of the Republic of Srpska, Vidovdanska 43, Banja Luka, Republic of Srpska, Bosnia and Herzegovina*

*Corresponding author.
E-mail: tariqslamat@buaa.edu.cn

20.1 Introduction

Toxicology, the scientific study of the nature of chemicals and their adverse effects on living organisms, has significantly been transformed over the years. Initially, the focus was primarily on individual substances and their isolated impacts. However, real-world conditions often involve exposure to a complex mixture of chemicals, posing new and intricate challenges still under researchers' active investigation [1, 2]. A cornerstone in this field is the practice of toxicological assessment (TA), a comprehensive method used to evaluate the potential risks and hazards presented by chemicals. When evaluating mixed compounds, often known as "chemical mixtures," the evaluation procedure becomes increasingly difficult. These mixtures can exhibit a range of interactions, including additive, synergistic, or antagonistic effects, that defy straightforward predictions based on the assessment of individual chemicals [3, 4].

Researchers have drawn upon various analytical and modeling techniques to overcome the challenges of chemical mixture toxicology. One common approach integrates environmental fate analysis with epidemiological data and toxic kinetic dynamic models to estimate the internal exposure levels of organisms [5]. This integration allows a more nuanced understanding of how chemical mixtures interact within biological systems. Building upon this, the Concentration Addition Approach has been frequently employed to predict the cumulative toxicity of multiple chemicals on nontarget organisms, substantiating its relevance in contemporary risk assessments [6]. With advancements in computational biology, novel methodologies such as adverse outcome pathway footprinting have emerged. These new techniques offer unique insights into the mechanistic bases of toxicological interactions and broaden the scope of chemical mixture assessment [7].

Other RAMs have also been developed to quantify human health risks associated with environmental chemical mixtures. These include joint effect indices, concentration addition, independent action, two-step prediction, integrated addition model, median-effect equation, hazard index, point of departure index, margin of exposure (MOE), relative potency factor, toxic equivalence quantity, cumulative risk index, and toxic unit summation [8]. Each model has its own advantages and limitations, indicating the need for tailored approaches depending on the context of exposure and chemical composition. Yet, despite these methodological advancements, substantial gaps in our understanding persist. The toxicity of many environmental mixtures remains underexplored, largely due to the complex and often unpredictable

Toxicological Assessment of Combined Chemicals in the Environment, First Edition. Edited by De-Sheng Pei and Yiyun Liu.

synergistic or antagonistic effects between chemical components [9]. These interactions can complicate the assessment process and necessitate ongoing research to refine existing models and develop new evaluative frameworks.

Several key scientific terms are central to the discussions that will unfold in this chapter. "Combined Toxicity" refers to the cumulative harmful effects that arise when an organism is exposed to multiple chemicals simultaneously. "RAMs" are the computational frameworks or algorithms that quantify these potential risks. The term "Accreditation" implies that a particular RAM has been formally validated and is recognized as reliable and effective. In light of emerging challenges, cutting-edge techniques such as "multivariate toxicogenomics (MTG)" are being employed. These advanced methodologies combine biological assays and computational algorithms to delve into the molecular interactions of chemicals [10]. Even with these developments, there are still a lot of unanswered questions, especially regarding the long-term or chronic consequences of exposure to chemical combinations [11].

This chapter comprehensively explores the current state of knowledge in the TA of combined chemical exposures (CCE). It discusses the various RAMs, analyzes their advantages and disadvantages, and looks at the best ways to optimize the legal frameworks that control these assessments. Additionally, this chapter will identify areas where future research could enhance the reliability and precision of these toxicological evaluations. In conclusion, the field of toxicology has evolved from a focus on individual substances to a broader investigation of complex chemical mixtures. Numerous challenges and questions remain, and this chapter serves as a guide to understanding the existing complexities, the tools developed for their evaluation, and the future direction in this vital area of study.

20.2 General Terms of TA

It is essential to delve into specialized terms at the core of understanding TAs in the context of CCE, and some are presented in Table 20.1. These terms serve as the lexicon for this scientific discourse and embody intricate concepts integral to toxicology and risk assessment. Starting with TA, this term signifies a comprehensive scientific approach to identifying the dangers inherent in chemicals or mixtures. The scope of TA is vast, incorporating both observational and computational methods. This range enables researchers to evaluate risks from multiple perspectives, providing a nuanced understanding crucial for effective regulation and policy formulation.

Table 20.1 General terms of toxicological assessment.

Term	Definition
Aggregate exposure	Exposure to the same substance from multiple sources or routes
Combined exposure	Exposure to multiple chemicals by a single route and exposure to multiple chemicals by multiple routes
Combined hazard	Hazard from multiple substances by a single route or multiple routes
Risk from combined exposures	Risk from exposure to multiple substances by a single route or multiple routes
Risk assessment of combined exposures	Risk assessment of exposure to multiple substances by a single route or multiple substances by multiple routes
Cumulative exposure or cumulative risk	Evaluation of chemical exposure to more than one pesticide at a time from a group that shares an identified common mechanism of toxicity
Point of departure (POD)	A selected measure of effect may be a no-observed effect level or a lowest observed (adverse) effect level.
Benchmark dose (BMD10)	Dose or concentration associated with a specified increase in the incidence of an effect (e.g. a benchmark dose [BMD10])
Benchmark dose concentration (BMC10)	Associated with a 10% increase in the incidence of an effect.
Mode of action (MOA)	It is a biologically plausible sequence of key events leading to an observed effect, supported by robust experimental observations and mechanistic data.
Margin of exposure (MOE)	The ratio of the selected measure of effect to the estimated exposure dose or concentration
Substance	Chemical elements and their compounds in the natural state or obtained by any manufacturing process

Term	Definition
Mixture	Group of two or more substances in which they do not react
Intentional mixtures	Manufactured formulations, e.g., commercial mixtures of industrial substances, technical mixtures, and product formulations
Discharge mixtures	Substance combinations that are emitted by a single industrial site, e.g., effluent of a production site
Coincidental mixture	Substances from different sources occur in a medium, such as the combination of substances applied dermally from the use of two or more product formulations.
Environmental mixtures:	Substance combinations in one environmental compartment such as substances found in soil from various exposure sources (application of product formulation, deposition from air, water run-off, etc.).

Next, the combined chemical exposure (CCE) concept introduces complexity into TAs. In real-world scenarios, organisms often encounter multiple chemicals either simultaneously or sequentially. The interactions among these chemicals can lead to a variety of effects, including additive, synergistic, or antagonistic outcomes. Accurately capturing these complexities is vital for both risk assessment and subsequent regulatory actions. Focusing on the risk assessment model (RAM) encompasses a set of validated scientific frameworks used for quantifying the risks associated with chemical exposure. RAM integrates a myriad of data, including but not limited to epidemiological studies, animal testing, and in vitro experiments. The result is a holistic risk profile, a critical tool for public health and environmental protection decision-making.

Advanced methodologies, such as MTG, represent the frontier in toxicological research. MTG employs a blend of genomic information and computational analysis to decode the complex molecular and cellular interactions triggered by chemical mixtures. This approach offers a depth of insight that is often beyond the reach of traditional toxicological methods. The term "Bioaccumulation" addresses the tendency for certain chemicals to accumulate within biological systems over time. This can lead to chronic health issues and poses a significant concern for long-term risk assessment. It serves as a reminder that the impacts of chemical exposure can extend far beyond immediate or acute effects, underscoring the need for comprehensive risk assessments.

The terms synergistic and antagonistic effects (SAEs) elucidate the potential for chemical mixtures to exhibit toxicity that is either greater or less than the sum of individual chemical components. These effects are not static but can vary depending on multiple factors such as dosage, timing of exposure, and the physiological condition of the organism being studied. In summary, a thorough understanding of these general terms is imperative for anyone engaged in studying, assessing, or regulating CCE risks. Equipped with this foundational knowledge, we are better prepared to explore the complex topics discussed in this chapter's subsequent sections.

20.3 Exposure to Combined Chemicals or Mixtures

In both natural and human-made environments, exposure to a mixture is often far more prevalent than exposure to individual chemicals. The dynamics that shape CCE are intricate and influenced by various contributing sources and biological responses. Among the significant contributors to CCE are industrial operations. Emissions from manufacturing facilities, particularly those that deal with chemicals, release complex mixtures into air and water resources. This poses an environmental hazard and elevates the risk levels for human health.

In light of this, Panizzi et al. [6] focus on the ecotoxicological risk assessment of pesticide mixtures, emphasizing using the concentration addition approach to predict multiple toxicity to nontarget organisms. This complements the need for nuanced approaches, especially given the complexities of risk assessment for mixtures. Such complexities often include reliance on multiple assumptions and the potential for interactions among mixture components [12]. The field has seen developments in hazard identification and risk assessment of chemical mixtures, stressing the significance of both empirical and mechanistic approaches [13].

Agricultural activities similarly contribute to CCE, with runoffs carrying a blend of fertilizers and pesticides into water bodies, further complicating the chemical landscape. Concurrently, there is a pressing need for a shift towards assessing mixtures as they occur in the environment and consumer products, employing integrated approaches that account for exposure, toxicokinetics, and biological responses [5]. Household exposures offer another layer to CCE. The concurrent use of various cleaning agents can result in complex chemical interactions, leading to unpredictable health risks. Pharmaceuticals are also major contributors. Residues from medicines, ranging from antidepressants to antibiotics, find their way into natural water and irrigation systems, adding to the existing milieu of chemicals and complicating TA efforts. At this point, it is important to note that toxicological interactions vary depending on exposure order, duration, and route [14]. The effects of mixtures are generally additive or antagonistic, yet their risk assessment for regulatory purposes usually overlooks the "real-life" exposure to multiple substances [15]. The toxicity of inorganic and organic mixtures derived from analytical laboratory waste presents another important dimension in their TA [16].

Biological processes, such as bioaccumulation, are pivotal in the toxicological landscape. This mechanism allows certain chemicals to accumulate in living tissues, leading to potential long-term or chronic health implications. The effects can be synergistic or antagonistic, emphasizing the complexity of assessing CCE's risks. Another layer of complexity comes from metabolic interactions. Enzymatic processes that typically metabolize one chemical can be altered by another, affecting the overall metabolic pathway. One major drawback is the absence of a harmonized approach for conducting a mixture of risk assessments and management across different regulatory sectors [17]. These biochemical complexities are often the focus of MTG studies, which aim to decode the molecular interactions between chemicals and their subsequent effects on metabolic pathways. Understanding the multifaceted nature of CCE is crucial for developing and validating RAM. Moreover, it forms the basis for legislative actions aimed at regulating chemical exposures.

20.3.1 Framework for Risk Assessment of Combined Exposures to Multiple Chemicals

The framework was developed with an evaluation strategy tailored to the specific objective, utilizing a hierarchical approach and allocating resources only as necessary to facilitate informed decision-making. Significant advancements have been achieved by the World Health Organization (WHO) in addressing the issue of chemical mixtures present in both drinking water and source water. This encompasses formulating diverse approaches and pragmatic suggestions to evaluate and control potential risks to human well-being linked to contact with mixtures of chemicals in drinking water and its origins [18].

20.3.1.1 European Chemicals Agency

The European Chemicals Agency (ECHA) has developed guidance sheets intending to facilitate the deliberate utilization of biocidal products. These materials have been developed per the Biocidal Products Regulation to support the cumulative risk assessment of the environment. The present regulation pertains to the allocation of deliberate mixes.

ECHA has prepared a guidance sheet for conducting a risk assessment of exposure to numerous biocidal compounds in combination, in alignment with the WHO and International Programme on Chemical Safety (IPCS) framework on combined exposures [19]. They examine the approach employed in conducting a risk assessment, considering several components, such as compounds with different danger levels and fate-related information.

20.3.1.2 The European Food Safety Authority

The European Food Safety Authority (EFSA) has made efforts to develop hierarchical approaches for evaluating risks related to concurrent exposures. EFSA has put forth a range of techniques aimed at assessing the effects of pesticides on human health, considering both similar and unique modes of action. Furthermore, EFSA has put out several approaches to evaluate the potential harm to human health from concurrent chemical exposure [20, 21]. EFSA emphasized the possible uses of physiologically based models, OMICs, and in silico methods for evaluating the risk associated with exposure to numerous chemicals in combination [20]. Following Nutrition and Allergies (NDA), EFSA [22] is actively endorsing the advancement of technical tools, such as the Monte Carlo risk assessment software, which may be flexibly utilized for extensive cumulative assessment cohorts.

Furthermore, EFSA has made significant advancements in enhancing worldwide communication and has provided suggestions for future undertakings in this domain to establish standardized techniques. This has been accomplished through the organization's facilitation of workshops and dissemination of published reports [23, 24]. EFSA commenced consultations in 2018 about the draft documents on MIXTOX (MIXTOX is an R package that provides toxicologists with a comprehensive framework for curve fitting, mixture investigation methodology, and mixture toxicity prediction) [25]. The provided materials encompass preliminary guidance that presents standardized approaches for evaluating the risks associated with concurrent exposure to various chemicals. This guidance places particular emphasis on assessing human health, animal health, and ecological factors. During the consultation period, EFSA also made available for evaluation of a proposed statement titled "Genotoxicity assessment of chemical mixtures."

20.3.1.3 Joint Research Centre of the European Commission

The committees underlined the significance of screening, wherein filters are employed to aid in the concentration on combinations that may raise concerns. A report has been released by the European Commission's Joint Research Centre (JRC), which offers a thorough examination of methodologies and existing legislation within the EU that are relevant to the assessment of chemicals in various matrices. Furthermore, the report provides a comprehensive analysis of the degree to which existing legislation effectively tackles the potential hazards posed by combinations. Additionally, it summarizes various global initiatives on ecological and human health, as outlined by Kienzler et al. [15].

20.3.1.4 EuroMix

The EuroMix project commenced in May 2015 as a collaborative initiative with multiple partners from the EU. The objective of the project is to formulate a hierarchical approach for test strategy, bioassays, and models that can be employed in subsequent endeavors pertaining to the evaluation of chemical mixes' risk. Numerous endeavors have been undertaken in this domain, one of which is EDCMix Risk, an initiative by the EU that aims to examine the consequences of chemical combinations that interfere with the endocrine system, specifically in relation to children. Another program, developed by Demeneix emphasizes studying the impact of mixes on neurotoxicity and its effects on the developmental processes of children. The Acropolis program is centered around the comprehensive assessment and aggregation of pesticide-related risks.

Bunke et al. [26] presented an alternative way to assess mixtures' danger, which the German Federal Environment Agency suggested. This approach adopts a stratified component-based methodology. This study establishes a correlation between the existing methodology for evaluating the hazards associated with mixtures and the data requirements and evaluation principles outlined in the Registration, Evaluation, Authorization, and Restriction of Chemicals (REACH) regulation. Furthermore, it offers various alternatives for assessing technical combinations within the REACH framework, including priority procedures.

The European Chemical Industry Council launched the construction of a decision tree for mixes. This decision tree is founded on the layered frameworks of the WHO/IPCS and the committees of the European Commission. Likewise, Price et al. [27] outlined it in their publications from 2012. The Health and Environmental Sciences Institute (HESI) has devised a methodology for evaluating cumulative hazards in the twenty-first century as a component of its RISK21 initiative [28]. Furthermore, Solomon et al. [29] have offered guidelines on issue formulation in the evaluation of combined exposure hazards, as outlined by the HESI. The nations of the Americas possess contiguous boundaries. In summary, the complexities surrounding CCE underscore the need for specialized approaches in TA, supported by validated RAM and guided by legislative frameworks. This understanding is foundational for ensuing discussions on risk assessment and legislative considerations, which will be the focus of the subsequent sections of this chapter.

20.4 Risk Exposure Assessment

Accurate risk evaluations related to CCE form a cornerstone in ensuring both public health and environmental safety [30]. The credibility of these assessments hinges on the reliability and effectiveness of RAM. Ensuring the quality and accreditation of RAM involves a multilayered process, encompassing scientific validation and formal recognition by regulatory bodies [31]. The accreditation journey typically initiates with the peer review of the RAM under consideration. Panels of experts in toxicology, statistics, and other relevant fields rigorously scrutinize the model to assess its scientific validity. Among the evaluated aspects is the model's capability to integrate diverse types of data, including epidemiological findings, results from animal testing, and in vitro experiments [32].

Following this, the RAM undergoes a series of validation tests. These involve juxtaposing the model's predictive outputs against observations derived from controlled environmental settings and field studies [33]. The RAM's performance is quantified using statistical metrics like sensitivity, specificity, and predictive values. Moreover, the model's aptitude in accounting for SAEs is assessed, given that CCE often entails such interactive scenarios [34]. Upon successful navigation through these initial steps, the RAM is presented to regulatory agencies for possible accreditation [35]. These agencies may be governmental or international bodies that examine the model against predetermined criteria. Criteria often pertain to the model's robustness, versatility across varied exposure scenarios, and alignment with existing legislative parameters [36].

Accreditation is not a stagnant milestone but rather a dynamic process requiring periodic updates [37]. As ongoing research frequently uncovers new insights into chemical interactions and long-term health effects, RAM must adapt to integrate these findings [38]. MTG studies have been particularly impactful, revealing previously unknown molecular pathways that CCE can affect [39]. Challenges in RAM accreditation are numerous, ranging from the absence of standardized validation protocols across different agencies to the rapid advancements in toxicological science that can render existing models obsolete [40]. The complexity is further heightened by the sheer diversity of chemical properties and biological responses involved, making a universal model an ambitious endeavor [41].

In summary, the accreditation of RAM is an intricate yet indispensable component in CCE risk assessments [42]. The complex process involves layers of scientific validation, regulatory scrutiny, and continuous updates, each with unique challenges. This foundational understanding will be crucial for the subsequent segments of this chapter, focusing on legislative frameworks and future perspectives in toxicology related to CCE.

20.4.1 Risk Assessment: Potential of Small Concentrations of Different Substance

A thorough legislative framework has been established to ensure the maintenance of safe limits for human, animal, and plant exposure to chemical compounds. Given adherence to relevant regulatory obligations, it is imperative to guarantee that notwithstanding exposure to minuscule amounts of various chemicals, encompassing those derived from fauna and flora, none of these compounds surpass their individual allowed thresholds. To adequately respond to the issue mentioned above, it is imperative to thoroughly examine the potential negative impacts that may arise from the amalgamation of chemical compounds, specifically concerning human health and ecological systems. The origins and paths of these chemicals can vary, and their quantities are extremely low. Under specific circumstances, the chemical constituents present in a combination will demonstrate a synergistic phenomenon, leading to an alteration in the overall hazardous properties of the mixture. Chemicals possessing analogous mechanisms of action can manifest synergistic effects upon the combination, yielding results surpassing their individual consequences' cumulative impact.

In relation to substances characterized by unique mechanisms of action, it is important to note that the available evidence is inadequate to support the notion that exposure to a combination of these substances presents a health hazard. This holds if the individual chemicals are present at or below their respective thresholds where no adverse effects are observed. The inquiry into the potential health risks associated with human exposure to chemical combinations is now being studied. It might be postulated that the degree of apprehension for mixtures of substances with divergent modes of action is negligible if the intended level of safeguarding is attained for each constituent component.

20.4.2 Data Needs, Limitations, and Uncertainties

The classification of strata is contingent upon the accessibility of data, the specific aim of the risk assessment, and the resources at hand. Examination pertaining to exposure factors, constraints, and uncertainty in single chemical exposure evaluations should be equally applied to multiple chemical exposure assessments.

A hierarchical methodology incorporating the integration and iterative assessment of exposure and risk across all stages is known as a tiered approach (Figure 20.1). The primary aim of employing a tiered technique is to optimize the efficiency of screening and prioritization processes by allocating resources to minimize expenditure while achieving conclusive outcomes.

The subsequent layer exhibits greater refinement compared to its antecedent, leading to an outcome characterized by increased realism and less uncertainty. However, according to Meek et al. [42], this approach necessitates additional human effort and data.

In the lower tiers, developers focus on creating simple and fundamental generic scenarios, whereas in the upper tiers, they concentrate on formulating intricate and illustrative receptor-oriented scenarios. The receptors being discussed encompass a range of entities, including individual species or populations within the environment and human individuals, groups, or communities [42, 43]. At lower levels, the conceptualizations of exposure may rely on deterministic techniques, whereas at higher levels, the utilization of probabilistic approaches may become more intricate. The latter approach produces more precise and definitive exposure assessments, albeit with the drawback of requiring additional effort and data.

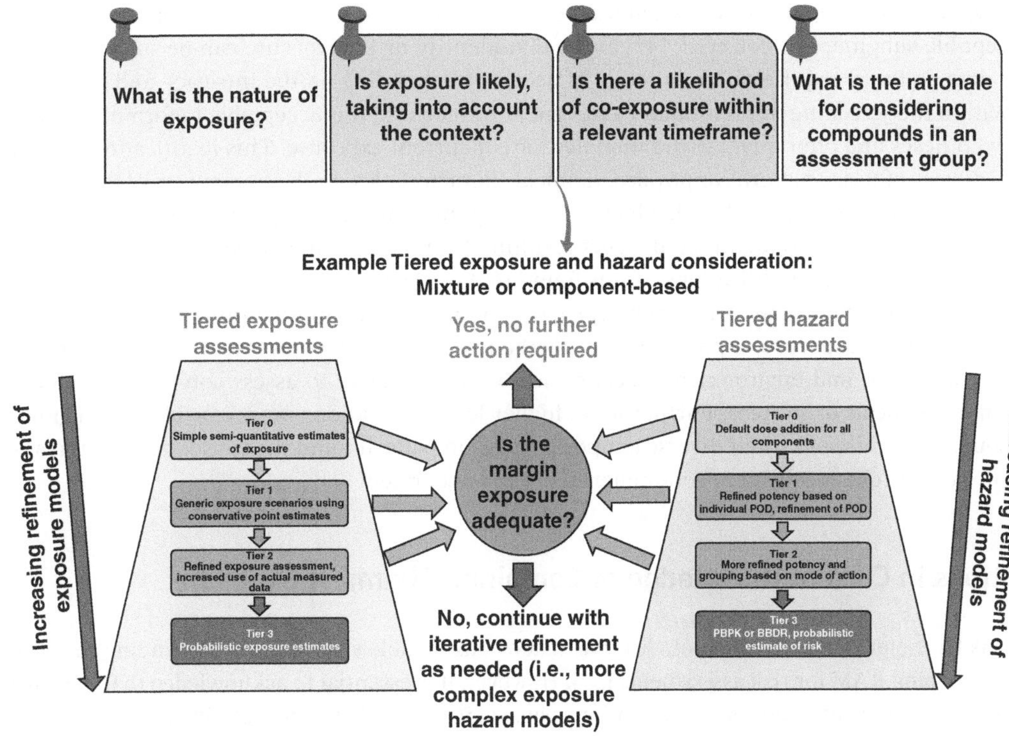

Figure 20.1 A conceptual representation of the framework.

Generic exposure scenarios are generated within the framework of Tier 1 exposure assessment. Instead, relying on empirical evidence, the circumstances are substantiated by hypotheses and simulated data. The subject of human health warrants careful examination of various applications, including cosmetic items. It would be advisable to use a cautious approach by assuming full absorption through the skin. According to Gosens et al. [44], the evaluations being discussed demonstrate a significant level of conservatism, leading to a negative outcome that may not accurately reflect a feasible estimation of combined exposure. Environmental scenarios can consider contaminants that have similar characteristics in terms of how they are transported and their fate in the environment. These scenarios can also include conservative assumptions, such as assuming that effluent treatment does not remove the pollutant. In order to mitigate the likelihood of overestimating exposure in the context of higher levels of human risk assessment, it is advisable to use a person-oriented approach, as suggested by Delmaar and Van Engelen [45]. This approach involves representing scenarios that closely mirror the real-life conditions experienced by the individual.

The process of Tier 2 exposure assessment involves the integration of customized exposure scenarios, leading to a more precise and advanced methodology. These scenarios may include drugs or contaminants that have comparable mechanisms of action. Reliance on precise and reliable information, such as data about assessments of cutaneous absorption, can be beneficial to mitigate the occurrence of unwarranted assumptions. The improvement of evaluating concurrent exposures to multiple chemicals could be achieved by employing a person-centered methodology, as discussed in previous studies by Price and Chaisson [46], Van Raaij [47], Dudzina et al. [48], and the European Centre for Ecotoxicology and Toxicology of Chemicals (ECETOC) [49]. As stated by Dudzina et al. [48], the individual under consideration could serve as a theoretical construct to symbolize a specific fraction within a larger population. This may entail selecting a regular user who can serve as a representative for a worst-case situation or an ordinary customer who utilizes various goods. In the framework of a probabilistic assessment, it is conceivable that the individual in issue could be considered a plausible member of a given population. By removing impractical combinations of product exposures, such as the simultaneous exposure to numerous chemicals found in hair dyes and infant care products, the use of a person-oriented approach ensures consistency in the assessment of exposure [47, 49].

According to Meek et al. [42], Tier 3 exposure evaluations can be fully probabilistic by relying on a restricted number of assumptions, specifically those generated from empirical evidence. As a result, there has been a rise in the formulation of increasingly complex exposure scenarios. Delmaar and Van Engelen [45] argue that a thorough assessment of the combined exposure to consumer products may need the analysis of multiple subscenarios. These subscenarios encompass a diverse range of product categories, simultaneous usage of items, seasonal usage patterns, and usage patterns among susceptible subgroups. Meek et al. [42] argue that identifying co-exposure can be substantiated through different types of data and may require different levels of accuracy depending on the intricacy and characteristics of the assessment. Within the screening or preliminary assessment framework, it is acceptable to formulate logical, negative, or extreme hypotheses and offer qualitative rationales for concurrent exposure. This justification may have been obtained from an analysis of usage patterns or physicochemical characteristics. In the framework of these assessments, standard sources of exposure data are employed, which may incorporate unclear or approximate data and depend on a range of assumptions. If there are no identified dangers resulting from the combined exposures in this specific case, there is no need for any additional modifications. If it is found that further refinement is necessary; it may be imperative to get further empirical data on the co-occurrence, such as biomonitoring or environmental monitoring data. In the context of environmental exposure, a proportionate methodology might be employed. In the realm of physicochemical qualities, emissions, and environmental occurrence, it is imperative to assess comparable factors that are applicable to distinct chemicals or sources. Furthermore, higher level evaluations must consider the regional dispersion of different sources of emissions, their diverse temporal emission patterns, and the consequent spatial and temporal distributions of multiple chemicals more comprehensively and accurately.

20.5 Limitations in Current Knowledge of Combined Chemical Exposure

As previous sections have elucidated, the field of CCE has made considerable strides in understanding the toxicological implications and developing RAM for risk assessments [37]. However, it is essential to acknowledge that the current body of knowledge is far from exhaustive, and several limitations persist [50]. One of the most glaring limitations is the inadequacy of data. Despite the presence of a multitude of studies, the coverage is far from comprehensive [32]. Most existing research focuses on high-profile chemicals, often overlooking less-studied but potentially hazardous substances [31]. This selectivity skews the data, presenting challenges in developing holistic RAM representative of real-world conditions [51].

Another significant limitation is simplifying complex biological responses in the current RAM [34]. Most models predominantly employ linear algorithms that inadequately capture the intricacies of biological systems, such as synergistic or antagonistic interactions among chemicals [33]. These simplifications can lead to inaccurate risk projections, compromising the efficacy of preventive measures and policy decisions [36]. Likewise, methodological inconsistencies present another obstacle. Research protocols vary widely across studies, making it challenging to compile and analyze data cohesively [35]. These inconsistencies impede the development of universally accepted RAM, necessitating labor-intensive efforts to harmonize different datasets for meaningful analysis [40].

Even though limitations exist in the realm of MTG, it has provided invaluable insights into molecular mechanisms. However, it remains largely a laboratory-based method with limited field data [39]. Its predictive power is yet to be extensively validated in real-world scenarios, which curtails its immediate applicability in policy formulation [38]. The rate at which new chemicals are introduced into the environment also outpaces the speed of scientific research [41]. This disparity creates a perpetual state of "catch-up," where regulatory frameworks and RAM are often reactive rather than proactive [37].

In summary, while rich in insights, the current landscape of CCE research is fraught with limitations ranging from data inadequacies and methodological inconsistencies to the simplification of biological complexities. Recognizing these limitations is critical as it informs the current state of risk assessments and highlights areas that require urgent attention. It serves as a prelude to the forthcoming discussions on available information, data limitations, and legislative challenges, topics that will be covered in the subsequent sections of this chapter.

20.6 Data Limitations

The limitations in our current understanding of CCE, as outlined in above section, are intrinsically tied to data availability and quality. Data limitations manifest in various ways and have far-reaching implications on risk assessments, policy formulations, and public health interventions [52]. First, data sparsity is a significant issue [53]. While considerable research has been conducted on certain chemicals, data for many others are either scarce or entirely lacking [37]. This absence of comprehensive data hampers the development of inclusive RAM, limiting the scope and effectiveness of regulatory frameworks [34].

Another critical limitation pertains to the quality of existing data [31]. Research studies often employ varying methodologies, leading to data quality and reliability inconsistencies. This heterogeneity poses challenges in data aggregation, making it difficult to derive universally applicable insights for RAM development [32]. Temporal limitations further complicate the issue [36]. Many studies are short-term, focusing on immediate or acute effects. The lack of longitudinal studies impairs our understanding of chronic exposure risks and long-term health impacts, creating a significant gap in the existing body of knowledge [33].

Data accessibility is also important to thoroughly understand and compare the different findings [40]. While many studies exist, not all are readily available for public or scientific scrutiny, often due to proprietary restrictions or publication behind paywalls [39]. This limitation hinders the free flow of information, restricting collaborative efforts to advance the field [38]. Even with advancements in MTG, the area suffers from data limitations, primarily due to the specialized nature of the tests and the high costs associated with them [35]. These factors often result in smaller sample sizes, reducing the statistical power of the studies and limiting their applicability in broader contexts [41].

In brief, data availability and quality limitations are significant roadblocks in advancing our understanding of CCE. These constraints impact the development and accreditation of RAM and the efficacy of regulatory actions and public health measures. Recognizing and addressing these data limitations is crucial and will serve as a segue into the ensuing discussions on legislative challenges and future perspectives in CCE research.

20.6.1 Legislation of Combined Chemical Exposure

Regulatory frameworks governing CCE are as complex as the scientific issues they aim to address. While significant progress has been made over the years, legislative initiatives often lag behind scientific advancements, resulting in frameworks that are not always up-to-date or comprehensive [50]. This section aims to dissect the existing CCE legislation, highlighting its strengths and weaknesses and examining the areas that require urgent attention. One of the primary challenges in legislating CCE is the variance between national and international laws [37]. While countries have regulations, such as the Toxic Substances Control Act (TSCA) in the United States or REACH in the EU, there is a lack of uniform international standards [54]. This disparity makes it difficult to harmonize regulatory practices across borders, posing challenges for multinational corporations and complicating global efforts to manage CCE [34].

Another critical aspect is the issue of data transparency and availability [31]. Current legislation often does not mandate full disclosure of all data, particularly regarding proprietary chemicals [32]. This lack of openness hampers the development of effective RAM and limits the scope of independent scientific scrutiny. Legislation related to CCE also involves setting risk assessment guidelines and mitigation strategies [33]. While these are typically based on the best available science, the limitations discussed in Sections 20.5 and 20.6, such as data sparsity and quality issues, often trickle down to these legislative frameworks, affecting their effectiveness [36].

20.6.1.1 Public Involvement and Awareness

Public awareness and involvement are crucial yet often overlooked components in legislation [40]. Most legislative frameworks do not involve public consultations or adequately inform the public about the risks associated with CCE [39]. This lack of public engagement limits the effectiveness of these laws and prevents a broader societal shift towards safer practices [38].

20.6.1.2 Enforcement and Compliance

Enforcement mechanisms are another vital but often weak link in the legislative chain [41]. Monitoring and ensuring compliance is a mammoth task that requires substantial resources in terms of workforce and technology [35]. Weak enforcement often results in violations going unpunished, undermining the legislation's very purpose [37].

20.6.1.3 Future Perspectives and Adaptability

Legislation must be dynamic and adaptable to remain effective [50]. However, the bureaucratic nature of legislative processes often results in slow updates, making the laws less responsive to new scientific findings and emerging risks.

In short, whereas current regulatory frameworks have made considerable progress in regulating CCE, serious limitations still need to be addressed. These range from a lack of international harmonization and data transparency to weak enforcement mechanisms and limited public engagement. Future legislative efforts must focus on these areas to develop more robust, effective, and dynamic frameworks for managing the risks associated with CCE.

20.7 Improvements and Future Perspectives

As we navigate through the complex and multifaceted landscape of CCE, it becomes increasingly evident that the field is at a critical juncture. While considerable advances have been made in understanding the toxicological implications, risk assessment, and legislative frameworks, significant challenges and limitations remain [37]. This final section outlines prospective improvements and future directions that could enhance our understanding and management of CCE.

Some parts require improvements, as listed in the following:

- First and foremost, there is an urgent need to advance scientific research in the field [53]. This involves expanding the scope of existing research to include lesser studied chemicals and employing more advanced methodologies to capture the complex biological responses associated with CCE [34]. More longitudinal studies are needed to understand the long-term effects, and enhanced data-sharing mechanisms can facilitate more robust and comprehensive RAM [31].
- Given the global nature of chemical exposures, international harmonization of regulatory frameworks is crucial [32]. Joint efforts could be made to standardize risk assessment protocols, data requirements, and enforcement mechanisms. This will facilitate more effective global management of CCE and ease the burden on industries that operate across borders [33].
- Public awareness and participation are fundamental to effectively managing CCE risks [36]. Legislative frameworks should incorporate mechanisms for public consultations and ensure that risk information is readily accessible and understandable to the general public [40].
- Rapid technological advancements offer promising avenues for improving research and regulation [39]. Advanced analytical tools can provide more nuanced insights into chemical interactions, while big data analytics can aid in developing more accurate RAM [38].
- Finally, scientific research and legislative frameworks must be more adaptable and responsive [41]. As new information becomes available, it should be rapidly integrated into risk assessments and regulatory guidelines to ensure they remain up-to-date and effective [35].

In conclusion, the field of CCE stands at a crossroads, with significant opportunities for advancement and improvement [37]. By focusing on these key areas, we can develop a more robust, comprehensive, and dynamic approach to managing the complexities and risks associated with CCE [55].

20.7.1 Logical Assessment in Combined Chemical Risk Assessment

In light of the extensive range of chemical compositions that humans and organisms in the environment may come into contact with, it is crucial to establish a preliminary evaluation procedure to identify combinations that could potentially provide a hazard.

The following criteria are recommended for consideration:

- There is a potential for human and/or environmental exposure to occur at considered substantial levels, such as those close to the human biomonitoring guidance values, derived no effect levels, or predicted no effect concentrations for various components.
- The substances referred to in this context are those that are produced and/or marketed as multicomponent chemicals or commercial blends comprising various constituents, including active ingredients and substances that are considered of concern according to EU regulations such as REACH, CLP, pesticides and biocidal products legislation, as well as food law, among other relevant legislations. The potential exposure to one or more chemical substances may lead to notable detrimental consequences.
- There is a possibility that the human population or the ecology can experience repeated exposure or widespread distribution of chemicals.
- The persistent existence of chemical substances within the human body and/or the surrounding environment.
- The characteristic of possessing high persistence and bioaccumulation is of considerable significance.

This concerns the current understanding of the potential interaction between individuals and environmental factors across different levels of exposure. One method for forecasting chemical behavior involves utilizing quantitative structure-activity relationships and structural alerts, which enable the identification of commonalities in chemical activity. It is imperative to allocate particular attention to blends that consist of one or more ingredients that are believed to have no discernible threshold for their effects, such as genotoxic carcinogens. In instances of this nature, utilizing a MOE or a lifetime cancer risk technique may be deemed suitable. If a mixture is subjected to one or more components that come close to the threshold levels for producing negative effects, it is imperative to prioritize its evaluation. The potential exclusion of irrelevant combinations can be achieved by applying a method based on the Threshold of Toxicological Concern (TTC). To obtain additional insights into the appropriateness of utilizing a TTC framework for assessing chemical mixtures, we recommend consulting the publications authored by Boobis et al. [56] and Price et al. [57].

20.7.2 Rectification of Knowledge Gaps

There is a significant lack of assessing chemical combinations, as outlined in the mandate, in terms of identifying the precise location, frequency, and magnitude of human and environmental exposure to specific chemical mixtures. Moreover, the uncertainty around the potential fluctuations in exposure during a given period remains unclear. According to Tornero-Velez et al. [58], a thorough understanding of the complex relationship between environmental exposures and human beings requires the implementation of monitoring and modeling techniques. Insufficient data are available concerning the mechanism of action for many chemicals. Currently, there is a lack of consensus on a comprehensive inventory of modes of action, established criteria for identifying or predicting modes of action for compounds with limited evidence, and a framework for classifying substances into evaluation groups. Predicting compound interactions within mixtures, particularly long-term effects, might present a significant challenge. The necessity of doing research lies in the identification of characteristics that possess the potential to act as predictors of potentiation or synergy. The subject under consideration exhibits a heightened degree of intricacy within the discipline of ecotoxicology. Gaining a comprehensive understanding of the various operational modes that may exist among the extensive range of organisms found in a complex biological community is challenging, if not elusive.

20.7.3 Prospective to Improve Qualitative and Quantitative Data

A comprehensive understanding of the mode of action (MOA) for each relevant taxonomic group found in aquatic and terrestrial ecosystems is crucial in studying ecotoxicology. The utilization of the dose/concentration addition model is imperative in forecasting effects at the population level due to the scarcity of information and the potential influence of diverse individual-level actions on endpoints important to the population.

However, choosing the independent action strategy may lead to underestimating the cumulative effects of drugs that exhibit comparable modes of action. Hence, when confronted with insufficient MOA data, giving precedence to the dose/concentration addition technique rather than the independent action approach is recommended. The need to assess and evaluate individualized knowledge arises from expecting prospective interaction. In the future, there is potential for the increased viability of prospective toxicity assessments that utilize pathway analysis, employing in silico and in vitro methodologies to evaluate several pathways, such as inflammation, oxidative stress, and genotoxicity. These methodologies provide the capability to ascertain common impacts across various drugs. The recent meeting of the US National Academic's Standing Committee on the Use of Emerging Science for Environmental Health Decision has concluded that several challenges must be addressed before considering the outcomes derived from high-throughput screens and in silico models as robust and informative. The report was authored by Rusyn and Daston in the year 2010 [59]. The field of ecotoxicology addresses the significant issue of potential cumulative effects that might have adverse repercussions on reproductive processes, population dynamics, and overall ecological health. In specific cases, the observed occurrence of these effects may persist even after exposure to particular substances is discontinued.

The determination of "significant" exposure is contingent upon the frequency, duration, and amount of the exposure. In the context of environmental assessment, it is deemed inadequate to rely on an exposure-driven approach without incorporating a preliminary risk characterization and using the TTC model. Hence, it is imperative to acknowledge the significance of any exposure resulting from emissions that have the potential to alter the prevailing natural background circumstances. Evidence of interaction can be observed at different stages of the decision tree, for as when comparing product information with compound-based assessment. To effectively prioritize chemical mixes for potential evaluation, it is imperative to initially evaluate the extent of human or environmental exposure to the mixture or its individual components. In the absence of substantial evidence suggesting a notable interaction, it is appropriate to employ a dose/concentration addition model when the constituents of the mixture elicit their biological effects through an identical or similar mode/mechanism of action. The independent action model would be employed if the components of the mixture exhibit dissimilar behavior. It is reasonable to argue that employing a dose/concentration addition technique as the default method is appropriate when there is a lack of knowledge regarding both the MOA and the dose–response relationship. This approach ensures a sufficient level of conservatism in the assessment.

20.7.4 Working Under Current Limited Legislation

The existing legislation, specifically in the context of European countries, lacks provisions for a thorough and cohesive evaluation of the cumulative impacts of various chemicals while also considering the diverse pathways by which individuals may be exposed. In situations where a combination of apprehension is detected, which comprises chemical substances subject to several EU regulations, there is presently no established procedure to facilitate a comprehensive and harmonized evaluation spanning the numerous legislations. In instances where chemicals possess distinct modes of action, determining "safe levels" through assessing individual components seems to offer an adequate means of protecting human health against potential adverse effects arising from mixtures or combinations. Nevertheless, in cases where chemicals exhibit comparable modes of action, there exists the possibility of cumulative effects when these chemicals coexist in a mixture, even if the concentration of each substance falls below its designated "safe level." Consequently, employing the concentration/dose addition approach is recommended to ensure sufficient protection.

Regarding the impacts on wild species and ecosystems, the matter is less certain, and it is important to consider the potential for combined or mixed effects. This applies to both substances that operate independently and those with comparable mechanisms of action. Various methodologies exist for identifying and assessing chemical combinations that may provide potential concerns. Nevertheless, significant gaps exist in knowledge and data, particularly about the mode of action and exposure statistics, which impose limitations on the comprehensive assessment of mixes. The collection of information within the framework of EU legislation, including the REACH Regulation, will significantly mitigate existing uncertainty. Despite knowledge and data shortages, it is feasible to evaluate the toxicity of mixtures more methodically within the framework of EU legislation. In cases where there is a lack of information or inconclusive evidence on the mode of action and dose/response relationship, adopting a default assumption of dose/concentration addition can offer a heightened level of protection. However, it is important to note that this approach may also overestimate adverse effects. The potential constraints and associated financial implications should be duly considered when evaluating potential management strategies. Although not explicitly discussed in the Scientific Committee's opinion, the evaluation of chemical mixtures must adhere to the principles of minimizing, refining, and substituting the use of vertebrate animals in testing.

20.7.4.1 Operational Follow-up

Forming an Ad Hoc working group consisting of pertinent services, associated agencies, and authorities such as EFSA, ECHA, EMEA, and EEA is recommended. This group would enhance coordination among various legislative components and facilitate the comprehensive evaluation of priority mixtures while considering the potential risks associated with human and environmental exposure. The Ad Hoc working group will coordinate the efforts to consolidate the data and supervise the comprehensive evaluation of priority mixtures. Any subsequent course of action would be undertaken within the framework of the existing legislation governing each respective chemical's regulation. By June 2014, it is proposed to develop technical recommendations that consider the input of the Scientific Committees. These guidelines aim to foster a uniform approach in evaluating priority combinations within the various EU legislations. These principles should neither serve as a substitute for preexisting regulations nor should they impose any additional responsibilities or costs on economic entities. The ad-hoc working group will oversee the coordination of the development of these guidelines. Following actions should be taken to enhance comprehension of the chemical mixtures to which both human populations and the natural environment are exposed: (1) Conduct a comprehensive review, in collaboration with relevant agencies, of the existing monitoring data collected under EU legislation or generated within the framework of EU-funded research projects. (2) Facilitate a more cohesive approach towards generating, collecting, storing, and utilizing chemical monitoring data pertaining to humans and the environment. This can be achieved by establishing a dedicated platform specifically designed for chemical monitoring data. This approach would facilitate the identification of associations between exposure and epidemiological data, enabling the investigation of potential biological impacts and ultimately contributing to enhanced health outcomes.

Acknowledgment

Dr. Tariq Mehmood thanks the Alexander von Humboldt Foundation for granting him the AvH Postdoc fellowship and the Helmholtz Centre for Environmental Research-UFZ, Leipzig, for enabling him to conduct his experimental work.

References

1 Cleuvers M. Mixture toxicity of the anti-inflammatory drugs diclofenac, ibuprofen, naproxen, and acetylsalicylic acid. Ecotoxicol Environ Saf. 2004, 59(3): 309–315.

2 Yang RSH. 1.24 - Toxicology and risk assessment of chemical mixtures and multiple stressors. In: McQueen CA (Ed.), Comprehensive Toxicology (Third Edition). Elsevier, Oxford, 2018, pp. 489–518.

3 Cedergreen N. Quantifying synergy: a systematic review of mixture toxicity studies within environmental toxicology. PLoS One. 2014, 9(5): e96580.

4 Ding K, Lu L, Wang J, Wang J, Zhou M, Zheng C, Liu J, Zhang C, Zhuang S. In vitro and in silico investigations of the binary-mixture toxicity of phthalate esters and cadmium (II) to Vibrio qinghaiensis sp.-Q67. Sci Total Environ. 2017, 580: 1078–1084.

5 Sarigiannis D, Gotti A, Cimino Reale G, Marafante E. Reflections on new directions for risk assessment of environmental chemical mixtures. Int J Risk Assess Manag. 2009, 13 (3–4): 216–241.

6 Panizzi S, Suciu NA, Trevisan M. Combined ecotoxicological risk assessment in the frame of European authorization of pesticides. Sci Total Environ. 2017, 580: 136–146.

7 Lambert JC. An integrated approach to testing and assessment of chemical mixtures in the environment: the advent of adverse outcome pathway footprinting. American Society for Cellular and Computational Toxicology, 2019.

8 Ge HL, Su BX, Zheng XH, Lin JL, Liu CH, Li PP. Research progress on assessment and prediction models of joint toxicity of chemical mixtures. Appl Mech Mater. 2014, 448: 287–290.

9 Baderna D, Golbamaki N, Maggioni S, Vaccari M, Colacci A, Benfenati E. Chapter 9 – Toxicological characterization of waste-related products using alternative methods: three case studies. In: B Bilitewski (Ed.), Global Risk-Based Management of Chemical Additives II: Risk-Based Assessment and Management Strategies. Springer, Berlin, Heidelberg, 2013, pp. 171–205.

10 Altenburger R, Scholz S, Schmitt-Jansen M, Busch W, Escher BI. Mixture toxicity revisited from a toxicogenomic perspective. Environ Sci Technol. 2012, 46(5): 2508–2522.

11 Barata C, Baird DJ, Nogueira AJ, Soares AM, Riva MC. Toxicity of binary mixtures of metals and pyrethroid insecticides to Daphnia magna Straus. Implications for multi-substance risks assessment. Aquat Toxicol. 2006, 78(1): 1–14.

12 Rider Simmons JE. Risk assessment strategies and techniques for combined exposures. In: Torres JA, S Bobst (Eds.), Toxicological Risk Assessment for Beginners. Springer International Publishing, Cham, 2015, pp. 111–134.

13 Jonker D, Freidig AP, Groten JP, De Hollander AE, Stierum RH, Woutersen RA, Feron VJ. Safety evaluation of chemical mixtures and combinations of chemical and non-chemical stressors. Rev Environ Health. 2004, 19(2): 83–140.

14 Schaaper JE, Hertzberg RC, Simmons JE, Mumtaz MM, Rice GE. Chemical mixtures: toxicologic interactions and risk assessment. In: Information Resources in Toxicology. Elsevier. 2020, pp. 401–414.

15 Kienzler A, Berggren E, Bessems J, Bopp S, van der Linden S, Worth A. Assessment of mixtures-review of regulatory requirements and guidance. JRC Science for Policy Report, EUR 26675EN. Publications Office of the European Union, Luxembourg, 2014.

16 Scanarotti C, Vernazza S, Tirendi S, Calzia D, Alloisio S, Novellino A, Panfoli I, Bassi AM. Evaluation of potential risks to human health and ecosystems during exposure to discarded laboratory chemical mixtures by in vitro multimodel approach. Appl Vitro Toxicol. 2020, 6(4): 131–143.

17 Bopp SK, Barouki R, Brack W, Dalla Costa S, Dorne JL, Drakvik PE, Faust M, Karjalainen TK, Kephalopoulos S, van Klaveren J, Kolossa-Gehring M, Kortenkamp A, Lebret E, Lettieri T, Nørager S, Rüegg J, Tarazona JV, Trier X, van de Water B, van Gils J, Bergman Å. Current EU research activities on combined exposure to multiple chemicals. Environ Int. 2018, 120: 544–562.

18 World Health Organization. Chemical Mixtures in Source Water and Drinking-Water. World Health Organization, 2017.

19 Meek ST, Greathouse JA, Allendorf MD. Metal-organic frameworks: a rapidly growing class of versatile nanoporous materials. Adv Mater. 2011, 23(2): 249–267.

20 European Food Safety Authority. EFSA Guidance Document on the risk assessment of plant protection products on bees (Apis mellifera, Bombus spp. and solitary bees). EFSA J. 2013, 11(7): 3295.

21 EFSA Panel on Biological Hazards (BIOHAZ). Scientific Opinion on the maintenance of the list of QPS biological agents intentionally added to food and feed (2013 update). EFSA J. 2013, 11(11): 3449.

22 EFSA Panel on Dietetic Products, Nutrition and Allergies (NDA). FHI LFC 24, a bovine milk-derived casein hydrolysate, and a reduction of post-prandial blood glucose responses: evaluation of a health claim pursuant to Article 13 (5) of Regulation (EC) No 1924/2006. EFSA J. 2016, 14(7): e04540.

23 European Food Safety Authority. EFSA Guidance Document for evaluating laboratory and field dissipation studies to obtain DegT50 values of active substances of plant protection products and transformation products of these active substances in soil. EFSA J. 2014, 12(5): 3662.

24 Ludden V, Godfrey E, Kobilsky A, Hahn F, Jansen L. The 3rd independent external evaluation of EFSA 2011–2016. Final report. Coffey International Development Ltd, 2018.

25 Zhu X-W, Chen J-Y. mixtox: an R package for mixture toxicity assessment. R J. 2016, 8(2): 421.

26 Bunke M, Alexander ME, Dick JT, Hatcher MJ, Paterson R, Dunn AM. Eaten alive: cannibalism is enhanced by parasites. Royal Soc Open Sci. 2015, 2(3): 140369.

27 Price NL, Gomes AP, Ling AJ, Duarte FV, Martin-Montalvo A, North BJ, Agarwal B, Ye L, Ramadori G, Teodoro JS, Hubbard BP. SIRT1 is required for AMPK activation and the beneficial effects of resveratrol on mitochondrial function. Cell Metab. 2012, 15(5): 675–690.

28 Moretto A, Bachman A, Boobis A, Solomon KR, Pastoor TP, Wilks MF, Embry MR. A framework for cumulative risk assessment in the 21st century. Crit Rev Toxicol. 2017, 47(2): 85–97.

29 Solomon KR, Wilks MF, Bachman A, Boobis A, Moretto A, Pastoor TP, Phillips R, Embry MR. Problem formulation for risk assessment of combined exposures to chemicals and other stressors in humans. Crit Rev Toxicol. 2016, 46(10): 835–844.

30 European Food Safety Authority, Hart A, Maxim L, Siegrist M, Von Goetz N, da Cruz C, Merten C, Mosbach-Schulz O, Lahaniatis M, Smith A, Hardy A. Guidance on communication of uncertainty in scientific assessments. EFSA J. 2019, 17(1): e05520.

31 Jones ML, Gibbons PM, Roussel AJ, Dominguez BJ. Mineral composition of uroliths obtained from sheep and goats with obstructive urolithiasis. J Vet Intern Med. 2017, 31(4): 1202–1208.

32 Clark J, Glasziou P, Del Mar C, Bannach-Brown A, Stehlik P, Scott AM. A full systematic review was completed in 2 weeks using automation tools: a case study. J Clin Epidemiol. 2020, 121: 81–90.

33 Nguyen VM, Young N, Brownscombe JW, Cooke SJ. Collaboration and engagement produce more actionable science: quantitatively analyzing uptake of fish tracking studies. Ecol Appl. 2019, 29(6): e01943.

34 Martinez-Martinez A, Cegarra-Navarro JG, Garcia-Perez A, Vicentini F. Extending structural capital through pro-environmental behaviour intention capital: an outlook on Spanish hotel industry. J Intellect Cap. 2021, 22(3): 633–652.

35 Harrison J, Vanbaelen R. Engineering education accreditation: a look at communication and language. In: 2015 IEEE International Professional Communication Conference (IPCC). Limerick, Ireland, 12-15 July 2015, IEEE.

36 Turner C, Kadiyala S, Aggarwal A, Coates J, Drewnowski A, Hawkes C, Herforth A, Kalamatianou S, Walls H. Concepts and methods for food environment research in low and middle income countries. Agriculture, Nutrition and Health Academy Food Environments Working Group (ANH-FEWG) Innovative Methods and Metrics for Agriculture and Nutrition Actions (IMMANA) Programme, London, UK, 2017.

37 Smith B, Hernandez M, Gordon J. Competency-Based Learning in 2018. Advanced Distributed Learning Initiative, 2019.

38 Hernández AF, Tsatsakis AM. Human exposure to chemical mixtures: challenges for the integration of toxicology with epidemiology data in risk assessment. Food Chem Toxicol. 2017, 103: 188–193.

39 Li Y, Barregard L, Xu Y, Scott K, Pineda D, Lindh CH, Jakobsson K, Fletcher T. Associations between perfluoroalkyl substances and serum lipids in a Swedish adult population with contaminated drinking water. Environ Health. 2020, 19(1): 1–11.

40 Edwards SW, Tan YM, Villeneuve DL, Meek ME, McQueen CA. Adverse outcome pathways—organizing toxicological information to improve decision making. J Pharmacol Exp Ther. 2016, 356(1): 170–181.

41 García-Rodríguez Y, Bravo-Monzón AE, Espinosa-García FJ. Growth response of maize seed-borne fungi to cereal phenolic acid mixtures. Biochem Syst Ecol. 2021, 98: 104321.

42 Meek AME, Boobis AR, Crofton KM, Heinemeyer G, Van Raaij M, Vickers C. Risk assessment of combined exposure to multiple chemicals: a WHO/IPCS framework. Regul Toxicol Pharmacol. 2011, 60(2): S1–S14.

43 EFSA Panel on Plant Protection Products and their Residues (PPR). Scientific Opinion on the identification of pesticides to be included in cumulative assessment groups on the basis of their toxicological profile. EFSA J. 2013, 11(7): 3293.

44 Gosens I, Delmaar C, ter Burg W, Schuur G. Aggregate consumer assessment of chemicals in consumer products. Paraben exposure in children as a case study. 2012.

45 Delmaar J, van Engelen V. Aggregating human exposure to chemicals: an overview of tools and methodologies. RIVM Rapport 630700001, 2007.

46 Price PS, Chaisson CF. A conceptual framework for modeling aggregate and cumulative exposures to chemicals. J Expo Sci Environ Epidemiol. 2005, 15(6): 473–481.

47 Meek MB, Boobis AR, Crofton KM, Heinemeyer G, Kleiner J, Lund BO, Olin S, Pavittranon S, Rodriguez C, Van Raaij M, Vickers C. Assessment of combined exposures to multiple chemicals: report of a WHO/IPCS international workshop on aggregate/cumulative risk assessment. In Assessment of Combined Exposures to Multiple Chemicals: Report of a WHO/IPCS International Workshop on Aggregate/Cumulative Risk Assessment. 2009. Citeseer.

48 Dudzina T, Delmaar CJ, Biesterbos JW, Bakker MI, Bokkers BG, Scheepers PT, van Engelen JG, Hungerbuehler K, Von Goetz N. The probabilistic aggregate consumer exposure model (PACEM): validation and comparison to a lower-tier assessment for the cyclic siloxane D5. Environ Int. 2015, 79: 8–16.

49 Ramón BS. ECETOC-European Centre for Ecotoxicology and Toxicology of Chemicals. 2024.

50 Altenburger R, Scholze M, Busch W, Escher BI, Jakobs G, Krauss M, Krüger J, Neale PA, Ait-Aissa S, Almeida AC, Seiler TB. Mixture effects in samples of multiple contaminants – An inter-laboratory study with manifold bioassays. Environ Int. 2018, 114: 95–106.

51 Browne P, Delrue N, Gourmelon A. Regulatory use and acceptance of alternative methods for chemical hazard identification. Curr Opin Toxicol. 2019, 15: 18–25.

52 Price JH, Murnan J. Research limitations and the necessity of reporting them. Am J Health Educ. 2004, 35(2): 66.

53 Yang P, Wang J, Hong AB, Huang LL, Xie QT, Wang YX, Xiong CL, Meng TQ, Pan A, Chen D. Exposure profiles and predictors of a cocktail of environmental chemicals in Chinese men of reproductive age. Chemosphere. 2022, 299: 134337.

54 Botos Á, Graham JD, Illés Z. Industrial chemical regulation in the European Union and the United States: a comparison of REACH and the amended TSCA. J Risk Res. 2019, 22(10): 1187–1204.

55 Drakvik E, Altenburger R, Aoki Y, Backhaus T, Bahadori T, Barouki R, Brack W, Cronin MT, Demeneix B, Bennekou SH, van Klaveren J. Statement on advancing the assessment of chemical mixtures and their risks for human health and the environment. Environ Int. 2020, 134: 105267.

56 Boobis A, Budinsky R, Collie S, Crofton K, Embry M, Felter S, Hertzberg R, Kopp D, Mihlan G, Mumtaz M, Price P, Solomon K, Teuschler L, Yang R, Zaleski R. Critical analysis of literature on low-dose synergy for use in screening chemical mixtures for risk assessment. Crit Rev Toxicol. 2011. 41(5): 369–383.

57 Price PS, Hollnagel HM, Zabik JM. Characterizing the noncancer toxicity of mixtures using concepts from the TTC and quantitative models of uncertainty in mixture toxicity. Risk Anal. 2009, 29(11): 1534–1548.

58 Tornero-Velez R, Egeghy PP, Cohen Hubal EA. Biogeographical analysis of chemical co-occurrence data to identify priorities for mixtures research. Risk Anal. 2012, 32(2): 224–236.

59 Rusyn I, Daston GP. Computational toxicology: realizing the promise of the toxicity testing in the 21st century. Environ Health Perspect. 2010, 118(8): 1047–1050.

21

Combined Toxicity of Chemicals: Final Thoughts and Concluding Remarks

De-Sheng Pei[a,], Marriya Sultan[b], Yiyun Liu[a]*

[a] School of Public Health, Chongqing Medical University, Chongqing, China
[b] Chongqing Institute of Green and Intelligent Technology, Chinese Academy of Sciences, Chongqing School, University of Chinese Academy of Sciences, Chongqing, China

*Corresponding author.
E-mail: peids@cqmu.edu.cn (D.S.P)

21.1 Summary of Key Findings of This Book

Chapter 1 in this book has shed light on the complexities of combined toxicity assessment, offering valuable insights and contributions. The toxicokinetic and toxicodynamic processes involved in chemical mixtures exposure were elucidated (Chapters 2 and 3), highlighting the intricate interplay between various components and their potential for synergistic or antagonistic effects. Additionally, the principal sources, fate, and mechanisms governing the distribution and transformation of chemical mixtures in the environment were explored (Chapters 4, 5, and 6), along with the analytical techniques employed for their detection (Chapter 7). Common toxicological experimental models were introduced (Chapter 8), providing a foundation for understanding the combined molecular toxicity mechanisms of diverse chemical classes, including heavy metals, pesticides, persistent organic pollutants (POPs), emerging pollutants (PPCPs), phthalates (PAEs), microplastics, and flame retardants (Chapters 9–15). The concept of adverse outcome pathways (AOPs) was also explored, offering a framework for understanding the toxicogenetic endpoints associated with combined pollutant mixtures (Chapter 16). In silico approaches played a crucial role, with mathematical models for predicting combined toxicity in both in vivo and in vitro settings (Chapter 17), as well as novel techniques such as quantitative structure–activity relationships (QSARs) and Tox21 methods (Chapter 18), demonstrating their potential for advancing our understanding of combined toxicity assessment. Despite the significant progress in combined toxicology, several challenges and limitations persist. Experimental, analytical, and predictive modeling approaches face inherent complexities, including the vast number of potential chemical combinations, varying exposure scenarios, and the intricacies of mixture interactions (Chapter 19). Addressing these challenges will require continued interdisciplinary collaboration, innovative methodologies, and robust data integration strategies. Looking ahead, several promising avenues for future research can be envisioned (Chapter 20). The development of high-throughput screening techniques and advanced computational approaches, such as machine learning and artificial intelligence, could revolutionize our ability to rapidly assess and predict the combined toxicity of chemical mixtures [1]. Additionally, the integration of "omics technologies," including genomics, proteomics, and metabolomics, holds the potential to provide a more comprehensive understanding of the molecular mechanisms underpinning combined toxicity [2, 3].

21.2 Final Thoughts

The wide range of chemical mixtures in ecosystems resulting from inputs from domestic, agricultural, and industrial sources raises concerns about exposure to living organisms [4, 5]. Understanding the consequences of combined exposure to these chemicals presents one of the major challenges in ecotoxicology [6] since these mixtures usually have various exposure pathways that change over time, affecting both the overall exposure level and the relative concentrations of

Toxicological Assessment of Combined Chemicals in the Environment, First Edition. Edited by De-Sheng Pei and Yiyun Liu.

constituent chemicals [7]. The concern for health risks related to combined chemical exposure has increased over the past few decades. Previously, dose–response models, such as concentration addition (CA) and independent action (IA) models, have been used to study the combined effects of chemical mixtures [8]. These models involve the same mode of action for chemicals with a similar mode of action and independent action for chemicals with different modes of action, respectively [4]. While these models were applied to predict mixture effects [9], deviations were observed for many chemical mixtures in different species due to synergistic or antagonistic responses of the chemicals, resulting in underestimation or overestimation of effects [10, 11].

In chronic and subchronic toxicity assessments using models that involve repeated administration of chemicals to different laboratory organisms in single or binary combinations, effects are assessed by many biochemical and morphometric indices. However, these primarily reflect the mathematical description of experimental data rather than differences in the toxicity mechanism between different chemical combinations. In a chemical mixture, if each chemical is below its threshold concentration and has a different mode of action, it produces a low or negligible combined effect. However, if individual chemicals are present in concentrations above their respective thresholds, toxicokinetic and/or toxicodynamic interactions occur, and subsequently, new toxicities emerge because of synergism or potentiation [12]. As such interactions are hard to predict, assessments are based on the analysis of relevant chemical mixtures using a wide range of molecular and toxic response biomarkers, such as DNA damage and oxidative stress [13]. Toxicokinetic–toxicodynamic (TKTD) models simulate the time course of progressions leading to toxicity [14] and use biological parameters that can deal with concentrations that vary over time [4, 15]. A wide range of mixture effects could be determined by toxicokinetic and toxicodynamic factors [16], including cellular and molecular targets and metabolic pathways of individual chemicals. However, for chemical mixtures, unpredictable effects occur due to the complexity of toxicological interaction. The system toxicology approach helps predict the biological responses elicited by the interaction of different chemicals in a mixture using molecular toxicity assessment, such as gene expression profiles [12].

Over the past decade, novel approaches have been integrated to assess the toxicity of mixture compounds, including structure–activity (QSAR) [17], omics technologies [18], and Adverse Outcome Pathway (AOP) approaches [19] for chemical and biological information relevant to risk assessment. The AOP framework assists in interpreting molecular or biochemical data from field-sampled organisms exposed to chemical mixtures to endpoints that are useful in inferring possible risks. Similarly, toxicological-based QSARs allow the formulation of multipurpose predictive tools capable of apprehending the activities of possible chemical mixtures. Predictions are made via correlation estimation between measurable chemical properties and the endpoint of concern, allowing the assessment of missing data using structural information [20]. In addition to these traditional and novel approaches, integrated dose-response-based in vitro and in vivo bioassays and in silico mathematical modeling approaches have also been developed, integrating laboratory data and computational modeling for a more robust evaluation of risk [21]. These emerging tools provide significant improvements in the hazard assessment of multifaceted mixtures of heavy metals [22], organochlorine pesticides [23], flame retardants [24], pharmaceuticals [25], and microplastics [26–28]. Moreover, the assessment of combined effects of emerging pollutants, such as perfluorinated compounds, nanoparticles, plasticizers, pharmaceuticals, and endocrine-disrupting chemicals using multiplexed toxicogenomic techniques has become a hot research topic [29]. This holistic approach to integrated toxicological assessment is significantly useful in prioritizing lethal chemical mixtures for safer research [30].

Appraising the risk from chemical mixtures is a complex task that involves multiple challenges, such as the consideration of cumulative risk from exposure to multiple agents, the aggregate risk from exposure to a particular chemical from different sources, and the interaction of exposures to multiple chemicals. Consequently, in the case of chemical mixtures, combined toxicity assessment requires information on the toxicity profiles and epidemiological investigations of the chemicals in question, from which hazard quotients (HQs) may be generated [31]. However, in cases where the mode of action of a mixture is unidentified, information on the mode of action of each chemical in the mixture must be collected, considering the structure–activity similarities and possible interactions.

21.3 Concluding Remarks

The assessment of the combined toxicity of chemicals in the environment is a complex and multifaceted challenge that necessitates a concerted effort from the scientific community, policymakers, and stakeholders [32]. This book has provided a comprehensive overview of the current state of knowledge, highlighting the significant progress made in understanding the intricate mechanisms, environmental distribution, and assessment methodologies associated with chemical mixtures. However, much work remains to be done. Continued research, interdisciplinary collaboration, and a commitment to translating scientific findings into practical solutions are essential for mitigating the potential risks posed by combined chemical exposures. By embracing a holistic and integrated approach, we can work towards a future where the combined toxicity of chemicals is effectively assessed, regulated, and managed, safeguarding both human health and the integrity of our natural environments. It is our collective responsibility to address this pressing challenge, fostering a sustainable and resilient world for present and future generations.

Acknowledgment

The authors are grateful for the support from the Chongqing Technology Innovation and Application Development Sichuan-Chongqing Special Key Project (CSTB2024TIAD-CYKJCXX0017), High-level Talents Project of Chongqing Medical University (No. R4014), Chongqing Postdoctoral Innovation Mentor Studio (X7928 D.S.P.), and the CAS-TWAS Scholarship (No. 2019A8018537003 to M.S).

Conflict of Interest

The authors declare no conflict of interest.

References

1 Singh AV, Varma M, Laux P, Choudhary S, Datusalia AK, Gupta N, Luch A, Gandhi A, Kulkarni P, Nath B. Artificial intelligence and machine learning disciplines with the potential to improve the nanotoxicology and nanomedicine fields: a comprehensive review. Arch Toxicol. 2023, 97: 963–979.
2 Zhang X, Zhou Q, Li X, Zou W, Hu X. Integrating omics and traditional analyses to profile the synergistic toxicity of graphene oxide and triphenyl phosphate. Environ Pollut. 2020, 263: 114473.
3 Ma Y, Liu H, Du X, Shi Z, Liu X, Wang R, Zhang S, Tian Z, Shi L, Guo H, Zhang H. Advances in the toxicology research of microcystins based on Omics approaches. Environ Int. 2021, 154: 106661.
4 Bart S, Short S, Jager T, Eagles EJ, Robinson A, Badder C, Lahive E, Spurgeon DJ, Ashauer R. How to analyse and account for interactions in mixture toxicity with toxicokinetic-toxicodynamic models. Sci Total Environ. 2022, 843: 157048.
5 Bopp SK, Barouki R, Brack W, Dalla Costa S, Dorne J-LC, Drakvik PE, Faust M, Karjalainen TK, Kephalopoulos S, van Klaveren J. Current EU research activities on combined exposure to multiple chemicals. Environ Int. 2018, 120: 544–562.
6 Escher BI, Stapleton HM, Schymanski EL. Tracking complex mixtures of chemicals in our changing environment. Science. 2020, 367(6476): 388–392.
7 Schaaper JE, Hertzberg RC, Simmons JE, Mumtaz MM, Rice GE. Chemical mixtures: toxicologic interactions and risk assessment. In: Information Resources in Toxicology. Elsevier, 2020, pp. 401–414.
8 Kim S, Kang K, Kim H, Seo M. In vitro toxicity screening of fifty complex mixtures in HepG2 cells. Toxics. 2024, 12: 126.
9 Cedergreen N. Quantifying synergy: a systematic review of mixture toxicity studies within environmental toxicology. PLOS ONE. 2014, 9(5): e96580.
10 Gottardi M, Cedergreen N. The synergistic potential of azole fungicides does not directly correlate to the inhibition of cytochrome P450 activity in aquatic invertebrates. Aquat Toxicol. 2019, 207: 187–196.
11 Schäfer RB, Piggott JJ. Advancing understanding and prediction in multiple stressor research through a mechanistic basis for null models. Glob Change Biol. 2018, 24(5): 1817–1826.
12 Hernández AF, Gil F, Lacasaña M. Toxicological interactions of pesticide mixtures: an update. Arch Toxicol. 2017, 91: 3211–3223.
13 Hernández AF, Gil F, Tsatsakis AM. Biomarkers of chemical mixture toxicity. In: Biomarkers in toxicology. Elsevier, 2019, pp. 569–585.
14 Jager T, Albert C, Preuss TG, Ashauer R. General unified threshold model of survival: a toxicokinetic-toxicodynamic framework for ecotoxicology. Environ Sci Technol. 2011, 45(7): 2529–2540.

15 Ashauer R, Escher BI. Advantages of toxicokinetic and toxicodynamic modelling in aquatic ecotoxicology and risk assessment. J Environ Monit. 2010, 12(11): 2056–2061.

16 Bloch D, Diel P, Epe B, Hellwig M, Lampen A, Mally A, Marko D, Villar Fernández MA, Guth S, Roth A, Marchan R, Ghallab A, Cadenas C, Nell P, Vartak N, Van Thriel C, Luch A, Schmeisser S, Herzler M, Landsiedel R, Leist M, Marx-Stoelting P, Tralau T, Hengstler JG. Basic concepts of mixture toxicity and relevance for risk evaluation and regulation. Arch Toxicol. 2023, 97: 3005–3017.

17 Zhang F, Wang Z, Peijnenburg WJGM, Vijver MG. Machine learning-driven QSAR models for predicting the mixture toxicity of nanoparticles. Environ Int. 2023, 177: 108025.

18 Luan Y, Zhao J, Chen Y, Shen J, Cheng L. Toxicologic, transcriptomic, and metabolomic insights into the effect of a mixture of 26 veterinary antimicrobials on rat liver. Chemosphere. 2023, 315: 137752.

19 Yang L, Zeng J, Gao N, Zhu L, Feng J. Predicting the metal mixture toxicity with a toxicokinetic–toxicodynamic model considering the time-dependent adverse outcome pathways. Environ Sci Technol. 2024, 58: 3714–3725.

20 Belfield SJ, Firman JW, Enoch SJ, Madden JC, Tollefsen KE, Cronin MT. A review of quantitative structure-activity relationship modelling approaches to predict the toxicity of mixtures. Comput Toxicol. 2022, 25: 100251.

21 Villeneuve DL, Coady K, Escher BI, Mihaich E, Murphy CA, Schlekat T, Garcia-Reyero N. High-throughput screening and environmental risk assessment: state of the science and emerging applications. Environ Toxicol Chem. 2019, 38(1): 12–26.

22 Su C, Hua Y, Liu Y, Tao S, Jia F, Zhao W, Lin W. Identification of the single and combined acute toxicity of Cr and Ni with Heterocypris sp. and the quantitative structure-activity relationship (QSAR) model. PLOS ONE. 2024, 19: e0300800.

23 Liu Q, Fan G, Bi J, Qin X, Fang Q, Wu M, Mei S, Wan Z, Lv Y, Song L, Wang Y. Associations of polychlorinated biphenyls and organochlorine pesticides with metabolic dysfunction-associated fatty liver disease among Chinese adults: effect modification by lifestyle. Environ Res. 2024, 240: 117507.

24 Du J, Li H, Xu S, Zhou Q, Jin M, Tang J. A review of organophosphorus flame retardants (OPFRs): occurrence, bioaccumulation, toxicity, and organism exposure. Environ Sci Pollut Res. 2019, 26: 22126–22136.

25 Kidd KA, Backhaus T, Brodin T, Inostroza PA, McCallum ES. Environmental risks of pharmaceutical mixtures in aquatic ecosystems: reflections on a decade of research. Environ Toxicol Chem. 2024, 43: 549–558.

26 Goutam Mukherjee A, Ramesh Wanjari U, Eladl MA, El-Sherbiny M, Elsherbini DMA, Sukumar A, Kannampuzha S, Ravichandran M, Renu K, Vellingiri B. Mixed contaminants: occurrence, interactions, toxicity, detection, and remediation. Molecules. 2022, 27(8): 2577.

27 Zhang S-Q, Li P, He S-W, Xing S-Y, Cao Z-H, Zhao X-L, Sun C, Li Z-H. Combined effect of microplastic and triphenyltin: insights from the gut-brain axis. Environ Sci Ecotechnol. 2023, 16: 100266.

28 Chen H, Chen X, Ding P, Gu Y, Jiang Y, Li X, Hu G, Li L, Wang C, Yu J, Li H. Photoaging enhances combined toxicity of microplastics and tetrabromobisphenol A by inducing intestinal damage and oxidative stress in Caenorhabditis elegans. Sci Total Environ. 2024, 912: 169259.

29 Hayes AW, Li R, Hoeng J, Iskandar A, Peistch MC, Dourson ML. New approaches to risk assessment of chemical mixtures. Toxicol Res Appl. 2019, 3: 2397847318820768.

30 Hamid N, Junaid M, Pei D-S. Combined toxicity of endocrine-disrupting chemicals: a review. Ecotoxicol Environ Saf. 2021, 215: 112136.

31 Neale PA, Neelamraju C, Warne MSJ. Derivation of species sensitivity distributions and ecotoxicity threshold values for 66 pesticide active ingredients and the hazard and risk they pose to freshwater waterways that discharge to the Great Barrier Reef, Australia. Sci Total Environ. 2024, 920: 170988.

32 Zhang J, Zhang J, Huang X, Xie F, Dai B, Ma T, Zeng J. Combined toxicity and adverse outcome pathways of common pesticides on *Chlorella pyrenoidosa*. Environ Sci Proces Impact. 2024, 26: 611–621.

Index